Oscillation Theory for Second Order
Linear, Half-Linear, Superlinear and Sublinear Dynamic
Equations

T0215768

Oscillation Theory for Second Order Linear, Half-Linear, Superlinear and Sublinear Dynamic Equations

by

Ravi P. Agarwal

Florida Institute of Technology,
Melbourne, Florida, U.S.A.

Said R. Grace

Cairo University,
Orman, Giza, Egypt

and

Donal O'Regan

National University of Ireland,
Galway, Ireland

KLUWER ACADEMIC PUBLISHERS
DORDRECHT / BOSTON / LONDON

A C.I.P. Catalogue record for this book is available from the Library of Congress.

ISBN 978-90-481-6095-2

Published by Kluwer Academic Publishers,
P.O. Box 17, 3300 AA Dordrecht, The Netherlands.

Sold and distributed in North, Central and South America
by Kluwer Academic Publishers,
101 Philip Drive, Norwell, MA 02061, U.S.A.

In all other countries, sold and distributed
by Kluwer Academic Publishers,
P.O. Box 322, 3300 AH Dordrecht, The Netherlands.

Printed on acid-free paper

Contents

Chapter 4 Oscillation Theory for Superlinear Differential Equations

Chapter 5 Oscillation Theory for Sublinear Differential Equations

Chapter 6 Further Results on the Oscillation of Differential Equations

viii

Chapter 7   Oscillation Results for Differential Systems

Chapter 8   Asymptotic Behavior of Solutions of Certain Differential Equations

Chapter 9   Miscellaneous Topics

Chapter 10   Nonoscillation Theory for Multivalued Differential Equations

Subject Index

Preface

This book is devoted to the qualitative theory of second order dynamic equations. In the last 50 years the *Oscillation Theory* of ordinary, functional, neutral, partial, and impulsive differential equations, and their discrete versions has attracted many researchers. This has resulted in hundreds of research papers in every major mathematical journal. There are many books which deal exclusively with oscillation of solutions of differential equations. However, most of those books appeal only to researchers who already are familiar with the subject. Thus, in an effort to bring oscillation theory to a new and wider audience, in this book we present a compact, thorough, and self–contained account for second order dynamic equations. An important feature of this book is that many examples of current interest are given to illustrate the theory.

In Chapter 1, we introduce oscillation theory of second order differential equations. Here also we provide conditions which guarantee that all solutions of a particular class of second order nonlinear differential equations are continuable, bounded, and converge to zero. Our final section in this chapter states several fixed point theorems which play an important role in establishing existence criteria of nonoscillatory solutions.

It is natural to begin our discussion with second order linear differential equations. The interest in second order linear oscillations is due, in a large part, to the fact that many physical systems are modelled by such equations. We begin Chapter 2 with some of the most basic results in the theory of oscillations of linear ordinary differential equations of second order. In particular, we present Sturm and Sturm–Picone comparison theorems. Then, we provide some necessary and sufficient conditions for nonoscillation as well as some comparison theorems of Sturm type. Also we present sufficiency criteria for nonoscillation of solutions. Next, we establish sufficient conditions for oscillation of second order differential equations with alternating coefficients. Integral averaging techniques and interval criteria are two of the most important concepts in oscillation theory and both are discussed in Chapter 2. Criteria for oscillation of linear second order differential equations with integrable coefficients are also given. We conclude this chapter by discussing the problem of forced oscillations.

In recent years the study of half–linear differential equations has become an important area of research. This is largely due to the fact that half–linear differential equations occur in a variety of real world problems such as in the study of p–Laplace equations, non–Newtonian fluid theory and the turbulent flow of a polytrophic gas in a porous medium. In Chapter 3, we present oscillation and nonoscillation criteria for second order half–linear differential equations. We begin with some preliminaries on half–linear differential equations, and then we present Sturm and Levin type comparison theorems and a Liapunov type inequality. Next, we provide an oscillation criterion for almost periodic Sturm–Liouville equations. In this chapter we also present a systematic study of the zeros of solutions of singular half–linear differential equations, nonoscillation characterizations (necessary and sufficient conditions), comparison results and sufficiency criteria for nonoscillation. In addition we study oscillation by employing integral and weighted averaging techniques. Here, interval criteria for the oscillation of half–linear equations will also be provided. Next, we discuss oscillation of half–linear equations with integrable coefficients, and damped and forced equations. We also derive lower bounds for the distance between consecutive zeros of an oscillatory solution. Finally in this chapter, we present a systematic study of the oscillation and nonoscillation of half–linear equations with a deviating argument. Here, classifications of nonoscillatory solutions, and existence results which guarantee that the solutions have prescribed asymptotic behavior will be given.

In Chapter 4, we present a oscillation and nonoscillation theory for second order nonlinear differential equations of superlinear type. We begin with the oscillation of superlinear equations with sign changing coefficients. In particular, we discuss results which involve integrals and weighted integrals of the alternating coefficients, and then provide several criteria which use average behaviors of these integrals. More general averages such as 'weighted averages' and 'iterated averages' will also be employed. Also, we provide sufficient conditions which guarantee the existence of nonoscillatory solutions, and then we present necessary and sufficient conditions for oscillation of superlinear equations. Oscillation results via comparison of nonlinear equations of the same form as well as with linear ones of the same order will also be presented. We conclude this chapter by discussing oscillation and nonoscillation of forced and damped superlinear differential equations.

In Chapter 5, we present oscillation and nonoscillation criteria for all solutions of second order nonlinear differential equations of sublinear type with alternating coefficients. Our results will involve integrals and weighted integrals of alternating coefficients and in some results we use integral aver-

$F(x)$ is bounded from below, $F(x) \geq -K$ for some constant $K > 0$.
Define

$$V(x, y, t) = \frac{1}{q(t)} G(y) + \frac{1}{a(t)} [F(x) + K],$$

then

$$\begin{aligned}
V'(x, y, t) &= -\frac{q'(t)}{q^2(t)} G(y) - \frac{a'(t)}{a(t)q(t)} \frac{y^2}{g(y)} - \frac{1}{a(t)q(t)} h(t, x, y) \left(\frac{y}{g(y)} \right) \\
&\quad + \frac{1}{a(t)q(t)} e(t, x, y) \left(\frac{y}{g(y)} \right) - \frac{a'(t)}{a^2(t)} [F(x) + K] \\
&\leq \frac{[q'(t)]^-}{q^2(t)} G(y) + \frac{r(t)}{a(t)q(t)} \frac{|y|}{g(y)} \\
&\leq \frac{[q'(t)]^-}{q^2(t)} G(y) + m \frac{r(t)}{a(t)q(t)} + n \frac{r(t)}{a(t)q(t)} G(y).
\end{aligned}$$

Integrating the above inequality from t_0 to t and noting that $r(t)/(a(t)q(t))$
is bounded on $[t_0, T]$, we obtain for $t \in [t_0, T]$ that

$$\frac{1}{q(t)} G(y(t)) \leq V(t) \leq K_1 + \int_{t_0}^t \left[\frac{[q'(s)]^-}{q(s)} + n \frac{r(s)}{a(s)} \right] \left(\frac{1}{q(s)} \right) G(y(s)) ds$$

for some constant $K_1 > 0$. Now, from Gronwall's inequality, we find

$$\begin{aligned}
\frac{1}{q(t)} G(y(t)) &\leq K_1 \exp \left(\int_{t_0}^t \left[\frac{[q'(s)]^-}{q(s)} + n \frac{r(s)}{a(s)} \right] ds \right) \\
&\leq K_1 \exp \left(\int_{t_0}^T \left[\frac{[q'(s)]^-}{q(s)} + n \frac{r(s)}{a(s)} \right] ds \right) \leq K_2 < \infty,
\end{aligned}$$

where K_2 is a constant. Thus, $G(y(t))$ is bounded on $[t_0, T)$, and
hence $y(t) = x'(t)$ is bounded on $[t_0, T)$. Finally, an integration shows
that $x(t)$ is also bounded on $[t_0, T)$, and so we have a contradiction to
the assumption that $(x(t), y(t))$ is a solution of (1.1.2) with finite escape
time. ∎

Remark 1.1.1.

1. If $e(t, x, y) \equiv 0$ in Theorem 1.1.1, then condition (1.1.3) can be dropped.

2. We can drop the condition on $a'(t)$ by requiring a stronger condition
on $g(y)$, namely, that there are positive constants M and k such that

$$\frac{y^2}{g(y)} \leq MG(y) \quad \text{for} \quad |y| \geq k. \tag{1.1.6}$$

The proof of this result involves more details than the proof of Theorem
1.1.1, so we omit it here. We also note that (1.1.6) implies (1.1.5).

Now, in what follows by the term 'solution' we shall mean a continuously continuable solution on $[t_0, \infty)$.

In the following results, we shall present sufficient conditions which guarantee that all solutions of (1.1.1) are bounded. These results also illustrate the interplay between $a(t)$ and $g(x')$ in equation (1.1.1).

Theorem 1.1.2. Suppose (1.1.4) and (1.1.5) hold with $n > 0$,

$$a'(t) \geq 0 \quad \text{and} \quad a(t) \leq a_2, \quad a_2 \text{ is a constant}, \quad t \geq t_0 \qquad (1.1.7)$$

and

$$|e(t, x, y)| \leq \frac{1}{n} a(t) \frac{q'(t)}{q(t)}.$$

If

$$F(x) \rightarrow \infty \quad \text{as} \quad |x| \rightarrow \infty, \qquad (1.1.8)$$

then all solutions of equation (1.1.1) are bounded.

Proof. Since $F(x) \rightarrow \infty$ as $|x| \rightarrow \infty$, $F(x)$ is bounded from below, say, $F(x) \geq -K$ for some constant K. Letting

$$V(x, y, t) = \frac{q(t)}{a(t)}[F(x) + K] + G(y),$$

we have

$$\begin{aligned} V'(t) &\leq [F(x) + K]\frac{q'(t)}{a(t)} + \frac{1}{a(t)}e(t, x, y)\left(\frac{y}{g(y)}\right) \\ &\leq \frac{q'(t)}{q(t)}\left(\frac{q(t)}{a(t)}[F(x) + K] + \frac{1}{n}\frac{|y|}{g(y)}\right). \end{aligned}$$

Integrating the above inequality from t_0 to t and using (1.1.5), we find

$$\frac{|y(t)|}{g(y(t))} + n\frac{q(t)}{a(t)}[F(x(t)) + K] \leq m + nV(t)$$

$$\leq m + nV(t_0) + \int_{t_0}^{t} \frac{q'(s)}{q(s)}\left(n\frac{q(s)}{a(s)}[F(x(s)) + K] + \frac{|y(s)|}{g(y(s))}\right) ds.$$

Now, applying Gronwall's inequality, we get

$$\frac{|y(t)|}{g(y(t))} + n\frac{q(t)}{a(t)}[F(x(t)) + K] \leq K_1 \exp\left(\int_{t_0}^{t} \frac{q'(s)}{q(s)} ds\right) = K_1 \frac{q(t)}{q(t_0)},$$

where $K_1 > 0$ is a constant. Hence, $nq(t)F(x(t))/a(t) \leq K_1 q(t)/q(t_0)$, so $F(x(t))$ is bounded for $t \geq t_0$. The conclusion of the theorem now follows from (1.1.8). ∎

Theorem 1.1.3. Suppose (1.1.4), (1.1.6) and (1.1.8) hold, $a(t) \leq a_2$, where a_2 is a constant, and

$$\int_{t_0}^{\infty} \frac{[a'(s)]^-}{a(s)} ds < \infty \qquad (1.1.9)$$

and $|e(t, x, y)| \leq a(t)q'(t)/(Mq(t))$. Then all solutions of equation (1.1.1) are bounded.

Proof. Condition (1.1.6) implies that there exists a constant $A > 0$ such that $y^2/g(y) \leq A + MG(y)$ for all y. Notice also that if $|y| \leq 1$, then $|y|/g(y) \leq B$ for some constant $B > 0$, and if $|y| \geq 1$, then $|y|/g(y) \leq y^2/g(y)$, so

$$\frac{|y|}{g(y)} \leq B + \frac{y^2}{g(y)} \quad \text{for all} \quad y.$$

By condition (1.1.8), $F(x) \geq -K$ for some constant $K > 0$. If $M \geq 1$, we define

$$V(x, y, t) = \frac{q(t)}{a(t)}[F(x) + K] + G(y) + A + B.$$

Then since $M \geq 1$, we have

$$
\begin{aligned}
V'(t) &= \frac{q'(t)}{q(t)} \left(\frac{q(t)}{a(t)}[F(x) + K] + \frac{1}{M}\frac{|y|}{g(y)} \right) \\
&\quad + \frac{[a'(t)]^-}{a(t)} \left(\frac{q(t)}{a(t)}[F(x) + K] + \frac{y^2}{g(y)} \right) \\
&\leq \frac{q'(t)}{q(t)} \left(\frac{q(t)}{a(t)}[F(x) + K] + \frac{1}{M}(A + B) + G(y) \right) \\
&\quad + \frac{[a'(t)]^-}{a(t)} \left(\frac{q(t)}{a(t)}[F(x) + K] + A + MG(y) \right) \\
&\leq \left(\frac{q'(t)}{q(t)} + M\frac{[a'(t)]^-}{a(t)} \right) \left(\frac{q(t)}{a(t)}[F(x) + K] + A + B + G(y) \right).
\end{aligned}
$$

Integrating $V'(t)$ from t_0 to t, we get

$$V(t) \leq V(t_0) + \int_{t_0}^{t} \left(\frac{q'(s)}{q(s)} + M\frac{[a'(s)]^-}{a(s)} \right) V(s)ds,$$

and hence

$$
\begin{aligned}
V(t) &\leq V(t_0) \exp\left(\int_{t_0}^{t} \left(\frac{q'(s)}{q(s)} + M\frac{[a'(s)]^-}{a(s)} \right) ds \right) \\
&\leq K_1 \exp\left(\int_{t_0}^{t} \frac{q'(s)}{q(s)} ds \right) = K_1 \frac{q(t)}{q(t_0)}, \quad K_1 \text{ is a constant.}
\end{aligned}
$$

The boundedness of $x(t)$ now follows as in Theorem 1.1.2.

If $M < 1$, we define

$$V(x, y, t) = \frac{q(t)}{a(t)}[F(x) + K] + G(y) + \frac{1}{M}(A + B).$$

Then since $M < 1$, we have

$$
\begin{aligned}
V'(t) &\leq \frac{q'(t)}{q(t)}\left(\frac{q(t)}{a(t)}[F(x) + K] + \frac{1}{M}(A + B) + G(y)\right) \\
&\quad + \frac{[a'(t)]^-}{a(t)}\left(\frac{q(t)}{a(t)}[F(x) + K] + A + MG(y)\right) \\
&\leq \left(\frac{q'(t)}{q(t)} + \frac{[a'(t)]^-}{a(t)}\right)V(t).
\end{aligned}
$$

The remainder of the proof follows as before.   ■

Now, from Theorems 1.1.2 and 1.1.3 we deduce the following corollary.

Corollary 1.1.1. If in addition to the hypotheses of either Theorem 1.1.2 or 1.1.3, $q(t) \leq q_2$, q_2 is a constant, and $G(y) \to \infty$ as $|y| \to \infty$, then all solutions of the system (1.1.2) are bounded.

Proof. From the proof of Theorem 1.1.2, we have

$$V'(t) \leq \frac{q'(t)}{q(t)}\left[\frac{m}{n} + V(t)\right],$$

so

$$V(t) \leq V(t_0) + \int_{t_0}^{t} \frac{q'(s)}{q(s)}V(s)ds + \frac{m}{n}\ln\left(\frac{q(t)}{q(t_0)}\right).$$

Hence, it follows that

$$V(t) \leq K_2 \exp\left(\int_{t_0}^{t} \frac{q'(s)}{q(s)}ds\right) \leq K_2\frac{q(t)}{q(t_0)} < \infty, \quad K_2 \text{ is a constant.}$$

The boundedness of $y(t)$ now follows from that of $G(y(0))$.   ■

The following lemma will be used to prove the convergence to zero of the nonoscillatory solutions of equation (1.1.1). For this, we need the following additional assumptions:

(I_1) $xf(x) > 0$ if $x \neq 0$ and $f(x)$ is bounded away from zero if x is bounded away from zero,

(I_2) condition (1.1.3) holds and $r(t)/q(t) \to 0$ as $t \to \infty$,

(I_3) if x is bounded, then there exist a continuous function $k(t)$ and $t_1 \geq t_0$ such that $|h(t, x, y)| \leq k(t)g(y)$ for $(t, x, y) \in [t_1, \infty) \times \mathbb{R}^2$ and $k(t)/q(t) \to 0$ as $t \to \infty$,

(I_4) $g(y) \geq c > 0$, c is a constant, $\int_{t_0}^{\infty} q(s)ds = \infty$ and $\int_{t_0}^{\infty} ds/a(s) = \infty$.

Lemma 1.1.2. If conditions (I_1) – (I_4) hold and $x(t)$ is a bounded nonoscillatory solution of equation (1.1.1), then $\liminf_{t \to \infty} |x(t)| = 0$.

Proof. Let $x(t)$ be a bounded nonoscillatory solution of equation (1.1.1), say, $0 < x(t) < B$ for $t \geq T \geq t_0$, B is a constant, and let $k(t)$ and $t_1 \geq T$ be determined by (I_3). Suppose that $\liminf_{t \to \infty} x(t) \neq 0$. Then there exists $t_2 \geq t_1$ such that $x(t)$ is bounded away from zero for $t \geq t_2$. Hence by condition (I_1), $f(x(t)) \geq A > 0$ for $t \geq t_2$, A is a constant. Choose $t_3 \geq t_2$ so that

$$\frac{r(t)}{cq(t)} < \frac{1}{4}A \quad \text{and} \quad \frac{k(t)}{q(t)} < \frac{1}{4}A \quad \text{for} \quad t \geq t_3.$$

From equation (1.1.1), we have

$$\begin{aligned}
\frac{(a(t)x')'}{g(x')} &= \frac{e(t, x, x')}{g(x')} - \frac{h(t, x, x')}{g(x')} - q(t)f(x) \\
&\leq \frac{1}{c}r(t) - k(t) - Aq(t) \\
&\leq q(t)\left[\frac{1}{c}\frac{r(t)}{q(t)} - A\right] \leq -\frac{1}{2}Aq(t) \quad \text{for} \quad t \geq t_3.
\end{aligned}$$

Thus, $(a(t)x')' \leq -Acq(t)/2 < 0$ for $t \geq t_3$. Integrating this inequality from t_3 to t, we find

$$a(t)x'(t) \leq a(t_3)x'(t_3) - \frac{1}{2}Ac\int_{t_3}^{t} q(s)ds \to -\infty \quad \text{as} \quad t \to \infty,$$

so there exists a $t_4 \geq t_3$ such that $x'(t) < 0$ for $t \geq t_4$. Hence,

$$x(t) < x(t_4) + a(t_4)x'(t_4)\int_{t_4}^{t} \frac{1}{a(s)}ds \to -\infty \quad \text{as} \quad t \to \infty$$

contradicting the fact that $x(t) > 0$ for $t \geq T$. A similar argument holds if $x(t) < 0$ for $t \geq T$. ∎

The following two examples illustrate that condition (I_3) above is essential.

Example 1.1.1. Consider the equation

$$x''(t) + tx'(t) + \frac{1}{t}x(t) = \frac{1}{t^2} + \frac{2}{t^3}, \quad t > 0. \tag{1.1.10}$$

Here $g(x') = 1$ and $h(t, x, x') = tx'$, and it is clear that all the hypotheses of Lemma 1.1.2 are satisfied except (I_3) since we do not have $|h(t, x, x')| \leq k(t)g(x')$. Equation (1.1.10) has a bounded nonoscillatory solution $x(t) = (t + 1)/t$ which does not have $\liminf_{t \to \infty} x(t) = 0$.

Example 1.1.2. The equation

$$x''(t) + tx'(t) + \frac{1}{t}x(t)[1 + (x'(t))^2] = \frac{1}{t^6}[t^4 + 2t^3 + t + 1], \quad t > 0 \quad (1.1.11)$$

satisfies all the conditions of Lemma 1.1.2 except (I_3). Here $|h(t, x, x')| = t|x'| \leq t[1 + (x')^2] = tg(x')$ but $k(t)/q(t) = t^2 \not\to 0$ as $t \to \infty$. Again $x(t) = (t + 1)/t$ is a bounded nonoscillatory solution of equation (1.1.11).

In what follows it will be convenient to have the following notation.

Condition C. If $x(t)$ is a nonoscillatory solution of equation (1.1.1), then $\lim_{t \to \infty} x(t) = 0$.

Also, we define

$$p(t) = \exp\left(-\int_{t_0}^t \frac{[q'(s)]^-}{q(s)} ds\right) \quad \text{and} \quad b(t) = \exp\left(-\int_{t_0}^t \frac{[a'(s)]^-}{a(s)} ds\right).$$

Note that $p(t) \leq 1$ and $b(t) \leq 1$ for $t \geq t_0$.

Theorem 1.1.4. Suppose conditions (1.1.3), (1.1.4), (1.1.8) and (1.1.9) hold,

$$\int_{t_0}^\infty \frac{[q'(s)]^-}{q(s)} ds < \infty, \qquad (1.1.12)$$

$$\int_{t_0}^\infty \frac{r(s)}{q(s)} ds < \infty \qquad (1.1.13)$$

and there is a positive constant N such that

$$\frac{y^2}{g(y)} \leq N. \qquad (1.1.14)$$

Then all solutions of equation (1.1.1) are bounded. If, in addition (I_1) – (I_4) are satisfied, then condition C holds.

Proof. From (1.1.8), $F(x) \geq -K$ for some constant $K > 0$. Let

$$V(x, y, t) = b(t)p(t)\left(\frac{1}{a(t)}[F(x) + K] + \frac{1}{q(t)}G(y)\right),$$

then

$$V'(t) = b(t)p(t) \left\{ \frac{-a'(t)}{a^2(t)} [F(x) + K] - \frac{q'(t)}{q^2(t)} G(y) - \frac{a'(t)}{a(t)q(t)} \frac{y^2}{g(y)} - \right.$$

$$- \frac{1}{a(t)q(t)} h(t, x, y) \left(\frac{y}{g(y)} \right) + \frac{1}{a(t)q(t)} e(t, x, y) \left(\frac{y}{g(y)} \right)$$

$$\left. - \left(\frac{1}{a(t)} [F(x) + K] + \frac{1}{q(t)} G(y) \right) \left(\frac{[a'(t)]^-}{a(t)} + \frac{[q'(t)]^-}{q(t)} \right) \right\}$$

$$\le b(t)p(t) \left\{ -\frac{[a'(t)]^+}{a^2(t)} [F(x) + K] - \frac{[q'(t)]^+}{q^2(t)} G(y) - \frac{[q'(t)]^-}{a(t)q(t)} [F(x) + K] \right.$$

$$\left. - \frac{[a'(t)]^-}{a(t)q(t)} G(y) + \frac{[a'(t)]^-}{a(t)q(t)} \left(\frac{y^2}{g(y)} \right) + \frac{1}{a(t)q(t)} e(t, x, y) \left(\frac{y}{g(y)} \right) \right\}$$

$$\le b(t)p(t) \left\{ \frac{[a'(t)]^-}{a(t)q(t)} \left(\frac{y^2}{g(y)} \right) + \frac{r(t)}{a(t)q(t)} \frac{|y|}{g(y)} \right\}.$$

Now conditions (1.1.9) and (1.1.12) imply that $q(t) \ge q_1 > 0$, $p(t) \ge p_1 > 0$, $a(t) \ge a_1 > 0$ and $b(t) \ge b_1 > 0$ where q_1, p_1, a_1 and b_1 are constants. Also, $|y|/g(y) \le N_1$ for all y and some constant $N_1 > 0$. Thus, on integrating $V'(t)$ from t_0 to t, we get

$$V(t) \le V(t_0) + \frac{N}{q_1} \int_{t_0}^t \frac{[a'(s)]^-}{a(s)} ds + \frac{N_1}{a_1} \int_{t_0}^t \frac{r(s)}{q(s)} ds \le K_1 < \infty$$

for all $t \ge t_0$, K_1 is a constant. Hence,

$$F(x(t)) \le K_1 \frac{a(t)}{b(t)p(t)} \le K_1 \frac{a_2}{b_1 p_1} \quad \text{for} \quad t \ge t_0,$$

and so by condition (1.1.8), $x(t)$ is bounded.

Next, let $x(t)$ be a nonoscillatory solution of equation (1.1.1). Note that by (I_1) we can choose $K = 0$. Since $\liminf_{t \to \infty} x(t) = 0$ by Lemma 1.1.2, if $x(t)$ is ultimately monotonic, we are done. If $x(t)$ is not ultimately monotonic, let $\epsilon > 0$ be given and choose $t_1 \ge t_0$ so that (I_3) is satisfied for $t \ge t_1$, $y(t_1) = 0$,

$$F(x(t_1)) < \frac{1}{3a_2} a_1 b_1 p_1 \epsilon, \qquad \int_{t_1}^\infty \frac{[a'(s)]^-}{a(s)} ds \le \frac{1}{3a_2 N} b_1 p_1 q_1 \epsilon$$

and

$$\int_{t_1}^\infty \frac{r(s)}{q(s)} ds < \frac{1}{3a_2 N_1} a_1 b_1 p_1 \epsilon.$$

Then integrating $V'(t)$ from t_1 to t, we find for $t \geq t_1$,

$$F(x(t)) \leq \frac{q(t)}{b(t)p(t)} V(t) \leq \frac{a_2}{b_1 p_1} V(t)$$

$$\leq \frac{a_2}{b_1 p_1} V(t_1) + \frac{a_2 N}{b_1 p_1 q_1} \int_{t_1}^{t} \frac{[a'(s)]^-}{a(s)} ds + \frac{a_2 N_1}{a_1 b_1 p_1} \int_{t_1}^{t} \frac{r(s)}{q(s)} ds < \epsilon.$$

This implies that $\lim_{t \to \infty} x(t) = 0$, since by (I_1), $F(x(t)) \to 0$ if and only if $x(t) \to 0$. ∎

Finally, we state the following two theorems.

Theorem 1.1.5.   Suppose conditions (1.1.3), (1.1.4), (1.1.7), (1.1.8), (1.1.12) and (1.1.13) hold, and there is a positive constant L such that $|y|/g(y) \leq L$. Then all solutions of equation (1.1.1) are bounded. If in addition $(I_1) - (I_4)$ hold, then condition C holds.

Theorem 1.1.6.   Suppose conditions (1.1.3), (1.1.4), (1.1.6), (1.1.8), (1.1.9) and (1.1.12) hold, $g(y) \geq c > 0$ and $\int_{t_0}^{\infty} r(s) ds / q^{1/2}(s) < \infty$. Then all solutions of equation (1.1.1) are bounded. If in addition $(I_1) - (I_4)$ hold, then condition C holds.

1.2. Some Useful Results from Analysis and Fixed Point Theorems

In this section, we state some fixed point theorems and other results from analysis which are needed throughout this book. We recall that a vector space X equipped with a norm $\|.\|$ is called a normed vector space. A subset E of a normed space X is said to be bounded if there exists a number K such that $\|x\| \leq K$ for all $x \in E$. A subset E of a normed vector space X is called convex if for any $x, y \in E$, $ax + (1 - a)y \in E$ for all $a \in [0, 1]$. A sequence $\{x_n\}$ in a normed vector space X is said to converge to the vector x in X if and only if the sequence $\|x_n - x\|$ converges to zero as $n \to \infty$. A sequence $\{x_n\}$ in a normed vector space X is called a Cauchy sequence if for every $\epsilon > 0$ there exists an $N = N(\epsilon)$ such that for all $n, m \geq N(\epsilon)$, $\|x_n - x_m\| < \epsilon$. Clearly a convergent sequence is also a Cauchy sequence, but the converse may not be true. A space X where every Cauchy sequence of elements of X converges to an element of X is called a complete space. A complete normed vector space is said to be a Banach space. Let E be a subset of a Banach space X. A point $x \in X$ is said to be a limit point of E if there exists a sequence of vectors in E which converges to x. We say a subset E is closed if E contains all of its limit points. The union of E and its limit points is called the closure of E and will be denoted by $\overline{E}$. Let X, F be normed vector spaces, and E be a subset of X. An operator $T : E \to F$ is continuous at a point $x \in E$ if and only if for any $\epsilon > 0$ there is a $\delta > 0$ such that $\|Tx - Ty\| < \epsilon$ for all $y \in E$ such that $\|x - y\| < \delta$. Further, T is continuous on E, or simply continuous, if it is continuous at all points of E.

We say that a subset E of a Banach space X is compact if every sequence of vectors in E contains a subsequence which converges to a vector in E. We say that E is relatively compact if every sequence of vectors in E contains a subsequence which converges to a vector in X, i.e., E is relatively compact if $\overline{E}$ is compact.

A family E in $C([a, b], \mathbb{R})$ is called uniformly bounded if there exists a positive constant K such that $|f(t)| \leq K$ for all $t \in [a, b]$ and all $f \in E$. Further, E is called equicontinuous, if for every $\epsilon > 0$ there exists a $\delta = \delta(\epsilon) > 0$ such that $|f(t_1) - f(t_2)| < \epsilon$ for all t_1, $t_2 \in [a, b]$ with $|t_1 - t_2| < \delta$ and all $f \in E$. A set E in $C([a, \infty), \mathbb{R})$ is said to be equiconvergent at ∞ if all $f \in E$ are convergent in $\mathbb{R}$ at the point ∞, and moreover, for every $\epsilon > 0$ there exists a $T \geq a$ such that for all $f \in E$, $t \geq T \Rightarrow |f(t) - \lim_{s \to \infty} f(s)| < \epsilon$.

Theorem 1.2.1 (Arzela–Ascoli Theorem). A subset E in $C([a, b], \mathbb{R})$ with norm $\|f\| = \sup_{t \in [a,b]} |f(t)|$ is relatively compact if and only if it is uniformly bounded and equicontinuous on $[a, b]$.

Let $B([T, \infty), \mathbb{R})$ be the Banach space of all continuous and bounded real valued functions on the interval $[T, \infty)$, endowed with the usual supnorm $\|.\|$. The following result provides a compactness criterion for subsets of $B([T, \infty), \mathbb{R})$.

Compactness Criterion. Let E be an equicontinuous and uniformly bounded subset of the Banach space $B([T, \infty), \mathbb{R})$. If E is equiconvergent at ∞, it is also relatively compact.

Theorem 1.2.2 (Schauder's Fixed Point Theorem). Let E be a closed, convex and nonempty subset of a Banach space X. Let $T : E \to E$ be a continuous mapping such that TE is a relatively compact subset of X. Then, T has at least one fixed point in E, i.e., there exists an $x \in E$ such that $Tx = x$.

Theorem 1.2.3 (Schauder–Tychonov Fixed Point Theorem). Let X be a locally convex linear space, S be a compact convex subset of X and let $T : S \to S$ be a continuous mapping with $T(S)$ compact. Then, T has a fixed point in S.

Finally, we state the Lebesgue dominated convergence theorem.

Theorem 1.2.4 (Lebesgue's Dominated Convergence Theorem). Let E be a measurable set and let $\{f_n\}$ be a sequence of measurable functions such that $\lim_{n\to\infty} f_n(x) = f(x)$ a.e. in E, and for every $n \in \mathbb{N}$, $|f_n(x)| \leq g(x)$ a.e. in E, where g is integrable on E. Then

$$\lim_{n\to\infty} \int_E f_n(x)dx = \int_E f(x)dx.$$

1.3. Notes and General Discussions

1. The results of Section 1.1 are taken from Graef and Spikes [8]. For several other related works, we refer to Burton and Grimmer [2], Grace and Lalli [3–5], and Graef and Spikes [6–8].

2. For recent contributions in fixed point theorems and their applications see the recent monographs by Agarwal, Meehan and O'Regan [1], and O'Regan and Precup [9].

1.4. References

1. **R.P. Agarwal, M. Meehan and D. O'Regan,** *Fixed Point Theory and Applications,* Cambridge University Press, Cambridge, 2001.

2. **T.A. Burton and R. Grimmer,** On the asymptotic behavior of solutions of $x'' + a(t)f(x) = e(t)$, Pacific J. Math. **41**(1972), 43–55.

3. **S.R. Grace and B.S. Lalli,** On boundedness and asymptotic behavior of certain second order integro–differential equations, J. Math. Phys. Sci. **14**(1980), 191–203.

4. **S.R. Grace and B.S. Lalli,** On continuability and boundedness of solutions of certain second order integro–differential equations, Indian J. Pure Appl. Math. **12**(1981), 950–963.

5. **S.R. Grace and B.S. Lalli,** Oscillation and convergence to zero of solutions of damped second order nonlinear differential equations, J. Math. Anal. Appl. **102**(1984), 539–548.

6. **J.R. Graef and P.W. Spikes,** Continuability, boundedness and asymptotic behavior of solutions of $x'' + q(t)f(x) = r(t)$, Ann. Mat. Pura Appl. **101**(1974), 307–320.

7. **J.R. Graef and P.W. Spikes,** Asymptotic behavior of solutions of a second order nonlinear differential equation, J. Differential Equations **17**(1975), 461–476.

8. **J.R. Graef and P.W. Spikes,** Boundedness and convergence to zero of solutions of a forced second order nonlinear differential equation, J. Math. Anal. Appl. **62**(1978), 295–309.

9. **D. O'Regan and R. Precup,** *Theorems of Leray–Schauder Type and Applications,* Gordon & Breach, Amsterdam, 2001.

Chapter 2

Oscillation and Nonoscillation of Linear Ordinary Differential Equations

2.0. Introduction

The oscillation and nonoscillation property of solutions of second order linear differential equations is of special interest, and therefore, it has been the subject of many investigations. The interest in second order linear oscillations is due, in a large part, to the fact that many physical systems are modelled by such equations. In this chapter we shall discuss some of the most basic results in the theory of oscillations of linear ordinary differential equations of second order. In Section 2.1, we shall present Sturm and Sturm–Picone comparison theorems which are useful in oscillation theory. In Section 2.2, we shall provide some necessary and sufficient conditions for the nonoscillation as well as some comparison theorems of Sturm's type. Sufficiency criteria for the nonoscillation are given in Section 2.3. In Section 2.4, we shall establish sufficient conditions for the oscillation of second order differential equations with alternating coefficients. Integral averaging techniques as well as interval criteria for the oscillation are discussed in Section 2.5. In Section 2.6, several criteria for oscillation of linear second order differential equations with integrable coefficients are established. Finally, in Section 2.7 we shall discuss the problems of forced oscillations.

2.1. Sturm Comparison Theorem

This basic result is usually stated as follows:

Theorem 2.1.1. Let $x(t)$ and $y(t)$ be nontrivial solutions of

$$x''(t) + q(t)x(t) = 0 \quad \text{and} \quad y''(t) + q_1(t)y(t) = 0$$

respectively in an interval $[t_1, t_2] \subseteq \mathbb{R}$. Suppose $x(t_1) = x(t_2) = 0$, $q(t)$, $q_1(t) \in C([t_1, t_2], \mathbb{R})$, and satisfy $q_1(t) \geq q(t)$, $(q_1(t) \neq q(t))$ in $[t_1, t_2]$. Then, $y(t)$ must change sign in (t_1, t_2).

The standard method of proof is based on a 'Wronskian argument' [4,29], i.e., since

$$\frac{d}{dt}(y(t)x'(t) - x(t)y'(t)) = y(t)x''(t) - x(t)y''(t) = (q_1(t) - q(t))x(t)y(t),$$

it follows that

$$y(t)x'(t) - x(t)y'(t)\Big|_{t_1}^{t_2} = \int_{t_1}^{t_2} [q_1(s) - q(s)]x(s)y(s)ds,$$

from which one derives the desired contradiction to the assumption that $x(t) > 0$ and $y(t) > 0$ in (t_1, t_2).

This result extends readily to the solutions of

$$(a(t)x'(t))' + q(t)x(t) = 0 \quad \text{and} \quad (a(t)y'(t))' + q_1(t)y(t) = 0,$$

where $a(t) \in C^1([t_1, t_2], \mathbb{R}^+)$ and $q(t)$, $q_1(t) \in C([t_1, t_2], \mathbb{R})$. In fact, to prove this we only need the Sturmian identity

$$\frac{d}{dt}(y(t)a(t)x'(t) - x(t)a(t)y'(t)) = (q_1(t) - q(t))x(t)y(t).$$

If however, we wish to compare the solutions of

$$(a(t)x'(t))' + q(t)x(t) = 0 \tag{2.1.1}$$

and

$$(a_1(t)y'(t))' + q_1(t)y(t) = 0, \tag{2.1.2}$$

where $a(t)$, $a_1(t) \in C^1([t_1, t_2], \mathbb{R}^+)$ and $q(t)$, $q_1(t) \in C([t_1, t_2], \mathbb{R})$ under the hypotheses

$$q_1(t) \geq q(t) \ (q_1(t) \neq q(t)) \ \text{in} \ [t_1, t_2], \quad a(t) \geq a_1(t), \tag{2.1.3}$$

then the Sturmian identity which takes the form

$$\frac{d}{dt}(y(t)a(t)x'(t) - x(t)a_1(t)y'(t)) = (q_1(t) - q(t))x(t)y(t)$$
$$+ (a(t) - a_1(t))x'(t)y'(t),$$

is not useful since the signs of $x'(t)$ and $y'(t)$ are not known. However, if $y(t) \neq 0$, then this identity modified by Picone [26] becomes

$$\frac{d}{dt}\left(\frac{x(t)}{y(t)}(y(t)a(t)x'(t) - x(t)a_1(t)y'(t))\right) = (q_1(t) - q(t))x(t)y(t)$$
$$+ (a(t) - a_1(t))(x'(t))^2 + a_1(t)\left(x'(t) - \frac{x(t)}{y(t)}y'(t)\right)^2,$$

$$\tag{2.1.4}$$

which immediately yields a more general result. Indeed, if $x(t_1) = x(t_2) = 0$, $y(t) > 0$ in $[t_1, t_2]$ and (2.1.3) holds in $[t_1, t_2]$, we can integrate (2.1.4) between t_1 and t_2 to obtain the desired contradiction.

Thus the Picone identity (2.1.4) provides an elementary, but rigorous proof of Sturm's original theorem. This result is known as Sturm–Picone theorem which we state in the following:

Theorem 2.1.2. If there exists a nontrivial real–valued solution $x(t)$ of equation (2.1.1) in (t_1, t_2) such that $x(t_1) = x(t_2) = 0$, and

$$\int_{t_1}^{t_2} \left[(a(s) - a_1(s))(x'(s))^2 + (q_1(s) - q(s))x^2(s) \right] ds \geq 0$$

then every solution $y(t)$ of (2.1.2) has one of the following properties:

(i) $y(t)$ has at least one zero in (t_1, t_2), or

(ii) $y(t)$ is a constant multiple of $x(t)$ on $[t_1, t_2]$.

2.2. Nonoscillatory Characterizations and Comparison Theorems

In this section, we shall establish necessary and sufficient conditions for the nonoscillation of equation (2.1.1) when $a(t) \in C([t_0, \infty), \mathbb{R}^+)$ and $q(t) \in C([t_0, \infty), \mathbb{R})$ for some $t_0 \geq 0$. For this, we shall need the following lemma.

Lemma 2.2.1. Equation (2.1.1) is nonoscillatory if and only if there exists a function $u(t) \in C^1([t_0, \infty), \mathbb{R})$ which satisfies the inequality

$$u'(t) + \frac{u^2(t)}{a(t)} + q(t) \leq 0 \quad \text{eventually.}$$

Consider the more general equation

$$(a(t)x'(t))' + p(t)x'(t) + q(t)f(x(t)) = 0, \qquad (2.2.1)$$

where $a(t) \in C^1([t_0, \infty), \mathbb{R}^+)$, $p(t) \in C^1([t_0, \infty), \mathbb{R})$, $q(t) \in C([t_0, \infty), \mathbb{R})$ and $f \in C(\mathbb{R}, \mathbb{R})$. We shall assume that

(i_1) $xf(x) > 0$ for $x \neq 0$, and

(i_2) $f'(x) \geq k > 0$ for $x \neq 0$, where k is a real constant.

Now, we are in the position to prove the following nonoscillation result.

Theorem 2.2.1. Assume that there exists a function $\rho(t) \in C^2([t_0, \infty), \mathbb{R}^+)$. If (2.2.1) is nonoscillatory, then the equation

$$(a(t)\rho(t)y'(t))' + kQ^*(t)y(t) = 0 \qquad (2.2.2)$$

is also nonoscillatory, where

$$Q^*(t) = \rho(t)\left[q(t) + ka(t)h^{*2}(t) - (a(t)h^*(t))' - (p(t)h^*(t))\right] \qquad (2.2.3)$$

and

$$h^*(t) = \frac{p(t)\rho(t) - a(t)\rho'(t)}{2\,k\,a(t)\rho(t)}. \qquad (2.2.4)$$

Proof. Let $x(t)$ be a nonoscillatory solution of (2.2.1). Define

$$w(t) = \rho(t)a(t)\left[\frac{x'(t)}{f(x(t))} + h^*(t)\right], \quad t \geq t_0 \geq 0.$$

Then, we have

$$w'(t) = \rho(t)\left[\frac{(a(t)x'(t))'}{f(x(t))} + (a(t)h^*(t))' - \frac{a(t)f'(x(t))(x'(t))^2}{f^2(x(t))}\right] + \frac{\rho'(t)}{\rho(t)}w(t)$$

$$\leq -\rho(t)q(t) - p(t)\rho(t)\frac{x'(t)}{f(x(t))} + \rho(t)(a(t)h^*(t))' - ka(t)\rho(t)\left(\frac{x'(t)}{f(x(t))}\right)^2$$

$$+ \left(\frac{p(t)}{a(t)} - 2kh^*(t)\right)(a(t)\rho(t))\left(\frac{x'(t)}{f(x(t))} + h^*(t)\right)$$

$$= -Q^*(t) - ka(t)\rho(t)\left(\frac{x'(t)}{f(x(t))} + h^*(t)\right)^2 = -Q^*(t) - \frac{k}{a(t)\rho(t)}w^2(t),$$

or

$$w'(t) + \frac{k}{a(t)\rho(t)}w^2(t) + Q^*(t) \leq 0, \quad \text{eventually.}$$

Applying Lemma 2.2.1, we arrive at the desired conclusion. ∎

Corollary 2.2.1. If there exists a function $\rho(t) \in C^2([t_0, \infty), \mathbb{R}^+)$ such that equation (2.2.2) is oscillatory, where the functions $Q^*(t)$, $h^*(t)$ are defined in (2.2.3) and (2.2.4) respectively, then (2.2.1) is oscillatory.

Remark 2.2.1. From (2.2.4), it is clear that the function $\rho(t)$ can be defined as

$$\rho(t) = \exp\left(-2k\int^t\left[h^*(s) - \frac{p(s)}{2ka(s)}\right]ds\right). \qquad (2.2.5)$$

For the special case when $f(x) = x$ and $p(t) = 0$, i.e., when (2.2.1) is reduced to equation (2.1.1), we have the following interesting result.

Lemma 2.2.2. Let $\rho(t) \in C^2([t_0, \infty), \mathbb{R}^+)$ be a given function. Then, (2.1.1) is oscillatory if and only if the equation

$$(a(t)\rho(t)w'(t))' + Q(t)w(t) = 0 \qquad (2.2.6)$$

is oscillatory, where

$$Q(t) = \rho(t)\left[q(t) + a(t)h^2(t) - (a(t)h(t))'\right] \qquad (2.2.7)$$

and

$$h(t) = -\frac{\rho'(t)}{2\rho(t)}. \qquad (2.2.8)$$

Combining Theorem 2.1.2 and Lemma 2.2.2, we obtain the following result.

Theorem 2.2.2. Let $a(t)$, $a_1(t) \in C^1([t_1, t_2], \mathbb{R}^+)$ and $q(t)$, $q_1(t) \in C([t_1, t_2], \mathbb{R})$. Suppose the boundary value problem

$$(a(t)x'(t))' + q(t)x(t) = 0, \qquad x(t_1) = x(t_2) = 0$$

has a solution $x(t)$ with $x(t) \neq 0$ on (t_1, t_2) and there exist two functions $\rho(t)$, $\rho_1(t) \in C^2([t_1, t_2], \mathbb{R}^+)$ such that

$$\int_{t_1}^{t_2} \left\{ [a(s)\rho(s) - a_1(s)\rho_1(s)](x'(s))^2 + (Q_1(s) - Q(s))x^2(s) \right\} ds \geq 0,$$

where the functions Q, h are defined in (2.2.7) and (2.2.8) respectively,

$$Q_1(t) = \rho_1(t)\left[q_1(t) + a_1(t)h_1^2(t) - (a_1(t)h_1(t))'\right] \qquad (2.2.9)$$

and

$$h_1(t) = -\frac{\rho_1'(t)}{2\rho_1(t)}. \qquad (2.2.10)$$

Then every solution $y(t)$ of equation (2.1.2) must have a zero in (t_1, t_2) unless $x(t)$ and $y(t)$ are proportional.

Corollary 2.2.2. Let $a(t)$, $a_1(t) \in C^1([t_0, \infty), \mathbb{R}^+)$ and $q(t)$, $q_1(t) \in C([t_0, \infty), \mathbb{R})$. Suppose there exist two functions $\rho(t)$, $\rho_1(t) \in C^2([t_0, \infty), \mathbb{R}^+)$ such that $a_1(t)\rho_1(t) \leq a(t)\rho(t)$ and $Q_1(t) \geq Q(t)$ on $[t_0, \infty)$, where the functions Q, Q_1 are defined in (2.2.7) and (2.2.9) respectively. If equation (2.1.1) is oscillatory, then (2.1.2) is also oscillatory.

Example 2.2.1. Consider the Euler differential equation

$$x''(t) + \frac{c}{t^2}x(t) = 0 \qquad (2.2.11)$$

and the differential equation

$$(t^2 y'(t))' + c_1 y(t) = 0 \qquad (2.2.12)$$

on the interval $[1, \infty)$, where c and c_1 are real constants. We know that the Euler equation (2.2.11) is oscillatory if $c > 1/4$ and nonoscillatory if $c \leq 1/4$. Let $\rho(t) = 1$ and $\rho_1(t) = 1/t^2$ for equations (2.2.11) and (2.2.12) respectively. Then it follows from Corollary 2.2.2 that (2.2.12) is oscillatory if $c_1 \geq c > 1/4$ and nonoscillatory if $c_1 \leq c \leq 1/4$.

Now, in what follows we shall assume that the functions $\rho(t)$, $\rho_1(t) \in C([t_0, \infty), \mathbb{R}^+)$, Q, h, Q_1 and h_1 are defined in (2.2.7), (2.2.8), (2.2.9) and (2.2.10) respectively,

$$\int^{\infty} \frac{1}{a(s)\rho(s)} ds \; = \; \infty \; = \; \int^{\infty} \frac{1}{a_1(s)\rho_1(s)} ds \qquad (2.2.13)$$

and for fixed $t \geq t_0$,

$$\beta(t) \; = \; \int_t^{\infty} Q(s) ds \; < \; \infty, \quad \beta_1(t) \; = \; \int_t^{\infty} Q_1(s) ds \; < \; \infty. \quad (2.2.14)$$

Theorem 2.2.3. The following four statements are equivalent:

(i)   Equation (2.1.1) is nonoscillatory.

(ii) There exist a number $T \geq t_0$ and a function $v(t) \in C([T_1, \infty), \mathbb{R})$ such that

$$v(t) \; = \; \beta(t) + \int_t^{\infty} \frac{v^2(s)}{a(s)\rho(s)} ds \quad \text{for } t \geq T. \qquad (2.2.15)$$

In particular, if $w(t)$ is a nonoscillatory solution of equation (2.2.6), then $v(t)$ can be taken as $v(t) = a(t)\rho(t)w'(t)/w^2(t)$ for $t \geq T$.

(iii) There exist a number $T \geq t_0$ and a function $v(t) \in C([T, \infty), \mathbb{R})$ such that

$$|v(t)| \; \geq \; \left| \beta(t) + \int_t^{\infty} \frac{v^2(s)}{a(s)\rho(s)} ds \right| \quad \text{for } t \geq T. \qquad (2.2.16)$$

(iv) There exists a function $y(t) \in C^1([T, \infty), \mathbb{R})$ for some $T \geq t_0$ such that for $t \geq T$

$$y'(t) + Q(t) + \frac{y^2(t)}{a(t)\rho(t)} \; \leq \; 0. \qquad (2.2.17)$$

Proof (i) $\Rightarrow$ (ii). Since equation (2.1.1) is nonoscillatory, (2.2.6) is also nonoscillatory by Lemma 2.2.2. Let $w(t)$ be a solution of (2.2.6). Without any loss of generality, we may assume that $w(t) > 0$ for $t \geq T \geq t_0$. Define $v(t) = a(t)\rho(t)w'(t)/w(t)$ for $t \geq T$. Then, we have

$$v'(t) + Q(t) + \frac{v^2(t)}{a(t)\rho(t)} \; = \; 0 \quad \text{for } t \geq T.$$

Integrating the above equation from t to u, $u \geq t \geq T$, we find

$$v(u) - v(t) + \int_t^u Q(s)ds + \int_t^u \frac{v^2(s)}{a(s)\rho(s)}ds = 0. \tag{2.2.18}$$

We claim for fixed $t \geq T$ that

$$\int_t^\infty \frac{v^2(s)}{a(s)\rho(s)}ds < \infty. \tag{2.2.19}$$

For this, assume that $\int_t^\infty v^2(s)/(a(s)\rho(s))ds = \infty$ for some $t \geq T$, then there is a $T_1 \geq T$ such that

$$v(u) - \int_{T_1}^u \frac{v^2(s)}{a(s)\rho(s)}ds = v(t) - \int_t^u Q(s)ds - \int_t^{T_1} \frac{v^2(s)}{a(s)\rho(s)}ds \leq -1$$

for $u \geq T_1$, or equivalently,

$$-v(u) \geq 1 + \int_{T_1}^u \frac{v^2(s)}{a(s)\rho(s)}ds \quad \text{for } u \geq T_1. \tag{2.2.20}$$

It follows that $w'(u) < 0$ for $u \geq T_1$ and

$$\frac{v^2(u)}{a(u)\rho(u)[1 + \int_{T_1}^u \frac{v^2(s)}{a(s)\rho(s)}ds]} \geq -\frac{v(u)}{a(u)\rho(u)} = -\frac{w'(u)}{w(u)} \quad \text{for } u \geq T_1.$$

Integrating the above inequality from T_1 to u, we get

$$\ln\left(1 + \int_{T_1}^u \frac{v^2(s)}{a(s)\rho(s)}ds\right) \geq \ln\left(\frac{w(T_1)}{w(u)}\right) \quad \text{for } u \geq T_1.$$

Using (2.2.20) in the above inequality, we obtain $-w'(u) \geq w(T_1)(a(u)\rho(u))$ for $u \geq T_1$. Finally, integrating this last inequality, we find that $w(u) \to -\infty$ as $u \to \infty$, which contradicts the fact that $w(t) > 0$ for $t \geq T$. Thus, (2.2.19) holds.

Now, letting $u \to \infty$ in (2.2.18), it follows that $v(u) \to v(\infty) < \infty$. Hence, in view of (2.2.19), $v(\infty) = 0$. This in (2.2.18) now implies (2.2.15), i.e., (ii) holds.

(ii) $\Rightarrow$ (iii). It is obvious.

(iii) $\Rightarrow$ (iv). Suppose there exists a function $v(t) \in C([T, \infty), \mathbb{R})$ satisfying (2.2.16). Define $y(t) = \beta(t) + \int_t^\infty v^2(s)/(a(s)\rho(s))ds$. Then, $|v(t)| \geq |y(t)|$, and

$$y'(t) = -Q(t) - \frac{v^2(t)}{a(t)\rho(t)} \leq -Q(t) - \frac{y^2(t)}{a(t)\rho(t)}.$$

Hence, (iv) holds.

(iv) $\Rightarrow$ (i). By Lemma 2.2.2, equation (2.2.6) is nonoscillatory and hence (2.1.1) is nonoscillatory. This completes the proof. ∎

Next, we present two nonoscillatory characterizations of (2.1.1).

Theorem 2.2.4. The following three statements are equivalent:

(i)   Equation (2.1.1) is nonoscillatory.

(ii) There exist $T \geq t_0$ and a function $y(t) \in C([T, \infty), \mathbb{R})$ such that

$$y(t) \geq \xi(t) + \int_t^\infty \frac{\mu[s,t]}{a(s)\rho(s)} y^2(s)ds \quad \text{for} \quad t \geq T. \tag{2.2.21}$$

(iii) There exist $T \geq t_0$ and a function $z(t) \in C([T, \infty), \mathbb{R})$ such that

$$z(t) = \xi(t) + \int_t^\infty \frac{\mu[s,t]}{a(s)\rho(s)} z^2(s)ds \quad \text{for} \quad t \geq T, \tag{2.2.22}$$

where

$$\xi(t) = \int_t^\infty \frac{\beta^2(s)}{a(s)\rho(s)} \mu[s,t]ds \tag{2.2.23}$$

and

$$\mu[s,t] = \exp\left(2 \int_t^s \frac{\beta(\tau)}{a(\tau)\rho(\tau)} d\tau \right). \tag{2.2.24}$$

Proof (i) $\Rightarrow$ (ii). Assume that equation (2.1.1) is nonoscillatory. It follows from Theorem 2.2.3 that (2.2.15) has a solution $v(t)$. Define $y(t) = \int_t^\infty v^2(s)/(a(s)\rho(s))ds$. Then, $y'(t) = -[y(t) + \beta(t)]^2/(a(t)\rho(t))$, or

$$y'(t) + \frac{2\beta(t)}{a(t)\rho(t)} y(t) = -\frac{y^2(t)}{a(t)\rho(t)} - \frac{\beta^2(t)}{a(t)\rho(t)}. \tag{2.2.25}$$

Using $\mu[s,t]$ as an integrating factor and integrating (2.2.25) from t to $u \geq t \geq T$, we get

$$\begin{aligned}
y(t) &= \mu[u,t]y(u) + \int_t^u \mu[s,t]\frac{\beta^2(s)}{a(s)\rho(s)}ds + \int_t^u \mu[s,t]\frac{y^2(s)}{a(s)\rho(s)}ds \\
&\geq \int_t^u \mu[s,t]\frac{\beta^2(s)}{a(s)\rho(s)}ds + \int_t^u \mu[s,t]\frac{y^2(s)}{a(s)\rho(s)}ds.
\end{aligned} \tag{2.2.26}$$

The left–hand side of inequality (2.2.26) is independent of u, and hence we may let $u \to \infty$ in (2.2.26). This immediately leads to (2.2.21).

(ii) $\Rightarrow$ (iii). Suppose there exists a function $y(t) \in C([T, \infty), \mathbb{R})$ satisfying (2.2.21). Define $z_0(t) = y(t)$, and inductively for $n = 1, 2, \cdots$

$$z_n(t) = \xi(t) + \int_t^\infty \mu[s,t]\frac{z_{n-1}^2(s)}{a(s)\rho(s)}ds. \tag{2.2.27}$$

Using (2.2.21) and (2.2.27), it is easy to show by induction that $0 \leq \xi(t) \leq z_n(t) \leq z_{n-1}(t) \leq y(t)$. Thus, the sequence $\{z_n(t)\}$ has a pointwise limiting function $z(t) = \lim_{n \to \infty} z_n(t)$. Since the integrand in (2.2.27) is nonnegative, it follows from the Lebesgue dominated convergence theorem that $z(t)$ satisfies the equation (2.2.22).

(iii) $\Rightarrow$ (i). Assume that (2.2.22) has a solution $z(t)$, then an easy computation shows that $v(t) = z(t) + \beta(t)$ defines a solution of (2.2.17). As a consequence of Theorem 2.2.3, we conclude that equation (2.1.1) is nonoscillatory. ∎

Theorem 2.2.5. The following three statements are equivalent:

(i) Equation (2.1.1) is nonoscillatory.

(ii) There exist $T \geq t_0$ and a function $y(t) \in C([T, \infty), \mathbb{R})$ such that

$$y(t) \geq \xi^*(t) + \int_t^\infty \mu^*[s, t] \frac{y^2(s)}{a(s)\rho(s)} ds, \quad t \geq T. \tag{2.2.28}$$

(iii) There exist $T \geq t_0$ and a function $z(t) \in C([T, \infty), \mathbb{R})$ such that

$$z(t) = \xi^*(t) + \int_t^\infty \mu^*[s, t] \frac{z^2(s)}{a(s)\rho(s)} ds, \quad t \geq T \tag{2.2.29}$$

where

$$\xi^*(t) = \int_t^\infty \mu^*[s, t] \frac{\xi^2(s)}{a(s)\rho(s)} ds \tag{2.2.30}$$

and

$$\mu^*[s, t] = \exp\left(2 \int_t^s \frac{\beta(\tau) + \xi(\tau)}{a(\tau)\rho(\tau)} d\tau\right). \tag{2.2.31}$$

Proof (i) $\Rightarrow$ (ii). Assume that equation (2.1.1) is nonoscillatory. It follows from Theorem 2.2.4 that (2.2.22) has a solution $z(t)$. Define $y(t) = \int_t^\infty \mu[s, t] z^2(s)/(a(s)\rho(s)) ds$. Then, we have

$$y'(t) = -\frac{1}{a(t)\rho(t)} \left[y^2(t) + \xi^2(t) + 2(\beta(t) + \xi(t))y(t)\right]. \tag{2.2.32}$$

Using $\mu^*[s, t]$ as an integrating factor and integrating (2.2.32) from t to $u \geq t \geq T$, we get

$$\begin{aligned} y(t) &= \mu^*[u, t] y(u) + \int_t^u \mu^*[s, t] \frac{\xi^2(s)}{a(s)\rho(s)} ds + \int_t^u \mu^*[s, t] \frac{y^2(s)}{a(s)\rho(s)} ds \\ &\geq \int_t^u \mu^*[s, t] \frac{\xi^2(s)}{a(s)\rho(s)} ds + \int_t^u \mu^*[s, t] \frac{y^2(s)}{a(s)\rho(s)} ds. \end{aligned} \tag{2.2.33}$$

Letting $u \to \infty$ in (2.2.33), we obtain (2.2.28).

(ii) $\Rightarrow$ (iii). Suppose there exists a function $y(t) \in C([T, \infty), \mathbb{R})$ satisfying (2.2.28). Define $z_0(t) = y(t)$, and inductively for $n = 1, 2, \cdots$

$$z_n(t) = \xi^*(t) + \int_t^\infty \mu^*[s, t] \frac{z_{n-1}^2(s)}{a(s)\rho(s)} ds. \tag{2.2.34}$$

Using (2.2.27) and (2.2.34), it is easy to show by induction that $0 \le \xi^*(t) \le z_n(t) \le z_{n-1}(t) \le y(t)$. Thus, the sequence $\{z_n(t)\}$ has a pointwise limiting function $z(t) = \lim_{n \to \infty} z_n(t)$. Since the integrand in (2.2.34) is nonnegative, it follows from the Lebesgue dominated convergence theorem that $z(t)$ satisfies (2.2.29).

(iii) $\Rightarrow$ (i). Assume that (2.2.29) has a solution $z(t)$, then an easy computation shows that $v(t) = z(t) + \beta(t) + \xi(t)$ defines a solution of (2.2.17). As a consequence of Theorem 2.2.3, we conclude that equation (2.1.1) is nonoscillatory. ∎

Remark 2.2.2. It is clear from Theorems 2.2.4 and 2.2.5 that the process of generating higher order iterated Riccati integral equations can be continued.

Now, using above nonoscillatory characterizations of solutions of equation (2.1.1), we shall derive the following comparison results.

Theorem 2.2.6. With respect to equations (2.1.1) and (2.1.2) assume that there exist two functions $\rho(t)$, $\rho_1(t) \in C^2([t_0, \infty), \mathbb{R}^+)$ such that

$$a(t)\rho(t) \ge a_1(t)\rho_1(t) \quad \text{and} \quad |\beta(t)| \le \beta_1(t) \quad \text{for} \quad t \ge t_0. \tag{2.2.35}$$

If (2.1.2) is nonoscillatory, then equation (2.1.1) is also nonoscillatory.

Proof. Since equation (2.1.2) is nonoscillatory, it follows from Theorem 2.2.3 that there exist $T \ge t_0$ and a function $v(t) \in C([T, \infty), \mathbb{R})$ such that

$$|v(t)| \ge \left| \beta_1(t) + \int_t^\infty \frac{v^2(s)}{a_1(s)\rho_1(s)} ds \right| \quad \text{for} \quad t \ge T. \tag{2.2.36}$$

Using (2.2.35) in inequality (2.2.36), we obtain

$$|v(t)| \ge \left| \beta(t) + \int_t^\infty \frac{v^2(s)}{a(s)\rho(s)} ds \right| \quad \text{for} \quad t \ge T.$$

Now, it follows from Theorem 2.2.3 that (2.1.1) is nonoscillatory. ∎

Example 2.2.2. Consider the differential equation

$$x''(t) + \frac{\alpha \sin \gamma t}{t} x(t) = 0, \quad t \ge 1 \tag{2.2.37}$$

where α and $\gamma \neq 0$ are constants, and a nonoscillatory Euler's equation

$$y''(t) + \frac{1}{4t^2} y(t) = 0, \quad t \geq 1. \tag{2.2.38}$$

Let

$$\rho(t) = \exp\left(\frac{2\alpha}{\gamma} \int_\eta^t \frac{\cos \gamma s}{s} ds\right), \quad \eta = \frac{\pi}{2|\gamma|}\left(2 + \operatorname{sgn} \frac{\alpha}{\gamma}\right) \text{ and } \rho_1(t) = 1.$$

Then as $t \to \infty$, $\rho(t) = 1 + O(1/t) \geq 1 = \rho_1(t)$ and

$$Q(t) = \left[1 + O\left(\frac{1}{t}\right)\right]\left(\frac{\alpha^2}{2\gamma^2 t^2} + \frac{\alpha^2 \cos 2\gamma t}{2\gamma^2 t^2} - \frac{\alpha \cos \gamma t}{\gamma t^2}\right) \quad \text{as} \quad t \to \infty.$$

Thus, we have

$$\beta(t) = \frac{\alpha^2}{2\gamma^2 t} + O\left(\frac{1}{t^2}\right) \quad \text{as} \quad t \to \infty \quad \text{and} \quad \beta_1(t) = \frac{1}{4t}.$$

It follows from Theorem 2.2.6 that equation (2.2.37) is nonoscillatory if $|\alpha/\gamma| < 1/\sqrt{2}$.

Theorem 2.2.7. With respect to equations (2.1.1) and (2.1.2) assume that there exist two functions $\rho(t), \rho_1(t) \in C^2([t_0, \infty), \mathbb{R}^+)$ such that

$$a(t)\rho(t) \geq a_1(t)\rho_1(t), \quad \frac{\beta^2(t)}{a(t)\rho(t)} \leq \frac{\beta_1^2(t)}{a_1(t)\rho_1(t)} \quad \text{for} \quad t \geq t_0 \tag{2.2.39}$$

and

$$\frac{\beta(t)}{a(t)\rho(t)} \leq \frac{\beta_1(t)}{a_1(t)\rho_1(t)} \quad \text{for} \quad t \geq t_0. \tag{2.2.40}$$

If (2.1.2) is nonoscillatory, then equation (2.1.1) is also nonoscillatory.

Proof. Since equation (2.1.2) is nonoscillatory, it follows from Theorem 2.2.4 that there exist $T \geq t_0$ and a function $y(t) \in C([T, \infty), \mathbb{R})$ such that

$$y(t) = \xi_1(t) + \int_t^\infty \frac{\mu_1[s, t]}{a_1(s)\rho_1(s)} y^2(s) ds,$$

where

$$\xi_1(t) = \int_t^\infty \frac{\beta_1^2(s)}{a_1(s)\rho_1(s)} \mu_1[s, t] ds \tag{2.2.41}$$

and

$$\mu_1[s, t] = \exp\left(2 \int_t^s \frac{\beta_1(\tau)}{a_1(\tau)\rho_1(\tau)} d\tau\right). \tag{2.2.42}$$

From (2.2.39) and (2.2.40), we have $\xi_1(t) \geq \xi(t)$ and $\mu_1[s,t] \geq \mu[s,t]$, where $\xi(t)$ and $\mu[s,t]$ are defined in (2.2.23) and (2.2.24) respectively. Hence,

$$y(t) \geq \xi(t) + \int_t^\infty \frac{\mu[s,t]}{a(s)\rho(s)} y^2(s)\, ds.$$

Now, it follows from Theorem 2.2.4 that (2.1.1) is nonoscillatory.   ∎

Theorem 2.2.8. With respect to equations (2.1.1) and (2.1.2) assume that there exist two functions $\rho(t)$, $\rho_1(t) \in C^2([t_0, \infty), \mathbb{R}^+)$ such that

$$a(t)\rho(t) \geq a_1(t)\rho_1(t), \qquad \frac{\xi^2(t)}{a(t)\rho(t)} \leq \frac{\xi_1^2(t)}{a_1(t)\rho_1(t)} \qquad (2.2.43)$$

and

$$\frac{\beta(t) + \xi(t)}{a(t)\rho(t)} \leq \frac{\beta_1(t) + \xi_1(t)}{a_1(t)\rho_1(t)}, \qquad (2.2.44)$$

where $\xi(t)$, $\xi_1(t)$ and $\mu_1[s,t]$ are defined in (2.2.23), (2.2.41) and (2.2.42) respectively. If (2.1.2) is nonoscillatory, then equation (2.1.1) is also nonoscillatory.

Proof. Since equation (2.1.2) is nonoscillatory, it follows from Theorem 2.2.5 that there exist $T \geq t_0$ and a function $y(t) \in C([T, \infty), \mathbb{R})$ such that

$$y(t) = \xi_1^*(t) + \int_t^\infty \mu_1^*[s,t] \frac{y^2(s)}{a_1(s)\rho_1(s)}\, ds,$$

where

$$\xi_1^*(t) = \int_t^\infty \frac{\xi_1^2(s)}{a_1(s)\rho_1(s)} \mu_1^*[s,t]\, ds$$

and

$$\mu_1^*[s,t] = \exp\left(2 \int_t^s \left(\frac{\beta_1(\tau) + \xi_1(\tau)}{a_1(\tau)\rho(\tau)} \right) d\tau \right).$$

From (2.2.43) and (2.2.44), we have $\xi_1^*(t) \geq \xi^*(t)$ and $\mu_1^*[s,t] \geq \mu^*[s,t]$, where $\xi^*(t)$ and $\mu^*[s,t]$ are defined in (2.2.30) and (2.2.31) respectively. Hence,

$$y(t) \geq \xi^*(t) + \int_t^\infty \frac{\mu^*[s,t]}{a(s)\rho(s)} y^2(s)\, ds, \quad t \geq T.$$

Now, it follows from Theorem 2.2.5 that (2.1.1) is nonoscillatory.   ∎

As an immediate consequence of the above theorems, we state the following result.

Corollary 2.2.3. With respect to equations (2.1.1) and (2.1.2) assume that there exist two functions $\rho(t)$, $\rho_1(t) \in C^2([t_0, \infty), \mathbb{R}^+)$ such that $a(t)\rho(t) \geq a_1(t)\rho_1(t)$, $\xi(t) \leq \xi_1(t)$ and $\beta(t) + \xi(t) \leq \beta_1(t) + \xi_1(t)$, where

$\xi(t)$ and $\xi_1(t)$ are defined in (2.2.23) and (2.2.41) respectively. If (2.1.2) is nonoscillatory, then equation (2.1.1) is also nonoscillatory.

Remark 2.2.3. It is clear from Theorems 2.2.7 and 2.2.8 that we can establish the higher order iterated comparison theorems by using the nonoscillatory characterizations.

2.3. Nonoscillation Criteria

In this section, we shall present easily verifiable nonoscillation criteria for equation (2.1.1).

Theorem 2.3.1. Equation (2.1.1) is nonoscillatory if and only if there exist $T \geq t_0$ and a function $h(t) \in C^1([T, \infty), \mathbb{R})$ satisfying

$$q(t) + a(t)h^2(t) - (a(t)h(t))' \leq 0 \quad \text{for} \quad t \geq T.$$

Proof. Let $x(t)$ be a nonoscillatory solution of equation (2.1.1) such that $x(t) \neq 0$ for $t \geq T \geq t_0$. Define $h(t) = -x'(t)/x(t)$ for $t \geq T$. Then it follows from equation (2.1.1) that

$$
\begin{aligned}
&q(t) + a(t)h^2(t) - (a(t)h(t))' \\
&= q(t) + a(t)\left(\frac{x'(t)}{x(t)}\right)^2 + \frac{(a(t)x'(t))'\, x(t) - a(t)\,(x'(t))^2}{x^2(t)} \\
&= q(t) - q(t)\frac{x^2(t)}{x^2(t)} = 0.
\end{aligned}
$$

Conversely, if $q(t) + a(t)h^2(t) - (a(t)h(t))' \leq 0$, then it follows from Corollary 2.2.2 that (2.2.6) is nonoscillatory because equation $(a(t)\rho(t)w'(t))' = 0$ is nonoscillatory, where $\rho(t) = \exp\left(-2\int_T^t h(s)ds\right)$. Thus, equation (2.1.1) is nonoscillatory. ∎

Example 2.3.1. Consider the Euler differential equation (2.2.11). Let $h(t) = -1/(2t)$. Then, $h'(t) = 1/(2t^2)$ and

$$q(t) + h^2(t) - h'(t) = \frac{c}{t^2} + \frac{1}{4t^2} - \frac{1}{2t^2} = \frac{4c-1}{4t^2}.$$

Thus, by Theorem 2.3.1, equation (2.2.11) is nonoscillatory if $c \leq 1/4$.

Example 2.3.2. Consider the differential equation

$$x''(t) + \left[\frac{1}{4t^2} + \frac{c}{(t \ln t)^2}\right] x(t) = 0, \tag{2.3.1}$$

where c is a real constant. Let

$$h(t) = -\frac{1}{2}\left(\frac{1}{t} + \frac{1}{t \ln t}\right).$$

Then, we have

$$h'(t) = \frac{1}{2}\left(\frac{1}{t^2} + \frac{1}{t^2 \ln t} + \frac{1}{t^2 \ln^2 t}\right)$$

and

$$\begin{aligned}
q(t) + h^2(t) - h'(t) &= \frac{1}{4t^2} + \frac{c}{(t \ln t)^2} + \frac{1}{4}\left(\frac{1}{t} + \frac{1}{t \ln t}\right)^2 \\
&\quad - \frac{1}{2}\left(\frac{1}{t^2} + \frac{1}{t^2 \ln t} + \frac{1}{(t \ln t)^2}\right) \\
&= \frac{4c-1}{4(t \ln t)^2} \quad \text{when } c \le \frac{1}{4}.
\end{aligned}$$

Thus, by Theorem 2.3.1, equation (2.3.1) is nonoscillatory if $c \le 1/4$.

As applications to Theorem 2.3.1, we present the following two corollaries.

Corollary 2.3.1. If $a'(t) \le 0$ for $t \ge t_0$, and

$$\limsup_{t \to \infty} \frac{t^2 q(t)}{a(t)} < \frac{1}{4}, \tag{2.3.2}$$

then equation (2.1.1) is nonoscillatory.

Proof. It follows from (2.3.2) that there exist two numbers $T \ge t_0$ and $c < 1/4$ such that $q(t) \le ca(t)/t^2$ for $t \ge T$. Let $h(t) = -1/(2t)$. Then, $h'(t) = 1/(2t^2)$ and

$$q(t) + a(t)h^2(t) - (a(t)h(t))' \le \frac{4c-1}{4t^2}a(t) - a'(t)h(t) \le 0 \quad \text{for } t \ge T.$$

This and Theorem 2.3.1 imply that equation (2.1.1) is nonoscillatory. ∎

Corollary 2.3.2. If $a'(t) \le 0$ for $t \ge t_0$, and

$$\limsup_{t \to \infty} (t \ln t)^2 \left(\frac{q(t)}{a(t)} - \frac{1}{4t^2}\right) < \frac{1}{4}, \tag{2.3.3}$$

then equation (2.1.1) is nonoscillatory.

Proof. By (2.3.3), there exist two numbers $T \ge t_0$ and $c < 1/4$ such that

$$q(t) < a(t)\left(\frac{1}{4t^2} + \frac{c}{(t \ln t)^2}\right) \quad \text{for } t \ge T.$$

Let

$$h(t) \;=\; -\frac{1}{2}\left(\frac{1}{t} + \frac{1}{t\ln t}\right).$$

Then for $t \geq T$, we have

$$h'(t) \;=\; \frac{1}{2}\left(\frac{1}{t^2} + \frac{1}{t^2\ln t} + \frac{1}{(t\ln t)^2}\right)$$

and

$$q(t) + a(t)h^2(t) - (a(t)h(t))'$$

$$< a(t)\left[\frac{1}{4t^2} + \frac{c}{(t\ln t)^2} + \frac{1}{4}\left(\frac{1}{t} + \frac{1}{t\ln t}\right)^2 \right.$$

$$\left. - \frac{1}{2}\left(\frac{1}{t^2} + \frac{1}{t^2\ln t} + \frac{1}{(t\ln t)^2}\right)\right] - a'(t)h(t)$$

$$= a(t)\frac{(4c-1)}{(t\ln t)^2} - a'(t)h(t) \;\leq\; 0 \quad \text{for} \quad t \geq T.$$

Now, it follows from Theorem 2.3.1 that equation (2.1.1) is nonoscillatory. $\blacksquare$

In what follows, we shall assume that $\rho(t) \in C^2([t_0, \infty), \mathbb{R}^+)$ and $h(t) = -\rho'(t)/(2\rho(t))$. Clearly, from Lemma 2.2.2 and Theorem 2.3.1 the following result is immediate.

Theorem 2.3.2. Equation (2.1.1) is nonoscillatory if and only if there exist $T \geq t_0$ and a function $v(t) \in C^1([T, \infty), \mathbb{R})$ satisfying

$$Q(t) + a(t)\rho(t)v^2(t) - (a(t)\rho(t)v(t))' \;\leq\; 0 \quad \text{for} \quad t \geq T,$$

where Q is defined in (2.2.7).

As applications to Theorem 2.3.2, we have the following corollaries.

Corollary 2.3.3. Suppose there exist two functions $g(t) \in C^1([t_0, \infty), \mathbb{R}^+)$ and $\psi(t) \in C^1([t_0, \infty), \mathbb{R})$ such that $g'(t) \geq 1/(a(t)\rho(t))$ and $\psi'(t) \leq -Q(t)$ for $t \geq t_0$. If

$$\limsup_{t\to\infty} g(t)|\psi(t)| \;<\; \frac{1}{4}, \tag{2.3.4}$$

then equation (2.1.1) is oscillatory.

Proof. By (2.3.4), there are two numbers $T \geq t_0$ and $k \in (0, 1/4)$ such that $|\psi(t)| < k/g(t)$ for $t \geq T$. Let

$$v(t) \;=\; -\frac{1}{2a(t)\rho(t)}\left[2\psi(t) + \frac{1-2k}{g(t)}\right].$$

Then, we have

$$
\begin{aligned}
(a(t)\rho(t)v(t))' &= -\psi'(t) + \frac{(1-2k)g'(t)}{2g^2(t)} \\
&\geq \rho(t)\left[q(t) + a(t)h^2(t) - (a(t)h(t))'\right] + \frac{1-2k}{2a(t)\rho(t)g^2(t)}
\end{aligned}
$$

and

$$
\begin{aligned}
\rho(t)&\left[q(t) + a(t)h^2(t) - (a(t)h(t))'\right] + a(t)\rho(t)v'(t) - (a(t)\rho(t)v(t))' \\
&\leq \frac{1}{4a(t)\rho(t)}\left[2\psi(t) + \frac{1-2k}{g(t)}\right]^2 - \left(\frac{1-2k}{2a(t)\rho(t)g^2(t)}\right) \\
&= \frac{1}{4a(t)\rho(t)}\left[4\psi^2(t) + \frac{4(1-2k)}{g(t)}\psi(t) + \left(\frac{1-2k}{g(t)}\right)^2 - \frac{2(1-2k)}{g^2(t)}\right] \\
&\leq \frac{1}{4a(t)\rho(t)g^2(t)}\left[4k^2 + 4k(1-2k) + (1-2k)^2 - 2(1-2k)\right] \\
&= \frac{4k-1}{4a(t)\rho(t)g^2(t)} \leq 0 \quad \text{for} \quad t \geq T.
\end{aligned}
$$

Now, it follows from Theorem 2.3.2 that (2.1.1) is nonoscillatory.   ∎

For the special case of (2.1.1), namely, the equation

$$x''(t) + q(t)x(t) = 0, \tag{2.3.5}$$

where $q(t) \in C([t_0, \infty), \mathbb{R}^+)$, Corollary 2.3.3 reduces to the following result.

Corollary 2.3.4. If $\int_t^\infty q(s)ds < \infty$ and $\limsup_{t\to\infty} t\int_t^\infty q(s)ds < 1/4$, then equation (2.3.5) is nonoscillatory.

Proof. Apply Corollary 2.3.3 with $a(t) = 1$, $\rho(t) = 1$, $g(t) = t$ and $\psi(t) = \int_t^\infty q(s)ds$.   ∎

Similarly, we have the following corollary.

Corollary 2.3.5. Let $g(t) \in C^1([t_0, \infty), \mathbb{R}^+)$ and $\psi(t) \in C^1([t_0, \infty), \mathbb{R})$. Suppose $g(t)$ and $\psi(t)$ satisfy $g'(t) \leq -1/(a(t)\rho(t))$ and $\psi'(t) \geq -Q(t)$ for $t \geq t_0$. If $\limsup_{t\to\infty} g(t)|\psi(t)| < 1/4$, then equation (2.1.1) is nonoscillatory.

Corollary 2.3.6. Let $\rho(t) \in C^2([t_0, \infty), \mathbb{R}^+)$, $g(t) \in C^1([t_0, \infty), \mathbb{R}^+)$ and $\psi(t) \in C^1([t_0, \infty), \mathbb{R})$. Suppose $g(t)$ and $\psi(t)$ satisfy $g'(t) \geq 1/(a(t)\rho(t))$ and $\psi'(t) \leq -Q(t)$ for $t \geq t_0$. If there exists a number $c > 0$ such that $-\sqrt{c} - c \leq g(t)\psi(t) \leq \sqrt{c} - c \leq 1/4$, then equation (2.1.1) is nonoscillatory.

Proof. Let
$$v(t) = -\frac{1}{a(t)\rho(t)}\left(\psi(t) + \frac{c}{g(t)}\right).$$

Then, we have
$$(a(t)\rho(t)v(t))' = -\psi'(t) + c\frac{g'(t)}{g^2(t)} \geq Q(t) + \frac{c}{a(t)\rho(t)g^2(t)}$$

and
$$Q(t) + a(t)\rho(t)v^2(t) - (a(t)\rho(t)v(t))'$$
$$\leq \frac{1}{a(t)\rho(t)}\left(\psi(t) + \frac{c}{g(t)}\right)^2 - \frac{c}{a(t)\rho(t)g^2(t)}$$
$$\leq \frac{1}{a(t)\rho(t)}\left(\frac{c}{g^2(t)} - \frac{c}{g^2(t)}\right) = 0 \quad \text{for} \quad t \geq t_0.$$

Now, it follows from Theorem 2.3.2 that (2.1.1) is nonoscillatory.   ∎

Corollary 2.3.7. If there exists a number $c > 0$ such that
$$-\sqrt{c} - c \leq \left(1 + \int_{t_0}^t \frac{1}{a(s)}ds\right)\left(\int_t^\infty q(s)ds\right) \leq \sqrt{c} - c \leq \frac{1}{4}$$

and $\int_t^\infty q(s)ds < \infty$, then equation (2.1.1) is nonoscillatory.

Proof. Apply Corollary 2.3.6 with $\rho(t) = 1$, $g(t) = 1 + \int_{t_0}^t 1/a(s)ds$ and $\psi(t) = \int_t^\infty q(s)ds$.   ∎

Example 2.3.3. Consider equation (2.3.1). Let $\rho(t) = t(\ln t)^\lambda$, where $\lambda < 1$ is a constant. Then, we find
$$h(t) = -\frac{1}{2}\left(\frac{1}{t} + \frac{\lambda}{t \ln t}\right) \quad \text{and} \quad Q(t) = (\lambda^2 - 2\lambda + 4c)\frac{(\ln t)^{\lambda-2}}{4t}.$$

Let
$$g(t) = \frac{1}{1-\lambda}(\ln t)^{1-\lambda} \quad \text{and} \quad \psi(t) = \frac{\lambda^2 - 2\lambda + 4c}{4(1-\lambda)}(\ln t)^{\lambda-1}.$$

Then, we have $g'(t) = 1/(a(t)\rho(t))$, $\psi'(t) = -Q(t)$ and
$$g(t)\psi(t) = \frac{\lambda^2 - 2\lambda + 4c}{4(1-\lambda)^2} \leq \frac{1}{4} \quad \text{if} \quad c \leq \frac{1}{4}.$$

Now, it follows from Corollary 2.3.6 that (2.3.1) is nonoscillatory if $c \leq 1/4$.   ∎

A result similar to that of Corollary 2.3.6 is the following:

Corollary 2.3.8. Let $\rho(t) \in C^2([t_0, \infty), \mathbb{R}^+)$, $g(t) \in C^1([t_0, \infty), \mathbb{R}^+)$ and $\psi(t) \in C^1([t_0, \infty), \mathbb{R})$. Suppose $g(t)$ and $\psi(t)$ satisfy $g'(t) \leq -1/a(t)\rho(t))$ and $\psi'(t) \geq -Q(t)$ for $t \geq t_0$. If there exists a number $c > 0$ such that $-\sqrt{c} - c \leq g(t)\psi(t) \leq \sqrt{c} - c \leq 1/4$, then equation (2.1.1) is nonoscillatory.

Corollary 2.3.9. Let $\rho(t) \in C^2([t_0, \infty), \mathbb{R}^+)$, $g(t) \in C^1([t_0, \infty), \mathbb{R}^+)$ and $\psi(t) \in C^1([t_0, \infty), \mathbb{R})$. If $\lim_{t \to \infty} \psi(t)$ exists, $g'(t) \geq 1/(a(t)\rho(t))$ and $\psi'(t) \leq -g(t)Q(t)$ for $t \geq t_0$, then equation (2.1.1) is nonoscillatory.

Proof. Since $\lim_{t \to \infty} \psi(t)$ exists, there are numbers $T \geq t_0$ and m such that $0 < m + \psi(t) \leq 1$ for $t \geq T$. Let $v(t) = (m + \psi(t))/(a(t)\rho(t)g(t))$ for $t \geq T$. Then, we have

$$(a(t)\rho(t)v(t))' = \frac{g'(t)(m + \psi(t))}{g^2(t)} - \frac{\psi'(t)}{g(t)} \geq Q(t) + \frac{m + \psi(t)}{a(t)\rho(t)g^2(t)}$$

and for $t \geq T$,

$$Q(t) + a(t)\rho(t)v^2(t) - (a(t)\rho(t)v(t))'$$
$$\leq \frac{(m + \psi(t))^2}{a(t)\rho(t)g^2(t)} - \frac{m + \psi(t)}{a(t)\rho(t)g^2(t)} \leq \frac{m + \psi(t)}{a(t)\rho(t)g^2(t)} - \frac{m + \psi(t)}{a(t)\rho(t)g^2(t)} = 0.$$

Now, it follows from Theorem 2.3.2 that (2.1.1) is nonoscillatory. ∎

Corollary 2.3.10. Let $\rho(t) \in C^2([t_0, \infty), \mathbb{R}^+)$, $g(t) \in C^1([t_0, \infty), \mathbb{R}^+)$ and $\psi(t) \in C^1([t_0, \infty), \mathbb{R})$. If $\lim_{t \to \infty} \psi(t)$ exists, $g'(t) \leq -1/(a(t)\rho(t))$ and $\psi'(t) \geq -g(t)Q(t)$ for $t \geq t_0$, then equation (2.1.1) is nonoscillatory.

Next, we present the following results.

Corollary 2.3.11. Let $\rho(t) \in C^2([t_0, \infty), \mathbb{R}^+)$ and $g(t) \in C^1([t_0, \infty), \mathbb{R}^+)$ satisfy $Q(t) \leq 1/g^2(t)$ for $t \geq t_0$. If either $g'(t) - (1/(a(t)\rho(t))) \geq 1$ for $t \geq t_0$, or

$$\lim_{t \to \infty} \left(g'(t) - \frac{1}{a(t)\rho(t)} \right) = m \quad \text{exists}, \ m > 1 \ \text{ is a constant,}$$

then equation (2.1.1) is nonoscillatory.

Proof. It follows from the hypothesis that there is a number $T \geq t_0$ such that $g'(t) - (1/(a(t)\rho(t))) \geq 1$ for $t \geq T$. Let $v(t) = -1/(a(t)\rho(t)g(t))$, $t \geq T$. Then for $t \geq T$, we have

$$Q(t) + a(t)\rho(t)v(t) - (a(t)\rho(t)v(t))'$$
$$= \frac{1}{g^2(t)} + \frac{1}{a(t)\rho(t)g^2(t)} - \frac{g'(t)}{g^2(t)} = \frac{1}{g^2(t)} \left(1 + \frac{1}{a(t)\rho(t)} - g'(t) \right) \leq 0.$$

Now, by Theorem 2.3.2 equation (2.1.1) is nonoscillatory. ∎

Example 2.3.4. Consider the differential equation (2.3.1) with $c \le 1/4$. Let $\rho(t) = t \ln t$ for $t \ge t_0 > 1$. Then for $t \ge t_0$, we have

$$h(t) = -\frac{\rho'(t)}{2\rho(t)} = -\frac{1}{2}\left(\frac{1}{t} + \frac{1}{t \ln t}\right) \quad \text{and} \quad Q(t) = \frac{4c-1}{4t \ln t} \le 0.$$

If $g(t) = t + \ln \ln t$, $t > e$, then $Q(t) \le 1/g^2(t)$ and $g'(t) - (1/(t \ln t)) = 1$. Thus, by Corollary 2.3.11 equation (2.3.1) is nonoscillatory.

Corollary 2.3.12. Let $\rho(t) \in C^2([t_0, \infty), \mathbb{R}^+)$ and $\psi(t) \in C^1([t_0, \infty), \mathbb{R})$. Suppose $\psi(t)$ satisfies $\psi'(t) \le -Q(t)$ for $t \ge t_0$. If

$$\int_t^\infty \frac{\psi^2(s)}{a(s)\rho(s)}\,ds \le \frac{1}{4}\psi(t),$$

then equation (2.1.1) is nonoscillatory.

Proof. Let

$$v(t) = -\frac{1}{a(t)\rho(t)}\left[\psi(t) + 4\int_t^\infty \frac{\psi^2(s)}{a(s)\rho(s)}\,ds\right].$$

Then, we have $(a(t)\rho(t)v(t))' \ge Q(t) + 4\psi^2(t)/(a(t)\rho(t))$ and

$$Q(t) + a(t)\rho(t)v^2(t) - (a(t)\rho(t)v(t))'$$
$$\le \frac{1}{a(t)\rho(t)}\left[\psi(t) + 4\int_t^\infty \frac{\psi^2(s)}{a(s)\rho(s)}\,ds\right]^2 - \frac{4\psi^2(t)}{a(t)\rho(t)}$$
$$= \frac{1}{a(t)\rho(t)}\left[\psi^2(t) + 8\psi(t)\int_t^\infty \frac{\psi^2(s)}{a(s)\rho(s)}\,ds + 16\left(\int_t^\infty \frac{\psi^2(s)}{a(s)\rho(s)}\,ds\right)^2 - 4\psi^2(t)\right]$$
$$\le \frac{1}{a(t)\rho(t)}\left[\psi^2(t) + 2\psi^2(t) + \psi^2(t) - 4\psi^2(t)\right] = 0 \quad \text{for } t \ge t_0.$$

Now, it follows from Theorem 2.3.2 that (2.1.1) is nonoscillatory. ∎

Example 2.3.5. Consider the differential equation (2.3.1) with $c \le 1/4$. Let $\rho(t) = t(\ln t)^\lambda$, $t \ge t_0 > 1$ where $\lambda < 1 - \sqrt{1 - 4c}$ is a constant. Then, we have

$$h(t) = -\frac{\rho'(t)}{2\rho(t)} = -\frac{1}{2}\left(\frac{1}{t} + \frac{\lambda}{t \ln t}\right) \quad \text{and} \quad Q(t) = \frac{(\lambda^2 - 2\lambda + 4c)}{4t}(\ln t)^{\lambda-2}.$$

Let

$$\psi(t) = \frac{(\lambda^2 - 2\lambda + 4c)}{4(1 - \lambda)}(\ln t)^{\lambda-1}.$$

Then, $\psi'(t) = -Q(t)$ and

$$\int_t^\infty \frac{\psi^2(s)}{a(s)\rho(s)} ds = \frac{\lambda^2 - 2\lambda + 4c}{4(1-\lambda)^2} \psi(t) \le \frac{1}{4}\psi(t).$$

It follows from Corollary 2.3.12 that equation (2.3.1) is nonoscillatory if $c \le 1/4$.

Corollary 2.3.13. Let $\rho(t) \in C^2([t_0,\infty), \mathbb{R}^+)$ and $\psi(t) \in C^1([t_0,\infty), \mathbb{R})$. Suppose $\psi'(t) \le -Q(t)$ for $t \ge t_0$, and

$$\psi_1(t) = \int_t^\infty \frac{\psi^2(s)}{a(s)\rho(s)} \exp\left(2\int_t^s \frac{\psi(\tau)}{a(\tau)\rho(\tau)} d\tau\right) ds.$$

If $\psi_1(t) \le (1/2)\psi(t)$, then equation (2.1.1) is nonoscillatory.

Proof. Let $v(t) = -(\psi(t) + 2\psi_1(t))/(a(t)\rho(t))$, $t \ge t_0$. Then, we have

$$(a(t)\rho(t)v(t))' \ge Q(t) + \frac{4\psi(t)\psi_1(t) + 2\psi^2(t)}{a(t)\rho(t)}$$

and

$$Q(t) + a(t)\rho(t)v^2(t) - (a(t)\rho(t)v(t))'$$
$$\le \frac{(\psi(t) + 2\psi_1(t))^2}{a(t)\rho(t)} - \left(\frac{4\psi(t)\psi_1(t) + 2\psi^2(t)}{a(t)\rho(t)}\right)$$
$$= \frac{1}{a(t)\rho(t)} \left[\psi^2(t) + 4\psi(t)\psi_1(t) + 4\psi_1^2(t) - 4\psi(t)\psi_1(t) - 2\psi^2(t)\right]$$
$$\le \frac{1}{a(t)\rho(t)} \left[4\left(\frac{1}{2}\psi(t)\right)^2 - \psi^2(t)\right] = 0 \quad \text{for } t \ge t_0.$$

Now, it follows from Theorem 2.3.2 that (2.1.1) is nonoscillatory. ∎

Corollary 2.3.14. Let the functions $\rho(t)$, $\psi(t)$ and $\psi_1(t)$ be as in Corollary 2.3.13. If $\psi_1(t) < \infty$ for $t \ge t_0$, and

$$\psi_2(t) = \int_t^\infty \frac{\psi_1^2(s)}{a(s)\rho(s)} \exp\left(2\int_t^s \frac{\psi(\tau)}{a(\tau)\rho(\tau)} d\tau\right) ds \le \frac{1}{4}\psi_1(t), \quad t \ge t_0$$

then equation (2.1.1) is nonoscillatory.

Proof. Let $v(t) = -\left[\psi(t) + \psi_1(t) + 4\psi_2(t)\right]/(a(t)\rho(t))$. Then, we have

$$(a(t)\rho(t)v(t))' \ge Q(t) + \frac{1}{a(t)\rho(t)} \left[2\psi(t)\psi_1(t) + \psi^2(t) + 8\psi(t)\psi_2(t) + 4\psi_1^2(t)\right]$$

and for $t \geq t_0$,

$$Q(t) + a(t)\rho(t)v^2(t) - (a(t)\rho(t)v(t))'$$

$$\leq \frac{1}{a(t)\rho(t)} \left\{ [\psi(t) + \psi_1(t) + 4\psi_2(t)]^2 - [2\psi(t)\psi_1(t) + \psi^2(t) \right.$$

$$\left. + 8\psi(t)\psi_2(t) + 4\psi_1^2(t)] \right\}$$

$$= \frac{1}{a(t)\rho(t)} \left[\psi_1^2(t) + 8\psi_1(t)\psi_2(t) + 16\psi_2^2(t) - 4\psi_1^2(t) \right]$$

$$\leq \frac{1}{a(t)\rho(t)} \left[-3\psi_1^2(t) + 8\psi_1(t) \left(\frac{1}{4}\psi_1(t) \right) + 16 \left(\frac{1}{16}\psi_1^2(t) \right) \right] = 0.$$

Now, it follows from Theorem 2.3.2 that (2.1.1) is nonoscillatory.   ■

The following theorem provides a variant of inequality (2.2.16) in Theorem 2.2.3.

Theorem 2.3.3.   Equation (2.1.1) is nonoscillatory if and only if there exists a function $c(t) \in C([T, \infty), \mathbb{R})$ for some $T \geq t_0$ such that

$$|\beta(t) + c(t)| \geq \left| \beta(t) + \int_t^\infty \frac{[\beta(s) + c(s)]^2}{a(s)\rho(s)} ds \right| \quad \text{for } t \geq T, \qquad (2.3.6)$$

where the function $\beta(t)$ is defined in (2.2.14).

As an immediate consequence of this result, we have the following corollary.

Corollary 2.3.15.   If $\beta(t) \geq 0$ for $t \geq t_0$ and there exists a function $c(t) \in C([T, \infty), \mathbb{R})$, $T \geq t_0$ such that

$$\int_t^\infty \frac{[\beta(s) + c(s)]^2}{a(s)\rho(s)} ds \leq c(t) \quad \text{for } t \geq T,$$

then equation (2.1.1) is nonoscillatory.

Now, we let $c_1(t) \in C([t_0, \infty), \mathbb{R})$ and $c(t) = c_1^+(t) - \beta(t)$, where $c_1^+(t) = \max\{0, c(t)\}$. Then by Theorem 2.3.3 the following corollary is immediate.

Corollary 2.3.16.   If there exists a function $c_1(t) \in C([T, \infty), \mathbb{R})$ for some $T \geq t_0$ such that

$$c_1^+(t) \geq \left| \beta(t) + \int_t^\infty \frac{(c_1^+(s))^2}{a(s)\rho(s)} ds \right| \quad \text{for } t \geq T,$$

then equation (2.1.1) is nonoscillatory.

Corollary 2.3.17. If

$$\int_t^\infty \frac{\beta^2(s)}{a(s)\rho(s)}\,ds \;\leq\; \frac{1}{4}|\beta(t)|, \quad t \geq t_0$$

then equation (2.1.1) is nonoscillatory.

Proof. It follows from Corollary 2.3.16 by taking $c_1(t) = 2|\beta(t)|$. ∎

Now, let $\phi(t) \in C^1([t_0,\infty),\mathbb{R})$ and $\rho(t) \in C^2([t_0,\infty),\mathbb{R}^+)$ satisfy $\phi'(t) = 1/(a(t)\rho(t))$, and let $c(t) = (\sqrt{k}/\phi(t)) - \beta(t)$, where $k > 0$ is a constant. Then, $\lim_{t\to\infty}\phi(t) = \infty$, $\beta(t) + c(t) = \sqrt{k}/\phi(t)$ and

$$\left| \beta(t) + \int_t^\infty \frac{[\beta(s) + c(s)]^2}{a(s)\rho(s)}\,ds \right| \;=\; \left| \beta(t) + \frac{k}{\phi(t)} \right|.$$

Thus, we have the following corollary.

Corollary 2.3.18. If there exists a constant $k > 0$ such that $-\sqrt{k} - k \leq \phi(t)\beta(t) \leq \sqrt{k} - k \leq 1/4$, then equation (2.1.1) is nonoscillatory.

Corollary 2.3.19. If there exists a constant $k > 0$ such that

$$-\sqrt{k} - k \;\leq\; \left(1 + \int_{t_0}^t \frac{1}{a(s)}ds\right)\left(\int_t^\infty q(s)ds\right) \;\leq\; \sqrt{k} - k \;\leq\; \frac{1}{4},$$

then equation (2.1.1) is nonoscillatory.

Proof. It follows from Corollary 2.3.18 by letting $\rho(t) = 1$ and $\phi(t) = 1 + \int_{t_0}^t 1/a(s)ds$. ∎

Corollary 2.3.20. Let

$$\xi(t) \;=\; \int_t^\infty \frac{\beta^2(s)}{a(s)\rho(s)} \exp\left(\lambda \int_t^s \frac{\beta(\tau)}{a(\tau)\rho(\tau)}d\tau\right) ds,$$

where λ is a constant. If one of the following statements holds:

(i) $\lambda > 1$ and $\dfrac{\beta(t)}{\xi(t)} \geq \dfrac{\lambda}{\lambda - 1}$,

(ii) $\lambda > 1$ and $\dfrac{\beta(t)}{\xi(t)} \leq -\lambda$,

(iii) $\lambda = 1$ and $\beta(t) \leq -\xi(t)$,

(iv) $\lambda < 1$ and $\dfrac{\lambda}{\lambda - 1} \leq \dfrac{\beta(t)}{\xi(t)} \leq -\lambda$,

then equation (2.1.1) is nonoscillatory.

Proof. Let $c(t) = (\lambda\beta(t)[\lambda\xi(t) + \beta(t)])^{1/2} - \beta(t)$. Then, we have

$$\left| \beta(t) + \int_t^\infty \frac{[\beta(s) + c(s)]^2}{a(s)\rho(s)} ds \right| = |\beta(t) + \lambda\xi(t)|$$

and

$$[\beta(t) + c(t)]^2 = \lambda\beta(t)[\lambda\xi(t) + \beta(t)] \geq [\lambda\xi(t) + \beta(t)]^2,$$

which implies (2.3.6). Now, it follows from Theorem 2.3.3 that (2.1.1) is nonoscillatory. ∎

Finally, in this section we state the following results.

Theorem 2.3.4. Let $\xi(t)$ and $\mu[s,t]$ be as in Theorem 2.2.4. Then equation (2.1.1) is nonoscillatory if and only if there exists a function $c(t) \in C([T,\infty), \mathbb{R})$ for some $T \geq t_0$ such that

$$c(t) \geq \int_t^\infty \frac{[\xi(s) + c(s)]^2}{a(s)\rho(s)}\mu[s,t]ds.$$

Theorem 2.3.5. Let $\xi^*(t)$ and $\mu^*[s,t]$ be as in Theorem 2.2.5. Then equation (2.1.1) is nonoscillatory if and only if there exists a function $c^*(t) \in C([T,\infty), \mathbb{R})$ for some $T \geq t_0$ such that

$$c^*(t) \geq \int_t^\infty \frac{[\xi^*(s) + c^*(s)]^2}{a(s)\rho(s)}\mu^*[s,t]ds.$$

Taking $c(t) = \xi(t)$ and $c^*(t) = \xi^*(t)$ in Theorems 2.3.4 and 2.3.5 respectively, we obtain the following corollary.

Corollary 2.3.21. Equation (2.1.1) is nonoscillatory if

$$\frac{1}{4}\xi(t) \geq \int_t^\infty \frac{\xi^2(s)}{a(s)\rho(s)}\mu[s,t]ds, \quad \text{or} \quad \frac{1}{4}\xi^*(t) \geq \int_t^\infty \frac{\xi^{*2}(s)}{a(s)\rho(s)}\mu^*[s,t]ds.$$

2.4. Oscillation Criteria

In this section, we shall begin with some well–known oscillation criteria for second order differential equations of the type

$$(a(t)x'(t))' + q(t)f(x(t)) = 0, \tag{2.4.1}$$

where $a(t) \in C([t_0,\infty), \mathbb{R}^+)$, $q(t) \in C([t_0,\infty), \mathbb{R})$, $t_0 \geq 0$ and $f \in C(\mathbb{R}, \mathbb{R})$. In what follows we shall assume that

(i) $\int^\infty ds/a(s) = \infty$, and

(ii) $xf(x) > 0$ and $f'(x) \geq 0$ for $x \neq 0$.

Theorem 2.4.1. If (i) and (ii) hold, and

$$\int^{\infty} q(s)ds = \infty. \tag{2.4.2}$$

then equation (2.4.1) is oscillatory.

Proof. Suppose $x(t)$ is a nonoscillatory solution of (2.4.1), say, $x(t) \neq 0$ for $t \geq t_1 \geq t_0$. Define $w(t) = a(t)x'(t)/f(x(t))$ for $t \geq t_1$. Then for $t \geq t_1$, we have

$$w'(t) = -q(t) - \frac{f'(x(t))}{a(t)}w^2(t) \leq -q(t). \tag{2.4.3}$$

Integrating (2.4.3) from t_1 to t, we obtain $w(t) \leq w(t_1) - \int_{t_1}^t q(s)ds$. Now, assume that $x(t) > 0$ for $t \geq t_1$. The proof in the case $x(t) < 0$ for $t \geq t_1$ is similar and hence omitted. By condition (2.4.2) there exists $t_2 \geq t_1$ such that $x'(t) < 0$ for $t \geq t_2$. Condition (2.4.2) also implies that there exists a $T \geq t_2$ such that $\int_{t_2}^T q(s)ds = 0$ and $\int_T^t q(s)ds \geq 0$ for $t \geq T$. Integrating equation (2.4.1) by parts from T to t, we find

$$\begin{aligned} a(t)x'(t) &= a(T)x'(T) - \int_T^t q(s)f(x(s))ds \\ &= a(T)x'(T) - f(x(t))\int_T^t q(s)ds \\ &\quad + \int_T^t f'(x(s))x'(s)\left(\int_T^s q(u)du\right)ds \leq a(T)x'(T). \end{aligned}$$

Thus, it follows that

$$x(t) \leq x(T) + a(T)x'(T)\int_T^t \frac{1}{a(s)}ds \quad \to -\infty \quad \text{as } t \to \infty,$$

which is a contradiction. ∎

Remark 2.4.1. Theorem 2.4.1 is applicable to the Emden–Fowler equation, namely,

$$(a(t)x'(t))' + q(t)|x(t)|^\gamma \text{sgn } x(t) = 0, \tag{2.4.4}$$

where $\gamma > 0$ is a real constant.

When $\gamma = 1$ in (2.4.4), we can employ Lemma 2.2.2 to obtain the following improvement of Theorem 2.4.1.

Theorem 2.4.2. Suppose there exists a function $\rho(t) \in C^2([t_0, \infty), \mathbb{R})$ such that

$$\int^\infty \frac{1}{a(s)\rho(s)} ds = \infty \tag{2.4.5}$$

and

$$\int^\infty Q(s) ds = \infty, \tag{2.4.6}$$

where $Q(t)$ is defined in (2.2.7), then equation (2.1.1) is oscillatory.

Remark 2.4.2. We can employ Theorem 2.2.1 to extend Theorem 2.4.2 to more general equations of type (2.2.1). The details are left to the reader.

Example 2.4.1. Consider the differential equation (2.3.1). It is easy to check that Theorem 2.4.1 is not applicable to (2.3.1), in fact, we have

$$\int^\infty \left[\frac{1}{4s^2} + \frac{c}{(t \ln t)^2} \right] ds < \infty, \quad \text{if } c > \frac{1}{4}.$$

However, if we take $\rho(t) = t \ln t$, then

$$h(t) = -\frac{\rho'(t)}{2\rho(t)} = -\left[\frac{1}{2t} + \frac{1}{2t \ln t} \right], \quad t > 1$$

$$\int^\infty \frac{1}{\rho(s)} ds = \int^\infty \frac{1}{s \ln s} ds = \infty$$

and

$$\int^\infty Q(s) ds = \int^\infty \rho(s) \left[q(s) + h^2(s) - h'(s) \right] ds$$

$$= \int^\infty \frac{4c - 1}{4s \ln s} ds = \infty \quad \text{if } c > \frac{1}{4}.$$

Thus all conditions of Theorem 2.4.2 are satisfied if $c > 1/4$, and hence (2.3.1) is oscillatory if $c > 1/4$.

As an immediate consequence of Theorems 2.2.4 and 2.2.5, we have the following interesting criterion for the oscillation of (2.1.1).

Corollary 2.4.1. Let $\xi(t)$ be defined in (2.2.23). If

$$\int^\infty \frac{\beta^2(s)}{a(s)\rho(s)} \exp \left(2 \int^s \frac{\beta(\tau)}{a(\tau)\rho(\tau)} d\tau \right) ds = \infty,$$

or

$$\int^\infty \frac{\xi^2(s)}{a(s)\rho(s)} \exp \left(2 \int^s \frac{\beta(\tau) + \xi(\tau)}{a(\tau)\rho(\tau)} d\tau \right) ds = \infty,$$

then equation (2.1.1) is oscillatory.

Example 2.4.2. Consider the Euler differential equation (2.2.11) for $c > 1/4$. We let $\rho(t) = t^\alpha$, where $1 - \sqrt{4c-1} \le \alpha < 1$. Then, we have

$$\beta(t) = \frac{\alpha^2 - 2\alpha + 4c}{1-\alpha} t^{\alpha-1} \quad \text{and} \quad \int^\infty \frac{\beta^2(s)}{\rho(s)} \exp\left(2\int^s \frac{\beta(\tau)}{\rho(\tau)} d\tau\right) ds = \infty.$$

It follows from Corollary 2.4.1 that (2.2.11) is oscillatory if $c > 1/4$.

Example 2.4.3. Consider the following differential equation

$$x''(t) + \left(\frac{\alpha \sin \gamma t}{t^\sigma} + \frac{c}{t^2}\right) x(t) = 0, \tag{2.4.7}$$

where $\alpha, \gamma \ne 0$, $\sigma > 0$ and c are constants. Let

$$\rho(t) = \exp\left(-2\int^t \left[\frac{\lambda}{s} - \frac{\alpha \cos \gamma s}{\gamma s^\sigma}\right] ds\right),$$

where $\lambda > \max\{-1/2, (1-2\sigma)/2\}$. Then, $\rho(t) = t^{-2\lambda} + O\left(t^{-2\lambda-\sigma}\right)$ as $t \to \infty$,

$$h(t) = -\frac{\rho'(t)}{2\rho(t)} = \frac{\lambda}{t} - \frac{\alpha \cos \gamma t}{t^\sigma}$$

and as $t \to \infty$,

$$Q(t) = \left[t^{-2\lambda} + O\left(t^{-2\lambda-\sigma}\right)\right]\left(\frac{\lambda^2 + \lambda + c}{t^2} + \frac{\alpha^2}{2\gamma^2 t^{2\sigma}} + \frac{\alpha^2 \cos 2\gamma t}{2\gamma^2 t^{2\sigma}}\right.$$
$$\left. - \frac{\alpha(2\lambda+\gamma)\cos \gamma t}{\gamma t^{\sigma+1}}\right).$$

We consider the following three cases:

Case (i). If $0 < \sigma < 1$, we let $(1-2\sigma)/2 < \lambda \le 1/2$. Then as $t \to \infty$,

$$\beta(t) = \frac{\alpha^2}{2\gamma^2(2\lambda + 2\sigma - 1)} t^{-2\sigma-2\lambda+1} + O\left(t^{-2\lambda-1}\right),$$

$$\exp\left(2\int^t \frac{\beta(s)}{\rho(s)} ds\right) = \exp\left(\frac{\alpha^2}{2\gamma^2(2\lambda+2\sigma-1)(2-2\sigma)} t^{2-2\sigma}\right)\left[1 + O\left(t^{-2\lambda}\right)\right]$$

and

$$\int^\infty \frac{\beta^2(s)}{\rho(s)} \exp\left(2\int^s \frac{\beta(\tau)}{\rho(\tau)} d\tau\right) ds = \infty.$$

It follows from Corollary 2.4.1 that (2.4.7) is oscillatory if $0 < c < 1$.

Case (ii). If $\sigma = 1$ and $c > (1/4) - (\alpha^2/2\gamma^2)$, we let

$$-\frac{1}{2} < \lambda \le \min\left\{\frac{1}{2}, \ -\frac{1}{2} + \sqrt{c + \frac{\alpha^2}{2\gamma^2} - \frac{1}{4}}\right\}.$$

Then,

$$\beta(t) = \frac{2\gamma^2(\lambda^2 + \lambda + c) + \alpha^2}{2\gamma^2(1 + 2\lambda)}t^{-2\lambda-1} + O\left(t^{-2\lambda-2}\right) \quad \text{as} \quad t \to \infty,$$

$$\exp\left(2\int^t \frac{\beta(s)}{\rho(s)}ds\right) = t^{(2\gamma^2(\lambda^2+\lambda+c)+\alpha^2)/\gamma^2(1+2\lambda)}\left[1 + O\left(t^{-2\lambda-1}\right)\right] \text{ as } t \to \infty$$

and

$$\int^\infty \frac{\beta^2(s)}{\rho(s)}\exp\left(2\int^s \frac{\beta(\tau)}{\rho(\tau)}d\tau\right)ds = \infty.$$

It follows from Corollary 2.4.1 that (2.4.7) is oscillatory if $\sigma = 1$ and $c > (1/4) - (\alpha^2/2\gamma^2)$.

Case (iii). If $\sigma > 1$ and $c > 1/4$, we let

$$-\frac{1}{2} < \lambda \le \min\left\{\frac{1}{2}, \ -\frac{1}{2} + \sqrt{c - \frac{1}{4}}\right\}.$$

Then,

$$\beta(t) = \frac{\lambda^2 + \lambda + c}{1 + 2\lambda}t^{-2\lambda-1} + O\left(t^{-2\lambda-2}\right) \quad \text{as} \quad t \to \infty,$$

$$\exp\left(2\int^t \frac{\beta(s)}{\rho(s)}ds\right) = t^{2(\lambda^2+\lambda+c)/(1+2\lambda)}\left[1 + O\left(t^{-2\lambda-1}\right)\right] \quad \text{as} \quad t \to \infty$$

and

$$\int^\infty \frac{\beta^2(s)}{\rho(s)}\exp\left(2\int^s \frac{\beta(\tau)}{\rho(\tau)}d\tau\right)ds = \infty.$$

It follows from Corollary 2.4.1 that (2.4.7) is oscillatory if $\sigma > 1$ and $c > (1/4)$.

Under an additional condition on $q(t)$, (2.4.2) can be replaced by

$$\limsup_{t\to\infty} \int_T^t q(s)ds = \infty \quad \text{for all large} \ T \ge t_0. \tag{2.4.8}$$

In fact, we have the following result.

Theorem 2.4.3. Assume that (i) and (ii) are satisfied. If

$$\liminf_{t\to\infty} \int_T^t q(s)ds \ge 0 \quad \text{for all large} \ T \ge t_0 \tag{2.4.9}$$

and condition (2.4.8) holds, then equation (2.4.1) is oscillatory.

Proof. Let $x(t)$ be a nonoscillatory solution of equation (2.4.1), say, $x(t) > 0$ for $t \ge t_1 \ge t_0$. As in Theorem 2.4.1, we obtain (2.4.3). Next, we consider the following three cases:

Case 1. Suppose $x'(t)$ changes signs. Then there exists a sequence $\{T_n\}_{n=1}^{\infty}$ with $\lim_{n\to\infty} T_n = \infty$ such that $x'(T_n) = 0$. Choose N large enough so that (2.4.9) holds. We then have

$$\frac{a(t)x'(t)}{f(x(t))} \leq \frac{a(T_N)x'(T_N)}{f(x(T_N))} - \int_{T_N}^{t} q(s)ds,$$

so

$$\limsup_{t\to\infty} \frac{a(t)x'(t)}{f(x(t))} \leq \frac{a(T_N)x'(T_N)}{f(x(T_N))} + \limsup_{t\to\infty} \left(-\int_{T_N}^{t} q(s)ds \right) < 0,$$

which contradicts the fact that $x'(t)$ oscillates.

Case 2. Suppose $x'(t) > 0$ for all $t \geq t_2 \geq t_1$, then from (2.4.3), we have

$$\frac{a(t)x'(t)}{f(x(t))} - \frac{a(t_2)x'(t_2)}{f(x(t_2))} + \int_{t_2}^{t} q(s)ds \leq 0,$$

and from (2.4.8), we obtain $\liminf_{t\to\infty} a(t)x'(t)/f(x(t)) = -\infty$, which is a contradiction.

Case 3. Suppose $x'(t) < 0$ for $t \geq t_3 \geq t_0$. Condition (2.4.9) implies that for any $t_4 \geq t_3$, there exists a $T \geq t_4$ such that $\int_T^t q(s)ds \geq 0$ for all $t \geq T$. Choosing $T \geq t_3$ as indicated, and then integrating (2.4.1), we obtain

$$a(t)x'(t) \leq a(T)x'(T) - \int_T^t q(s)f(x(s))ds$$

$$= a(T)x'(T) - f(x(t))\int_T^t q(s)ds + \int_T^t f'(x(s))x'(s)\int_T^s q(u)du\,ds$$

$$\leq a(T)x'(T).$$

Thus, $x(t) \leq x(T) + a(T)x'(T) \int_T^t 1/a(s)ds \to -\infty$ as $t \to \infty$. But, this contradicts the fact that $x(t) > 0$ for $t \geq t_1$.

2.5. Oscillation Criteria–Integral Averaging

In this section we shall employ *averaging techniques* to study the oscillatory behavior of equations (2.1.1) and (2.2.1), so that the criteria we shall present will involve the average behavior of the integral of the alternating coefficients. The motivation of this criteria comes from the disconjugate property of (2.1.1) on $[t_0, \infty)$, $t_0 \geq 0$.

Theorem 2.5.1. If

$$\int^{\infty} \left(\int_{t_0}^{s} a(u)du \right)^{-1} ds = \infty \tag{2.5.1}$$

and

$$\lim_{t \to \infty} \frac{1}{t} \int_{t_0}^t \int_{t_0}^s q(u) du ds = \infty \qquad (2.5.2)$$

then equation (2.1.1) is oscillatory.

Proof. Let $x(t)$ be a nonoscillatory solution of equation (2.1.1), say, $x(t) > 0$ for $t \ge t_0 > 0$. Define $w(t) = a(t)x'(t)/x(t)$ for $t \ge t_0$. Then, we have

$$w'(t) = -q(t) - \frac{1}{a(t)} w^2(t) \qquad \text{for} \quad t \ge t_0. \qquad (2.5.3)$$

Integrating (2.5.3) twice from t_0 to t, we obtain

$$\int_{t_1}^t w(s) ds + \int_{t_0}^t \int_{t_0}^s \frac{w^2(u)}{a(u)} du du = t \left(k - \frac{1}{t} \int_{t_0}^t \int_{t_0}^s q(u) du ds \right),$$

where k is a constant. By (2.5.2), there exists a $t_1 \ge t_0$ such that

$$\int_{t_0}^t w(s) ds + \int_{t_0}^t \int_{t_0}^s \frac{w^2(u)}{a(u)} du du < 0 \quad \text{for} \quad t \ge t_1.$$

Thus for $t \ge t_1$ it follows that

$$F(t) = \int_{t_0}^t \int_{t_0}^s \frac{w^2(u)}{a(u)} du du < -\int_{t_0}^t w(s) ds.$$

Since $F(t)$ is nonnegative, we get

$$F^2(t) < \left(\int_{t_0}^t w(s) ds \right)^2. \qquad (2.5.4)$$

By Schwartz's inequality, we have

$$\left(\int_{t_0}^t w(s) ds \right)^2 \le \left(\int_{t_0}^t a(s) ds \right) \left(\int_{t_0}^t \frac{w^2(s)}{a(s)} ds \right) \qquad \text{for} \quad t \ge t_1.$$

It is easy to check that there exists a $t_2 \ge t_1$ such that $F(t) > 0$ for $t \ge t_2$, and hence

$$\left(\int_{t_0}^t a(s) ds \right)^{-1} \le F^{-2}(t) F'(t) \qquad \text{for} \quad t \ge t_2. \qquad (2.5.5)$$

Now, integrating (2.5.5) from t_2 to t, we get

$$\int_{t_2}^t \left(\int_{t_0}^s a(u) du \right)^{-1} ds \le \frac{1}{F(t_2)} < \infty,$$

which contradicts condition (2.5.1).   ∎

From Lemma 2.2.2 and Theorem 2.5.1 the following corollary is immediate.

Corollary 2.5.1. If there exists a function $\rho(t) \in C^2([t_0, \infty), \mathbb{R}^+)$ such that

$$\int^\infty \left(\int_{t_0}^s a(u)\rho(u)du \right)^{-1} ds = \infty \qquad (2.5.6)$$

and

$$\lim_{t \to \infty} \frac{1}{t} \int_{t_0}^t \int_{t_0}^s Q(u)duds = \infty, \qquad (2.5.7)$$

where $Q(t)$ is defined in (2.2.7), then equation (2.1.1) is oscillatory.

Similarly, from Theorems 2.2.1 and 2.5.1 the following corollary is immediate.

Corollary 2.5.2. If there exists a function $\rho(t) \in C^2([t_0, \infty), \mathbb{R}^+)$ such that condition (2.5.6) holds and

$$\lim_{t \to \infty} \frac{1}{t} \int_{t_0}^t \int_{t_0}^s Q^*(u)duds = \infty,$$

where $Q^*(t)$ is defined in (2.2.3), then equation (2.2.1) is oscillatory.

Example 2.5.1. Consider the differential equation

$$\left(\frac{1}{t}x'(t) \right)' + \frac{1}{t^2}x'(t) + \frac{c}{t^2}x(t) = 0, \quad t \geq 1 \qquad (2.5.8)$$

where c is any positive constant. Let $\rho(t) = t$. Then, $a(t)\rho'(t) - p(t)\rho(t) = 0$ and $Q^*(t) = \rho(t)q(t) = c/t$. Thus, all conditions of Corollary 2.5.2 are satisfied, and hence (2.5.8) is oscillatory for all $c > 0$.

Taking $\rho(t) = t$ in Corollary 2.5.1, it follows that the undamped equation associated with (2.5.8), namely, the equation

$$\left(\frac{1}{t}x'(t) \right)' + \frac{c}{t^2}x(t) = 0 \qquad (2.5.9)$$

is oscillatory, and hence the damping term in (2.5.8) preserves oscillations.

Remark 2.5.1. In Theorem 2.5.1 and Corollaries 2.5.1 and 2.5.2 if we let

$$a(t) \leq a_1, \qquad (2.5.10)$$

From equation (2.5.26), we have

$$v(\tau) = v(T) - \int_T^\tau Q(s)ds - \int_T^\tau \frac{v^2(s)}{a(s)\rho(s)}ds.$$

Since $g(t) \in \Omega$, (2.5.23) holds, and hence

$$A_g(\tau, t) = \frac{\int_T^t g(s)ds}{\int_\tau^t g(s)ds} A_g(T, t) - \int_T^\tau Q(s)ds - \frac{\int_T^\tau g(s) \int_T^s Q(u)duds}{\int_\tau^t g(s)ds}$$

$$= A_g(T, t) - \int_T^\tau Q(s)ds + o(1) \quad \text{as} \quad t \to \infty.$$

Thus, it follows that

$$v(\tau) - A_g(\tau, t) = v(T) - A_g(T, t) - \int_T^\tau \frac{v^2(s)}{a(s)\rho(s)}ds + o(1) \quad \text{as} \quad t \to \infty.$$
$$(2.5.31)$$

Since $g(t) \in \Omega$, there exists a positive number $\lambda > 0$ such that

$$\frac{1}{\lambda} < (1-k)\limsup_{t\to\infty} \left(\int^t g(s)ds\right)^{1-k} [G_k(\infty) - G_k(t)], \qquad (2.5.32)$$

where k is as in (2.5.21). It follows from (2.5.27), (2.5.29) and (2.5.31) that there exist two numbers t_1 and t_2 with $t_2 \geq t_1 \geq T$ such that

$$v(t_1) - A_g(t_1, t) \leq -\lambda \quad \text{for all} \quad t \geq t_2. \qquad (2.5.33)$$

Let $y(t) = \int_{t_1}^t g(s)v(s)ds$. Then Schwartz's inequality gives

$$\int_{t_1}^s \frac{v^2(u)}{a(u)\rho(u)}du \geq \frac{y^2(s)}{\int_{t_1}^s a(u)\rho(u)g^2(u)du}.$$

It follows from (2.5.30) and (2.5.33) that

$$y(t) \leq -\lambda \int_{t_1}^t g(s)ds - \int_{t_1}^t \frac{g(s)y^2(s)}{\int_{t_1}^s a(u)\rho(u)g^2(u)du}ds = -F(t). \quad (2.5.34)$$

Thus, we find

$$F'(t) = \lambda g(t) + \frac{g(t)y^2(t)}{\int_{t_1}^t a(s)\rho(s)g^2(s)ds} \qquad (2.5.35)$$

and

$$0 \leq \lambda \int_{t_1}^t g(s)ds \leq F(t) \leq |y(t)|. \qquad (2.5.36)$$

Now, for $t \geq t_1$ from (2.5.34) – (2.5.36), we obtain

$$F^{k-2}(t)F'(t) \geq \frac{F^k(t)F'(t)}{y^2(t)} \geq \lambda^k g(t) \left(\int_{t_1}^t g(s)ds \right)^k \left(\int_{t_1}^t a(s)\rho(s)g^2(s)ds \right)^{-1}.$$

Integrating the above inequality from $t \geq t_2$ to z and letting $z \to \infty$, and noting that $k \in [0, 1)$, we get $F^{k-1}(t) \geq (1 - k)\lambda^k[G_k(\infty) - G_k(t)]$. Finally, it follows from (2.5.36) that

$$\frac{1}{\lambda} \geq (1 - k) \left(\int_{t_1}^t g(s)ds \right)^{1-k} [G_k(\infty) - G_k(t)],$$

which contradicts (2.5.32). This completes the proof. ∎

Lemma 2.5.2. If equation (2.5.25) has a solution $v(t)$ satisfying (2.5.28), then for any $g(t) \in \Omega_0 = \Omega_0(g)$, $\lim_{t\to\infty} A_g(,.t)$ exists, and

$$\lim_{t\to\infty} A_g(\tau, t) = v(\tau) - \int_\tau^\infty \frac{v^2(s)}{a(s)\rho(s)}ds. \qquad (2.5.37)$$

Proof. As in Lemma 2.5.1, (2.5.30) holds and this implies that

$$A(\tau, t) = v(\tau) - \frac{\int_\tau^t g(s)v(s)ds}{\int_\tau^t g(s)ds} - \frac{\int_\tau^t g(s) \int_\tau^s \frac{v^2(u)}{a(u)\rho(u)}duds}{\int_\tau^t g(s)ds}. \qquad (2.5.38)$$

Since $g \in \Omega_0$, (2.5.23) holds. Thus, it follows that

$$\lim_{t\to\infty} \frac{\int_\tau^t g(s) \int_\tau^s \frac{v^2(u)}{a(u)\rho(u)}duds}{\int_\tau^t g(s)ds} < \infty.$$

Next, by Schwartz's inequality, we have

$$0 \leq \lim_{t\to\infty} \frac{\left| \int_\tau^t g(s)v(s)ds \right|}{\int_\tau^t g(s)ds}$$

$$\leq \lim_{t\to\infty} \frac{\left(\int_\tau^t a(s)\rho(s)g^2(s)ds \right)^{1/2} \left(\int_\tau^t \frac{v^2(s)}{a(s)\rho(s)}ds \right)^{1/2}}{\int_\tau^t g(s)ds} = 0.$$

Hence, from (2.5.38), $\lim_{t\to\infty} A_g(\tau, t)$ exists and (2.5.37) holds. ∎

From Lemmas 2.5.1 and 2.5.2 the following theorem is immediate.

Theorem 2.5.4. Equation (2.1.1) is oscillatory if there exists a function $g(t) \in \Omega_0$ such that

$$-\infty < \liminf_{t\to\infty} A_g(., t) < \limsup_{t\to\infty} A_g(., t) \leq \infty. \qquad (2.5.39)$$

Further, following Lemma 2.5.1 we can easily prove the following result.

Theorem 2.5.5. Assume that there exist two functions $\rho(t) \in C^2([t_0, \infty), \mathbb{R}^+)$ and $g(t) \in \Omega$ such that $\lim_{t\to\infty} G_k(t) = \infty$ for some $k \in [0, 1)$ and $\lim_{t\to\infty} A_g(., t) = \infty$, then equation (2.1.1) is oscillatory.

Example 2.5.2. Consider (2.3.1) with $c > 0$. If $\rho(t) = t \ln t$, then

$$h(t) = -\frac{1}{2t} - \frac{1}{2t \ln t} \quad \text{and} \quad Q(t) = \frac{4c - 1}{4t \ln t} \quad \text{for} \quad t \geq T > 1.$$

Let $g(t) = 1/(t \ln t)$, $t \geq T$. Then, we have

$$\lim_{t\to\infty} G_k(t) = \lim_{t\to\infty} \int_T^t \frac{[\ln \ln s - \ln \ln T]^{k-1}}{s \ln s} ds = \lim_{t\to\infty} [\ln \ln t - \ln \ln T] = \infty$$

and

$$
\begin{aligned}
\lim_{t\to\infty} A_g(T, t) &= \lim_{t\to\infty} \frac{\int_T^t g(s) \int_T^s Q(u) du \, ds}{\int_T^t g(s) ds} \\
&= \lim_{t\to\infty} \left(\frac{4c - 1}{4}\right) \frac{\int_T^t (s \ln s)^{-1} \int_T^s (u \ln u)^{-1} du \, ds}{\int_T^t (s \ln s)^{-1} ds} \\
&= \lim_{t\to\infty} \left(\frac{4c - 1}{4}\right) \int_T^t \frac{1}{s \ln s} ds \\
&= \lim_{t\to\infty} \left(\frac{4c - 1}{4}\right) [\ln \ln t - \ln \ln T] \\
&\to \infty \quad \text{as} \quad t \to \infty \quad \text{when} \quad c > \frac{1}{4},
\end{aligned}
$$

where $T > 1$ is large enough. Hence, by Theorem 2.5.5 equation (2.3.1) is oscillatory if $c > 1/4$.

Remark 2.5.4. Theorem 2.5.3 is an application of Theorem 2.5.4. In fact, if we let $g(t) = 1$, then from (2.5.17) and (2.5.18) it follows that $g(t) \in \Omega_0$ and $\lim_{t\to\infty} A_g(., t)$ does not exist. Thus, by Theorem 2.5.3, equation (2.1.1) is oscillatory.

As further applications of Theorem 2.5.4, we have the following results:

Corollary 2.5.3. Let $k(t) \in C^1([t_0, \infty), \mathbb{R}^+)$ be such that $k'(t) = 1/(a(t)\rho(t))$ and $\lim_{t\to\infty} k(t) = \infty$. If

$$
\begin{aligned}
-\infty &< \liminf_{t\to\infty} \frac{1}{k(t)} \int_{t_0}^t \int_{t_0}^s \frac{Q(u)}{a(u)\rho(u)} du \, ds \\
&< \limsup_{t\to\infty} \frac{1}{k(t)} \int_{t_0}^t \int_{t_0}^s \frac{Q(u)}{a(u)\rho(u)} du \, ds \leq \infty,
\end{aligned}
\tag{2.5.40}
$$

then equation (2.1.1) is oscillatory.

Proof. Let $g(t) = 1/(a(t)\rho(t)) = k'(t)$. It follows from (2.5.40) that $g(t) \in \Omega_0$ and $\lim_{t \to \infty} A_g(.,t)$ does not exist. Thus, by Theorem 2.5.4, equation (2.1.1) is oscillatory. ∎

Corollary 2.5.4. Suppose there exist two nonnegative bounded functions $g_1(t)$ and $g_2(t)$ on $[T, \infty)$, $T \geq t_0$ satisfying $\int^\infty g_1(s)ds = \infty = \int^\infty g_2(s)ds$, and $a(t)\rho(t)g_1(t)$ and $a(t)\rho(t)g_2(t)$ are bounded. If $\lim_{t \to \infty} A_{g_1}(T,t) < \lim_{t \to \infty} A_{g_2}(T,t)$, then equation (2.1.1) is oscillatory.

Proof. Let α_1 and α_2 be two numbers satisfying $\lim_{t \to \infty} A_{g_1}(T,t) < \alpha_1 < \alpha_2 < \lim_{t \to \infty} A_{g_2}(T,t)$. Let $r(t) = g_2(t)$ for $T \leq t < t_1$, where t_1 is such that $A_r(T,t_1) > \alpha_2$ and $\int_T^{t_1} r(s)ds \geq 1$. Let $r(t) = g_1(t)$ for $t_1 \leq t < t_2$, where t_2 is such that $A_r(T,t_2) \leq \alpha_1$ and $\int_T^{t_2} r(s)ds > 2$. This is possible because

$$
\begin{aligned}
A_r(T,t_2) &= \frac{\int_T^{t_2} r(s) \int_T^s Q(u)du\,ds}{\int_T^{t_2} r(s)ds} \\
&= \frac{\int_T^{t_1} [g_2(s) - g_1(s)] \int_T^s Q(u)du\,ds}{\int_T^{t_1} g_2(s)ds + \int_{t_1}^{t_2} g_1(s)ds} \\
&\quad + \left(\frac{\int_T^{t_2} g_1(s) \int_T^s Q(u)du\,ds}{\int_T^{t_2} g_1(s)ds} \right) \left(\frac{\int_T^{t_2} g_1(s)ds}{\int_T^{t_1} g_2(s)ds + \int_{t_1}^{t_2} g_1(s)ds} \right) \\
&= A_{g_1}(T,t_2)[1 + o(1)] + o(1) \quad \text{as} \quad t \to \infty.
\end{aligned}
$$

Continuing in this manner, we obtain a nonnegative, nonintegrable and bounded function $r(t)$ defined on $[T, \infty)$ such that $\liminf_{t \to \infty} A_r(T,t) \leq \alpha_1 < \alpha_2 \leq \limsup_{t \to \infty} A_r(T,t)$ and

$$
\lim_{t \to \infty} \frac{\int_T^t a(s)\rho(s)r^2(s)ds}{(\int_T^t r(s)ds)^2} = 0.
$$

This implies that $r(t) \in \Omega_0$ and $\lim_{t \to \infty} A_r(T,t)$ does not exist. Thus, by Theorem 2.5.4, equation (2.1.1) is oscillatory. ∎

Remark 2.5.5. It follows from the results presented above and Theorem 2.2.1 that in the hypotheses of each of the Theorems 2.5.2 – 2.5.5 the function $Q(t)$ can be replaced by $Q^*(t)$ defined in (2.2.3).

Example 2.5.3. Consider the differential equation

$$
(t^2 x'(t))' - 3tx'(t) + \frac{1}{2}t^2 \sqrt{t}(2 + \cos t - 2t \sin t)f(x(t)) = 0, \quad t \geq t_0 \geq \frac{\pi}{2} \tag{2.5.41}
$$

where the function $f(x)$ is one of the following:

(i) $f(x) = mx$, $x \in \mathbb{R}$, m is a positive constant,

(ii) $f(x) = |x|^\gamma + mx$, $x \in \mathbb{R}$, m and γ are positive constants,

(iii) $f(x) = x \ln^2(\mu + |x|)$, $x \in \mathbb{R}$, $\mu > 1$ is a constant,

(iv) $f(x) = x \exp(\lambda x)$, $x \in \mathbb{R}$, $\lambda > 0$ is a constant,

(v) $f(x) = \sinh x$, $x \in \mathbb{R}$.

We take $\rho(t) = 1/t^3$ and $g(t) = t$, $t \geq \pi/2$. Then, we have $h^*(t) = (p(t)\rho(t) - a(t)\rho'(t))/(2k\ a(t)\rho(t)) = 0$, and hence $Q^*(t) = \rho(t)q(t) = (2 + \cos t - 2t \sin t)/(2\sqrt{t})$. Now, since

$$\int_{t_0}^t \rho(s)q(s)ds = \int_{\pi/2}^t \left[\sqrt{s}\sin s + \frac{1}{2\sqrt{s}}(2 + \cos s) \right] ds$$
$$= \int_{\pi/2}^t d(\sqrt{s}(2 + \cos s))$$
$$= \sqrt{t}(2 + \cos t) - 2(\pi/2)^{1/2} \geq \sqrt{t} - 2(\pi/2)^{1/2}$$

and

$$\frac{1}{\int_{t_0}^t g(s)ds} \int_{t_0}^t g(s) \left(\int_{t_0}^s \rho(u)q(u)du \right) ds$$
$$\geq \frac{1}{t^2 - (\pi/2)^2} \int_{\pi/2}^t (s\sqrt{s} - 2(\pi/2)^{1/2}s)ds \quad \to \quad \infty \quad \text{as} \quad t \to \infty$$

all conditions of Theorem 2.5.5 with the function Q replaced by Q^* are satisfied and hence equation (2.5.41) is oscillatory.

We note that in Theorems 2.5.2 – 2.5.5 condition $\int^\infty ds/a(s) = \infty$ is not required. We also note that condition (2.5.2) is equivalent to

$$\lim_{t \to \infty} \frac{1}{t} \int_{t_0}^t (t - s)q(s)ds = \infty, \tag{2.5.42}$$

which in turn implies that for any integer $n > 1$

$$\lim_{t \to \infty} \frac{1}{t^n} \int_{t_0}^t (t - s)^n q(s)ds = \infty. \tag{2.5.43}$$

It is to be further noted that (cf. Remark 2.5.3) in (2.5.42) the limit cannot be replaced by an upper bound.

Now, we state the following oscillation criterion for the linear differential equation

$$x''(t) + q(t)x(t) = 0, \tag{2.5.44}$$

where $q(t) \in C([t_0, \infty), \mathbb{R})$.

Theorem 2.5.6. If for some integer $n > 1$,

$$\limsup_{t \to \infty} \frac{1}{t^n} \int_{t_0}^t (t - s)^n q(s) ds = \infty, \qquad (2.5.45)$$

then equation (2.5.44) is oscillatory.

Further, in the case when condition (2.5.45) is violated, we have

Theorem 2.5.7. Let

$$\limsup_{t \to \infty} \frac{1}{t^n} \int_{t_0}^t (t - s)^n q(s) ds < \infty \quad \text{for some} \quad n > 1.$$

If there exists a function $\Omega(t) \in C([t_0, \infty), \mathbb{R})$ such that

$$\liminf_{t \to \infty} \frac{1}{t^n} \int_T^t (t - s)^n q(s) ds \geq \Omega(T) \quad \text{for every} \quad T \geq t_0$$

and $\int_{t_0}^\infty (\Omega^+(s))^2 ds = \infty$, where $\Omega^+(t) = \max\{\Omega(t), 0\}$, $t \geq t_0$ then equation (2.5.44) is oscillatory.

An extension of Theorem 2.5.6 is as follows:

Theorem 2.5.8. Suppose $H : \mathcal{D} = \{(t, s) : t \geq s \geq t_0\} \to \mathbb{R}$ is a continuous function such that

$$H(t, t) = 0 \quad \text{for} \quad t \geq t_0, \quad H(t, s) > 0 \quad \text{for} \quad t > s \geq t_0 \qquad (2.5.46)$$

and has a continuous and nonnegative partial derivative on $\mathcal{D}$ with respect to the second variable. Let $h : \mathcal{D} \to \mathbb{R}$ be a continuous function with

$$-\frac{\partial}{\partial s} H(t, s) = h(t, s) \sqrt{H(t, s)} \quad \text{for all} \quad (t, s) \in \mathcal{D}. \qquad (2.5.47)$$

If

$$\limsup_{t \to \infty} \frac{1}{H(t, t_0)} \int_{t_0}^t \left[H(t, s)q(s) - \frac{1}{4} h^2(t, s) \right] ds = \infty,$$

then equation (2.5.44) is oscillatory.

Remark 2.5.6. Oscillation criteria stated in Theorems 2.5.6 – 2.5.8 for (2.5.44) are of very limited applications. In fact, these fail to deduce the oscillatory character of Euler's equation (2.2.11).

These criteria can be extended rather easily to equation (2.1.1) by applying Lemma 2.2.2. In fact, we only need to assume that $p(t) \in$

$C^1([t_0, \infty), \mathbb{R}^+)$, and replace the functions $a(t)$ and $q(t)$ by $a(t)\rho(t)$ and $Q(t) = \rho(t)\left[q(t) + a(t)h^2(t) - (a(t)h(t))'\right]$ respectively, where $h(t) = -\rho'(t)/(2\rho(t))$.

Further extensions and improvements of these criteria to more general equations of type (2.2.1) can be obtained by applying Theorem 2.2.1. For this, we let $\rho(t) \in C^1([t_0, \infty), \mathbb{R}^+)$, and replace the functions $a(t)$ and $q(t)$ by $a(t)\rho(t)$ and $kQ^*(t)$ where $Q^*(t) = \rho(t)[q(t) + a(t)h^{2*}(t) - (a(t)h^*(t))']$, $h^*(t) = (p(t)\rho(t) - a(t)\rho'(t))/(2ka(t)\rho(t))$ and the constant k is as in (i_2), i.e., $f'(x) \geq k > 0$ for $x \neq 0$.

In the following result we extend Theorem 2.5.8 to equation (2.1.1).

Theorem 2.5.9. Let the functions H and h be as in Theorem 2.5.8 such that (2.5.46) and (2.5.47) hold. If

$$\limsup_{t \to \infty} \frac{1}{H(t, t_0)} \int_{t_0}^t \left[H(t, s)q(s) - \frac{1}{4}h^2(t, s)a(s) \right] ds = \infty, \qquad (2.5.48)$$

then equation (2.1.1) is oscillatory.

Proof. Let $x(t)$ be a nonoscillatory solution of (2.1.1), and let $T_0 \geq t_0$ be such that $x(t) \neq 0$ for $t \geq T_0$. Define $v(t) = a(t)x'(t)/x(t)$ for $t \geq T_0$. Then from equation (2.1.1) it follows that $q(t) = -v'(t) - (v^2(t)/a(t))$ for $t \geq T_0$. Thus, for every t, T with $t \geq T \geq T_0$, we obtain

$$
\begin{aligned}
\int_T^t H(t, s)q(s)ds &= -\int_T^t H(t, s)v'(s)ds - \int_T^t H(t, s)\frac{v^2(s)}{a(s)}ds \\
&= H(t, T)v(T) - \int_T^t \left[-\frac{\partial}{\partial s}H(t, s)v(s) + H(t, s)\frac{v^2(s)}{a(s)} \right]ds \\
&= H(t, T)v(T) - \int_T^t \left[h(t, s)\sqrt{H(t, s)} + H(t, s)\frac{v^2(s)}{a(s)} \right]ds \\
&= H(t, T)v(T) - \int_T^t \left[\left(\frac{H(t, s)}{a(s)}\right)^{1/2}v(s) + \frac{1}{2}\sqrt{a(s)}h(t, s) \right]^2 ds \\
&\quad + \frac{1}{4}\int_T^t h^2(t, s)a(s)ds.
\end{aligned}
$$

Hence, for every $t \geq T_0$, we have

$$\int_{T_0}^t \left[H(t, s)q(s) - \frac{1}{4}h^2(t, s)a(s) \right]ds \leq H(t, T_0)v(T_0)$$

$$\leq H(t, T_0)|v(T_0)| \leq H(t, t_0)|v(T_0)|.$$

Therefore, for all $t \geq T_0$ it follows that

$$\int_{t_0}^t \left[H(t,s)q(s) - \frac{1}{4}h^2(t,s)a(s) \right] ds$$

$$= \int_{t_0}^{T_0} \left[H(t,s)q(s) - \frac{1}{4}h^2(t,s)a(s) \right] ds + \int_{T_0}^t \left[H(t,s)q(s) - \frac{1}{4}h^2(t,s)a(s) \right] ds$$

$$\leq \int_{t_0}^{T_0} H(t,s)|q(s)|ds + H(t,t_0)|v(T_0)| \leq H(t,t_0) \left[\int_{t_0}^{T_0} |q(s)|ds + |v(T_0)| \right].$$

This gives

$$\limsup_{t \to \infty} \frac{1}{H(t,t_0)} \int_{t_0}^t \left[H(t,s)q(s) - \frac{1}{4}h^2(t,s)a(s) \right] ds \leq |v(T_0)| + \int_{t_0}^{T_0} |q(s)|ds,$$

which contradicts (2.5.48). ∎

Now, in view of Remark 2.5.6 we can state the following more general results.

Theorem 2.5.10. Let H, h be as in Theorem 2.5.8 and let there exist a function $\rho(t) \in C^1([t_0, \infty), \mathbb{R}^+)$ such that

$$\limsup_{t \to \infty} \frac{1}{H(t,t_0)} \int_{t_0}^t \left[H(t,s)Q(s) - \frac{1}{4}h^2(t,s)a(s)\rho(s) \right] ds = \infty, \quad (2.5.49)$$

where $Q(t)$ is defined in (2.2.7), then equation (2.1.1) is oscillatory.

Theorem 2.5.11. Let H, h be as in Theorem 2.5.8 and let there exist a function $\rho(t) \in C^1([t_0, \infty), \mathbb{R}^+)$ such that

$$\limsup_{t \to \infty} \frac{1}{H(t,t_0)} \int_{t_0}^t \left[H(t,s)Q^*(s) - \frac{1}{4k}h^2(t,s)a(s)\rho(s) \right] ds = \infty,$$
$$(2.5.50)$$

where $Q^*(t)$ and k are defined in Remark 2.5.6, then equation (2.2.1) is oscillatory.

Corollary 2.5.5. Let $R(t) = \int_{t_0}^t ds/a(s)$. If there exists a function $\rho(t) \in C^1([t_0, \infty), \mathbb{R}^+)$ such that for some $n > 1$,

$$\limsup_{t \to \infty} \frac{1}{R^n(t)} \int_{t_0}^t (R(t)-R(s))^{n-2} \left[(R(t)-R(s))^{n-2}Q^*(s) - \frac{n^2}{4k}\frac{\rho(s)}{a(s)} \right] ds = \infty,$$
$$(2.5.51)$$

then equation (2.2.1) is oscillatory.

Proof. Let $H(t,s) = (R(t) - R(s))^n$ for $t \geq s \geq t_0$. Then, $h(t,s) = (n/a(s))(R(t) - R(s))^{(n-2)/2}$ for $t > s \geq t_0$, and this implies

$$\limsup_{t \to \infty} \frac{1}{H(t,t_0)} \int_{t_0}^t \left[H(t,s) Q^*(s) - \frac{1}{4k} a(s) \rho(s) h^2(t,s) \right] ds$$

$$= \limsup_{t \to \infty} \frac{1}{R^n(t)} \int_{t_0}^t \left[(R(t) - R(s))^n Q^*(s) - \frac{n^2}{4k} \frac{\rho(s)}{a(s)} (R(t) - R(s))^{n-2} \right] ds$$

$$= \infty.$$

Now, it follows from Theorem 2.5.11 that (2.2.1) is oscillatory.   ∎

If $\rho(t) = 1$, then we have

$$\int_{t_0}^t a(s) \rho(s) h^2(t,s) ds = \frac{n^2}{n-1} (R(t) - R(t_0))^{n-1} \quad \text{for} \quad t \geq t_0$$

and hence the following corollary is immediate.

Corollary 2.5.6. If

$$\limsup_{t \to \infty} \frac{1}{R^n(t)} \int_{t_0}^t (R(t) - R(s))^n Q^*(s) ds = \infty \quad \text{for some} \quad n > 1, \quad (2.5.52)$$

then equation (2.2.1) is oscillatory.

Remark 2.5.7. Condition (2.5.52) for equation (2.1.1) reduces to

$$\limsup_{t \to \infty} \frac{1}{R^n(t)} \int_{t_0}^t (R(t) - R(s))^n q(s) ds = \infty \quad \text{for some} \quad n > 1.$$

Corollary 2.5.7. Let $R_1(t) = \int_t^\infty ds/a(s) < \infty$, $t \geq t_0$. If for some $n > 1$,

$$\limsup_{t \to \infty} (-\ln R_1(t))^{-n} \int_{t_0}^t \left(\ln \left(\frac{R_1(s)}{R_1(t)} \right) \right)^n \left[R_1(s) Q(s) - \frac{1}{4a(s) R_1(s)} \right] ds = \infty,$$

$$(2.5.53)$$

then equation (2.1.1) is oscillatory.

Proof. Let $\rho(t) = R_1(t)$ for $t \geq t_0$ and $H(t,s) = (\ln(R_1(s)/R_1(t)))^n$ for $t \geq s \geq t_0$. Then, we have $h(t) = -\rho'(t)/(2\rho(t)) = 1/(2a(t)R_1(t))$,

$$Q(t) = \rho(t) \left[q(t) + a(t) h^2(t) - (a(t)h(t))' \right] = q(t) R_1(t) - \frac{1}{4a(t) R_1(t)}$$

and

$$h(t,s) = \frac{n}{a(s) R_1(s)} \left(\ln \frac{R_1(s)}{R_1(t)} \right)^{(n-2)/2} \quad \text{for} \quad t > s \geq t_0.$$

Thus, for $t \geq t_0$ it follows that

$$
\begin{aligned}
\int_{t_0}^{t} a(s)\rho(s)h^2(t,s)ds &= \int_{t_0}^{t} n^2 \left(\ln \frac{R_1(s)}{R_1(t)} \right)^{n-2} \frac{1}{a(s)R_1(s)} ds \\
&= \frac{n^2}{n-1} \left(\ln \frac{R_1(t_0)}{R_1(t)} \right)^{n-1} < \infty.
\end{aligned}
$$

It is easy to check that (2.5.49) implies condition (2.5.53), and hence equation (2.1.1) is oscillatory. ∎

Next, we let $H(t,s) = (t-s)^n$ for $t \geq s \geq t_0$, where $n > 1$ is a constant. Then from Theorem 2.5.11 the following corollary is immediate.

Corollary 2.5.8. Let $n > 1$ be a constant and there exist a function $\rho(t) \in C^1([t_0, \infty), \mathbb{R}^+)$ such that

$$
\limsup_{t \to \infty} \frac{1}{t^n} \int_{t_0}^{t} (t-s)^{n-2} \left[(t-s)^2 Q^*(s) - \frac{n^2}{4k} a(s)\rho(s) \right] ds = \infty,
$$

then equation (2.2.1) is oscillatory.

Example 2.5.4. Consider the differential equation

$$
(t^2 x'(t))' - tx'(t) + cf(x(t)) = 0, \quad t \geq t_0 \geq 1 \tag{2.5.54}
$$

where $c > 0$ is a constant and $f(x)$ is as in Example 2.5.3 with $f'(x) \geq k > 0$ for $x \neq 0$. Let $\rho(t) = 1/t$, $t \geq 1$ and $n > 1$ a number. Then, we find $a(t)\rho'(t) - p(t)\rho(t) = 0$, and hence $h^*(t) = 0$ and $Q^*(t) = c/t$, $t \geq 1$. Now,

$$
\begin{aligned}
\limsup_{t \to \infty} \frac{1}{t^n} &\int_{t_0}^{t} \left[(t-s)^n Q^*(s) - \frac{n^2}{4k} a(s)\rho(s)(t-s)^{n-2} \right] ds \\
&= \limsup_{t \to \infty} \frac{1}{t^n} \int_{1}^{t} \left[(t-s)^n \left(\frac{c}{s} \right) - \frac{n^2}{4k} (t-s)^{n-2} s \right] ds.
\end{aligned}
$$

Since $n > 1$ it follows from Hardy et. al. [10] that $(t-s)^n \geq t^n - nst^{n-1}$ for $t \geq s \geq 1$. Thus,

$$
\begin{aligned}
\limsup_{t \to \infty} \frac{c}{t^n} \int_{1}^{t} \frac{1}{s}(t-s)^n ds &\geq c \limsup_{t \to \infty} \frac{1}{t^n} \int_{1}^{t} \frac{t^n - nst^{n-1}}{s} ds \\
&= c \limsup_{t \to \infty} \left[\ln t - \frac{n(t-1)}{t} \right] = \infty.
\end{aligned}
$$

Also, we have

$$
\frac{n^2}{4} \int_{1}^{t} s(t-s)^{n-2} ds = \frac{n^2}{4(n-1)}(t-1)^{n-1} + \frac{n}{4(n-1)}(t-1)^n,
$$

and hence

$$\limsup_{t\to\infty} \frac{1}{t^n} \int_1^t \left[\frac{c}{s}(t-s)^n - \frac{n^2}{4k}s(t-s)^{n-2} \right] ds$$

$$= c \limsup_{t\to\infty} \left[\ln t - \frac{n(t-1)}{t} \right] - \frac{n}{4(n-1)k} = \infty \quad \text{if} \quad c > 0.$$

Now, it follows from Corollary 2.5.8 that (2.5.54) is oscillatory if $c > 0$.

Example 2.5.5. Consider the differential equation

$$(tx'(t))' + x'(t) + \frac{c}{t}f(x(t)) = 0, \quad t \geq 1 \tag{2.5.55}$$

where $c > 0$ is a constant and $f(x)$ is as in Example 2.5.4. Let $\rho(t) = 1$. Then, $h^*(t) = 0$ and $Q^*(t) = c/t$, $t \geq 1$. Now, proceeding exactly as in Example 2.5.4, we find that (2.5.55) is oscillatory by Corollary 2.5.8 if $c > 0$.

In Theorems 2.5.6 – 2.5.11 no assumption has been made on $\int^\infty ds/a(s)$. Therefore, the conclusions of these criteria hold in either of the following two cases:

(I) $\displaystyle\int^\infty ds/a(s) = \infty$, (II) $\displaystyle\int^\infty ds/a(s)ds < \infty$.

For illustration see Examples 2.5.4 and 2.5.5.

We note that if (II) is satisfied, then the condition (2.5.51) is equivalent to the following

$$\limsup_{t\to\infty} \int_{t_0}^t (R(t) - R(s))^{n-2} \left[(R(t) - R(s))^2 Q^*(t) - \frac{n^2}{4}\frac{\rho(s)}{a(s)} \right] ds = \infty$$

for some $n > 1$.

Also, it is easy to show that the condition

$$\liminf_{t\to\infty} R_1(t) \int_{t_0}^t Q(s)ds > \frac{1}{4} \tag{2.5.56}$$

implies (2.5.53) with $n = 2$. Thus, we have the following corollary.

Corollary 2.5.9. Let condition (2.5.53) of Corollary 2.5.7 be replaced by (2.5.56), then equation (2.1.1) is oscillatory.

We are now in the position to prove the following oscillation criteria which extend and improve Theorem 2.5.7.

Theorem 2.5.12. Let H and h be as in Theorem 2.5.8 and conditions (2.5.46) and (2.5.47) hold, and let

$$0 \; < \; \inf_{s \geq t_0} \left\{ \liminf_{t \to \infty} \frac{H(t,s)}{H(t,t_0)} \right\} \; \leq \; \infty. \tag{2.5.57}$$

Suppose there exist two functions $\rho \in C^1([t_0, \infty), \mathbb{R}^+)$ and $\Omega \in C([t_0, \infty), \mathbb{R})$ such that

$$\limsup_{t \to \infty} \frac{1}{H(t,t_0)} \int_{t_0}^t a(s)\rho(s)h^2(t,s)ds \; < \; \infty, \tag{2.5.58}$$

$$\int_{t_0}^{\infty} \frac{(\Omega^+(s))^2}{a(s)\rho(s)} ds \; = \; \infty \tag{2.5.59}$$

and for every $T \geq t_0$,

$$\limsup_{t \to \infty} \frac{1}{H(t,T)} \int_T^t \left[H(t,s)Q(s) - \frac{1}{4}h^2(t,s)a(s)\rho(s) \right] ds \; \geq \; \Omega(T), \tag{2.5.60}$$

where Q is defined in (2.2.7) and $\Omega^+(t) = \max\{\Omega(t), 0\}$, then equation (2.1.1) is oscillatory.

Proof. Assume to the contrary that equation (2.1.1) is nonoscillatory. Let the function $v(t)$ be as in Theorem 2.2.3. Then, we have

$$v'(t) + Q(t) + \frac{v^2(t)}{a(t)\rho(t)} \; = \; 0 \quad \text{for} \quad t \geq T_0 \geq t_0.$$

Multiplying both sides of the above equation by $H(t,s)$ and proceeding as in Theorem 2.5.9, we get

$$\frac{1}{H(t,T)} \int_T^t \left[H(t,s)Q(s) - \frac{1}{4}a(s)\rho(s)h^2(t,s) \right] ds$$

$$= \; v(T) - \frac{1}{H(t,T)} \int_T^t \left[\left(\frac{H(t,s)}{a(s)\rho(s)} \right)^{1/2} v(s) + \frac{1}{2}\sqrt{a(s)\rho(s)}h(t,s) \right]^2 ds$$

for $t \geq T \geq T_0$. Consequently,

$$\limsup_{t \to \infty} \frac{1}{H(t,T)} \int_T^t \left[H(t,s)Q(s) - \frac{1}{4}a(s)\rho(s)h^2(t,s) \right] ds$$

$$= \; v(T) - \liminf_{t \to \infty} \frac{1}{H(t,T)} \int_T^t \left[\left(\frac{H(t,s)}{a(s)\rho(s)} \right)^{1/2} v(s) + \frac{1}{2}\sqrt{a(s)\rho(s)}h(t,s) \right]^2 ds$$

for all $T \geq T_0$. Thus, by (2.5.60), we obtain

$$v(T) \geq \Omega(T) + \liminf_{t \to \infty} \frac{1}{H(t,T)} \int_T^t \left[\left(\frac{H(t,s)}{a(s)\rho(s)} \right)^{1/2} v(s) + \frac{1}{2} \sqrt{a(s)\rho(s)} h(t,s) \right]^2 ds$$

for all $T \geq T_0$. This shows that for $T \geq T_0$,

$$v(T) \geq \Omega(T) \tag{2.5.61}$$

and

$$\liminf_{t \to \infty} \frac{1}{H(t,T)} \int_T^t \left[\left(\frac{H(t,s)}{a(s)\rho(s)} \right)^{1/2} v(s) + \frac{1}{2} \sqrt{a(s)\rho(s)} h(t,s) \right]^2 ds < \infty.$$

Hence,

$$\liminf_{t \to \infty} \left[\frac{1}{H(t,T_0)} \int_{T_0}^t H(t,s) \frac{v^2(s)}{a(s)\rho(s)} ds + \frac{1}{H(t,T_0)} \int_{T_0}^t h(t,s) \sqrt{H(t,s)} v(s) ds \right]$$

$$\leq \liminf_{t \to \infty} \frac{1}{H(t,T_0)} \int_{T_0}^t \left[\left(\frac{H(t,s)}{a(s)\rho(s)} \right)^{1/2} v(s) + \frac{1}{2} \sqrt{a(s)\rho(s)} h(t,s) \right]^2 ds < \infty,$$

$$\tag{2.5.62}$$

i.e., we have

$$\liminf_{t \to \infty} [U(t) + W(t)] < \infty, \tag{2.5.63}$$

where

$$U(t) = \frac{1}{H(t,T_0)} \int_{T_0}^t H(t,s) \frac{v^2(s)}{a(s)\rho(s)} ds, \quad t \geq T_0$$

and

$$W(t) = \frac{1}{H(t,T_0)} \int_{T_0}^t h(t,s) \sqrt{H(t,s)} v(s) ds, \quad t \geq T_0.$$

Now, we claim that

$$\int_{T_0}^\infty \frac{v^2(s)}{a(s)\rho(s)} ds < \infty. \tag{2.5.64}$$

For this, suppose that

$$\int_{T_0}^\infty \frac{v^2(s)}{a(s)\rho(s)} ds = \infty. \tag{2.5.65}$$

By condition (2.5.57) there is a positive constant η such that

$$\inf_{s \geq t_0} \left\{ \liminf_{t \to \infty} \frac{H(t,s)}{H(t,t_0)} \right\} > \eta > 0. \tag{2.5.66}$$

Let μ be an arbitrary positive number. Then in view of (2.5.65) there exists a $T_1 > T_0$ such that

$$\int_{T_0}^t \frac{v^2(s)}{a(s)\rho(s)} ds \geq \frac{\mu}{\eta} \quad \text{for all } t \geq T_1.$$

Therefore, for $t \geq T_1$,

$$
\begin{aligned}
U(t) &= \frac{1}{H(t,T_0)} \int_{T_0}^t H(t,s) d\left(\int_{T_0}^s \frac{v^2(\tau)}{a(\tau)\rho(\tau)} d\tau\right) \\
&= \frac{1}{H(t,T_0)} \int_{T_0}^t -\frac{\partial}{\partial s} H(t,s) \left(\int_{T_0}^s \frac{v^2(\tau)}{a(\tau)\rho(\tau)} d\tau\right) ds \\
&\geq \frac{1}{H(t,T_0)} \int_{T_1}^t -\frac{\partial}{\partial s} H(t,s) \left(\int_{T_0}^s \frac{v^2(\tau)}{a(\tau)\rho(\tau)} d\tau\right) ds \\
&\geq \left(\frac{\mu}{\eta}\right) \frac{1}{H(t,T_0)} \int_{T_1}^t -\frac{\partial}{\partial s} H(t,s) ds = \left(\frac{\mu}{\eta}\right) \frac{H(t,T_1)}{H(t,T_0)}.
\end{aligned}
$$

By (2.5.66), there exists a $T_2 \geq T_1$ such that $H(t,T_1)/H(t,t_0) \geq \eta$ for all $t \geq T_2$, which implies that $U(t) \geq \mu$ for all $t \geq T_2$. Since μ is arbitrary

$$\lim_{t \to \infty} U(t) = \infty. \tag{2.5.67}$$

Next, consider a sequence $\{t_k\}_{k=1}^\infty$ in (t_0, ∞) with $\lim_{k \to \infty} t_k = \infty$ satisfying $\lim_{k \to \infty}[U(t_k) + W(t_k)] = \liminf_{t \to \infty}[U(t) + W(t)]$. In view of (2.5.63), there exists a constant M such that

$$U(t_k) + W(t_k) \leq M \quad \text{for } k = 1, 2, \cdots. \tag{2.5.68}$$

It follows from (2.5.67) that

$$\lim_{k \to \infty} U(t_k) = \infty, \tag{2.5.69}$$

and hence (2.5.68) gives

$$\lim_{k \to \infty} W(t_k) = -\infty. \tag{2.5.70}$$

Taking into account (2.5.67), from (2.5.69), we get

$$1 + \frac{W(t_k)}{U(t_k)} \leq \frac{M}{U(t_k)} < \frac{1}{2}$$

provided k is sufficiently large. Thus, $W(t_k)/U(t_k) < -1/2$ for all large k, which by (2.5.70) ensures that

$$\lim_{k \to \infty} \frac{W^2(t_k)}{U(t_k)} = \infty. \tag{2.5.71}$$

On the other hand, by Schwartz's inequality, we have

$$W^2(t_k) = \left[\frac{1}{H(t_k, T_0)} \int_{T_0}^{t_k} h(t_k, s) \sqrt{H(t_k, s)} v(s) ds \right]^2$$

$$\leq \left[\frac{1}{H(t_k, T_0)} \int_{T_0}^{t_k} a(s)\rho(s)h^2(t_k, s)ds \right] \left[\frac{1}{H(t_k, T_0)} \int_{T_0}^{t_k} H(t_k, s) \frac{v^2(s)}{a(s)\rho(s)} ds \right]$$

$$\leq U(t_k) \left[\frac{1}{H(t_k, T_0)} \int_{t_0}^{t_k} a(s)\rho(s)h^2(t_k, s)ds \right] \quad \text{for all large } k.$$

But (2.5.66) guarantees that $\liminf_{t \to \infty} H(t, T_0)/H(t, t_0) > \eta$. This means that there exists a $T_3 \geq T_0$ such that $H(t, T_0)/H(t, t_0) \geq \eta$ for every $t \geq T_3$. Thus, $H(t_k, T_0)/H(t_k, t_0) \geq \eta$ for all large k, and therefore,

$$\frac{W^2(t_k)}{U(t_k)} \leq \frac{1}{\eta \, H(t_k, t_0)} \int_{t_0}^{t_k} a(s)\rho(s)h^2(t_k, s)ds \quad \text{for all large } n.$$

It follows from (2.5.71) that

$$\lim_{k \to \infty} \frac{1}{H(t_k, t_0)} \int_{t_0}^{t_k} a(s)\rho(s)h^2(t_k, s)ds = \infty. \tag{2.5.72}$$

This gives $\limsup_{t \to \infty} (1/H(t, t_0)) \int_{t_0}^{t} a(s)\rho(s)h^2(t, s)ds = \infty$, which contradicts (2.5.58). Thus, (2.5.64) holds. Hence, by (2.5.61),

$$\int_{T_0}^{\infty} \frac{(\Omega^+(s))^2}{a(s)\rho(s)} ds \leq \int_{T_0}^{\infty} \frac{v^2(s)}{a(s)\rho(s)} ds < \infty,$$

which contradicts condition (2.5.59). This completes the proof.   ∎

Consider the function $H(t, s) = (t - s)^n$ for $t \geq s \geq t_0$, where $n > 1$ is a constant. Then from Theorem 2.5.12 the following corollary is immediate.

Corollary 2.5.10. Let $n > 1$ be a constant and there exist two functions $\rho(t) \in C^1([t_0, \infty), \mathbb{R}^+)$ and $\Omega(t) \in C([t_0, \infty), \mathbb{R})$ such that condition (2.5.59) hold,

$$\limsup_{t \to \infty} \frac{1}{t^n} \int_{t_0}^{t} (t - s)^{n-2} a(s)\rho(s)ds < \infty$$

and for all $T \geq t_0$,

$$\limsup_{t \to \infty} \frac{1}{t^n} \int_{t_0}^{t} \left[(t - s)^n Q(s) - \frac{n^2}{4}(t - s)^{n-2} a(s)\rho(s) \right] ds \geq \Omega(T).$$

Then equation (2.1.1) is oscillatory.

Theorem 2.5.13. Let H and h be as in Theorem 2.5.8 and conditions (2.5.46) and (2.5.47) hold. Suppose there exist two functions $\rho(t) \in C^1([t_0, \infty), \mathbb{R}^+)$ and $\Omega(t) \in C([t_0, \infty), \mathbb{R})$ such that condition (2.5.59) hold,

$$\liminf_{t \to \infty} \frac{1}{H(t, t_0)} \int_{t_0}^t H(t, s) Q(s) ds < \infty \qquad (2.5.73)$$

and for every $T \ge t_0$,

$$\liminf_{t \to \infty} \frac{1}{H(t, T)} \int_T^t \left[H(t, s) Q(s) - \frac{1}{4} h^2(t, s) a(s) \rho(s) \right] ds \ge \Omega(T), \quad (2.5.74)$$

where $Q(t)$ is defined in (2.2.7), then equation (2.1.1) is oscillatory.

Proof. Assume to the contrary that equation (2.1.1) is nonoscillatory. As in Theorem 2.5.12, we obtain for $t > T \ge T_0$,

$$\frac{1}{H(t, T)} \int_T^t \left[H(t, s) Q(s) - \frac{1}{4} a(s) \rho(s) h^2(t, s) \right] ds$$

$$= v(T) - \frac{1}{H(t, T)} \int_T^t \left[\left(\frac{H(t, s)}{a(s) \rho(s)} \right)^{1/2} v(s) + \frac{1}{2} \sqrt{a(s) \rho(s)} h(t, s) \right]^2 ds.$$

Consequently,

$$\liminf_{t \to \infty} \frac{1}{H(t, T)} \int_T^t \left[H(t, s) Q(s) - \frac{1}{4} a(s) \rho(s) h^2(t, s) \right] ds$$

$$= v(T) - \limsup_{t \to \infty} \frac{1}{H(t, T)} \int_T^t \left[\left(\frac{H(t, s)}{a(s) \rho(s)} \right)^{1/2} v(s) + \frac{1}{2} \sqrt{a(s) \rho(s)} h(t, s) \right]^2 ds$$

for all $T \ge T_0$. It follows from (2.5.73) that

$$v(T) \ge \Omega(T) + \limsup_{t \to \infty} \frac{1}{H(t, T)} \int_T^t \left[\left(\frac{H(t, s)}{a(s) \rho(s)} \right)^{1/2} v(s) + \frac{1}{2} \sqrt{a(s) \rho(s)} h(t, s) \right]^2 ds$$

for all $T \ge T_0$. Hence, (2.5.61) holds for all $T \ge T_0$ and

$$\limsup_{t \to \infty} \frac{1}{H(t, T_0)} \int_{T_0}^t \left[\left(\frac{H(t, s)}{a(s) \rho(s)} \right)^{1/2} v(s) + \frac{1}{2} \sqrt{a(s) \rho(s)} h(t, s) \right]^2 ds$$

$$\le v(T_0) - \Omega(T_0) < \infty.$$

This implies that

$$\limsup_{t \to \infty} [U(t) + W(t)]$$

$$\le \limsup_{t \to \infty} \frac{1}{H(t, T_0)} \int_{T_0}^t \left[\left(\frac{H(t, s)}{a(s) \rho(s)} \right)^{1/2} v(s) + \frac{1}{2} \sqrt{a(s) \rho(s)} h(t, s) \right]^2 ds < \infty,$$

$$(2.5.75)$$

where $U(t)$ and $W(t)$ are defined in the proof of Theorem 2.5.12. By condition (2.5.74),

$$
\begin{aligned}
\Omega(t_0) \;\leq\; & \liminf_{t\to\infty} \frac{1}{H(t,t_0)} \int_{t_0}^{t} \left[H(t,s)Q(s) - \frac{1}{4}h^2(t,s)a(s)\rho(s) \right] ds \\
\leq\; & \liminf_{t\to\infty} \frac{1}{H(t,t_0)} \int_{t_0}^{t} H(t,s)Q(s)\,ds \\
& - \frac{1}{4} \liminf_{t\to\infty} \frac{1}{H(t,t_0)} \int_{t_0}^{t} h^2(t,s)a(s)\rho(s)\,ds.
\end{aligned}
$$

From the above inequality and (2.5.73), we have

$$
\liminf_{t\to\infty} \frac{1}{H(t,t_0)} \int_{t_0}^{t} h^2(t,s)a(s)\rho(s)\,ds < \infty.
$$

Thus there exists a sequence $\{t_k\}_{k=1}^{\infty}$ in (t_0,∞) with $\lim_{k\to\infty} t_k = \infty$ satisfying

$$
\begin{aligned}
\lim_{k\to\infty} \frac{1}{H(t_k,t_0)} \int_{t_0}^{t_k} h^2(t_k,s)a(s)\rho(s)\,ds & \\
= \liminf_{t\to\infty} \frac{1}{H(t,t_0)} \int_{t_0}^{t} h^2(t,s)a(s)\rho(s)\,ds & < \infty.
\end{aligned}
\tag{2.5.76}
$$

Now, suppose (2.5.65) holds. Using the procedure of Theorem 2.5.12, we conclude that (2.6.69) is satisfied. It follows from (2.5.75) that there exists a constant M such that (2.5.68) is fulfilled. Then as in Theorem 2.5.12, we find that (2.5.72) holds, which contradicts (2.5.76). Hence, (2.5.65) fails. Since the remainder of the proof is similar to that of Theorem 2.5.12, we omit the details. ∎

Example 2.5.6. Consider the differential equation

$$
\left(t^\lambda x'(t)\right)' + (t^\mu \cos t)x(t) = 0 \quad \text{for} \quad t \geq t_0 > 0, \tag{2.5.77}
$$

where λ and μ are constants, $\lambda \leq 1$, $-1 < \mu \leq 1$ and $2\mu + 1 \geq \lambda$. Let $\rho(t) = 1$ and $H(t,s) = (t-s)^2$ for $t \geq s \geq t_0$. Then, we have $\limsup_{t\to\infty}(1/t^2)\int_{t_0}^{t}(t-s)^2 s^\mu \cos s\,ds = -t_0^\mu \cos t_0 < \infty$, and

$$
\liminf_{t\to\infty} \frac{1}{t^2} \int_{T}^{t} \left[(t-s)^2 s^\mu \cos s - s^\lambda \right] ds \geq -T^\mu \sin T - k_1 \quad \text{for} \quad T \geq t_0,
$$

where k_1 is a positive small constant. Set $\Omega(T) = -T^\mu \sin T - k_1$. Then there exists an integer N such that $(2N+1)\pi + (\pi/4) > t_0$ and if $n \geq N$, $(2n+1)\pi + (\pi/4) \leq T \leq 2(n+1)\pi - (\pi/4)$, and

$\Omega(T) = -T^\mu \sin T - k_1 \geq \delta T^\mu$, where δ is a small constant. Noting that $2\mu + 1 \geq \lambda$, we obtain

$$\int_{t_0}^t \frac{(\Omega^+(s))^2}{a(s)\rho(s)} ds \;\geq\; \sum_{N=n}^\infty \delta^2 \int_{(2n+1)\pi+(\pi/4)}^{2(n+1)-(\pi/4)} s^{2\mu-\lambda} ds$$

$$\geq\; \sum_{N=n}^\infty \delta^2 \int_{(2n+1)\pi+(\pi/4)}^{2(n+1)-(\pi/4)} \frac{ds}{s} \;=\; \infty.$$

Thus, all conditions of Theorem 2.3.13 are satisfied, and hence equation (2.5.77) is oscillatory.

The above oscillation criteria as well as most of the known results in the literature require information of equations (2.1.1) and (2.2.1) on the entire half-line $[t_0, \infty)$. Now, motivated by the Sturm separation theorem, we will present oscillation results for equations (2.1.1) and (2.2.1) which depend on the behavior of the coefficients of these equations only on a sequence of subintervals $\{[a_i, b_i]\}_{i=1}^\infty$ of $[t_0, \infty)$, no matter how 'bad' the coefficients are on the remaining parts of $[t_0, \infty)$.

In what follows, we say that the function $H = H(t, s) \in \Gamma$ if H is as in Theorem 2.5.8, i.e., if $H \in C(\mathcal{D}, \mathbb{R}^+)$, where $\mathcal{D} = \{(t, s) : t \geq s \geq t_0\}$ which satisfies (2.5.46) and has partial derivatives $\partial H/\partial t$ and $\partial H/\partial s$ on $\mathcal{D}$ such that

$$\frac{\partial}{\partial t} H(t, s) = h_1(t, s)\sqrt{H(t, s)} \quad \text{and} \quad \frac{\partial}{\partial s} H(t, s) = -h_2(t, s)\sqrt{H(t, s)},$$

$$(2.5.78)$$

where $h_1, h_2 \in \mathcal{L}_{Loc}(\mathcal{D}, \mathbb{R})$.

From the results of Section 2.2, since equation (2.1.1) is equivalent to (2.2.6), equation (2.2.6) is nonoscillatory if and only if (2.1.1) is nonoscillatory. As earlier, let $w(t)$ be a solution of (2.2.6) and assume that $w(t) > 0$ for $t \geq T \geq t_0$. For any $\rho(t) \in C^1([t_0, \infty), \mathbb{R}^+)$, we define $v(t) = a(t)\rho(t)w'(t)/w(t)$, $t \geq T$. Then, it follows that

$$v'(t) + Q(t) + \frac{v^2(t)}{a(t)\rho(t)} = 0 \quad \text{for} \quad t \geq T, \qquad (2.5.79)$$

where $Q(t)$ is defined in (2.2.7).

The following lemma plays a fundamental role in establishing the interval oscillation criteria for equation (2.1.1).

Lemma 2.5.3. Let $w(t)$, $v(t)$ and T be defined as above.

(i)   Assume $[c, b] \subset [T, \infty)$. Then for any $H \in \Gamma$,

$$\int_c^b H(b, s)Q(s)ds \;\leq\; H(b, c)v(c) + \frac{1}{4}\int_c^b a(s)\rho(s)h_2^2(b, s)ds. \qquad (2.5.80)$$

(ii)   Assume $[a, c] \subset [T, \infty)$. Then for any $H \in \Gamma$,

$$\int_a^c H(s, a)Q(s)ds \leq -H(c, a)v(c) + \frac{1}{4}\int_a^c a(s)\rho(s)h_1^2(s, a)ds. \quad (2.5.81)$$

Proof. (i) Multiplying equation (2.5.79) by $H(b, s)$, integrating it with respect to s from c to b and using (2.5.46) and (2.5.78), we get

$$\int_c^b H(b, s)Q(s)ds = -\int_c^b H(b, s)v'(s)ds - \int_c^b H(b, s)\frac{v^2(s)}{a(s)\rho(s)}ds$$

$$= H(b, c)v(c) - \int_c^b h_2(b, s)\sqrt{H(b, s)}v(s)ds - \int_c^b H(b, s)\frac{v^2(s)}{a(s)\rho(s)}ds$$

$$= H(b, c)v(c) + \frac{1}{4}\int_c^b h_2^2(b, s)a(s)\rho(s)ds$$

$$\quad - \int_c^b \left[\left(\frac{H(b, s)}{a(s)\rho(s)}\right)^{1/2} v(s) - \frac{1}{2}\sqrt{a(s)\rho(s)}h_2(b, s)\right]^2 ds$$

$$\leq H(b, c)v(c) + \frac{1}{4}\int_c^b h_2^2(b, s)a(s)\rho(s)ds.$$

(ii) Similar to part (i), we multiply (2.5.79) by $H(s, a)$, integrate it with respect to s from a to c and use (2.5.46) and (2.5.78), to get

$$\int_a^c H(s, a)Q(s)ds = -\int_a^c H(s, a)v'(s)ds - \int_a^c H(s, a)\frac{v^2(s)}{a(s)\rho(s)}ds$$

$$= -H(c, a)v(c) + \int_a^c h_1(s, a)\sqrt{H(s, a)}v(s)ds - \int_a^c H(s, a)\frac{v^2(s)}{a(s)\rho(s)}ds$$

$$= -H(c, a)v(c) + \frac{1}{4}\int_a^c h_1^2(s, a)a(s)\rho(s)ds$$

$$\quad - \int_a^c \left[\left(\frac{H(s, a)}{a(s)\rho(s)}\right)^{1/2} v(s) + \frac{1}{2}\sqrt{a(s)\rho(s)}h_1(s, a)\right]^2 ds$$

$$\leq -H(c, a)v(c) + \frac{1}{4}\int_a^c h_1^2(s, a)a(s)\rho(s)ds. \quad \blacksquare$$

Corollary 2.5.11. Let $w(t)$, $v(t)$ and T be defined as above. Then for any $H \in \Gamma$ and a, b, $c \in \mathbb{R}$ such that $T \leq a < c < b$,

$$\frac{1}{H(c, a)}\int_a^c H(s, a)Q(s)ds + \frac{1}{H(b, c)}\int_c^b H(b, s)Q(s)ds$$

$$\leq \frac{1}{4}\left[\frac{1}{H(c, a)}\int_a^c h_1^2(s, a)a(s)\rho(s)ds + \frac{1}{H(b, c)}\int_c^b h_2^2(b, s)a(s)\rho(s)ds\right].$$

$$(2.5.82)$$

Proof. Dividing (2.5.80) and (2.5.81) by $H(b,c)$ and $H(c,a)$ respectively, and adding them, we get (2.5.82). ∎

Now, we shall prove the main interval oscillation theorems.

Theorem 2.5.14. If for each $T \geq t_0$ there exist $H \in \Gamma$, $\rho \in C^1([t_0, \infty), \mathbb{R}^+)$ and a, b, $c \in \mathbb{R}$ such that $T \leq a < c < b$, and

$$
\frac{1}{H(c,a)} \int_a^c H(s,a)Q(s)ds + \frac{1}{H(b,c)} \int_c^b H(b,s)Q(s)ds
$$
$$
> \frac{1}{4}\left[\frac{1}{H(c,a)}\int_a^c h_1^2(s,a)a(s)\rho(s)ds + \frac{1}{H(b,c)}\int_c^b h_2^2(b,s)a(s)\rho(s)ds\right],
$$
$$
\tag{2.5.83}
$$

then equation (2.1.1) is oscillatory.

Proof. Without any loss of generality we assume that $w(t)$ is an eventually positive solution of equation (2.2.6). Then from Corollary 2.5.11 there exists $T \geq t_0$ such that inequality (2.5.82) holds for any $H \in \Gamma$ and a, b, $c \in \mathbb{R}$ satisfying $T \leq a < c < b$. But, this contradicts (2.5.83). ∎

Theorem 2.5.15. If there exists $H \in \Gamma$ such that for any $r \geq t_0$,

$$
\limsup_{t \to \infty} \frac{\int_r^t H(s,r)Q(s)ds}{\int_r^t h_1^2(s,r)a(s)\rho(s)ds} = \infty \tag{2.5.84}
$$

and

$$
\limsup_{t \to \infty} \frac{\int_r^t H(t,s)Q(s)ds}{\int_r^t h_2^2(t,s)a(s)\rho(s)ds} = \infty, \tag{2.5.85}
$$

then equation (2.1.1) is oscillatory.

Proof. For any $T \geq t_0$ let $a = T$. In (2.5.84) we choose $r = a$. Then there exists $c > a$ such that

$$
\int_a^c H(s,a)Q(s)ds > \frac{1}{4}\int_a^c h_1^2(s,a)a(s)\rho(s)ds. \tag{2.5.86}
$$

In (2.5.85) we choose $r = c$. Then there exists $b > c$ such that

$$
\int_c^b H(b,s)Q(s)ds > \frac{1}{4}\int_c^b h_2^2(b,s)a(s)\rho(s)ds. \tag{2.5.87}
$$

Combining (2.5.86) and (2.5.87), we obtain (2.5.83). The conclusion now follows from Theorem 2.5.14. ∎

The following result is an equivalent version of Theorem 2.5.15.

Theorem 2.5.16. If there exists $H \in \Gamma$ such that for any $r \geq t_0$,

$$\limsup_{t \to \infty} \int_r^t \left[H(s,r)Q(s) - \frac{1}{4}h_1^2(s,r)a(s)\rho(s) \right] ds > 0 \qquad (2.5.88)$$

and

$$\limsup_{t \to \infty} \int_r^t \left[H(t,s)Q(s) - \frac{1}{4}h_2^2(t,s)a(s)\rho(s) \right] ds > 0, \qquad (2.5.89)$$

then equation (2.1.1) is oscillatory.

Proof. For any $T \geq t_0$ let $a = T$. In (2.5.88) we choose $r = a$. Then there exists $c > a$ such that

$$\int_a^c \left[H(s,a)Q(s) - \frac{1}{4}h_1^2(s,a)a(s)\rho(s) \right] ds > 0. \qquad (2.5.90)$$

In (2.5.89) we choose $r = c$. Then there exists $b > c$ such that

$$\int_c^b \left[H(b,s)Q(s) - \frac{1}{4}h_2^2(b,s)a(s)\rho(s) \right] ds > 0. \qquad (2.5.91)$$

Combining (2.5.90) and (2.5.91), we obtain (2.5.83). The conclusion now follows from Theorem 2.5.14. ∎

In the case $H = H(t - s) \in \Gamma$, we have $h_1(t - s) = h_2(t - s)$ and denote them by $h(t - s)$. The subclass of Γ containing such $H(t - s)$ is denoted by Γ_0.

Theorem 2.5.14 when applied to Γ_0 leads to the following result.

Theorem 2.5.17. If for each $T \geq t_0$ there exist two functions $H \in \Gamma_0$ and $\rho(t) \in C^1([t_0, \infty), \mathbb{R}^+)$ and $a, c \in \mathbb{R}$ such that $T \leq a < c$, and

$$\int_a^c H(s-a)[Q(s)+Q(2c-s)]ds > \frac{1}{4}\int_a^c h^2(s-a)[a(s)\rho(s)+a(2c-s)\rho(2c-s)]ds, \qquad (2.5.92)$$

then equation (2.1.1) is oscillatory.

Proof. Let $b = 2c - a$. Then, $H(b - c) = H(c - a) = H((b - a)/2)$, and for any $y \in \mathcal{L}[a, b]$, we have $\int_c^b y(s)ds = \int_a^c y(2c - s)ds$. Hence, $\int_c^b H(b - s)Q(s)ds = \int_a^c H(s - a)Q(2c - s)ds$ and

$$\int_c^b a(s)\rho(s)h^2(b - s)ds = \int_a^c a(2c - s)\rho(2c - s)h^2(s - a)ds.$$

Thus (2.5.92) implies that (2.5.83) holds for $H \in \Gamma_0$, and therefore equation (2.1.1) is oscillatory. ∎

Now, let $H(t-s) = (t-s)^n$ where $n > 1$. It is clear that $H \in \Gamma_0$ and $h(t-s) = n(t-s)^{(n-2)/2}$. Then from the above results the following oscillation criterion is immediate.

Corollary 2.5.12. Equation (2.1.1) is oscillatory provided for some $n > 1$ and for any $r \geq t_0$ either

(I)   the following inequalities hold

$$\limsup_{t \to \infty} \int_r^t \left[(s-r)^n Q(s) - \frac{n^2}{4}(s-r)^{n-2} a(s)\rho(s) \right] ds > 0$$

and

$$\limsup_{t \to \infty} \int_r^t \left[(t-s)^n Q(s) - \frac{n^2}{4}(t-s)^{n-2} a(s)\rho(s) \right] ds > 0,$$

or

(II)   the following inequality holds

$$\limsup_{t \to \infty} \int_r^t \left[(s-r)^n (Q(s) + Q(2t-s)) - \frac{n^2}{4}(s-r)^{n-2} \right.$$
$$\left. \times\, (a(s)\rho(s) + a(2t-s)\rho(2t-s)) \right] ds > 0.$$

Again, we define $R(t) = \int_r^t du/a(u)$ for $t \geq r \geq t_0$, and let $H(t,s) = (R(t) - R(s))^n$ for $t \geq s \geq t_0$, where $n > 1$ is a real constant. Then by Theorem 2.5.16, we have the following result.

Corollary 2.5.13. Equation (2.1.1) is oscillatory provided $\lim_{t \to \infty} R(t) = \infty$ for each $r \geq t_0$, and for some $n > 1$ the following inequalities hold

$$\limsup_{t \to \infty} \frac{1}{R^{n-1}(t)} \int_r^t (R(s) - R(r))^n q(s) ds > \frac{n^2}{4(n-1)} \qquad (2.5.93)$$

and

$$\limsup_{t \to \infty} \frac{1}{R^{n-1}(t)} \int_r^t (R(t) - R(s))^n q(s) ds > \frac{n^2}{4(n-1)}. \qquad (2.5.94)$$

Proof. Let $\rho(t) = 1$. Then, $Q(t) = q(t)$, $t \geq t_0$,

$$h_1(t,s) = \frac{n}{a(t)}(R(t) - R(s))^{(n-2)/2}$$

and

$$h_2(t,s) = \frac{n}{a(s)}(R(t) - R(s))^{(n-2)/2}.$$

Next, since

$$\limsup_{t\to\infty} \frac{1}{R^{n-1}(t)} \int_r^t \left[(R(s) - R(r))^{n-1} q(s) - \frac{1}{4} a(s) h_1^2(s,r) \right] ds$$

$$= \limsup_{t\to\infty} \frac{1}{R^{n-1}(t)} \int_r^t \left[(R(s) - R(r))^n q(s) - \frac{n^2}{4} (R(s) - R(r))^{n-1} \frac{1}{a(s)} \right] ds$$

$$= \limsup_{t\to\infty} \frac{1}{R^{n-1}(t)} \int_r^t (R(s) - R(r))^n q(s) ds$$

$$- \lim_{t\to\infty} \frac{1}{R^{n-1}(t)} \left[\frac{n^2}{4(n-1)} (R(t) - R(r))^{n-1} \right]$$

$$= \limsup_{t\to\infty} \frac{1}{R^{n-1}(t)} \int_r^t (R(s) - R(r))^n q(s) ds - \frac{n^2}{4(n-1)}$$

in view of (2.5.93) inequality (2.5.88) holds. Similarly, (2.5.94) implies that (2.5.89) holds. Now, by Theorem 2.5.16 equation (2.1.1) is oscillatory.   ∎

Remark 2.5.8.

1. Special cases of the above oscillatory interval criteria for equation (2.1.1) when $a(t) = 1$ and/or $\rho(t) = 1$ can be easily formulated. The details are left to the reader.

2. Extensions of the above results to more general equations of type (2.2.1) can be easily formulated (see Remark 2.5.6).

The following examples illustrate how the theory can be applied in practice.

Example 2.5.7.   Consider the differential equation

$$x''(t) + q(t)x(t) = 0, \tag{2.5.95}$$

where $q(t)$ is defined by

$$q(t) = \begin{cases} 5(t - 3m) & \text{for} \quad 3m \le t \le 3m + 1 \\ 5(-t + 3m + 2) & \text{for} \quad 3m + 1 < t \le 3m + 2 \\ -m & \text{for} \quad 3m + 2 < t < 3m + 3, \end{cases}$$

where $m \in \mathbb{N}$. For any $T \ge 0$ there exists $m \in \mathbb{N}$ such that $3m \ge T$. Let $a = 3m$, $c = 3m + 1$, $H(t - s) = (t - s)^2$, $t \ge s \ge T$ and $\rho(t) = 1$ for $t \ge T$. Then, $Q(t) = q(t)$ for $t \ge T$. It is easy to check that inequality (2.5.92) holds, and hence every solution of equation (2.5.95) is oscillatory by Theorem 2.5.14. We note that in this equation $\int_0^\infty q(s)ds = -\infty$.

Example 2.5.8.   Consider the Euler differential equation (2.2.11) with $t \ge 1$ and $c > 1/4$. Clearly, $R(t) = \int_1^t ds = t - 1$ and $\lim_{t\to\infty} R(t) = \infty$.

Since for each $t \geq r \geq 1$ and $n > 1$,

$$\lim_{t\to\infty} \frac{1}{R^{n-1}(t)} \int_r^t (R(s) - R(r))^n q(s) ds$$

$$= \lim_{t\to\infty} \frac{1}{R^{n-1}(t)} \int_r^t (s-r)^n \frac{c}{s^2} ds \geq c \lim_{t\to\infty} \frac{1}{t^{n-1}} \int_r^t \frac{s^n - nrs^{n-1}}{s^2} ds$$

$$= c \lim_{t\to\infty} \frac{1}{t^{n-1}} \int_r^t \left[s^{n-2} - nrs^{n-3} \right] ds = \frac{c}{n-1}$$

and

$$\lim_{t\to\infty} \frac{1}{R^{n-1}(t)} \int_r^t (R(t) - R(s))^n q(s) ds$$

$$= \lim_{t\to\infty} \frac{1}{R^{n-1}(t)} \int_r^t (t-s)^n \frac{c}{s^2} ds \geq c \lim_{t\to\infty} \frac{1}{t^{n-1}} \int_r^t \frac{t^n - nst^{n-1}}{s^2} ds$$

$$= c \lim_{t\to\infty} \frac{1}{t^{n-1}} \int_r^t t^{n-1} \left[\frac{t}{s^2} - \frac{n}{s} \right] ds = c \lim_{t\to\infty} \left[\frac{t}{r} - 1 - n \ln \frac{t}{r} \right] = \infty.$$

condition (2.5.93) holds if $c/(n-1) > n^2/[4(n-1)]$. Also condition (2.5.94) is automatically fulfilled. Thus, Corollary 2.5.13 implies that the equation (2.2.11) is oscillatory when $c > 1/4$.

2.6. Oscillation Criteria-Integrable Coefficients

In this section we shall present oscillation criteria for equation (2.1.1) when the coefficient is integrably small. For this, we shall assume that $z_0(t) \in C([t_0, \infty), \mathbb{R})$ is a given function and there exists a function $\rho(t) \in C^2([t_0, \infty), \mathbb{R}^+)$ such that

$$\int^\infty \frac{1}{a(s)\rho(s)} ds = \infty \tag{2.6.1}$$

and

$$\beta(t) = \int_t^\infty Q(s) ds < \infty, \tag{2.6.2}$$

where once more $Q(t) = \rho(t)[q(t) + a(t)h^2(t) - (a(t)h(t))']$ and $h(t) = -\rho'(t)/(2\rho(t))$.

We define a sequence $\{z_n(t)\}_{n=0}^\infty$ for $t \geq t_0$ as follows (if it exists)

$$z_n(t) = z_0(t) + \int_t^\infty \frac{(z_{n-1}^+(s))^2}{a(s)\rho(s)} ds, \quad n = 1, 2, \cdots$$

where $z^+(t) = (z(t) + |z(t)|)/2$. Clearly, $z_1(t) \geq z_0(t)$ and this implies that $z_1^+(t) \geq z_0^+(t)$. Thus, by induction, we have

$$z_{n+1}(t) \geq z_n(t), \quad n = 1, 2, \cdots, \quad t \geq t_0 \tag{2.6.3}$$

i.e., the sequence $\{z_n(t)\}$ is nondecreasing on $[t_0, \infty)$.

Theorem 2.6.1. Suppose $z_0(t) \leq \beta(t)$ for $t \geq t_0$. If equation (2.1.1) is nonoscillatory, then there exists a $t_1 \geq t_0$ such that

$$\lim_{n \to \infty} z_n(t) = z(t) < \infty \quad \text{for} \quad t \geq t_1. \tag{2.6.4}$$

Proof. Suppose equation (2.1.1) is nonoscillatory. Then in view of Theorem 2.2.3 there exists $v(t) \in C([t_1, \infty), \mathbb{R})$ such that

$$v(t) = \beta(t) + \int_t^\infty \frac{v^2(s)}{a(s)\rho(s)} ds$$

on $[t_1, \infty)$ for some $t_1 \geq t_0$. Thus, $v(t) \geq z_0(t)$ for $t \geq t_1$, and hence $v^+(t) \geq z_0^+(t)$ for $t \geq t_1$. This implies

$$
\begin{aligned}
v(t) &= \beta(t) + \int_t^\infty \frac{v^2(s)}{a(s)\rho(s)} ds \geq z_0(t) + \int_t^\infty \frac{(v^+(s))^2}{a(s)\rho(s)} ds \\
&\geq z_0(t) + \int_t^\infty \frac{(z_0^+(s))^2}{a(s)\rho(s)} ds = z_1(t) \quad \text{for} \quad t \geq t_1.
\end{aligned}
$$

Hence, by induction, we have

$$v(t) \geq z_n(t), \quad n = 0, 1, 2, \cdots \quad \text{for} \quad t \geq t_1. \tag{2.6.5}$$

Now, it follows from (2.6.3) and (2.6.5) that the sequence $\{z_n(t)\}$ is bounded above on $[t_1, \infty)$. Therefore, (2.6.4) holds. ∎

From Theorem 2.6.1 the following oscillation result is immediate.

Theorem 2.6.2. Suppose $z_0(t) \leq \beta(t)$. If either

(i)   there exists a positive integer m such that $z_n(t)$ is defined for $n = 1, 2, \cdots, m - 1$, but $z_m(t)$ does not exist, or

(ii) $z_n(t)$ is defined for $n = 1, 2, \cdots$, but for arbitrarily large $T^* \geq t_0$ there is $t^* \geq T^*$ such that

$$\lim_{n \to \infty} z_n(t^*) = \infty, \tag{2.6.6}$$

then equation (2.1.1) is oscillatory.

The following result guarantees the nonoscillation of (2.1.1).

Theorem 2.6.3. Suppose $z_0(t) \geq |\beta(t)|$ for $t \geq t_0$. If there exists a $t_1 \geq t_0$ such that (2.6.4) holds, then equation (2.1.1) is nonoscillatory.

Proof. From (2.6.3) and (2.6.4) it follows that $z_n(t) \leq z(t)$, $n = 0, 1, 2, \cdots$ for $t \geq t_1$. Applying the monotone convergence theorem, we find

$$z(t) = z_0(t) + \int_t^\infty \frac{(z^+(s))^2}{a(s)\rho(s)} ds \geq 0 \quad \text{for} \quad t \geq t_1.$$

Thus, we have

$$z^+(t) = z(t) = \int_t^\infty \frac{(z^+(s))^2}{a(s)\rho(s)} ds + z_0(t) \geq \int_t^\infty \frac{(z^+(s))^2}{a(s)\rho(s)} ds + |\beta(t)|$$

$$\geq \left| \int_t^\infty \frac{(z^+(s))^2}{a(s)\rho(s)} ds + \beta(t) \right| \quad \text{for} \quad t \geq t_1.$$

Now, it follows from Corollary 2.3.16 that (2.1.1) is nonoscillatory. ∎

The next result provides necessary conditions for the oscillation of equation (2.1.1).

Theorem 2.6.4. Suppose $z_0(t) \geq |\beta(t)|$ for $t \geq t_0$. If equation (2.1.1) is oscillatory, then either '

(i)   there exists a positive integer m such that $z_n(t)$ is defined for $n = 1, 2, \cdots, m - 1$, but $z_m(t)$ does not exist, or

(ii) $z_n(t)$ is defined for $n = 1, 2, \cdots$, but for arbitrarily large $T^* \geq t_0$ there is $t^* \geq T^*$ such that (2.6.6) holds.

If $\beta(t) \geq 0$ for $t \geq t_0$, then from the above results the following corollaries are immediate.

Corollary 2.6.1. Suppose $z_0(t) = \beta(t) \geq 0$ for $t \geq t_0$. Then equation (2.1.1) is nonoscillatory if and only if there exists a $t_1 \geq t_0$ such that (2.6.4) holds.

Corollary 2.6.2. Suppose $z_0(t) = \beta(t) \geq 0$ for $t \geq t_0$. Then equation (2.1.1) is oscillatory if and only if either

(i)   there exists a positive integer m such that $z_n(t)$ is defined for $n = 1, 2, \cdots, m - 1$, but $z_m(t)$ does not exist, or

(ii) $z_n(t)$ is defined for $n = 1, 2, \cdots$, but for arbitrarily large $T^* \geq t_0$ there is $t^* \geq T^*$ such that (2.6.6) holds.

Example 2.6.1. Consider the differential equation

$$x''(t) + \frac{1}{3} t^{-4/3} (2 + \cos t + 3t \sin t) x(t) = 0, \quad t > 0. \qquad (2.6.7)$$

Let $\rho(t) = 1$. Then, $q(t) = Q(t) = (1/3)t^{-4/3}(2 + \cos t + 3t \sin t)$, $t > 0$. Now, we let

$$z_0(t) = \beta(t) = \int_t^\infty q(s)ds = t^{-1/3}(2 + \cos t) \geq t^{-1/3}, \quad t > 0.$$

Then, $z_0(t) = z_0^+(t)$ for all large t, and we have

$$\int_t^\infty \frac{(z_0^+(s))^2}{a(s)\rho(s)}ds \geq \int_t^\infty s^{-2/3}ds = \infty,$$

and hence $m = 1$. The oscillation of (2.6.7) now follows from Theorem 2.6.2.

Example 2.6.2. Consider the differential equation

$$x''(t) + \frac{1}{2}\left(t^{-3/4}\sin\sqrt{t} + \frac{1}{2}t^{-5/4}\cos\sqrt{t}\right)x(t) = 0, \quad t > 0. \qquad (2.6.8)$$

Here, we take $\rho(t) = 1$. Then,

$$q(t) = Q(t) = \frac{1}{2}\left(t^{-3/4}\sin\sqrt{t} + \frac{1}{2}t^{-5/4}\cos\sqrt{t}\right), \quad t > 0.$$

Now, we let

$$z_0(t) = \beta(t) = \int_t^\infty q(s)ds$$

$$= \frac{1}{2}\int_t^\infty\left[s^{-3/4}\sin\sqrt{s} + \frac{1}{2}s^{-5/4}\cos\sqrt{s}\right]ds = t^{-1/4}\cos\sqrt{t}.$$

Thus, $z_0^+(t) = (1/2)t^{-1/4}[\cos\sqrt{t} + |\cos\sqrt{t}|]$, and for $k = [[t]] + 1$,

$$z_1(t) = \frac{1}{2}\int_t^\infty \frac{1}{\sqrt{s}}\left[\cos^2\sqrt{s} + \cos\sqrt{s}|\cos\sqrt{s}|\right]ds + z_0(t)$$

$$\geq \frac{1}{2}\sum_{n=k}^\infty \int_{[(2n+1)\pi/2]^2}^{[(2n+3)\pi/2]^2}\left[\frac{1}{\sqrt{s}}\left(\cos^2\sqrt{s} + \cos\sqrt{s}|\cos\sqrt{s}|\right)\right]ds + z_0(t)$$

$$= \sum_{n=k}^\infty \int_{[(4n-1)\pi/2]^2}^{[(4n+1)\pi/2]^2}\frac{\cos^2\sqrt{s}}{\sqrt{s}}ds + z_0(t)$$

$$= 2\sum_{n=k}^\infty \int_{3\pi/2}^{5\pi/2}\cos^2 s\, ds + z_0(t) = \infty.$$

Hence, all conditions of Theorem 2.6.2(i) are satisfied with $m = 1$, and therefore (2.6.8) is oscillatory.

Now, we shall consider equation (2.1.2). Suppose $\alpha_0(t) \in C([t_0, \infty), \mathbb{R})$ is a given function and there exists a function $\rho(t) \in C^2([t_0, \infty), \mathbb{R}^+)$ such that

$$\beta_1(t) = \int_t^\infty Q_1(s)ds \quad \text{and} \quad \int^\infty \frac{1}{a_1(s)\rho_1(s)}ds = \infty,$$

where $Q_1(t)$ is defined in (2.2.9).

Again, we define the sequence $\{\alpha_n(t)\}_{n=0}^\infty$ for $t \geq t_0$ as follows (if it exists)

$$\alpha_n(t) = \alpha_0(t) + \int_t^\infty \frac{(\alpha_{n-1}^+(s))^2}{a_1(s)\rho_1(s)}ds, \quad n = 1, 2, \cdots, \quad t \geq t_0. \quad (2.6.9)$$

Clearly, $\alpha_1(t) \geq \alpha_0(t)$ and this implies that $\alpha_1^+(t) \geq \alpha_0^+(t)$, $t \geq t_0$. Now, by induction, we have

$$\alpha_{n+1}(t) \geq \alpha_n(t), \quad n = 1, 2, \cdots, \quad t \geq t_0 \quad (2.6.10)$$

i.e., the sequence $\{\alpha_n(t)\}$ defined by (2.6.9) is nondecreasing on $[t_0, \infty)$.

From Theorems 2.6.1 and 2.6.3, we obtain the following comparison result.

Theorem 2.6.5. Assume that

$$0 < a(t)\rho(t) \leq a_1(t)\rho_1(t), \quad |\beta_1(t)| \leq \beta(t) \quad \text{for all large} \quad t. \quad (2.6.11)$$

If equation (2.1.1) is nonoscillatory, then (2.1.2) is nonoscillatory, or equivalently, if equation (2.1.2) is oscillatory, then (2.1.1) is oscillatory.

Proof. Let $z_0(t) = \beta(t)$ and $\alpha_0(t) = |\beta_1(t)|$ for $t \geq t_0$. Suppose equation (2.1.1) is nonoscillatory. It follows from Theorem 2.6.1 that there exists a $t_1 \geq t_0$ such that (2.6.4) holds. Clearly, by (2.6.11), $\alpha_0(t) = |\beta_1(t)| \leq \beta(t) = z_0(t)$, and hence $\alpha_0^+(t) \leq z_0^+(t)$ for $t \geq t_1$. Thus, for $t \geq t_1$,

$$\begin{aligned}
\alpha_1(t) &= \alpha_0(t) + \int_t^\infty \frac{(\alpha_0^+(s))^2}{a_1(s)\rho_1(s)}ds \\
&\leq \alpha_0(t) + \int_t^\infty \frac{(\alpha_0^+(s))^2}{a(s)\rho(s)}ds = \alpha_1(t).
\end{aligned}$$

Now, by induction, we have

$$\alpha_n(t) \leq z_n(t), \quad n = 0, 1, 2, \cdots, \quad t \geq t_1. \quad (2.6.12)$$

Therefore, by (2.6.4), (2.6.10) and (2.6.12), we get

$$\alpha(t) = \lim_{n\to\infty} \alpha_n(t) \leq \lim_{n\to\infty} z_n(t) = z(t) < \infty, \quad t \geq t_1.$$

Hence, by Theorem 2.6.3, equation (2.1.2) is nonoscillatory. ∎

Theorem 2.6.6. Suppose $z_0(t) \leq \beta(t)$, $t \geq t_0$. If equation (2.1.1) is nonoscillatory, then

$$\limsup_{t \to \infty} (z(t) - \beta(t)) \exp \left(4 \int^t \frac{\beta^+(s)}{a(s)\rho(s)} ds \right) < \infty, \qquad (2.6.13)$$

where $z(t)$ is defined in (2.6.4).

Proof. Assume that equation (2.1.1) is nonoscillatory, then it follows from Theorem 2.2.3 that $v(t) = \beta(t) + y(t)$ for $t \geq T \geq t_0$, where $y(t) = \int_t^\infty v^2(s)/(a(s)\rho(s))ds$. Now,

$$y'(t) = -\frac{v^2(t)}{a(t)\rho(t)} = -\frac{(\beta(t) + y(t))^2}{a(t)\rho(t)} \leq 0 \quad \text{for} \quad t \geq T.$$

We claim that

$$y'(t) + 4\frac{\beta^+(t)}{a(t)\rho(t)}y(t) \leq 0, \qquad t \geq T. \qquad (2.6.14)$$

Since $y(t) > 0$ and $y'(t) < 0$, (2.6.14) holds if $\beta(t) \leq 0$. If $\beta(t) \geq 0$, then $[y(t) + \beta(t)]^2 \geq 4\beta^+(t)y(t)$. This implies that (2.6.14) holds.

Clearly (2.6.14) implies that

$$y(t) \leq y(T) \exp \left(-4 \int_T^t \frac{\beta^+(s)}{a(s)\rho(s)} ds \right), \qquad t \geq T. \qquad (2.6.15)$$

On the other hand, we have $v(t) = \beta(t) + y(t) \geq \beta(t) \geq z_0(t)$, and hence

$$y(t) = \int_t^\infty \frac{v^2(s)}{a(s)\rho(s)} ds \geq \int_t^\infty \frac{(z_0^+(s))^2}{a(s)\rho(s)} ds.$$

This gives

$$v(t) = \beta(t) + y(t) \geq z_0(t) + \int_t^\infty \frac{(z_0^+(s))^2}{a(s)\rho(s)} ds = z_1(t) \quad \text{for} \quad t \geq T$$

and by induction

$$v(t) = \beta(t) + y(t) \geq z_n(t), \quad n = 0, 1, 2, \cdots, \quad t \geq T. \qquad (2.6.16)$$

Therefore, from (2.6.15) and (2.6.16), we find

$$z_n(t) - \beta(t) \leq y(t) \leq y(T) \exp \left(-4 \int_T^t \frac{\beta^+(s)}{a(s)\rho(s)} ds \right),$$

and hence

$$(z_n(t) - \beta(t))\exp\left(4\int_T^t \frac{\beta^+(s)}{a(s)\rho(s)}ds\right) \le y(T), \quad n = 0, 1, 2, \cdots, \quad t \ge T.$$

This and (2.6.4) lead to

$$(z(t) - \beta(t))\exp\left(4\int_T^t \frac{\beta^+(s)}{a(s)\rho(s)}ds\right)$$

$$= \lim_{n\to\infty}(z_n(t) - \beta(t))\exp\left(4\int_T^t \frac{\beta^+(s)}{a(s)\rho(s)}ds\right) \le y(T) < \infty \text{ for } t \ge T.$$

Therefore, (2.6.13) holds.   ■

The following result is now immediate.

Theorem 2.6.7. Suppose $z_0(t) \le \beta(t)$. If either

(i) $z_n(t)$ exists for $n = 1, 2, \cdots, m$ and

$$\limsup_{t\to\infty}(z_m(t) - \beta(t))\exp\left(4\int^t \frac{\beta^+(s)}{a(s)\rho(s)}ds\right) = \infty,$$

or

(ii)   (2.6.4) holds and

$$\limsup_{t\to\infty}(z(t) - \beta(t))\exp\left(4\int^t \frac{\beta^+(s)}{a(s)\rho(s)}ds\right) = \infty,$$

then equation (2.1.1) is oscillatory.

Theorem 2.6.8. Suppose $z_0(t) \le \beta(t)$, $t \ge t_0$. If

$$\int^\infty \exp\left(-4\int^s \frac{\beta^+(u)}{a(u)\rho(u)}du\right)ds < \infty, \qquad (2.6.17)$$

and there exists a nonnegative integer m such that

$$\int^\infty z_m(s)ds = \infty, \qquad (2.6.18)$$

then equation (2.1.1) is oscillatory.

Proof.   Suppose to the contrary that equation (2.1.1) is nonoscillatory. As in Theorem 2.6.6, we obtain

$$(z_n(t) - \beta(t)) \le y(T)\exp\left(-4\int_T^t \frac{\beta^+(s)}{a(s)\rho(s)}ds\right), \quad n = 0, 1, 2, \cdots, \quad t \ge T.$$

$$(2.6.19)$$

Integrating (2.6.19) from T to t and letting $t \to \infty$, we find

$$\int_T^\infty z_n(s)ds \leq y(T) \int_T^\infty \exp\left(-4 \int_T^s \frac{\beta^+(u)}{a(u)\rho(u)} du\right) ds + \int_T^\infty \beta(s)ds < \infty,$$

which contradicts (2.6.18). This completes the proof. ∎

To prove our next result we shall need the following lemma.

Lemma 2.6.1. Every nonoscillatory equation (2.1.1) has a solution $x(t)$ such that $\int^\infty ds/(a(s)x^2(s)) < \infty$ and a nontrivial solution $y(t)$ such that $\int^\infty ds/(a(s)y^2(s)) = \infty$.

Theorem 2.6.9. If

$$\int^\infty \frac{1}{a(s)\rho(s)} \exp\left(-4 \int^s \frac{\beta^+(\tau)}{a(\tau)\rho(\tau)} d\tau\right) ds < \infty, \qquad (2.6.20)$$

then equation (2.1.1) is oscillatory.

Proof. If (2.1.1) is nonoscillatory, then from Theorem 2.2.3 there exists a $T \geq t_0$ such that

$$v(t) = \beta(t) + \int_t^\infty \frac{v^2(s)}{a(s)\rho(s)} ds \quad \text{for} \quad t \geq T,$$

where $v(t) = a(t)\rho(t)w'(t)/w(t)$ and $w(t)$ is a nonoscillatory solution of (2.2.6) satisfying $\int^\infty ds/(a(s)\rho(s)w^2(s)) < \infty$. Now, as in Theorem 2.6.6, we define $y(t) = \int_t^\infty v^2(s)/(a(s)\rho(s))ds$, and obtain (2.6.15). It follows from condition (2.6.20) that $\int^\infty y(s)/(a(s)\rho(s))ds < \infty$, which implies $\int^\infty k(s)v^2(s)/(a(s)\rho(s))ds < \infty$, where $k'(t) = 1/(a(t)\rho(t))$. Next, since

$$\left(\ln \frac{w(t)}{w(T)}\right)^2 = \left(\int_T^t \frac{v(s)}{a(s)\rho(s)} ds\right)^2$$

$$\leq \left(\int_T^t \frac{ds}{k(s)a(s)\rho(s)}\right) \left(\int_T^t \frac{k(s)v^2(s)}{a(s)\rho(s)} ds\right),$$

there exist positive constants c_0, c_1 such that $|w(t)| \leq c_0 \exp\left(c_1 \sqrt{\ln k(t)}\right)$ for all large t. Thus, we have

$$\int^\infty \frac{ds}{a(s)\rho(s)w^2(s)} \geq \frac{1}{c_0^2} \int^\infty \frac{1}{a(s)\rho(s)} \exp\left(-2c_1 \sqrt{\ln k(s)}\right) ds = \infty,$$

which is a contradiction. ∎

Corollary 2.6.3. If

$$\int^\infty \frac{1}{a(s)\rho(s)} \exp\left(-\delta \int^s \frac{\beta(\tau)}{a(\tau)\rho(\tau)} d\tau\right) ds < \infty \quad \text{for some} \quad \delta, \ 0 < \delta \leq 4,$$

$$\qquad (2.6.21)$$

then equation (2.1.1) is oscillatory.

Example 2.6.3. Consider (2.3.1) with $c > 1/4$. Let $\rho(t) = t \ln^{2\lambda} t$, where the constant λ is such that $(1 - \sqrt{4c - 1})/2 < \lambda < 1/2$. Then, we have

$$\frac{2c - 2\lambda^2}{2\lambda - 1} < -1, \qquad \int^\infty \frac{1}{a(s)\rho(s)} ds = \infty,$$

$$Q(t) = (\lambda^2 - \lambda + c)\frac{\ln^{2(\lambda - 1)} t}{t}, \qquad t > 0$$

$$\beta(t) = \int_t^\infty Q(s) ds = \frac{\lambda^2 - \lambda + c}{1 - 2\lambda} \ln^{2\alpha - 1} t < \infty$$

and for $T > 0$,

$$\int_T^\infty \frac{1}{a(s)\rho(s)} \exp\left(-2 \int_T^s \frac{\beta(\tau)}{a(\tau)\rho(\tau)} d\tau\right) ds$$

$$= \left(\ln^{2(c+\lambda^2-\lambda)/(2\lambda-1)} T\right) \int_T^\infty \frac{\ln^{(2c-2\lambda^2)/(2\lambda-1)} s}{s} ds < \infty.$$

Now, it follows from Corollary 2.6.3 with $\delta = 2$ that (2.3.1) is oscillatory if $c > 1/4$.

Corollary 2.6.4. Let $k(t) \in C^1([t_0, \infty), \mathbb{R}^+)$ satisfy $k'(t) = 1/(a(t)\rho(t))$. If

$$\liminf_{t \to \infty} k(t)\beta(t) > \frac{1}{4}, \qquad (2.6.22)$$

then equation (2.1.1) is oscillatory.

Proof. In view of (2.6.22) there exist two numbers $\lambda > 1/4$ and $T \geq t_0$ such that $k(t)\beta(t) > \lambda$ for $t \geq T$. Then, we have

$$\int_T^\infty \frac{1}{a(s)\rho(s)} \exp\left(-4 \int_T^s \frac{\beta^+(\tau)}{a(\tau)\rho(\tau)} d\tau\right) ds$$

$$< \int_T^\infty \frac{1}{a(s)\rho(s)} \exp\left(-4\lambda \int_T^s \frac{1}{a(\tau)\rho(\tau)k(\tau)} d\tau\right) ds$$

$$= (k(T))^{4\lambda} \int_T^\infty \frac{1}{a(s)\rho(s)(k(s))^{4\lambda}} ds < \infty.$$

Now, it follows from Theorem 2.6.9 that (2.1.1) is oscillatory. ∎

Example 2.6.4. Consider the differential equation

$$((t \ln^2 t)x'(t))' + \frac{c}{t}x(t) = 0 \quad \text{for} \quad t \geq 2, \qquad (2.6.23)$$

where $c > 1/4$ is a constant. Let $\rho(t) = \ln^\lambda t$ and $k(t) = -\ln^{-1-\lambda} t/(1+\lambda)$, where $\sqrt{4c-1} < \lambda < -1$. Then, we have

$$\beta(t) = -\frac{\lambda^2 + 2\lambda + 4c}{4(1+\lambda)} \ln^{1+\lambda} t \quad \text{and} \quad \liminf_{t\to\infty} k(t)\beta(t) = \frac{\lambda^2 + 2\lambda + 4c}{4(1+\lambda)^2} > \frac{1}{2}.$$

Now, it follows from Corollary 2.6.4 that (2.1.1) is oscillatory if $c > 1/4$.

Corollary 2.6.5. Let $k(t) \in C^1([t_0, \infty), \mathbb{R}^+)$ satisfy $k'(t) = 1/(a(t)\rho(t))$. If

$$\liminf_{t\to\infty} \frac{t}{4a(t)\rho(t)} \left[4\beta^+(t) + (a(t)\rho(s))' \right] > \frac{1}{4}, \tag{2.6.24}$$

then equation (2.1.1) is oscillatory.

Proof. In view of (2.6.24) there exist two constants $T \geq t_0$ and $c > 1/4$ such that

$$\frac{\beta^+(t)}{a(t)\rho(t)} > \frac{c}{t} - \frac{(a(t)\rho(t))'}{4a(t)\rho(t)} \quad \text{for} \quad t \geq T.$$

Then, we have

$$\int_T^\infty \frac{1}{a(s)\rho(s)} \exp\left(-4 \int_T^s \frac{\beta^+(\tau)}{a(\tau)\rho(\tau)} d\tau \right) ds$$

$$< \int_T^\infty \frac{1}{a(s)\rho(s)} \exp\left(\int_T^t \left[-\frac{4c}{\tau} + \frac{(a(\tau)\rho(\tau))'}{a(\tau)\rho(\tau)} \right] d\tau \right) ds$$

$$= \frac{T^{4c}}{a(T)\rho(T)} \int_T^\infty s^{-4c} ds < \infty.$$

Now, it follows from Theorem 2.6.9 that (2.1.1) is oscillatory. ∎

Example 2.6.5. Consider the Euler equation (2.2.11) with $c > 1/4$. Let $\rho(t) = t^\lambda$, where $\lambda < 1$ is a constant. Then, we have

$$\beta(t) = \frac{\lambda^2 - 2\lambda + 4c}{4(1-\lambda)} t^{\lambda-1}, \quad t \geq T > 0$$

and

$$\liminf_{t\to\infty} \frac{t}{4a(t)\rho(t)} \left[4\beta^+(t) + (a(t)\rho(t))' \right] = \frac{4c - \lambda}{4(1-\lambda)}.$$

Now, it follows from Corollary 2.6.5 that (2.2.11) is oscillatory if $(c - (\lambda/4))/(1-\lambda) > 1/4$.

In equation (2.2.6) if $Q(t) > 0$ for $t \geq t_0$, then we have the following result.

Theorem 2.6.10. Suppose $Q(t) > 0$ for $t \geq t_0$ and let $k(t) \in C^1([t_0, \infty), \mathbb{R}^+)$ satisfy $k'(t) = 1/(a(t)\rho(t))$. If

$$\limsup_{t\to\infty} k(t)\beta(t) > 1, \tag{2.6.25}$$

then equation (2.1.1) is oscillatory.

Proof. Suppose (2.1.1) is nonoscillatory. It follows from Theorem 2.2.3 that there exists a $T \geq t_0$ such that $v(t) = \beta(t) + \int_t^\infty v^2(s)/(a(s)\rho(s))ds$, where

$$v(t) = a(t)\rho(t)\frac{w'(t)}{w(t)} \qquad (2.6.26)$$

and $w(t)$ is a solution of equation (2.2.6) with $w(t) > 0$ for $t \geq T$. If $Q(t) > 0$ for $t \geq t_0$, then the function $a(t)\rho(t)w'(t)$ is strictly decreasing and $v(t) \geq \beta(t)$ for $t \geq T$. Also, it is easy to check that $w'(t) > 0$ for $t \geq T$. Now, since

$$\frac{w(t)}{k(t)w'(t)} = \frac{w(T) + \int_T^t \frac{a(s)\rho(s)}{a(s)\rho(s)}w'(s)ds}{k(t)w'(t)}$$

$$\geq \frac{w(T) + a(t)\rho(t)w'(t)[k(t) - k(T)]}{k(t)w'(t)} \geq a(t)\rho(t)\left(\frac{k(t) - k(T)}{k(t)}\right)$$

it follows that

$$\limsup_{t\to\infty} k(t)\beta(t) \leq \limsup_{t\to\infty} k(t)v(t) \leq \limsup_{t\to\infty} \frac{k(t)}{k(t) - k(T)} = 1,$$

which contradicts (2.6.25). ∎

Theorem 2.6.11. If

$$\int^\infty \frac{\beta(s)}{a(s)\rho(s)}ds < \infty \qquad (2.6.27)$$

and $\xi(t)$ defined in (2.2.23) satisfies

$$\int^\infty \frac{1}{a(s)\rho(s)} \exp\left(-4\int^s \frac{\xi(\tau)}{a(\tau)\rho(\tau)}d\tau\right)ds < \infty, \qquad (2.6.28)$$

then equation (2.1.1) is oscillatory.

Proof. Assume to the contrary that (2.1.1) is nonoscillatory. Then, by Lemmas 2.2.2 and 2.6.1, equation (2.2.6) has a nonoscillatory solution $w(t)$ satisfying $\int^\infty ds/(a(s)\rho(s)w^2(s)) < \infty$. As in the proof of Theorem 2.2.4, the function $z(t) = \int_t^\infty v^2(s)/(a(s)\rho(s))ds$ satisfies (2.2.22), where $v(t)$ is defined in (2.6.26). Define $y(t) = \int_t^\infty z^2(s)/(a(s)\rho(s))ds$. Then, we have

$$y'(t) = -\frac{z^2(t)}{a(t)\rho(t)} - \frac{2\beta(t)y(t)}{a(t)\rho(t)}$$

$$= -\frac{1}{a(t)\rho(t)}\left[z^2(t) + 2(\xi(t) + \beta(t))y(t) + \xi^2(t)\right] \qquad (2.6.29)$$

$$\leq -\frac{1}{a(t)\rho(t)}[4\xi(t) + 2\beta(t)]y(t).$$

Using (2.6.27) and (2.6.28) in (2.6.29), we obtain $\int^\infty y(s)/(a(s)\rho(s))ds < \infty$. Observe that

$$\int_T^\eta \frac{y(s)}{a(s)\rho(s)}ds$$

$$= \int_T^\eta \left(\int_T^s \frac{1}{a(t)\rho(t)} \exp\left(-2\int_{t_0}^t \frac{\beta(u)}{a(u)\rho(u)}du\right) dt\right) \frac{z^2(s)}{a(s)\rho(s)}\mu[s,t_0]ds$$

$$\geq c_1 \int_T^\eta \frac{k(s)z^2(s)}{a(s)\rho(s)}\mu[s,t_0]ds \geq c_2 \int_T^\eta \frac{k(s)z^2(s)}{a(s)\rho(s)}ds$$

for some positive constants c_1 and c_2, $k(s) = \int_T^s dt/(a(t)\rho(t))$ and $\mu[s,t]$ as in (2.2.24). Now, since

$$\left(\int_T^\eta \frac{z(s)}{a(s)\rho(s)}ds\right)^2 \leq \left(\int_T^\eta \frac{k(s)z^2(s)}{a(s)\rho(s)}ds\right)\left(\int_T^\eta \frac{1}{k(s)a(s)\rho(s)}ds\right)$$

it follows that

$$\int_T^\eta \frac{z(s)}{a(s)\rho(s)}ds \leq c_3\sqrt{\ln k(\eta)} \quad \text{for some} \quad c_3 > 0. \tag{2.6.30}$$

By Schwartz's inequality and integration by parts, we have

$$\left(\int_T^\eta \frac{v(s)}{a(s)\rho(s)}ds\right)^2 \leq \left(\int_T^\eta \frac{k(s)v^2(s)}{a(s)\rho(s)}ds\right)\left(\int_T^\eta \frac{1}{k(s)a(s)\rho(s)}ds\right) \tag{2.6.31}$$

and

$$\int_T^\eta \frac{k(s)v^2(s)}{a(s)\rho(s)}ds = -k(\eta)z(\eta) + k(T)z(T) + \int_T^\eta \frac{z(s)}{a(s)\rho(s)}ds. \tag{2.6.32}$$

From (2.6.30) – (2.6.32) the following estimate on the nonoscillatory solution $w(t)$ is immediate

$$\ln\frac{w(\eta)}{w(T)} = \int_T^\eta \frac{v(s)}{a(s)\rho(s)}ds \leq c_4 \ln^{3/4} k(\eta), \tag{2.6.33}$$

where c_4 is an appropriate positive constant depending on T. As a consequence of (2.6.33), we have

$$\frac{1}{a(\eta)\rho(\eta)w^2(\eta)} \geq \frac{c_5 \exp\left(-2c_4 \ln^{3/4} k(\eta)\right)}{a(\eta)\rho(\eta)}$$

$$\geq c_6 \frac{\exp(-\ln k(\eta))}{a(\eta)\rho(\eta)} = \frac{c_6}{k(\eta)a(\eta)\rho(\eta)}$$

for some positive constants c_5 and c_6. Thus,

$$\int^\infty \frac{1}{a(\eta)\rho(\eta)w^2(\eta)}\, d\eta \geq c_6 \int^\infty \frac{1}{a(\eta)\rho(\eta)k(\eta)}\, d\eta = \infty,$$

which is a contradiction. ∎

Finally, we present the following result.

Theorem 2.6.12. If equation (2.1.1) is nonoscillatory and

$$\int^\infty a(s)\rho(s) \exp\left(2\int^s \frac{\beta(\tau)}{a(\tau)\rho(\tau)}\, d\tau\right) ds = \infty, \qquad (2.6.34)$$

then (2.1.1) cannot have a solution $u(t)$ satisfying $\int^\infty a(s)u^2(s)ds < \infty$.

Proof. Since equation (2.1.1) is nonoscillatory, it follows from Theorem 2.2.3 that there exist a $T \geq t_0$ and a function $v(t) \in C([T,\infty), \mathbb{R})$ such that $v(t) = \beta(t) + \int_t^\infty v^2(s)/(a(s)\rho(s))ds$, where $v(t)$ is defined in (2.6.26) and $w(t)$ is a solution of (2.2.6). Hence, for all $t \geq T$, we have

$$\frac{d}{dt}\ln w(t) = \frac{w'(t)}{w(t)} = \frac{v(t)}{a(t)\rho(t)}$$

$$= \frac{1}{a(t)\rho(t)}\left[\beta(t) + \int_t^\infty \frac{v^2(s)}{a(s)\rho(s)}ds\right] \geq \frac{\beta(t)}{a(t)\rho(t)}.$$

This implies

$$\int^\infty a(s)u^2(s)ds = \int^\infty a(s)\rho(s)w^2(s)ds$$

$$\geq \int^\infty a(s)\rho(s)\exp\left(2\int^s \frac{\beta(\tau)}{a(\tau)\rho(\tau)}d\tau\right)ds = \infty. \quad ∎$$

2.7. Forced Oscillations

In this section we are concerned with the oscillatory behavior of forced equations of the form

$$(a(t)x'(t))' + p(t)x'(t) + q(t)x(t) = e(t), \qquad (2.7.1)$$

where $a(t) \in C([t_0,\infty), \mathbb{R}^+)$, and $e(t), p(t), q(t) \in C([t_0,\infty), \mathbb{R})$. In what follows we shall assume that the unforced equation

$$(a(t)z'(t))' + p(t)z'(t) + q(t)z(t) = 0 \qquad (2.7.2)$$

is nonoscillatory. Let $z(t)$ be a nonprincipal solution of (2.7.2), i.e., $z(t)$ satisfies

$$\int^{\infty} \frac{1}{a(s)\rho(s)z^2(s)} ds < \infty, \tag{2.7.3}$$

where

$$\rho(t) = \exp\left(\int^t \frac{p(s)}{a(s)} ds\right) \tag{2.7.4}$$

(see Lemma 2.6.1). Define the function

$$\psi(t) = \int^t \frac{1}{a(s)\rho(s)z^2(s)} \left(\int^s e(u)\rho(u)z(u)du\right) ds. \tag{2.7.5}$$

Theorem 2.7.1. Suppose equation (2.7.2) is nonoscillatory and let $z(t)$ be a nonprincipal solution. Then, (2.7.1) is oscillatory if

$$\limsup_{t\to\infty} \psi(t) = -\liminf_{t\to\infty} \psi(t) = \infty. \tag{2.7.6}$$

Proof. The change of variable $x(t) = y(t)z(t)$ transforms (2.7.1) into

$$(a(t)\rho(t)z^2(t)y'(t))' + z(t)\left[(a(t)z'(t))' + p(t)z'(t) + q(t)z(t)\right]\rho(t)y(t)$$
$$= e(t)\rho(t)z(t), \tag{2.7.7}$$

where $\rho(t)$ is defined in (2.7.4). We can express $y(t)$ by integration of (2.7.7) as follows

$$y(t) = c_1 + c_2 \int_{t_0}^t \frac{1}{a(s)\rho(s)z^2(s)} ds + \int_{t_0}^t \frac{1}{a(s)\rho(s)z^2(s)} \left(\int_{t_0}^s e(u)\rho(u)z(u)du\right) ds,$$

where c_1 and c_2 are constants depending on the initial conditions $y(t_0)$ and $y'(t_0)$. Since $z(t)$ is a nonprincipal solution, (2.7.3) and (2.7.6) imply that $y(t)$ satisfies

$$\limsup_{t\to\infty} y(t) = -\liminf_{t\to\infty} y(t) = \infty. \tag{2.7.8}$$

Because $z(t)$ is nonoscillatory, $z(t) \neq 0$ for $t \geq t_1 \geq t_0$. Now, (2.7.8) implies that $y(t)$ is oscillatory, and hence $x(t) = y(t)z(t)$ is also oscillatory. ∎

Example 2.7.1. Consider the forced differential equation

$$x''(t) + \frac{\sin kt}{t} x(t) = t^\alpha \sin \beta t, \quad t \geq t_0 > 0 \tag{2.7.9}$$

where α, β, k are constants. For $|k| > \sqrt{2}$, the linear unforced equation

$$z''(t) + \frac{\sin kt}{t} z(t) = 0 \tag{2.7.10}$$

is nonoscillatory. Equation (2.7.10) has two linearly independent solutions $z_1(t)$ and $z_2(t)$ with the following asymptotic behavior

$$z_1(t) \sim t^{\gamma + (1/2)} = t^{\gamma + (1/2)}[1 + o(1)] \quad \text{as} \quad t \to \infty$$

and

$$z_2(t) \sim t^{-\gamma + (1/2)} = t^{-\gamma + (1/2)}[1 + o(1)] \quad \text{as} \quad t \to \infty,$$

where $\gamma = \left((k^2 - 2)/(4k^2)\right)^{1/2} > 0$. Hence, $z_1(t)$ is a nonprincipal solution. Substituting $z_1(t)$ and $t^{\alpha} \sin \beta t$ in (2.7.5), we find that (2.7.6) is satisfied if $\alpha > \gamma + (1/2)$ and any $\beta \neq 0$. Because $\gamma < 1/2$, this shows that equation (2.7.9) is oscillatory for any $\alpha > 1$ and for all $\beta, k \neq 0$.

Example 2.7.2. Consider the forced differential equation

$$x''(t) - x(t) = e(t), \quad t \geq t_0 \geq 0 \tag{2.7.11}$$

where $e(t) \in C([t_0, \infty), \mathbb{R})$. Here, $a(t) = 1$, $p(t) = 0$ and $\rho(t) = 1$. The linear unforced equation $z''(t) - z(t) = 0$ is nonoscillatory. It has two linearly independent solutions e^{-t} and e^t. Clearly, $z(t) = e^t$ is a nonprincipal solution.

Now, we consider the following two cases:

(i)   Let $e(t) = \sin t$. Then, we have $\psi(t) = \int^t e^{-2s} \left(\int^s e^{\tau} \sin \tau d\tau \right) ds$. Clearly, $|\psi(t)| < \infty$ for all $t \geq 0$, and hence condition (2.7.6) is violated. In this case the general solution of equation (2.7.11) is $x(t) = c_1 e^{-t} + c_2 e^t - (1/2) \sin t$, where c_1 and c_2 are arbitrary constants. It is nonoscillatory for all sufficiently large t and $c_2 \neq 0$.

(ii)   Let $e(t) = t^{\delta} e^t \sin t$ or $e^{\epsilon t} \sin t$, where $\delta > 0$ and $\epsilon > 1$ are constants. Substituting $z(t) = e^t$ and $e(t)$ in (2.7.5), we find that (2.7.6) is satisfied. It follows from Theorem 2.7.1 that equation (2.7.11) is oscillatory. Thus, the large size of the forcing term generates oscillation.

Example 2.7.3. Consider the forced differential equation

$$x''(t) + 2x'(t) + x(t) = \sin t, \quad t \geq t_0 \geq 0. \tag{2.7.12}$$

Here, $a(t) = 1$, $p(t) = 2$ and $\rho(t) = e^{2t}$. The linear unforced equation $z''(t) + 2z'(t) + z(t) = 0$ is nonoscillatory. It has two linearly independent solutions e^{-t} and te^{-t}. Clearly, $z(t) = te^{-t}$ is a nonprincipal solution. Now, since $\psi(t) = \int^t (1/s^2) \left(\int^s \tau e^{\tau} \sin \tau d\tau \right) ds$ condition (2.7.6) is satisfied. Hence, by Theorem 2.7.1, equation (2.7.12) is oscillatory.

The general solution of (2.7.12) is $x(t) = c_1 e^{-t} + c_2 te^{-t} - (1/2) \sin t$, where c_1 and c_2 are arbitrary constants. It is oscillatory for all large t.

When condition (2.7.3) is violated, i.e., $z(t)$ is not a nonprincipal solution of equation (2.7.2), we have the following result.

Theorem 2.7.2. If there exists a positive solution $z(t)$ of equation (2.7.2) such that for each $t_0 > 0$ and for some constant $m > 0$,

$$\liminf_{t \to \infty} \int_{t_0}^{t} e(s)\rho(s)z(s)ds = -\infty \quad \text{and} \quad \limsup_{t \to \infty} \int_{t_0}^{t} e(s)\rho(s)z(s)ds = \infty,$$
$$(2.7.13)$$

$$|\psi(t)| \leq m \int_{t_0}^{t} \frac{1}{a(s)\rho(s)z^2(s)} ds \qquad (2.7.14)$$

and

$$\lim_{t \to \infty} \int_{t_0}^{t} \frac{1}{a(s)\rho(s)z^2(s)} ds = \infty, \qquad (2.7.15)$$

where $\rho(t)$ and $\psi(t)$ are defined in (2.7.4) and (2.7.5) respectively, then (2.7.1) is oscillatory.

Proof. Let $x(t)$ be an eventually positive solution of (2.7.1), say, $x(t) > 0$ for $t \geq t_0 > 0$. The function $y(t)$ defined by $x(t) = y(t)z(t)$ is a nonoscillatory solution of equation (2.7.7). Integrating (2.7.7) from $T > t_0$ to t, where T is sufficiently large, we get

$$a(t)\rho(t)z^2(t)y'(t) = a(T)\rho(T)z^2(T)y'(T) + \int_{T}^{t} e(s)\rho(s)z(s)ds \quad (2.7.16)$$

and hence by (2.7.13), we find $\liminf_{t \to \infty} a(t)\rho(t)z^2(t)y'(t) = -\infty$. Next, we choose a sufficiently large $t_1 \geq T$ such that

$$a(t_1)\rho(t_1)z^2(t_1)y'(t_1) < -2m. \qquad (2.7.17)$$

Replacing T by t_1 in (2.7.16), integrating it from t_1 to t and using (2.7.14) and (2.7.17), we obtain

$$y(t) \leq y(t_1) - m \int_{t_1}^{t} \frac{1}{a(s)\rho(s)z^2(s)} ds$$

and thus by condition (2.7.15), the solution $y(t)$ is eventually negative. This contradicts the fact that $x(t)$ is eventually positive. ∎

Example 2.7.4. Consider the forced differential equation

$$x''(t) - 3x'(t) + 2x(t) = e^{3t} \sin t, \quad t \geq 0. \qquad (2.7.18)$$

The unforced equation $z''(t) - 3z'(t) + 2z(t) = 0$ is nonoscillatory. It has two linearly independent solutions e^t and e^{2t}. Here, we let $z(t) = e^t$. Since $a(t) = 1$ and $p(t) = -3$, we have $\rho(t) = e^{-3t}$. Thus, conditions (2.7.14) and (2.7.15) are satisfied. It is easy to see that (2.7.13) also holds, and hence, by Theorem 2.7.2 equation (2.7.18) is oscillatory. The same conclusion follows from Theorem 2.7.1 if we let $z(t) = e^{2t}$ as a nonprincipal solution.

The following corollary is now immediate.

Corollary 2.7.1. If there exists a positive solution $y(t)$ of (2.7.2) such that for each $t_0 > 0$ conditions (2.7.6) and (2.7.13) are satisfied, then equation (2.7.1) is oscillatory.

Next, we consider a special case of (2.7.1), namely, the equation

$$(a(t)x'(t))' + q(t)x(t) = e(t) \tag{2.7.19}$$

and state the following result.

Theorem 2.7.3. Let there exist two positive increasing divergent sequences $\{p_n^+\}$, $\{p_n^-\}$, and two sequences $\{c_n^+\}$ and $\{c_n^-\}$ such that c_n^+, c_n^- are positive numbers, and

$$V^\pm = \int_{p_n^\pm}^{p_n^\pm + \pi/\sqrt{c_n^\pm}} \left[c_n^\pm (1 - q(s)) \cos^2 \left(\sqrt{c_n^\pm}(t - p_n^\pm) \right) \right. \tag{2.7.20}$$
$$\left. + (q(s) - c_n^\pm) \sin^2 \left(\sqrt{c_n^\pm}(t - p_n^\pm) \right) \right] ds \geq 0$$

for all $n \geq n_0$, where n_0 is a fixed positive integer. Further, let the function $e(t)$ satisfy

$$e(t) \begin{cases} \geq 0 & \text{for } t \in [p_n^+, p_n^+ + \pi/\sqrt{c_n^+}] \\ \leq 0 & \text{for } t \in [p_n^-, p_n^- + \pi/\sqrt{c_n^-}] \end{cases} \tag{2.7.21}$$

for all $n \geq n_0$. Then equation (2.7.19) is oscillatory.

The following result extends Theorem 2.7.3.

Theorem 2.7.4. Suppose for any $T \geq 0$ there exist $T \leq s_1 < t_1 \leq s_2 < t_2$ such that

$$e(t) \begin{cases} \leq 0 & \text{for } t \in [s_1, t_1] \\ \geq 0 & \text{for } t \in [s_2, t_2]. \end{cases} \tag{2.7.22}$$

If there exist

$$u(t) \in D(s_i, t_i) = \left\{ u(t) \in C^1([s_i, t_i], \mathbb{R}): u(t) \neq 0, \ u(s_i) = u(t_i) = 0 \right\},$$

such that

$$Q_i(u) = \int_{s_i}^{t_i} \left[q(s)u^2(s) - a(s)(u'(s))^2 \right] ds \geq 0 \tag{2.7.23}$$

for $i = 1, 2$ then equation (2.7.19) is oscillatory.

Now, we shall consider a more general forced equation

$$(a(t)x'(t))' + p(t)x'(t) + q(t)f(x(t)) = e(t), \qquad (2.7.24)$$

where the functions a, e, p and q are as in (2.7.1) and $f(x) \in C(\mathbb{R}, \mathbb{R})$, $xf(x) > 0$ for $x \neq 0$.

The following oscillation criterion for equation (2.7.24) extends and unifies Theorems 2.7.3 and 2.7.4. For this, as in Theorem 2.7.4 for each $T \geq 0$, $T \leq s_1 < t_1 \leq s_2 < t_2$, and $i = 1, 2$ we shall need the set $D(s_i, t_i)$.

Theorem 2.7.5. Let

$$f'(x) \geq k > 0 \quad \text{for} \quad x \neq 0, \quad \text{where } k \text{ is a real constant,} \quad (2.7.25)$$

and suppose for any $T \geq 0$, there exist $T \leq s_1 < t_1 \leq s_2 < t_2$ such that condition (2.7.22) holds. If there exist $\rho(t) \in C^1([s_i, t_i], \mathbb{R}^+)$ and $u(t) \in D(s_i, t_i)$ such that

$$Q_i(u) = \int_{s_i}^{t_i} \rho(s) \left[q(s)u^2(s) - \frac{a(s)}{k} \left(u'(s) + \frac{\gamma(s)}{2a(s)\rho(s)} \right)^2 \right] ds \geq 0$$

$$(2.7.26)$$

for $i = 1, 2$ where

$$\gamma(t) = a(t)\rho'(t) - p(t)\rho(t), \qquad (2.7.27)$$

then equation (2.7.24) is oscillatory.

Proof. Suppose $x(t)$ is a nonoscillatory solution of equation (2.7.24), say, $x(t) > 0$ for $t \geq t_0 \geq 0$. Define $w(t) = -\rho(t)a(t)x'(t)/f(x(t))$ for $t \geq t_0$. Then for $t \geq t_0$, we have

$$w'(t) = \rho(t)q(t) - \frac{\rho(t)e(t)}{f(x(t))} + \frac{\gamma(t)}{a(t)\rho(t)}w(t) + f'(x(t))\frac{w^2(t)}{a(t)\rho(t)}. \quad (2.7.28)$$

By hypotheses, we can choose s_1, $t_1 \geq t_0$, $s_1 < t_1$ so that $e(t) \leq 0$ on $I = [s_1, t_1]$. Thus, on the interval I, $w(t)$ satisfies the differential inequality

$$w'(t) \geq \rho(t)q(t) + \frac{\gamma(t)}{a(t)\rho(t)}w(t) + f'(x(t))\frac{w^2(t)}{a(t)\rho(t)} \quad \text{on} \quad I. \quad (2.7.29)$$

Let $u(t) \in D(s_1, t_1)$ be as in the hypothesis. Multiplying (2.7.29) by $u^2(t)$ and integrating over I, we find

$$\int_I u^2(s)w'(s)ds \geq \int_I u^2(s)\rho(s)q(s)ds + \int_I \frac{\gamma(s)}{a(s)\rho(s)}u^2(s)w(s)ds$$

$$+ \int_I f'(x(s))u^2(s)\frac{w^2(s)}{a(s)\rho(s)}ds.$$

$$(2.7.30)$$

Integrating the left–hand side of (2.7.30) by parts and using the fact that $u(s_1) = u(t_1) = 0$, we obtain

$$-2\int_I u(s)u'(s)w(s)ds \geq \int_I u^2(s)\rho(s)q(s)ds + \int_I \frac{\gamma(s)}{a(s)\rho(s)}u^2(s)w(s)ds$$
$$+ \int_I f'(x(s))u^2(s)\frac{w^2(s)}{a(s)\rho(s)}ds,$$

which is equivalent to

$$0 \geq \int_I \left[\sqrt{\frac{f'(x(s))}{a(s)\rho(s)}}u(s)w(s) + \left(u'(s) + \frac{\gamma(s)u(s)}{2a(s)\rho(s)}\right)\sqrt{\frac{a(s)\rho(s)}{f'(x(s))}}\right]^2 ds + Q_1(u).$$

$$(2.7.31)$$

Since $Q_1(u) \geq 0$ inequality (2.7.31) yields

$$\frac{f'(x(t))}{a(t)\rho(t)}u(t)\left(-\rho(t)\frac{a(t)x'(t)}{f(x(t))}\right) + u'(t) + \frac{\gamma(t)u(t)}{2a(t)\rho(t)} = 0,$$

i.e.,

$$\left[-\frac{f'(x(t))x'(t)}{f(x(t))}u(t) + u'(t)\right] + \frac{\gamma(t)u(t)}{2a(t)\rho(t)} = 0,$$

or

$$f(x(t))\frac{d}{dt}\left(\frac{u(t)}{f(x(t))}\right) + \frac{\gamma(t)u(t)}{2a(t)\rho(t)} = 0,$$

or

$$\frac{d}{dt}y'(t) + \frac{\gamma(t)}{2a(t)\rho(t)}y(t) = 0 \quad \text{on} \quad I, \qquad (2.7.32)$$

where $y(t) = u(t)/f(x(t))$, $t \in I$. It follows from (2.7.32) that $(c(t)y(t))' = 0$ on I, where

$$c(t) = \exp\left(\int_{s_1}^t \frac{\gamma(s)}{2a(s)\rho(s)}ds\right), \quad t \in I.$$

Thus, $c(t)y(t) = C$ for some constant C, and hence $u(t) = Cf(x(t))/c(t)$ on I. Since $u(t) \in D(s_1, t_1)$ and $u(t) \neq 0$, this is incompatible to the fact that $x(t) > 0$ on I. This contradiction proves that $x(t)$ is oscillatory.

When $x(t) < 0$ eventually, we use $u(t) \in D(s_2, t_2)$ and $e(t) \geq 0$ on $[s_2, t_2]$ to reach a similar contradiction. This completes the proof. ∎

Theorem 2.7.5 when specialized to (2.7.19) takes the following form:

Corollary 2.7.2. Suppose for any $T \geq 0$ there exist $T \leq s_1 < t_1 \leq s_2 < t_2$ such that condition (2.7.22) holds. If there exist $\rho(t) \in C^1([s_i, t_i], \mathbb{R}^+)$

and $u(t) \in D(s_i, t_i)$ such that

$$Q_i(u) = \int_{s_i}^{t_i} \rho(s) \left[q(s)u^2(s) - a(s) \left(u'(s) + \frac{\rho'(s)}{2\rho(s)} u(s) \right)^2 \right] ds \geq 0$$

(2.7.33)

for $i = 1, 2$ then equation (2.7.19) is oscillatory.

Remark 2.7.1. Theorem 2.7.4 follows from Corollary 2.7.2 by letting $\rho(t) = 1$. Also, if the weight function

$$\rho(t) = \exp \left(\int^t \frac{p(s)}{a(s)} ds \right)$$

(2.7.34)

then $\gamma(t)$ defined in (2.7.27) will be identically zero. Thus, the condition (2.7.26) in Theorem 2.7.5 takes the form

$$Q_i(u) = \int_{s_i}^{t_i} \rho(s) \left[q(s)u^2(s) - \frac{a(s)}{k} (u'(s))^2 \right] ds \geq 0$$

(2.7.35)

for $i = 1, 2$.

Example 2.7.5. Consider the forced equation

$$(\sqrt{t}x'(t))' + \frac{1}{\sqrt{t}}x'(t) + \frac{1}{\sqrt{t}}x(t) = \sin\sqrt{t}, \quad t > 0.$$

(2.7.36)

Here, the zeros of the forcing term $\sin\sqrt{t}$ are $(n\pi)^2$.

For any $T > 0$ choose n sufficiently large so that $(n\pi)^2 \geq T$, and set $s_1 = (n\pi)^2$ and $t_1 = (n+1)^2\pi^2$ and take $u(t) = \sin\sqrt{t}$, $\rho(t) = t$ and $k = 1$. Then, we have

$$\begin{aligned}
Q_1(u) &= \int_{s_1}^{t_1} \left[\rho(s)q(s)u^2(s) - a(s)\rho(s)(u'(s))^2 \right] ds \\
&= \int_{n^2\pi^2}^{(n+1)^2\pi^2} \sqrt{s} \left(\sin^2\sqrt{s} - \frac{1}{4}\cos^2\sqrt{s} \right) ds \\
&= \int_{n\pi}^{(n+1)\pi} 2v^2 \left(\sin^2 v - \frac{1}{4}\cos^2 v \right) dv \\
&= \int_{n\pi}^{(n+1)\pi} \frac{v^2}{4} (3 - 5\cos 2v) dv \\
&= \frac{3n^2 + 3n + 1}{4}\pi^3 - \frac{5}{16}\pi > 0.
\end{aligned}$$

Similarly, for $s_2 = (n+1)^2\pi^2$ and $t_2 = (n+2)^2\pi^2$ we can show that $Q_2(u) > 0$. Now, it follows from Theorem 2.7.5 that equation (2.7.36) is oscillatory.

2.8. Notes and General Discussions

1. The results of Section 2.1 are due to Kreith [17,18] and Swanson [30].

2. Lemma 2.2.1 is theorem 2.15 of Swanson [30], Lemma 2.2.2 can be proved by Kummer transformation (see Willett [31]), i.e., for (2.1.1) we let $w(t) = x(t)/\sqrt{\rho(t)}$, where $\rho(t) \in C^2([t_0, \infty), \mathbb{R}^+)$. Theorem 2.2.6 improves a result of Hartman [12], Theorem 2.2.7 improves the comparison theorem due to Hille [13] and Wintner [34], while Theorem 2.2.8 improves a result of Wong [35]. In Theorem 2.2.1, we can let the function $\rho(t) \in C^2([t_0, \infty), \mathbb{R}^+)$ to be $\rho(t) = \exp\left(\int^t (p(s)/a(s)) ds\right)$. In this case $Q^*(t)$ and $h^*(t)$ defined in (2.2.3) and (2.2.4), respectively, become $h^*(t) = 0$ and $Q^*(t) = \rho(t)q(t)$.

3. Theorem 2.3.1, Corollaries 2.3.1, 2.3.4 and 2.3.7 are due to Wintner [33], Kneser [15], Hille [13] and Moore [22], respectively. Corollary 2.3.6 with $a(t) = 1$ and $c = 1/4$ reduces to Wintner's criterion [34]. Also Corollary 2.3.11 with $a(t) = \rho(t) = 1$ reduces to a result of Potter [27]. Theorem 2.3.3 improves a result of Wong [35], whereas Corollary 2.3.17 improves a result of Wintner [33]. Corollary 2.3.19 is due to Moore [22]. Corollary 2.3.20 with $\lambda = 2$, $\rho(t) = 1$ and $\beta(t) \geq 0$, reduces to Willett's criterion [31].

4. Theorems 2.4.1 and 2.4.3 are taken from Graef et. al. [9]. Theorem 2.4.1 extends Leighton's oscillation criterion established in [20].

5. Theorem 2.5.1 is extracted from Grace and Lalli [8], and it includes Wintner result [32]. Theorem 2.5.2 is due to Graef et. al. [9]. Theorem 2.5.3 includes Hartman's oscillation criterion [11] as a special case. Lemmas 2.5.1 and 2.5.2 and Theorem 2.5.4 extend Willett's oscillation result [31]. Theorems 2.5.6 – 2.5.8 are taken from Kamenov [14], Yan [37] and Philos [25], respectively. Corollaries 2.5.5 – 2.5.7 are due to Philos [23,24]. Theorems 2.5.12 and 2.5.13 are the extensions of Philos results [25] and are borrowed from Garce [7]. Lemma 2.5.3 and Theorems 2.5.14 – 2.5.17 are the generalizations of the work originated by Kong [16].

6. Theorems 2.6.1 – 2.6.4 extend the results of Yan [38]. Theorem 2.6.5 is due to Li [21]. Lemma 2.6.1 is taken from Coppel [5]. Corollary 2.6.3 includes Wintner's criterion [33]. Theorem 2.6.11 generalizes theorem 2 of Wong [35], and Theorem 2.6.12 improves Wintner result [32].

7. Theorems 2.7.1 and 2.7.2 are the extensions of Rankin results [28]. Theorem 2.7.3 is due to El–Sayed [6], and Theorem 2.7.4 is taken from Wong [36]. Theorem 2.7.5 improves Theorem 2.7.4, and is taken from Agarwal and Grace [2].

8. For some other related results of this chapter see Agarwal et. al. [1,3].

2.9. References

1. **R.P. Agarwal and S.R. Grace,** On the oscillation of certain second order differential equations, *Georgian Math. J.* **7**(2000), 201–213.

2. **R.P. Agarwal and S.R. Grace,** Second order nonlinear forced oscillations, *Dyn. Sys. Appl.*, to appear.

3. **R.P. Agarwal, S.R. Grace and D. O'Regan** , *Oscillation Theory for Second Order Dynamic Equations*, to appear.

4. **R.P. Agarwal and R.C. Gupta,** *Essentials of Ordinary Differential Equations*, McGraw–Hill, Singapore, 1991.

5. **W.A. Coppel,** *Disconjugacy, Lecture Notes in Math.*, 220, Springer–Verlag, New York, 1971.

6. **M.A. El–Sayed,** An oscillation criterion for a forced second order linear differential equation, *Proc. Amer. Math. Soc.* **118**(1993), 813–817.

7. **S.R. Grace,** Oscillation theorems for nonlinear differential equations of second order, *J. Math. Anal. Appl.* **171**(1992), 220–241.

8. **S.R. Grace and B.S. Lalli,** Integral averaging and the oscillation of second order nonlinear differential equations, *Ann. Mat. Pura Appl.* **151**(1988), 149–159.

9. **J.R. Graef, S.M. Rankin and P.W. Spikes,** Oscillation theorems for perturbed nonlinear differential equations, *J. Math. Anal. Appl.* **65**(1978), 375–390.

10. **G.H. Hardy, J.E. Littlewood and G. Polya,** *Inequalities*, 2nd ed. *Cambridge Univ. Press*, 1988.

11. **P. Hartman,** On nonoscillatory linear differential equations of second order, *Amer. J. Math.* **74**(1952), 389–400.

12. **P. Hartman,** *Ordinary Differential Equations*, John Wiley, New York, 1964.

13. **E. Hille,** Nonoscillation theorems, *Trans. Amer. Math. Soc.* **64**(1948), 234–252.

14. **I.V. Kamenev,** An integral criterion for oscillation of linear differential equations of second order, *Mat. Zametki* **23**(1978), 249–251.

15. **A. Kneser,** Untersuchungen über die reellen Nullstellen der Integrale linearer Differentialgleichungen, *Math. Annalen* **42**(1893), 409–435.

16. **Q. Kong,** Interval criteria for oscillation of second order linear ordinary differential equations, *J. Math. Anal. Appl.* **239**(1999), 285–270.

17. **K. Kreith,** *Oscillation Theory, Lecture Notes in Math.*, 324, Springer Verlag, New York, 1973.

18. **K. Kreith,** PDE generalization of Sturm comparison theorem, *Memoirs Amer. Math. Soc.* **48**(1984), 31–46.

19. **M.K. Kwong and J.S.W. Wong,** An application of integral inequality to second order nonlinear oscillation, *J. Differential Equations* **46**(1982), 63–77.

20. **W. Leighton,** The detection of the oscillation of solutions of a second order linear differential equation, *Duke J. Math.* **17**(1950), 57–62.

21. **H.J. Li,** Oscillation criteria for second order linear differential equations, *J. Math. Anal. Appl.* **194**(1995), 217–234.

22. **R.A. Moore,** The behavior of solutions of a linear differential equation of second order, *Pacific J. Math.* **5**(1955), 125–145.

23. **Ch.G. Philos,** Oscillation of second order linear ordinary differential equations with alternating coefficients, *Bull. Austral. Math. Soc.* **27**(1983), 307–313.

24. **Ch.G. Philos,** On a Kamenev's integral criterion for oscillation of linear differential equations of second order, *Utilitas Math.* **24**(1983), 277–289.

25. **Ch.G. Philos,** Oscillation theorems for linear differential equations of second order, *Arch. Math.* **53**(1989), 482–492.

26. **M. Picone,** Sui valorieccezionali di un parametro de cui dipends un equazioni differenziale lineare ordinaria del second ordine, *Ann. Scuola Norm. Pisa* **11**(1909), 1–141.

27. **R.L. Potter,** On self–adjoint differential equations of second order, *Pacific J. Math.* **3**(1953), 467–491.

28. **S.M. Rankin,** Oscillation results for a nonhomogeneous equation, *Pacific J. Math.* **80**(1979), 237–243.

29. **B. Sturm,** Sur les équations différentielles linéaires due second ordré, *J. Math. Pures Appl.* **1**(1836), 106–186.

30. **C.A. Swanson,** *Comparison and Oscillation Theory of Linear Differential Equations*, *Academic Press*, New York, 1968.

31. **D. Willet,** On the oscillatory behavior of the solutions of second order linear differential equations, *Ann. Polon. Math.* **21**(1969), 175–194.

32. **A. Wintner,** A criterion for oscillatory stability, *Quart. Appl. Math.* **7**(1949), 115–117.

33. **A. Wintner,** On the nonexistence of conjugate points, *Amer. J. Math.* **73**(1951), 368–380.

34. **A. Wintner,** On the comparison theorem of Kneser–Hille, *Math. Scand.* **5**(1957), 255–260.

35. **J.S.W. Wong,** Oscillation and nonoscillation of solutions of second order linear differential equations with integrable coefficients, *Trans. Amer. Math. Soc.* **144**(1969), 197–215.

36. **J.S.W. Wong,** Oscillation criteria for a forced second order linear differential equation, *J. Math. Anal. Appl.* **231**(1999), 235–240.

37. **J. Yan,** Oscillation theorems for second order linear differential equations with damping, *Proc. Amer. Math. Soc.* **98**(1986), 276–282.

38. **J. Yan,** Oscillation property for second order differential equations with an 'integrably small' coefficient, *Acta Math. Sinica* **30**(1987), 206–215.

Chapter 3

Oscillation and Nonoscillation of Half–Linear Differential Equations

3.0. Introduction

In this chapter we shall present oscillation and nonoscillation criteria for second order half–linear differential equations. In recent years these equations have attracted considerable attention. This is largely due to the fact that half–linear differential equations occur in a variety of real world problems; moreover, these are the natural generalizations of second order linear differential equations. In Section 3.1, we shall provide some preliminaries for the study of half–linear differential equations. In Sections 3.2 and 3.3, respectively, Sturm's and Levin's type comparison theorems are developed. In Section 3.4, we shall establish a Liapunov type inequality. Section 3.5 presents an oscillation criterion for almost periodic Sturm–Liouville equations. A systematic study on the zeros of solutions of singular half–linear equations is made in Section 3.6. Nonoscillation characterizations (necessary and sufficient conditions), comparison results as well as several sufficient criteria for the nonoscillation are presented in Section 3.7. Section 3.8 is devoted to the study of oscillation of half–linear equations. In Section 3.9, we shall establish oscillation criteria by employing integral and weighted averaging techniques. Here, interval criteria for the oscillation of half–linear equations are also provided. Section 3.10 deals with the oscillation of half–linear equations with integrable coefficients. Section 3.11 addresses the oscillation of damped and forced equations. In Section 3.12, we shall derive lower bounds for the distance between consecutive zeros of an oscillatory solution. Finally, in Section 3.13, we shall present a systematic study of the oscillation and nonoscillation of half–linear equations with a deviating argument. Here, classifications of the nonoscillatory solutions, and the existence results which guarantee that the solutions have prescribed asymptotic behavior are also presented.

3.1. Preliminaries

In this chapter we shall study half–linear differential equations of the form

$$\frac{d}{dt}\left(a(t)\psi(x'(t))\right) + q(t)\psi(x(t)) = 0, \tag{3.1.1}$$

where $a(t) \in C([t_0, \infty), \mathbb{R}^+)$, $q(t) \in C([t_0, \infty), \mathbb{R})$ and $\psi : \mathbb{R} \to \mathbb{R}$ is defined by $\psi(y) = |y|^{\alpha-1}y$ with $\alpha > 0$ is a fixed number. If $\alpha = 1$, then (3.1.1) reduces to the linear differential equation

$$\frac{d}{dt}\left(a(t)x'(t)\right) + q(t)x(t) = 0. \tag{3.1.2}$$

A solution of (3.1.1) is a real valued function $x(t) \in C^1([t_0, \infty), \mathbb{R})$. If $x(t)$ is a positive solution of (3.1.1) and either increasing, or decreasing, then equation (3.1.1) reduces to the Euler–Lagrange equation

$$\frac{d}{dt}\left(a(t)[x'(t)]^\alpha\right) + q(t)x^\alpha(t) = 0, \tag{3.1.3}$$

or

$$\frac{d}{dt}\left(a(t)[-x'(t)]^\alpha\right) - q(t)x^\alpha(t) = 0 \tag{3.1.4}$$

respectively.

The existence of solutions to initial value problems for (3.1.1) on $[T, \infty)$, $T \geq 0$ has been addressed in [15]. We note that any constant multiple of a solution of (3.1.1) is also a solution. A solution $x(t)$ of (3.1.1) is called oscillatory if for each $t_1 > 0$ there exists a $t > t_1$ such that $x(t) = 0$; otherwise, it is said to be nonoscillatory. Equation (3.1.1) is called oscillatory if all its solutions are oscillatory, and nonoscillatory if all its solutions are nonoscillatory. The main objective of this chapter is to discuss the oscillation and nonoscillation behavior of all solutions of equation (3.1.1).

3.2. Sturm–Type Comparison Theorems

In what follows $AC(I)$ denotes the set of all absolutely continuous functions on an interval $I \subseteq \mathbb{R}$. To prove our main results, we shall need the following five lemmas.

Lemma 3.2.1 [23]. If A and B are nonnegative constants, then

$$A^\lambda + (\lambda - 1)B^\lambda - \lambda AB^{\lambda-1} \geq 0 \quad \text{for} \quad \lambda > 1$$

and

$$A^\lambda + (\lambda - 1)B^\lambda - \lambda AB^{\lambda-1} \leq 0 \quad \text{for} \quad 0 < \lambda < 1,$$

where equality holds if and only if $A = B$.

Lemma 3.2.2. Let $a(t) \in C^1((a,b), \mathbb{R}^+)$ and $q(t) \in C((a,b), \mathbb{R})$. Suppose equation (3.1.3) has a positive increasing solution $x(t)$ on (a,b). If $y(t) \in AC((a,b), \mathbb{R}_0)$ satisfies

$$\liminf_{t \to b^-} y^{\alpha+1}(t) \left[\frac{x'(t)}{x(t)}\right]^\alpha a(t) \geq \limsup_{t \to a^+} y^{\alpha+1}(t) \left[\frac{x'(t)}{x(t)}\right]^\alpha a(t), \quad (3.2.1)$$

then

$$\liminf_{\substack{c \to a^+ \\ d \to b^-}} \int_c^d \left[a(s)|y'(s)|^{\alpha+1} - q(s)y^{\alpha+1}(s)\right] ds \geq 0, \quad (3.2.2)$$

where equality holds if and only if x and y are proportional.

Proof. Define $g(t) = x'(t)/x(t)$ on (a,b). It follows from (3.1.3) that

$$(a(t)g^\alpha(t))' + \alpha a(t)g^{\alpha+1}(t) = -q(t) \quad \text{on} \quad (a,b). \quad (3.2.3)$$

Let

$$A = a^{1/(\alpha+1)}(t)|y'(t)|, \quad B = a^{1/(\alpha+1)}(t)y(t)g(t) \quad \text{and} \quad \lambda = \alpha + 1$$

in Lemma 3.2.1, to obtain

$$a(t)|y'(t)|^{\alpha+1} + \alpha a(t)y^{\alpha+1}(t)g^{\alpha+1}(t) \geq (\alpha+1)a(t)y^\alpha(t)y'(t)g^\alpha(t),$$

where equality holds if and only if $g(t) = y'(t)/y(t)$. Thus, we have

$$a(t)|y'(t)|^{\alpha+1} - q(t)y^{\alpha+1}(t) \geq \left(a(t)y^{\alpha+1}(t)g^\alpha(t)\right)', \quad (3.2.4)$$

where equality holds if and only if x and y are proportional. Now, integrating (3.2.4) from $c(>a)$ to $d(<b)$, we obtain

$$\int_c^d \left[a(s)|y'(s)|^{\alpha+1} - q(s)y^{\alpha+1}(s)\right] ds \geq a(d)y^{\alpha+1}(d)g^\alpha(d) - a(c)y^{\alpha+1}(c)g^\alpha(c).$$

It follows from (3.2.1) that (3.2.2) holds. ∎

Using an argument similar to that in Lemma 3.2.2, we can prove the following lemma.

Lemma 3.2.3. Let $a(t) \in C^1((a,b), \mathbb{R}^+)$ and $q(t) \in C((a,b), \mathbb{R})$. Suppose equation (3.1.4) has a positive decreasing solution $x(t)$ on (a,b). If $y(t) \in AC((a,b), \mathbb{R}_0)$ satisfies

$$\liminf_{t \to a^+} y^{\alpha+1}(t) \left[\frac{x'(t)}{x(t)}\right]^\alpha a(t) \geq \limsup_{t \to b^-} y^{\alpha+1}(t) \left[\frac{x'(t)}{x(t)}\right]^\alpha a(t),$$

then (3.2.2) holds with equality only if x and y are proportional.

Lemma 3.2.4.   Let $a(t) \in C^1((a,b),\mathbb{R}^+)$ and $q(t) \in C((a,b),\mathbb{R})$. Suppose $x(t)$ is a solution of equation (3.1.1) on $[a,b]$ satisfying $x(a) = x(b) = 0$, then $x'(a) \neq 0$, or $x'(b) \neq 0$.

Proof.   We shall prove only the case $x'(a) \neq 0$. Assume to the contrary that $x'(a) = 0$. Since any constant multiple of a solution of (3.1.1) is also a solution of (3.1.1), we can assume without loss of generality that $x(t) > 0$ on (a,b). It follows from equation (3.1.1) that $a(t)\psi(x'(t)) = -\int_a^t q(s)\psi(x(s))ds$ on $[a,b]$, which implies

$$x'(t) \; = \; \psi^{-1}\left(-\frac{1}{a(t)}\int_a^t q(s)\psi(x(s))ds\right),$$

where ψ^{-1} is the inverse function of ψ. Since $x(a) = 0$ and $a \in C^1([a,b],\mathbb{R}^+)$, we have

$$\begin{aligned} x(t) \;&= \; \int_a^t \psi^{-1}\left(-\frac{1}{a(\tau)}\int_a^\tau q(s)\psi(x(s))ds\right)d\tau \\ &\leq \; (t-a)\psi^{-1}\left(M\int_a^t |q(s)|\psi(x(s))ds\right) \end{aligned}$$

for $a \leq t \leq b$, where $M = \max\{1/a(t) : a \leq t \leq b\}$. Hence,

$$\psi(x(t)) \; \leq \; (b-a)^\alpha M \int_a^t |q(s)|\psi(x(s))ds \quad \text{for} \quad a \leq t \leq b.$$

Thus, it follows from Lemma 1.1.1 that $\psi(x(t)) = 0$ for each $t \in [a,b]$. This implies that $x(t) = 0$ on $[a,b]$, which contradicts the hypothesis $x(t) > 0$ on (a,b). This contradiction completes the proof. ∎

From Lemmas 3.2.2 – 3.2.4, we shall establish the following Wirtinger–type inequality.

Lemma 3.2.5.   Let $x(t)$ be a solution of equation (3.1.1) on $[a,b]$ satisfying $x(t) \neq 0$ on (a,b). Denote by Y the family

$$Y \; = \; \{y(t) \in C^1([a,b],\mathbb{R}) : y(a) = y(b) = 0 \text{ and } y(t) \neq 0 \text{ on } (a,b)\}.$$

Then for every $y \in Y$,

$$\int_a^b a(s)|y'(s)|^{\alpha+1}ds \; \geq \; \int_a^b q(s)|y(s)|^{\alpha+1}ds, \tag{3.2.5}$$

where equality holds if and only if x and y are proportional.

Proof. Let $y \in Y$. We shall prove only the case $y(t) > 0$ on (a, b), since the proof of the case $y(t) < 0$ on (a, b) is similar. Since $x(t)$ is a solution of equation (3.1.1) on $[a, b]$ satisfying $x(t) \neq 0$ on (a, b), we may assume without loss of generality that $x(t) > 0$ on (a, b). Now, we separate our proof into the following six cases:

Case 1. Either (i). there is a number c, $a < c < b$ such that $x'(c) = 0$ and $x'(t) \geq 0$ on (a, c), or (ii). $x'(t) > 0$ on $(a, b]$.

(i). It follows from $y(a) = 0$, $x'(c) = 0$ and Lemma 3.2.4 that

$$\lim_{t \to a^+} y^{\alpha+1}(t) \left[\frac{x'(t)}{x(t)} \right]^\alpha a(t) = \lim_{t \to c^-} y^{\alpha+1}(t) \left[\frac{x'(t)}{x(t)} \right]^\alpha a(t) = 0. \quad (3.2.6)$$

The proof of (3.2.6) is as follows:

(I) If $x(a) = 0$, then $x'(a) \neq 0$ by Lemma 3.2.4. This and $y(a) = 0$ imply that

$$\lim_{t \to a^+} y^{\alpha+1}(t) \left[\frac{x'(t)}{x(t)} \right]^\alpha a(t) = \left(\lim_{t \to a^+} \frac{y(t)}{x(t)} \right)^\alpha \lim_{t \to a^+} y(t) a(t)$$

$$= \left(\lim_{t \to a^+} \frac{y(t)}{x(t)} \right)^\alpha \times 0 = \left[\frac{y'(a)}{x'(a)} \right]^\alpha \times 0 = 0.$$

(II) If $x'(a) = 0$, then $x(a) \neq 0$ by Lemma 3.2.4. This and $y(a) = 0$ imply

$$\lim_{t \to a^+} y^{\alpha+1}(t) \left[\frac{x'(t)}{x(t)} \right]^\alpha a(t) = 0.$$

(III) It follows from $x'(c) = 0$ and $x(t) \neq 0$ on (a, b) that

$$\lim_{t \to c^-} y^{\alpha+1}(t) \left[\frac{x'(t)}{x(t)} \right]^\alpha a(t) = 0.$$

From (I) – (III), (3.2.6) is clear.

Now, by (3.2.6) and Lemma 3.2.2, we have

$$\int_a^c a(s) |y'(s)|^{\alpha+1} ds \geq \int_a^c q(s) y^{\alpha+1}(s) ds. \quad (3.2.7)$$

(ii). It follows from $y(a) = y(b) = 0$ and Lemma 3.2.4 that

$$\lim_{t \to a^+} y^{\alpha+1}(t) \left[\frac{x'(t)}{x(t)} \right]^\alpha a(t) = \lim_{t \to b^-} y^{\alpha+1}(t) \left[\frac{x'(t)}{x(t)} \right]^\alpha a(t) = 0. \quad (3.2.8)$$

This and Lemma 3.2.2 now imply that

$$\int_a^b a(s) |y'(s)|^{\alpha+1} ds \geq \int_a^b q(s) y^{\alpha+1}(s). \quad (3.2.9)$$

Case 2. Either (iii). there is a number c, $a < c < b$ satisfying $x'(c) = 0$ and $x'(t) \leq 0$ on (a, c), or (iv). $x'(t) < 0$ on $(a, b]$.

(iii). It follows from $x(a) = 0$, $x'(c) = 0$ and Lemma 3.2.4 that

$$\lim_{t \to a^+} y^{\alpha+1}(t) \left[-\frac{x'(t)}{x(t)} \right]^\alpha a(t) = \lim_{t \to c^-} y^{\alpha+1}(t) \left[-\frac{x'(t)}{x(t)} \right]^\alpha a(t) = 0.$$
(3.2.10)

This and Lemma 3.2.3 imply that (3.2.7) holds.

(iv). It follows from $y(a) = y(b) = 0$ and Lemma 3.2.4 that

$$\lim_{t \to a^+} y^{\alpha+1}(t) \left[-\frac{x'(t)}{x(t)} \right]^\alpha a(t) = \lim_{t \to b^-} y^{\alpha+1}(t) \left[-\frac{x'(t)}{x(t)} \right]^\alpha a(t) = 0.$$

This and Lemma 3.2.3 imply that (3.2.9) holds.

Case 3. If there are two numbers c_0 and c_1, $a < c_0 < c_1 < b$ such that $x'(c_0) = x'(c_1) = 0$ and $x'(t) > 0$ on (c_0, c_1), then

$$\lim_{t \to c_0^+} y^{\alpha+1}(t) \left[\frac{x'(t)}{x(t)} \right]^\alpha a(t) = \lim_{t \to c_1^-} y^{\alpha+1}(t) \left[\frac{x'(t)}{x(t)} \right]^\alpha a(t) = 0.$$

It follows from Lemma 3.2.2 that

$$\int_{c_0}^{c_1} a(s)|y'(s)|^{\alpha+1} ds \geq \int_{c_0}^{c_1} q(s) y^{\alpha+1}(s) ds.$$
(3.2.11)

Case 4. If there are two numbers c_0 and c_1, $a < c_0 < c_1 < b$ such that $x'(c_0) = x'(c_1) = 0$ and $x'(t) < 0$ on (c_0, c_1), then

$$\lim_{t \to c_0^+} y^{\alpha+1}(t) \left[-\frac{x'(t)}{x(t)} \right]^\alpha a(t) = \lim_{t \to c_1^-} y^{\alpha+1}(t) \left[-\frac{x'(t)}{x(t)} \right]^\alpha a(t) = 0.$$

It follows from Lemma 3.2.2 that (3.2.11) holds.

Case 5. Either (v). there is a number c, $a < c < b$ satisfying $x'(c) = 0$ and $x'(t) > 0$ on $(c, b]$, or (vi). $x'(t) > 0$ on $[a, b)$.

(v). It follows from $y(b) = 0$, $x'(c) = 0$ and Lemma 3.2.4 that

$$\lim_{t \to c^+} y^{\alpha+1}(t) \left[\frac{x'(t)}{x(t)} \right]^\alpha a(t) = \lim_{t \to b^-} y^{\alpha+1}(t) \left[\frac{x'(t)}{x(t)} \right]^\alpha a(t) = 0.$$

This and Lemma 3.2.2 imply that

$$\int_{c}^{b} a(s)|y'(s)|^{\alpha+1} ds \geq \int_{c}^{b} q(s) y^{\alpha+1}(s) ds.$$
(3.2.12)

(vi). It follows from $y(a) = y(b) = 0$ and Lemma 3.2.4 that (3.2.8) holds. This and Lemma 3.2.2 imply that (3.2.9) holds.

Case 6. Either (vii). there is a number c, $a < c < b$ satisfying $x'(c) = 0$ and $x'(t) < 0$ on $(c, b]$, or (viii). $x'(t) < 0$ on $(a, b]$.

(vii). It follows from $y(b) = 0$, $x'(c) = 0$ and Lemma 3.2.4 that

$$\lim_{t \to c^+} y^{\alpha+1}(t) \left[-\frac{x'(t)}{x(t)} \right]^\alpha a(t) \geq \lim_{t \to b^-} y^{\alpha+1}(t) \left[-\frac{x'(t)}{x(t)} \right]^\alpha a(t) = 0.$$

This and Lemma 3.2.2 imply that (3.2.12) holds.

(viii). It follows from $y(a) = y(b) = 0$ and Lemma 3.2.4 that (3.2.10) holds. This and Lemma 3.2.3 imply that (3.2.8) holds.

Combining the above six cases, we obtain the inequality (3.2.5). ∎

Remark 3.2.1. If α is a quotient of two positive odd integers, then the condition '$y(t) \neq 0$ on (a, b)' in Y can be omitted.

Now, we shall use Lemma 3.2.5, to prove the following Sturm–type comparison theorem for equation (3.1.1).

Theorem 3.2.1. Let $a(t)$, $a_1(t) \in C^1([a, b], \mathbb{R}^+)$ and $q(t)$, $q_1(t) \in C([a, b], \mathbb{R})$. If the boundary value problem

$$\begin{aligned}
(a_1(t)\psi(y'(t)))' + q_1(t)\psi(y(t)) &= 0 \\
y(a) = y(b) &= 0,
\end{aligned} \tag{3.2.13}$$

has a solution $y(t)$ with $y(t) \neq 0$ on (a, b) satisfying

$$\int_a^b \left[(a(s) - a_1(s))|y'(s)|^{\alpha+1} - (q(s) - q_1(s))|y(s)|^{\alpha+1} \right] ds \leq 0, \tag{3.2.14}$$

then every solution $x(t)$ of (3.1.1) must have a zero in (a, b) unless $x(t)$ and $y(t)$ are proportional.

Proof. Suppose to the contrary that $x(t)$ is a solution of equation (3.1.1) such that $x(t) \neq 0$ on (a, b). It follows from Lemma 3.2.5 that (3.2.5) holds. Multiplying equation (3.2.13) by $y(t)$ and integrating it by parts from a to b, we get $\int_a^b a_1(s)|y'(s)|^{\alpha+1} ds = \int_a^b q_1(s)|y(s)|^{\alpha+1} ds$. Combining this with (3.2.5), we find

$$\int_a^b \left[(a(s) - a_1(s))|y'(s)|^{\alpha+1} - (q(s) - q_1(s))|y(s)|^{\alpha+1} \right] ds \geq 0,$$

which contradicts (3.2.14) unless equality holds. In view of Lemma 3.2.5 the later situation occurs if and only if $x(t)$ and $y(t)$ are proportional. This completes the proof. ∎

Remark 3.2.2. If $\alpha = 1$, then Theorem 3.2.1 reduces to the well–known Sturm comparison theorem.

If $a(t) = a_1(t)$ and $q(t) = q_1(t)$, then Theorem 3.2.1 yields the following corollary.

Corollary 3.2.1. The zeros of two linearly independent solutions of equation (3.1.1) separate each other.

Remark 3.2.3. Corollary 3.2.1 implies that the oscillatory and nonoscillatory equations (3.1.1) are mutually exclusive.

Corollary 3.2.2. Let $a(t)$, $a_1(t) \in C^1([t_0, \infty), \mathbb{R}^+)$ and $q(t)$, $q_1(t) \in C([t_0, \infty), \mathbb{R})$. Suppose $a_1(t) \geq a(t)$ and $q_1(t) \leq q(t)$ on $[t_0, \infty)$. If the differential equation

$$(a_1(t)\psi(u'(t)))' + q_1(t)\psi(u(t)) = 0 \qquad (3.2.15)$$

is oscillatory, then so is (3.1.1); or equivalently, the nonoscillation of (3.1.1) implies that of equation (3.2.15).

3.3. Levin–Type Comparison Theorems

In this section we shall extend Levin's comparison theorem which is known for the differential equation

$$x''(t) + q_1(t)x(t) = 0 \quad \text{and} \quad y''(t) + q_2(t)y(t) = 0$$

to the half–linear differential inequality

$$x(t)\{(a_1(t)\psi(x'(t)))' + q_1(t)\psi(x(t))\} \leq 0 \qquad (3.3.1)$$

and the half–linear differential equation

$$(a_2(t)\psi(y'(t)))' + q_2(t)\psi(y(t)) = 0, \qquad (3.3.2)$$

where $a_1(t)$, $a_2(t) \in C^1([T, \infty), \mathbb{R}^+)$, $T \geq t_0 \geq 0$, $q_1(t)$, $q_2(t) \in C([T, \infty), \mathbb{R})$ and q_1, $q_2 \in \mathcal{L}^1_{Loc}([T, \infty), \mathbb{R})$, and $a_1(t) \leq a_2(t)$ for all $t \geq T$.

Theorem 3.3.1. Let $x(t)$ and $y(t)$ be nontrivial solutions of (3.3.1) and (3.3.2) respectively, on a closed subinterval $[a, b]$ of $[T, \infty)$ satisfying either

(i) $y(a) \geq x(a) > 0$ and $x(t) > 0$ on $[a, b]$, or

(ii) $y(a) \leq x(a) < 0$ and $x(t) < 0$ on $[a, b]$.

If

$$-a_1(a)\frac{\psi(x'(a))}{\psi(x(a))} + \int_a^t q_1(s)ds > \left| -a_2(a)\frac{\psi(y'(a))}{\psi(y(a))} + \int_a^t q_2(s)ds \right| \qquad (3.3.3)$$

on $[a, b]$, then $y(t)$ does not vanish on $[a, b]$, and

$$x(t)x'(t) < 0, \qquad y(t) \geq x(t) > 0 \quad \text{if (i) holds,}$$
$$y(t) \leq x(t) < 0 \quad \text{if (ii) holds}$$

also

$$-a_1(t)\frac{\psi(x'(t))}{\psi(x(t))} > \left|a_2(t)\frac{\psi(y'(t))}{\psi(y(t))}\right| \quad \text{for all} \quad t \in [a, b]. \tag{3.3.4}$$

If $>$ in (3.3.3) is replaced by $\geq$, then $>$ in (3.3.4) should be replaced by $\geq$.

Proof. Since the proofs of (i) and (ii) are similar, we shall prove only case (i). Since $x(t) > 0$ on $[a, b]$, the continuous function

$$w(t) = -a_1(t)\frac{\psi(x'(t))}{\psi(x(t))} \quad \text{on} \quad [a, b] \tag{3.3.5}$$

satisfies

$$w'(t) \geq q_1(t) + \alpha a_1^{-1/\alpha}(t)|w(t)|^{(\alpha+1)/\alpha} \geq q_1(t).$$

Thus, we find

$$w(t) \geq w(a) + \int_a^t q_1(s)ds + \int_a^t \alpha a^{-1/\alpha}(s)|w(s)|^{(\alpha+1)/\alpha}ds$$
$$\geq w(a) + \int_a^t q_1(s)ds. \tag{3.3.6}$$

Hence, $x'(t) < 0$ for all $t \in [a, b]$. Since $y(a) > 0$ the function

$$z(t) = -a_2(t)\frac{\psi(y'(t))}{\psi(y(t))} \tag{3.3.7}$$

is continuous on some interval $[a, c]$, where $a < c < b$. Clearly, $z(t)$ satisfies the integral equation

$$z(t) = z(a) + \int_a^t q_2(s)ds + \int_a^t \alpha a_2^{-1/\alpha}(s)|w(s)|^{(\alpha+1)/\alpha}ds \tag{3.3.8}$$

for all $t \in [a, c]$. Moreover, we have

$$z(t) \geq z(a) + \int_a^t q_2(s)ds. \tag{3.3.9}$$

It follows from (3.3.3), (3.3.6) and (3.3.9) that

$$z(t) \geq z(a) + \int_a^t q_2(s)ds > -w(a) - \int_a^t q_1(s)ds \geq -w(t) \quad \text{on} \quad [a, c].$$
$$\tag{3.3.10}$$

Hence, $w(t) > -z(t)$ on $[a, c]$. Next, we claim that $w(t) > z(t)$ on $[a, c]$. For this, assume that there exists a $t_1 \in (a, c)$ such that $w(t_1) = z(t_1)$ and $z(t) < w(t)$ on $[a, t_1]$. By (3.3.10), we have $0 < |z(t)| < w(t)$ on $[a, t_1]$. Thus, it follows from (3.3.3), (3.3.6) and (3.3.8) that

$$
\begin{aligned}
z(t_1) \;=\; & z(a) + \int_a^{t_1} q_2(s)ds + \int_a^{t_1} \alpha a_2^{-1/\alpha}(s)|z(s)|^{(\alpha+1)/\alpha}ds \\
< \;& w(a) + \int_a^{t_1} q_1(s)ds + \int_a^{t_1} \alpha a_1^{-1/\alpha}|w(s)|^{(\alpha+1)/\alpha}ds \;\leq\; w(t_1),
\end{aligned}
$$

which is a contradiction. Therefore,

$$
|z(t)| \;<\; w(t) \quad \text{on} \quad [a, c]. \tag{3.3.11}
$$

Next, we shall show that $y(t)$ cannot vanish on $[a, b]$. Suppose the first point to the right of a at which $y(t)$ vanishes is $t = t_2 \leq b$, i.e., $y(t) > 0$ on $[a, t_2)$ and $y(t_2) = 0$. Then from Lemma 3.2.4 it follows that $y'(t_2) \neq 0$. This means that the solutions of equation (3.3.2) have only simple zeros. However, since $|z(t)| < w(t)$ on $[a, t_2)$ and $w(t)$ is bounded on $[a, b]$, we get $\infty = \limsup_{t \to t_2^-} |z(t)| \leq \lim_{t \to t_2^-} w(t) = w(t_2) < \infty$, which is a contradiction. Hence, $y(t)$ cannot vanish on $[a, b]$. Thus, (3.3.11) holds on $[a, c] \subset [a, b]$ on which $z(t)$ is continuous. But, this implies that $z(t)$ is continuous on the entire interval $[a, b]$, since $w(t)$ is bounded on $[a, b]$ and $y(t)$ cannot vanish on $[a, b]$. Therefore, (3.3.11) holds on the interval $[a, b]$.

Now, from (3.3.5), (3.3.7) and (3.3.11) it follows that $y(t) \geq x(t)$ on $[a, b]$. This completes the proof. ∎

Theorem 3.3.2. Let $x(t)$ and $y(t)$ be nontrivial solutions of (3.3.1) and (3.3.2) respectively, on $[a, b] \subset [T, \infty)$ satisfying either

(i) $x(t) > 0$ on $[a, b]$ and $y(b) \geq x(b) > 0$, or

(ii) $x(t) < 0$ on $[a, b]$ and $y(b) \leq x(b) < 0$.

If

$$
a_1(b)\frac{\psi(x'(b))}{\psi(x(b))} + \int_t^b q_1(s)ds \;>\; \left| a_2(b)\frac{\psi(y'(b))}{\psi(y(b))} + \int_t^b q_2(s)ds \right| \tag{3.3.12}
$$

for all $t \in [a, b]$, then $y(t)$ does not vanish on $[a, b]$, and

$$
x(t)x'(t) \;<\; 0, \qquad
\begin{aligned}
& y(t) \;\geq\; x(t) > 0 \quad \text{if (i) holds,} \\
& y(t) \;\leq\; x(t) < 0 \quad \text{if (ii) holds}
\end{aligned}
$$

also

$$
a_1(t)\frac{\psi(x'(t))}{\psi(x(t))} \;>\; \left| a_2(t)\frac{\psi(y'(t))}{\psi(y(t))} \right| \quad \text{for all} \quad t \in [a, b]. \tag{3.3.13}
$$

If $>$ in (3.3.12) is replaced by $\geq$, then $>$ in (3.3.13) should be replaced by $\geq$.

Proof. Define

$$
\begin{aligned}
x_1(t) &= x(a+b-t), & y_1(t) &= y(a+b-t) \\
A_1(t) &= a_1(a+b-t), & A_2(t) &= a_2(a+b-t) \\
Q_1(t) &= q_1(a+b-t), & Q_2(t) &= q_2(a+b-t).
\end{aligned}
$$

Then, $x_1(t)$ does not vanish on $[a,b]$, $A_1(a) = a_1(b)$, $y_1(a) = y(b) \geq x(b) = x_1(a) > 0$, or $y_1(a) = y(b) \leq x(b) = x_1(a) < 0$, and

$$
-A_1(a)\frac{\psi(x_1'(a))}{\psi(x_1(a))} + \int_a^{a+b-t} Q_1(s)ds = a_1(b)\frac{\psi(x'(b))}{\psi(x(b))} + \int_a^b q_1(s)ds,
$$

$$
\left| a_2(b)\frac{\psi(y'(b))}{\psi(y(b))} + \int_t^b q_2(s)ds \right| = \left| -A_2(a)\frac{\psi(y_1'(a))}{\psi(y_1(a))} + \int_a^{a+b-t} Q_2(s)ds \right|.
$$

Thus, (3.3.12) is equivalent to the assumption (3.3.3), since $t \in [a,b]$ if and only if $a+b-t \in [a,b]$. Hence the conclusion (3.3.13) follows from Theorem 3.3.1. ∎

Corollary 3.3.1. If the hypotheses of Theorems 3.3.1 and 3.3.2 satisfied, then the following inequalities hold for $t \in [a,b]$,

$$
\begin{aligned}
& a_2(t)\frac{\psi(y'(t))}{\psi(y(t))} - a_1(t)\frac{\psi(x'(t))}{\psi(x(t))} \\
& > a_2(a)\frac{\psi(y'(a))}{\psi(y(a))} - a_1(a)\frac{\psi(x'(a))}{\psi(x(a))} + \int_a^t (q_1(s) - q_2(s))ds
\end{aligned} \tag{3.3.14}
$$

and

$$
\begin{aligned}
& a_1(t)\frac{\psi(x'(t))}{\psi(x(t))} - a_2(t)\frac{\psi(y'(t))}{\psi(y(t))} \\
& > a_1(b)\frac{\psi(x'(b))}{\psi(x(b))} - a_2(b)\frac{\psi(y'(b))}{\psi(y(b))} - \int_t^b (q_1(s) - q_2(s))ds.
\end{aligned} \tag{3.3.15}
$$

Proof. It follows from the proof of Theorem 3.3.1 that

$$
\begin{aligned}
z(t) - z(a) - \int_a^t q_2(s)ds &= \int_a^t \alpha a_2^{-1/\alpha}(s)|z(s)|^{(\alpha+1)/\alpha}ds \\
&< \int_a^t \alpha a_1^{-1/\alpha}(s)|w(s)|^{(\alpha+1)/\alpha}ds = w(t) - w(a) - \int_a^t q_1(s)ds.
\end{aligned}
$$

This in view of

$$
z(t) = -a_2(t)\frac{\psi(y'(t))}{\psi(y(t))} \quad \text{and} \quad w(t) = -a_1(t)\frac{\psi(x'(t))}{\psi(x(t))}
$$

is equivalent to (3.3.14). Similarly, (3.3.15) can be obtained from (3.3.14) by the substitution used in the proof of Theorem 3.3.2.   ∎

As applications of Theorem 3.3.1, we shall prove the following results.

Theorem 3.3.3. Suppose for some $a \geq T$,

(i) $x(t)$ and $y(t)$ are nontrivial solutions of (3.3.1) and (3.3.2), respectively, and either

(I) $y'(a) = 0$, $x'(a) \leq 0$, $y(a) \geq x(a) > 0$ and $x(t) > 0$ on $[a, \infty)$, or

(II) $y'(a) = 0$, $x'(a) \geq 0$, $y(a) \leq x(a) < 0$ and $x(t) < 0$ on $[a, \infty)$,

(ii) $\displaystyle\int_a^t q_1(s)ds \geq \left|\int_a^t q_2(s)ds\right|$ for all $t \geq a$.

Then, $y(t)$ does not vanish for $t \geq a$, and

$$x(t)x'(t) < 0, \qquad \begin{aligned} y(t) &\geq x(t) > 0 \quad \text{if (I) holds,} \\ y(t) &\leq x(t) < 0 \quad \text{if (II) holds} \end{aligned}$$

also

$$-a_1(t)\frac{\psi(x'(t))}{\psi(x(t))} \geq \left|a_2(t)\frac{\psi(y'(t))}{\psi(y(t))}\right| \qquad \text{for } t \geq a.$$

Proof. Since (i) implies $-a_1(t)\psi(x'(a))/\psi(x(a)) \geq 0$, (3.3.3) holds on every closed interval $[a, b]$. Hence, the conclusion of Theorem 3.3.1 holds on every such interval. This in turn implies that the conclusion of Theorem 3.3.3 holds on every such interval. This completes the proof.   ∎

Theorem 3.3.4. Suppose there exists a nontrivial solution $y(t)$ of (3.3.2) which satisfies the conditions $y(a) = y(b) = 0$, $y'(c) = 0$, $a < c < b$. If the inequalities

$$\int_t^c q_1(s)ds \geq \left|\int_t^c q_2(s)ds\right| \quad \text{and} \quad \int_c^t q_1(s)ds \geq \left|\int_c^t q_2(s)ds\right| \quad (3.3.16)$$

hold on $[a, c]$ and $[c, b]$ respectively, then every solution $x(t)$ of the differential equation

$$(a_1(t)\psi(x'(t)))' + q_1(t)\psi(x(t)) = 0 \qquad (3.3.17)$$

has at least one zero in $[a, b]$.

Proof. Let $x(t)$ be a nontrivial solution of (3.3.17) satisfying $x'(c) = 0$. We assert that $x(t)$ has at least one zero in each of the intervals $[a, c)$ and $(c, b]$. From Lemma 3.2.4 it follows that $x(c) \neq 0$ and $y(c) \neq 0$. Without loss of generality we can assume that $x(c) = y(c)$. If $x(t)$ has no zero in $(c, b]$, and hence no zero on $[c, b]$, (3.3.16) ensures that all the hypotheses of Theorem 3.3.1 are satisfied. Therefore, $y(t)$ has no zero in $[c, b]$. This contradicts the hypothesis $y(b) = 0$. Similarly, if $x(t)$ has no zero in $[a, c)$ an application of Theorem 3.3.2 yields the contradiction that $y(t)$ has no zero in $[a, c]$. Hence, $x(t)$ has at least two zeros in $[a, b]$, and therefore by Corollary 3.2.1 every solution of equation (3.3.17) has at least one zero in $[a, b]$. ■

3.4. Liapunov's Inequality

The well–known Liapunov inequality can be stated as follows:

Theorem 3.4.1. Let $q(t) \in C((a, b), \mathbb{R}^+)$ be not identically zero on any open subset of (a, b). Suppose $x(t)$ is a nontrivial solution of equation

$$x''(t) + q(t)x(t) \ = \ 0$$

having consecutive zeros at $t = a$ and $t = b$. Then the following inequality holds

$$(b - a) \int_a^b q(s)ds \ > \ 4.$$

In this section, first we shall extend Liapunov's inequality to the half–linear differential equation

$$(\psi(x'(t)))' + q(t)\psi(x(t)) \ = \ 0 \tag{3.4.1}$$

and then present an application.

Theorem 3.4.2. Let $q(t) \in C([a, b], \mathbb{R}_0)$ be not identically zero on any open subset of $[a, b]$. Suppose $x(t)$ is a nontrivial solution of equation (3.4.1) having consecutive zeros at $t = a$ and $t = b$. Then the following inequality holds

$$2^{\alpha+1} \ < \ (b - a)^\alpha \int_a^b q(s)ds.$$

Proof. Since for the equation (3.4.1) any constant multiple of a solution $x(t)$ is also a solution, we may assume without loss of generality that $x(t) > 0$ on (a, b). Let M be the maximum of $x(t)$ on $[a, b]$ which is attained at some point $t_0 = a + \xi(b - a)$, $0 < \xi < 1$. Then, we have

$$\psi(x'(a)) - \psi(x'(b)) \ = \ -\int_a^b \frac{d}{dt}\psi(x'(t))dt$$
$$= \ \int_a^b q(t)\psi(x(t))dt \ < \ M^\alpha \int_a^b q(t)dt. \tag{3.4.2}$$

It follows from $x(t) > 0$ and $q(t) \geq 0$ that $x(t)$ is concave. Hence, we find $x'(a) \geq M/[\xi(b - a)]$ and $x'(b) \leq -M/[(1 - \xi)(b - a)]$. This and (3.4.2) imply that $\xi^{-\alpha} + (1 - \xi)^{-\alpha} < (b - a)^\alpha \int_a^b q(t)dt$. On the interval $(0,1)$ the function $\xi^{-\alpha} + (1 - \xi)^{-\alpha}$ has a maximum $2^{\alpha+1}$ which is attainded at $\xi = 1/2$. Therefore, $2^{\alpha+1} < (b - a)^\alpha \int_a^b q(t)dt$. This completes the proof. ∎

Theorem 3.4.3. Let $q(t) \in C([a, b], \mathbb{R})$. Suppose $x(t)$ is a solution of equation (3.4.1) having consecutive zeros at $t = a$ and $t = b$. Then the following inequality holds

$$2^{\alpha+1} < (b - a)^\alpha \int_a^b q^+(t)dt, \tag{3.4.3}$$

where $q^+(t) = \max\{q(t), 0\}$.

Proof. Consider the differential equation

$$(\psi(y'(t)))' + q^+(t)\psi(y(t)) = 0. \tag{3.4.4}$$

The existence theory ensures that the initial value problem (3.4.4), $y(a) = 0$, $y'(a) = 1$ has a solution $y(t)$. Then from Theorem 3.2.1 it follows that $y(t)$ has at least one zero in $(a, b]$. Let $\xi = \min\{t \in (a, b] : y(t) = 0\}$. Thus, from Theorem 3.4.2, we have $(b - a)^\alpha \int_a^b q^+(t)dt \geq (\xi - a)^\alpha \int_a^\xi q^+(t)dt > 2^{\alpha+1}$. This completes the proof. ∎

Suppose $x(t)$ is a nontrivial solution of (3.4.1) having $N+1$ consecutive zeros at $a = a_0 < a_1 < \cdots < a_N = b$. Then from (3.4.3), we have

$$\int_{a_{k-1}}^{a_k} q^+(s)ds > 2^{\alpha+1}(a_k - a_{k-1})^{-\alpha}, \quad k = 1, 2, \cdots, N. \tag{3.4.5}$$

Thus, by Hölder's inequality it follows that

$$(b - a)^{\alpha/(\alpha+1)} \left(\sum_{k=1}^N (a_k - a_{k-1})^{-\alpha} \right)^{1/(\alpha+1)}$$

$$= \left(\sum_{k=1}^N (a_k - a_{k-1})^{-\alpha} \right)^{1/(\alpha+1)} \left(\sum_{k=1}^N (a_k - a_{k-1}) \right)^{\alpha/(\alpha+1)}$$

$$= \left(\sum_{k=1}^N \left[(a_k - a_{k-1})^{-\alpha/(\alpha+1)} \right]^{\alpha+1} \right)^{1/(\alpha+1)}$$

$$\times \left(\sum_{k=1}^N \left[(a_k - a_{k-1})^{\alpha/(\alpha+1)} \right]^{(\alpha+1)/\alpha} \right)^{\alpha/(\alpha+1)}$$

$$\geq \sum_{k=1}^N (a_k - a_{k-1})^{-\alpha/(\alpha+1)}(a_k - a_{k-1})^{\alpha/(\alpha+1)} = N.$$

Therefore, we have $\sum_{k=1}^{N}(a_k - a_{k-1})^{-\alpha} \geq N^{\alpha+1}(b - a)^{-\alpha}$. This and (3.4.5) imply that $\int_a^b q^+(t)dt > (2N)^{\alpha+1}(b - a)^{-\alpha}$. From this inequality the following corollary is immediate. ∎

Corollary 3.4.1. Let $q(t) \in C([a,b], \mathbb{R})$ be not identically zero on $[a, b]$. Suppose $x(t)$ is a nontrivial solution of equation (3.4.1) having consecutive zeros at $a = a_0 < a_1 < \cdots < a_N = b$. Then the following inequality holds

$$\int_a^b q^+(t)dt > (2N)^{\alpha+1}(b - a)^{-\alpha}.$$

Definition 3.4.1. A function $f \in C(\mathbb{R}, \mathbb{R})$ is said to be *quickly oscillatory* if it is defined in a neighborhood of ∞ and there exists a sequence $\{t_n\}_{n=1}^{\infty}$ with $\lim_{n\to\infty} t_n = \infty$ such that $f(t_n) = 0$, $n = 1, 2, \cdots$, $t_{n+1} > t_n$ and $\lim_{n\to\infty}(t_{n+1} - t_n) = 0$.

As an application of Theorem 3.4.3, we shall prove the following result.

Theorem 3.4.4. If equation (3.4.1) has a quickly oscillating solution, then

$$\int^{\infty} q^+(t)dt = \infty \tag{3.4.6}$$

and $\limsup_{t\to\infty} q(t) = \infty$.

Proof. Let $x(t)$ be a quickly oscillating solution of equation (3.4.1) with zeros t_n such that $t_n \to \infty$ and $t_{n+1} - t_n \to 0$ as $n \to \infty$. Consider the consecutive zeros $t_{n+1} > t_n > t_0$ and the interval $[t_n, t_{n+1}]$. Then it follows from Theorem 3.4.3 that

$$\int_{t_0}^{\infty} q^+(t)dt > \int_{t_n}^{t_{n+1}} q^+(t)dt > 2^{\alpha+1}(t_{n+1} - t_n)^{-\alpha} \to \infty \quad \text{as} \quad n \to \infty.$$

Hence (3.4.6) holds. Now, from the mean value theorem for integrals, we have $\int_{t_n}^{t_{n+1}} q^+(t)dt = q^+(\xi_n)(t_{n+1} - t_n) > 2^{\alpha+1}(t_{n+1} - t_n)^{-\alpha}$, where $t_n < \xi_n < t_{n+1}$. This implies that $q^+(\xi_n) > 2^{\alpha+1}(t_{n+1} - t_n)^{-\alpha-1}$ and hence $\limsup_{t\to\infty} q(t) = \infty$. This completes the proof. ∎

3.5. An Oscillation Criterion for Almost–Periodic Sturm–Liouville Equations

The class $\Omega \subset \mathcal{L}^1_{Loc}(\mathbb{R})$ of Besicovitch almost periodic functions is the closure of the set of all finite trigonometric polynomials with the Besicovitch seminorm $\| \cdot \|_B$ defined by

$$\|q\|_B = \limsup_{t\to\infty} \frac{1}{2t} \int_{-t}^{t} |q(s)|ds,$$

where $q \in \Omega$. The mean value, $M\{q\}$ of $q \in \Omega$ always exists, is finite and is uniform with respect to β for $\beta \in \mathbb{R}$, where

$$M\{q\} = \lim_{t \to \infty} \frac{1}{t} \int_{t_0}^{t} q(s + \beta)ds = 0 \quad \text{for some} \quad t_0 \geq 0.$$

Consider the following half–linear differential equation

$$(\psi(x'(t)))' - \lambda q(t)\psi(x(t)) = 0, \tag{3.5.1}$$

where $q(t) \in \Omega$, $\lambda \in \mathbb{R} - \{0\}$ and $\psi(x)$ is as in equation (3.1.1).

Definition 3.5.1. We say equation (3.5.1) is oscillatory at $+\infty$ and $-\infty$ if every solution of (3.5.1) has an infinity of zeros clustering only at $+\infty$ and $-\infty$, respectively.

When $\alpha = 1$ equation (3.5.1) reduces to the linear differential equation

$$x''(t) - \lambda q(t)x(t) = 0. \tag{3.5.2}$$

From Levin's comparison theorem with $\alpha = 1$ the following result is immediate.

Theorem 3.5.1. Let $q(t) \in \Omega$ and $M\{|q|\} > 0$. Then, $M\{q\} = 0$ if and only if equation (3.5.2) is oscillatory at $+\infty$ and $-\infty$ for every $\lambda \in \mathbb{R} - \{0\}$.

In what follows we shall extend the sufficient condition of Theorem 3.5.1 to equation (3.5.1). For this, we need the following two lemmas.

Lemma 3.5.1. If $\gamma > 1$, a, $b \in \mathbb{R}^+$ then $(a + b)^\gamma > a^\gamma + b^\gamma$.

Proof. Without loss of generality we can assume that $a \geq b$. Then by Lemma 1.1.1, we have

$$(a + b)^\gamma \geq \gamma a^{\gamma-1}(a + b) - (\gamma - 1)a^\gamma = a^\gamma + \gamma a^{\gamma-1}b > a^\gamma + b^\gamma. \quad \blacksquare$$

Lemma 3.5.2. Suppose

$$\begin{cases} q : [t_0, \infty) \to \mathbb{R} \text{ is locally Lebesgue integrable,} \\ t_0 \geq 0 \text{ and has the mean value } M\{q\} = 0. \end{cases} \tag{3.5.3}$$

If $x(t) \neq 0$ is a solution of equation

$$(\psi(x'(t)))' - q(t)\psi(x(t)) = 0 \tag{3.5.4}$$

on $[t_0, \infty)$, then

$$\lim_{t \to \infty} \frac{1}{t} \int_{t_0}^{t} \left(\frac{|x'(s)|}{|x(s)|} \right)^{\alpha+1} ds = 0.$$

Proof. Define $w(t) = -\psi(x'(t))/\psi(x(t))$ for $t \geq t_0$. It follows from (3.5.4) that $w(t)$ is a solution of the equation

$$w'(t) - \alpha|w(t)|^{(\alpha+1)/\alpha} + q(t) = 0 \quad \text{for} \quad t \geq t_0. \qquad (3.5.5)$$

It suffices to show that $\lim \sup_{t\to\infty}(1/t)\int_{t_0}^t |w(s)|^{(\alpha+1)/\alpha}ds = 0$. Assume to the contrary that

$$\lim_{t\to\infty} \sup \frac{1}{t}\int_{t_0}^t |w(s)|^{(\alpha+1)/\alpha}ds > 0. \qquad (3.5.6)$$

Integrating (3.5.5) from t_0 to t and dividing it by t, we obtain

$$\frac{w(t)}{t} = \frac{w(t_0)}{t} + \frac{1}{t}\int_{t_0}^t q(s)ds + \frac{\alpha}{t}\int_{t_0}^t |w(s)|^{(\alpha+1)/\alpha}ds \text{ for } t > t_0. \quad (3.5.7)$$

From (3.5.6), (3.5.7) and (3.5.3) there exist a positive constant m and an increasing sequence $\{t_n\}_{n=1}^\infty$ of $[t_0,\infty)$ with $\lim_{n\to\infty} t_n = \infty$ such that

$$\frac{w(t_n)}{t_n} > \alpha m^{\alpha+1} \quad \text{for all sufficiently large } n. \qquad (3.5.8)$$

It follows from (3.5.3) that there exists t^* large enough so that

$$\left|\int_{t_0}^t q(s)ds\right| < \frac{\alpha}{2^{\alpha+1}}m^{\alpha+1}t \quad \text{for all} \quad t \geq t^*. \qquad (3.5.9)$$

Using (3.5.9) for $t \geq t_n \geq t^*$, we find

$$\int_{t_n}^t q(s)ds = \int_{t_0}^t q(s)ds - \int_{t_0}^{t_n} q(s)ds < \frac{\alpha}{2^{\alpha+1}}m^{\alpha+1}(t + t_n). \qquad (3.5.10)$$

From (3.5.8) and (3.5.10), we get

$$w(t_n) - \int_{t_n}^t q(s)ds > \alpha m^{\alpha+1}t_n - \frac{\alpha}{2^{\alpha+1}}m^{\alpha+1}(t + t_n)$$

$$\geq \alpha m^{\alpha+1}t_n - \frac{\alpha}{2^{\alpha+1}}m^{\alpha+1}\left[(2^{\alpha+1} - 1)t_n + t_n\right] = 0$$

$$(3.5.11)$$

for all $t \in I = [t_n, (2^{\alpha+1} - 1)t_n] \subset [t^*, \infty)$.

From the existence theory the differential equation

$$\frac{d}{dt}(\psi(x_n'(t))) - \alpha\left(\frac{m}{2}\right)^{\alpha+1}\psi(x_n(t)) = 0 \qquad (3.5.12)$$

has a solution $x_n(t)$ on I satisfying $x_n(t_n) = x(t_n)$, and

$$-\frac{\psi(x_n'(t_n))}{\psi(x_n(t_n))} = w(t_n) - \frac{\alpha\, m^{\alpha+1}}{2^\alpha}t_n.$$

Now, from (3.5.10) and (3.5.11), we obtain

$$-\frac{\psi(x'(t_n))}{\psi(x(t_n))} - \int_{t_n}^t q(s)ds = w(t_n) - \int_{t_n}^t q(s)ds$$

$$\geq w(t_n) - \alpha\left(\frac{m}{2}\right)^{\alpha+1}(t+t_n) = \left[w(t_n) - \frac{\alpha m^{\alpha+1}}{2^\alpha}t_n\right] - \alpha\left(\frac{m}{2}\right)^{\alpha+1}(t-t_n)$$

$$= -\frac{\psi(x_n'(t_n))}{\psi(x_n(t_n))} - \int_{t_n}^t \alpha\left(\frac{m}{2}\right)^{\alpha+1}ds \geq 0 \quad \text{on} \quad I.$$

Thus, from Theorem 3.3.1 it follows that

$$-\frac{\psi(x(t_n))}{\psi(x(t_n))} > \left|-\frac{\psi(x_n'(t_n))}{\psi(x_n(t_n))}\right| \quad \text{on} \quad I. \tag{3.5.13}$$

Now, define $w_n(t) = -\psi(x_n'(t_n))/\psi(x_n(t_n))$ on I. It follows that $w_n(t)$ is a solution of the differential equation

$$w_n'(t) - \alpha|w_n(t)|^{(\alpha+1)/\alpha} + \alpha(m/2)^{\alpha+1} = 0 \quad \text{on} \quad I, \tag{3.5.14}$$

where $w_n(t_n) = w(t_n) - \alpha\left(m^{\alpha+1}/2^m\right)t_n$. Let

$$c_n = [w_n(t_n) - (m/2)^\alpha]^{-1/\alpha} \tag{3.5.15}$$

and

$$y_n(t) = (m/2)^\alpha + (t_n - t + c_n)^{-\alpha} \tag{3.5.16}$$

on $I_1 = [t_n, t_n + c_n] \subset [t^*, \infty)$, where n is large enough so that $w_n(t_n) > (m/2)^\alpha$. Then, $y_n(t_n) = w_n(t_n)$, and it follows from Lemma 3.5.1 that

$$
\begin{aligned}
y_n'(t) &= \alpha(t_n - t + c_n)^{-(\alpha+1)} \\
&= \alpha\left[(t_n - t + c_n)^{-(\alpha+1)} + (m/2)^{\alpha+1}\right] - \alpha(m/2)^{\alpha+1} \\
&< \alpha\left[(t_n - t + c_n)^{-\alpha} + (m/2)^\alpha\right]^{(\alpha+1)/\alpha} - \alpha(m/2)^{\alpha+1} \\
&= \alpha|y_n(t)|^{(\alpha+1)/\alpha} - \alpha(m/2)^{\alpha+1} \quad \text{on} \quad I_1.
\end{aligned}
$$

Thus, we have

$$
\begin{aligned}
y_n'(t) - \alpha|y_n(t)|^{(\alpha+1)/\alpha} + \alpha(m/2)^{\alpha+1} &< 0 \\
&= w_n'(t) - \alpha|w_n(t)|^{(\alpha+1)/\alpha} + \alpha(m/2)^{\alpha+1}
\end{aligned}
$$

for all $t \in I_2 = I \cap I_1 \subset [t^*, \infty)$. Now, a comparison argument gives $y_n(t) \leq x_n(t)$ on I_2. Hence, it follows from

$$w_n(t_n) = w(t_n) - 2\alpha(m/2)^{\alpha+1}t_n > \alpha\left(1 - 2^{-\alpha}\right)m^{\alpha+1}t_n$$

that $t_n + c_n \in I$ for all n large enough.

By the definition of $y_n(t)$ we find that for all sufficiently large n, $\lim_{t \to (t_n + c_n) -} y_n(t) = \infty$, and hence

$$\lim_{t \to (t_n + c_n) -} w_n(t) = \infty. \tag{3.5.17}$$

Now, we take k large enough so that $t_k + c_k \in I_k = [t_k, (2^{\alpha+1} - 1)t_k]$. Clearly, there exists a positive constant M such that

$$-\frac{\psi(x'(t_n))}{\psi(x(t_n))} \leq M < \infty \quad \text{on} \quad I_k.$$

Finally, it follows from (3.5.13) and (3.5.17) that

$$\infty = \lim_{t \to (t_n + c_n) -} w_n(t) \leq \lim_{t \to (t_n + c_n) -} \frac{\psi(x'(t_n))}{\psi(x(t_n))} \leq M < \infty,$$

which is a contradiction. This completes the proof. ∎

Theorem 3.5.2. If $q(t) \in \Omega$ satisfies (3.5.3) and $M\{|q|\} > 0$, then equation (3.5.1) is oscillatory at $+\infty$ and $-\infty$ for every $\lambda \in \mathbb{R} - \{0\}$.

Proof. Without loss of generality we shall only show that equation (3.5.4) is oscillatory at $+\infty$. Assume to the contrary that (3.5.4) has a solution $x(t)$ which is nonoscillatory at $+\infty$. Thus, we can assume that there exists $t_0 > 0$ such that $x(t) > 0$ on $[t_0, \infty)$. Define $w(t) = -\psi(x'(t))/\psi(x(t))$ for $t \geq t_0$. Then, $w(t)$ is a solution of (3.5.5) on $[t_0, \infty)$. Hence, for any fixed $\delta > 0$, we have

$$\frac{1}{\delta} \int_t^{t+\delta} q(s)ds = -\frac{w(t+\delta)}{\delta} + \frac{w(t)}{\delta} + \frac{\alpha}{\delta} \int_t^{t+\delta} |w(s)|^{(\alpha+1)/\alpha} ds \tag{3.5.18}$$

on $[t_0, \infty)$. Applying the Besicovitch semi–norm $\| \cdot \|_{B'}$, essentially a restriction of $\| \cdot \|_B$ to the interval $[t_0, \infty)$ defined by

$$\|f\|_{B'} = \limsup_{t \to \infty} \frac{1}{t} \int_{t_0}^t |f(s)| ds$$

to (3.5.18), we find for all $\delta > 0$,

$$0 \leq \left\| \frac{1}{\delta} \int_t^{t+\delta} q(s)ds \right\|_{B'} \leq \left\| \frac{\alpha}{\delta} \int_t^{t+\delta} |w(s)|^{(\alpha+1)/\alpha} ds \right\|_{B'} \tag{3.5.19}$$
$$+ \left\| \frac{w(t+\delta)}{\delta} \right\|_{B'} + \left\| \frac{w(t)}{\delta} \right\|_{B'}.$$

It follows from Lemma 3.5.2 that $M\left\{|w|^{(\alpha+1)/\alpha}\right\} = 0$. Thus, $\|w\|_{B'} = \|w(t+\delta)\|_{B'}$ for all $\delta > 0$. Using Fubini's theorem, we obtain

$$\frac{1}{\delta t}\int_{t_0}^{t}\int_{s}^{s+\delta}|w(\xi)|^{(\alpha+1)/\alpha}d\xi ds$$

$$= \frac{1}{\delta t}\int_{t_0}^{t}\int_{0}^{\delta}|w(\xi+s)|^{(\alpha+1)/\alpha}d\xi ds = \frac{1}{\delta t}\int_{0}^{\delta}\int_{t_0}^{t}|w(\xi+s)|^{(\alpha+1)/\alpha}ds d\xi$$

$$\leq \frac{1}{\delta t}\int_{0}^{\delta}\int_{t_0}^{t+\delta}|w(s)|^{(\alpha+1)/\alpha}ds d\xi = \frac{1}{t}\int_{t_0}^{t_0+\delta}|w(s)|^{(\alpha+1)/\alpha}ds,$$

$$(3.5.20)$$

for any fixed $\delta > 0$. Now, from (3.5.20) and Lemma 3.5.2, we get

$$\left\|\frac{\alpha}{\delta}\int_{t}^{t+\delta}|w(s)|^{(\alpha+1)/\alpha}ds\right\|_{B'} = 0 \quad \text{for any fixed} \quad \delta > 0. \qquad (3.5.21)$$

Applying (3.5.21) and $\|w\|_{B'} = \|w(t+\delta)\|_{B'} = 0$ to (3.5.20), we find

$$\left\|\frac{1}{\delta}\int_{t}^{t+\delta}q(s)ds\right\|_{B'} = 0 \quad \text{for all} \quad \delta > 0. \qquad (3.5.22)$$

Finally, since $q(t)$ is Besicovitch almost periodic, it follows from [9] that

$$\lim_{\delta\to 0}\left\|q(t) - \frac{1}{\delta}\int_{t}^{t+\delta}q(s)ds\right\|_{B'} = 0.$$

This and (3.6.22) imply that $M\{|q|\} = \|q\|_{B'} = 0$, which is a contradiction. This completes the proof. ∎

Example 3.5.1. Consider the half–linear differential equation

$$\left(\psi(x'(t))\right)' - \lambda\cos t\,\psi(x(t)) = 0, \quad t \geq t_0 \geq 0. \qquad (3.5.23)$$

Then, $q(t) = \cos t$, and hence

$$M\{q\} = \lim_{t\to\infty}\frac{1}{t}\int_{0}^{t}q(s)ds = \lim_{t\to\infty}\frac{\sin t}{t} = 0$$

and

$$M\{|q|\} = \lim_{t\to\infty}\frac{1}{t}\int_{0}^{t}|q(s)|ds = \lim_{t\to\infty}\frac{1}{2(n+1)\pi}\int_{0}^{2(n+1)\pi}|\cos s|ds$$

$$= \lim_{n\to\infty}\frac{2}{\pi}\int_{0}^{\pi/2}|\cos s|ds = \frac{2}{\pi} > 0.$$

Thus, it follows from Theorem 3.5.2 that for each $\lambda \in \mathbb{R} - \{0\}$ equation (3.5.23) is oscillatory at $+\infty$ and $-\infty$.

3.6. On Zeros of Solutions of a Singular Half–linear Differential Equation

In this section, we shall consider the differential equation

$$x''(t) = q(t)|x(t)|^{\alpha}|x'(t)|^{1-\alpha} \operatorname{sgn} x(t), \qquad (3.6.1)$$

where $\alpha \in (0,1]$ is a constant, the function $q : (a,b) \to \mathbb{R}$ is locally integrable and

$$\int_a^b (s-a)^{\alpha}(b-s)^{\alpha}|q(s)|ds < \infty. \qquad (3.6.2)$$

Sufficient conditions for the existence of a solution of (3.6.1) having at least two zeros in the interval $[a,b]$ are presented.

In what follows we will use the following notation: $\mathcal{L}_{loc}((a,b))$ is the set of functions $q : (a,b) \to \mathbb{R}$ which are Lebesgue integrable on each segment contained in (a,b). $AC([a,b])$ is the set of functions $u : [a,b] \to \mathbb{R}$ absolutely continuous on the segment $[a,b]$; $AC_{Loc}(I)$, where $I \subset \mathbb{R}$ is the set of functions $u : I \to \mathbb{R}$ absolutely continuous on each segment contained in I; $AC^1_{loc}(I)$, where $I \subset \mathbb{R}$ is the set of functions $u \in AC_{loc}(I)$ for which $u' \in AC_{loc}(I)$; $u(s+)$ and $u(s-)$ respectively are the right and left limits of the function u at the point s; and $q^-(t) = (|q(t)| - q(t))/2$.

By a solution of equation (3.6.1), where $q \in \mathcal{L}_{loc}((a,b))$ we mean a function $x \in AC^1_{loc}((a_1,b_1))$, where $a_1 \in [a,b)$ and $b_1 \in (a_1,b]$ which satisfies (3.6.1) almost everywhere in (a_1,b_1). In the case of linear differential equation, i.e., when $\alpha = 1$, the number of zeros of two arbitrary non–trivial solutions differ from each other by not more than 1. This fact in general does not hold for (3.6.1) when $\alpha \neq 1$, since any constant function is also its solution; however, it remains valid for a definite subset of the set of solutions, which in what follows will be called the set of proper solutions.

Definition 3.6.1. A solution $x(t)$ of (3.6.1) is said to be *proper* if there exists $A \subset \mathbb{R}$ such that *mes* $A = 0$ and $\{t \in (a,b) : x'(t) = 0\} \subset \{t \in (a,b) : q(t) = 0\} \cup A$.

Definition 3.6.2. The function $q(t)$ is said to belong to the set $O_{\alpha}((a,b))$ if there exists a proper solution of (3.6.1) having at least two zeros on the segment $[a,b]$.

In other words, $q \notin O_\alpha((a,b))$ if and only if there is no proper solution $x(t)$ of (3.6.1) satisfying for some $a_1 \in [a,b)$ the conditions

$$x(a_1+) = 0, \quad x(b_1-) = 0. \tag{3.6.3}$$

Definition 3.6.3. The function $q(t)$ is said to belong to the set $U_\alpha((a,b))$ if for any $a_1 \in [a,b)$ and $b_1 \in (a_1,b]$ the problem (3.6.1), (3.6.3) has no non–zero (not necessarily proper) solution.

It is clear that if $q \in U_\alpha((a,b))$, then $q \notin O_\alpha((a,b))$. In the case when $\alpha = 1$, or $\alpha \in (0,1)$ and $q(t) \leq 0$ for $a < t < b$, the converse is also valid, i.e., if $q \notin U_\alpha((a,b))$, then $q \in O_\alpha((a,b))$.

For the convenience of reference we state the following proposition.

Proposition 3.6.1. The following equalities hold

$$\lim_{t\to a^+} (t-a)^\alpha \int_t^{(a+b)/2} |q(s)|ds = 0 \quad \text{and} \quad \lim_{t\to b^-} (b-t)^\alpha \int_{(a+b)/2}^t |q(s)|ds = 0.$$

Lemma 3.6.1. There exist two solutions $x_1(t)$ and $x_2(t)$ of the equation (3.6.1) satisfying respectively the initial conditions

$$x_1(a+) = 0, \quad x_1'(a+) = 1 \tag{3.6.4}$$

$$x_2(b-) = 0, \quad x_2'(b-) = -1. \tag{3.6.5}$$

Moreover, all noncontinuable solutions of the problems (3.6.1), (3.6.4) and (3.6.1), (3.6.5) are defined on the entire segment (a,b).

Proof. We will prove only the existence of x_1. The existence of x_2 can be proved similarly. Let $T \in (a,b)$. Denote by Ω_T the set of all noncontinuable to the right solutions of (3.6.1) satisfying the initial conditions

$$x(T) = 0, \quad x'(T) = 1. \tag{3.6.6}$$

Let $x(\cdot,T) \in \Omega_T$. We shall show that this solution is defined in the interval $[T,b)$. Suppose that

$$I_T = \{t \in (T,b) : |x(s,T)| < \infty, \ |x'(s,T)| < \infty \quad \text{for} \quad T \leq s \leq t\}.$$

Integrating (3.6.1) and taking into account (3.6.6), we obtain

$$x'(t,T) = 1 + \int_T^t \cdot q(s)|x(s,T)|^\alpha |x'(s,T)|^{1-\alpha} \operatorname{sgn} x(s,T)ds \quad \text{for} \quad t \in I_T \tag{3.6.7}$$

and

$$x(t,T) = (t-T) + \int_T^t (t-s)|x(s,T)|^\alpha |x'(s,T)|^{1-\alpha} \operatorname{sgn} x(s,T)ds \quad \text{for} \quad t \in I_T. \tag{3.6.8}$$

Thus, it follows that

$$|x'(t,T)| \leq 1 + \int_T^t (s-a)^\alpha |q(s)| \left(\left| \frac{x(s,T)}{s-T} \right|^\alpha \right) |x'(s,T)|^{1-\alpha} ds \quad \text{for} \quad t \in I_T,$$

$$\left| \frac{x(t,T)}{t-T} \right| \leq 1 + \int_T^t (s-a)^\alpha |q(s)| \left(\left| \frac{x(s,T)}{s-T} \right|^\alpha \right) |x'(s,T)|^{1-\alpha} ds \quad \text{for} \quad t \in I_T.$$

Adding the above inequalities and using the fact that

$$y^\alpha < 1 + y \quad \text{for} \quad y > 0, \tag{3.6.9}$$

we find for $t \in I_T$,

$$\left| \frac{x(t,T)}{t-T} \right| + |x'(t,T)| \leq 2 + 2 \int_T^t (s-a)^\alpha |q(s)| \left[\left| \frac{x(s,T)}{s-T} \right| + |x'(s,T)| \right] ds.$$

Applying Lemma 1.1.1, we get

$$\left| \frac{x(t,T)}{t-T} \right| + |x'(t,T)| \leq 2 \exp \left(2 \int_T^t (s-a)^\alpha |q(s)| ds \right) \quad \text{for} \quad t \in I_T. \tag{3.6.10}$$

Consequently, $\sup I_T = b$. Moreover, from (3.6.7) and (3.6.10), we also obtain

$$|x(t,T)| \leq 2(t-a) \exp \left(2 \int_a^t (s-a)^\alpha |q(s)| ds \right) \quad \text{for} \quad T \leq t < b, \tag{3.6.11}$$

$$|x'(t,T)| \leq 2 \exp \left(2 \int_a^t (s-a)^\alpha |q(s)| ds \right) \quad \text{for} \quad T \leq t < b \tag{3.6.12}$$

and for $T \leq t < b$,

$$|x'(t,T) - 1| \leq 2 \left(\int_a^t (s-a)^\alpha |q(s)| ds \right) \exp \left(2 \int_a^t (s-a)^\alpha |q(s)| ds \right). \tag{3.6.13}$$

Let $T_k \in (a,b)$, $T_{k+1} < T_k$ for $k = 1, 2, \cdots$, $\lim_{k \to \infty} T_k = a$. Suppose $u_k(t) = x(t, T_k)$ for $T_k \leq t < b$ for $k = 1, 2, \cdots$. From (3.6.11) and (3.6.12) the sequences $\{u_k\}_{k=1}^\infty$ and $\{u'_k\}_{k=1}^\infty$ are uniformly bounded and equicontinuous in (a,b), (i.e., on each segment contained in (a,b)). Hence, without loss of generality, by the Arzela–Ascoli theorem, we can assume that $\lim_{k \to \infty} u_k^{(i)}(t) = x_1^{(i)}(t)$, $i = 0, 1$ uniformly in (a,b).

It is easy to see that the function $x_1(t)$ is a solution of (3.6.1). Further, by (3.6.11) and (3.6.13) the function $x_1(t)$ satisfies condition (3.6.4) also. Thus, we have proved that the set of solutions Ω of the problem (3.6.1), (3.6.4) is non–empty.

Now, we shall show that all functions in Ω are defined on the entire segment $[a, b]$. For this, let $x_1 \in \Omega$ be a noncontinuable solution. Then analogous to what has been done above, we find

$$x_1'(t) = 1 + \int_a^t (s-a)^\alpha q(s) \left| \frac{x_1(s)}{s-a} \right|^\alpha |x_1'(s)|^{1-\alpha} \operatorname{sgn} x_1(s) ds \quad \text{for} \quad t \in I,$$

$$\frac{x_1(t)}{t-a} = 1 + \frac{1}{t-a} \int_a^t (s-a)^\alpha q(s) \left| \frac{x_1(s)}{s-a} \right|^\alpha |x_1'(s)|^{1-\alpha} \operatorname{sgn} x_1(s) ds \quad \text{for} \quad t \in I,$$

$$(3.6.14)$$

where

$$I = \{ t \in (a, b) : |x_1(s)| < \infty, \ |x_1'(s)| < \infty \quad \text{for} \quad a \le s \le t \}.$$

These two inequalities on I immediately yield

$$(b-t)|x_1'(t)| \le (b-a) + \int_a^t (s-a)^\alpha (b-s)^\alpha q(s) \left| \frac{x_1(s)}{s-a} \right|^\alpha |(b-s)x'(s)|^{1-\alpha} ds,$$

$$\frac{|x_1(t)|}{t-a} \le 1 + \frac{1}{b-a} \int_a^t (s-a)^\alpha (b-s)^\alpha |q(s)| \left| \frac{x_1(s)}{s-a} \right|^\alpha |(b-s)x_1'(s)|^{1-\alpha} ds.$$

Adding these inequalities, using (3.6.9) and applying Lemma 1.1.1, we obtain

$$\frac{|x_1(t)|}{t-a} + (b-t)|x_1'(t)| \le (1+b-a) \exp \left(\frac{1+b-a}{b-a} \right.$$

$$\left. \times \int_a^b (s-a)^\alpha (b-s)^\alpha |q(s)| ds \right) \quad \text{for} \quad t \in I.$$

$$(3.6.15)$$

The inequality (3.6.14) for $a < t < b$ results in

$$x_1(t) = t - a - (b-t) \int_a^t \frac{[(s-a)(b-s)]^\alpha}{b-s} q(s) \left| \frac{x_1(s)}{s-a} \right|^\alpha$$

$$\times |(b-s)x_1'(s)|^{1-\alpha} \operatorname{sgn} x_1(s) ds$$

$$+ \int_a^t [(s-a)(b-s)]^\alpha q(s) \left| \frac{x_1(s)}{s-a} \right|^\alpha |(b-s)x_1'(s)|^{1-\alpha} \operatorname{sgn} x_1(s) ds$$

from which in view of (3.6.15) and Proposition 3.6.1, we conclude that there exists a finite limit $x_1(b-)$. ∎

The proof of the following lemma is similar to that of Lemma 3.6.1.

Lemma 3.6.2. All noncontinuable solutions of (3.6.1) are defined on the entire segment $[a, b]$.

Remark 3.6.1. If $c \in (a, b)$ and $x(t)$ is a solution of (3.6.1) satisfying the conditions $x(c) = 0$, $x'(c) = 0$ then $x(t)$ is identically zero.

Lemma 3.6.3. If $x(t)$ is a solution of equation (3.6.1), then

$$\lim_{t \to a+} (t - a)|x'(t)| = 0 \quad \left(\lim_{t \to b-} (b - t)|x'(t)| = 0 \right). \tag{3.6.16}$$

Proof. We claim that

$$\liminf_{t \to a+} (t - a)|x'(t)| = 0 \quad \left(\liminf_{t \to b-} (b - t)|x'(t)| = 0 \right). \tag{3.6.17}$$

For this, assume that (3.6.17) does not hold. Then there exist $c \in (a, b)$ and $\epsilon > 0$ such that

$$|x'(t)| > \frac{\epsilon}{t - a} \quad \text{for} \quad a < t < c \quad \left(|x'(t)| > \frac{\epsilon}{b - t} \quad \text{for} \quad c < t < b \right).$$

An integration of this inequality from t to c (from c to t) yields

$$|x(t) - x(c)| > \epsilon \ln \frac{c - a}{t - a} \quad \text{for} \quad a < t < c$$

$$\left(|x(t) - x(c)| > \epsilon \ln \frac{b - c}{b - t} \quad \text{for} \quad c < t < b \right),$$

which is impossible because the function $x(t)$ is bounded. Thus, (3.6.17) holds.

Multiplying both sides of (3.6.1) by $(t-a)$ (by $(b-t)$) and integrating it from τ to t (from t to τ), we obtain

$$(t - a)x'(t) = (\tau - a)x'(\tau) + x(t) - x(\tau)$$

$$+ \int_\tau^t (s - a)q(s)|x(s)|^\alpha |x'(s)|^{1-\alpha} \operatorname{sgn} x(s)ds \quad \text{for} \quad a < \tau < t < b \tag{3.6.18}$$

$$\left(\begin{array}{l} (b - t)x'(t) = (b - \tau)x'(\tau) + x(\tau) - x(t) \\ - \int_t^\tau (b - s)q(s)|x(s)|^\alpha |x'(s)|^{1-\alpha} \operatorname{sgn} x(s)ds \quad \text{for} \quad a < t < \tau < b \end{array} \right)$$

from this and (3.6.9), we conclude that

$$(t - a)|x'(t)| \leq (\tau - a)|x'(\tau)| + |x(t) - x(\tau)|$$

$$+ \int_\tau^t (s - a)^\alpha |q(s)|(|x(s)| + (s - a)|x'(s)|)ds \quad \text{for} \quad a < \tau < t < b \tag{3.6.19}$$

$$\left(\begin{array}{l} (b-t)|x'(t)| \ \le \ (b-\tau)|x'(\tau)| + |x(t) - x(\tau)| \\[2mm] \quad + \int_t^\tau (b-s)^\alpha |q(s)|(|x(s)| + (b-s)|x'(s)|)ds \quad \text{for} \ \ a < t < \tau < b. \end{array}\right)$$

Suppose

$$\begin{aligned} c(T) \ &= \ 2\max\{|x(t) - x(a+)| : \ \text{for} \ \ a \le t \le T\} \\[2mm] &\quad + \int_a^T (s-a)^\alpha |q(s)||x(s)|ds \quad \text{for} \ \ a < T < b \end{aligned}$$

$$\left(\begin{array}{l} c(T) \ = \ 2\max\{|x(t) - x(b-)| : \ \text{for} \ \ T \le t \le b\} \\[2mm] \quad + \int_T^b (b-s)^\alpha |q(s)||x(s)|ds \quad \text{for} \ \ a < T < b \end{array}\right),$$

and

$$\begin{aligned} M(\tau, T) \ &= \ (\tau - a)|x'(\tau)| + c(T) \quad \text{for} \ \ a < \tau \le T < b \\ (M(\tau, T) \ &= \ (b - \tau)|x'(\tau)| + c(T) \quad \text{for} \ \ a < T \le \tau < b). \end{aligned} \qquad (3.6.20)$$

Then from (3.6.19), we find

$$\begin{aligned} (t-a)|x'(t)| \ &\le \ M(\tau, T) + \int_\tau^t (s-a)^\alpha |q(s)|((s-a)|x'(s)|)ds \\ &\qquad\qquad\qquad \text{for} \ \ a < \tau < t < T < b \end{aligned}$$

$$\left(\begin{array}{l} (b-t)|x'(t)| \ \le \ M(\tau, T) + \int_t^\tau (b-s)^\alpha |q(s)|((b-s)|x'(s)|)ds \\[2mm] \qquad\qquad\qquad \text{for} \ \ a < T < t < \tau < b \end{array}\right).$$

Applying Lemma 1.1.1, we get

$$(t-a)|x'(t)| \ \le \ M(\tau, T)\exp\left(\int_a^T (s-a)^\alpha |q(s)|ds\right) \quad \text{for} \ a < \tau < t < T < b$$

$$\left((b-t)|x'(t)| \le M(\tau, T)\exp\left(\int_T^b (b-s)^\alpha |q(s)|ds\right) \quad \text{for} \ a < T < t < \tau < b.\right)$$

Now, in view of (3.6.17) and (3.6.20), we find

$$(t-a)|x'(t)| \ \le \ c(T)\exp\left(\int_a^T (s-a)^\alpha |q(s)|ds\right) \quad \text{for} \ \ a < t < T < b$$

$$\left((b-t)|x'(t)| \ \le \ c(T)\exp\left(\int_T^b (b-s)^\alpha |q(s)|ds\right) \quad \text{for} \ \ a < T < t < b\right).$$

Since $\lim_{T \to a^+} c(T) = 0$ $(\lim_{T \to b^-} c(T) = 0)$, the above inequality implies that (3.6.16) holds. ∎

Lemma 3.6.4. Let $a_1 \in (a, b)$, $b_1 \in (a_1, b)$ and let $x(t)$ and $y(t)$ be solutions of equation (3.6.1) satisfying the conditions

$$x(a_1) = 0, \quad x'(a_1) > 0 \quad (x(b_1) = 0, \quad x'(b_1) < 0)$$
$$y(t) > 0 \quad \text{for} \quad a_1 \le t \le b_1, \quad y'(b_1) = 0 \quad (y'(a_1) = 0).$$

(3.6.21)

Furthermore, let $u \in AC^1([a_1, b_1])$ be such that

$$u(t) > 0, \ u'(t) > 0 \ \text{for} \ a_1 \le t \le b_1 \ \ (u(t) > 0, \ u'(t) < 0 \ \text{for} \ a_1 \le t \le b_1)$$
$$u''(t) \le q(t)|u(t)|^\alpha |x'(t)|^{1-\alpha} \quad \text{for} \quad a_1 < t < b_1.$$

Then the following hold

$$x'(t) > 0 \ \text{for} \ a_1 < t < b_1 \quad (x'(b) < 0 \ \text{for} \ a_1 \le t \le b) \tag{3.6.22}$$

and

$$\frac{y'(t)}{y(t)} < \frac{u'(t)}{u(t)} \ \text{for} \ a_1 \le t \le b_1 \ \left(\frac{y'(t)}{y(t)} > \frac{u'(t)}{u(t)} \ \text{for} \ a_1 \le t \le b_1 \right).$$

(3.6.23)

Proof. Suppose to the contrary that (3.6.22) does not hold. Then by (3.6.21) there exists $c \in (a_1, b_1)$ such that $x'(t) > 0$ for $a_1 < t < c$ $(x'(t) < 0$ for $c < t \le b_1)$ and $x'(c) = 0$. Define

$$\rho(t) = \left| \frac{u'(t)}{u(t)} \right|^\alpha \ \text{sgn} \ u'(t) \quad \text{for} \quad a_1 < t < b_1$$

and

$$\sigma(t) = \left| \frac{x'(t)}{x(t)} \right|^\alpha \ \text{sgn} \ x'(t) \quad \text{for} \quad a < t \le c \ (\text{for} \ c \le t < b_1).$$

Clearly, it follows that

$$\rho'(t) \le \alpha q(t) - \alpha |\rho(t)|^{(\alpha+1)/\alpha} \quad \text{for} \quad a_1 < t < b_1 \tag{3.6.24}$$

$$\sigma'(t) = \alpha q(t) - \alpha |\sigma(t)|^{(\alpha+1)/\alpha} \quad \text{for} \quad a_1 < t < c \ (\text{for} \ c < t < b_1) \tag{3.6.25}$$

$$\sigma(a_1+) = +\infty, \ \sigma(c) = 0 \ (\sigma(b_1-) = -\infty, \ \sigma(c) = 0). \tag{3.6.26}$$

From (3.6.26) there exist $c_1 \in (a_1, c)$ and $c_2 \in (c_1, c)$ $(c_1 \in (c_1, b_1)$ and $c_2 \in (c, c_1))$ such that

$$\rho(t) > \sigma(t) > 0 \quad \text{for} \quad c_1 < t < c_2, \quad \rho(c_1) = \sigma(c_1)$$
$$(\rho(t) < \sigma(t) < 0 \quad \text{for} \quad c_2 < t < c_1, \quad \rho(c_1) = \sigma(c_1)).$$

(3.6.27)

Thus from (3.6.24) and (3.6.25), we have

$$\sigma(t) = \sigma(c_1) + \alpha \int_{c_1}^{t} q(s)ds - \alpha \int_{c_1}^{t} |\sigma(s)|^{(\alpha+1)/\alpha}ds$$

$$> \rho(c_1) + \alpha \int_{c_1}^{t} q(s)ds - \alpha \int_{c_1}^{t} |\rho(s)|^{(\alpha+1)/\alpha}ds \geq \rho(t) \quad \text{for} \quad c_1 < t < c_2$$

$$\left(\begin{array}{l} \sigma(t) = \sigma(c_1) - \alpha \int_{t}^{c_1} q(s)ds + \alpha \int_{t}^{c_1} |\sigma(s)|^{(\alpha+1)/\alpha}ds \\[2mm] < \rho(c_1) - \alpha \int_{t}^{c_1} q(s)ds + \alpha \int_{t}^{c_1} |\rho(s)|^{(\alpha+1)/\alpha}ds \leq \rho(t) \quad \text{for} \quad c_2 < t < c_1 \end{array} \right),$$

which contradicts (3.6.27). Hence, (3.6.22) holds. Similarly, we can prove (3.6.23). This completes the proof. ∎

Lemma 3.6.5. Let $x(t)$ be a nontrivial solution of equation (3.6.1) satisfying the condition

$$x(a+) = 0 \quad (x(b-) = 0) \tag{3.6.28}$$

and let $u \in AC^1_{Loc}((a,c])$ $(u \in AC^1_{loc}([c,b)))$, $c \in (a,b)$ have a finite limit $u(a+) \geq 0$ $(u(b-) \geq 0)$ and satisfy the conditions

$$u'(t) > 0 \quad \text{for} \quad a < t \leq c \quad (u'(t) < 0 \quad \text{for} \quad c \leq t < b),$$

$$u''(t) \leq q(t)|u(t)|^{\alpha}|u'(t)|^{1-\alpha} \quad \text{for} \quad a < t < c \quad (\text{for} \quad c < t < b).$$

Then the following hold

$$x'(t) \neq 0 \quad \text{for} \quad a < t \leq c \quad (\text{for} \quad c \leq t < b). \tag{3.6.29}$$

Proof. Let $x_1(t)$ be a solution of the problem (3.6.1), (3.6.4). First we claim that

$$x_1'(t) > 0 \quad \text{for} \quad a < t \leq c. \tag{3.6.30}$$

If false, then there exists $c_1 \in (a,c)$ such that $x_1'(t) > 0$ for $a < t < c_1$, $x_1'(c_1) = 0$.

By Lemma 3.6.4, we have

$$\frac{x_1'(t)}{x_1(t)} < \frac{u'(t)}{u(t)} \quad \text{for} \quad a < t < c_1. \tag{3.6.31}$$

Define

$$y(t) = u'(t)x_1(t) - x_1'(t)u(t) \quad \text{for} \quad a < t < c_1,$$

$$h(t) = \frac{|u'(t)x_1(t)|^{1-\alpha} - |x_1'(t)u(t)|^{1-\alpha}}{u'(t)x_1(t) - x_1'(t)u(t)} \quad \text{for} \quad a < t < c_1,$$

$$g(t) = q(t)h(t)(x_1(t)u(t))^\alpha \quad \text{for} \quad a < t < c_1. \tag{3.6.32}$$

Since the function $f(x) = (|x|^{1-\alpha} - 1)/(x-1)$ is bounded, there exists $M_0 > 0$ such that $|h(t)| \leq M_0/(u(t)x_1'(t))^\alpha)$ for $a < t < c_1$. From (3.6.32), we find

$$|g(t)| \leq M_0(t-a)^\alpha |q(t)| \left(\frac{x_1(t)}{(t-a)x_1'(t)} \right)$$

$$\leq M(t-a)^\alpha |q(t)| \quad \text{for} \quad a < t < (a+c_1)/2,$$

where

$$M = M_0 \sup \left\{ \left(\frac{x_1(t)}{(t-a)x_1'(t)} \right)^\alpha : a < t < \frac{a+c_1}{2} \right\}.$$

Thus, the function $g(t)$ is integrable on the segment $[a, (a+c_1)/2]$.

Since $y'(t) \leq g(t)y(t)$ for $a < t < (a+c_1)/2$ it follows that

$$y(t) \geq y \left(\frac{a+c_1}{2} \right) \exp \left(-\int_t^{(a+c_1)/2} g(s)ds \right)$$

$$\geq y \left(\frac{a+c_1}{2} \right) \exp \left(-\int_a^{(a+c_1)/2} |g(s)|ds \right) \quad \text{for} \quad a < t \leq \frac{a+c_1}{2}.$$

This inequality and (3.6.31) imply that

$$\liminf_{t \to a^+} y(t) > 0. \tag{3.6.33}$$

However, from the proof of Lemma 3.6.3 and condition (3.6.4), we have

$$\liminf_{t \to a^+} y(t) = \liminf_{t \to a^+} \left((t-a)u'(t) \left(\frac{x_1(t)}{t-a} \right) - x_1'(t)u(t) \right) \leq 0,$$

which contradicts (3.6.33). Thus, (3.6.30) holds.

Now, we shall show that (3.6.29) holds. By Lemma 3.6.4 there exists $c_0 \in (a, c)$ such that $x(t) \neq 0$ for $a < t < c_0$. Without loss of generality assume that

$$x(t) > 0 \quad \text{for} \quad a < t < c_0. \tag{3.6.34}$$

We claim that

$$x'(t) > 0 \quad \text{for} \quad a < t < c_0. \tag{3.6.35}$$

If false, then there exists $c_2 \in (a, c_0)$ such that $x'(c_2) = 0$. By Lemma 3.6.4, we have $x'(t)/x(t) < x_1'(t)/x_1(t)$ for $a < t < c_2$. Thus, as above it follows that $y_1'(t) = g_1(t)y_1(t)$ for $a < t < c_2$, where

$$y_1(t) = x'(t)x_1(t) - x_1'(t)x(t) \quad \text{for} \quad a < t < c_2,$$

$$g_1(t) = q(t)h_1(t)(x(t)x_1(t))^\alpha \quad \text{for} \quad a < t < c_2,$$

$$h_1(t) = \frac{|x'(t)x_1(t)|^{1-\alpha} - |x_1'(t)x(t)|^{1-\alpha}}{x'(t)x_1(t) - x_1'(t)x(t)} \quad \text{for} \quad a < t < c_2,$$

which as before leads to a contradiction.

Hence, if for some $c_0 \in (a, c)$ inequality (3.6.34) holds, then (3.6.35) also holds. Therefore, $x(t) > 0$ for $a < t \le c$ and (3.6.29) is satisfied. ∎

Remark 3.6.2. Let $x_1(t)$ $(x_2(t))$ be a solution of the problem (3.6.1), (3.6.4) ((3.6.5)). It is clear that there exists $c \in (a, b)$ such that $x_1'(t) > 0$ for $a < t < c$ $(x_2'(t) < 0$ for $c < t < b)$. Thus, in view of Lemma 3.6.5 for the case $u(t) = x_1(t)$, $a < t \le c$ $(u(t) = x_2(t)$, $c \le t < b)$, if $x(t)$ is a nontrivial solution of equation (3.6.1) satisfying (3.6.28), then there exists $c \in (a, b)$ such that (3.6.29) holds.

Lemma 3.6.6. Let $x(t)$ be a nontrivial solution of equation (3.6.1) satisfying the conditions

$$x(a+) = 0, \quad \liminf_{t \to a^+} (t - a) \left| \frac{x'(t)}{x(t)} \right| < \infty$$

$$\left(x(b-) = 0, \quad \liminf_{t \to b^-} (b - t) \left| \frac{x'(t)}{x(t)} \right| < \infty \right). \tag{3.6.36}$$

Then the following hold

$$\lim_{t \to a^+} (t - a) \frac{x'(t)}{x(t)} = 1 \quad \left(\lim_{t \to b^-} (b - t) \frac{x'(t)}{x(t)} = -1 \right). \tag{3.6.37}$$

Proof. By Remark 3.6.2, without loss of generality, we can assume that for some $c_0 \in (a, b)$,

$$x'(t) > 0 \quad \text{for} \quad a < t < c_0 \quad (x'(t) < 0 \quad \text{for} \quad c_0 < t < b). \tag{3.6.38}$$

Multiplying both sides of (3.6.1) by $(t - a)$ (by $(b - t)$) and integrating it from τ to t (from t to τ), we obtain equality (3.6.18). Using (3.6.9) and (3.6.38) in (3.6.18), we obtain

$$(t - a) \frac{x'(t)}{x(t)} \le 1 + (\tau - a) \frac{x'(\tau)}{x(\tau)} + \int_\tau^t (s - a)^\alpha |q(s)|$$
$$\times \left[1 + (s - a) \frac{x'(s)}{x(s)} \right] ds \quad \text{for} \quad a < \tau < t < c_0$$

$$\left(\begin{array}{l} (b - t) \frac{|x'(t)|}{x(t)} \le 1 + (b - \tau) \frac{|x'(\tau)|}{x(\tau)} + \int_t^\tau (b - s)^\alpha |q(s)| \\ \qquad\qquad \times \left[1 + (b - s) \frac{|x'(s)|}{x(s)} \right] ds \quad \text{for} \quad c_0 < t < \tau < b \end{array} \right)$$

$$\tag{3.6.39}$$

and

$$\left| (t-a)\frac{x'(t)}{x(t)} - 1 \right| \leq \frac{(\tau-a)x'(\tau)+x(\tau)}{x(t)}$$

$$+ \int_\tau^t (s-a)^\alpha |q(s)| \left[1+(s-a)\frac{x'(s)}{x(s)} \right] ds \text{ for } a < \tau < t < c_0$$

$$\left(\begin{array}{l} \left| (b-t)\dfrac{x'(t)}{x(t)} + 1 \right| \leq \dfrac{(b-\tau)|x'(\tau)|+x(\tau)}{x(t)} \\[3mm] + \displaystyle\int_t^\tau (b-s)^\alpha |q(s)| \left[1+(b-s)\dfrac{x'(s)}{x(s)} \right] ds \text{ for } c_0 < t < \tau < b \end{array} \right) .$$

$$(3.6.40)$$

From (3.6.39), we find

$$(t-a)\frac{x'(t)}{x(t)} \leq M(\tau) + \int_\tau^t (s-a)^\alpha |q(s)| \left((s-a)\frac{x'(s)}{x(s)} \right) ds \text{ for } a < \tau < t < c_0$$

$$\left((b-t)\frac{|x'(t)|}{x(t)} \leq M(\tau) + \int_t^\tau (b-s)^\alpha |q(s)| \left((b-s)\frac{|x'(s)|}{x(s)} \right) ds \right),$$
$$\text{for } c_0 < t < \tau < b$$

where

$$M(\tau) = 1+(\tau-a)\frac{x'(\tau)}{x(\tau)} + \int_a^{c_0} (s-a)^\alpha |q(s)| ds \quad \text{for} \quad a < \tau < c_0$$

$$\left(M(\tau) = 1+(b-\tau)\frac{|x'(\tau)|}{x(\tau)} + \int_{c_0}^b (b-s)^\alpha |q(s)| ds \quad \text{for} \quad c_0 < \tau < b \right).$$

Applying Lemma 1.1.1, we get

$$(t-a)\frac{x'(t)}{x(t)} \leq M(\tau) \exp \left(\int_a^t (s-a)^\alpha |q(s)| ds \right) \quad \text{for} \quad a < \tau < t < c_0$$

$$\left((b-t)\frac{|x'(t)|}{x(t)} \leq M(\tau) \exp \left(\int_t^b (b-s)^\alpha |q(s)| ds \right) \quad \text{for } c_0 < t < \tau < b \right).$$

Thus, in view of condition (3.6.36), there exists a constant $M_0 > 0$ such that

$$(t-a)\frac{x'(t)}{x(t)} \leq M_0 \text{ for } a < t < c_0 \quad \left((b-t)\frac{|x'(t)|}{x(t)} \leq M_0 \text{ for } c_0 < t < b \right).$$

From (3.6.40), we find

$$\left| (t-a)\frac{x'(t)}{x(t)} - 1 \right| \leq \frac{(\tau-a)x'(\tau)+x(\tau)}{x(t)} + (1+M_0) \int_a^t (s-a)^\alpha |q(s)| ds$$
$$\text{for } a < \tau < t < c_0$$

$$\left(\left|(b-t)\frac{x'(t)}{x(t)}+1\right| \le \frac{(b-\tau)|x'(\tau)|+x(\tau)}{x(t)} + (1+M_0)\int_t^b (b-s)^\alpha |q(s)|ds\right).$$
$$\text{for } c_0 < t < \tau < b$$

Thus, by Lemma 3.6.3, we conclude that

$$\left|(t-a)\frac{x'(t)}{x(t)}-1\right| \le (1+M_0)\int_a^t (s-a)^\alpha |q(s)|ds \quad \text{for} \quad a < t < c_0$$

$$\left(\left|(b-t)\frac{x'(t)}{x(t)}+1\right| \le (1+M_0)\int_t^b (b-s)^\alpha |q(s)|ds \quad \text{for} \quad c_0 < t < b\right).$$

Hence, (3.6.37) holds. ∎

Lemma 3.6.7. Let $x(t)$ be a nontrivial solution of equation (3.6.1) satisfying the condition (3.6.28). Then, (3.6.37) holds.

Proof. By Remark 3.6.2, without loss of generality, we can assume that (3.6.38) holds for some $c_0 \in (a,b)$. Multiplying both sides of (3.6.1) by $(t-a)$ (by $(b-t)$), integrating from a to t (from t to b) and taking Lemma 3.6.3 into account, we obtain

$$(t-a)x'(t) = x(t) + \int_a^t (s-a)q(s)|x(s)|^\alpha |x'(s)|^{1-\alpha}ds \quad \text{for} \quad a < t < c_0$$

$$\left((b-t)|x'(t)| = x(t) + \int_t^b (b-s)q(s)|x(s)|^\alpha |x'(s)|^{1-\alpha}ds \quad \text{for} \quad c_0 < t < b\right).$$
$$(3.6.41)$$

By Lemma 3.6.6, we have

$$\liminf_{t\to a^+} (t-a)\frac{x'(t)}{x(t)} > 0 \quad \left(\liminf_{t\to b^-} (b-t)\frac{|x'(t)|}{x(t)} > 0\right).$$

Hence, there exist $\epsilon > 0$ and $c_1 \in (a,c_0)$ $(c_1 \in (c_0,b))$ such that

$$x(t) < \epsilon(t-a)x'(t) \text{ for } a < t < c_1 \quad (x(t) < \epsilon(b-t)|x'(t)| \text{ for } c_1 < t < b).$$
$$(3.6.42)$$

Using (3.6.38) and (3.6.42) in (3.6.41), we get

$$(t-a)x'(t) \le x(T) + \epsilon^\alpha \int_a^t (s-a)^\alpha |q(s)|[(s-a)|x'(s)|]ds, \quad a < t \le T < c_1$$

$$\left((b-t)|x'(t)| \le x(T) + \epsilon^\alpha \int_t^b (b-s)^\alpha |q(s)|[(b-s)|x'(s)|]ds, \quad c_1 < T \le t < b\right).$$

From this and Lemma 1.1.1, we find

$$(T-a)\frac{x'(T)}{x(T)} \le \exp\left(\epsilon^\alpha \int_a^T (s-a)^\alpha |q(s)| ds\right) \quad \text{for } a < T < c_1$$

$$\left((b-T)\frac{|x'(T)|}{x(T)} \le \exp\left(\epsilon^\alpha \int_T^b (b-s)^\alpha |q(s)| ds\right) \quad \text{for } c_1 < T < b\right).$$

Thus, (3.6.36) holds, and hence by Lemma 3.6.6, (3.6.37) follows. ∎

Lemma 3.6.8. Let $x(t)$ be a solution of equation (3.6.1) satisfying the conditions

$$x(a+) = 0, \quad \liminf_{t\to a^+} |x'(t)| = 0 \quad \left(x(b-) = 0, \quad \liminf_{t\to b^-} |x'(t)| = 0\right).$$

$$(3.6.43)$$

Then, $x(t)$ is identically equal to zero.

Proof. Assume to the contrary that $x(t)$ is a nontrivial solution of (3.6.1) satisfying the condition (3.6.43). By Remark 3.6.2, without loss of generality we can assume that (3.6.38) holds for some $c_0 \in (a,b)$. From equation (3.6.1) for $a < \tau < t < b$, we find

$$x'(t) = x'(\tau) + \int_\tau^t q(s)|x(s)|^\alpha |x'(s)|^{1-\alpha} \text{ sgn } x(s) ds$$

and

$$x(t) = x(\tau) + (t-\tau)x'(\tau) + \int_\tau^t (t-s)q(s)|x(s)|^\alpha |x'(s)|^{1-\alpha} \text{ sgn } x(s) ds.$$

Thus, in view of (3.6.9) for $a < \tau < t < b$, we get

$$|x'(t)| \le |x'(\tau)| + \int_\tau^t (s-a)^\alpha |q(s)| \left[\frac{|x(s)|}{s-a} + |x'(s)|\right] ds,$$

$$\frac{|x(t)|}{t-a} \le \frac{|x(\tau)|}{\tau-a} + |x'(\tau)| + \int_\tau^t (s-a)^\alpha |q(s)| \left[\frac{|x(s)|}{s-a} + |x'(s)|\right] ds.$$

Hence, by Lemma 3.6.7 there exist $\epsilon > 0$ and $c_1 \in (a, c_0)$ so that (3.6.42) holds. Combining the last two inequalities, using Lemma 1.1.1 and condition (3.6.42), we obtain for $a < \tau < t < b$, $\tau < c_1$,

$$\frac{|x(t)|}{t-a} + |x'(t)| \le (2+\epsilon)x'(\tau) \exp\left(\int_a^t (s-a)^\alpha |q(s)| ds\right).$$

Now, from (3.6.43) and the above inequality it follows that $|x(t)| + (t - a)|x'(t)| = 0$ for $a < t < b$, which is a contradiction. This completes the proof. ∎

Lemma 3.6.9. The set of proper solutions of equation (3.6.1) is nonempty.

Proof. Let $a_1 \in [a, b)$ and $b_1 \in (a_1, b]$. Denote by $\mathcal{B}([a_1, b_1])$ the set of all nontrivial solutions of equation (3.6.1) satisfying the condition

$$\text{mes}\,\{\{t \in (a_1, b_1) : x'(t) = 0\} \setminus \{t \in (a_1, b_1) : q(t) = 0\}\} \;=\; 0.$$

We will show that $\mathcal{B}([a, b]) \neq \emptyset$.

First we shall show that if $x_0 \in \mathcal{B}([a_1, b_1])$, where $b_1 < b$ then there exist $b_2 \in (b_1, b)$ and a nontrivial solution $\bar{x}_0$ of (3.6.1) such that

$$\bar{x}_0(t) \;=\; x_0(t) \quad \text{for} \quad a_1 \leq t \leq b, \quad \bar{x}_0 \in \mathcal{B}([a_1, b_2]). \tag{3.6.44}$$

Indeed, by Remark 3.6.1, either

$$x_0(b) \;\neq\; 0, \tag{3.6.45}$$

or

$$x_0(b) \;=\; 0, \quad x_0'(b) \neq 0. \tag{3.6.46}$$

Suppose (3.6.45) holds. Denote by ρ_0 the solution of the Cauchy problem

$$\rho'(t) \;=\; \alpha q(t) - \alpha |\rho(t)|^{(\alpha+1)/\alpha}, \quad \rho(b_1) \;=\; \left|\frac{x_0'(b_1)}{x_0(b_1)}\right|^{\alpha} \, \text{sgn}\, x_0'(b_1).$$

Assume $\rho_0(t)$ is defined on $[b_1, b_2]$. Let

$$x(t) \;=\; \begin{cases} x_0(t) \quad \text{for} \quad a_1 \leq t \leq b_1 \\ x_0(b_1) \exp\left(\displaystyle\int_{b_1}^{t} |\rho_0(s)|^{1/\alpha} \, \text{sgn}\, \rho_0(s) ds\right) \quad \text{for} \quad b_1 \leq t \leq b_2. \end{cases}$$

The function $x(t)$ clearly satisfies equation (3.6.1) for $t \in (a, b_2)$. By Lemma 3.6.2 there exists a solution $\bar{x}_0(t)$ of (3.6.1) such that $\bar{x}_0(t) = x(t)$ for $t \in [a_1, b_2]$ and satisfies (3.6.44).

Suppose now that (3.6.46) is satisfied. Denote by $y_0(t)$ a solution of the Cauchy problem

$$x''(t) \;=\; q(t)|x(t)|^{\alpha}|x'(t)|^{1-\alpha} \, \text{sgn}\, x(t), \quad x(b_1) = 0, \quad x'(b_1) = x_0'(b_1).$$

Obviously, there exists $b_2 \in (b_1, b)$ such that $y_0'(t) \neq 0$ for $t \in [b_1, b_2]$. Suppose

$$x(t) \;=\; \begin{cases} x_0(t) \quad \text{for} \quad a_1 \leq t \leq b_1 \\ y_0(t) \quad \text{for} \quad b_1 \leq t \leq b_2. \end{cases}$$

Again, it follows that there exists a solution $\bar{x}_0(t)$ of (3.6.1) satisfying (3.6.44).

Now, let $x_1(t)$ be a solution of the problem (3.6.1), (3.6.4). It is easy to see that there exists $\bar{a}_1 \in (a, b)$ such that $x_1 \in \mathcal{B}([a, \bar{a}_1])$. Denote by I the set of those $b_1 \in [\bar{a}_1, b]$ for which there exists a solution $\bar{x}(t)$ of the equation (3.6.1) such that $\bar{x}(t) = x_1(t)$ for $a \le t \le \bar{a}_1$, $\bar{x} \in \mathcal{B}([a, b])$. Let $\bar{b}_1 = \sup I$. By Lemma 3.6.2, to prove the result it suffices to show that $\bar{b}_1 = b$. Assume to the contrary that $\bar{b}_1 < b$. Then as is shown above if $\bar{b}_1 \in I$, there exists $b_2 \in (\bar{b}_1, b)$ such that $b_2 \in I$. Consequently, $I \ne \emptyset$, and I is a connected set whose every point, except the point $\bar{a}_1$ is an interior point. In view of this fact, we have

$$b \notin I. \tag{3.6.47}$$

Choose a sequence of points $\{t_k\}_{k=1}^{\infty}$ and a sequence of functions $\{y_k(t)\}_{k=1}^{\infty}$ such that $t_k < t_{k+1}$, $k = 1, 2, \cdots$, $\bar{a}_1 < t_1$, $\lim_{k \to \infty} t_k = \bar{b}_1$

$$y_k \in \mathcal{B}([a, t_k]), \quad k = 1, 2, \cdots$$
$$y_1(t) = x_1(t) \quad \text{for} \quad a < t < \bar{a}_1 \tag{3.6.48}$$
$$y_{k+1}(t) = y_k(t) \quad \text{for} \quad a < t < t_k, \quad k = 1, 2, \cdots.$$

Define

$$x(t) = \begin{cases} y_1(t) & \text{for} \quad a \le t < t_1 \\ y_k(t) & \text{for} \quad t_{k-1} \le t < t_k, \quad k = 2, 3, \cdots. \end{cases} \tag{3.6.49}$$

It can be seen that $x \in AC_{loc}^1((a, \bar{b}_1))$ and $x(t)$ satisfies (3.6.1) for $t \in (a, \bar{b}_1)$. Hence by Lemma 3.6.2 there exists a solution $\bar{x}(t)$ of (3.6.1) such that $\bar{x}(t) = x(t)$ for $a \le t \le \bar{b}_1$. From (3.6.48) and (3.6.49), we find

$$\{t \in (a, t_k) : y_k'(t) = 0\} \subset \{t \in (a, t_k) : q(t) = 0\} \cup A_k,$$
$$\text{mes } A_k = 0, \quad k = 1, 2, \cdots$$
$$\{t \in (a, \bar{b}_1) : (\bar{x}(t))' = 0\} \subset \{t \in (a, \bar{b}_1) : q(t) = 0\} \cup A,$$
$$A = \cup_{k=1}^{\infty} A_k.$$

Since mes $A = \sum_{k=1}^{\infty} \text{mes } A_k = 0$, this implies that $\bar{x} \in \mathcal{B}([a, \bar{b}_1])$, and consequently $\bar{b}_1 \in I$, which contradicts (3.6.47). ∎

Remark 3.6.3. In the above lemma, we proved that the problem (3.6.1), (3.6.4) has at least one proper solution. Similarly, we can show that the problem (3.6.1), (3.6.5) has at least one proper solution, and for any $t_0 \in (a, b)$ and $c_1, c_2 \in \mathbb{R}$, $|c_1| + |c_2| \ne 0$ the Cauchy problem

$$x''(t) = q(t)|x(t)|^{\alpha}|x'(t)|^{1-\alpha} \text{ sgn } x(t)$$
$$x(t_0) = c_1, \quad x'(t_0) = c_2$$

has at least one proper solution.

Lemma 3.6.10. Let there exist $c \in (a, b)$ and a continuous function $y \in AC^1_{loc}((a, c) \cup (c, b))$ having finite limit

$$y(a+) \geq 0, \quad y(b-) \geq 0, \quad y'(c-) > 0, \quad y'(c+) < 0$$

and satisfying the conditions

$$y''(t) \leq q(t)|y(t)|^{\alpha}|y'(t)|^{1-\alpha} \quad \text{for} \quad a < t < b,$$
$$y(t) > 0 \quad \text{for} \quad a < t < b, \quad y'(t) > 0 \quad \text{for} \quad a < t < c,$$
$$y'(t) < 0 \quad \text{for} \quad c < t < b.$$

Then, $q \in U_{\alpha}((a, b))$.

Proof. Assume to the contrary. Let $q \notin U_{\alpha}((a, b))$. Then there exist $a_1 \in [a, b)$, $b_1 \in (a_1, b]$ and a solution $x_0(t)$ of (3.6.1) satisfying the conditions

$$x_0(t) > 0 \quad \text{for} \quad a_1 < t < b_1, \quad x_0(a_1+) = 0, \quad x_0(b_1-) = 0. \quad (3.6.50)$$

By Lemma 3.6.8 and Remark 3.6.1 there exist $\tau_1 \in (a_1, b_1)$ and $\tau_2 \in [\tau_1, b_1)$ such that

$$x'_0(t) > 0 \quad \text{for} \quad a_1 < t < \tau_1, \quad x'_0(\tau_1) = 0$$
$$x'_0(t) < 0 \quad \text{for} \quad \tau_1 < t < b_1, \quad x'_0(\tau_2) = 0.$$

Let $x_1(t)$ and $x_2(t)$ be some solutions of problems (3.6.1), (3.6.4) and (3.6.1), (3.6.5) respectively. Then by Lemmas 3.6.4 and 3.6.5 (with $y(t) = x_1(t)$, $x(t) = x_0(t)$ and $y(t) = x_2(t)$, $x(t) = x_0(t)$, respectively), there exist $t_1 \in (a, \tau_1]$ and $t_2 \in [\tau_2, b)$ such that

$$x'_1(t) > 0 \quad \text{for} \quad a < t < t_1, \quad x'_1(t_1) = 0$$
$$x'_2(t) < 0 \quad \text{for} \quad t_2 < t < b, \quad x'_2(t_2) = 0.$$

By Lemma 3.6.4 (with $x(t) = x_1(t)$ and $x(t) = x_2(t)$, respectively), we have $t_1 > c$ and $t_2 < c$ which is impossible, since $t_2 \geq t_1$. ∎

Lemma 3.6.11. Let there exist $c \in (a, b)$ and a function $\sigma \in AC_{Loc}((a, c) \cup (c, b))$ having finite limits $\sigma(c-) \geq \sigma(c+)$ and satisfying the conditions

$$\sigma'(t) \leq \alpha q(t) - \alpha|\sigma(t)|^{(\alpha+1)/\alpha} \quad \text{for} \quad a < t < b \quad (3.6.51)$$

$$\liminf_{t \to a^+} (t - a)^{\alpha}\sigma(t) < 1, \quad \limsup_{t \to b^-} (b - t)^{\alpha}\sigma(t) > -1. \quad (3.6.52)$$

Then, $q \notin O_{\alpha}((a, b))$.

Proof. Assume to the contrary that $q \in O_\alpha((a,b))$. Then there exist $a_1 \in [a,b)$, $b_1 \in (a_1, b]$ and a proper solution $x_0(t)$ of (3.6.1) satisfying the condition (3.6.50). Suppose $a_1 = a$ and $b_1 = b$ (the lemma for the other cases can be proved analogously). First, we introduce the function ρ by

$$\rho(t) = \left| \frac{x_0'(t)}{x_0(t)} \right|^\alpha \operatorname{sgn} x_0'(t) \quad \text{for} \quad a < t < b.$$

Clearly, we have

$$\rho'(t) = \alpha q(t) - \alpha |\rho(t)|^{(\alpha+1)/\alpha} \quad \text{for} \quad a < t < b \tag{3.6.53}$$

and by Lemma 3.6.6,

$$\lim_{t \to a^+} (t-a)^\alpha \rho(t) = 1, \quad \lim_{t \to b^-} (b-t)^\alpha \rho(t) = -1. \tag{3.6.54}$$

From (3.6.52) and (3.6.54) we conclude that there exist $t_1 \in (a,b)$ and $\epsilon \in (0, b - t_1)$ such that $\sigma \in AC([t_1, t_1 + \epsilon])$, and

$$\sigma(t) > \rho(t) \quad \text{for} \quad t_1 < t < t_1 + \epsilon, \quad \sigma(t_1) = \rho(t_1). \tag{3.6.55}$$

Suppose $u(t) = \sigma(t) - \rho(t)$ for $t_1 \leq t \leq t_1 + \epsilon$. Then there exists an integrable function $h : (t_1, t_1 + \epsilon) \to \mathbb{R}$ such that

$$|\sigma(t)|^{(\alpha+1)/\alpha} - |\rho(t)|^{(\alpha+1)/\alpha} = (\sigma(t) - \rho(t)) h(t) \quad \text{for} \quad t_1 < t < t_1 + \epsilon.$$

It follows from this fact, (3.6.51) and (3.6.53) that $u'(t) \leq -h(t) u(t)$ for $t_1 < t < t_1 + \epsilon$, and hence we conclude that $u(t) < 0$ for $t_1 < t < t_1 + \epsilon$, which contradicts (3.6.55). ∎

Before we present the main results of this section, we introduce the notation

$$Q(t, t_0, \alpha) = \begin{cases} \alpha (t-a)^\alpha \displaystyle\int_t^{t_0} q(s) ds & \text{for} \quad a < t < t_0 \\[2mm] \alpha (b-t)^\alpha \displaystyle\int_{t_0}^t q(s) ds & \text{for} \quad t_0 < t < b, \end{cases}$$

$$Q_*(t_0, \alpha) = \inf\{Q(t, t_0, \alpha) : a < t < b\},$$
$$Q^*(t_0, \alpha) = \sup\{Q(t, t_0, \alpha) : a < t < b\}.$$

Theorem 3.6.1. Let there exist $\lambda \in (a, b)$ and $\mu \in (\lambda, b)$ such that

$$-\alpha \int_\lambda^\mu q(s) ds \geq \frac{1}{(\lambda - a)^\alpha} + \frac{1}{(b - \mu)^\alpha} + \frac{\alpha}{(\lambda - a)^{\alpha+1}} \int_a^\lambda (s-a)^{\alpha+1} q(s) ds$$

$$+ \frac{\alpha}{(b - \mu)^{\alpha+1}} \int_\mu^b (b-s)^{\alpha+1} q(s) ds.$$

$$\tag{3.6.56}$$

Then, $q \in O_\alpha((a, b))$.

Proof. Let $x_0(t)$ be a proper solution of (3.6.1). We claim that $x_0(t)$ has at least one zero in the interval (a, b). Assume to the contrary that

$$x_0(t) > 0 \quad \text{for} \quad a < t < b. \tag{3.6.57}$$

Let

$$\rho(t) = \left| \frac{x_0'(t)}{x_0(t)} \right|^\alpha \operatorname{sgn} x_0'(t) \quad \text{for} \quad a < t < b \tag{3.6.58}$$

so that

$$\rho'(t) = \alpha q(t) - \alpha |\rho(t)|^{(\alpha+1)/\alpha} \quad \text{for} \quad a < t < b. \tag{3.6.59}$$

Thus, it follows that

$$\{ t \in (a, \lambda] : (t - a)^\alpha \rho(t) = 1 \} \cup \{ t \in [\mu, b) : (b - t)^\alpha \rho(t) = -1 \}$$
$$\cup \{ t \in [\lambda, \mu] : \rho(t) = 0 \} \ne (a, b). \tag{3.6.60}$$

Multiplying both sides of (3.6.59) by $(t - a)^{\alpha+1}$, integrating from a to λ and taking into account Lemmas 3.6.3 and 3.6.7, we obtain

$$(\lambda - a)^\alpha \rho(\lambda) = \frac{\alpha}{\lambda - a} \int_a^\lambda (s - a)^{\alpha+1} q(s) ds$$
$$+ \frac{1}{\lambda - a} \int_a^\lambda \left[(\alpha + 1)(s - a)^\alpha \rho(s) - \alpha(s - a)^{\alpha+1} |\rho(s)|^{(\alpha+1)/\alpha} \right] ds. \tag{3.6.61}$$

Multiplying both sides of (3.6.59) by $(b - t)^{\alpha+1}$, integrating from μ to b and taking into account Lemmas 3.6.3 and 3.6.7, we get

$$(b - \mu)^\alpha \rho(\mu) = \frac{\alpha}{b - \mu} \int_\mu^b (b - s)^{\alpha+1} q(s) ds$$
$$+ \frac{1}{b - \mu} \int_\mu^b \left[(\alpha + 1)(b - s)^\alpha \rho(s) + \alpha(b - s)^{\alpha+1} |\rho(s)|^{(\alpha+1)/\alpha} \right] ds. \tag{3.6.62}$$

Finally, integrating (3.6.59) from λ to μ, we find

$$-\alpha \int_\lambda^\mu q(s) ds = \rho(\lambda) - \rho(\mu) - \alpha \int_\lambda^\mu |\rho(s)|^{(\alpha+1)/\alpha} ds. \tag{3.6.63}$$

Clearly, we have

$$(\alpha + 1)T - \alpha |T|^{(\alpha+1)/\alpha} < 1 \quad \text{for} \quad T \in \mathbb{R} \backslash \{1\}$$
$$(\alpha + 1)T + \alpha |T|^{(\alpha+1)/\alpha} > -1 \quad \text{for} \quad T \in \mathbb{R} \backslash \{-1\}. \tag{3.6.64}$$

Now, in view of (3.6.60) and (3.6.63), from (3.6.61) – (3.6.63) we obtain the following three inequalities

$$(\lambda - a)^\alpha \rho(\lambda) \leq 1 + \frac{\alpha}{\lambda - a} \int_a^\lambda (s - a)^{\alpha+1} q(s) ds, \tag{3.6.65}$$

$$(b - \mu)^\alpha \rho(\mu) \geq -1 - \frac{\alpha}{b - \mu} \int_\mu^b (b - s)^{\alpha+1} q(s) ds \tag{3.6.66}$$

$$-\alpha \int_\lambda^\mu q(s) ds \leq \rho(\lambda) - \rho(\mu) \tag{3.6.67}$$

of which at least one is strictly satisfied. Using (3.6.65) and (3.6.66) in (3.6.67), we find that condition (3.6.56) is violated. Thus, every proper solution of (3.6.1) has at least one zero in the interval (a, b). In particular, the proper solution of the problem (3.6.1), (3.6.4) (see Remark 3.6.3) has also at least one zero in the interval (a, b), and hence $q \in O_\alpha((a, b))$. ∎

Corollary 3.6.1. Let

$$-(\alpha + 1) \int_a^b Q\left(s, \frac{a+b}{2}, \alpha\right) ds > b - a. \tag{3.6.68}$$

Then, $q \in O_\alpha((a, b))$.

Proof. From condition (3.6.68) one can choose $\lambda \in (a, (a + b)/2]$ and $\mu \in [(a + b)/2, b)$ such that

$$-\alpha(\alpha + 1) \left[\frac{1}{(\lambda - a)^{\alpha+1}} \int_a^\lambda Q\left(s, \frac{a+b}{2}, \alpha\right) ds \right.$$
$$\left. + \frac{1}{(b - \mu)^{\alpha+1}} \int_\mu^b Q\left(s, \frac{a+b}{2}, \alpha\right) ds \right] \geq \frac{1}{(\lambda - a)^{\alpha+1}} + \frac{1}{(b - \mu)^{\alpha+1}}. \tag{3.6.69}$$

By Proposition 3.6.1, we can easily verify that

$$\int_a^\lambda Q\left(s, \frac{a+b}{2}, \alpha\right) ds$$
$$= \frac{1}{\alpha + 1} \left[(\lambda - a)^{\alpha+1} \int_\lambda^{(a+b)/2} q(s) ds + \int_a^\lambda (s - a)^{\alpha+1} q(s) ds \right]$$

and

$$\int_\mu^b Q\left(s, \frac{a+b}{2}, \alpha\right) ds$$
$$= \frac{1}{\alpha + 1} \left[(b - \mu)^{\alpha+1} \int_{(a+b)/2}^\mu q(s) ds + \int_\mu^b (b - s)^{\alpha+1} q(s) ds \right].$$

In view of these facts, we find from (3.6.69) that the inequality (3.6.56) is satisfied. Hence, by Theorem 3.6.1, $q \in O_\alpha((a,b))$. ■

Corollary 3.6.2. Let $t_0 \in (a,b)$,

$$q(t) \leq 0 \quad \text{for} \quad a < t < b \tag{3.6.70}$$

$$Q_*(t_0, \alpha) \leq -1 - \max\left\{ \left(\frac{t_0 - a}{b - t_0}\right)^\alpha, \left(\frac{b - t_0}{t_0 - a}\right)^\alpha \right\}. \tag{3.6.71}$$

Then, $q \in O_\alpha((a,b))$.

Proof. By (3.6.71) there exists $c \in (a,b) \setminus \{t_0\}$ such that

$$-Q(c, t_0, \alpha) \geq 1 + \max\left\{ \left(\frac{t_0 - a}{b - t_0}\right)^\alpha, \left(\frac{b - t_0}{t_0 - a}\right)^\alpha \right\}. \tag{3.6.72}$$

Suppose

$$\lambda = \begin{cases} c & \text{if } c < t_0 \\ t_0 & \text{if } c > t_0, \end{cases} \qquad \mu = \begin{cases} t_0 & \text{if } c < t_0 \\ c & \text{if } c > t_0. \end{cases}$$

From (3.6.72), we conclude that

$$-\alpha \int_\lambda^\mu q(s)ds > \frac{1}{(\lambda - a)^\alpha} + \frac{1}{(b - \mu)^\alpha}.$$

This and inequality (3.6.70) show that (3.6.56) holds, and hence by Theorem 3.6.1, $q \in O_\alpha((a,b))$. ■

   In the case when condition (3.6.70) holds Theorem 3.6.1 can be improved. In fact, the following result holds.

Theorem 3.6.2. Let condition (3.6.70) hold and there exist $\lambda \in (a,b)$, $\mu \in (\lambda, b)$ and $n \in \mathbb{N}$ such that

$$\alpha \int_\lambda^\mu |q(s)|ds > a_n^{-\alpha}(\lambda) + b_n^{-\alpha}(\mu), \tag{3.6.73}$$

where

$$a_1(t) = t - a, \quad a_{k+1}(t) = t - a + \int_a^t a_k^\alpha(s)|q(s)|ds, \quad k \geq 1 \quad \text{for} \quad a < t < b,$$

$$b_1(t) = b - t, \quad b_{k+1}(t) = b - t + \int_t^b b_k^\alpha(s)|q(s)|ds, \quad k \geq 1 \quad \text{for} \quad a < t < b.$$

Then, $q \in O_\alpha((a,b))$.

Proof. Let $x_1(t)$ be a proper solution of the problem (3.6.1), (3.6.4). We claim that $x_1(t)$ has at least one zero in the interval (a, b). Assume the contrary. Then by (3.6.70) either

$$x_1'(t) > 0 \quad \text{for} \quad a < t < b \tag{3.6.74}$$

or there exist $c_1 \in (a, b)$ and $c_2 \in [c_1, b)$ such that

$$x_1'(t) > 0, \ a < t < c_1, \quad x_1'(t) < 0, \ c_2 < t < b, \quad x_1'(t) = 0, \ c_1 \le t \le c_2. \tag{3.6.75}$$

Suppose

$$\rho(t) \;=\; \left| \frac{x_1'(t)}{x_1(t)} \right|^{\alpha} \ \operatorname{sgn} x_1'(t) \quad \text{for} \quad a < t < b$$

so that (3.6.59) is satisfied.

Assume that (3.6.75) holds. Multiplying both sides of (3.6.59) by $(t - a)^{\alpha+1}$ and integrating from a to t, we obtain for $a < t < b$,

$$(t - a)^{\alpha}\rho(t) \;=\; \frac{\alpha}{t - a} \int_a^t (s - a)^{\alpha+1}|q(s)|ds$$
$$+ \frac{1}{t - a} \int_a^t \left[(\alpha + 1)(s - a)^{\alpha+1}\rho(s) - \alpha(s - a)^{\alpha+1}|\rho(s)|^{(\alpha+1)/\alpha} \right] ds,$$

which in view of (3.6.64) results in $\rho(t) \le a_1^{-\alpha}(t)$ for $a < t < b$. Assume now that for some $k \in \mathbb{N}$,

$$\rho(t) \;\le\; a_k^{-\alpha}(t) \quad \text{for} \quad a < t < b. \tag{3.6.76}$$

From (3.6.59) with regard to (3.6.76), we have for $a < t < c_1$,

$$\left(\rho^{-1/\alpha}(t) \right)' \;=\; |q(t)|(\rho(t))^{-(\alpha+1)/\alpha} + 1 \;\ge\; a_k^{\alpha+1}(t)|q(t)| + 1.$$

Integrating the above inequality from a to t, we arrive at

$$\rho^{-1/\alpha}(t) \;\ge\; t - a + \int_0^t a_k^{\alpha+1}(s)|q(s)|ds \quad \text{for} \quad a < t < c_1.$$

From (3.6.75), we find that $\rho(t) \le a_{k+1}^{-\alpha}(t)$ for $a < t < b$. Thus, it is proved by induction that for any $k \in \mathbb{N}$ the inequality (3.6.76) holds. Analogously, we can show that for any $k \in \mathbb{N}$,

$$\rho(t) \;\ge\; -b_k^{-\alpha}(t) \quad \text{for} \quad a < t < b. \tag{3.6.77}$$

Suppose now that (3.6.74) is satisfied. Then, as above, we find that (3.6.76) holds and the inequality (3.6.77) is trivial in this case.

Integrating (3.6.59) from λ to μ and using (3.6.76) and (3.6.77) (for $k = n$), we obtain $-\alpha \int_\lambda^\mu |q(s)| ds \le a_n^{-\alpha}(\lambda) + b_n^{-\alpha}(\mu)$, which contradicts (3.6.73). ∎

Theorem 3.6.3. Let there exist $c \in (a, b)$ and the functions $f \in AC([a, c])$ and $g \in AC([c, b])$ such that

$$f(a) = 0, \quad g(b) = 0, \quad f(t) > 0 \quad \text{for} \quad a < t < c,$$

$$g(t) > 0 \quad \text{for} \quad c < t < b, \quad \frac{|f'(t)|^{\alpha+1}}{f^\alpha(t)} \quad \text{and} \quad \frac{|g'(t)|^{\alpha+1}}{g^\alpha(t)}$$

be integrable on $[a, c]$ and $[c, b]$ respectively, and

$$-\alpha \left[g(c) \int_a^c f(s)q(s)ds + f(c) \int_c^b g(s)q(s)ds \right]$$

$$> (\alpha + 1)^{-(\alpha+1)} \left[g(c) \int_a^c \frac{|f'(s)|^{\alpha+1}}{f^\alpha(s)} ds + f(c) \int_c^b \frac{|g'(s)|^{\alpha+1}}{g^\alpha(s)} ds \right].$$

$$\tag{3.6.78}$$

Then, $q \in O_\alpha((a, b))$.

Proof. Let $x_0(t)$ be a proper solution of (3.6.1). We claim that $x_0(t)$ has at least one zero in the interval (a, b). Assume to the contrary that (3.6.57) holds. Define the function $\rho(t)$ by (3.6.58) and obtain (3.6.59). We can easily verify that for $a < t < c$,

$$(\alpha + 1)f^{1/(\alpha+1)}(t) \le \int_a^t \frac{|f'(s)|}{f^{\alpha/(\alpha+1)}(s)} ds \le (t - a)^{\frac{\alpha}{\alpha+1}} \left[\int_a^t \frac{|f'(s)|^{\alpha+1}}{f^\alpha(s)} ds \right]^{\frac{1}{\alpha+1}}$$

and for $c < t < b$,

$$(\alpha + 1)g^{1/(\alpha+1)}(t) \le \int_t^b \frac{|g'(s)|}{g^{\alpha/(\alpha+1)}(s)} ds \le (b - t)^{\frac{\alpha}{\alpha+1}} \left[\int_t^b \frac{|g'(s)|^{\alpha+1}}{g^\alpha(s)} ds \right]^{\frac{1}{\alpha+1}}.$$

From Lemmas 3.6.3 and 3.6.7, we have

$$\lim_{t \to a^+} f(t)\rho(t) = 0, \quad \lim_{t \to b^-} g(t)\rho(t) = 0. \tag{3.6.79}$$

Multiplying both sides of (3.6.59) by $f(t)$ and integrating from $a + \epsilon$ to c, where $\epsilon \in (0, c - a)$, we obtain

$$-\alpha \int_{a+\epsilon}^c f(s)q(s)ds + f(c)\rho(c) - f(a + \epsilon)\rho(a + \epsilon)$$

$$= \int_{a+\epsilon}^c \left[f'(s)\rho(s) - \alpha f(s)|\rho(s)|^{(\alpha+1)/\alpha} \right] ds \tag{3.6.80}$$

$$\le \int_{a+\epsilon}^c f(s) \left[\frac{|f'(s)|}{f(s)} |\rho(s)| - \alpha|\rho(s)|^{(\alpha+1)/\alpha} \right] ds.$$

Similarly, multiplying both sides of (3.6.59) by $g(t)$ and integrating from c to $b - \eta$, where $\eta \in (0, b - c)$, we get

$$-\alpha \int_c^{b-\eta} g(s)q(s)ds + g(b - \eta)\rho(b - \eta) - g(c)\rho(c)$$
$$\leq \int_c^{b-c} g(s) \left[\frac{|g'(s)|}{g(s)} |\rho(s)| - \alpha|\rho(s)|^{(\alpha+1)/\alpha} \right] ds. \tag{3.6.81}$$

It can be easily verified that

$$\lambda|T| - \alpha|T|^{(\alpha+1)/\alpha} \leq \left(\frac{\lambda}{\alpha+1} \right)^{\alpha+1} \quad \text{for} \quad T \in \mathbb{R}.$$

Thus, we can conclude that

$$-\alpha g(c) \int_a^c f(s)q(s)ds + g(c)f(c)\rho(c) \leq \frac{g(c)}{(\alpha+1)^{\alpha+1}} \int_a^c \frac{|f'(s)|^{\alpha+1}}{f^\alpha(s)} ds$$

and

$$-\alpha f(c) \int_c^b g(s)q(s)ds - g(c)f(c)\rho(c) \leq \frac{f(c)}{(\alpha+1)^{\alpha+1}} \int_c^b \frac{|g'(s)|^{\alpha+1}}{g^\alpha(s)} ds.$$

Addition of the last two inequalities gives a contradiction to (3.6.78). ∎

Corollary 3.6.3. Let one of the following conditions be satisfied

$$-\alpha \int_a^b (s - a)^{\alpha+1}(b - s)^{\alpha+1}q(s)ds > \frac{(b - a)^{\alpha+2}}{\alpha+2}, \tag{3.6.82}$$

$$-\alpha \left[\int_a^{(a+b)/2} (s - a)^\lambda q(s)ds + \int_{(a+b)/2}^b (b - s)^\lambda q(s)ds \right]$$
$$> \frac{2\lambda^{\alpha+1}}{(\lambda - \alpha)(\alpha + 1)^{\alpha+1}} \left(\frac{b - a}{2} \right)^{\lambda-\alpha}, \quad \text{where} \quad \lambda > \alpha \tag{3.6.83}$$

or

$$-\int_a^b \sin^{\alpha+1}\left(\frac{\pi(s - a)}{b - a} \right) q(s)ds > \frac{2(\alpha + 2)^\alpha}{(\alpha + 1)^{\alpha+1}} \left(\frac{\pi}{b - a} \right)^\alpha. \tag{3.6.84}$$

Then, $q \in O_\alpha((a, b))$.

Proof. Condition (3.6.82) follows from (3.6.81) in the case when $f(t) = [(t - a)(b - t)]^{\alpha+1} = g(t)$ for $a < t < b$; condition (3.6.83) is obtained in the case when $f(t) = (t - a)^\lambda$ and $g(t) = (b - t)^\lambda$ for $a < t < b$, and condition (3.6.84) in the case when $f(t) = \sin(\pi(t - a)/b - a) = g(t)$ for $a < t < b$. ∎

Finally, we shall present results which ensure that the function $q(t) \in U_\alpha((a,b))$, and when $q(t) \notin O_\alpha((a,b))$.

Theorem 3.6.4. Let $\sup\{A^\alpha(t)|B(t)|^{1-\alpha} : a < t < b\} < b - a$, where

$$
A(t) = \frac{(b-t)^{1-\lambda}}{(t-a)^\lambda} \int_a^t (s-a)^{1+\alpha\lambda}(b-s)^{\alpha\lambda}[q(s)]^- ds
$$
$$
+ \frac{(t-a)^{1-\lambda}}{(b-t)^\lambda} \int_t^b (s-a)^{\alpha\lambda}(b-s)^{1+\alpha\lambda}[q(s)]^- ds,
$$

$$
B(t) = \int_a^t (s-a)^{1+\alpha\lambda}(b-s)^{\alpha\lambda}[q(s)]^- ds - \int_t^b (s-a)^{\alpha\lambda}(b-s)^{1+\alpha\lambda}[q(s)]^- ds
$$

and $\lambda \in [0,1]$. Then, $q \in U_\alpha((a,b))$.

Proof. Suppose $[q(t)]^- \not\equiv 0$, since otherwise $q \in U_\alpha((a,b))$. Choose $\epsilon > 0$ such that $\sup\{(A(t)+\epsilon)^\alpha|B(t)|^{1-\alpha} : a < t < b\} < b - a$. Let

$$
y_0(t) = \epsilon + A(t) \quad \text{for} \quad a < t < b. \tag{3.6.85}
$$

Then, we have $y_0(t) > 0$ for $a < t < b$, $y_0(a+) = 0$, $y_0(b-) = \epsilon$, and there exists $c_0 \in (a,b)$ such that $y_0'(t)\,\mathrm{sgn}\,(c_0 - t) > 0$ for $a < t < b$ and

$$
y_0''(t) = -[q(t)]^-((t-a)(b-t))^{\alpha\lambda} \le q(t)|y_0(t)|^\alpha|y_0'(t)|^{1-\alpha} \quad \text{for} \quad a < t < b. \tag{3.6.86}
$$

Let $x_1(t)$ and $x_2(t)$ be proper solutions of the problems (3.6.1), (3.6.4) and (3.6.1), (3.6.5) respectively. By Lemma 3.6.5 (in this case $u(t) = y_0(t)$ for $a < t < b$) there exist $t_1 \in [c_0, b)$ and $t_2 \in (a, c_0]$ such that

$$
\begin{aligned}
x_1'(t) &> 0 \quad \text{for} \quad a < t < t_1, \quad x_1'(t_1) = 0 \\
x_2'(t) &< 0 \quad \text{for} \quad t_2 < t < b, \quad x_2'(t_2) = 0.
\end{aligned} \tag{3.6.87}
$$

Suppose $t_1 = t_2$. Then by (3.6.87) the function

$$
x(t) = \begin{cases} x_1(t) & \text{for} \quad a < t < c_0 \\ x_2(t)\dfrac{x_1(c_0)}{x_2(c_0)} & \text{for} \quad c_0 < t < b \end{cases}
$$

will be a proper solution of equation (3.6.1), and hence

$$
q \in O_\alpha((a,b)). \tag{3.6.88}
$$

Define

$$
\sigma(t) = \left|\frac{y_0'(t)}{y_0(t)}\right|^\alpha \mathrm{sgn}\, y_0'(t) \quad \text{for} \quad a < t < b
$$

then by (3.6.85), (3.6.86) and Proposition 3.6.1, we find that $\sigma(t)$ satisfies the conditions of Lemma 3.6.11, and hence $q \notin O_\alpha((a, b))$, which contradicts (3.6.88).

Consequently, $t_1 > t_2$. Then from (3.6.87) there exists $c \in (t_1, t_2)$ such that $x_1'(c) > 0 > x_2'(c)$. It can be easily verified that the function

$$u(t) = \begin{cases} x_1(t) & \text{for } a < t \le c \\ x_2(t)\dfrac{x_1(c)}{x_2(c)} & \text{for } c < t < b \end{cases}$$

satisfies the conditions of Lemma 3.6.10 and hence $q \in U_\alpha((a, b))$. ∎

Corollary 3.6.4. Let

$$C^{(1-\alpha)/\alpha} \int_a^b (s - a)^\alpha (b - s)^\alpha [q(s)]^- ds < (b - a)^{1/\alpha},$$

where

$$C = \max \left\{ \int_a^b (s-a)^{1+\alpha}(b-s)^\alpha [q(s)]^- ds, \ \int_a^b (s-a)^\alpha (b-s)^{\alpha+1} [q(s)]^- ds \right\}.$$

Then, $q \in U_\alpha((a, b))$.

In particular, this corollary implies that if

$$\int_a^b (s - a)^\alpha (b - s)^\alpha [q(s)]^- ds \ \le \ (b - a)^\alpha,$$

then $q \in U_\alpha((a, b))$.

Denote by $\xi(\lambda)$, where $\lambda > -\dfrac{1}{\alpha + 1}\left(\dfrac{\alpha}{\alpha + 1}\right)^\alpha$, the largest positive root of the equation $T - |T|^{\alpha/(\alpha+1)} - \lambda = 0$.

Theorem 3.6.5. Let

$$Q_*(t_0, \alpha) \ \ge \ -\dfrac{1}{\alpha + 1}\left(\dfrac{\alpha}{\alpha + 1}\right)^\alpha$$

and

$$Q^*(t_0, \alpha) \ < \ \xi(Q_*(t_0, \alpha)). \tag{3.6.89}$$

Then, $q \in U_\alpha((a, b))$.

Proof. Let $k = \xi(Q_*(t_0, \alpha))$,

$$\sigma(t) = \begin{cases} -\dfrac{Q(t, t_0, \alpha) - k}{(t - a)^\alpha} & \text{for } a < t < t_0 \\ \dfrac{Q(t, t_0, \alpha) - k}{(b - t)^\alpha} & \text{for } t_0 < t < b. \end{cases} \tag{3.6.90}$$

Then from (3.6.89), we have

$$k - k^{\alpha/(\alpha+1)} \leq Q(t, t_0, \alpha) < k \quad \text{for} \quad a < t < b \qquad (3.6.91)$$

from which it follows that $\sigma(t) \operatorname{sgn}(t_0 - t) > 0$ for $a < t < b$, $t \neq t_0$ and

$$|\sigma(t)|^{(\alpha+1)/\alpha} \leq k\, h(t) \quad \text{for} \quad a < t < b, \qquad (3.6.92)$$

where

$$h(t) = \begin{cases} \dfrac{1}{(t-a)^{\alpha+1}} & \text{for} \quad a < t \leq t_0 \\[2mm] \dfrac{1}{(b-t)^{\alpha+1}} & \text{for} \quad t_0 < t < b. \end{cases} \qquad (3.6.93)$$

Define

$$y(t) = \exp\left(\operatorname{sgn}(t_0 - t) \int_{t_0}^{t} |\sigma(s)|^{1/\alpha} ds \right) \quad \text{for} \quad a < t < b.$$

From (3.6.90) – (3.6.92) it follows that $y \in AC^1_{loc}((a, t_0) \cup (t_0, b))$,

$$y'(t) \operatorname{sgn}(t_0 - t) > 0 \quad \text{for} \quad a < t < b, \ t \neq t_0,$$
$$y'(t_0+) > 0 > y'(t_0-), \quad y(a+) = 0, \quad y(b-) = 0$$

and

$$y''(t) = q(t)|y(t)|^\alpha |y'(t)|^{1-\alpha} + y(t)|\sigma(t)|^{(1-\alpha)/\alpha}\left(|\sigma(t)|^{(\alpha+1)/\alpha} - kh(t) \right)$$
$$\leq q(t)|y(t)|^\alpha |y'(t)|^{1-\alpha} \quad \text{for} \quad a < t < b.$$

Hence, by Lemma 3.6.10, we have $q \in U_\alpha((a, b))$. ∎

Corollary 3.6.5. Let

$$-\frac{1}{\alpha+1}\left(\frac{\alpha}{\alpha+1}\right)^\alpha \leq Q_*(t_0, \alpha) \quad \text{and} \quad Q^*(t_0, \alpha) < \left(\frac{\alpha}{\alpha+1}\right)^{\alpha+1}.$$

Then, $q \in U_\alpha((a, b))$.

Theorem 3.6.6. Let

$$Q_*(t_0, \alpha) \geq -\frac{1}{\alpha+1}\left(\frac{\alpha}{\alpha+1}\right)^\alpha$$

and either

$$Q_*(t_0, \alpha) \neq 0 \quad \text{and} \quad Q^*(t_0, \alpha) \leq 2\xi(Q_*(t_0, \alpha)) - Q_*(t_0, \alpha), \qquad (3.6.94)$$

or

$$Q_*(t_0, \alpha) = 0 \quad \text{and} \quad Q^*(t_0, \alpha) < 2. \qquad (3.6.95)$$

Then, $q \notin O_\alpha((a, b))$.

Proof. Let

$$k = \begin{cases} \xi(Q_*(t_0, \alpha)) & \text{for} \quad Q_*(t_0, \alpha) \neq 0 \\ k_1 & \text{for} \quad Q_*(t_0, \alpha) = 0, \end{cases}$$

where $k_1 \in ((\alpha/(\alpha + 1))^{\alpha+1}, 1)$ is chosen in such a way that $Q^*(t_0, \alpha) < k_1 + k_1^{\alpha/(\alpha+1)}$. It follows that $k \in (0, 1)$.

Define the function $\sigma(t)$ by (3.6.90). Clearly,

$$\sigma \in AC_{loc}((a, t_0) \cup (t_0, b)), \quad \sigma(t_0-) = \frac{k}{(t_0 - a)^\alpha},$$

$$\sigma(t_0+) = -\frac{k}{(b - t_0)^\alpha}, \quad \lim_{t \to a^+} (t - a)^\alpha \sigma(t) = k < 1$$

and $\lim_{t \to b-} (b - t)^\alpha \sigma(t) = -k > -1$. In view of (3.6.94) and (3.6.95), we can conclude that

$$k - k^{\alpha/(\alpha+1)} \leq \alpha Q(t, t_0, \alpha) \leq k + k^{\alpha/(\alpha+1)} \quad \text{for} \quad a < t < b.$$

Thus, (3.6.92), where $h(t)$ is the function defined by (3.6.93) is satisfied. In view of this fact for $a < t < b$, we have

$$\sigma'(t) = \alpha q(t) - \alpha k h(t) \leq \alpha q(t) - \alpha |\sigma(t)|^{(\alpha+1)/\alpha}.$$

By Lemma 3.6.11, we now have $q \in O_\alpha((a, b))$. ∎

Corollary 3.6.6. Let

$$-\frac{1}{\alpha+1} \left(\frac{\alpha}{\alpha+1}\right)^\alpha \leq Q_*(t_0, \alpha) \quad \text{and} \quad Q^*(t_0, \alpha) \leq \frac{2\alpha + 1}{\alpha + 1} \left(\frac{\alpha}{\alpha + 1}\right)^\alpha.$$

Then, $q \notin O_\alpha((a, b))$.

We note that Corollaries 3.6.5 and 3.6.6 follow from the fact that

$$\xi(Q_*(t_0, \alpha)) \geq \left(\frac{\alpha}{\alpha + 1}\right)^{\alpha+1}.$$

Corollary 3.6.7. Let $\alpha = 1$, $Q_*(t_0, 1) \geq -1/4$ and either

$$Q_*(t_0, 1) \neq 0, \quad Q^*(t_0, 1) \leq 1 + Q_*(t_0, 1) + \sqrt{1 + 4Q_*(t_0, 1)},$$

or

$$Q_*(t_0, 1) = 0, \quad Q^*(t_0, 1) < 2.$$

Then, $q \notin O_1((a, b))$ (i.e., $q \in U_1((a, b))$).

3.7. Nonoscillation Criteria

We begin with the following nonoscillatory characterization of (3.1.1).

3.7.1. Nonoscillatory Characterization

Theorem 3.7.1. Equation (3.1.1) is nonoscillatory if and only if there exist a number $T \geq t_0$ and a function $w(t) \in C^1([T, \infty), \mathbb{R})$ such that

$$w'(t) + \alpha a^{-1/\alpha}(t)|w(t)|^{(\alpha+1)/\alpha} + q(t) \leq 0 \quad \text{on} \quad [T, \infty). \tag{3.7.1}$$

Proof. Let $x(t)$ be a nonoscillatory solution of (3.1.1). Then there is a number $T \geq t_0$ such that $x(t) \neq 0$ for all $t \geq T$. Define $w(t) = a(t)\psi(x'(t))/\psi(x(t))$ on $[T, \infty)$. Clearly, $w(t) \in C^1([T, \infty), \mathbb{R})$ and satisfies (3.7.1). Conversely, let inequality (3.7.1) have a solution $w(t)$ on $[T, \infty)$. Define $-q_1(t) = w'(t) + \alpha a^{-1/\alpha}(t)|w(t)|^{(\alpha+1)/\alpha}$ on $[T, \infty)$. It follows from (3.7.1) that

$$q_1(t) \geq q(t) \quad \text{for all} \quad t \geq T. \tag{3.7.2}$$

A simple calculation shows that

$$u(t) = \exp\left(\int_T^t \psi^{-1}\left(\frac{w(s)}{a(s)}\right)\right) ds$$

is a solution of $(a(t)\psi(x'(t)))' + q_1(t)\psi(x(t)) = 0$, where $\psi^{-1}(s) = |s|^{(1-\alpha)/\alpha}s$.

Since $u(t)$ does not vanish on $[T, \infty)$, (3.7.2) and Corollary 3.2.2 imply that equation (3.1.1) is nonoscillatory. ∎

Now, we shall study the nonoscillatory behavior of (3.1.1) under the conditions

$$\int_{t_0}^{\infty} a^{-1/\alpha}(s)ds = \infty \quad \text{and} \quad \int^{\infty} q(s)ds < \infty. \tag{3.7.3}$$

In what follows we shall frequently use the functions $R(t) = \int_{t_0}^t a^{-1/\alpha}(s)ds$ and $Q(t) = \int_t^{\infty} q(s)ds$.

Theorem 3.7.2. The following three statements are equivalent:

(i)   Equation (3.1.1) is nonoscillatory.

(ii)   There exists a function $w(t) \in C([T, \infty), \mathbb{R})$, $T \geq t_0$ such that for $t \geq T$,

$$w(t) = \int_t^{\infty} q(s)ds + \alpha \int_t^{\infty} a^{-1/\alpha}(s)|w(s)|^{(\alpha+1)/\alpha}ds. \tag{3.7.4}$$

(iii) There exists a function $w(t) \in C([T, \infty), \mathbb{R})$, $T \geq t_0$ such that for $t \geq T$,

$$|w(t)| \geq \left| \int_t^\infty q(s)ds + \alpha \int_t^\infty a^{-1/\alpha}(s)|w(s)|^{(\alpha+1)/\alpha}ds \right|. \qquad (3.7.5)$$

Proof. (i) $\Rightarrow$ (ii). Let $x(t)$ be a nonoscillatory solution of (3.1.1). Without loss of generality, we can assume that $x(t) > 0$ on $[T, \infty)$ for some $T \geq t_0$. Define

$$w(t) = a(t)\frac{\psi(x'(t))}{\psi(x(t))} \quad \text{on} \quad [T, \infty). \qquad (3.7.6)$$

From equation (3.1.1), we find

$$w'(t) + \alpha a^{-1/\alpha}(t)|w(t)|^{(\alpha+1)/\alpha} + q(t) = 0 \quad \text{on} \quad [T, \infty). \qquad (3.7.7)$$

Let $t \geq T$ be fixed arbitrarily. Integrating (3.7.7) over $[t, \xi]$, we get

$$w(\xi) - w(t) + \alpha \int_t^\xi a^{-1/\alpha}(s)|w(s)|^{(\alpha+1)/\alpha}ds + \int_t^\xi q(s)ds = 0 \quad \text{for } \xi \geq t \geq T.$$

$$\qquad (3.7.8)$$

We claim that

$$\int_t^\infty a^{-1/\alpha}(s)|w(s)|^{(\alpha+1)/\alpha}ds < \infty. \qquad (3.7.9)$$

Suppose to the contrary that $\int_t^\infty a^{-1/\alpha}(s)|w(s)|^{(\alpha+1)/\alpha}ds = \infty$. Then in view of (3.7.8) and (3.7.3) there is a $T_0 \geq t$ large enough so that for $\xi \geq T_0$,

$$w(\xi) + \alpha \int_{T_0}^\xi a^{-1/\alpha}(s)|w(s)|^{(\alpha+1)/\alpha}ds$$

$$= w(t) - \int_t^\xi q(s)ds - \alpha \int_t^{T_0} a^{-1/\alpha}(s)|w(s)|^{(\alpha+1)/\alpha}ds \leq -1,$$

or equivalently,

$$-w(\xi) \geq 1 + \alpha \int_{T_0}^\xi a^{-1/\alpha}(s)|w(s)|^{(\alpha+1)/\alpha}ds \quad \text{for } t \geq T_0. \qquad (3.7.10)$$

Thus, it follows that $x'(\xi) < 0$ and

$$\frac{a^{-1/\alpha}(\xi)|w(\xi)|^{(\alpha+1)/\alpha}}{1 + \alpha \int_{T_0}^\xi a^{-1/\alpha}(s)|w(s)|^{(\alpha+1)/\alpha}ds} \geq a^{-1/\alpha}(\xi)(-w(\xi))^{1/\alpha} \quad \text{for } \xi \geq T_0.$$

$$\qquad (3.7.11)$$

Integrating (3.7.11) from T_0 to ξ, we obtain

$$\frac{1}{\alpha} \ln \left(1 + \alpha \int_{T_0}^{\xi} a^{-1/\alpha}(s) |w(s)|^{(\alpha+1)/\alpha} ds \right) \geq \ln \frac{x(T_0)}{x(\xi)} \quad \text{for} \quad \xi \geq T_0,$$

which when combined with (3.7.10) implies that $-x'(\xi) \geq a^{-1/\alpha}(\xi) x(T_0)$ for $\xi \geq T_0$. Integrating the last inequality and using (3.7.3), we find that $x(\xi) \to \infty$ as $\xi \to \infty$, which contradicts the assumption that $x(t) > 0$ for $t \geq T$. Therefore, (3.7.9) must hold.

We now let $\xi \to \infty$ in (3.7.8). Using (3.7.9) and (3.7.3), we observe that $w(\xi)$ tends to a finite limit $w(\infty)$. But, $x(\infty)$ must be zero, otherwise, in view of (3.7.3), (3.7.9) must fail. Thus, (3.7.4) holds.

(ii) $\Rightarrow$ (iii). The proof is clear.

(iii) $\Rightarrow$ (i). Let

$$w(t) = \int_t^{\infty} q(s) ds + \alpha \int_t^{\infty} a^{-1/\alpha}(s) |w(s)|^{(\alpha+1)/\alpha} ds.$$

Then, (3.7.1) holds. Now, it follows from Theorem 3.7.1 that equation (3.1.1) is nonoscillatory. ∎

To prove our next result, we need the following lemma.

Lemma 3.7.1. The function

$$h(x) = |a + \alpha x|^{(\alpha+1)/\alpha} - (\alpha+1)|a|^{1/\alpha} x \, \text{sgn} \, a,$$

where a is a real number, is nondecreasing on $[0, \infty)$.

Proof. Since

$$h'(x) = \begin{cases} (\alpha+1)[a + \alpha x]^{1/\alpha} - (\alpha+1)|a|^{1/\alpha} \, \text{sgn} \, a \geq 0 & \text{if } a + \alpha x \geq 0 \\ -(\alpha+1)[-a - \alpha x]^{1/\alpha} + (\alpha+1)|a|^{1/\alpha} \geq 0 & \text{if } a + \alpha x < 0, \end{cases}$$

$h(x)$ is nondecreasing on $[0, \infty)$. ∎

In what follows we shall let by

$$\mu[s, t] = \exp \left((\alpha+1) \int_t^s |Q(\xi)|^{1/\alpha} a^{-1/\alpha}(\xi) \, \text{sgn} \, Q(\xi) d\xi \right). \qquad (3.7.12)$$

Theorem 3.7.3. Let condition (3.7.3) hold. Then the following three statements are equivalent:

(i) Equation (3.1.1) is nonoscillatory.

(ii)   There exist a number $T \geq t_0$ and a function $y(t) \in C([T, \infty), \mathbb{R})$ such that

$$
\begin{aligned}
y(t) &\geq \int_t^\infty \left[|Q(s) + \alpha y(s)|^{(\alpha+1)/\alpha} - (\alpha + 1)|Q(s)|^{1/\alpha} y(s) \, \text{sgn} \, Q(s) \right] \\
&\qquad\qquad\qquad\qquad\qquad\qquad\qquad\qquad \times a^{-1/\alpha}(s)\mu[s, t]ds \\
&\geq \int_t^\infty |Q(s)|^{(\alpha+1)/\alpha} a^{-1/\alpha}(s)\mu[s, t] = Q_1(t).
\end{aligned} \tag{3.7.13}
$$

(iii)   There exist a number $T \geq t_0$ and a function $z(t) \in C([T, \infty), \mathbb{R})$ such that

$$
\begin{aligned}
z(t) &= \int_t^\infty \left[|Q(s) + \alpha z(s)|^{(\alpha+1)/\alpha} - (\alpha + 1)|Q(s)|^{1/\alpha} z(s) \, \text{sgn} \, Q(s) \right] \\
&\qquad\qquad\qquad\qquad\qquad\qquad\qquad \times a^{-1/\alpha}(s)\mu[s, t]ds.
\end{aligned} \tag{3.7.14}
$$

Proof.   (i) $\Rightarrow$ (ii).   Assume that equation (3.1.1) is nonoscillatory. It follows from Theorem 3.7.2 that (3.7.4) has a solution $w(t)$. Define $y(t) = \int_t^\infty a^{-1/\alpha}(s)|w(s)|^{(\alpha+1)/\alpha}ds \geq 0$. Clearly, $y'(t) = -a^{-1/\alpha}(t)|w(t)|^{(\alpha+1)/\alpha}$. Using $\mu[s, t]$ as an integrating factor and integrating from t to T $(\geq t)$, we obtain

$$
\begin{aligned}
y(t) &= \mu[T, t]y(T) \\
&+ \int_t^T \left[|Q(s)+\alpha y(s)|^{\frac{\alpha+1}{\alpha}} - (\alpha+1)|Q(s)|^{\frac{1}{\alpha}} y(s) \, \text{sgn} \, Q(s) \right] a^{-\frac{1}{\alpha}}(s)\mu[s, t]ds \\
&\geq \int_t^T |Q(s)|^{\frac{\alpha+1}{\alpha}} a^{-\frac{1}{\alpha}}(s)\mu[s, t]ds.
\end{aligned} \tag{3.7.15}
$$

The left hand side of (3.7.15) is independent of T, and hence as $T \to \infty$, (3.7.13) follows.

(ii) $\Rightarrow$ (iii).   Suppose there exists a function $y(t) \in C([T, \infty), \mathbb{R})$ which satisfies (3.7.13). Define

$$
z_0(t) = y(t)
$$
$$
z_n(t) = \int_t^\infty \left[|Q(s)+\alpha z_{n-1}(s)|^{(\alpha+1)/\alpha} - (\alpha+1)|Q(s)|^{1/\alpha} z_{n-1}(s) \, \text{sgn} \, Q(s) \right] \\
\times a^{-1/\alpha}(s)\mu[s, t]ds
$$

for $n = 1, 2, \cdots$. Then from Lemma 3.7.1 and (3.7.13), it is easy to show by induction that $0 \leq Q_1(t) \leq z_n(t) \leq z_{n-1}(t) \leq y(t)$. Thus, the sequence of functions $\{z_n(t)\}_{n=0}^\infty$ has a pointwise limiting function $z(t) = \lim_{n\to\infty} z_n(t)$. Now, it follows from (3.7.13) and the Lebesgue dominated convergence theorem that $z(t)$ satisfies (3.7.14).

(iii) $\Rightarrow$ (i).   Assume that (3.7.14) has a solution $z(t)$,   then an easy computation shows that $w(t) = -[Q(t) + \alpha z(t)]/a(t)$ defines a solution of (3.7.1). Now, as a consequence of Theorem 3.7.1, we conclude that equation (3.1.1) is nonoscillatory.   ∎

To establish a nonoscillation characterization of equation (3.1.1) when $q(t) \in C([t_0, \infty), \mathbb{R}^+)$,   and

$$\pi(t) = \int_t^\infty a^{-1/\alpha}(s)ds \quad \text{and} \quad \pi(t_0) < \infty \qquad (3.7.16)$$

we need the following lemma.

Lemma 3.7.2.   Let λ be a constant.   Then the function $g(x) = \alpha|x|^{(\alpha+1)/\alpha} + \lambda x$ is nondecreasing for $x \geq -(\lambda/(\alpha+1))^\alpha$.

Proof.   Since

$$g'(x) = (\alpha+1)|x|^\alpha \operatorname{sgn} x + \lambda \geq 0 \quad \text{for} \quad x \geq -(\lambda/(\alpha+1))^\alpha,$$

$g(x)$ is nondecreasing for $x \geq -(\lambda/(\alpha+1))^\alpha$.   ∎

Lemma 3.7.3.   Assume that $q(t) \in C([t_0, \infty), \mathbb{R}^+)$ and $\pi(t_0) < \infty$. Let $x(t)$ be a solution of equation (3.1.1) such that $x(t) \neq 0$ for $t \geq T \geq t_0$. Then, $x(t)$ as well as the function

$$\pi^\alpha(t)a(t)\frac{\psi(x'(t))}{\psi(x(t))}. \qquad (3.7.17)$$

are bounded on $[T, \infty)$.   Furthermore,

$$\pi^\alpha(t)a(t)\frac{\psi(x'(t))}{\psi(x(t))} \geq -1 \quad \text{for} \quad t \geq T \qquad (3.7.18)$$

and

$$\limsup_{t\to\infty} \pi^\alpha(t)a(t)\frac{\psi(x'(t))}{\psi(x(t))} \leq 0. \qquad (3.7.19)$$

Proof.   Without loss of generality, we assume that $x(t) > 0$ for $t \geq T \geq t_0$. Since $a(t)\psi(x'(t))$ is nonincreasing on $[T, \infty)$,   we find that $x'(t)$ is eventually of constant sign, i.e., $x'(t) > 0$ for $t \geq T$,   or there is a $T_1 > T$ such that $x'(t) < 0$ for $t \geq T_1$ and

$$a^{1/\alpha}(s)x'(s) \leq a^{1/\alpha}(t)x'(t) \quad \text{for} \quad s \geq t \geq T. \qquad (3.7.20)$$

Dividing both sides of (3.7.20) by $a^{1/\alpha}(s)$ and integrating it over $[t, \xi]$ gives

$$x(\xi) \leq x(t) + a^{1/\alpha}(t)x'(t)\int_t^\xi a^{-1/\alpha}(s)ds \quad \text{for} \quad \xi \geq t \geq T. \qquad (3.7.21)$$

If $x'(t) > 0$ for $t \geq T$, then from (3.7.21) we have $x(\xi) \leq x(t) + a^{1/\alpha}(t)x'(t)\pi(t)$ for $\xi \geq t \geq T$, which shows that $x(t)$ is bounded on $[T, \infty)$. If $x'(t) < 0$ for $t \geq T_1$, then $x(t)$ is clearly bounded. Letting $\xi \to \infty$ in (3.7.21), we get $0 \leq x(t) + a^{1/\alpha}(t)x'(t)\pi(t)$ for $t \geq T$. In either case, we obtain $\pi(t)a^{1/\alpha}(t)x'(t)/x(t) \geq -1$ for $t \geq T$ from which (3.7.18) is an immediate consequence.

The relation (3.7.19) trivially holds if $x'(t) < 0$ for $t \geq T_1$, since in this case the function defined in (3.7.17) itself is negative for $t \geq T_1$. If $x'(t) > 0$ for $t \geq T$, then there exist two positive constants k_1 and k_2 such that $x(t) \geq k_1$ and $a(t)\psi(x'(t)) \leq k_2$ for $t \geq T$, which imply

$$a(t)\frac{\psi(x'(t))}{\psi(x(t))} \leq \frac{k_2}{k_1^\alpha} \quad \text{for} \quad t \geq T.$$

Now, since $\pi(t) \to 0$ as $t \to \infty$, we conclude that

$$\lim_{t \to \infty} \pi^\alpha(t)a(t)\frac{\psi(x'(t))}{\psi(x(t))} \leq 0.$$

This proves (3.7.19) and completes the proof.   ■

Theorem 3.7.4. Let $\lambda \geq \alpha + 1$ be a constant. Then equation (3.1.1) is nonoscillatory if and only if

$$\int^\infty \pi^\lambda(s)q(s)ds < \infty, \tag{3.7.22}$$

and there exists a function $y(t) \in C([T, \infty), \mathbb{R})$, $T \geq t_0$ such that

$$\pi^\alpha(t)y(t) \text{ is bounded on } [T, \infty), \quad \pi^\alpha(t)y(t) \geq -1 \quad \text{for} \quad t \geq T \tag{3.7.23}$$

and

$$\pi^\alpha(t)y(t) \geq \int_t^\infty \pi^\lambda(s)q(s)ds + \lambda \int_t^\infty a^{-1/\alpha}(s)\pi^{\lambda-1}(s)y(s)ds$$
$$+\alpha \int_t^\infty a^{-1/\alpha}(s)\pi^\lambda(s)|y(s)|^{(\alpha+1)/\alpha}ds \quad \text{for} \quad t \geq T. \tag{3.7.24}$$

Proof. (The 'only if' part). Let $x(t)$ be a solution of equation (3.1.1) such that $x(t) \neq 0$ on $[T, \infty)$ for some $T \geq t_0$. Define $w(t) = a(t)\psi(x'(t))/\psi(x(t))$ on $[T, \infty)$. Then, we have

$$w'(t) + \alpha a^{-1/\alpha}(t)|w(t)|^{(\alpha+1)/\alpha} + q(t) = 0 \quad \text{on} \quad [T, \infty). \tag{3.7.25}$$

Multiplying (3.7.25) by $\pi^\lambda(t)$ and integrating it over $[t, \xi]$, $\xi \geq t \geq T$, we obtain

$$\pi^\lambda(\xi)w(\xi) - \pi^\lambda(t)w(t) + \lambda \int_t^\xi a^{-1/\alpha}(s)\pi^{\lambda-1}(s)w(s)ds$$

$$+ \int_t^\xi \pi^\lambda(s)q(s)ds + \alpha \int_t^\xi a^{-1/\alpha}(s)\pi^\lambda(s)|w(s)|^{(\alpha+1)/\alpha}ds = 0, \quad \xi \geq t \geq T.$$

$$(3.7.26)$$

Clearly, (3.7.24) follows from (3.7.26). Also $\pi^\alpha(t)y(t) \geq -1$ for $t \geq T$ is trivial from Lemma 3.7.3.

(The 'if' part). Let

$$y(t) = \pi^{-\lambda}(t)\left[\int_t^\infty \pi^\lambda(s)q(s)ds + \lambda \int_t^\infty a^{-1/\alpha}(s)\pi^{\lambda-1}(s)w(s)ds\right.$$

$$\left. + \alpha \int_t^\infty a^{-1/\alpha}(s)\pi^\lambda(s)|w(s)|^{(\alpha+1)/\alpha}ds\right].$$

Then by (3.7.24), $w(t) \geq y(t)$ for $t \geq T$. This and Lemma 3.7.2 imply that

$$\pi^\alpha(t)w(t) \geq \pi^\alpha(t)y(t)$$

$$\geq \pi^{\alpha-\lambda}(t)\int_t^\infty a^{-1/\alpha}(s)\pi^{\lambda-\alpha-1}(s)\left[\lambda\pi^\alpha(s)w(s) + \alpha|\pi^\alpha(s)w(s)|^{(\alpha+1)/\alpha}\right]ds$$

$$\geq -\left(\frac{\lambda}{\alpha+1}\right)^{\alpha+1}\pi^{\alpha-\lambda}(t)\int_t^\infty a^{-1/\alpha}(s)\pi^{\lambda-\alpha-1}(s)ds$$

$$= -\frac{1}{\lambda-\alpha}\left(\frac{\lambda}{\alpha+1}\right)^{\alpha+1}\pi^{\alpha-\lambda}(t)\pi^{\lambda-\alpha}(t) = -\frac{1}{\lambda-\alpha}\left(\frac{\lambda}{\alpha+1}\right)^{\alpha+1},$$

and $g(\pi^\alpha(t)w(t)) \geq g(\pi^\alpha(t)y(t))$ for $t \geq T$, where the function g is defined in Lemma 3.7.2.

Since

$$y'(t) = \lambda\pi^{-1}(t)a^{-1/\alpha}(t)w(t) - c(t) - \lambda\pi^{-1}(t)a^{-1/\alpha}(t)w(t)$$

$$-\alpha a^{-1/\alpha}(t)|w(t)|^{(\alpha+1)/\alpha}$$

for $t \geq T$, we have

$$y'(t) + q(t) + \alpha a^{-1/\alpha}(t)|y(t)|^{(\alpha+1)/\alpha}$$

$$= \lambda\pi^{-1}(t)a^{-1/\alpha}(t)y(t) - \lambda\pi^{-1}(t)a^{-1/\alpha}(t)w(t) - \alpha a^{-1/\alpha}(t)|w(t)|^{(\alpha+1)/\alpha}$$

$$+ \alpha a^{-1/\alpha}(t)|y(t)|^{(\alpha+1)/\alpha}$$

$$= a^{-1/\alpha}(t)\pi^{-\alpha-1}(t)\left\{\left[\alpha|\pi^\alpha(t)y(t)|^{(\alpha+1)/\alpha} + \lambda\pi^\alpha(t)y(t)\right]\right.$$

$$\left. - \left[\alpha|\pi^\alpha(t)w(t)|^{(\alpha+1)/\alpha} + \lambda\pi^\alpha(t)w(t)\right]\right\}$$

$$= a^{-1/\alpha}(t)\pi^{-\alpha-1}(t)\left[g(\pi^\alpha(t)w(t)) - g(\pi^\alpha(t)y(t))\right] \leq 0 \text{ for } t \geq T.$$

Now, Theorem 3.7.1 guarantees that equation (3.1.1) is nonoscillatory. ■

3.7.2. Comparison Theorems

In this subsection, we shall consider (3.1.1) and the equation

$$(a_1(t)\psi(y'(t)))' + q_1(t)\psi(y(t)) = 0, \tag{3.7.27}$$

where $a_1(t) \in C^1([t_0, \infty), \mathbb{R}^+)$ and $q_1(t) \in C([t_0, \infty), \mathbb{R})$.

We shall employ the nonoscillation characterization presented in Theorem 3.7.2 to prove the following comparison theorem.

Theorem 3.7.5. Let $a(t) \geq a_1(t)$, $\lim_{t \to \infty} R(t) = \infty$, and

$$\left| \int_t^\infty q(s)ds \right| \leq \int_t^\infty q_1(s)ds \quad \text{for} \quad t \geq t_0.$$

If (3.7.27) is nonoscillatory, then so is the equation (3.1.1).

Proof. Since equation (3.7.27) is nonoscillatory it follows from Theorem 3.7.1 that there exists $w(t) \in C^1([T, \infty), \mathbb{R})$ for some $T \geq t_0$ such that

$$w'(t) + q_1(t) + \alpha a_1^{-1/\alpha}(t)|w(t)|^{(\alpha+1)/\alpha} \leq 0 \quad \text{for} \quad t \geq T.$$

Using the fact that $a(t) \geq a_1(t)$ in the above inequality, we obtain

$$w'(t) + q_1(t) + \alpha a^{-1/\alpha}(t)|w(t)|^{(\alpha+1)/\alpha} \leq 0 \quad \text{for} \quad t \geq T.$$

Thus, by Theorem 3.7.1, we find that the differential equation

$$(a(t)\psi(x'(t)))' + q_1(t)\psi(x(t)) = 0 \tag{3.7.28}$$

is nonoscillatory. Now, it follows from Theorem 3.7.2 that there exists a function $y(t) \in C([T^*, \infty), \mathbb{R})$ for some $T^* \geq T$ such that

$$|y(t)| \geq \left| \int_t^\infty q_1(s)ds + \alpha \int_t^\infty a^{-1/\alpha}(s)|y(s)|^{(\alpha+1)/\alpha}ds \right|$$

$$\geq \left| \int_t^\infty q(s)ds + \alpha \int_t^\infty a^{-1/\alpha}(s)|y(s)|^{(\alpha+1)/\alpha}ds \right|.$$

Once again, we can apply Theorem 3.7.2 to conclude that equation (3.1.1) is nonoscillatory. This completes the proof. ■

Next, we shall consider equations (3.1.1) and (3.7.27) where $q(t)$, $q_1(t) \in C([t_0, \infty), \mathbb{R}^+)$, and

$$\pi_1(t_0) < \infty \quad \text{where} \quad \pi_1(t) = \int_t^\infty \frac{1}{a_1(s)}ds. \tag{3.7.29}$$

Using the nonoscillation characterization established in Theorem 3.7.4, we shall prove the following comparison theorem.

Theorem 3.7.6. Assume that $a(t) \geq a_1(t)$ and (3.7.29) holds, and let

$$\int_t^\infty \pi^\lambda(s)q(s)ds \;\leq\; \int_t^\infty \pi^\lambda(s)q_1(s)ds \qquad (3.7.30)$$

for all sufficiently large t, where $\lambda \geq \alpha + 1$ is a constant. If (3.7.27) is nonoscillatory, then so is the equation (3.1.1).

Proof. Since (3.7.27) is nonoscillatory it follows from Theorem 3.7.1 that there exists a function $w(t) \in C^1([T,\infty), \mathbb{R})$ for some $T \geq t_0$ such that $w'(t) + q_1(t) + \alpha a_1^{-1/\alpha}(t)|w(t)|^{(\alpha+1)/\alpha} \leq 0$ for $t \geq T$. Hence, we have $w'(t) + q_1(t) + \alpha a^{-1/\alpha}(t)|w(t)|^{(\alpha+1)/\alpha} \leq 0$ for $t \geq T$. By Theorem 3.7.1, equation (3.7.28) is nonoscillatory. Now, from Theorem 3.7.4 and (3.7.30), we conclude that (3.1.1) is nonoscillatory. This completes the proof. ∎

3.7.3. Sufficient Conditions for Nonoscillation of Equation (3.1.1)

Another version of Theorem 3.7.1 is the following result.

Theorem 3.7.7. Equation (3.1.1) is nonoscillatory if and only if there exist $T \geq t_0$ and a function $w(t) \in C^1([T,\infty), \mathbb{R})$ such that

$$q(t) + \alpha a(t)|w(t)|^{(\alpha+1)/\alpha} - (a(t)w(t))' \;\leq\; 0 \quad \text{on} \quad [T,\infty). \qquad (3.7.31)$$

The proof of Theorem 3.7.7 is similar to that of Theorem 3.7.1, except that now we let $w(t) = \psi(x'(t))/\psi(x(t))$, $t \geq T$. The details are left to the reader.

As applications of Theorem 3.7.7, now we present the following nonoscillation criteria for the equation (3.1.1). In what follows, we shall let by $\gamma = (1/\alpha)(\alpha/(\alpha+1))^{\alpha+1}$.

Theorem 3.7.8. Let $h(t) \in C^1([t_0,\infty), \mathbb{R}^+)$ and $\phi(t) \in C^1([t_0,\infty), \mathbb{R})$ satisfy

$$h'(t) \;\geq\; a^{-1/\alpha}(t), \quad t \geq t_0 \qquad (3.7.32)$$

and

$$\phi'(t) \;\leq\; -q(t), \quad t \geq t_0. \qquad (3.7.33)$$

If

$$\limsup_{t\to\infty} \; h^\alpha(t)|\phi(t)| \;<\; \gamma, \qquad (3.7.34)$$

then equation (3.1.1) is nonoscillatory.

Proof. Inequality (3.7.34) implies that there are two numbers $T \geq t_0$ and $k \in (0, \gamma)$ such that $|\phi(t)| \leq kh^{-\alpha}(t)$ for $t \geq T$. Let

$$w(t) = -\frac{1}{\lambda a(t)}\left(\lambda\phi(t) + \frac{1-\lambda k}{h^\alpha(t)}\right),$$

where $\lambda = ((\alpha+1)/\alpha)^\alpha$. Then, we have

$$\lambda k \leq \lambda\gamma = \left(\frac{\alpha+1}{\alpha}\right)^\alpha \frac{1}{\alpha}\left(\frac{\alpha}{\alpha+1}\right)^{\alpha+1} = \frac{1}{\alpha+1} < 1, \qquad (3.7.35)$$

$$(a(t)w(t))' = -\phi'(t) + \frac{\alpha(1-\lambda k)}{\lambda}\frac{h'(t)}{h^{\alpha+1}(t)} \geq q(t) + \frac{\alpha(1-\lambda k)}{\lambda}\frac{a^{-1/\alpha}(t)}{h^{\alpha+1}(t)}$$

and

$$q(t) + \alpha a(t)|w(t)|^{(\alpha+1)/\alpha} - (a(t)w(t))'$$

$$\leq q(t) + \alpha a(t)\left|\frac{1}{\lambda a(t)}\left(\lambda\phi(s) + \frac{1-\lambda k}{h^\alpha(t)}\right)\right|^{(\alpha+1)/\alpha} - q(t) - \frac{\alpha(1-\lambda k)}{\lambda}\frac{a^{-1/\alpha}(t)}{h^{\alpha+1}(t)}$$

$$= \alpha a^{-1/\alpha}(t)h^{-(\alpha+1)}(t)\left[\left|\phi(t)h^\alpha(t) + \frac{1-\lambda k}{\lambda}\right|^{(\alpha+1)/\alpha} - \frac{1-\lambda k}{\lambda}\right]$$

$$\leq \alpha a^{-1/\alpha}(t)h^{-(\alpha+1)}(t)\left[\left|k + \frac{1-\lambda k}{\lambda}\right|^{(\alpha+1)/\alpha} - \frac{1-\lambda k}{\lambda}\right]$$

$$= \alpha a^{-1/\alpha}(t)h^{-(\alpha+1)}(t)\left(\frac{1}{\lambda^{(\alpha+1)/\alpha}} - \frac{1}{\lambda} + k\right)$$

$$= \alpha a^{-1/\alpha}(t)h^{-(\alpha+1)}(t)(k-\gamma) \leq 0.$$

Now, it follows from Theorem 3.7.7 that (3.1.1) is nonoscillatory. ∎

Theorem 3.7.9. Conditions (3.7.32) and (3.7.33) of Theorem 3.7.8 can be replaced by

$$h'(t) \leq -a^{-1/\alpha}(t) \quad \text{for} \quad t \geq t_0 \qquad (3.7.36)$$

and

$$\phi'(t) \geq q(t) \quad \text{for} \quad t \geq t_0, \qquad (3.7.37)$$

respectively.

The proof of Theorem 3.7.9 is the same as that of Theorem 3.7.8 except that now, we let

$$w(t) = \frac{1}{\lambda a(t)}\left(\lambda\phi(t) + \frac{1-\lambda k}{h^\alpha(t)}\right).$$

Theorem 3.7.10. Let the functions $h(t)$ and $\phi(t)$ be as in Theorem 3.7.8 such that conditions (3.7.32) and (3.7.33) hold. If there exists a number $k > 0$ such that

$$-k^{\alpha/(\alpha+1)} - k \leq h^\alpha(t)|\phi(t)| \leq k^{\alpha/(\alpha+1)} - k < \gamma, \qquad (3.7.38)$$

then equation (3.1.1) is nonoscillatory.

Proof. Let

$$w(t) = -\frac{1}{a(t)}\left(\phi(t) + \frac{k}{h^\alpha(t)}\right).$$

Then, we have

$$(a(t)w(t))' = -\phi'(t) + \alpha k\frac{h'(t)}{h^{\alpha+1}(t)} \geq q(t) + \alpha k\frac{a^{-1/\alpha}(t)}{h^{\alpha+1}(t)}$$

and

$$q(t) + \alpha a(t)|w(t)|^{(\alpha+1)/\alpha} - (a(t)w(t))'$$

$$\leq q(t) + \alpha a(t)\left|\frac{1}{a(t)}\left(\phi(t) + \frac{k}{h^\alpha(t)}\right)\right|^{(\alpha+1)/\alpha} - q(t) - \alpha k\frac{a^{-1/\alpha}(t)}{h^{\alpha+1}(t)}$$

$$= \alpha a^{-1/\alpha}(t)h^{-(\alpha+1)}(t)\left[|\phi(t)h^\alpha(t) + k|^{(\alpha+1)/\alpha} - k\right]$$

$$\leq \alpha a^{-1/\alpha}(t)h^{-(\alpha+1)}(t)(k - k) = 0.$$

Now, it follows from Theorem 3.7.7 that (3.1.1) is nonoscillatory. ∎

Theorem 3.7.11. Conditions (3.7.32) and (3.7.33) of Theorem 3.7.10 can be replaced by (3.7.36) and (3.7.37), respectively.

The proof of Theorem 3.7.11 is the same as that of Theorem 3.7.10 except that now, we let

$$w(t) = \frac{1}{a(t)}\left(\phi(t) + \frac{k}{h^\alpha(t)}\right).$$

Example 3.7.1. Consider the half–linear equation

$$\left(t^\beta|x'(t)|^{\alpha-1}x'(t)\right)' + \theta t^{\beta-\alpha-1}|x(t)|^{\alpha-1}x(t) = 0, \qquad (3.7.39)$$

where $\beta > \alpha$ and $\theta > 0$ are constants. Let $h(t) = \int_t^\infty s^{-\beta/\alpha}ds$ and $\phi(t) = \int^t \theta s^{\beta-\alpha-1}ds$. Then for $t \geq t_0 > 0$,

$$h(t) = \frac{\alpha}{\beta-\alpha}t^{(\alpha-\beta)/\alpha}, \quad \phi(t) = \frac{\theta}{\beta-\alpha}t^{\beta-\alpha} \text{ and } h^\alpha(t)\phi(t) = \frac{\theta}{\beta-\alpha}\left(\frac{\alpha}{\beta-\alpha}\right)^\alpha.$$

If $\theta \leq ((\beta - \alpha)/(\alpha+1))^{\alpha+1}$, then $h^\alpha(t)\phi(t) \leq \gamma$. Now, it follows from Theorem 3.7.11 that (3.7.39) is nonoscillatory.

Theorem 3.7.12. Let $h(t) \in C^1([t_0,\infty), \mathbb{R}^+)$ be such that condition (3.7.32) hold. If there exists $\phi(t) \in C^1([t_0,\infty), \mathbb{R})$ such that

$$\lim_{t\to\infty} \phi(t) \text{ exists and } \phi'(t) \leq -h^\alpha(t)q(t), \tag{3.7.40}$$

then equation (3.1.1) is nonoscillatory.

Proof. Since $\lim_{t\to\infty} \phi(t)$ exists, there are two numbers $T \geq t_0$ and m such that $0 < m + \phi(t) \leq 1$ for $t \geq T$. Let $w(t) = -(m + \phi(t))/(a(t)h^\alpha(t))$, $t \geq T$. Then for $t \geq T$, we have

$$(a(t)w(t))' = \alpha(m+\phi(t))\frac{h'(t)}{h^{\alpha+1}(t)} - \frac{\phi'(t)}{h^\alpha(t)} \geq q(t) + \alpha(m+\phi(t))\frac{a^{-1/\alpha}(t)}{h^{\alpha+1}(t)},$$

which implies

$$q(t) + \alpha a(t)|w(t)|^{(\alpha+1)/\alpha} - (a(t)w(t))'$$
$$\leq q(t) + \alpha a(t)\left|\frac{m+\phi(t)}{a(t)h^\alpha(t)}\right|^{(\alpha+1)/\alpha} - q(t) - \alpha(m+\phi(t))\frac{a^{-1/\alpha}(t)}{h^{\alpha+1}(t)}$$
$$\leq \alpha a^{-1/\alpha}(t)h^{-(\alpha+1)}(t)[(m+\phi(t)) - (m+\phi(t))] = 0.$$

Now, it follows from Theorem 3.7.7 that (3.1.1) is nonoscillatory. ∎

Theorem 3.7.13. Condition (3.7.32) in Theorem 3.7.12 can be replaced by (3.7.36).

Proof. Since $\lim_{t\to\infty} \phi(t)$ exists, there are two numbers $T \geq t_0$ and m such that $0 < m - \phi(t) \leq 1$ for $t \geq T$. Let $w(t) = (m - \phi(t))/(a(t)h^\alpha(t))$. The rest of the proof is similar to that of Theorem 3.7.12 and hence omitted. ∎

Theorem 3.7.14. Let $h(t) \in C^1([t_0,\infty), \mathbb{R}^+)$ be such that $q(t)h^{\alpha+1}(t) \leq 1$. If either

$$h'(t) - a^{-1/\alpha}(t) \geq \frac{1}{\alpha} \text{ for all sufficiently large } t, \tag{3.7.41}$$

or

$$\lim_{t\to\infty}\left(h'(t) - a^{-1/\alpha}(t)\right) = \ell \text{ exists and } \ell \geq 1/\alpha, \tag{3.7.42}$$

then equation (3.1.1) is nonoscillatory.

Proof. It follows from (3.7.41) or (3.7.42) that there is a number $T \geq t_0$ such that $h'(t) - a^{-1/\alpha}(t) \geq 1/\alpha$ for all $t \geq T$. Let $w(t) = -1/(a(t)h^\alpha(t))$.

Then, we have

$$q(t) + \alpha a(t)|w(t)|^{(\alpha+1)/\alpha} - (a(t)w(t))'$$

$$\leq \frac{1}{h^{\alpha+1}(t)} + \alpha\frac{a^{-1/\alpha}(t)}{h^{\alpha+1}(t)} - \alpha\frac{h'(t)}{h^{\alpha+1}(t)}$$

$$= \alpha h^{-(\alpha+1)}(t)\left(\frac{1}{\alpha} + a^{-1/\alpha}(t) - h'(t)\right) \leq 0 \quad \text{for} \quad t \geq T.$$

Hence, by Theorem 3.7.7, equation (3.1.1) is nonoscillatory. ∎

Theorem 3.7.15. Let $\phi(t) \in C^1([t_0, \infty), \mathbb{R}_0)$ be such that condition (3.7.33) hold. If

$$\int_t^\infty a^{-1/\alpha}(s)\phi^{(\alpha+1)/\alpha}(s)ds < \left(\frac{1}{\alpha+1}\right)^{(\alpha+1)/\alpha}\phi(t), \quad t \geq t_0$$

then equation (3.1.1) is nonoscillatory.

Proof. Let

$$w(t) = -\frac{1}{a(t)}\left[\phi(t) + \alpha(\alpha+1)^{(\alpha+1)/\alpha}\int_t^\infty a^{-1/\alpha}(s)\phi^{(\alpha+1)/\alpha}(s)ds\right].$$

Then, we have

$$(a(t)w(t))' \geq q(t) + \alpha(\alpha+1)^{(\alpha+1)/\alpha}a^{-1/\alpha}(t)\phi^{(\alpha+1)/\alpha}(t),$$

which implies

$$q(t) + \alpha a(t)|w(t)|^{(\alpha+1)/\alpha} - (a(t)w(t))'$$

$$\leq q(t) + \alpha a(t)\left|\frac{1}{a(t)}\left(\phi(t) + \alpha(\alpha+1)^{(\alpha+1)/\alpha}\int_t^\infty a^{-1/\alpha}(s)\phi^{(\alpha+1)/\alpha}(s)ds\right)\right|^{\frac{\alpha+1}{\alpha}}$$

$$- q(t) - \alpha(\alpha+1)^{(\alpha+1)/\alpha}a^{-1/\alpha}(t)\phi^{(\alpha+1)/\alpha}(t)$$

$$= \alpha a^{-1/\alpha}(t)\left[\left|\phi(t) + \alpha(\alpha+1)^{(\alpha+1)/\alpha}\int_t^\infty a^{-1/\alpha}(s)\phi^{(\alpha+1)/\alpha}(s)ds\right|^{(\alpha+1)/\alpha}\right.$$

$$\left. -(\alpha+1)^{(\alpha+1)/\alpha}\phi^{(\alpha+1)/\alpha}(t)\right]$$

$$\leq \alpha a^{-1/\alpha}(t)\left[\left|\phi(t) + \alpha(\alpha+1)^{(\alpha+1)/\alpha}(\alpha+1)^{-(\alpha+1)/\alpha}\phi(t)\right|^{(\alpha+1)/\alpha}\right.$$

$$\left. -(\alpha+1)^{(\alpha+1)/\alpha}\phi^{(\alpha+1)/\alpha}(t)\right] = 0.$$

Now, it follows from Theorem 3.7.7 that (3.1.1) is nonoscillatory. ∎

Theorem 3.7.16. Let $\phi(t) \in C^1([t_0, \infty), \mathbb{R}_0)$ satisfy condition (3.7.33), and let

$$\phi_1(t) = \int_t^\infty a^{-1/\alpha}(s)\phi^{(\alpha+1)/\alpha}(s) \exp\left[\alpha(\alpha+1)^{1/\alpha} \int_t^s a^{-1/\alpha}(\xi)\phi^{1/\alpha}(\xi)d\xi\right] ds.$$

If $\phi_1(t) \le (\alpha+1)^{-1/\alpha}\phi(t)$, $t \ge t_0$ then equation (3.1.1) is nonoscillatory.

Proof. Let $w(t) = -\left[\phi(t) + \alpha(\alpha+1)^{1/\alpha}\phi_1(t)\right]/a(t)$. Then, we have

$$(a(t)w(t))' = -\phi'(t) - \alpha(\alpha+1)^{1/\alpha}\phi_1'(t)$$
$$\ge q(t) + \alpha(\alpha+1)^{1/\alpha}a^{-1/\alpha}(t)\left[\alpha(\alpha+1)^{1/\alpha}\phi^{1/\alpha}(t)\phi_1(t) + \phi^{(\alpha+1)/\alpha}(t)\right],$$

which implies

$$q(t) + \alpha a(t)|w(t)|^{(\alpha+1)/\alpha} - (a(t)w(t))'$$
$$\le q(t) + \alpha a^{-1/\alpha}(t)\left[\phi(t) + \alpha(\alpha+1)^{1/\alpha}\phi_1(t)\right]^{(\alpha+1)/\alpha} - q(t)$$
$$\quad - \alpha(\alpha+1)^{1/\alpha}a^{-1/\alpha}(t)\left[\alpha(\alpha+1)^{1/\alpha}\phi^{1/\alpha}(t)\phi_1(t) + \phi^{(\alpha+1)/\alpha}(t)\right]$$
$$= \alpha a^{-1/\alpha}(t)\left[\phi(t) + \alpha(\alpha+1)^{1/\alpha}\phi_1(t)\right]$$
$$\quad \times \left\{\left[\phi(t) + \alpha(\alpha+1)^{1/\alpha}\phi_1(t)\right]^{1/\alpha} - (\alpha+1)^{1/\alpha}\phi^{1/\alpha}(t)\right\}$$
$$\le \alpha a^{-1/\alpha}(t)\left[\phi(t) + \alpha(\alpha+1)^{1/\alpha}\phi_1(t)\right]$$
$$\quad \times \left\{\left[\phi(t) + \alpha(\alpha+1)^{1/\alpha}(\alpha+1)^{-1/\alpha}\phi(t)\right]^{1/\alpha} - (\alpha+1)^{1/\alpha}\phi^{1/\alpha}(t)\right\}$$
$$= \alpha a^{-1/\alpha}(t)\left[\phi(t) + \alpha(\alpha+1)^{1/\alpha}\phi_1(t)\right]\left[(\alpha+1)^{1/\alpha}\phi^{1/\alpha}(t)\right.$$
$$\quad \left. -(\alpha+1)^{1/\alpha}\phi^{1/\alpha}(t)\right] = 0.$$

Now, it follows from Theorem 3.7.7 that (3.1.1) is nonoscillatory. ∎

Theorem 3.7.17. Let $\phi(t) \in C^1([t_0, \infty), \mathbb{R}^+)$ be such that $\phi'(t) = a^{-1/\alpha}(t)$. If there exists a $T \ge t_0$ such that

$$\left|\int_t^\infty \phi^\alpha(s)q(s)ds\right| < \infty \quad \text{on} \quad [T, \infty), \tag{3.7.43}$$

then equation (3.1.1) is nonoscillatory.

Proof. Let

$$y(t) = \frac{1}{2}\phi^{-\alpha}(t)\left[1 + 2\int_t^\infty \phi^\alpha(s)q(s)ds\right].$$

Then, we have

$$y'(t) = -\frac{\alpha}{2}a^{-1/\alpha}(t)\phi^{-(\alpha+1)}(t)\left[1 + 2\int_t^\infty \phi^\alpha(s)q(s)ds\right] - q(t). \quad (3.7.44)$$

In view of (3.7.43) there exists a $T_1 \geq T$ such that

$$0 < 1 + 2\int_t^\infty \phi^\alpha(s)q(s)ds \leq 2 \quad \text{for all} \quad t \geq T_1.$$

Thus, we find

$$\begin{aligned}
|y(t)|^{(\alpha+1)/\alpha} &= 2^{-(\alpha+1)/\alpha}\phi^{-(\alpha+1)}(t)\left[1 + 2\int_t^\infty \phi^\alpha(s)q(s)ds\right] \\
&\leq \frac{1}{2}\phi^{-(\alpha+1)}(t)\left[1 + 2\int_t^\infty \phi^\alpha(s)q(s)ds\right].
\end{aligned}$$

Consequently, (3.7.44) implies that (3.7.1) holds on $[T_1, \infty)$. Now, Theorem 3.7.1 guarantees that equation (3.1.1) is nonoscillatory. ∎

3.7.4. Further Results on Nonoscillation of Equation (3.1.1)

Here, we shall consider the following half–linear differential equation

$$\left(a(t)|x'(t)|^{\alpha-1}x'(t)\right)' + \lambda a^{-1/\alpha}(t)R^{-(\alpha+1)}(t)|x(t)|^{\alpha-1}x(t) = 0, \quad (3.7.45)$$

where $R(t) \in C^1([t_0,\infty),\mathbb{R}^+)$ satisfies $R'(t) = a^{-1/\alpha}(t)$ and $\lim_{t\to\infty} R(t) = \infty$.

Clearly, (3.7.45) has a solution $x(t) = R^m(t)$ on $[t_0,\infty)$ if m satisfies the indicial equation $f(m) = \alpha m(m-1)|m|^{\alpha-1} + \lambda = 0$. If $\lambda = (\alpha/(\alpha+1))^{\alpha+1}$, then (3.7.45) is nonoscillatory because it has a nonoscillatory solution $y(t) = R^{\alpha/(\alpha+1)}(t)$. Equation (3.7.45) is oscillatory if $\lambda > (\alpha/(\alpha+1))^{\alpha+1}$ because then $f(m) = 0$ has no real roots. Further, (3.7.45) is nonoscillatory if $\lambda \leq (\alpha/(\alpha+1))^{\alpha+1}$ because then $f(m) = 0$ has a real root.

As an application of Theorem 3.7.5, we have the following result.

Theorem 3.7.18. Let $\int_t^\infty q(s)ds < \infty$ and $R(t) \in C^1([t_0,\infty),\mathbb{R}^+)$ satisfy $R'(t) = a^{-1/\alpha}(t)$ and $\lim_{t\to\infty} R(t) = \infty$. Then,

(i)   equation (3.1.1) is nonoscillatory if

$$R^\alpha(t)\int_t^\infty q(s)ds \leq \frac{1}{\alpha}\left(\frac{\alpha}{\alpha+1}\right)^{\alpha+1} \quad \text{for all sufficiently large} \quad t,$$

(ii)   equation (3.1.1) is oscillatory if there exists a number λ^* such that

$$R^\alpha(t) \int_t^\infty q(s)ds \geq \lambda^* > \frac{1}{\alpha}\left(\frac{\alpha}{\alpha+1}\right)^{\alpha+1} \qquad \text{for all sufficiently large } t.$$

Proof.   (i)   Let $q_1(t) = (\alpha/(\alpha+1))^{\alpha+1} a^{-1/\alpha}(t) R^{-(\alpha+1)}(t)$. Then, we have

$$\int_t^\infty q_1(s)ds = \frac{1}{\alpha}\left(\frac{\alpha}{\alpha+1}\right)^{\alpha+1} R^{-\alpha}(t).$$

Now, it follows from Theorem 3.7.5 that (3.1.1) is nonoscillatory.

(ii)   By assumption there exists a $T \geq t_0$ such that

$$R^\alpha(t) \int_t^\infty q(s)ds \geq \lambda^* = \alpha\lambda^* R^\alpha(t) \int_t^\infty a^{-1/\alpha}(s) R^{-(\alpha+1)}(s)ds \text{ for } t \geq T.$$

Since equation (3.7.45) with $\lambda = \alpha\lambda^*$ is oscillatory, it follows from Theorem 3.7.5 that (3.1.1) is oscillatory.   ∎

Next, we present another useful version of Theorem 3.7.2.

Theorem 3.7.19.   Equation (3.1.1) is nonoscillatory if and only if there exists a function $c(t) \in C([T,\infty),\mathbb{R})$ for some $T \geq t_0$ such that

$$|c(t)+Q(t)| \geq \left|Q(t) + \alpha \int_t^\infty a^{-1/\alpha}(s)|c(s) + Q(s)|^{(\alpha+1)/\alpha}ds\right| \qquad \text{for } t \geq T. \tag{3.7.46}$$

Corollary 3.7.1.   If $Q(t) \geq 0$ and there exists a function $c(t) \in C([T,\infty),\mathbb{R}^+)$ for some $T \geq t_0$ such that

$$c(t) \geq \alpha \int_t^\infty a^{-1/\alpha}(s)|c(s) + Q(s)|^{(\alpha+1)/\alpha}ds \quad \text{for } t \geq T, \tag{3.7.47}$$

then equation (3.1.1) is nonoscillatory.

Let $b(t) \in C([t_0,\infty),\mathbb{R})$ and $c(t) = b^+(t) - Q(t)$, where $b^+(t) = \max\{b(t),0\}$. Then by Theorem 3.7.19, we have the following corollary.

Corollary 3.7.2.   If there exists a function $b(t) \in C([T,\infty),\mathbb{R})$ for some $T \geq t_0$ such that

$$b^+(t) \geq \left|Q(t) + \alpha \int_t^\infty a^{-1/\alpha}(s)(b^+(s))^{(\alpha+1)/\alpha}ds\right| \quad \text{for } t \geq T, \tag{3.7.48}$$

then equation (3.1.1) is nonoscillatory.

Next, we let $b(t) = (\alpha+1)|Q(t)|$ for $t \geq T \geq t_0$ in Corollary 3.7.2, to obtain the following result.

Corollary 3.7.3. If

$$\int_t^\infty a^{-1/\alpha}(s)|Q(s)|^{(\alpha+1)/\alpha}ds \; < \; \left(\frac{1}{\alpha+1}\right)^{(\alpha+1)/\alpha} |Q(t)|,$$

then equation (3.1.1) is nonoscillatory.

To obtain the next result, we let $c(t) = k^{\alpha/(\alpha+1)}R^\alpha(t) - Q(t)$ where $k > 0$ is a constant. Then, we have $|c(t) + Q(t)| = k^{\alpha/(\alpha+1)}R^{-\alpha}(t)$ and

$$\left|Q(t) + \alpha\int_t^\infty a^{-1/\alpha}(s)|c(s) + Q(s)|^{(\alpha+1)/\alpha}ds\right| \; = \; |Q(t) + kR^{-\alpha}(t)|.$$

Corollary 3.7.4. If there exists $k > 0$ such that

$$-k^{\alpha/(\alpha+1)} - k \; \leq \; R^\alpha(t)Q(t) \; \leq \; k^{\alpha/(\alpha+1)} - k \; \leq \; \frac{1}{\alpha}\left(\frac{\alpha}{\alpha+1}\right)^{\alpha+1},$$

then equation (3.1.1) is nonoscillatory.

Now, taking $y(t) = Q_1(t) + b(t)$ in (3.7.13) for some $b(t) \in C([T,\infty),\mathbb{R})$ where $T \geq t_0$, we have the following nonoscillatory characterization result.

Theorem 3.7.20. Let $\mu[s,t]$, $Q(t)$ and $Q_1(t)$ be as in Theorem 3.7.3. Then equation (3.1.1) is nonoscillatory if and only if there exists a function $b(t) \in C([T,\infty),\mathbb{R}_0)$ for some $T \geq t_0$ such that

$$b(t) \; \geq \; \int_t^\infty \Big\{|Q(s)+\alpha[Q_1(s)+b(s)]|^{(\alpha+1)/\alpha} - (\alpha+1)|Q(s)|^{1/\alpha}[Q_1(s)+b(s)]$$
$$\times \text{ sgn } Q(s) - |Q(s)|^{(\alpha+1)/\alpha}\Big\} a^{-1/\alpha}(s)\mu[s,t]ds.$$

$$(3.7.49)$$

Let $b(t) = (\lambda - 1)Q_1(t)$, $\lambda > 1$ in (3.7.49), to obtain the following corollary.

Corollary 3.7.5. Let $\mu[s,t]$, $Q(t)$ and $Q_1(t)$ be as in Theorem 3.7.3. If there exists a constant $\lambda > 1$ such that

$$(\lambda - 1)Q_1(t) \; \geq \; \int_t^\infty \Big\{|Q(s) + \alpha\lambda Q_1(s)|^{(\alpha+1)/\alpha} - (\alpha+1)\lambda|Q(s)|^{1/\alpha}$$
$$\times Q_1(s) \text{ sgn } Q(s) - |Q(s)|^{(\alpha+1)/\alpha}\Big\} a^{-1/\alpha}(s)\mu[s,t]ds,$$

then equation (3.1.1) is nonoscillatory.

If we let $\lambda = (\alpha+1)/\alpha$ in Corollary 3.7.5, then an easy computation gives the following result.

Corollary 3.7.6. If $Q(t) \geq 0$ for $t \geq t_0$ and

$$Q_1(t) \leq \left(\frac{1}{\alpha+1}\right)\left[\left(\frac{\alpha+1}{\alpha}\right)^\alpha - 1\right]Q(t),$$

then equation (3.1.1) is nonoscillatory.

Corollary 3.7.7. If $Q(t) \geq 0$ for $t \geq t_0$ and

$$\int_t^\infty Q^{1/\alpha}(s)a^{-1/\alpha}(s)ds \leq \left(\frac{1}{\alpha+1}\right)\left[1 - \left(\frac{\alpha}{\alpha+1}\right)^\alpha\right]Q(t), \qquad (3.7.50)$$

then equation (3.1.1) is nonoscillatory.

Proof. Let

$$\gamma = \frac{1}{\alpha+1}\left[1 - \left(\frac{\alpha}{\alpha+1}\right)^\alpha\right].$$

Multiplying both sides of (3.7.50) by

$$\frac{Q^{1/\alpha}(t)a^{-1/\alpha}(t)}{\int_t^\infty Q^{(\alpha+1)/\alpha}(s)a^{-1/\alpha}(s)ds} \quad (\geq 0 \text{ by assumption}),$$

and integrating it over $[s, t]$, we obtain

$$\mu[s,t] = \exp\left((\alpha+1)\int_t^s |Q(\xi)|^{1/\alpha}a^{-1/\alpha}(\xi) \operatorname{sgn} Q(\xi)d\xi\right)$$

$$\leq \left(\frac{\int_t^\infty Q^{(\alpha+1)/\alpha}(\xi)a^{-1/\alpha}(\xi)d\xi}{\int_s^\infty Q^{(\alpha+1)/\alpha}(\xi)a^{-1/\alpha}(\xi)d\xi}\right)^{\gamma(\alpha+1)} = \mu_1[s,t] \text{ for } s \geq t \geq T.$$

Thus, we have

$$Q_1(t) \leq \int_t^\infty Q^{(\alpha+1)/\alpha}(s)a^{-1/\alpha}(s)\mu_1[s,t]ds$$

$$= \frac{1}{1-\gamma(\alpha+1)}\int_t^\infty Q^{(\alpha+1)/\alpha}(s)ds \leq \frac{\gamma}{1-\gamma(\alpha+1)}Q(t)$$

$$= \frac{1}{\alpha+1}\left(\left(\frac{\alpha+1}{\alpha}\right)^\alpha - 1\right)Q(t).$$

Now, it follows from Corollary 3.7.6 that (3.1.1) is nonoscillatory. ∎

The following result improves Theorem 3.7.20.

Theorem 3.7.21. Let $\phi(t) \in C^1([t_0, \infty), \mathbb{R})$ be such that $\phi'(t) \leq -q(t)$ for $t \geq t_0$, and let

$$\beta(t) = \int_t^\infty a^{-1/\alpha}(s)|\phi(s)|^{\frac{\alpha+1}{\alpha}} \exp\left(\alpha\lambda\int_t^s a^{-1/\alpha}(\xi)|\phi(\xi)|^{1/\alpha}\operatorname{sgn} \phi(\xi)d\xi\right)ds,$$

where λ is a constant. If one of the following conditions holds

(i) $\lambda > 1$ and $\phi(t) \geq \dfrac{\alpha\lambda}{\lambda^\alpha - 1}\beta(t)$,

(ii) $\lambda \geq 1$ and $\phi(t) \leq -\alpha\lambda\beta(t)$,

(iii) $\lambda < 1$ and $\dfrac{\alpha\lambda}{-1 + |\lambda|^\alpha \, \text{sgn} \, \lambda}\beta(t) \leq \phi(t) \leq -\alpha\lambda\beta(t)$,

then equation (3.1.1) is nonoscillatory.

Proof. Since one of the conditions (i), (ii) and (iii) holds,

$$|\phi(t) + \alpha\lambda\beta(t)|^{(\alpha+1)/\alpha} - \lambda\left\{|\phi(t)|^{(\alpha+1)/\alpha} + \alpha\lambda|\phi(t)|^{1/\alpha}\beta(t) \, \text{sgn} \, \phi(t)\right\} \leq 0.$$

Let $w(t) = -[\phi(t) + \alpha\lambda\beta(t)]/a(t)$. Then, we have

$$
\begin{aligned}
(a(t)w(t))' &= -\phi'(t) - \alpha\lambda\beta'(t)\\
&\geq q(t) + \alpha\lambda a^{-1/\alpha}(t)\left\{|\phi(t)|^{(\alpha+1)/\alpha} + \alpha\lambda|\phi(t)|^{1/\alpha}\beta(t) \, \text{sgn} \, \phi(t)\right\},
\end{aligned}
$$

which implies

$$
\begin{aligned}
q(t) + \alpha a(t)|w(t)|^{\frac{\alpha+1}{\alpha}} - (a(t)w(t))' &= \alpha a^{-1/\alpha}(t)\left\{|\phi(t) + \alpha\lambda\beta(t)|^{\frac{\alpha+1}{\alpha}}\right.\\
&\left. -\lambda\left[|\phi(t)|^{\frac{\alpha+1}{\alpha}} + \alpha\lambda|\phi(t)|^{1/\alpha}\beta(t) \, \text{sgn} \, \phi(t)\right]\right\} \leq 0.
\end{aligned}
$$

Now, it follows from Theorem 3.7.7 that (3.1.1) is nonoscillatory. ∎

Similar to Corollary 3.7.7, we have the following result.

Corollary 3.7.8. Let $\phi \in C^1([t_0, \infty), \mathbb{R})$ be such that $\phi'(t) \leq -q(t)$, $t \geq t_0$. If there exists a constant $\lambda > 1$ such that

$$\int_t^\infty a^{-1/\alpha}(s)|\phi(s)|^{(\alpha+1)/\alpha}ds \leq \frac{\lambda^\alpha - 1}{\alpha\lambda^{\alpha+1}}\phi(t), \quad t \geq t_0$$

then equation (3.1.1) is nonoscillatory.

Remark 3.7.1. Let $Q(t) = \phi(t) \geq 0$ for $t \geq t_0$. If $\lambda = (\alpha+1)/\alpha$, then Theorem 3.7.21 and Corollary 3.7.8 reduce to Corollaries 3.7.6 and 3.7.7 respectively.

Now, we shall consider the following half–linear differential equation

$$(a(t)|x'(t)|^{\alpha-1}x'(t))' + \lambda a^{-1/\alpha}(t)\pi^{-(\alpha+1)}(t)|x(t)|^{\alpha-1}x(t) = 0, \quad t \geq t_0 \tag{3.7.51}$$

where λ is a constant, $\pi(t) = \int_t^\infty a^{-1/\alpha}(s)ds$ and $\pi(t_0) < \infty$.

Clearly, (3.7.51) has a solution $x(t) = \pi^{-m}(t)$ for $t \geq t_0$ if m satisfies the indicial equation $g(m) = \alpha m(m - 1)|m|^{\alpha-1} + \lambda$. If $\lambda = (\alpha/(\alpha + 1))^{\alpha+1}$, then (3.7.51) is nonoscillatory because it has a nonoscillatory solution $x(t) = \pi^{\alpha/(\alpha+1)}(t)$. Equation (3.7.51) is oscillatory if $\lambda > (\alpha/(\alpha + 1))^{\alpha+1}$ because then $g(m) = 0$ has no real roots. Further, (3.7.51) is nonoscillatory if $\lambda \leq (\alpha/(\alpha + 1))^{\alpha+1}$ because then $g(m) = 0$ has a real root.

By Theorem 3.7.6 and a technique similar to the one used in Theorem 3.7.16, we have the following result.

Theorem 3.7.22. Let $q(t) \in C([t_0, \infty), \mathbb{R}^+)$ and $\pi(t_0) < \infty$. Then,

(i) equation (3.1.1) is nonoscillatory if

$$\frac{1}{\pi(t)} \int_t^\infty \pi^{\alpha+1}(s)q(s)ds \leq \left(\frac{\alpha}{\alpha + 1}\right)^{\alpha+1} \quad \text{for all sufficiently large } t,$$

(ii) equation (3.1.1) is oscillatory if there exists a number λ^* such that

$$\frac{1}{\pi(t)} \int_t^\infty \pi^{\alpha+1}(s)q(s)ds \geq \lambda^* > \left(\frac{\alpha}{\alpha + 1}\right)^{\alpha+1} \quad \text{for all sufficiently large } t.$$

The following corollary is an immediate consequence of Theorem 3.7.22.

Corollary 3.7.9. Consider the following half–linear differential equation

$$\left(t^\gamma |x'(t)|^{\alpha-1}x'(t)\right)' + \lambda t^{\gamma-\alpha-1}|x(t)|^{\alpha-1}x(t) = 0, \quad t \geq t_0 \quad (3.7.52)$$

where $\gamma > \alpha$ and $\lambda > 0$ are constants. Then,

(i) equation (3.7.52) is nonoscillatory if $\lambda \leq ((\gamma - \alpha)/(\alpha + 1))^{\alpha+1}$,

(ii) equation (3.7.52) is oscillatory if $\lambda > ((\gamma - \alpha)/(\alpha + 1))^{\alpha+1}$.

Proof. From equation (3.7.52) it follows that

$$\pi(t) = \int_t^\infty s^{-\gamma/\alpha}ds = \frac{\alpha}{\gamma - \alpha}t^{(\alpha-\gamma)/\alpha}$$

and

$$\frac{1}{\pi(t)} \int_t^\infty \pi^{\alpha+1}(s)\lambda s^{\gamma-\alpha-1}ds = \lambda\left(\frac{\alpha}{\gamma - \alpha}\right)^{\alpha+1}.$$

Now, the result follows from Theorem 3.7.22. ∎

In an analogous way we can extend Corollary 3.7.9, to obtain

Corollary 3.7.10. Consider the following half–linear differential equation

$$\left(t^\gamma |x'(t)|^{\alpha-1}x'(t)\right)' + q(t)|x(t)|^{\alpha-1}x(t) = 0, \quad t \geq t_0 \quad (3.7.53)$$

where $\gamma > \alpha$ is a constant. Then,

(i)   equation (3.7.53) is nonoscillatory if

$$t^{(\gamma-\alpha)/\alpha} \int_t^\infty s^{-(\gamma-\alpha)(\alpha+1)/\alpha} q(s)ds \ \leq\ \left(\frac{\alpha}{\alpha+1}\right)\left(\frac{\gamma-\alpha}{\alpha+1}\right)^\alpha$$

for all sufficiently large t,

(ii)   equation (3.7.53) is oscillatory if there exists a number λ^* such that

$$t^{(\gamma-\alpha)/\alpha} \int_t^\infty s^{-(\gamma-\alpha)(\alpha+1)/\alpha} q(s)ds \ \geq\ \lambda^* \ > \ \left(\frac{\alpha}{\alpha+1}\right)\left(\frac{\gamma-\alpha}{\alpha+1}\right)^\alpha$$

for all sufficiently large t.

Next, by using an approach different than those given above, we shall present nonoscillation criteria for the more general equations

$$\left(|x'(t)|^{\alpha-1}x'(t)\right)' + F(t, x(t)) \ = \ 0 \qquad\qquad (3.7.54)$$

and

$$\left(|x'(t)|^{\alpha-1}x'(t)\right)' + q(t)f(x(t)) \ = \ 0, \qquad\qquad (3.7.55)$$

where

(i) $\alpha > 0$ is a constant,
(ii) $F \in C([t_0, \infty) \times \mathbb{R}, \mathbb{R})$, sgn $F(t, x) =$ sgn x for each $t \in [t_0, \infty)$,
(iii) $f \in C(\mathbb{R}, \mathbb{R})$, sgn $f(x) =$ sgn x, and
(iv) $q(t) \in C([t_0, \infty), \mathbb{R}^+)$.

Theorem 3.7.23. Suppose $F(t, x)$ is nondecreasing in x for each fixed t and has the partial derivative with respect to t such that

$$\text{sgn } \frac{\partial F}{\partial t}(t, x) \ = \ - \text{ sgn } x \quad \text{for} \quad t \geq t_0. \qquad\qquad (3.7.56)$$

Furthermore, suppose that

$$\int^\infty |F(s, cs)|ds \ < \ \infty \quad \text{for all nonzero constant } c \qquad\qquad (3.7.57)$$

and

$$\int^\infty \limsup_{\lambda \to 0} \frac{F(s, \lambda s)}{\lambda^\alpha} ds \ < \ \infty, \qquad\qquad (3.7.58)$$

then equation (3.7.54) is nonoscillatory.

Proof. Suppose equation (3.7.54) has an oscillatory solution $x(t)$. Let $\{t_k\}_{k=1}^\infty$ be an increasing sequence of zeros of $x(t)$ such that $x'(t_k) > 0$

for $k = 1, 2, \cdots$. Let τ_k be the first zero of $x'(t)$ which is located to the right of t_k. Since $x'(t) > 0$ on (t_k, τ_k) and $x'(t)$ is decreasing there, we have

$$x(t) = x(t_k) + \int_{t_k}^t x'(s)ds \leq x'(t_k)(t - t_k) \leq ty'(t_k) \quad \text{for} \quad t \in (t_k, \tau_k).$$
$$(3.7.59)$$

Define

$$V(t) = \frac{\alpha}{\alpha+1}|x'(t)|^{\alpha+1} + \int_0^{x(t)} F(t, u)du \quad \text{for} \quad t \geq t_1. \qquad (3.7.60)$$

Clearly, $V(t) \geq 0$ for $t \geq t_1$, and

$$V'(t) = \left(|x'(t)|^{\alpha-1}x'(t)\right)' x'(t) + F(t, x(t))x'(t) + \int_0^{x(t)} \frac{\partial F}{\partial t}(t, u)du$$

$$= \int_0^{x(t)} \frac{\partial F}{\partial t}(t, u)du \leq 0 \quad \text{for} \quad t \geq t_1.$$

This implies that $V(t)$ is bounded for $t \geq t_1$. Thus, in view of (3.7.60), $x'(t)$ remains bounded as $t \to \infty$. It follows that there exists a constant $M > 0$ such that

$$|x'(t)| \leq M \quad \text{for} \quad t \geq t_1. \qquad (3.7.61)$$

Now, we integrate (3.7.54) from t_k to τ_k, and use (3.7.59) and (3.7.61), to obtain

$$0 < (x'(t_k))^\alpha = \int_{t_k}^{\tau_k} F(s, x(s))ds \leq \int_{t_k}^{\tau_k} F(s, x'(t_k))ds$$

$$\leq \int_{t_k}^{\tau_k} F(s, Ms)ds \leq \int_{t_k}^\infty F(s, Ms)ds,$$

which implies that $x'(t_k) \to 0$ as $k \to \infty$, and

$$1 \leq \int_{t_k}^{\tau_k} \frac{F(s, sx'(t_k))}{(x'(t_k))^\alpha}ds \quad \text{for} \quad k = 1, 2, \cdots. \qquad (3.7.62)$$

Let $T > t_1$ be fixed arbitrarily and choose t_k so that $t_k \geq T$. Then, it follows from (3.7.62) that

$$1 \leq \int_T^{\tau_k} \frac{F(s, sx'(t_k))}{(x'(t_k))^\alpha}ds \leq \int_T^\infty \frac{F(s, sx'(t_k))}{(x'(t_k))^\alpha}ds$$

and hence by Fatou's lemma as $T \to \infty$,

$$1 \leq \int_T^\infty \limsup_{k\to\infty} \frac{F(s, sx'(t_k))}{(x'(t_k))^\alpha}ds \leq \int_T^\infty \limsup_{\lambda\to 0} \frac{F(s, \lambda s)}{\lambda^\alpha}ds \to 0,$$

which is a contradiction. Hence, equation (3.7.54) cannot have an oscillatory solution. ∎

Theorem 3.7.24. Suppose $f(x)$ is nondecreasing, $q(t) \in C^1([t_0, \infty),$ $\mathbb{R}^+)$ and satisfies

$$\int_0^\infty \frac{[q'(s)]^+}{q(s)} ds < \infty, \quad [q'(t)]^+ = \max\{q'(t), 0\}. \tag{3.7.63}$$

Furthermore, suppose that

$$\int^\infty q(s)|f(cs)|ds < \infty \quad \text{for all nonzero constant } c \tag{3.7.64}$$

and

$$\int^\infty q(s) \limsup_{\lambda \to 0} \frac{|f(\lambda s)|}{|\lambda|^\alpha} ds < \infty. \tag{3.7.65}$$

Then equation (3.7.55) is nonoscillatory.

Proof. Let $x(t)$ be an oscillatory solution of (3.7.55) on $[t_0, \infty)$. We will show that $x'(t)$ remains bounded as $t \to \infty$. Define the function $V(t)$ as follows

$$V(t) = \frac{\alpha}{\alpha+1}|x'(t)|^{\alpha+1} + q(t) \int_0^{x(t)} f(u)du, \quad t \geq t_0.$$

Differentiation of $V(t)$ gives

$$V'(t) = q'(t) \int_0^{x(t)} f(u)du \leq ([q'(t)]^+) \int_0^{x(t)} f(u)du,$$

which implies that $V'(t)$ satisfies the first order differential inequality

$$V'(t) \leq \frac{[q'(t)]^+}{q(t)} V(t) \quad \text{for } t \geq t_0.$$

Hence, we have

$$V(t) \leq V(t_0) \exp\left(\int_{t_0}^t \frac{[q'(s)]^+}{q(s)} ds\right), \quad t \geq t_0.$$

Thus, $V(t)$ is bounded. The definition of $V(t)$ now shows that $x'(t)$ remains bounded as claimed. Proceeding as in Theorem 3.7.23, we get the required contradiction. ∎

Example 3.7.2. Consider the differential equation

$$\left(|x'(t)|^{\alpha-1}x'(t)\right)' + \frac{|x(t)|^{\beta-1}x(t)}{t^\gamma(t^2 + x^2(t))^\delta} = 0, \quad t \geq 1 \tag{3.7.66}$$

where α, β, γ and δ are positive constants. The function

$$F(t,x) = \frac{|x|^{\beta-1}x}{t^\gamma(t^2+x^2)^\delta}$$

is nondecreasing in x if $\beta \geq 2\delta$ and satisfies (3.7.56). The condition (3.7.57) is satisfied if $\beta - 2\delta + 1 < \gamma$, whereas (3.7.58) is satisfied if $\alpha \leq \beta$ and $\beta - 2\delta + 1 < \delta$. Consequently, Theorem 3.7.23 implies that all solutions of (3.7.66) are nonoscillatory if $\beta \geq \max\{\alpha, 2\delta\}$ and $\gamma > \beta - 2\delta + 1$.

Similarly, it can be shown that all solutions of the equation

$$\left(|x'(t)|^{\alpha-1}x'(t)\right)' + e^{-t^\beta} \sinh\left(|x(t)|^{\alpha-1}x(t)\right) = 0, \quad \beta > \alpha > 0 \quad (3.7.67)$$

are oscillatory.

3.8. Oscillation Criteria

Here, we shall consider (3.1.1) with $a(t) = 1$, i.e., the equation

$$\left(|x'(t)|^{\alpha-1}x'(t)\right)' + q(t)|x(t)|^{\alpha-1}x(t) = 0, \tag{3.8.1}$$

where $\alpha > 0$ and $q(t) \in C([t_0, \infty), \mathbb{R})$. When $\alpha = 1$, (3.8.1) reduces to the linear equation

$$x''(t) + q(t)x(t) = 0. \tag{3.8.2}$$

When $\alpha \neq 1$, although equation (3.8.1) is nonlinear it has the property that any constant multiple of a solution is also a solution. The term 'half–linear' is used to denote this property.

In the oscillatory theory of (3.8.1) there are several results which are similar to those known for the equation (3.8.2). We state some of these in the following theorems.

Theorem 3.8.1. Equation (3.8.1) is oscillatory if

$$\liminf_{t\to\infty} t^{\alpha+1}q(t) > \left(\frac{\alpha}{\alpha+1}\right)^{\alpha+1}$$

and is nonoscillatory if

$$\limsup_{t\to\infty} t^{\alpha+1}q(t) < \left(\frac{\alpha}{\alpha+1}\right)^{\alpha+1}.$$

Another form of Theorem 3.8.1 is the following result.

Theorem 3.8.1'. Equation (3.8.1) is oscillatory if there exists a constant $\epsilon > 0$ such that

$$t^{\alpha+1}q(t) \geq \left(\frac{\alpha}{\alpha+1}\right)^{\alpha+1} + \epsilon \quad \text{for all sufficiently large } t$$

and is nonoscillatory if

$$t^{\alpha+1}q(t) \leq \left(\frac{\alpha}{\alpha+1}\right)^{\alpha+1} \quad \text{for all sufficiently large } t.$$

Theorem 3.8.2. Suppose $q(t) > 0$ for $t \geq t_0 \geq 0$. Then equation (3.8.1) is oscillatory if either one of the following conditions holds

(i) $q(t) \notin \mathcal{L}[t_0, \infty)$,   or

(ii) $q(t) \in \mathcal{L}[t_0, \infty)$ and

$$\liminf_{t\to\infty} t^{\alpha} \int_t^{\infty} q(s)ds > \frac{\alpha^{\alpha}}{(\alpha+1)^{\alpha+1}}$$

and equation (3.8.1) is nonoscillatory if $q(t) \in \mathcal{L}[t_0, \infty)$ and

$$\limsup_{t\to\infty} t^{\alpha} \int_t^{\infty} q(s)ds < \frac{\alpha^{\alpha}}{(\alpha+1)^{\alpha+1}}.$$

Theorem 3.8.3 (Generalized Sine Function). Let $S(t)$ denote the solution of (3.8.1) with $q(t) = \alpha$ determined by the initial conditions $x(0) = 0$, $x'(0) = 1$. Then, $S(t)$ exists on $\mathbb{R}$ and for $t \in \mathbb{R}$ has the properties

$$S(t + \pi_{\alpha}) = S(t) \quad \text{and} \quad |S(t)|^{\alpha+1} + |S'(t)|^{\alpha+1} = 1,$$

where $\pi_{\alpha} = \dfrac{2\pi}{(\alpha+1)\,\sin(\pi/(\alpha+1))}.$

Theorem 3.8.4 (Generalized Prüfer Transform). Let $x(t)$ be a solution of (3.8.1) on $[t_0, \infty)$. Define the functions $\rho(t)$ and $\phi(t)$ by

$$x(t) = \rho(t)S(\phi(t)), \qquad x'(t) = \rho(t)S'(\phi(t)). \tag{3.8.3}$$

Then, $\rho(t) > 0$ and $\phi(t) \in C^1([t_0, \infty), \mathbb{R})$ and satisfy the following differential equations

$$\frac{\rho'(t)}{\rho(t)} = \left(1 - \frac{1}{\alpha}q(t)\right) S'(\phi(t))|S(\phi(t))|^{\alpha-1}S(\phi(t))$$

and

$$\phi'(t) = |S'(\phi(t))|^{\alpha+1} + \frac{1}{\alpha}q(t)|S(\phi(t))|^{\alpha+1}.$$

Now, we shall state a result for the equation

$$\left(|x'(t)|^{\alpha-1}x'(t)\right)' + F(t, x(t)) = 0, \tag{3.8.4}$$

where $\alpha > 0$ is a constant, $F \in C([t_0, \infty) \times \mathbb{R}, \mathbb{R})$ and sgn $F(t, x) =$ sgn x for $t \in [t_0, \infty)$, $t_0 \geq 0$.

Theorem 3.8.5. (i) Suppose there exists $q(t) \in C([t_0, \infty), \mathbb{R})$ such that

$$\inf_{x \neq 0} \frac{F(t, x)}{|x|^{\alpha-1}x} \geq q(t) \quad \text{for all sufficiently large } t. \tag{3.8.5}$$

If (3.8.1) is oscillatory with $q(t)$ as in (3.8.5), then equation (3.8.4) is oscillatory.

(ii) Suppose there exists $q(t) \in C([t_0, \infty), \mathbb{R})$ such that

$$\sup_{x \neq 0} \frac{F(t, x)}{|x|^{\alpha-1}x} \leq q(t) \quad \text{for all sufficiently large } t. \tag{3.8.6}$$

If (3.8.1) is nonoscillatory with $q(t)$ as in (3.8.6), then equation (3.8.4) is nonoscillatory.

Example 3.8.1. Consider the equation

$$\left(|x'(t)|^{\alpha-1}x'(t)\right)' + \frac{b + t^{\beta}|x(t)|^{\lambda}}{a + t^{\mu}|x(t)|^{\lambda}} t^{\alpha-1}|x(t)|^{\alpha-1}x(t) = 0 \tag{3.8.7}$$

for $t \geq 1$, where a, b, β, λ and μ are positive constants. Since

$$\inf_{x \neq 0} \frac{b + t^{\beta}|x|^{\lambda}}{a + t^{\mu}|x|^{\lambda}} \geq \frac{b}{a}, \quad \beta > \mu$$

and

$$\sup_{x \neq 0} \frac{b + t^{\beta}|x|^{\lambda}}{a + t^{\mu}|x|^{\lambda}} \leq \frac{b}{a}, \quad \beta < \mu$$

for all large t, it follows from Theorems 3.8.1' and 3.8.5 that

(i) equation (3.8.7) is oscillatory if $\beta > \mu$ and $b/a > (\alpha/(\alpha+1))^{\alpha+1}$,

(i) equation (3.8.7) is nonoscillatory if $\beta < \mu$ and $b/a \leq (\alpha/(\alpha+1))^{\alpha+1}$.

Finally, we shall prove two results for the more general equation, namely,

$$\left(a(t)|x'(t)|^{\alpha-1}x'(t)\right)' + F(t, x(t)) = 0, \tag{3.8.8}$$

where α and F are as in (3.8.4) and $a(t) \in C([t_0, \infty), \mathbb{R}^+)$.

Theorem 3.8.6. Suppose condition (3.8.5) holds and $\int^{\infty} a^{-1/\alpha}(s)ds = \infty$. If

$$\lim_{t\to\infty} \int_{t_0}^{t} q(s)ds = \infty, \tag{3.8.9}$$

then equation (3.8.8) is oscillatory.

Proof. Let $x(t)$ be a nonoscillatory solution of equation (3.8.8), say, $x(t) > 0$ for $t \geq t_0 \geq 0$. Define

$$w(t) = a(t)\frac{|x'(t)|^{\alpha-1}x'(t)}{x^{\alpha}(t)} \quad \text{for} \quad t \geq t_0. \tag{3.8.10}$$

Then, we have

$$w'(t) + \alpha a^{-1/\alpha}(t)|w(t)|^{(\alpha+1)/\alpha} + \frac{F(t, x(t))}{x^{\alpha}(t)} = 0, \quad t \geq t_0. \tag{3.8.11}$$

Thus, from (3.8.11) it follows that $w'(t) + q(t) \leq 0$ for $t \geq t_0$. Integrating this inequality from t_0 to t, we obtain $w(t) - w(t_0) + \int_{t_0}^{t} q(s)ds \leq 0, \ t \geq t_0$. Hence, $w(t) \to -\infty$ as $t \to \infty$, and therefore $x'(t) < 0$ for $t \geq t_1$ for some sufficiently large $t_1 \geq t_0$. Choose $t_2 \geq t_1$ so that $\int_{t_2}^{t} q(s)ds > 0$ for $t > t_2$. Integrating (3.8.8) from t_2 to t, we get

$$a(t)|x'(t)|^{\alpha-1}x'(t) - a(t_2)|x'(t_2)|^{\alpha-1}x'(t_2) + \int_{t_2}^{t} q(s)(x(s))^{\alpha}ds \leq 0. \tag{3.8.12}$$

Now, from the relation

$$\int_{t_2}^{t} q(s)(x^{\alpha}(s))ds = \left(\int_{t_2}^{t} q(s)ds\right)(x(t))^{\alpha} - \int_{t_2}^{t}\left(\int_{t_2}^{s} q(\tau)d\tau\right)\alpha x^{\alpha-1}(s)x'(s)ds,$$

we find that $\int_{t_2}^{t} q(s)(x^{\alpha}(s))ds \geq 0$ for $t \geq t_2$. Thus, from (3.8.12) we can conclude that $a(t)|x'(t)|^{\alpha-1}x'(t) \leq a(t_2)|x'(t_2)|^{\alpha-1}x'(t_2), \ t \geq t_2$ and hence

$$x'(t) \leq \left(\frac{a(t_2)}{a(t)}\right)^{1/\alpha} x'(t_2) \quad \text{for} \quad t \geq t_2. \tag{3.8.13}$$

An integration of (3.8.13) shows that $x(t) \to -\infty$ as $t \to \infty$, which contradicts the fact that $x(t) > 0$ eventually. This completes the proof. ∎

Theorem 3.8.7. Suppose condition (3.8.5) holds, $\alpha \geq 1$ and

$$\lim_{t\to\infty} \int_{t_0}^{t}\left(\int_{t_0}^{s} a(u)du\right)^{-1/\alpha} ds = \infty. \tag{3.8.14}$$

If

$$\lim_{t\to\infty} \frac{1}{t} \int_{t_0}^t \int_{t_0}^s q(u)\,du\,ds = \infty, \quad t_0 \geq 0 \tag{3.8.15}$$

then equation (3.8.8) is oscillatory.

Proof. Let $x(t)$ be a nonoscillatory solution of equation (3.8.8), say, $x(t) > 0$ for $t \geq t_0 \geq 0$. Define the function $w(t)$ as in Theorem 3.8.6 to obtain (3.8.11). Using condition (3.8.5) and integrating the resulting inequality twice from t_0 to $t \geq t_0$, we get

$$\frac{1}{t}\int_{t_0}^t w(s)\,ds + \frac{\alpha}{t}\int_{t_0}^t\int_{t_0}^s a^{-1/\alpha}(u)|w(u)|^{(\alpha+1)/\alpha}\,du\,ds + \frac{1}{t}\int_{t_0}^t\int_{t_0}^s q(u)\,du\,ds \leq c_0, \tag{3.8.16}$$

where $c_0 > 0$ is a constant independent of t.

Let $V(t) = \int_{t_0}^t |w(s)|\,ds$. Then by Höder's inequality, we have

$$V(t) \leq \left(\int_{t_0}^t a(s)\,ds\right)^{1/(\alpha+1)} \left(\int_{t_0}^t a^{-1/\alpha}(s)|w(s)|^{(\alpha+1)/\alpha}\,ds\right)^{\alpha/(\alpha+1)},$$

or

$$\left(\int_{t_0}^t a(s)\,ds\right)^{-1/\alpha} V^{(\alpha+1)/\alpha}(t) \leq \int_{t_0}^t a^{-1/\alpha}(s)|w(s)|^{(\alpha+1)/\alpha}\,ds \quad \text{for } t \geq t_0. \tag{3.8.17}$$

From (3.8.15) and (3.8.16) it follows that there exists a $t_1 \geq t_0$ such that

$$-\frac{1}{t}\int_{t_0}^t |w(s)|\,ds + \frac{\alpha}{t}\int_{t_0}^t\int_{t_0}^s a^{-1/\alpha}(u)|w(u)|^{(\alpha+1)/\alpha}\,du\,ds \leq 0 \quad \text{for } t \geq t_1. \tag{3.8.18}$$

Using (3.8.17) in (3.8.18), we get

$$-\frac{V(t)}{t} + \frac{\alpha}{t}\int_{t_0}^t \frac{V^{(\alpha+1)/\alpha}(s)}{\left(\int_{t_0}^s a(u)\,du\right)^{1/\alpha}}\,ds \leq 0,$$

or for $t \geq t_1$,

$$\frac{\alpha^{(\alpha+1)/\alpha}}{\left(\int_{t_0}^t a(s)\,ds\right)^{1/\alpha}}\left[\int_{t_0}^t \frac{V^{(\alpha+1)/\alpha}(s)}{\left(\int_{t_0}^s a(u)\,du\right)^{1/\alpha}}\,ds\right]^{(\alpha+1)/\alpha} \leq \frac{V^{(\alpha+1)/\alpha}(t)}{\left(\int_{t_0}^t a(u)\,du\right)^{1/\alpha}}.$$

Let

$$W(t) = \int_{t_0}^t \frac{V^{(\alpha+1)/\alpha}(s)}{\left(\int_{t_0}^s a(u)\,du\right)^{1/\alpha}}\,ds.$$

Then, we have

$$\frac{W'(t)}{W^{(\alpha+1)/\alpha}(t)} \geq \alpha^{(\alpha+1)/\alpha} \left(\int_{t_0}^t a(s)ds \right)^{-1/\alpha} \qquad \text{for} \quad t \geq t_2. \qquad (3.8.19)$$

Integrating (3.8.19) from t_1 to t, we obtain for $t \geq t_1$,

$$\alpha^{\alpha+1} \int_{t_1}^t \left(\int_{t_0}^s a(u)du \right)^{-1/\alpha} ds \leq \int_{t_1}^t \frac{W'(s)}{W^{(\alpha+1)/\alpha}(s)} ds$$

$$= \int_{W(t_1)}^{W(t)} \frac{dv}{v^{(\alpha+1)/\alpha}} = \alpha \left[W^{-1/\alpha}(t_1) - W^{-1/\alpha}(t) \right] \leq \alpha W^{-1/\alpha}(t_1).$$

Thus, it follows that

$$\lim_{t \to \infty} \int_{t_1}^t \left(\int_{t_0}^s a(u)du \right)^{-1/\alpha} ds \leq \alpha^{-\alpha} W^{-1/\alpha}(t_1) < \infty,$$

which contradicts condition (3.8.14). This completes the proof. ∎

3.9. Oscillation Criteria–Integral Averaging

3.9.1. Oscillation Theorems for Equation (3.1.1)

Here, we shall employ the integral averaging techniques to obtain oscillation criteria for equation (3.1.1). In what follows for convenience, we shall let by $\alpha_0 = (\alpha + 1)^{\alpha+1}$. Our results of this subsection extend Theorems 2.5.6 – 2.5.13.

Theorem 3.9.1. Suppose there exists a continuous function $H : \mathcal{D} = \{(t, s) : t \geq s \geq t_0\} \to \mathbb{R}$ such that

$$H(t,t) = 0 \quad \text{for} \quad t \geq t_0, \quad H(t,s) > 0 \quad \text{for} \quad (t,s) \in \mathcal{D}, \quad t > s \quad (3.9.1)$$

and

$$h(t,s) = -\partial H(t,s)/\partial s \text{ is nonnegative and continuous on } \mathcal{D}. \quad (3.9.2)$$

If

$$\limsup_{t \to \infty} \frac{1}{H(t,t_0)} \int_{t_0}^t \left[H(t,s)q(s) - \frac{a(s)h^{\alpha+1}(t,s)}{\alpha_0 H^\alpha(t,s)} \right] ds = \infty, \qquad (3.9.3)$$

then equation (3.1.1) is oscillatory.

Proof. Let $x(t)$ be a nonoscillatory solution of (3.1.1). Assume that $x(t) \neq 0$ for $t \geq t_0$. We define $w(t) = a(t)|x'(t)|^{\alpha-1}x'(t)/(|x(t)|^{\alpha-1}x(t))$ for $t \geq t_0$. Then for every $t \geq t_0$, we have

$$w'(t) = -q(t) - \alpha \frac{|w(t)|^{(\alpha+1)/\alpha}}{a^{1/\alpha}(s)}$$

and consequently,

$$\int_{t_0}^t H(t,s)w'(s)ds = -\int_{t_0}^t H(t,s)q(s)ds - \alpha \int_{t_0}^t H(t,s)\frac{|w(s)|^{(\alpha+1)/\alpha}}{a^{1/\alpha}(s)}ds. \tag{3.9.4}$$

Since

$$\int_{t_0}^t H(t,s)w'(s)ds = -w(t_0)H(t,t_0) - \int_{t_0}^t w(s)\frac{\partial}{\partial s}H(t,s)ds, \tag{3.9.5}$$

equation (3.9.4) gives

$$\int_{t_0}^t H(t,s)q(s)ds \leq H(t,t_0)w(t_0) + \int_{t_0}^t h(t,s)|w(s)|ds$$
$$-\alpha \int_{t_0}^t H(t,s)\frac{|w(s)|^{(\alpha+1)/\alpha}}{a^{1/\alpha}(s)}ds. \tag{3.9.6}$$

Letting

$$A = (\alpha H(t,s))^{\alpha/(\alpha+1)}\frac{|w(s)|}{a^{1/(\alpha+1)}(s)}, \qquad \lambda = \frac{\alpha+1}{\alpha}$$

and

$$B = \frac{\alpha^{\alpha/(\alpha+1)}}{(\alpha+1)^\alpha}\frac{a^{\alpha/(\alpha+1)}(s)h^\alpha(t,s)}{H^{\alpha^2/(\alpha+1)}(t,s)}$$

then from Lemma 3.2.1, we find for $t > s \geq t_0$,

$$|w(s)|h(t,s) - \alpha H(t,s)\frac{|w(s)|^{(\alpha+1)/\alpha}}{a^{1/\alpha}(s)} \leq \frac{a(s)h^{\alpha+1}(t,s)}{\alpha_0 H^\alpha(t,s)}.$$

Hence, (3.9.6) leads to

$$\frac{1}{H(t,t_0)}\int_{t_0}^t H(t,s)q(s)ds \leq w(t_0) + \frac{1}{H(t,t_0)}\int_{t_0}^t \frac{a(s)h^{\alpha+1}(t,s)}{\alpha_0 H^\alpha(t,s)}ds \tag{3.9.7}$$

for all $t \geq t_0$. Consequently, we obtain

$$\frac{1}{H(t,t_0)}\int_{t_0}^t \left[H(t,s)q(s) - \frac{a(s)h^{\alpha+1}(t,s)}{\alpha_0 H^\alpha(t,s)}\right]ds \leq w(t_0) \quad \text{for} \quad t \geq t_0.$$

Taking the upper limit as $t \to \infty$, we get a contradiction, which completes the proof. ■

A simpler version of Theorem 3.9.1 can be stated as follows.

Theorem 3.9.2. Let the function H be as in Theorem 3.9.1 such that (3.9.1) hold, and suppose that H has a continuous and nonnegative partial derivative with respect to the second variable. Let $h : \mathcal{D} \to \mathbb{R}$ be a continuous function with

$$-\frac{\partial}{\partial s}H(t,s) = h(t,s)H^{\alpha/(\alpha+1)}(t,s) \quad \text{for} \quad (t,s) \in \mathcal{D}. \qquad (3.9.8)$$

If

$$\limsup_{t\to\infty} \frac{1}{H(t,t_0)} \int_{t_0}^{t} \left[H(t,s)q(s) - \frac{a(s)}{\alpha_0}h^{\alpha+1}(t,s) \right] ds = \infty, \qquad (3.9.9)$$

then equation (3.1.1) is oscillatory.

Next, we shall prove the following result which extends and improves Theorem 3.9.1.

Theorem 3.9.3. Suppose the functions H and h are as in Theorem 3.9.1 such that conditions (3.9.1) and (3.9.2) hold. If there exists a nondecreasing function $\rho(t) \in C^1([t_0, \infty), \mathbb{R}^+)$ such that

$$\limsup_{t\to\infty} \frac{1}{H(t,t_0)} \int_{t_0}^{t} \left[H(t,s)\rho(s)q(s) \right.$$

$$\left. - \frac{a(s)\rho(s)}{\alpha_0 H^\alpha(t,s)} \left(h(t,s) + \frac{\rho'(s)}{\rho(s)}H(t,s) \right)^{\alpha+1} \right] ds = \infty,$$

$$(3.9.10)$$

then equation (3.1.1) is oscillatory.

Proof. Let $x(t)$ be a nonoscillatory solution of equation (3.1.1). Without loss of generality we assume that $x(t) \neq 0$ for $t \geq t_0$. Now, we define $w(t) = \rho(t)a(t)|x'(t)|^{\alpha-1}x'(t)/(|x(t)|^\alpha x(t))$ for $t \geq t_0$. Then for every $s \geq t_0$, we have

$$w'(s) = -\rho(s)q(s) + \frac{\rho'(s)}{\rho(s)}w(s) - \alpha\frac{|w(s)|^{(\alpha+1)/\alpha}}{(a(s)\rho(s))^{1/\alpha}}. \qquad (3.9.11)$$

Multiplying (3.9.11) by $H(t,s)$ for $t \geq s \geq t_0$ and integrating from t_0 to t, we get

$$\int_{t_0}^{t} H(t,s)w'(s)ds = -\int_{t_0}^{t} H(t,s)\rho(s)q(s)ds + \int_{t_0}^{t} \frac{\rho'(s)}{\rho(s)}w(s)ds$$

$$-\alpha \int_{t_0}^{t} H(t,s)\frac{|w(s)|^{(\alpha+1)/\alpha}}{(a(s)\rho(s))^{1/\alpha}}ds.$$

Thus, from (3.9.5), we find

$$\int_{t_0}^{t} H(t,s)\rho(s)q(s)ds \leq H(t,t_0)w(t_0) + \int_{t_0}^{t} \left(h(t,s) + \frac{\rho'(s)}{\rho(s)} H(t,s) \right) |w(s)| ds$$
$$- \alpha \int_{t_0}^{t} H(t,s) \frac{|w(s)|^{(\alpha+1)/\alpha}}{(a(s)\rho(s))^{1/\alpha}} ds.$$

$$(3.9.12)$$

Therefore, in view of Lemma 3.2.1 with

$$A = (\alpha H(t,s))^{\alpha/(\alpha+1)} \frac{|w(s)|}{(a(s)\rho(s))^{1/(\alpha+1)}}, \qquad \lambda = \frac{\alpha+1}{\alpha}$$

and

$$B = \left(\frac{\alpha}{\alpha+1} \right)^{\alpha} \left(\frac{a(s)\rho(s)}{(\alpha H(t,s))^{\alpha}} \right)^{\alpha/(\alpha+1)} \left(h(t,s) + \frac{\rho'(s)}{\rho(s)} H(t,s) \right)^{\alpha},$$

for $t > s \geq t_0$, we obtain

$$|w(s)| \left(h(t,s) + \frac{\rho'(s)}{\rho(s)} H(t,s) \right) - \alpha H(t,s) \frac{|w(s)|^{(\alpha+1)/\alpha}}{p^{1/\alpha}(s)}$$
$$\leq \frac{a(s)\rho(s)}{(\alpha+1)^{\alpha+1} H^{\alpha}(t,s)} \left(h(t,s) + \frac{\rho'(s)}{\rho(s)} H(t,s) \right)^{\alpha+1}.$$

$$(3.9.13)$$

From (3.9.12) and (3.9.13), we have

$$\frac{1}{H(t,t_0)} \int_{t_0}^{t} \left[H(t,s)\rho(s)q(s) - \frac{a(s)\rho(s)}{\alpha_0 H^{\alpha}(t,s)} \left(h(t,s) + \frac{\rho'(s)}{\rho(s)} H(t,s) \right)^{\alpha+1} \right] ds$$
$$\leq w(t_0).$$

$$(3.9.14)$$

Now, taking the upper limit as $t \to \infty$, we obtain a contradiction. ∎

Corollary 3.9.1. Let condition (3.9.10) in Theorem 3.9.3 be replaced by

$$\lim_{t \to \infty} \frac{1}{H(t,t_0)} \int_{t_0}^{t} \frac{a(s)\rho(s)}{H^{\alpha}(t,s)} \left(h(t,s) + \frac{\rho'(s)}{\rho(s)} H(t,s) \right)^{\alpha+1} ds < \infty \quad (3.9.15)$$

and

$$\limsup_{t \to \infty} \frac{1}{H(t,t_0)} \int_{t_0}^{t} H(t,s)\rho(s)q(s)ds = \infty, \qquad (3.9.16)$$

then the conclusion of Theorem 3.9.3 holds.

Example 3.9.1. Consider the differential equation

$$\left(t^{\nu} |x'(t)|^{\alpha-1} x'(t) \right)' + \left[\gamma t^{\gamma-3} (2 - \cos t) + t^{\gamma-2} \sin t \right] |x(t)|^{\alpha-1} x(t) = 0$$

$$(3.9.17)$$

for $t \geq t_0 > 0$, where γ is an arbitrary positive constant, α and ν are constants such that $\nu < \alpha - 2$, $\alpha < 1$.

Here, we choose $\rho(t) = t^2$ and $H(t,s) = (t-s)^2$ for $t \geq s \geq t_0$. Then since $\rho(t)q(t) = [t^\gamma(2 - \cos t)]'$, we find $\int_{t_0}^t \rho(s)q(s)ds \geq t^\gamma - k_0$ for some constant $k_0 > 0$, and therefore,

$$\frac{1}{t^2} \int_{t_0}^t (t-s)^2 \rho(s)q(s)ds \geq \frac{2t^\gamma}{(\gamma+1)(\gamma+2)} + \frac{k_1}{t^2} + \frac{k_2}{t} - k_0,$$

where

$$k_1 = \frac{2t_0^{\gamma+2}}{\gamma+2} - k_0 t_0^2 \quad \text{and} \quad k_2 = 2k_0 t_0 - \frac{2t_0^{\gamma+1}}{\gamma+1}.$$

Hence, condition (3.9.16) is satisfied. On the other hand, since

$$\frac{1}{t^2} \int_{t_0}^t \frac{s^{\nu+2}}{(t-s)^{2\alpha}} \left(2(t-s) + \frac{2}{s}(t-s)^2 \right)^{\alpha+1} ds$$

$$= 2^{\alpha+1} t^{\alpha-1} \int_{t_0}^t s^{\nu-\alpha+1}(t-s)^{1-\alpha} ds \leq 2^{\alpha+1} \left(1 - \frac{t_0}{t} \right)^{1-\alpha} \frac{t^{\nu-\alpha+2} - t_0^{\nu-\alpha+2}}{\nu-\alpha+2}$$

condition (3.9.15) is also satisfied. Thus, from Corollary 3.9.1 equation (3.9.17) is oscillatory.

Let $H(t,s) = (t-s)^n$ for $t \geq s \geq t_0$, where $n > \alpha$ is a constant. Then by Theorem 3.9.3 the following corollaries are immediate.

Corollary 3.9.2. If there exists a nondecreasing function $\rho \in C^1([t_0, \infty), \mathbb{R}^+)$ such that

$$\limsup_{t \to \infty} \frac{1}{t^n} \int_{t_0}^t \left[(t-s)^n \rho(s)q(s) \right.$$

$$\left. - \frac{1}{\alpha_0} a(s)\rho(s)(t-s)^{n-\alpha-1} \left(\frac{\rho'(s)}{\rho(s)}(t-s) + n \right)^{\alpha+1} \right] ds = \infty,$$
(3.9.18)

then equation (3.1.1) is oscillatory.

Corollary 3.9.3. If

$$\limsup_{t \to \infty} \frac{1}{t^n} \int_{t_0}^t \left[(t-s)^n q(s) - \frac{n^{\alpha+1}}{\alpha_0} a(s)(t-s)^{n-\alpha-1} \right] ds = \infty, \quad (3.9.19)$$

then equation (3.1.1) is oscillatory.

Theorem 3.9.4. Let the functions H, h and ρ be as in Theorem 3.9.1 such that conditions (3.9.1) and (3.9.2) hold, and let

$$0 < \inf_{s \geq t_0} \left\{ \liminf_{t \to \infty} \frac{H(t,s)}{H(t,t_0)} \right\} \leq \infty \qquad (3.9.20)$$

and

$$\limsup_{t\to\infty} \frac{1}{H(t,t_0)} \int_{t_0}^{t} \frac{a(s)\rho(s)}{H^\alpha(t,s)} \left(h(t,s) + \frac{\rho'(s)}{\rho(s)} H(t,s) \right)^{\alpha+1} ds \ < \ \infty.$$

(3.9.21)

If there exists a function $\Omega \in C([t_0,\infty),\mathbb{R})$ such that for every $T \ge t_0$,

$$\limsup_{t\to\infty} \frac{1}{H(t,T)} \int_{T}^{t} \left[H(t,s)\rho(s)q(s) \right.$$

$$\left. - \frac{a(s)\rho(s)}{\alpha_0 H^\alpha(t,s)} \left(h(t,s) + \frac{\rho'(s)}{\rho(s)} H(t,s) \right)^{\alpha+1} \right] ds \ \ge \ \Omega(T)$$

(3.9.22)

and

$$\int_{t_0}^{\infty} \frac{(\Omega^+(s))^{(\alpha+1)/\alpha}}{(a(s)\rho(s))^{1/\alpha}} ds \ = \ \infty,$$

(3.9.23)

where $\Omega^+(t) = \max\{0,\Omega(t)\}$, then equation (3.1.1) is oscillatory.

Proof. Let $x(t)$ be a nonoscillatory solution of equation (3.1.1), say, $x(t) \ne 0$ for $t \ge t_0$. Define the function $w(t)$ as in Theorem 3.9.3, to get (3.9.12) and (3.9.14). Then for $t \ge T \ge t_0$, we have

$$\limsup_{t\to\infty} \frac{1}{H(t,T)} \int_{T}^{t} \left[H(t,s)\rho(s)q(s) \right.$$

$$\left. - \frac{a(s)\rho(s)}{\alpha_0} \left(h(t,s) + \frac{\rho'(s)}{\rho(s)} H(t,s) \right)^{\alpha+1} \right] ds \ \le \ w(T).$$

Thus, from condition (3.9.22) it follows that

$$\Omega(T) \ \le \ w(T) \quad \text{for every} \quad T \ge t_0$$

(3.9.24)

and

$$\limsup_{t\to\infty} \frac{1}{H(t,t_0)} \int_{t_0}^{t} H(t,s)\rho(s)q(s)ds \ \ge \ \Omega(t_0).$$

(3.9.25)

Now, we define functions

$$U(t) \ = \ \frac{\alpha}{H(t,t_0)} \int_{t_0}^{t} H(t,s) \frac{|w(s)|^{(\alpha+1)/\alpha}}{(a(s)\rho(s))^{1/\alpha}} ds$$

and

$$W(t) \ = \ \frac{1}{H(t,t_0)} \int_{t_0}^{t} \beta(t,s)|w(s)|ds,$$

where

$$\beta(t,s) \ = \ h(t,s) + \frac{\rho'(s)}{\rho(s)} H(t,s), \quad t \ge s \ge t_0.$$

Then from (3.9.12) and (3.9.25), we find

$$\liminf_{t\to\infty}[U(t) - W(t)] \;\leq\; w(t_0) - \limsup_{t\to\infty} \frac{1}{H(t,t_0)} \int_{t_0}^{t} H(t,s)\rho(s)q(s)ds$$

$$\leq\; w(t_0) - \Omega(t_0) \;<\; \infty.$$

$$(3.9.26)$$

Next, we claim that

$$\int_{t_0}^{\infty} \frac{|w(s)|^{(\alpha+1)/\alpha}}{(a(s)\rho(s))^{1/\alpha}} ds \;<\; \infty. \qquad (3.9.27)$$

Suppose to the contrary that

$$\int_{t_0}^{\infty} \frac{|w(s)|^{(\alpha+1)/\alpha}}{(a(s)\rho(s))^{1/\alpha}} ds \;=\; \infty. \qquad (3.9.28)$$

By condition (3.9.20), there exists a positive constant ξ such that

$$\inf_{s\geq t_0}\left\{ \liminf_{t\to\infty} \frac{H(t,s)}{H(t,t_0)} \right\} \;>\; \xi \;>\; 0. \qquad (3.9.29)$$

Let μ be an arbitrary positive number. Then it follows from (3.9.28) that there exists a $t_1 > t_0$ such that

$$\int_{t_0}^{t} \frac{|w(s)|^{(\alpha+1)/\alpha}}{(a(s)\rho(s))^{1/\alpha}} ds \;>\; \frac{\mu}{\xi} \quad \text{for all} \;\; t \geq t_1.$$

Therefore, for all $t \geq t_1$,

$$\begin{aligned}
U(t) &= \frac{\alpha}{H(t,t_0)} \int_{t_0}^{t} H(t,s)d\left(\int_{t_0}^{s} \frac{|w(\tau)|^{(\alpha+1)/\alpha}}{(a(\tau)\rho(\tau))^{1/\alpha}} d\tau \right) \\
&= \frac{\alpha}{H(t,t_0)} \int_{t_0}^{t} \left(-\frac{\partial}{\partial s}H(t,s) \right) \left(\int_{t_0}^{s} \frac{|w(\tau)|^{(\alpha+1)/\alpha}}{(a(\tau)\rho(\tau))^{1/\alpha}} d\tau \right) ds \\
&\geq \frac{\alpha}{H(t,t_0)} \int_{t_1}^{t} \left(-\frac{\partial}{\partial s}H(t,s) \right) \left(\int_{t_0}^{s} \frac{|w(\tau)|^{(\alpha+1)/\alpha}}{(a(\tau)\rho(\tau))^{1/\alpha}} d\tau \right) ds \\
&\geq \frac{\mu\alpha}{\xi H(t,t_0)} \int_{t_1}^{t} -\frac{\partial}{\partial s}H(t,s)ds \;=\; \frac{\mu H(t,t_1)}{\xi H(t,t_0)}.
\end{aligned}$$

From (3.9.29), we find that there is a $t_2 > t_1$ such that $H(t,t_1)/H(t,t_0) \geq \xi$ for all $t \geq t_2$, which implies that $U(t) \geq \mu\alpha$ for all $t \geq t_2$. Since μ is arbitrary

$$\lim_{t\to\infty} U(t) \;=\; \infty. \qquad (3.9.30)$$

Further, consider a sequence $\{T_n\}_{n=1}^{\infty}$ in (t_0,∞) such that $\lim_{n\to\infty} T_n = \infty$ and

$$\lim_{n\to\infty} [U(T_n) - W(T_n)] \;=\; \liminf_{t\to\infty} [U(t) - W(t)] \;<\; \infty.$$

Then there exists a constant M such that

$$U(T_n) - W(T_n) \leq M \qquad (3.9.31)$$

for all sufficiently large n. Since (3.9.30) ensures that

$$\lim_{n \to \infty} U(T_n) = \infty, \qquad (3.9.32)$$

(3.9.31) implies that

$$\lim_{n \to \infty} W(T_n) = \infty. \qquad (3.9.33)$$

Combining (3.9.31) and (3.9.32), we derive for n sufficiently large

$$\frac{W(T_n)}{U(T_n)} - 1 \geq -\frac{M}{U(T_n)} > -\frac{1}{2}.$$

Therefore, we have

$$\frac{W(T_n)}{U(T_n)} > \frac{1}{2} \quad \text{for all large } n. \qquad (3.9.34)$$

Now from (3.9.33) and (3.9.33), we obtain

$$\lim_{n \to \infty} \frac{W^{\alpha+1}(T_n)}{U^\alpha(T_n)} = \infty. \qquad (3.9.35)$$

On the other hand by Hölder's inequality, we have

$$
\begin{aligned}
W(T_n) &= \frac{1}{H(T_n, t_0)} \int_{t_0}^{T_n} \beta(T_n, s) |w(s)| ds \\
&= \int_{t_0}^{T_n} \left(\frac{\alpha^{\alpha/(\alpha+1)}}{H^{\alpha/(\alpha+1)}(T_n, t_0)} \frac{|w(s)| H^{\alpha/(\alpha+1)}(T_n, s)}{(a(s)\rho(s))^{1/(\alpha+1)}} \right) \\
&\quad \times \left(\frac{\alpha^{-\alpha/(\alpha+1)}}{H^{1/(\alpha+1)}(T_n, t_0)} \frac{\beta(T_n, s)(a(s)\rho(s))^{1/(\alpha+1)}}{H^{\alpha/(\alpha+1)}(T_n, s)} \right) ds \\
&\leq \left(\frac{\alpha}{H(T_n, t_0)} \int_{t_0}^{T_n} \frac{|w(s)|^{(\alpha+1)/\alpha} H(T_n, s)}{(a(s)\rho(s))^{1/\alpha}} ds \right)^{\alpha/(\alpha+1)} \\
&\quad \times \left(\frac{1}{\alpha^\alpha H(T_n, t_0)} \int_{t_0}^{T_n} a(s)\rho(s) \frac{\beta^{\alpha+1}(T_n, s)}{H^\alpha(T_n, s)} ds \right)^{1/(\alpha+1)},
\end{aligned}
$$

and hence

$$\frac{W^{\alpha+1}(T_n)}{U^\alpha(T_n)} \leq \frac{1}{\alpha^\alpha H(T_n, t_0)} \int_{t_0}^{T_n} a(s)\rho(s) \frac{\beta^{\alpha+1}(T_n, s)}{H^\alpha(T_n, s)} ds.$$

Finally, it follows from (3.9.35) that

$$\lim_{n\to\infty} \frac{1}{H(T_n,t_0)} \int_{t_0}^{T_n} a(s)\rho(s)\frac{\beta^{\alpha+1}(T_n,s)}{H^\alpha(T_n,s)}ds = \infty,$$

which gives

$$\lim_{t\to\infty} \frac{1}{H(t,t_0)} \int_{t_0}^{t} a(s)\rho(s)\frac{\beta^{\alpha+1}(t,s)}{H^\alpha(t,s)}ds = \infty.$$

But, this contradicts condition (3.9.21) and hence (3.9.28) holds.

Now, from (3.9.24), we have

$$\int_{t_0}^{\infty} \frac{(\Omega^+(s))^{(\alpha+1)/\alpha}}{(a(s)\rho(s))^{1/\alpha}}ds < \int_{t_0}^{\infty} \frac{|w(s)|^{(\alpha+1)/\alpha}}{(a(s)\rho(s))^{1/\alpha}}ds < \infty,$$

which contradicts condition (3.9.23). This completes the proof. ∎

Theorem 3.9.5. Let the functions H, h and ρ be as in Theorem 3.9.4 such that conditions (3.9.1), (3.9.2) and (3.9.20) hold, and

$$\liminf_{t\to\infty} \frac{1}{H(t,t_0)} \int_{t_0}^{t} H(t,s)\rho(s)q(s)ds < \infty. \tag{3.9.36}$$

If there exists a function $\Omega \in C([t_0,\infty), \mathbb{R})$ such that for every $T \geq t_0$,

$$\liminf_{t\to\infty} \frac{1}{H(t,T)} \int_{T}^{t} \left[H(t,s)\rho(s)q(s) \right. \tag{3.9.37}$$
$$\left. -\frac{a(s)\rho(s)}{\alpha_0 H^\alpha(t,s)} \left(h(t,s) + \frac{\rho'(s)}{\rho(s)}H(t,s) \right)^{\alpha+1} \right] ds \geq \Omega(T)$$

and condition (3.9.23) holds, then equation (3.1.1) is oscillatory.

Proof. Let $x(t)$ be a nonoscillatory solution of equation (3.1.1), say, $x(t) \neq 0$ for $t \geq t_0$. As in Theorem 3.9.3, (3.9.12) and (3.9.14) are satisfied and proceeding as in Theorem 3.9.4, (3.9.24) holds for $t \geq T \geq t_0$. Using condition (3.9.36), we find

$$\limsup_{t\to\infty} [U(t)-W(t)] \leq w(t_0)-\liminf_{t\to\infty} \frac{1}{H(t,t_0)} \int_{t_0}^{t} H(t,s)\rho(s)q(s)ds < \infty,$$

where $U(t)$ and $W(t)$ are as in the proof of Theorem 3.9.4. Now, it follows from condition (3.9.37) that

$$\Omega(t_0) \leq \liminf_{t\to\infty} \frac{1}{H(t,t_0)} \int_{t_0}^{t} H(t,s)\rho(s)q(s)ds$$
$$- \liminf_{t\to\infty} \frac{1}{H(t,t_0)} \int_{t_0}^{t} \frac{a(s)\rho(s)}{\alpha_0 H^\alpha(t,s)} \left(h(t,s) + \frac{\rho'(s)}{\rho(s)}H(t,s) \right)^{\alpha+1} ds,$$

so that condition (3.9.36) implies that

$$\liminf_{t\to\infty} \frac{1}{\alpha_0 H(t,t_0)} \int_{t_0}^t \frac{a(s)\rho(s)}{H^\alpha(t,s)} \left(h(t,s) + \frac{\rho'(s)}{\rho(s)} H(t,s) \right)^{\alpha+1} ds < \infty.$$

Let $\{T_n\}_{n=1}^\infty$ be a sequence in (t_0,∞) with $\lim_{n\to\infty} T_n = \infty$ such that $\lim_{n\to\infty} [U(T_n) - W(T_n)] = \limsup_{t\to\infty} [U(t) - W(t)]$. Then using the procedure as in the proof of Theorem 3.9.4 we conclude that (3.9.27) is satisfied. The remainder of the proof proceeds as in Theorem 3.9.4 and hence omitted. ∎

Now, define $H(t,s) = (\pi(t) - \pi(s))^n$ for $t \geq s \geq t_0$, where $n > \alpha$ is a constant and $\pi(t) \in C^1([t_0,\infty), \mathbb{R}_0)$ with $\pi'(t) = 1/a^{1/\alpha}(t)$ for $t \geq t_0$ and $\lim_{t\to\infty} \pi(t) = \infty$. Then, $H(t,s)$ is continuous on $\mathcal{D} = \{(t,s) : t \geq s \geq t_0\}$ and satisfies $H(t,t) = 0$ for $t \geq t_0$, $H(t,s) > 0$ for $t > s \geq t_0$. Moreover, $H(t,s)$ has a continuous and nonpositive partial derivative on $\mathcal{D}$ with respect to the second variable. Furthermore, the function

$$h(t,s) = \frac{n}{a^{1/\alpha}(t)} (\pi(t) - \pi(s))^{(n-\alpha-1)/(\alpha+1)}, \quad t > s \geq t_0$$

is continuous and satisfies

$$-\frac{\partial}{\partial s} H(t,s) = h(t,s) H^{\alpha/(\alpha+1)}(t,s) \quad \text{for} \quad t > s \geq t_0.$$

We find that (3.9.20) holds, since for every $s \geq t_0$,

$$\lim_{t\to\infty} \frac{H(t,s)}{H(t,t_0)} = \lim_{t\to\infty} \frac{(\pi(t) - \pi(s))^n}{(\pi(t) - \pi(t_0))^n} = 1.$$

If $\rho(t) = 1$, then since

$$\limsup_{t\to\infty} \frac{1}{\pi^n(t)} \int_{t_0}^t a(s) \left[\frac{n}{a^{1/\alpha}(s)} (\pi(t) - \pi(s))^{(n-\alpha-1)/(\alpha+1)} \right]^{\alpha+1} ds$$

$$= \lim_{t\to\infty} \frac{1}{\pi^n(t)} \int_{t_0}^t n^{\alpha+1} a^{-1/\alpha}(s)(\pi(t) - \pi(s))^{n-\alpha-1} ds$$

$$= \lim_{t\to\infty} \frac{n^{\alpha+1}}{\pi^n(t)} \frac{(\pi(t) - \pi(t_0))^{n-\alpha}}{n-\alpha} = 0,$$

condition (3.9.21) is satisfied. Thus, Theorem 3.9.4 leads to the following corollary.

Corollary 3.9.4. Let $n > \alpha$ be a constant, and let $\pi(t) \in C^1([t_0,\infty), \mathbb{R}_0)$ with $\pi'(t) = a^{-1/\alpha}(t)$ and $\lim_{t\to\infty} \pi(t) = \infty$. Suppose there exists a function $\Omega(t) \in C([t_0,\infty), \mathbb{R})$ such that

$$\int_{t_0}^t \frac{(\Omega^+(s))^{(\alpha+1)/\alpha}}{a^{1/\alpha}(s)} ds = \infty \tag{3.9.38}$$

and

$$\limsup_{t\to\infty} \frac{1}{\pi^n(t)} \int_T^t [\pi(t) - \pi(s)]^n q(s)ds \geq \Omega(T), \quad \text{for all} \quad T \geq t_0 \ (3.9.39)$$

where $\Omega^+(t) = \max\{\Omega(t), 0\}$. Then equation (3.1.1) is oscillatory.

In the case when $H(t, s) = (t - s)^n$ for $t \geq s \geq t_0$ and some $n > \alpha$, we have the following corollary.

Corollary 3.9.5. Let $n > \alpha$ be a constant and suppose there exists a function $\Omega(t) \in C([t_0, \infty), \mathbb{R})$ such that condition (3.9.38) holds. If

$$\limsup_{t\to\infty} \frac{1}{t^n} \int_T^t (t - s)^{n-\alpha-1} \left[(t - s)^{\alpha+1} q(s) - \frac{n^{n+1}}{\alpha_0} a(s) \right] ds \geq \Omega(T)$$

$$(3.9.40)$$

for all $T \geq t_0$, then equation (3.1.1) is oscillatory.

Example 3.9.2. Consider the differential equation

$$\left(t^\nu |x'(t)|^{\alpha-1} x'(t) \right)' + (t^\lambda \cos t) |x(t)|^{\alpha-1} x(t) = 0 \quad \text{for} \quad t \geq t_0 > 0, \ (3.9.41)$$

where ν, λ, α are constants such that $-1 < \lambda \leq 1$, $\nu < \alpha$, $\alpha \neq 2$ and $\alpha^2 \lambda \geq (\alpha + 1)(\nu - \alpha)$. Taking $H(t, s) = (t - s)^2$ for $t \geq s \geq t_0$, we find

$$\frac{1}{t^2} \int_{t_0}^t s^\nu (t - s)^{1-\alpha} ds \leq \begin{cases} \dfrac{t^\nu}{t^2} \dfrac{(t - t_0)^{2-\alpha}}{2 - \alpha}, & \nu > 0 \\[2ex] \dfrac{t_0^\nu}{t^2} \dfrac{(t - t_0)^{2-\alpha}}{2 - \alpha}, & \nu < 0 \end{cases}$$

$$= \begin{cases} \dfrac{t^{\nu-\alpha}}{2 - \alpha} \left(1 - \dfrac{t_0}{t} \right)^{2-\alpha}, & \nu > 0 \\[2ex] \dfrac{t_0^\nu}{2 - \alpha} \dfrac{1}{t^\alpha} \left(1 - \dfrac{t_0}{t} \right)^{2-\alpha}, & \nu < 0. \end{cases}$$

Therefore, the condition $\limsup_{t\to\infty} (1/t^2) \int_{t_0}^t a(s)(t - s)^{n-\alpha-1} ds < \infty$ is satisfied, and for arbitrary small constant $\epsilon > 0$ there exists a $t_1 \geq t_0$ such that for $T \geq t_1$,

$$\limsup_{t\to\infty} \frac{1}{t^2} \int_T^t \left[(t - s)^2 s^\lambda \cos s - s^\nu \frac{(t - s)^{1-\alpha}}{(1 + \alpha)^{1+\alpha}} \right] ds \geq -T^\lambda \cos T - \epsilon.$$

Now, set $\Omega(t) = -T^\lambda \cos T - \epsilon$. Then there exists an integer N such that $(2N + 1)\pi - (\pi/4) > t_1$ and if $n \geq N$, $(2n + 1)\pi - (\pi/4) \leq T \leq (2n + 1)\pi + (\pi/4)$, $\Omega(T) \geq \delta T^\lambda$, where δ is a small constant. Now, since

$\alpha^2 \lambda \geq (\alpha + 1)(\nu - \alpha)$, we obtain

$$\int_{t_0}^{\infty} \frac{(\Omega^+(s))^{(\alpha+1)/\alpha}}{a^{1/\alpha}(s)} ds \geq \sum_{n=N}^{\infty} \delta^{(\alpha+1)/\alpha} \int_{(2n+1)\pi-(\pi/4)}^{(2n+1)\pi+(\pi/4)} s^{\lambda(\alpha+1)/\alpha-(\nu/\alpha)} ds$$

$$\geq \sum_{n=N}^{\infty} \delta^{(\alpha+1)/\alpha} \int_{(2n+1)\pi-(\pi/4)}^{(2n+1)\pi+(\pi/4)} \frac{ds}{s} = \infty.$$

Hence, all conditions of Corollary 3.9.5 are satisfied and therefore equation (3.9.41) is oscillatory.

3.9.2. Interval Criteria for the Oscillation of Equation (3.1.1)

It is well–known that equations (2.1.1) and (3.1.1) have several similar properties. For example, Sturmian comparison and separation theorems for equation (2.1.1) have been extended to (3.1.1) in Section 3.2. Thus, the zeros of two linearly independent solutions of (3.1.1) separate each other and all nontrivial solutions of (3.1.1) are either oscillatory or nonoscillatory. We note that most of the oscillation criteria obtained so far involve the integral of the function $q(t)$, and hence require the information of $q(t)$ on the entire half–line $[t_0, \infty)$. These results cannot be applied to the case when $q(t)$ has a 'bad' behavior on a large part of $[t_0, \infty)$, for example, when $\int_{t_0}^{\infty} q(s) ds = -\infty$. However, from Sturm separation theorem, we note that oscillation is only an interval property, i.e., if there exists a sequence of subintervals $[a_i, b_i]$ of $[t_0, \infty)$, $a_i \to \infty$ such that for each i there exists a solution of (3.1.1) that has at least two zeros in $[a_i, b_i]$, then every solution of (3.1.1) is oscillatory, no matter how 'bad' equation (3.1.1) is (or $a(t)$ and $q(t)$ are) on the remaining parts of $[t_0, \infty)$. In this section we shall use this idea to provide several oscillation criteria for (3.1.1). These results in particular extend and improve those given in Section 2.5 for the equation (2.1.1).

In what follows we say that the function $H = H(t, s)$ belongs to the class Γ, denoted by $H \in \Gamma$ if $H \in C(\mathcal{D}, \mathbb{R}^+)$ where $\mathcal{D} = \{(t, s) : -\infty < s \leq t < \infty\}$ which satisfies condition (3.9.1), and has partial derivatives $\partial H / \partial t$ and $\partial H / \partial s$ on $\mathcal{D}$ such that

$$\frac{\partial}{\partial t} H(t, s) = h_1(t, s) \quad \text{and} \quad \frac{\partial}{\partial s} H(t, s) = -h_2(t, s), \tag{3.9.42}$$

where h_1, h_2 are continuous and nonnegative on $\mathcal{D}$, h_1, $h_2 \in \mathcal{L}_{Loc}(\mathcal{D}, \mathbb{R})$.

Again, we shall let by $\alpha_0 = (\alpha + 1)^{\alpha+1}$. To prove our main results we need the following two lemmas.

Lemma 3.9.1. Assume that $x(t)$ is a solution of equation (3.9.1) such that $x(t) > 0$ on $[c, b)$. Let

$$w(t) = \rho(t)\frac{a(t)|x'(t)|^{\alpha-1}x'(t)}{|x(t)|^{\alpha-1}x(t)}, \tag{3.9.43}$$

where $\rho(t) \in C^1([t_0, \infty), \mathbb{R}^+)$ is any nondecreasing function. Then for any $H \in \Gamma$,

$$\int_c^b H(b,s)\rho(s)q(s)ds \leq H(b,c)w(c) + \int_c^b \frac{a(s)\rho(s)}{\alpha_0 H^\alpha(b,s)} \tag{3.9.44}$$
$$\times \left[H(b,s)\frac{\rho'(s)}{\rho(s)} + h_2(b,s)\right]^{\alpha+1} ds.$$

Proof. From equation (3.1.1) and (3.9.43), we have for $s \in [c, b)$,

$$\rho(s)q(s) = -w'(s) + \frac{\rho'(s)}{\rho(s)}w(s) - \alpha\frac{|w(s)|^{(\alpha+1)/\alpha}}{(a(s)\rho(s))^{1/\alpha}}. \tag{3.9.45}$$

Multiplying (3.9.45) by $H(b, s)$, integrating from c to b and using (3.9.1) and (3.9.42), we obtain

$$\int_c^b H(b,s)\rho(s)q(s)ds = -\int_c^b H(b,s)w'(s)ds + \int_c^b H(b,s)\frac{\rho'(s)}{\rho(s)}w(s)ds$$
$$-\alpha \int_c^b H(b,c)\frac{|w(s)|^{(\alpha+1)/\alpha}}{(a(s)\rho(s))^{1/\alpha}}ds$$
$$\leq H(b,c)w(c) + \int_c^b \left[H(b,s)\frac{\rho'(s)}{\rho(s)} + h_2(b,s)\right]|w(s)|ds$$
$$-\alpha \int_c^b H(b,s)\frac{|w(s)|^{(\alpha+1)/\alpha}}{(a(s)\rho(s))^{1/\alpha}}ds.$$

Thus, from Lemma 3.2.1 with

$$A = (\alpha H(b,s))^{\alpha/(\alpha+1)}\frac{|w(s)|}{(a(s)\rho(s))^{1/(\alpha+1)}}, \quad \lambda = \frac{\alpha+1}{\alpha}$$

and

$$B = \left(\frac{\alpha}{\alpha+1}\right)^\alpha \left(\frac{a(s)\rho(s)}{(\alpha H(b,s))^\alpha}\right)^{\alpha/(\alpha+1)} \left(h_2(b,s) + H(b,s)\frac{\rho'(s)}{\rho(s)}\right)^\alpha,$$

we get for $t > s \geq t_0$,

$$|w(s)| \left(h_2(b,s) + \frac{\rho'(s)}{\rho(s)}H(b,s)\right) - \alpha H(b,s)\frac{|w(s)|^{(\alpha+1)/\alpha}}{(a(s)\rho(s))^{1/\alpha}}$$
$$\leq \frac{a(s)\rho(s)}{\alpha_0 H^\alpha(b,s)}\left(h_2(b,s) + \frac{\rho'(s)}{\rho(s)}H(b,s)\right)^{\alpha+1}$$

and hence, we have

$$\int_c^b H(b,s)\rho(s)q(s)ds \;\le\; H(b,c)w(c) + \frac{1}{\alpha_0}\int_c^b \frac{a(s)\rho(s)}{H^\alpha(b,s)}$$
$$\times \left(h_2(b,s) + \frac{\rho'(s)}{\rho(s)} H(b,s)\right)^{\alpha+1} ds.$$

Therefore, (3.9.44) holds. ∎

Lemma 3.9.2. Assume that $x(t)$ is a solution of equation (3.1.1) such that $x(t) > 0$ on $(a,c]$. Let $\rho(t) \in C^1([t_0,\infty),\mathbb{R}^+)$ be a nondecreasing function and $w(t)$ be as in (3.9.43) on $(a,c]$. Then for any $H \in \Gamma$,

$$\int_a^c H(s,a)\rho(s)q(s)ds \;\le\; -H(c,a)w(c) + \int_a^c \frac{a(s)\rho(s)}{\alpha_0 H^\alpha(s,a)}$$
$$\times \left[H(s,a)\frac{\rho'(s)}{\rho(s)} + h_1(s,a)\right]^{\alpha+1} ds. \tag{3.9.46}$$

Proof. Following Lemma 3.9.1, we multiply (3.9.45) by $H(s,a)$ and integrate from a to c, to obtain

$$\int_a^c H(s,a)\rho(s)q(s)ds = -\int_a^c H(s,a)w'(s)ds + \int_a^c H(s,a)\frac{\rho'(s)}{\rho(s)}w(s)ds$$
$$-\alpha \int_a^c H(s,a)\frac{|w(s)|^{(\alpha+1)/\alpha}}{(a(s)\rho(s))^{1/\alpha}}ds$$
$$\le -H(c,a)w(c) + \int_a^c \left[H(s,a)\frac{\rho'(s)}{\rho(s)} + h_1(s,a)\right]|w(s)|ds$$
$$-\alpha \int_a^c H(s,a)\frac{|w(s)|^{(\alpha+1)/\alpha}}{(a(s)\rho(s))^{1/\alpha}}ds.$$

Thus, from Lemma 3.2.1 with

$$A = (\alpha H(s,a))^{\alpha/(\alpha+1)}\frac{|w(s)|}{(a(s)\rho(s))^{1/(\alpha+1)}}, \qquad \lambda = \frac{\alpha+1}{\alpha}$$

and

$$B = \left(\frac{\alpha}{\alpha+1}\right)^\alpha \left(\frac{a(s)\rho(s)}{(\alpha H(s,a))^\alpha}\right)^{\alpha/(\alpha+1)} \left(h_1(s,a) + H(s,a)\frac{\rho'(s)}{\rho(s)}\right)^\alpha,$$

we get for $t > s \ge t_0$,

$$|w(s)| \left(h_1(s,a) + \frac{\rho'(s)}{\rho(s)}H(s,a)\right) - \alpha H(s,a)\frac{|w(s)|^{(\alpha+1)/\alpha}}{(a(s)\rho(s))^{1/\alpha}}$$
$$\le \frac{a(s)\rho(s)}{\alpha_0 H^\alpha(s,a)} \left(h_1(s,a) + \frac{\rho'(s)}{\rho(s)}H(s,a)\right)^{\alpha+1}$$

and hence, we have

$$\int_a^c H(s,a)\rho(s)q(s)ds \;\leq\; -H(c,a)w(c) + \frac{1}{\alpha_0}\int_a^c \frac{a(s)\rho(s)}{H^\alpha(s,a)}$$
$$\times \left(h_1(s,a) + \frac{\rho'(s)}{\rho(s)}H(s,a)\right)^{\alpha+1} ds.$$

Therefore, (3.9.46) holds. ∎

Corollary 3.9.6. Let $x(t)$ be a solution of equation (3.1.1), $x(t) > 0$ for $t \geq T \geq t_0$ and let $w(t)$ and $\rho(t)$ be as in Lemma 3.9.1. Then for any $H \in \Gamma$ and $a,\ b,\ c \in \mathbb{R}$ such that $T \leq a < c < b$,

$$\frac{1}{H(c,a)}\int_a^c H(s,a)\rho(s)q(s)ds + \frac{1}{H(b,c)}\int_c^b H(b,s)\rho(s)q(s)ds$$
$$\leq \frac{1}{H(c,a)}\int_a^c \frac{a(s)\rho(s)}{\alpha_0 H^\alpha(s,a)}\left(H(s,a)\frac{\rho'(s)}{\rho(s)} + h_1(s,a)\right)^{\alpha+1} ds$$
$$+ \frac{1}{H(b,c)}\int_c^b \frac{a(s)\rho(s)}{\alpha_0 H^\alpha(b,s)}\left(H(b,s)\frac{\rho'(s)}{\rho(s)} + h_2(b,s)\right)^{\alpha+1} ds.$$
$$\tag{3.9.47}$$

Proof. Divide (3.9.44) and (3.9.46) by $H(b,c)$ and $H(c,a)$, respectively and add them together. ∎

Corollary 3.9.7. Assume that $\rho(t) \in C^1([t_0,\infty),\mathbb{R}^+)$, $\rho'(t) \geq 0$ for $t \geq t_0$ and for some $c \in (a,b)$ and $H \in \Gamma$,

$$\frac{1}{H(c,a)}\int_a^c H(s,a)\rho(s)q(s)ds + \frac{1}{H(b,c)}\int_c^b H(b,s)\rho(s)q(s)ds$$
$$> \frac{1}{H(c,a)}\int_a^c \frac{a(s)\rho(s)}{\alpha_0 H^\alpha(s,a)}\left(H(s,a)\frac{\rho'(s)}{\rho(s)} + h_1(s,a)\right)^{\alpha+1} ds$$
$$+ \frac{1}{H(b,c)}\int_c^b \frac{a(s)\rho(s)}{\alpha_0 H^\alpha(b,s)}\left(H(b,s)\frac{\rho'(s)}{\rho(s)} + h_2(b,s)\right)^{\alpha+1} ds.$$
$$\tag{3.9.48}$$

Then every solution of equation (3.1.1) has at least one zero in (a,b).

Proof. Assume to the contrary that $x(t)$ is an eventually positive solution of equation (3.1.1) on (a,b). From Corollary 3.9.6, there exists a $T \geq t_0$ such that inequality (3.9.47) holds for any $H \in \Gamma$ and $a,\ b,\ c \in \mathbb{R}$ satisfying $T \leq a < c < b$. This contradicts (3.9.48) and completes the proof. ∎

Now, we state and prove the main interval oscillation criterion for (3.1.1).

Theorem 3.9.6. If for each $T \geq t_0$ there exist $H \in \Gamma$, $\rho \in C^1([t_0, \infty), \mathbb{R}^+)$, $\rho'(t) \geq 0$ for $t \geq t_0$, and a, b, $c \in \mathbb{R}$ such that $T \leq a < c < b$ and condition (3.9.48) holds, then equation (3.1.1) is oscillatory.

Proof. Pick a sequence $\{T_i\} \subset [t_0, \infty)$ such that $T_i \to \infty$ as $i \to \infty$. By the hypotheses for each $i \in \mathbb{N}$ there exist a_i, b_i, $c_i \in \mathbb{R}$ such that $T_i \leq a_i < c_i < b_i$ and (3.9.48) holds with a, b, c replaced by a_i, b_i, c_i, respectively. From Corollary 3.9.7 every solution $x(t)$ has at least one zero, $t_i \in (a_i, b_i)$. Noting that $T_i \leq a_i < t_i$, $i \in \mathbb{N}$ we find that every solution has arbitrarily large zeros. Thus, every solution of equation (3.1.1) is oscillatory. ∎

Theorem 3.9.7. If there exist $H \in \Gamma$ and $\rho \in C^1([t_0, \infty), \mathbb{R}^+)$, $\rho'(t) \geq 0$ for $t \geq t_0$ such that for any $r \geq t_0$,

$$\limsup_{t \to \infty} \frac{\int_r^t H(s, r)\rho(s)q(s)ds}{\int_r^t \frac{a(s)\rho(s)}{H^\alpha(s, r)} \left(H(s, r)\frac{\rho'(s)}{\rho(s)} + h_1(s, r) \right)^{\alpha+1} ds} = \infty \qquad (3.9.49)$$

and

$$\limsup_{t \to \infty} \frac{\int_r^t H(t, s)\rho(s)q(s)ds}{\int_r^t \frac{a(s)\rho(s)}{H^\alpha(t, s)} \left(H(t, s)\frac{\rho'(s)}{\rho(s)} + h_2(t, s) \right)^{\alpha+1} ds} = \infty, \qquad (3.9.50)$$

then equation (3.1.1) is oscillatory.

Proof. In (3.9.49) let $r = a = T \geq t_0$. Then there exists a $c > a$ such that

$$\int_a^c H(s, a)\rho(s)q(s)ds > \frac{1}{\alpha_0} \int_a^c \frac{a(s)\rho(s)}{H^\alpha(s, a)} \left(H(s, a)\frac{\rho'(s)}{\rho(s)} + h_1(s, a) \right)^{\alpha+1} ds. \qquad (3.9.51)$$

In (3.9.50), we choose $r = c$. Then there exists a $b > c$ such that

$$\int_c^b H(b, s)\rho(s)q(s)ds > \frac{1}{\alpha_0} \int_c^b \frac{a(s)\rho(s)}{H^\alpha(b, s)} \left(H(b, s)\frac{\rho'(s)}{\rho(s)} + h_2(b, s) \right)^{\alpha+1} ds. \qquad (3.9.52)$$

Combining (3.9.51) and (3.9.52), we obtain (3.9.48). Now, it follows from Theorem 3.9.6 that equation (3.1.1) is oscillatory. ∎

The following result is equivalent to that of Theorem 3.9.6.

Theorem 3.9.8. If there exist $H \in \Gamma$ and a nondecreasing function

$\rho(t) \in C^1([t_0, \infty), \mathbb{R}^+)$ such that for any $r \geq t_0$,

$$\limsup_{t \to \infty} \int_r^t \left[H(s, r)\rho(s)q(s) - \frac{a(s)\rho(s)}{\alpha_0 H^\alpha(s, r)} \left(H(s, r)\frac{\rho'(s)}{\rho(s)} + h_1(s, r) \right)^{\alpha+1} \right] ds$$
$$> 0$$
$$(3.9.53)$$

and

$$\limsup_{t \to \infty} \int_r^t \left[H(t, s)\rho(s)q(s) - \frac{a(s)\rho(s)}{\alpha_0 H^\alpha(s, r)} \left(H(t, s)\frac{\rho'(s)}{\rho(s)} + h_2(t, s) \right)^{\alpha+1} \right] ds$$
$$> 0,$$
$$(3.9.54)$$

then equation (3.1.1) is oscillatory.

Proof. In (3.9.53) let $r = a = T \geq t_0$. Then there exists a $c > a$ such that

$$\int_a^c \left[H(s, a)\rho(s)q(s) - \frac{a(s)\rho(s)}{\alpha_0 H^\alpha(s, a)} \left(H(s, a)\frac{\rho'(s)}{\rho(s)} + h_1(s, a) \right)^{\alpha+1} \right] ds > 0.$$
$$(3.9.55)$$

In (3.9.54), we choose $r = c$. Then there exists a $b > c$ such that

$$\int_c^b \left[H(b, s)\rho(s)q(s) - \frac{a(s)\rho(s)}{\alpha_0 H^\alpha(b, s)} \left(H(b, s)\frac{\rho'(s)}{\rho(s)} + h_2(b, s) \right)^{\alpha+1} \right] ds > 0.$$
$$(3.9.56)$$

Combining (3.9.55) and (3.9.56), we obtain (3.9.48). Now, the conclusion follows from Theorem 3.9.6. ∎

The following result is a special case of Theorem 3.9.6. In fact, we let $\rho(t) = 1$ to obtain

Corollary 3.9.8. If for each $T \geq t_0$ there exist $H \in \Gamma$ and $a, b, c \in \mathbb{R}$ such that $T \leq a < c < b$ and

$$\frac{1}{H(c, a)} \int_a^c H(s, a)q(s)ds + \frac{1}{H(b, c)} \int_c^b H(b, s)q(s)ds$$
$$> \frac{1}{H(c, a)} \int_a^c \frac{a(s)h_1^{\alpha+1}(s, a)}{\alpha_0 H^\alpha(s, a)} ds + \frac{1}{H(b, c)} \int_c^b \frac{a(s)h_2^{\alpha+1}(b, s)}{\alpha_0 H^\alpha(b, s)} ds,$$
$$(3.9.57)$$

then equation (3.1.1) is oscillatory.

If condition (3.9.1) holds, and (3.9.42) is replaced by

$$\frac{\partial}{\partial t} H(t, s) = h_1(t, s)H^{\alpha/(\alpha+1)}(t, s) \quad \text{and} \quad \frac{\partial}{\partial s}(t, s) = -h_2(t, s)H^{\alpha/(\alpha+1)}(t, s),$$
$$(3.9.58)$$

where h_1, h_2 are nonnegative and continuous functions on $\mathcal{D}$, h_1, $h_2 \in \mathcal{L}_{Loc}(\mathcal{D}, \mathbb{R})$, then we shall say $H \in \Gamma^*$. In this case Corollary 3.9.8 is reduced to the following result.

Corollary 3.9.9. If for each $T \geq t_0$ there exist $H \in \Gamma^*$ and a, b, $c \in \mathbb{R}$ such that $T \leq a < c < b$ and

$$\frac{1}{H(c,a)} \int_a^c H(s,a)q(s)ds + \frac{1}{H(b,c)} \int_c^b H(b,s)q(s)ds$$

$$> \frac{1}{\alpha_0} \left[\frac{1}{H(c,a)} \int_a^c a(s)h_1^{\alpha+1}(s,a)ds + \frac{1}{H(b,c)} \int_c^b a(s)h_2^{\alpha+1}(b,s)ds \right],$$

$$\tag{3.9.59}$$

then equation (3.1.1) is oscillatory.

Next, we state the following interesting result.

Corollary 3.9.10. If for each $T \geq t_0$ there exist $H \in \Gamma$ and a, b, $c \in \mathbb{R}$ such that $T \leq a < c < b$, and either

$$\int_a^c H(s,a)q(s)ds > \frac{1}{\alpha_0} \int_a^c \frac{a(s)h_1^{\alpha+1}(s,a)}{H^\alpha(s,a)}ds, \tag{3.9.60}$$

or

$$\int_c^b H(b,s)q(s)ds > \frac{1}{\alpha_0} \int_c^b \frac{a(s)h_2^{\alpha+1}(b,s)}{H^\alpha(b,s)}ds, \tag{3.9.61}$$

then equation (3.1.1) is oscillatory.

In the case when condition (3.9.58) holds Corollary 3.9.10 reduces to

Corollary 3.9.11. If for each $T \geq t_0$ there exist $H \in \Gamma^*$ and a, b, $c \in \mathbb{R}$ such that $T \leq a < c < b$, and either

$$\int_a^c H(s,a)q(s)ds > \frac{1}{\alpha_0} \int_a^c a(s)h_1^{\alpha+1}(s,a)ds, \tag{3.9.62}$$

or

$$\int_c^b H(b,s)q(s)ds > \frac{1}{\alpha_0} \int_c^b a(s)h_2^{\alpha+1}(b,s)ds, \tag{3.9.63}$$

then equation (3.1.1) is oscillatory.

Example 3.9.3. Consider the half–linear differential equation

$$\left(a(t)|x'(t)|^{\alpha-1}x'(t) \right)' + q(t)|x(t)|^{\alpha-1}x(t) = 0, \tag{3.9.64}$$

where $\alpha > 0$ is a real constant,

$$a(t) = \begin{cases} e^{-2n(\alpha+1)} & \text{for} \quad 2n \leq t \leq 2n+1 \\ 1 & \text{for} \quad 2n+1 < t < 2n+2 \end{cases}$$

and

$$q(t) = \begin{cases} q_1 e^{t-2n(\alpha+2)} & \text{for} \quad 2n \le t \le 2n+1 \\ 0 & \text{for} \quad 2n+1 < t < 2n+2, \end{cases}$$

where $q_1 > 0$ is a constant and $n \in \mathbb{N}$. Obviously, $\int_0^t q(s)ds$ is convergent as $t \to \infty$. For $T \ge 0$ there exists $n \in \mathbb{N}$ such that $2n \ge T$. Let $a = 2n$, $b = 2n+1$ and $H(t,s) = (e^t - e^s)^{\alpha+1}$. Then, $H \in \Gamma^*$, $h_1(t,s) = (\alpha+1)e^t$ and $h_2(t,s) = (\alpha+1)e^s$. To show that equation (3.9.64) is oscillatory, by Corollary 3.9.11 it suffices to verify that for all $c \in (a,b)$ either condition (3.9.62), or (3.9.63) holds. In fact, by a simple computation, we find that condition (3.9.62) holds for each $\epsilon = c - 2n \in (0,1)$ if

$$q_1 \frac{(e^\epsilon - 1)^{\alpha+2}}{(e^{\epsilon(\alpha+1)} - 1)} > \frac{\alpha+2}{\alpha+1} \tag{3.9.65}$$

and condition (3.9.63) holds for each $\epsilon_1 = 2n+1-c \in (0,1)$ if

$$q_1 \frac{(e - e^{\epsilon_1})^{\alpha+2}}{(e - e^{\epsilon_1(\alpha+1)})} > \frac{\alpha+2}{\alpha+1}. \tag{3.9.66}$$

For an appropriate choice of q_1, we note that both the conditions (3.9.65) and (3.9.66) hold, and hence equation (3.9.64) is oscillatory.

For the case $H = H(t-s) \in \Gamma^*$, we have $h_1(t-s) = h_2(t-s)$ and denote them by $h(t-s)$. The subclass of Γ^* containing such $H(t-s)$ is denoted by Γ_0.

Applying Corollary 3.9.9 to Γ_0, we get the following result.

Theorem 3.9.9. If for each $T \ge t_0$ there exist $H \in \Gamma_0$ and $a, c \in \mathbb{R}$ such that $T \le a < c$, and

$$\int_a^c H(s,a)[q(s) + q(2c - s)]ds > \frac{1}{\alpha_0} \int_a^c h^{\alpha+1}(s,a)[a(s) + a(2c - s)]ds, \tag{3.9.67}$$

then equation (3.1.1) is oscillatory.

Proof. Let $b = 2c - a$. Then, $H(b - c) = H(c - a) = H((b - a)/2)$ and for any $y \in \mathcal{L}[a,b]$, we have $\int_c^b y(s)ds = \int_a^c y(2c - s)ds$. Hence, $\int_c^b H(b - s)q(s)ds = \int_a^c H(s - a)q(2c - s)ds$ and

$$\int_c^b a(s)h^{\alpha+1}(b - s)ds = \int_c^b a(2c - s)h^{\alpha+1}(s - a)ds.$$

Thus, (3.9.67) implies that (3.9.59) holds for each $H \in \Gamma_0$, and therefore, equation (3.1.1) is oscillatory. ∎

Example 3.9.4. Consider the half–linear differential equation

$$\left(|x'(t)|^{\alpha-1}x'(t)\right)' + q(t)|x(t)|^{\alpha-1}x(t) = 0, \tag{3.9.68}$$

where $\alpha > 0$ is a constant, and

$$q(t) = \begin{cases} 5(t-2n) & \text{for} \quad 3n \le t \le 3n+1 \\ 5(-t+3n+2) & \text{for} \quad 3n+1 < t \le 3n+2 \\ -n & \text{for} \quad 3n+2 < t < 3n+3, \quad n \in \mathbb{N}_0. \end{cases}$$

For any $T \ge 0$ there exists $n \in \mathbb{N}_0$ such that $3n \ge T$. Let $a = 3n$, $c = 3n+1$, and $H(t-s) = (t-s)^{\alpha+1}$. Then, $h(t-s) = \alpha+1$. It is easy to verify that condition (3.9.67) holds, and hence equation (3.9.68) is oscillatory by Theorem 3.9.9. Note that in equation (3.9.68), we have $\int_0^\infty q(s)ds = -\infty$.

Next, we choose $H(t-s) = (t-s)^n$ for $n > \alpha$. Then, $H \in \Gamma_0$ and $h(t-s) = n(t-s)^{(n-\alpha-1)/(\alpha+1)}$. Then, we have the following oscillation criterion for (3.1.1).

Theorem 3.9.10. Equation (3.1.1) is oscillatory if for some $n > \alpha$ and for any $r \ge t_0$ either

(I)   the following inequalities hold

$$\limsup_{t\to\infty} \int_r^t \left[(s-r)^n q(s) - \frac{n^{\alpha+1}}{\alpha_0}(s-r)^{n-\alpha-1}a(s)\right] ds > 0$$

and

$$\limsup_{t\to\infty} \int_r^t \left[(t-s)^n q(s) - \frac{n^{\alpha+1}}{\alpha_0}(t-s)^{n-\alpha-1}a(s)\right] ds > 0,$$

or

(II)   the following inequality holds

$$\limsup_{t\to\infty} \int_r^t \left[(s-r)^n(q(s)+q(2t-s)) - \frac{n^{\alpha+1}}{\alpha_0}(s-r)^{n-\alpha-1}(a(s)+a(2t-s))\right] ds$$
$$> 0.$$

The following corollary is an immediate consequence of Theorem 3.9.10.

Corollary 3.9.12. Let $a(t) = 1$. Then equation (3.1.1) is oscillatory if for any $r \ge t_0$ and some $n > \alpha$ either

(I)   the following two inequalities hold

$$\limsup_{t\to\infty} \frac{1}{t^{n-\alpha}} \int_r^t (s-r)^n q(s)ds > \frac{n^{\alpha+1}}{\alpha_0(n-\alpha)}$$

and

$$\limsup_{t\to\infty} \frac{1}{t^{n-\alpha}} \int_r^t (t-s)^n q(s)ds \ > \ \frac{n^{\alpha+1}}{\alpha_0(n-\alpha)},$$

or

(II) the following inequality holds

$$\limsup_{t\to\infty} \frac{1}{t^{n-\alpha}} \int_r^t (s-r)^n[q(s)+q(2t-s)]ds \ > \ \frac{n^{\alpha+1}}{2\alpha_0(n-\alpha)}.$$

For our next result, we define $R(t) = \int_r^t a^{-1/\alpha}(s)ds$ for $t \geq r \geq t_0$ and let $H(t,s) = (R(t) - R(s))^n$ for $t \geq s \geq t_0$, where $n > \alpha$ is a real constant. We let h_1 and h_2 be defined in such a way that condition (3.9.58) holds. In fact, we have

$$h_1(t,s) \ = \ na^{-1/\alpha}(t)(R(t)-R(s))^{(n-\alpha-1)/(\alpha+1)}$$

and

$$h_2(t,s) \ = \ na^{-1/\alpha}(s)(R(t)-R(s))^{(n-\alpha-1)/(\alpha+1)}.$$

Corollary 3.9.13. Equation (3.1.1) is oscillatory if $\lim_{t\to\infty} R(t) = \infty$ for some $n > \alpha$, and for each $r \geq t_0$ the following two inequalities hold

$$\limsup_{t\to\infty} \frac{1}{R^{n-\alpha}(t)} \int_r^t (R(s)-R(r))^n q(s)ds \ > \ \frac{n^{\alpha+1}}{\alpha_0(n-\alpha)}$$

and

$$\limsup_{t\to\infty} \frac{1}{R^{n-\alpha}(t)} \int_r^t (R(t)-R(s))^n q(s)ds \ > \ \frac{n^{\alpha+1}}{\alpha_0(n-\alpha)}.$$

Proof. The proof is easy and hence omitted. ∎

Example 3.9.5. Consider the half–linear differential equation

$$\left(t^\alpha |x'(t)|^{\alpha-1}x'(t)\right)' + \frac{c}{t\ln^\alpha t}|x(t)|^{\alpha-1}x(t) \ = \ 0, \tag{3.9.69}$$

where $\alpha > 0$ and $c > 0$ are constants. Let $R(t) = \int^t a^{-1/\alpha}(s)ds = \ln t,\ t > 1$. Then, we have $H(t,s) = (R(t)-R(s))^n = (\ln(t/s))^n$ for $t \geq s > 1$, where $n > \alpha$ is a constant. It is easy to check that all the conditions of Corollary 3.9.13 are satisfied and hence equation (3.9.69) is oscillatory for all $c > 0$ and $\alpha > 0$.

3.9.3. The Weighted–Average Oscillation Criteria

Here, we shall extend Lemmas 2.5.1 and 2.5.2, and Theorems 2.5.4 and 2.5.5 to equation (3.1.1). For this, let Ω be the set of all nonnegative locally integrable functions g on $[t_0, \infty)$, $t_0 \geq 0$ satisfying the condition

$$\limsup_{t \to \infty} \left(\int^t g(s)ds \right)^{(1/\alpha)-k} \{G_k(\infty) - G_k(t)\} > 0 \text{ for some } k \in [0, 1/\alpha),$$
(3.9.70)

where

$$G_k(t) = \int^t g(s) \frac{\left(\int^s g(\tau)d\tau \right)^k}{\left(\int^s a(\tau)g^{\alpha+1}(\tau)d\tau \right)^{1/\alpha}} ds.$$
(3.9.71)

If $G_k(\infty) = \infty$ in (3.9.70), then $g \in \Omega$. Let Ω_0 be the set of all nonnegative locally integrable function g on $[t_0, \infty)$ satisfying

$$\lim_{t \to \infty} \frac{\int^t a(s)g^{\alpha+1}(s)ds}{\int^t g(s)ds} = 0.$$
(3.9.72)

It is clear that for (3.9.70) or (3.9.72) to be satisfied by a nonnegative function g it is necessary that

$$\int^\infty g(s)ds = \infty.$$
(3.9.73)

On the other hand every bounded nonnegative locally integrable function g satisfying (3.9.73) belongs to Ω_0 and $\Omega_0 \subset \Omega$. Further, since all nonnegative polynomials are members of Ω_0, Ω_0 contains some unbounded functions. Members of the classes Ω and Ω_0 will be called weight functions.

If (3.1.1) is nonoscillatory and $x(t)$ is its nontrivial solution, then there exist a number $T \geq t_0$ and a function $w(t) = a(t)\psi(x'(t))/\psi(x(t)) \in C^1([T, \infty), \mathbb{R})$, which satisfies

$$w'(t) = -q(t) - \alpha a^{-1/\alpha}(t)|w(t)|^{(\alpha+1)/\alpha}, \quad t \geq T.$$
(3.9.74)

Clearly, (3.9.74) is equivalent to the integral equation

$$w(t) = w(s) - \alpha \int_s^t a^{-1/\alpha}(\tau)|w(\tau)|^{(\alpha+1)/\alpha}d\tau - \int_s^t q(\tau)d\tau$$
(3.9.75)

for $t \geq s \geq T$. In what follows for $g \in \Omega$, we shall denote by

$$A_g(s, t) = \frac{\int_s^t g(\tau) \int_s^\tau q(u)du d\tau}{\int_s^t g(\tau)d\tau}.$$

Lemma 3.9.3. Assume that $w(t)$ satisfies equation (3.9.74) on $[T, \infty)$ for some $T \geq t_0$. If there exists $g \in \Omega$ such that

$$\liminf_{t \to \infty} A_g(., t) > -\infty,$$
(3.9.76)

then

$$\int^\infty a^{-1/\alpha}(s)|w(s)|^{(\alpha+1)/\alpha}ds < \infty.$$

Proof. Let $A(s,t) = A_g(s,t)$ and assume that

$$\int^\infty a^{-1/\alpha}(s)|w(s)|^{(\alpha+1)/\alpha}ds = \infty. \qquad (3.9.77)$$

Multiplying both sides of equation (3.9.75) by $g(t)$ and integrating it from ξ to t, we obtain

$$\int_\xi^t g(s)w(s)ds$$

$$= w(\xi)\int_\xi^t g(s)ds - \int_\xi^t g(s)\int_\xi^s q(\tau)d\tau ds - \alpha\int_\xi^t g(s)\int_\xi^s a^{-\frac{1}{\alpha}}(\tau)|w(\tau)|^{\frac{\alpha+1}{\alpha}}d\tau ds$$

$$= w(\xi)\int_\xi^t g(s)ds - A(\xi,t)\int_\xi^t g(s)ds - \alpha\int_\xi^t g(s)\int_\xi^t a^{-\frac{1}{\alpha}}(\tau)|w(\tau)|^{\frac{\alpha+1}{\alpha}}d\tau ds$$

$$= [w(\xi) - A(\xi,t)]\int_\xi^t g(s)ds - \alpha\int_\xi^t g(s)\int_\xi^s a^{-\frac{1}{\alpha}}(\tau)|w(\tau)|^{\frac{\alpha+1}{\alpha}}d\tau ds,$$

$$(3.9.78)$$

where $t \geq \xi \geq T$. From equation (3.9.75), we have

$$w(\xi) = w(T) - \int_t^\xi q(s)ds - \alpha\int_T^\xi a^{-1/\alpha}(s)|w(s)|^{(\alpha+1)/\alpha}ds.$$

Since $g \in \Omega$, (3.9.73) holds. This implies

$$A(\xi,t) = \frac{\int_T^t g(s)ds}{\int_\xi^t g(s)ds}A(T,t) - \int_T^\xi q(s)ds - \frac{\int_T^\xi g(s)\int_T^s q(\tau)d\tau ds}{\int_\xi^t g(s)ds}$$

$$= \frac{\int_T^t g(s)ds}{\int_\xi^t g(s)ds}A(T,t) - \int_T^\xi q(s)ds + o(1) \quad \text{as} \quad t \to \infty.$$

Thus, we get

$$w(\xi) - A(\xi,t) = w(T) - \frac{\int_T^t g(s)ds}{\int_\xi^t g(s)ds}A(T,t)$$

$$-\alpha\int_T^\xi a^{-1/\alpha}(s)|w(s)|^{(\alpha+1)/\alpha}ds + o(1) \quad \text{as} \quad t \to \infty.$$

$$(3.9.79)$$

Since $g \in \Omega$ there exists a positive number $\lambda > 0$ such that

$$\frac{\lambda^{-1/\alpha}}{\alpha} < \left(\frac{1}{\alpha} - k\right)\limsup_{t\to\infty}\left(\int^t g(s)ds\right)^{(1/\alpha)-k}[G_k(\infty) - G_k(t)],$$

$$(3.9.80)$$

where k is as in (3.9.70). It follows from (3.9.76), (3.9.77) and (3.9.79) that there exist two numbers t_1 and t_2 with $t_2 \geq t_1 \geq T$ such that

$$w(t_1) - A(t_1, t) \leq -\lambda \quad \text{for all} \quad t \geq t_2. \tag{3.9.81}$$

Let $y(t) = \int_{t_1}^t g(s)w(s)ds$. Then Hölder's inequality for $s \geq t_1$ yields

$$|y(s)|^{(\alpha+1)/\alpha} \leq \left(\int_{t_1}^s a^{-1/\alpha}(\tau)|w(\tau)|^{(\alpha+1)/\alpha} d\tau \right) \left(\int_{t_1}^s a(\tau)g^{\alpha+1}(\tau)d\tau \right)^{1/\alpha}.$$

Thus, from (3.9.78) and (3.9.81), we find

$$y(t) \leq -\lambda \int_{t_1}^t g(s)ds - \alpha \int_{t_1}^t g(s)|y(s)|^{\frac{\alpha+1}{\alpha}} \left(\int_{t_1}^s a(\tau)g^{\alpha+1}(\tau)d\tau \right)^{-\frac{1}{\alpha}} ds = -F(t). \tag{3.9.82}$$

Hence,

$$F'(t) = \lambda g(t) + \alpha g(t)|y(t)|^{(\alpha+1)/\alpha} \left(\int_{t_1}^t a(s)g^{\alpha+1}(s)ds \right)^{-1/\alpha} \tag{3.9.83}$$

and

$$0 \leq \lambda \int_{t_1}^t g(s)ds \leq F(t) \leq |y(t)|. \tag{3.9.84}$$

Now, from (3.9.82), (3.8.83) and (3.9.84) it follows that

$$F'(t)F^{k-1-(1/\alpha)}(t) \geq F'(t)F^k(t)|y(t)|^{-(\alpha+1)/\alpha}$$
$$\geq \alpha\lambda^k g(t) \left(\int_{t_1}^t g(s)ds \right)^k \left(\int_{t_1}^t a(s)g^{\alpha+1}(s)ds \right)^{-1/\alpha}.$$

For $k < 1/\alpha$, we integrate the above inequality from $t \geq t_2$ to u and let $u \to \infty$, to obtain

$$\frac{1}{(1/\alpha) - k} F^{k-(1/\alpha)}(t) \geq \alpha\lambda^k[G_k(\infty) - G_k(t)].$$

Therefore, from (3.9.84), we get

$$\frac{\lambda^{-1/\alpha}}{\alpha} \geq \left(\frac{1}{\alpha} - k \right) \left(\int_{t_1}^t g(s)ds \right)^{(1/\alpha)-k} [G_k(\infty) - G_k(t)],$$

which contradicts (3.9.80). This completes the proof. ∎

Lemma 3.9.4. Assume that $w(t)$ satisfies equation (3.9.74) on $[T, \infty)$ for some $T \geq t_0$. If $\int^\infty a^{-1/\alpha}(s)|w(s)|^{(\alpha+1)/\alpha}ds < \infty$, then for any $g \in \Omega_0$, $\lim_{t\to\infty} A_g(., t)$ exists.

Proof. As in Lemma 3.9.3, (3.9.78) holds. This implies that

$$A(\xi,t) = w(\xi) - \frac{\int_\xi^t g(s)w(s)ds}{\int_\xi^t g(s)ds} - \alpha \frac{\int_\xi^t g(s)\int_\xi^s a^{-1/\alpha}(\tau)|w(\tau)|^{(\alpha+1)/\alpha}d\tau ds}{\int_\xi^t g(s)ds}.$$
(3.9.85)

Since $g \in \Omega_0$, (3.9.73) holds. Thus, we have

$$\lim_{t\to\infty} \frac{\int_\xi^t g(s)\int_\xi^s a^{-\frac{1}{\alpha}}(\tau)|w(\tau)|^{\frac{\alpha+1}{\alpha}}d\tau ds}{\int_\xi^t g(s)ds} = \int_\xi^\infty a^{-\frac{1}{\alpha}}(s)|w(s)|^{\frac{\alpha+1}{\alpha}}ds < \infty.$$

By Hölder's inequality, we find

$$0 \le \lim_{t\to\infty} \frac{\left|\int_\xi^t g(s)w(s)ds\right|}{\int_\xi^t g(s)ds}$$

$$\le \lim_{t\to\infty} \frac{\left(\int_\xi^t a(s)g^{\alpha+1}(s)ds\right)^{1/(\alpha+1)} \left(\int_\xi^t a^{-1/\alpha}(s)|w(s)|^{(\alpha+1)/\alpha}ds\right)^{\alpha/(\alpha+1)}}{\int_\xi^t g(s)ds}.$$

Hence, in view of (3.9.85), $\lim_{t\to\infty} A_g(\xi,t)$ exists and

$$\lim_{t\to\infty} A_g(\xi,t) = w(\xi) - \alpha \int_\xi^\infty a^{-1/\alpha}(s)|w(s)|^{(\alpha+1)/\alpha}ds. \qquad \blacksquare$$

From Lemmas 3.9.3 and 3.9.4 the following oscillation criterion for (3.1.1) is immediate.

Theorem 3.9.11. If there exists $g \in \Omega$ such that

$$\lim_{t\to\infty} G_k(t) = \infty \quad \text{for some} \quad k \in [0, 1/\alpha) \quad \text{and} \quad \lim_{t\to\infty} A_g(.,t) = \infty,$$
(3.9.86)

or, there exists $g \in \Omega_0$ such that

$$-\infty < \liminf_{t\to\infty} A_g(.,t) < \limsup_{t\to\infty} A_g(.,t) \le \infty,$$
(3.9.87)

then equation (3.1.1) is oscillatory.

Corollary 3.9.14. Suppose $g(t)$ and $h(t)$ are two nonnegative bounded functions on $[T,\infty)$ for some $T \ge t_0$ with $\int^\infty g(s)ds = \infty = \int^\infty h(s)ds$. If the functions $a(t)g^\alpha(t)$ and $a(t)h^\alpha(t)$ are bounded and

$$\lim_{t\to\infty} A_g(T,t) < \lim_{t\to\infty} A_h(T,t),$$
(3.9.88)

then equation (3.1.1) is oscillatory.

Proof. Let a and b be two numbers satisfying $\lim_{t\to\infty} A_g(T,t) < a < b < \lim_{t\to\infty} A_h(T,t)$. Let $f(t) = h(t)$ for $T \le t < t_1$, where t_1 is such that $A_h(T,t_1) \ge b$ and $\int_T^t h(s)ds \ge 1$. Further, let $f(t) = g(t)$ for $t_1 \le t < t_2$, where t_2 is such that $A_f(T,t_2) \le a$ and $\int_T^{t_2} f(s)ds \ge 2$. This is possible because

$$
A_f(T,t_2) = \frac{\int_T^{t_2} f(s) \int_T^s q(u)du\,ds}{\int_T^{t_2} f(s)ds} = \frac{\int_T^{t_1} [h(s) - g(s)] \int_T^s q(u)du\,ds}{\int_T^{t_1} h(s)ds + \int_{t_1}^{t_2} g(s)ds}
$$

$$
+ \frac{\int_T^{t_2} g(s) \int_T^s q(u)du\,ds}{\int_T^{t_2} g(s)ds} \cdot \frac{\int_T^{t_2} g(s)ds}{\int_T^{t_1} h(s)ds + \int_{t_1}^{t_2} g(s)ds}
$$

$$
= A_g(T,t_2)[1 + o(1)] + o(1) \quad \text{as} \quad t_2 \to \infty.
$$

Continuing in this manner, we obtain nonnegative, bounded and noninte-grable function $f(t)$ defined on $[T,\infty)$ such that $\limsup_{t\to\infty} A_f(T,t) \ge b > a \ge \liminf_{t\to\infty} A_f(T,t)$ and

$$
\lim_{t\to\infty} \frac{\int_T^t a(s) f^{\alpha+1}(s)ds}{\int_T^t f(s)ds} = 0.
$$

This implies that $f(t) \in \mathfrak{N}_0$ and $\lim A_f(T,t)$ does not exist. By Theorem 3.9.11 equation (3.1.1) is then oscillatory. ∎

Corollary 3.9.15. Let

$$
k(t) = \int_{t_0}^t a^{-1/\alpha}(s)ds \to \infty \quad \text{as} \quad t \to \infty. \tag{3.9.89}
$$

If

$$
-\infty < \liminf_{t\to\infty} \frac{1}{k(t)} \int^t \int^s a^{-1/\alpha}(u)q(u)du\,ds
$$
$$
< \limsup_{t\to\infty} \int^t \int^s a^{-1/\alpha}(u)q(u)du\,ds \le \infty, \tag{3.9.90}
$$

then equation (3.1.1) is oscillatory.

Proof. Let $g(t) = a^{-1/\alpha}(s)$. Then, $g(t) \in \mathfrak{N}_0$ and it follows from (3.9.90) that $\lim_{t\to\infty} A_g(.,t)$ does not exist. By Theorem 3.9.11 equation (3.1.1) is then oscillatory. ∎

Lemma 3.9.5. Assume that $c(s)$ and $P(s,t)$ are nonnegative continuous functions for $T \le s,\ t < \infty$. If

$$
\int_t^\infty P(s,t)c^{(\alpha+1)/\alpha}(s)ds \le (\alpha+1)^{-(\alpha+1)/\alpha}c(t) \quad \text{for} \quad t \ge T \tag{3.9.91}
$$

then the equation

$$y(t) = c(t) + \alpha \int_t^\infty P(s,t)|y(s)|^{(\alpha+1)/\alpha}ds \quad \text{for} \quad t \geq T \qquad (3.9.92)$$

has a continuous solution $y(t)$.

Proof. Let $y_1(t) = c(t)$ and define

$$y_{k+1}(t) = c(t) + \alpha \int_t^\infty P(s,t)|y_k(s)|^{(\alpha+1)/\alpha}ds, \quad k = 0,1,2,\cdots.$$

Then from (3.9.91), we find

$$\begin{aligned}
y_1(t) \leq y_2(t) &= c(t) + \alpha \int_t^\infty P(s,t)c^{(\alpha+1)/\alpha}(s)ds \\
&\leq c(t) + \alpha(\alpha+1)^{-(\alpha+1)/\alpha}c(t) \leq (\alpha+1)c(t)
\end{aligned}$$

and $y_1(t) \leq y_2(t) \leq (\alpha+1)c(t)$ for $t \geq T$. Suppose $y_1(t) \leq y_2(t) \leq \cdots \leq y_n(t) \leq (\alpha+1)c(t)$, $t \geq T$ where n is a positive integer. Then, we have

$$\begin{aligned}
y_n(t) &= c(t) + \alpha \int_t^\infty P(s,t)|y_{n-1}(s)|^{(\alpha+1)/\alpha}ds \\
&\leq c(t) + \alpha \int_t^\infty P(s,t)|y_n(s)|^{(\alpha+1)/\alpha}ds = y_{n+1}(t) \\
&\leq c(t) + \alpha(\alpha+1)^{(\alpha+1)/\alpha} \int_t^\infty P(s,t)c^{(\alpha+1)/\alpha}(s)ds \\
&\leq c(t) + \alpha(\alpha+1)^{(\alpha+1)/\alpha}(\alpha+1)^{-(\alpha+1)/\alpha}c(t) \\
&= (\alpha+1)c(t) \quad \text{for} \quad t \geq T.
\end{aligned}$$

Thus, the sequence $\{y_n(t)\}_{n=0}^\infty$ is monotonically nondecreasing and bounded above by $(\alpha+1)c(t)$. Hence, $\{y_n(t)\}$ converges uniformly to a continuous function $y(t)$, which is a solution of equation (3.9.92). ∎

Lemma 3.9.6. Assume that $c(t)$ and $P(s,t)$ are nonnegative continuous functions for $T \leq s$, $t < \infty$. If there exists $\epsilon > 0$ such that

$$\int_t^\infty P(s,t)c^{(\alpha+1)/\alpha}(s)ds \geq (\alpha+1)^{-(\alpha+1)/\alpha}(1+\epsilon)c(t) \not\equiv 0 \quad \text{for} \quad t \geq T,$$
$$(3.9.93)$$

then the inequality

$$y(t) \geq c(t) + \alpha \int_t^\infty P(s,t)|y(s)|^{(\alpha+1)/\alpha}ds \quad \text{for} \quad t \geq T \qquad (3.9.94)$$

does not have a continuous solution $y(t)$.

Proof.   Suppose to the contrary that $y(t)$ is a continuous function satisfying (3.9.94) for $t \geq T$. Then, $y(t) \geq c(t) \geq 0$ for $t \geq T$, which implies $y^{(\alpha+1)/\alpha}(t) \geq c^{(\alpha+1)/\alpha}(t) \geq 0$ for $t \geq T$. Thus, we have

$$
\begin{aligned}
y(t) &\geq c(t) + \alpha \int_t^\infty P(s,t) c^{(\alpha+1)/\alpha}(s) ds \\
&\geq \left[1 + \alpha(1+\epsilon)(\alpha+1)^{-(\alpha+1)/\alpha} \right] c(t) \quad \text{for} \quad t \geq T.
\end{aligned}
$$

Continuing this process, we obtain $y(t) \geq \beta_n c(t)$, where $\beta_1 = 1$, $\beta_n < \beta_{n+1}$ and

$$
\beta_{n+1} = 1 + \alpha \beta_n^{(\alpha+1)/\alpha} (\alpha+1)^{-(\alpha+1)/\alpha}(1+\epsilon), \quad n = 1, 2, \cdots. \quad (3.9.95)
$$

We claim that

$$
\lim_{n \to \infty} \beta_n = \infty. \quad (3.9.96)
$$

Assume to the contrary that $\lim_{n \to \infty} \beta_n = \lambda < \infty$, where λ is a constant. Clearly, $\lambda \geq 1$. Now, it follows from (3.9.95) that

$$
\lambda = 1 + \alpha(1+\epsilon)\lambda^{(\alpha+1)/\alpha}(\alpha+1)^{-(\alpha+1)/\alpha}. \quad (3.9.97)
$$

But, it is easy to verify that (3.9.97) is impossible for $\lambda \geq 1$. This contradiction proves (3.9.96). Then, $c(t) \equiv 0$ for $t \geq T$, which contradicts (3.9.93). This completes the proof.   ∎

Theorem 3.9.12. If for every $t \geq T > t_0$,

$$
c(t) = \int_t^\infty q(s) ds \geq 0 \quad (3.9.98)
$$

and

$$
\int_t^\infty a^{-1/\alpha}(s) c^{(\alpha+1)/\alpha}(s) ds \leq \left(\frac{1}{\alpha+1} \right)^{(\alpha+1)/\alpha} c(t),
$$

then equation (3.1.1) is nonoscillatory.

Proof. By Lemma 3.9.5 the equation

$$
y(t) = c(t) + \alpha \int_t^\infty a^{-1/\alpha}(s)|y(s)|^{(\alpha+1)/\alpha} ds
$$

has a continuous solution $y(t)$ on $[T, \infty)$. Then, we have

$$
y'(t) = -q(t) - \alpha a^{-1/\alpha}(t)|y(t)|^{(\alpha+1)/\alpha} \quad \text{for} \quad t \geq T.
$$

Now, from Theorem 3.7.1 equation (3.1.1) is nonoscillatory.   ∎

Theorem 3.9.13. Suppose there is a $T \geq t_0$ such that condition (3.9.98) holds. If there is an $\epsilon > 0$ such that

$$\int_t^\infty a^{-1/\alpha}(s)c^{(\alpha+1)/\alpha}(s)ds \geq \left(\frac{1}{\alpha+1}\right)^{(\alpha+1)/\alpha}(1+\epsilon)c(t) \quad \text{for} \quad t \geq T,$$

(3.9.99)

then equation (3.1.1) is oscillatory.

Proof. Suppose to the contrary that equation (3.1.1) is nonoscillatory. Then following the proof of Theorem 3.7.2 there exist a number T_1 and a function $w(t) \in C^1([T_1, \infty), \mathbb{R})$ satisfying (3.7.4) on $[T_1, \infty)$. Without loss of generality let $T = T_1$, then it follows from (3.9.98) that for $g \in \Omega$, $\liminf_{t \to \infty} A_g(T, t) = \int_T^\infty q(s)ds > -\infty$. Now, from Lemma 3.9.3, we have $\int_T^\infty a^{-1/\alpha}(s)|w(s)|^{(\alpha+1)/\alpha}ds < \infty$. Thus, in view of (3.9.98) and (3.9.75), $w(t)$ satisfies $w(t) = c(t) + \alpha \int_t^\infty a^{-1/\alpha}(s)|w(s)|^{(\alpha+1)/\alpha}ds$. But, by (3.9.99) and Lemma 3.9.6, the equation $y(t) = c(t) + \alpha \int_t^\infty a^{-1/\alpha}(s)|y(s)|^{(\alpha+1)/\alpha}ds$ does not have a continuous solution, which is a contradiction. ∎

Example 3.9.6. Consider the half–linear differential equation

$$\left(|x'(t)|^{\alpha-1}x'(t)\right)' + \left(\frac{\alpha+1}{\alpha}\right)^{-(\alpha+1)}(1+t)^{-(\alpha+1)}|x(t)|^{\alpha-1}x(t) = 0.$$

(3.9.100)

It is easy to check that equation (3.9.100) has a nonoscillatory solution $x(t) = (1+t)^{\alpha/(\alpha+1)}$. Since

$$q(t) = \left(\frac{\alpha}{\alpha+1}\right)^{\alpha+1}(t+1)^{-(\alpha+1)},$$

$$\int_t^\infty \left(\int_s^\infty q(\tau)d\tau\right)ds = \left(\frac{\alpha}{\alpha+1}\right)^\alpha a^{-(1+2\alpha)/\alpha}(1+t)^{-\alpha}$$

$$= \left(\frac{\alpha}{\alpha+1}\right)^{(\alpha+1)/\alpha}\left(\frac{1}{\alpha}\right)^{(\alpha+1)/\alpha}\int_t^\infty q(s)ds = \left(\frac{1}{\alpha+1}\right)^{(\alpha+1)/\alpha}\int_t^\infty q(s)ds.$$

This means that the number $(1/(\alpha+1))^{(\alpha+1)/\alpha}$ in Theorems 3.9.12 and 3.9.13 is the best possible.

3.10. Oscillation Criteria–Integrable Coefficients

In what follows, we shall assume that

$$R(t) = \int_{t_0}^t a^{-1/\alpha}(s)ds \to \infty \quad \text{as} \quad t \to \infty \quad \text{and} \quad Q(t) = \int_t^\infty q(s)ds < \infty.$$

(3.10.1)

To prove the main result of this section, we shall need the following lemma.

Lemma 3.10.1. If

$$\int^{\infty} R^{\alpha-1}(s)a^{-1/\alpha}(s)\exp\left(-(\alpha+1)^{\frac{\alpha+1}{\alpha}}\int^{s}a^{-1/\alpha}(\xi)(Q^{+}(\xi))^{1/\alpha}d\xi\right)ds < \infty,$$

$$(3.10.2)$$

then

$$\int^{\infty} a^{-1/\alpha}(s)R^{\alpha-1}(s)|x(s)|^{-\lambda}ds = \infty \qquad (3.10.3)$$

for any nonoscillatory solution $x(t)$ of (3.1.1) and for any constant $\lambda > 0$.

Proof. Let $x(t)$ be a nonoscillatory solution of equation (3.1.1). Without loss of generality, we assume that $x(t) \neq 0$ for $t \geq T \geq t_0$. It follows from Theorem 3.7.2 that (3.7.4) holds for $t \geq T$, where $w(t)$ is defined in (3.7.6). Let $y(t) = \int_t^{\infty} a^{-1/\alpha}(s)|w(s)|^{(\alpha+1)/\alpha}ds$ for $t \geq T$. Then for $t \geq T$, we have

$$y'(t) = -a^{-1/\alpha}(t)|w(t)|^{(\alpha+1)/\alpha} = -a^{-1/\alpha}(t)|Q(t)+\alpha y(t)|^{(\alpha+1)/\alpha} \leq 0.$$

Hence, $y'(t)+(\alpha+1)^{(\alpha+1)/\alpha}a^{-1/\alpha}(t)(Q^{+}(t))^{1/\alpha}y(t) \leq 0$ for $t \geq T$, which implies

$$y(t) \leq y(T)\exp\left(-(\alpha+1)^{(\alpha+1)/\alpha}\int_{T}^{t}a^{-1/\alpha}(s)(Q^{+}(s))^{1/\alpha}ds\right) \quad \text{for } t \geq T.$$

From (3.10.2), we find $\int^{\infty} R^{\alpha-1}(s)a^{-1/\alpha}(s)y(s)ds < \infty$, and hence $\int^{\infty} R^{\alpha}(s)a^{-1/\alpha}(s)|w(s)|^{(\alpha+1)/\alpha}ds < \infty$. Now, since

$$\ln\left(\frac{x(t)}{x(T)}\right) = \int_{T}^{t} a^{-1/\alpha}(s)|w(s)|^{(1-\alpha)/\alpha}w(s)ds,$$

by Hölder's inequality, we get

$$\ln\left(\frac{x(t)}{x(T)}\right) \leq \left(\int_{T}^{t} R^{\alpha}(s)a^{-1/\alpha}(s)|w(s)|^{(\alpha+1)/\alpha}ds\right)^{1/(\alpha+1)}$$
$$\times \left(\int_{T}^{t} \frac{a^{-1/\alpha}(s)}{R(s)}ds\right)^{\alpha/(\alpha+1)}, \quad t \geq T.$$

Thus, there exist constants k_1 and k_2 such that

$$|x(t)| \leq k_2\exp\left(k_1(\ln R(t))^{\alpha/(\alpha+1)}\right) \quad \text{for } t \geq T.$$

Therefore, we have

$$\int^{\infty} R^{\alpha-1}(s)a^{-1/\alpha}(s)|x(t)|^{-\lambda}ds$$

$$\geq k_2^{-\lambda}\int^{\infty} R^{\alpha-1}(s)a^{-1/\alpha}(s)\exp\left(-\lambda k_1(\ln R(s))^{\alpha/(\alpha+1)}\right)ds = \infty.$$

This completes the proof.   ■

Theorem 3.10.1. If there is a constant λ, $0 < \lambda \le (\alpha + 1)^{(\alpha+1)/\alpha}$ such that

$$\int^{\infty} R^{\alpha-1}(s)a^{-1/\alpha}(s)\exp\left(-\lambda \int^{s} a^{-1/\alpha}(\xi)|Q(\xi)|^{(1-\alpha)/\alpha}Q(\xi)d\xi\right)ds \;<\; \infty,$$

$$\text{(3.10.4)}$$

then equation (3.1.1) is oscillatory.

Proof. Let $x(t)$ be a nonoscillatory solution of equation (3.1.1), say, $x(t) > 0$ for $t \ge T \ge t_0$. From (3.10.4) we find that (3.10.2) holds, and this in view of Lemma 3.10.1 implies (3.10.3). Since $x(t)$ is nonoscillatory, it follows from Theorem 3.7.2 that (3.7.4) holds for $t \ge T$ where $w(t)$ is as in (3.7.6). Hence, for $t \ge T$,

$$\frac{d}{dt}(\ln x(t)) \;=\; a^{-1/\alpha}(t)|w(t)|^{(1-\alpha)/\alpha}w(t) \;\ge\; a^{-1/\alpha}(t)|Q(t)|^{(1-\alpha)/\alpha}Q(t),$$

which implies

$$x(t) \;\ge\; x(T)\exp\left(\int_{T}^{t}a^{-1/\alpha}(s)|Q(s)|^{(1-\alpha)/\alpha}Q(s)ds\right), \quad t \ge T.$$

Thus, from (3.10.4), we obtain $\int^{\infty} R^{\alpha-1}(s)a^{-1/\alpha}(s)|x(s)|^{-\lambda}ds < \infty$, which contradicts (3.10.3). This completes the proof.   ■

Corollary 3.10.1. If $\int^{\infty}\exp\left(-2\int^{s}Q(\xi)d\xi\right)ds < \infty$, then the linear equation $x''(t) + q(t)x(t) = 0$, where $q(t)$ is as in (3.1.1) is oscillatory.

Proof. Take $\alpha = 1$ and $\lambda = 2$ in Theorem 3.10.1.   ■

Example 3.10.1. Consider the half–linear equation

$$\left(|x'(t)|^{\alpha-1}x'(t)\right)' + ct^{-(\alpha+1)}|x(t)|^{\alpha-1}x(t) \;=\; 0, \quad t \ge t_0 > 0 \quad \text{(3.10.5)}$$

where $c > (1/(\alpha + 1))^{(\alpha+1)/\alpha}$ is a constant. Then, we have

$$R(t) \;=\; t, \quad Q(t) = \frac{c}{\alpha}t^{-\alpha}, \quad (\alpha+1)^{(\alpha+1)/\alpha}\left(\frac{c}{\alpha}\right)^{1/\alpha} \;>\; \alpha$$

and

$$\int^{\infty} R^{\alpha-1}(s)\exp\left(-(\alpha+1)^{(\alpha+1)/\alpha}\int^{s}Q^{\alpha}(\xi)d\xi\right)ds \;<\; \infty.$$

It follows from Theorem 3.10.1 that equation (3.10.5) is oscillatory. In fact, it has a nonoscillatory solution $x(t) = t^{\alpha/(\alpha+1)}$ if $c = (1/(\alpha+1))^{(\alpha+1)/\alpha}$.

Corollary 3.10.2. If

$$\liminf_{t\to\infty} R^\alpha(t)Q(t) > \frac{1}{\alpha}\left(\frac{\alpha}{\alpha+1}\right)^{\alpha+1}, \qquad (3.10.6)$$

then equation (3.1.1) is oscillatory.

Proof. It follows from (3.10.6) that there exist a constant $\gamma > (1/\alpha) \times (\alpha/(\alpha+1))^{\alpha+1}$ and a $T \geq t_0$ such that $R^\alpha(t)Q(t) > \gamma$ for $t \geq T$. Then, we have

$$\int^\infty R^{\alpha-1}(s)a^{-\frac{1}{\alpha}}(s)\exp\left(-\left(\frac{1}{\alpha+1}\right)^{\frac{\alpha+1}{\alpha}}\int^s a^{-\frac{1}{\alpha}}(\xi)|Q(\xi)|^{\frac{1-\alpha}{\alpha}}Q(\xi)d\xi\right)ds < \infty.$$

Thus, by Theorem 3.10.1, equation (3.1.1) is oscillatory. ■

Remark 3.10.1. From Example 3.10.1 it is clear that the inequality (3.10.6) is sharp, i.e., the constant $(1/\alpha)(\alpha/(\alpha+1))^{\alpha-1}$ is the best possible.

Corollary 3.10.3. If there exist a constant λ, $0 < \lambda < (\alpha+1)^{(\alpha+1)/\alpha}$ such that

$$\liminf_{t\to\infty} ta^{-1/\alpha}(t)\left[\lambda|Q(t)|^{(1-\alpha)/\alpha}Q(t) + \frac{1-\alpha}{R(t)} + \frac{1}{\alpha}a^{(1-\alpha)/\alpha}(t)a'(t)\right] > 1,$$
$$(3.10.7)$$

then equation (3.1.1) is oscillatory.

Proof. It follows from (3.10.7) that there exist a constant $\gamma > 1$ and a $T \geq t_0$ such that

$$ta^{-1/\alpha}(t)\left[\lambda|Q(t)|^{(1-\alpha)/\alpha}Q(t) + \frac{1-\alpha}{R(t)} + \frac{1}{\alpha}a^{(1-\alpha)/\alpha}(t)a'(t)\right] > \gamma \text{ for } t \geq T.$$

Then, we have

$$\int^\infty R^{\alpha-1}(s)a^{-1/\alpha}(s)\exp\left(-\lambda\int^s a^{-1/\alpha}(\xi)|Q(\xi)|^{(1-\alpha)/\alpha}Q(\xi)d\xi\right)ds < \infty.$$

Thus, by Theorem 3.10.1, equation (3.1.1) is oscillatory. ■

Next, we present the following result.

Theorem 3.10.2. Suppose equation (3.1.1) is nonoscillatory and

$$\int^\infty \exp\left((\alpha+1)\int^s a^{-1/\alpha}(\xi)|Q(\xi)|^{(1-\alpha)/\alpha}Q(\xi)d\xi\right)ds = \infty. \quad (3.10.8)$$

Then (3.1.1) has a solution $x(t)$ satisfying $\int^\infty |x(s)|^{\alpha+1}ds = \infty$.

Proof. Let $x(t)$ be a nonoscillatory solution of equation (3.1.1), say, $x(t) > 0$ for $t \geq T \geq t_0$. It follows from Theorem 3.10.1 that (3.10.4) holds for $t \geq T$, where $w(t)$ is as in (3.7.6). Then for $t \geq T$, we have

$$\frac{d}{dt} \ln x(t) = a^{-1/\alpha}(t)|w(t)|^{(1-\alpha)/\alpha} w(t) \geq a^{-1/\alpha}(t)|Q(t)|^{(1-\alpha)/\alpha} Q(t),$$

which implies

$$x(t) \geq x(T) \exp\left(\int_T^t a^{-1/\alpha}(s)|Q(s)|^{(1-\alpha)/\alpha} Q(s)ds \right).$$

Thus, from (3.10.8), we find that $\int^\infty |x(s)|^{\alpha+1} ds = \infty$. This completes the proof. ∎

Now, we define the sequence of functions $\{h_n(t)\}_{n=0}^\infty$ for $t \geq t_0$ as follows (if it exists):

$$h_0(t) = \int_t^\infty q(s)ds = Q(t)$$

$$h_n(t) = \alpha \int_t^\infty a^{-1/\alpha}(s) \left(h_{n-1}^+(s) \right)^{(\alpha+1)/\alpha} ds + h_0(t), \quad n = 1, 2, \cdots.$$

(3.10.9)

Clearly, $h_1(t) \geq h_0(t)$ and this implies that $h_1^+(t) \geq h_0^+(t)$ for $t \geq t_0$. Also, by induction

$$h_{n+1}(t) \geq h_n(t), \quad n = 1, 2, \cdots \quad \text{for} \quad t \geq t_0, \tag{3.10.10}$$

i.e., this sequence $\{h_n(t)\}$ is nondecreasing on $[t_0, \infty)$.

Theorem 3.10.3. If equation (3.1.1) is nonoscillatory, then there exists a $T \geq t_0$ such that

$$\lim_{t \to \infty} h_n(t) = h(t) < \infty \quad \text{for} \quad t \geq T. \tag{3.10.11}$$

Proof. Suppose equation (3.1.1) is nonoscillatory. Then it follows from Theorem 3.10.2 that there exists $w(t) \in C^1([T, \infty), \mathbb{R})$ such that

$$w(t) = \alpha \int_t^\infty a^{-1/\alpha}(s)|w(s)|^{(\alpha+1)/\alpha} ds + \int_t^\infty q(s)ds \quad \text{for} \quad t \geq T \geq t_0.$$

Thus, $w(t) \geq h_0(t)$, and hence $w^+(t) \geq h_0^+(t)$ for $t \geq T$. This implies that

$$w(t) = \alpha \int_t^\infty a^{-1/\alpha}(s)|w(s)|^{(\alpha+1)/\alpha} ds + \int_t^\infty q(s)ds$$

$$\geq \alpha \int_t^\infty a^{-1/\alpha}(s)[w^+(s)]^{(\alpha+1)/\alpha} ds + h_0(t)$$

$$\geq \alpha \int_t^\infty a^{-1/\alpha}(s)[h_0^+(s)]^{(\alpha+1)/\alpha} ds + h_0(t) = h_1(t) \quad \text{for} \quad t \geq T$$

and, by induction

$$w(t) \geq h_n(t), \quad n = 0, 1, \cdots, \quad t \geq T. \tag{3.10.12}$$

Now, from (3.10.10) and (3.10.12), we find that the sequence $\{h_n(t)\}$ is bounded above on $[T, \infty)$. Hence condition (3.10.11) holds. ∎

Corollary 3.10.4. Equation (3.1.1) is oscillatory, if either

(i) there exists a positive integer m such that $h_n(t)$ is defined for $n = 1, 2, \cdots, m - 1$, but $h_m(t)$ does not exist, or

(ii) $h_n(t)$ is defined for $n = 1, 2, \cdots$ but for arbitrarily large $T^* \geq t_0$ there is a $t^* \geq T^*$ such that $\lim_{n \to \infty} h_n(t^*) = \infty$.

Next, we define the sequence of functions $\{g_n(t)\}_{n=0}^{\infty}$ for $t \geq t_0$ as follows (if it exists):

$$g_0(t) \;=\; \int_t^{\infty} q_1(s)ds \;=\; Q_1(t)$$

$$g_n(t) \;=\; \alpha \int_t^{\infty} a_1^{-1/\alpha}(s)[g_{n-1}^+(s)]^{(\alpha+1)/\alpha}ds + |g_0(t)|, \quad n = 1, 2, \cdots.$$
$$\tag{3.10.13}$$

Clearly, $g_1(t) \geq |g_0(t)| \geq g_0^+(t)$ for $t \geq t_0$. Also, by induction

$$g_{n+1}(t) \;\geq\; g_n(t), \quad n = 1, 2 \cdots \quad \text{and} \quad t \geq t_0, \tag{3.10.14}$$

i.e., this sequence $\{g_n(t)\}$ is nondecreasing on $[t_0, \infty)$.

In what follows, we shall assume that

$$R_1(t) \;=\; \int_{t_0}^t a_1^{-1/\alpha}(s)ds \;\to\; \infty \quad \text{as} \quad t \to \infty \quad \text{and} \quad Q_1(t) < \infty. \tag{3.10.15}$$

Theorem 3.10.4. Equation (3.2.15) is nonoscillatory if there exists a $T \geq t_0$ such that

$$\lim_{n \to \infty} g_n(t) \;=\; g(t) \;<\; \infty \quad \text{for} \quad t \geq T. \tag{3.10.16}$$

Proof. If (3.10.16) holds, then from (3.10.14) and (3.10.16) it follows that $g_n(t) \leq g(t)$, $n = 0, 1, \cdots$ for $t \geq T$. Applying the monotone convergence theorem, we have

$$g(t) \;=\; \alpha \int_t^{\infty} a_1^{-1/\alpha}(s)[g^+(s)]^{(\alpha+1)/\alpha}ds + |g_0(t)| \;\geq\; 0.$$

Thus, we get

$$g^+(t) \;=\; g(t) \;=\; \alpha \int_t^{\infty} a_1^{-1/\alpha}(s)[g^+(s)]^{(\alpha+1)/\alpha}ds + |g_0(t)|$$

$$\geq\; \left| \alpha \int_t^{\infty} a_1^{-1/\alpha}(s)[g^+(s)]^{(\alpha+1)/\alpha}ds + g_0(t) \right|.$$

Clearly, $g^+(t) \in C([T,\infty), \mathbb{R}^+)$, and now it follows from Theorem 3.10.2 that equation (3.2.15) is nonoscillatory. This completes the proof. $\blacksquare$

Corollary 3.10.5. If equation (3.2.15) is oscillatory, then either

(i)   there exists a positive integer m such that $g_n(t)$ is defined for $n = 1, 2, \cdots, m-1$, but $g_m(t)$ does not exist, or

(ii) $g_n(t)$ is defined for $n = 1, 2, \cdots$, but for arbitrarily large $T^* \geq t_0$, there exists $t^* \geq T^*$ such that $\lim_{n\to\infty} g_n(t^*) = \infty$.

Now, if $a(t) = a_1(t)$ and $h_0(t) = g_0(t) = Q(t) = \int_t^\infty q(s)ds \geq 0$ for $t \geq t_0$, then $h_n(t) = g_n(t) \geq 0$, $n = 0, 1, 2, \cdots$. Thus, the following two corollaries hold:

Corollary 3.10.6. Suppose $Q(t) \geq 0$ for $t \geq t_0$. Then equation (3.1.1) is nonoscillatory if and only if there exists $T \geq t_0$ such that $\lim_{n\to\infty} h_n(t) = h(t) < \infty$ for $t \geq T$.

Corollary 3.10.7. Suppose $Q(t) \geq 0$ for $t \geq t_0$. Then equation (3.1.1) is oscillatory if and only if either

(i)   there exists a positive integer m such that $h_n(t)$ is defined for $n = 1, 2, \cdots, m-1$, but $h_m(t)$ does not exist, or

(ii) $h_n(t)$ is defined for $n = 1, 2, \cdots$, but for arbitrarily large $T^* \geq t_0$ there is a $t^* \geq T^*$ such that $\lim_{n\to\infty} h_n(t^*) = \infty$.

Now, using Theorems 3.10.3 and 3.10.4, we shall prove the following result.

Theorem 3.10.5. Assume that

$$0 < a_1(t) \leq a(t), \quad |Q(t)| \leq Q_1(t) \quad \text{for all sufficiently large } t. \quad (3.10.17)$$

If (3.2.15) is nonoscillatory, then equation (3.1.1) is nonoscillatory, or equivalently, if (3.1.1) is oscillatory, then equation (3.2.15) is oscillatory.

Proof. Suppose equation (3.2.15) is nonoscillatory. It follows from Theorem 3.10.3 that there exists a $T \geq t_0$ such that

$$\lim_{n\to\infty} g_n(t) = g(t) < \infty \quad \text{for } t \geq T. \quad (3.10.18)$$

Clearly, by (3.10.17), $h_0(t) \leq |h_0(t)| = |Q(t)| \leq Q_1(t) = g_0(t)$, and hence $h_0^+(t) \leq g_0^+(t)$ for $t \geq T$. Thus, for $t \geq T$,

$$
\begin{aligned}
h_1(t) &= \alpha \int_t^\infty a^{-1/\alpha}(s)[h_0^+(s)]^{(\alpha+1)/\alpha}ds + |h_0(t)| \\
&\leq \alpha \int_t^\infty a_1^{-1/\alpha}(s)[g_0^+(s)]^{(\alpha+1)/\alpha}ds + g_0(t) = g_1(t).
\end{aligned}
$$

Also, by induction

$$h_n(t) \leq g_n(t), \quad n = 0, 1, 2, \cdots \quad \text{and} \quad t \geq T. \tag{3.10.19}$$

Therefore, by (3.10.14), (3.10.18) and (3.10.19), we have

$$h(t) = \lim_{n \to \infty} h_n(t) \leq \lim_{n \to \infty} g_n(t) = g(t) < \infty \quad \text{for} \quad t \geq T.$$

Hence, by Theorem 3.10.4, equation (3.1.1) is nonoscillatory. This completes the proof. ∎

Next, we present the following criterion for the oscillation of equation (3.1.1) when condition (3.7.16) holds.

Theorem 3.10.6. Let $\lambda > \alpha$ be a constant. If there exists a number c such that

$$\limsup_{t \to \infty} \pi^{\alpha - \lambda}(t) \int_t^\infty \pi^\lambda(s)q(s)ds > c > \frac{1}{\lambda - \alpha} \left(\frac{\lambda}{\alpha + 1} \right)^{\alpha + 1}, \tag{3.10.20}$$

then equation (3.1.1) is oscillatory.

Proof. Assume to the contrary that equation (3.1.1) is nonoscillatory. We define the function $w(t)$ as in Theorem 3.7.4. Then there exists a $T \geq t_0$ such that (3.7.26) holds for $t \geq T$. In view of the boundedness of $\pi^\alpha(t)w(t)$ (cf. Lemma 3.7.3), we find that $\pi^\lambda(\xi)w(\xi) = \pi^{\lambda\alpha}(\xi)\pi^\alpha(\xi)w(\xi) \to 0$ as $\xi \to \infty$, and

$$\left| \int_t^\infty a^{-1/\alpha}(s)\pi^{\lambda - 1}(s)w(s)ds \right| \leq \int_t^\infty a^{-1/\alpha}(s)\pi^{\lambda - \alpha - 1}(s)|\pi^\alpha(s)w(s)|ds < \infty$$

and

$$\left| \int_t^\infty a^{-1/\alpha}(s)\pi^\lambda(s)|w(s)|^{(\alpha+1)/\alpha}ds \right|$$
$$\leq \int_t^\infty a^{-1/\alpha}(s)\pi^{\lambda - \alpha - 1}|\pi^\alpha(s)w(s)|^{(\alpha+1)/\alpha}ds < \infty,$$

for all $t \geq T$. Therefore, letting $\xi \to \infty$ in (3.7.26), we find

$$\pi^\lambda(t)w(t) = \int_t^\infty \pi^\lambda(s)q(s)ds + \lambda \int_t^\infty a^{-1/\alpha}(s)\pi^{\lambda - 1}(s)w(s)ds$$
$$+ \alpha \int_t^\infty a^{-1/\alpha}(s)\pi^\lambda(s)|w(s)|^{(\alpha+1)/\alpha}ds \quad \text{for} \quad t \geq T.$$

By Lemma 3.7.2, we have

$$|\pi^\alpha(s)w(s)|^{(\alpha+1)/\alpha} + \frac{\lambda}{\alpha}\pi^\alpha(s)w(s) + \frac{1}{\alpha}\left(\frac{\lambda}{\alpha + 1} \right)^{\alpha+1} \geq 0,$$

which implies

$$\pi^\lambda(t)w(t) \geq \int_t^\infty \pi^\lambda(s)q(s)ds - \frac{1}{\lambda-\alpha}\left(\frac{\lambda}{\alpha+1}\right)^{\alpha+1}\pi^{\lambda-\alpha}(t).$$

Hence, we get

$$\pi^\alpha(t)w(t) \geq \pi^{\alpha-\lambda}(t)\int_t^\infty \pi^\lambda(s)q(s)ds - \frac{1}{\lambda-\alpha}\left(\frac{\lambda}{\alpha+1}\right)^{\alpha+1}.$$

Now, it follows from (3.10.20) that $\limsup_{t\to\infty}\pi^\alpha(t)w(t) > 0$, which contradicts (3.7.19). This completes the proof. ∎

Corollary 3.10.8. Let $\lambda > \alpha$ be a constant. If

$$\int^\infty \pi^\lambda(s)q(s)ds = \infty,$$

then equation (3.1.1) is oscillatory.

Theorem 3.10.7. If

$$\limsup_{t\to\infty}\int_{t_0}^t \left[\pi^\alpha(s)q(s) - \left(\frac{\alpha}{\alpha+1}\right)^{\alpha+1}a^{-1/\alpha}(s)\pi^{-1}(s)\right]ds = \infty,$$
$$(3.10.21)$$

then equation (3.1.1) is oscillatory.

Proof. Assume to the contrary that equation (3.1.1) is nonoscillatory. We define the function $w(t)$ as in Theorem 3.7.4. Then there exists a $T \geq t_0$ such that (3.7.26) holds for $t \geq T$ with $\lambda = \alpha$. By Lemma 3.7.2, we have

$$|\pi^\alpha(s)w(s)|^{(\alpha+1)/\alpha} + \pi^\alpha(s)w(s) + \left(\frac{\alpha}{\alpha+1}\right)^{\alpha+1} \geq 0,$$

which implies

$$\pi^\alpha(\xi)w(\xi) \leq \pi^\alpha(T)w(T) - \int_T^\xi \left[\pi^\alpha(s)q(s) - \left(\frac{\alpha}{\alpha+1}\right)^{\alpha+1}a^{-1/\alpha}(s)\pi^{-1}(s)\right]ds.$$

Now, it follows from (3.10.21) that $\liminf_{t\to\infty}\pi^\alpha(t)w(t) = -\infty$, which contradicts the fact that $\pi^\alpha(t)w(t) \geq -1$ on $[T, \infty)$. Thus, equation (3.1.1) is oscillatory. ∎

Theorem 3.10.8. Let $\lambda < \alpha$ be a constant. If

$$\limsup_{t\to\infty}\pi^{\alpha-\lambda}(t)\int_{t_0}^t \pi^\lambda(s)q(s)ds > 1 + \frac{1}{\alpha-\lambda}\left(\frac{\lambda}{\alpha+1}\right)^{\alpha+1}, \quad (3.10.22)$$

then equation (3.1.1) is oscillatory.

Proof. Suppose equation (3.1.1) is nonoscillatory. As in Theorem 3.10.6, (3.7.26) holds for $t \geq T$ for some $T \geq t_0$. By Lemma 3.7.2, we have

$$|\pi^\alpha(s)w(s)|^{(\alpha+1)/\alpha} + \frac{\lambda}{\alpha}\pi^\alpha(s)w(s) + \frac{1}{\alpha}\left(\frac{\lambda}{\alpha+1}\right)^{\alpha+1} \geq 0,$$

which implies

$$\pi^\lambda(\xi)w(\xi) - \pi^\lambda(T)w(T) + \int_T^\xi \pi^\lambda(s)q(s)ds$$
$$+ \frac{1}{\lambda-\alpha}\left(\frac{\lambda}{\alpha+1}\right)^{\alpha+1}\left[\pi^{\lambda-\alpha}(\xi) - \pi^{\lambda-\alpha}(T)\right] \leq 0.$$

Then, we find

$$\pi^{\alpha-\lambda}(\xi)\int_T^\xi \pi^\lambda(s)q(s)ds \leq 1 + \frac{1}{\alpha-\lambda}\left(\frac{\lambda}{\alpha+1}\right)^{\alpha+1}$$
$$+ \pi^{\alpha-\lambda}(\xi)\left[\pi^\lambda(T)w(T) - \frac{1}{\alpha-\lambda}\left(\frac{\lambda}{\alpha+1}\right)^{\alpha+1}\pi^{\lambda-\alpha}(T)\right].$$

Hence, we obtain

$$\limsup_{\xi\to T}\ \pi^{\alpha-\lambda}(\xi)\int_T^\xi \pi^\lambda(s)q(s)ds \leq 1 + \frac{1}{\alpha-\lambda}\left(\frac{\lambda}{\alpha+1}\right)^{\alpha+1},$$

which contradicts (3.10.22). Thus, equation (3.1.1) is oscillatory. ∎

3.11. Oscillation of Damped and Forced Equations

In this section, we shall discuss the oscillatory behavior of the damped equation

$$\left(|x'(t)|^{\alpha-1}x'(t)\right)' + q(t)F(x(t), x'(t)) = 0 \tag{3.11.1}$$

and the forced equation

$$\left(|x'(t)|^{\alpha-1}x'(t)\right)' + q(t)|x(t)|^{\alpha-1}x(t) = e(t), \tag{3.11.2}$$

where

(i) $\alpha > 0$ is a constant,

(ii) $e(t),\ q(t) \in C([t_0, \infty), \mathbb{R})$, and

(iii) $F \in C(\mathbb{R}^2, \mathbb{R})$ and $\mathrm{sgn}\ F(x, y) = \mathrm{sgn}\ x$.

3.11.1. Oscillation of Equation (3.11.1)

Theorem 3.11.1. Suppose the function F satisfies the condition $(-1)^{\alpha+1} = 1$, $F(\lambda x, \lambda y) = \lambda^\alpha F(x, y)$ for all $(\lambda, x, y) \in \mathbb{R}^3$, and

$$\int_0^v \frac{dw}{F(1, w)} > c \quad \text{for every} \quad v \in \mathbb{R}. \tag{3.11.3}$$

If

$$\int_{t_0}^\infty q(s)ds = \infty, \quad t_0 \geq 0 \tag{3.11.4}$$

then equation (3.11.1) is oscillatory.

Proof. Suppose $x(t) \neq 0$ is a nonoscillatory solution of equation (3.11.1), then since $-F(-x, -y) = F(x, y)$, we can assume that $x(t) > 0$ for $t \geq t_0 \geq 0$. Define

$$w(t) = \frac{|x'(t)|^{\alpha-1} x'(t)}{x^\alpha(t)} \quad \text{for} \quad t \geq t_0.$$

Then for $t \geq t_0$, we have $w'(t) + \alpha |w(t)|^{(\alpha+1)/\alpha} + q(t) x^{-\alpha}(t) F(x(t), x'(t)) = 0$, or $w'(t) + \alpha |w(t)|^{(\alpha+1)/\alpha} + q(t) F(1, w(t)) = 0$, and hence

$$w'(t) \leq -q(t) F(1, w(t)) \quad \text{for} \quad t \geq t_0. \tag{3.11.5}$$

Dividing both sides of (3.11.5) by $F(1, w(t)) > 0$, we get

$$\int_{w(t_0)}^{w(t)} \frac{dv}{F(1, v)} \leq -\int_{t_0}^t q(s)ds \to -\infty \quad \text{as} \quad t \to \infty, \tag{3.11.6}$$

which contradicts condition (3.11.3). This completes the proof. ∎

Remark 3.11.1. When condition (3.11.3) does not hold the argument presented above shows that every solution of (3.11.1) is either oscillatory or tends to zero monotonically as $t \to \infty$. In fact, in this case inequality (3.11.6) yields

$$\lim_{t \to \infty} w(t) = \lim_{t \to \infty} \frac{|x'(t)|^{\alpha-1} x'(t)}{x^\alpha(t)} = -\infty,$$

so there exists a constant $k > 0$ such that $x'(t)/x(t) \leq -k$ for all t sufficiently large, say, $t \geq T \geq t_0$ which on integration gives $0 < x(t) < x(T) \exp(-(t - T))$ for all $t \geq T$.

Next, we shall prove the following result.

Theorem 3.11.2. Suppose $q(t) \geq 0$ eventually, and

$$F(\lambda x, \lambda y) = \lambda^\beta F(x, y) \quad \text{and} \quad (-1)^{\beta+1} = 1 \tag{3.11.7}$$

for all $(\lambda, x, y) \in \mathbb{R}^3$, where $\beta > \alpha$ is a constant. If

$$\int_{t_0}^{\infty} \left(\int_s^{\infty} q(u) du \right)^{1/\alpha} ds = \infty, \tag{3.11.8}$$

then equation (3.11.1) is oscillatory.

Proof. Let $x(t)$ be a nonoscillatory solution of equation (3.11.1), say, $x(t) > 0$ for $t \geq t_0 \geq 0$. Since $\left(|x'(t)|^{\alpha-1} x'(t) \right)' \leq 0$ for $t \geq t_0$, it is easy to check that $x'(t) > 0$ for $t \geq t_1 \geq t_0$, and

$$\lim_{t \to \infty} \frac{x'(t)}{x(t)} = 0, \tag{3.11.9}$$

and consequently, from the continuity of $F(x, y)$, given a fixed positive constant ϵ, $\epsilon < F(1, 0)$, there exists a $t_2 \geq t_1$ such that

$$k = F(1, 0) - \epsilon < F\left(1, \frac{x'(t)}{x(t)} \right) \leq F(1, 0) + \epsilon \tag{3.11.10}$$

for every $t \geq t_0$. Now, we have

$$
\begin{aligned}
x'(t) &\geq \left(\int_t^{\infty} q(s) F(x(s), x'(s)) ds \right)^{1/\alpha} \\
&= \left(\int_t^{\infty} q(s) x^{\beta}(s) F\left(1, \frac{x'(s)}{x(s)} \right) ds \right)^{1/\alpha} \\
&\geq (F(1, 0) - \epsilon)^{1/\alpha} \left(\int_t^{\infty} q(s) x^{\beta}(s) ds \right)^{1/\alpha},
\end{aligned}
$$

or

$$\frac{x'(t)}{x^{\beta/\alpha}(t)} \geq k^{1/\alpha} \left(\int_t^{\infty} q(s) ds \right)^{1/\alpha}. \tag{3.11.11}$$

Integrating (3.11.11) from t_2 to t, we find

$$k^{1/\alpha} \int_{t_2}^t \left(\int_s^{\infty} q(u) du \right)^{1/\alpha} ds \leq \int_{x(t_2)}^{x(t)} v^{-\beta/\alpha} dv < \infty,$$

which contradicts (3.11.8). This completes the proof. ∎

3.11.2. Oscillation of Equation (3.11.2)

Theorem 3.11.3. Let $q(t) \geq 0$ for $t \geq t_0$. If for every $T \geq t_0$,

$$\liminf_{t \to \infty} \int_T^t e(s) ds = -\infty, \quad \limsup_{t \to \infty} \int_T^t e(s) ds = \infty, \tag{3.11.12}$$

and

$$\liminf_{t\to\infty} \int_T^t a^{-1/\alpha}(s) \left(\int_T^s e(u)du\right)^{1/\alpha} ds = -\infty,$$

$$\limsup_{t\to\infty} \int_T^t a^{-1/\alpha}(s) \left(\int_T^s e(u)du\right)^{1/\alpha} ds = \infty,$$

(3.11.13)

then equation (3.11.2) is oscillatory.

Proof.   Assume the contrary. Then without loss of generality, we can assume that there is a nonoscillatory solution $x(t)$ of (3.11.2), say, $x(t) > 0$ for $t \geq t_1 \geq t_0$. From (3.11.2), we have $\left(a(t)|x'(t)|^{\alpha-1}x'(t)\right)' \leq e(t)$ for $t \geq t_1$. Thus, it follows that

$$a(t)|x'(t)|^{\alpha-1}x'(t) - a(t_1)|x'(t_1)|^{\alpha-1}x'(t_1) \leq \int_{t_1}^t e(s)ds. \qquad (3.11.14)$$

By (3.11.12) there exists a $T \geq t_0$ sufficiently large so that $x'(T) < 0$ and $x'(t) < 0$ for $t \geq T$. Replacing t_1 by T in (3.11.14), we get

$$x'(t) \leq a^{-1/\alpha}(t) \left(\int_T^t e(s)ds\right)^{1/\alpha}$$

and

$$x(t) \leq x(T) + \int_T^t a^{-1/\alpha}(s) \left(\int_T^s e(u)du\right)^{1/\alpha} ds.$$

Therefore, $\liminf_{t\to\infty} x(t) = -\infty$, which contradicts the fact that $x(t) > 0$ eventually.   ∎

Theorem 3.11.4.   Suppose for every $T \geq 0$ there exist $T \leq s_1 < t_1 \leq s_2 < t_2$ such that

$$e(t) \begin{cases} \leq 0 & \text{for} \quad t \in [s_1, t_1] \\ \geq 0 & \text{for} \quad t \in [s_2, t_2]. \end{cases} \qquad (3.11.15)$$

Denote by

$$D(s_i, t_i) = \{u \in C^1[s_i, t_i] : u(t) \neq 0, \ u(s_i) = u(t_i) = 0\}, \quad i = 1, 2.$$

If there exists $u \in D(s_i, t_i)$ such that

$$Q_i(u) = \int_{s_i}^{t_i} \left[q(s)u^{\alpha+1}(s) - a(s)(u'(s))^{\alpha+1}\right] ds > 0 \qquad (3.11.16)$$

for $i = 1, 2$ and $(-1)^{\alpha+1} = 1$, then equation (3.11.2) is oscillatory.

Proof. Suppose $x(t)$ is a nonoscillatory solution of equation (3.11.2), say, $x(t) > 0$ for $t \geq t_0$ for some t_0 depending on the solution $x(t)$. Define $w(t) = -a(t)|x'(t)|^{\alpha-1}x'(t)/x^\alpha(t)$ for $t \geq t_0$. It follows from (3.11.2) that $w(t)$ satisfies the first order nonlinear (generalized Riccati) equation

$$w'(t) = \alpha a^{-1/\alpha}(t)|w(t)|^{(\alpha+1)/\alpha} + q(t) - \frac{e(t)}{x^\alpha(t)} \quad \text{for} \quad t \geq t_0. \quad (3.11.17)$$

By assumption, we can choose s_1, $t_1 \geq t_0$ so that $e(t) \leq 0$ on the interval $I = [s_1, t_1]$ where $s_1 < t_1$. On the interval I, $w(t)$ in view of (3.11.15) satisfies the differential inequality

$$w'(t) \geq \alpha a^{-1/\alpha}(t)|w(t)|^{(\alpha+1)/\alpha} + q(t) \quad \text{on} \quad I. \quad (3.11.18)$$

Let $u(t) \in D(s_1, t_1)$ be as in the hypothesis. Multiplying (3.11.18) by $u^{\alpha+1}(t)$ and integrating over I, we find

$$\int_I u^{\alpha+1}(s)w'(s)ds \geq \int_I \alpha a^{-1/\alpha}(s)|w(s)|^{(\alpha+1)/\alpha}ds + \int_I u^{\alpha+1}(s)q(s)ds.$$
$$(3.11.19)$$

Integrating (3.11.19) by parts and using the fact that $u(s_1) = u(t_1) = 0$, we obtain

$$-(\alpha+1)\int_I u^\alpha(s)u'(s)w(s)ds \geq \int_I \alpha a^{-1/\alpha}(s)|w(s)|^{\frac{\alpha+1}{\alpha}}ds + \int_I u^{\alpha+1}(s)q(s)ds,$$

or

$$0 \geq \int_I u^{\alpha+1}(s)q(s)ds + \int_I \Big[\alpha a^{-1/\alpha}(s)u^{\alpha+1}(s)|w(s)|^{(\alpha+1)/\alpha}$$
$$-(\alpha+1)u^\alpha(s)|u'(s)||w(s)|\Big]\,ds.$$

Let $A^\lambda = \alpha a^{-1/\alpha}(s)u^{\alpha+1}(s)|w(s)|^{(\alpha+1)/\alpha}$, $\lambda = (\alpha+1)/\alpha$ and $B = \big((\alpha a(s))^{1/(\alpha+1)}|u'(s)|\big)^\alpha$, then from Lemma 3.2.1, we obtain for $t_1 > s \geq s_1 \geq t_0$,

$$\alpha a^{-1/\alpha}(s)u^{\alpha+1}(s)|w(s)|^{(\alpha+1)/\alpha} + \frac{1}{\alpha}\Big((\alpha a(s))^{1/(\alpha+1)}u'(s)\Big)^{\alpha+1}$$
$$-(\alpha+1)u^\alpha(s)|u'(s)||w(s)| \geq 0.$$

Thus, we have $0 \geq Q_1(t) > 0$, which contradicts (3.11.16).

When $x(t)$ is eventually negative, we use $u \in D(s_2, t_2)$ and $e(t) \geq 0$ on $[s_2, t_2]$ to reach a similar contradiction. This completes the proof. ∎

Remark 3.11.2. If $\alpha = 1$ in Theorem 3.11.4, then condition (3.11.16) can be replaced by

$$Q_i(u) = \int_{s_i}^{t_i} \big[q(s)u^2(s) - a(s)(u'(s))^2\big]\,ds \geq 0, \quad i = 1, 2.$$

Example 3.11.1. Consider the forced half–linear equation

$$\left(\sqrt{t}\,|x'(t)|^2 x'(t)\right)' + \frac{1}{2\sqrt{t}}|x(t)|^2 x(t) \;=\; \sin\sqrt{t}, \quad t > 0. \qquad (3.11.20)$$

Here, the zeros of the forcing term $\sin\sqrt{t}$ are $(n\pi)^2$. Let $u(t) = \sin\sqrt{t}$. For any $T \geq 0$ choose n sufficiently large so that $(n\pi)^2 \geq T$, and set $s_1 = (n\pi)^2$ and $t_1 = (n+1)^2\pi^2$ in (3.11.16). It is easy to verify that

$$
\begin{aligned}
Q_1(u) &= \int_{(n\pi)^2}^{(n+1)^2\pi^2} \left[\frac{1}{2\sqrt{s}}\sin^4\sqrt{s} - \frac{1}{16s\sqrt{s}}\cos^4\sqrt{s}\right] ds \\[2mm]
&= \int_{n\pi}^{(n+1)\pi} \left[\sin^4 v - \frac{1}{8v^2}\cos^4 v\right] dv \\[2mm]
&\geq \int_{n\pi}^{(n+1)\pi} \left[\frac{3}{4} - \frac{\cos 2v}{2} - \frac{\cos 4v}{2} - \frac{1}{8v^2}\right] dv \\[2mm]
&= \frac{3}{4}\pi - \frac{1}{8n(n+1)\pi} \;>\; 0.
\end{aligned}
$$

Similarly, with $s_2 = (n+1)^2\pi^2$ and $t_2 = (n+2)^2\pi^2$ we can show that $Q_2(u) > 0$. It follows from Theorem 3.11.4 that equation (3.11.20) is oscillatory.

3.12. Distance of Zeros of Oscillatory Solutions

Here, we shall derive lower bounds for the distance between consecutive zeros of solutions of the half–linear differential equation

$$\left(|x'(t)|^{\alpha-1}x'(t)\right)' + q(t)|x(t)|^{\alpha-1}x(t) \;=\; 0, \qquad \alpha > 0 \qquad (3.12.1)$$

where $q : [t_0, \infty) \to \mathbb{R}$ is locally integrable for some $t_0 \geq 0$.

To establish main results, we shall need the following lemmas.

Lemma 3.12.1. Suppose $p_1,\ p_2 \in C([a,b], \mathbb{R})$ satisfy $p_1(t) \leq p_2(t)$ for all $t \in [a,b]$. Let ϕ_1 and ϕ_2 defined on $[a,b]$ be respective solutions of the integral equations

$$\phi_1(t) \;=\; p_1(t) + k\int_a^t |\phi_1(s)|^\lambda ds \qquad (3.12.2)$$

and

$$\phi_2(t) \;=\; p_2(t) + k\int_a^t |\phi_2(t)|^\lambda ds, \qquad (3.12.3)$$

where $k > 0$ and $\lambda > 1$ are constants. If $\phi_1(t) \geq 0$ on $[a,b]$, then $\phi_1(t) \leq \phi_2(t)$ for all $t \in [a,b]$.

Proof. First, consider the case $p_1(t) < p_2(t)$ for all $t \in [a, b]$. We shall show that $\phi_1(t) < \phi_2(t)$ on $[a, b]$. Suppose to the contrary that $\phi_1(t^*) = \phi_2(t^*)$ for some $t^* \in [a, b]$. Since $\phi_2(a) = p_2(a) > p_1(a) = \phi_1(a)$, there must be a smallest t, say, $t_1 > a$, $t_1 \le t^*$ such that $\phi_2(t_1) = \phi_1(t_1)$ and $\phi_2(t) > \phi_1(t)$ on $[a, t_1)$.

Let $G(y) = |y|^\lambda$, then $G(y)$ is nondecreasing with respect to $y \ge 0$. It follows that

$$\phi_1(t_1) = p_1(t_1) + k \int_a^{t_1} G(\phi_1(s))ds$$

$$< p_2(t_1) + k \int_a^{t_1} G(\phi_2(s))ds = \phi_2(t_1),$$

which contradicts the fact that $\phi_1(t_1) = \phi_2(t_1)$. Hence, $\phi_1(t) < \phi_2(t)$ for all $t \in [a, b]$. Now, a continuity argument completes the proof. ∎

Lemma 3.12.2. If $x(t)$ is a solution of (3.12.1) satisfying $x'(c) = 0$, $x(b) = 0$, $x(t) > 0$ and $x'(t) \le 0$ for $t \in (c, b)$, then $\sup_{c \le t \le b} \int_c^t q(s)ds > 0$.

Proof. Suppose to the contrary that $\int_c^t q(s)ds \le 0$ for $t \in [c, b]$. Let $Q(t) = \int_c^t q(s)ds$, and define

$$w(t) = -\frac{|x'(t)|^{\alpha-1}x'(t)}{|x(t)|^{\alpha-1}x(t)}. \tag{3.12.4}$$

Then, we find

$$w'(t) = q(t) + \alpha|w(t)|^{(\alpha+1)/\alpha}, \quad t \in [c, b) \tag{3.12.5}$$

which implies

$$w(t) = Q(t) + \alpha \int_c^t |w(s)|^{(\alpha+1)/\alpha}ds, \quad t \in [c, b) \tag{3.12.6}$$

and clearly, we have

$$w(c) = 0, \quad \lim_{t\to b^-} w(t) = \infty \quad \text{and} \quad w(t) \ge 0, \quad t \in (c, b). \tag{3.12.7}$$

Similarly, if $z(t)$ is a nontrivial solution of the equation $(|z'(t)|^{\alpha-1}z'(t))' = 0$ with $z'(c) = 0$, we let $w_1(t) = -|z'(t)|^{\alpha-1}z'(t)/(|z(t)|^{\alpha-1}z(t))$, then $w_1(t) = 0$ for all $t \in [c, b]$. Since

$$w_1(t) = \alpha \int_c^t |w_1(s)|^{(\alpha+1)/\alpha}ds, \quad t \in [c, b)$$

it follows from Lemma 3.12.1 that $w(t) \le w_1(t) = 0$ for all $t \in [c, b)$. This contradicts the fact that $\lim_{t\to b^-} w(t) = \infty$. ∎

Theorem 3.12.1. Let $x(t)$ be a nontrivial solution of equation (3.12.1) satisfying $x'(c) = 0$, $x(b) = 0$ and $x(t) \neq 0$ for all $t \in [c, b)$. Then,

$$(b - c)^\alpha \sup_{c \leq t \leq b} \left| \int_c^t q(s)ds \right| > 1. \qquad (3.12.8)$$

Moreover, if there are no extreme values of $x(t)$ in (c, b), then

$$(b - c)^\alpha \sup_{c \leq t \leq b} \int_c^t q(s)ds > 1. \qquad (3.12.9)$$

Proof. Without loss of generality, we assume that $x(t) > 0$ for all $t \in [c, b)$. Define $w(t)$ as in (3.12.4), and let

$$\phi(t) = \alpha \int_c^t |w(s)|^{(\alpha+1)/\alpha} ds \quad \text{for} \quad t \in [c, b). \qquad (3.12.10)$$

It follows from (3.12.6) that

$$w(t) = \phi(t) + Q(t) \quad \text{for} \quad t \in [c, b), \qquad (3.12.11)$$

where $Q(t) = \int_c^t q(s)ds$. Thus, $w(c) = \phi(c) = 0$ and $\lim_{t \to b^-} w(t) = \lim_{t \to b^-} \phi(t) = \infty$.

Now set $Q^* = \sup_{c \leq t \leq b} |Q(t)|$, so that

$$\phi'(t) = \alpha |w(t)|^{(\alpha+1)/\alpha} \leq \alpha(Q^* + \phi(t))^{(\alpha+1)/\alpha} \quad \text{for} \quad t \in [c, b)$$

and

$$\frac{\phi'(t)}{\alpha(Q^* + \phi(t))^{(\alpha+1)/\alpha}} \leq 1 \quad \text{for} \quad t \in [c, b). \qquad (3.12.12)$$

Integrating the inequality (3.12.12) from c to b and using $\lim_{t \to b^-} \phi(t) = \infty$, we obtain

$$-\left. \frac{1}{(Q^* + \phi(t))^{1/\alpha}} \right|_{t=c}^{t=b} \leq b - c,$$

which implies that

$$(b - c)^\alpha Q^* \geq 1. \qquad (3.12.13)$$

We remark that the equality in (3.12.13) cannot hold, for otherwise, $|Q(t)| = \left| \int_c^t q(s)ds \right| = Q^*$ for $t \in [c, b)$, which contradicts the fact that $Q(t)$ is continuous and $Q(c) = 0$. Thus, inequality (3.12.8) holds.

If c is the largest extreme point of $x(t)$ on $[c, b)$, then $x'(t) \leq 0$ and thus $w(t) \geq 0$ for $t \in [c, b)$. Set $Q_* = \sup_{c \leq t \leq b} \int_c^t q(s)ds$. Since $x'(c) = 0$, $x(b) = 0$, $x(t) > 0$ and $x'(t) \leq 0$ on (c, b), it follows from

Lemma 3.12.2 that $Q_* > 0$. Hence, by (3.12.11), $0 \leq w(t) \leq Q_* + \phi(t)$. The rest of the proof is similar to that presented above, and hence we omit the details.   ∎

In a similar way, we can prove the following result.

Theorem 3.12.2. Let $x(t)$ be a nontrivial solution of equation (3.12.1) satisfying $x(a) = 0$, $x'(\tau) = 0$ and $x(t) \neq 0$ for all $t \in (a, \tau]$. Then,

$$(\tau - a)^\alpha \sup_{a \leq t \leq \tau} \left| \int_t^\tau q(s)ds \right| > 1. \tag{3.12.14}$$

Moreover, if there are no extreme values of $x(t)$ in (a, τ), then

$$(\tau - a)^\alpha \sup_{a \leq t \leq \tau} \int_t^\tau q(s)ds > 1. \tag{3.12.15}$$

Definition 3.12.1. We say that (3.12.1) is *right disfocal* (*left disfocal*) on $[a, b]$ if the solutions of equation (3.12.1) such that $x'(a) = 0$ ($x'(b) = 0$, respectively) have no zeros in $[a, b]$.

Corollary 3.12.1. If

$$(b - c)^\alpha \sup_{a \leq t \leq b} \left| \int_c^b q(s)ds \right| \leq 1,$$

then equation (3.12.1) is right disfocal on $[c, d]$, if

$$(c - a)^\alpha \sup_{a \leq t \leq c} \left| \int_a^c q(s)ds \right| \leq 1,$$

then equation (3.12.1) is left disfocal on $(a, c]$.

In what follows for convenience, we shall let $\eta = \min\{4, 4^\alpha\}$.

Theorem 3.12.3. Let a and b, $a < b$ denote consecutive zeros of a nontrivial solution $x(t)$ of equation (3.12.1). Then there exist two disjoint subintervals, say, I_1 and I_2 of $[a, b]$ satisfying

$$(b - a)^\alpha \int_{I_1 \cup I_2} q(s)ds > \eta \tag{3.12.16}$$

and

$$\int_{[a,b] \setminus (I_1 \cup I_2)} q(s)ds \leq 0. \tag{3.12.17}$$

Proof. Let c and τ denote the smallest and largest extreme points of $x(t)$ on $[a, b]$, respectively. (If there is only one zero of $x'(t)$ in (a, b), then c and τ coincide). Thus, $x'(c) = 0$, $x(b) = 0$ and $x'(t) \neq 0$ for all $t \in (c, b]$. By Theorem 3.12.1 inequality (3.12.9) holds. Thus, there exists $b_1 \in (c, b]$ such that

$$(b - c)^\alpha \int_c^{b_1} q(s)ds > 1 \quad \text{and} \quad \int_c^{b_1} q(s)ds \geq \int_c^b q(s)ds. \quad (3.12.18)$$

Similarly, it follows from Theorem 3.12.2 that there exists $a_1 \in [a, \tau)$ such that

$$(\tau - a)^\alpha \int_{a_1}^\tau q(s)ds > 1 \quad \text{and} \quad \int_{a_1}^\tau q(s)ds \geq \int_a^\tau q(s)ds. \quad (3.12.19)$$

Let $I_1 = [c, b_1]$ and $I_2 = [a_1, \tau]$. Now, we consider the following two cases:

Case 1. If $\alpha > 1$, then

$$(b-a)^\alpha \int_{I_1 \cup I_2} q(s)ds \geq [(b - c) + (\tau - a)]^\alpha \left(\int_c^{b_1} q(s)ds + \int_{a_1}^\tau q(s)ds \right)$$

$$> [(b - c)^\alpha + (\tau - a)^\alpha] \left[\frac{1}{(b - c)^\alpha} + \frac{1}{(\tau - a)^\alpha} \right]$$

$$\geq \left[(b-c)^{\alpha/2} \frac{1}{(b-c)^{\alpha/2}} + (\tau-a)^{\alpha/2} \frac{1}{(\tau-a)^{\alpha/2}} \right]^2 = 4.$$

Case 2. If $0 < \alpha < 1$, then

$$(b-a) \left(\int_{I_1 \cup I_2} q(s)ds \right)^{1/\alpha} = (b - a) \left(\int_c^{b_1} q(s)ds + \int_{a_1}^\tau q(s)ds \right)^{1/\alpha}$$

$$\geq (b-a) \left[\left(\int_c^{b_1} q(s)ds \right)^{1/\alpha} + \left(\int_{a_1}^\tau q(s)ds \right)^{1/\alpha} \right]$$

$$\geq [(b - c) + (\tau - a)] \left[\frac{1}{(b - c)} + \frac{1}{(\tau - a)} \right] \geq 4.$$

It follows from the cases 1 and 2 that $(b - a)^\alpha \int_{I_1 \cup I_2} q(s)ds > \eta$. Next, by (3.12.18) and (3.12.19), we have $\int_{b_1}^b q(s)ds \leq 0$ and $\int_a^{a_1} q(s)ds \leq 0$. Thus, to verify (3.12.17) it suffices to show that $\int_\tau^c q(s)ds \leq 0$. In fact, since $x'(\tau) = x'(c) = 0$, $w(\tau) = w(c) = 0$ from (3.12.6), we find

$$0 = w(c) - w(\tau) = \int_\tau^c q(s)ds + \alpha \int_\tau^c |w(s)|^{(\alpha+1)/\alpha}ds.$$

This means that $\int_\tau^c q(s)ds \leq 0$, which implies that (3.12.17) holds.   ∎

Corollary 3.12.2. Suppose for every two disjoint subintervals I_1 and I_2 of $[a, b]$,

$$(b - a)^\alpha \int_{I_1 \cup I_2} q(s)ds \leq \eta. \qquad (3.12.20)$$

Then each solution of equation (3.12.1) has at most one zero on $[a, b]$.

Proof. Suppose to the contrary, there exists a nontrivial solution $x(t)$ of equation (3.12.1) with $x(\tau) = x(c) = 0$ for $a \leq \tau < c \leq b$. Without loss of generality, we assume that $x'(t) \neq 0$ for $t \in (\tau, c)$. By Theorem 3.12.3 there exist two disjoint subintervals I_1 and I_2 on $[\tau, c) \subset [a, b]$ such that $(c - \tau)^\alpha \int_{I_1 \cup I_2} q(s)ds > \eta$. Thus, we have $(b - a)^\alpha \int_{I_1 \cup I_2} q(s)ds > \eta$, which is a contradiction. ∎

Corollary 3.12.3. Suppose a nontrivial solution $x(t)$ of equation (3.12.1) has $N \geq 2$ zeros in $[a, b]$. Then there exist $2(N - 1)$ disjoint subintervals of $[a, b]$, I_{ij} where $i = 1, 2, \cdots, N - 1$, $j = 1, 2$ such that

$$N < \left[\frac{(b - a)^\alpha}{\eta} \int_I q(s)ds \right]^{1/\alpha} + 1 \qquad (3.12.21)$$

and

$$\int_{[a,b] \setminus I} q(s)ds \leq 0, \qquad (3.12.22)$$

where $I = \cup_{i=1}^{N-1} \cup_{j=1}^{2} I_{ij}$.

Proof. Let t_i, $i = 1, 2, \cdots, N$ be the zeros of $x(t)$ in $[a, b]$. By Theorem 3.12.4 for each $i = 1, 2, \cdots, N - 1$ there are two disjoint subintervals, say, I_{i1} and I_{i2} of $[t_i, t_{i+1}]$ such that

$$\int_{I_{i1} \cup I_{i2}} q(s)ds > \frac{\eta}{(t_{i+1} - t_i)^\alpha} \qquad (3.12.23)$$

and

$$\int_{[t_i, t_{i+1}] \setminus (I_{i1} \cup I_{i2})} q(s)ds \leq 0. \qquad (3.12.24)$$

Summing (3.12.23) for i from 1 to $N - 1$, we get

$$\int_I q(s)ds > \eta \sum_{i=1}^{N-1} \frac{1}{(t_{i+1} - t_i)^\alpha}$$

$$\geq \eta(N - 1) \left[\frac{1}{(t_2 - t_1)^\alpha} \cdots \frac{1}{(t_N - t_{N-1})^\alpha} \right]^{1/(N-1)}$$

$$= \eta(N - 1) \left[\frac{1}{(t_2 - t_1)} \cdots \frac{1}{(t_N - t_{N-1})} \right]^{1/\alpha(N-1)}$$

$$\geq \eta(N - 1)^{\alpha+1} \frac{1}{(t_N - t_1)^\alpha},$$

which implies $(N-1)^{\alpha+1} < ((b-a)^\alpha)/\eta \int_I q(s)ds$. This implies that (3.12.21) holds. From (3.12.24) it is easy to deduce (3.12.22). This completes the proof. ∎

Example 3.12.1. Consider the differential equation

$$\left(|x'(t)|^{\alpha-1}x'(t)\right)' + (\sin kt)|x(t)|^{\alpha-1}x(t) = 0, \qquad (3.12.25)$$

where $k > 0$ is a constant. Then,

(i) equation (3.12.25) is right disfocal on $[0, b)$ and left disfocal on $(0, b]$ if $0 < b < k/2$, and

(ii) each solution of (3.12.25) has at most one zero in $[0, b]$ if $0 < b < (k\eta/4)^{1/\alpha}$.

(i) Since $q(t) = \sin kt$ and $0 < b < k/2$,

$$\sup_{0 \le t \le b} \left| \int_0^t q(s)ds \right| = \sup_{0 \le t \le b} \left| \int_0^t \sin ks\,ds \right| \le \frac{2}{k}.$$

Thus, $b\sup_{0 \le t \le b} \left| \int_0^t q(s)ds \right| \le 1$. Now, by Corollary 3.12.1 equation (3.12.25) is right disfocal on $[0, b)$. Similarly, (3.12.25) is also left disfocal on $(0, b]$.

(ii) It follows from $0 < b \le (k\eta/4)^{1/\alpha}$ that for any two disjoint subintervals I_1 and I_2 of $[0, b]$, $\int_{I_1 \cup I_2} q(s)ds = \int_{I_1 \cup I_2} \sin ks\,ds \le 4/k$, which implies $b^\alpha \int_{I_1 \cup I_2} q(s)ds \le \eta$. Now, by Corollary 3.12.2 each solution of equation (3.12.25) has at most one zero on $[0, b]$.

In what follows for any t and $\delta > 0$, we shall denote by $(I_1 \cup I_2)(t, \delta)$ the union of two disjoint subintervals I_1 and I_2 of $[t, t+\delta]$.

Theorem 3.12.4. Let $x(t)$ be an oscillatory solution of equation (3.12.1). If

$$\limsup_{t \to \infty} \left(\delta^\alpha \int_{(I_1 \cup I_2)(t,\delta)} q(s)ds \right) < \eta \qquad (3.12.26)$$

for all $\delta > 0$ and for every two disjoint subintervals I_1 and I_2 of $[t, t+\delta]$, then the distance between consecutive zeros of $x(t)$ is unbounded as $t \to \infty$.

Proof. Suppose to the contrary that equation (3.12.1) has an oscillatory solution $x(t)$ whose zeros $\{t_n\}_{n=1}^\infty$ contain a subsequence $\{t_{n_k}\}_{k=1}^\infty$ such that $0 < t_{n_{k+1}} - t_{n_k} < \delta$ for some $\delta > 0$ and all k. Then by Theorem 3.12.3 there are two disjoint subintervals $I_1(t_{n_k}, \delta)$ and $I_2(t_{n_k}, \delta)$ of $[t_{n_k}, t_{n_{k+1}}]$ satisfying

$$(t_{n_{k+1}} - t_{n_k})^\alpha \int_{(I_1 \cup I_2)(t_{n_k}, \delta)} q(s)ds > \eta \quad \text{for all} \quad k.$$

This implies that

$$\delta^\alpha \int_{(I_1 \cup I_2)(t_{n_k}, \delta)} q(s)ds > \eta \quad \text{for all} \quad k,$$

which contradicts (3.12.26). This completes the proof. ∎

Theorem 3.12.5. Let $x(t)$ be an oscillatory solution of equation (3.12.1). If there exists a $\delta_0 > 0$ such that

$$\lim_{t \to \infty} \int_{(I_1 \cup I_2)(t, \delta_0)} q(s)ds = 0 \tag{3.12.27}$$

for every two disjoint subintervals I_1 and I_2 of $[t, t + \delta_0]$, then the distance between consecutive zeros of $x(t)$ is unbounded as $t \to \infty$.

Proof. We will first show that for all $\delta > 0$ and any two disjoint subintervals I_1 and I_2 of $[t, t + \delta_0]$,

$$\lim_{t \to \infty} \int_{(I_1 \cup I_2)(t, \delta)} q(s)ds = 0. \tag{3.12.28}$$

Let k denote the least integer with $k\delta_0 \geq \delta$ and $t_i = t + i\delta_0$, $i = 0, 1, \cdots, k - 1$ and $t_k = t + \delta$, then

$$\int_{(I_1 \cup I_2)(t, \delta)} q(s)ds = \sum_{i=0}^{k-1} \int_{(I_{i1} \cup I_{i2})(t_i, \delta_0)} q(s)ds, \tag{3.12.29}$$

where $I_{ij}(t_i, \delta_0) = I_j(t, \delta) \cap [t_i, t_{i+1}]$ for $i = 0, 1, \cdots, k - 1$ and $j = 1, 2$. Noting that $t_i \to \infty$ as $t \to \infty$, it follows from (3.12.27) that

$$\lim_{t \to \infty} \int_{(I_{i1} \cup I_{i2})(t_i, \delta_0)} q(s)ds = 0, \quad i = 0, 1, \cdots, k - 1. \tag{3.12.30}$$

Combining (3.12.29) and (3.12.30), we obtain (3.12.28) which implies

$$\lim_{t \to \infty} \left(\delta^\alpha \int_{(I_1 \cup I_2)(t, \delta)} q(s)ds \right) = 0$$

for all $\delta > 0$ and any two disjoint subintervals I_1 and I_2 of $[t, t + \delta]$. The result now follows from Theorem 3.12.4. ∎

Remark 3.12.1. Consider the general half–linear equation (3.1.1). By direct computation it is easy to verify that the change of variables $(t, x) \to (\tau, X)$ given by $\tau = \tau(t) = \int_{t_0}^t a^{-1/\alpha}(s)ds$ for $t \geq t_0$ and $X(\tau) = x(t)$ transform (3.1.1) into the equation

$$\frac{d}{d\tau} \left(|\dot{X}(\tau)|^{\alpha-1} X'(\tau) \right) + Q(\tau)|X(\tau)|^{\alpha-1} X(\tau) = 0, \tag{3.12.31}$$

where $Q(\tau) = a^{1/\alpha}(t)q(t)$, $(\cdot = d/d\tau)$, $t = t(\tau)$ is the inverse function of $\tau = \tau(t)$ given above. Since equation (3.12.31) is of the form (3.12.1) it follows from Theorems 3.12.4 and 3.12.5 that the following oscillatory criteria for equation (3.1.1) hold.

Theorem 3.12.6. Let $x(t)$ denote an oscillatory solution of equation (3.1.1). If

$$\limsup_{t \to \infty} \left(\delta^\alpha \int_{(I_1 \cup I_2)(t,\delta)} q(s)ds \right) < \eta \qquad (3.12.32)$$

for all $\delta > 0$ and for every two disjoint subintervals I_1 and I_2 of $[t, t+\delta]$, then the distance between consecutive zeros of $x(t)$ is unbounded as $t \to \infty$.

Theorem 3.12.7. Let $x(t)$ denote an oscillatory solution of equation (3.1.1). If there exists a $\delta_0 > 0$ such that

$$\lim_{t \to \infty} \int_{(I_1 \cup I_2)(t,\delta_0)} q(s)ds = 0 \qquad (3.12.33)$$

for every two disjoint subintervals I_1 and I_2 of $[t, t + \delta_0]$, then the distance between consecutive zeros of $x(t)$ is unbounded as $t \to \infty$.

3.13. Oscillation and Nonoscillation of Half–Linear Equations with Deviating Arguments

This section is devoted to the study of the oscillatory behavior of half–linear functional differential equations of the type

$$\left(a(t)|x'(t)|^{\alpha-1}x'(t)\right)' + q(t)|x[g(t)]|^{\beta-1}x[g(t)] = 0 \qquad (3.13.1)$$

and

$$\left(a(t)|x'(t)|^{\alpha-1}x'(t)\right)' - q(t)|x[g(t)]|^{\alpha-1}x[g(t)] = 0, \qquad (3.13.2)$$

where

(i) $\alpha > 0$, $\beta > 0$ are positive constants,

(ii) $a(t) \in C([t_0, \infty), \mathbb{R}^+)$ and $\int^\infty a^{-1/\alpha}(s)ds = \infty$,

(iii) $q(t) \in C([t_0, \infty), \mathbb{R}_0)$, $q(t) \not\equiv 0$ for all large t,

(iv) $g(t) \in C^1([t_0, \infty), \mathbb{R})$, $g'(t) \geq 0$ for $t \geq t_0$ and $\lim_{t \to \infty} g(t) = \infty$.

The half–linear ordinary differential equation

$$\left(|x'(t)|^{\alpha-1}x'(t)\right)' = q(t)|x(t)|^{\alpha-1}x(t) \qquad (3.13.3)$$

which is the same as (3.13.2) with $a(t) \equiv 1$ and $g(t) \equiv t$ is nonoscillatory in the sense that all of its solutions are nonoscillatory, see Elbert [15]. However, the presence of a deviating argument $g(t) \not\equiv t$ in (3.13.2) may generate oscillation to some or all of its solutions. We illustrate this in the following example.

Example 3.13.1. Let $S_\alpha(t)$ denote the solution of the equation

$$\left(|x'(t)|^{\alpha-1}x'(t)\right)' + \alpha|x(t)|^{\alpha-1}x(t) \ = \ 0, \quad \alpha > 0$$

satisfying the initial conditions $x(0) = 0$, $x'(0) = 1$. Elbert [15] has shown that $S_\alpha(t)$ exists uniquely on $\mathbb{R}$ and it is periodic of period π_α, where $\pi_\alpha = \frac{2\pi}{\alpha+1} \Big/ \sin\left(\frac{\pi}{\alpha+1}\right)$. Furthermore, $S_\alpha(t)$ satisfies $S_\alpha(t - \pi_\alpha) = S_\alpha(t + \pi_\alpha) = -S_\alpha(t)$ for $t \in \mathbb{R}$. It follows that $S_\alpha(t)$ is an oscillatory solution of both the functional differential equations

$$\left(|x'(t)|^{\alpha-1}x'(t)\right)' \ = \ \alpha|x[t - \pi_\alpha]|^{\alpha-1}x[t - \pi_\alpha]$$

and

$$\left(|x'(t)|^{\alpha-1}x'(t)\right)' \ = \ \alpha|x[t + \pi_\alpha]|^{\alpha-1}x[t + \pi_\alpha].$$

3.13.1. Some Useful Lemmas

In what follows, we shall let $R[t, s] = \int_s^t a^{-1/\alpha}(u)du$ for $t \geq s \geq t$ and $R(t) = R[t, t_0]$. To prove our main results we shall need the following lemmas.

Lemma 3.13.1.   Assume that $a, g \in C([t_0, \infty), \mathbb{R}^+)$, $g(t) < t$ for $t \geq t_0$ and $\lim_{t\to\infty} g(t) = \infty$. Let $x \in C^1([t_0, \infty), \mathbb{R}^+)$ be such that $a^{1/\alpha}(t)x'(t) \in C^1([t_0, \infty), \mathbb{R}^+)$ and $(a(t)(x'(t))^\alpha)' \leq 0$ for $t \geq T \geq t_0$. Then,

(I)   for each constant k, $0 < k < 1$ there is a $T_k \geq T$ such that either

$$x[g(t)] \ \geq \ k\left(\frac{a(t)}{a(T)}\right)^{1/\alpha}\left(\frac{g(t)}{t}\right)x(t) \quad \text{for} \quad a'(t) \leq 0, \quad t \geq T_k \geq T,$$

(3.13.4)

or

$$x[g(t)] \ \geq \ k\left(\frac{g(t)}{t}\right)x(t) \quad \text{for} \quad a'(t) \geq 0, \quad t \geq T_k \geq T, \qquad (3.13.5)$$

(II) for all sufficiently large $T \geq t_0$,

$$x(t) \ \geq \ a^{1/\alpha}(t)R[t, T]x'(t), \quad t \geq T. \qquad (3.13.6)$$

Proof. (I) It suffices to consider only those t for which $g(t) < t$. Since $a^{1/\alpha}(t)x'(t)$ is decreasing, it follows from the mean–value theorem that for each $t > g(t) \geq T$, there exists a $t_1 \in (g(t), t)$ such that

$$x(t) - x[g(t)] = a^{-1/\alpha}(t_1)\left(a^{1/\alpha}(t_1)x'(t_1)\right)[t - g(t)]$$

$$\leq a^{-1/\alpha}(t_1)a^{1/\alpha}[g(t)]x'[g(t)][t - g(t)].$$

Since $x(t) > 0$, we get

$$\frac{x(t)}{x[g(t)]} \leq 1 + \left(\frac{a[g(t)]}{a(t_1)}\right)^{1/\alpha}\left(\frac{x'[g(t)]}{x[g(t)]}\right)[t - g(t)] \qquad (3.13.7)$$

for $t > g(t) \geq T$. Similarly,

$$x[g(t)] - x(T) = a^{-1/\alpha}(t_2)\left(a^{1/\alpha}(t_2)x'(t_2)\right)[g(t) - T]$$

$$\geq \left(\frac{a[g(t)]}{a(t_2)}\right)^{1/\alpha}x'[g(t)][g(t) - T],$$

for some $t_2 \in (T, g(t))$. Therefore,

$$x[g(t)] \geq \left(\frac{a[g(t)]}{a(t_2)}\right)^{1/\alpha}x'[g(t)][g(t) - T].$$

It follows from $x'(t) > 0$ that

$$\frac{x[g(t)]}{x'[g(t)]} \geq \left(\frac{a[g(t)]}{a(t_2)}\right)^{1/\alpha}[g(t) - T] = \left(\frac{a[g(t)]}{a(t_2)}\right)^{1/\alpha}k(t)g(t),$$

where $k(t) = [g(t) - T]/g(t)$. Since $\lim_{t\to\infty} g(t) = \infty$, it is clear that $\lim_{t\to\infty} k(t) = 1$. Thus, for each $k_1 \in (0, 1)$, there exists $T_{k_1} \geq T$ such that $k(t) \geq k_1$ for $t \geq T_{k_1}$. Then, we have

$$\frac{x[g(t)]}{x'[g(t)]} \geq \left(\frac{a[g(t)]}{a(t_2)}\right)^{1/\alpha}k_1g(t) \quad \text{for} \quad t \geq T_{k_1}. \qquad (3.13.8)$$

By (3.13.7) and (3.13.8), we find that for $t > g(t) \geq T_{k_1} \geq T$,

$$\frac{x(t)}{x[g(t)]} \leq 1 + \left(\frac{a[g(t)]}{a(t_1)}\right)^{1/\alpha}\left(\frac{x'[g(t)]}{x[g(t)]}\right)[t - g(t)]$$

$$\leq 1 + \left(\frac{a(t_2)}{a(t_1)}\right)^{1/\alpha}\left(\frac{t - g(t)}{k_1g(t)}\right) \leq \left(\frac{a(t_2)}{a(t_1)}\right)^{1/\alpha}\left(\frac{t}{k_1g(t)}\right),$$

or

$$x[g(t)] \geq k_1\left(\frac{a(t_1)}{a(t_2)}\right)^{1/\alpha}\left(\frac{g(t)}{t}\right)x(t) \quad \text{for} \quad t \geq T_{k_1}. \qquad (3.13.9)$$

Now, (3.13.4) and (3.13.5) easily follow from (3.13.9).

(II) Since $x'(t) > 0$ and $a(t)(x'(t))^\alpha$ is decreasing for $t \geq T$, we find that $a^{1/\alpha}(s)x'(s) \geq a^{1/\alpha}(t)x'(t)$ for $t \geq s \geq T$, and hence

$$x'(s) \geq a^{-1/\alpha}(s)a^{1/\alpha}(t)x'(t). \tag{3.13.10}$$

Integrating (3.13.10) from T to t, we obtain

$$x(t) \geq x(t) - x(T) \geq \left(a^{1/\alpha}(t)x'(t)\right)\int_T^t a^{-1/\alpha}(s)ds,$$

or $x(t) \geq a^{1/\alpha}(t)R[t, T]x'(t)$ for $t \geq T$. This completes the proof. ∎

Lemma 3.13.2. Let $x(t) \in C^1([t_0, \infty), \mathbb{R}^+)$ be such that $a^{1/\alpha}(t)x'(t) \in C^1([t_0, \infty), \mathbb{R})$, where α and $a(t)$ satisfy conditions (i) and (ii) respectively, and $\left(a(t)|x'(t)|^{\alpha-1}x'(t)\right)' \geq 0$ for $t \geq T \geq t_0$. Then,

(I$_1$) when $x'(t) < 0$,

$$x(\sigma) \geq a^{1/\alpha}(\tau)R[\tau, \sigma]x'(\tau), \quad \tau \geq \sigma \geq T \tag{3.13.11}$$

(I$_2$) when $x'(t) > 0$,

$$x(\tau) \geq a^{1/\alpha}(\sigma)R[\tau, \sigma]x'(\sigma) \quad \tau \geq \sigma \geq T. \tag{3.13.12}$$

Proof. (I$_1$) Since

$$x(\sigma) \geq x(\sigma) - x(\tau) \geq a^{1/\alpha}(\tau)\left(\int_\sigma^\tau a^{-1/\alpha}(s)ds\right)x'(\tau)$$

inequality (3.13.11) follows.

(I$_2$) Since

$$x(\tau) \geq x(\tau) - x(\sigma) \geq a^{1/\alpha}(\sigma)\left(\int_\sigma^\tau a^{-1/\alpha}(s)ds\right)x'(\sigma)$$

inequality (3.13.12) holds. ∎

Lemma 3.13.3. Consider the differential inequality

$$\left(a(t)|z'(t)|^{\alpha-1}z'(t)\right)' + q(t)|z[g(t)]|^{\beta-1}z[g(t)] \leq 0, \tag{3.13.13}$$

where conditions (i) – (iv) hold and $\beta > 0$ is a constant. Let $z(t)$ be an eventually positive solution of (3.13.13). Then for some t_0 and $0 < x_0 \leq z(t_0)$ there exists a solution $x(t)$ of

$$\left(a(t)|x'(t)|^{\alpha-1}x'(t)\right)' + q(t)|x[g(t)]|^{\beta-1}x[g(t)] = 0 \tag{3.13.14}$$

with $x(t_0) = x_0$ such that $0 < x^{(i)}(t) \leq z^{(i)}(t)$, $i = 0, 1$ for all $t \geq t_0$.

Proof. Since $z(t) > 0$ and $\left(a(t)|z'(t)|^{\alpha-1}z'(t)\right)' \leq -q(t)z^{\beta}[g(t)] < 0$ for all large t it follows that $z'(t)$ is eventually of one sign, i.e., either $x'(t) > 0$ for $t \geq t_0$, or there exists a $t_1 \geq t_0$ such that $z'(t) < 0$ for $t \geq t_1$. Suppose $z'(t) < 0$ for $t \geq t_1$. Then since $a(t)(z'(t))^{\alpha} = -a(t)(-x'(t))^{\alpha}$, we find that $-a(t)(-z'(t))^{\alpha} \leq -a(t_1)(-z'(t_1))^{\alpha}$ for $t \geq t_1$, or equivalently, $a^{1/\alpha}(t)z'(t) \leq a^{1/\alpha}(t_1)z'(t_1)$, and hence

$$z'(t) \leq \left(a^{1/\alpha}(t_1)z'(t_1)\right)a^{-1/\alpha}(t) \quad \text{for} \quad t \geq t_1. \tag{3.13.15}$$

Integrating (3.13.15) from t_1 to t, we get

$$z(t) \leq z(t_1) + \left(a^{1/\alpha}(t_1)z'(t_1)\right)R[t, t_1] \to -\infty \quad \text{as} \quad t \to \infty,$$

which contradicts the fact that $z(t) > 0$ eventually. Thus, $z'(t) > 0$ eventually. Now, let x_0 be such that $0 < x_0 < z(T)$ for some $T \geq t_0$. Then integrating (3.13.13) from t to u with $u \geq t \geq T$ and letting $u \to \infty$, we obtain for $t \geq T$,

$$z'(t) \geq a^{-1/\alpha}(t)\left(\int_t^{\infty} q(s)z^{\beta}[g(s)]ds\right)^{1/\alpha} = \Phi(t, z). \tag{3.13.16}$$

Integrating (3.13.16) from T to $t \geq T$, we obtain

$$z(t) \geq z(T) + \int_T^t \Phi(s, z)ds = z(T) + \Psi(t, z), \quad t \geq T. \tag{3.13.17}$$

Now, define

$$\begin{aligned} x_0(t) &= z(t) \\ x_{n+1}(t) &= x_0 + \Psi(t, x_n), \quad n = 1, 2, \cdots. \end{aligned} \tag{3.13.18}$$

Then from (3.13.17), we get by induction that $0 < x_n(t) \leq z(t)$ for $t \geq T$, $n = 0, 1, 2, \cdots$ and $x_{n+1}(t) \leq x_n(t)$ for $t \geq T$, $n = 0, 1, 2, \cdots$. Consequently, letting $\lim_{n\to\infty} x_n(t) = x(t)$ and applying Lebesgue's monotone convergence theorem we find $x(t) = x_0 + \Psi(t, x)$. Now, it follows easily that $x(t)$ has the desired properties. It is also clear that a corresponding result holds for the negative solution $z(t)$ of (2.13.13). ∎

Lemma 3.13.4. Let $q(t) \in C([t_0, \infty), \mathbb{R}^+)$, $\tau(t) \in C^1([t_0, \infty), \mathbb{R})$, $\tau(t) \leq t$ and $\tau'(t) \geq 0$ for $t \geq t_0$ and $\lim_{t\to\infty} \tau(t) = \infty$. If

$$\liminf_{t\to\infty} \int_{\tau(t)}^t q(s)ds > \frac{1}{e},$$

(a₁)   then the inequality $x'(t) + q(t)x[\tau(t)] \leq 0$ eventually, has no eventually positive solution,

(a₂)   then the inequality $x'(t) + q(t)x[\tau(t)] \geq 0$ eventually, has no eventually negative solution,

(a₃)   then the equation $x'(t) + q(t)x[\tau(t)] = 0$ is oscillatory.

Lemma 3.13.5.   Let $q(t) \in C([t_0, \infty), \mathbb{R}^+)$, $\tau(t) \in C^1([t_0, \infty), \mathbb{R})$, $\tau(t) \geq t$ and $\tau'(t) \geq 0$ for $t \geq t_0$. If

$$\liminf_{t \to \infty} \int_t^{\tau(t)} q(s)ds \; > \; \frac{1}{e},$$

(b₁)   then the inequality $x'(t) - q(t)x[\tau(t)] \geq 0$ eventually, has no eventually positive solution,

(b₂)   then the inequality $x'(t) - q(t)x[\tau(t)] \leq 0$ eventually, has no eventually negative solution,

(b₃)   then the equation $x'(t) - q(t)x[\tau(t)] = 0$ is oscillatory.

Lemma 3.13.6.   If $y(t)$ is a positive and strictly decreasing solution of the integral inequality $y(t) \geq \int_t^\infty H(s, y[g(s)])ds$, where $H \in C([t_0, \infty) \times \mathbb{R}, \mathbb{R})$, $xH(t, x) > 0$ for $x \neq 0$, $t \geq t_0 \geq 0$ and increasing in the second variable, $g \in C([t_0, \infty), \mathbb{R})$, $g(t) < t$ and $\lim_{t \to \infty} g(t) = \infty$, then there exists a positive solution $x(t)$ of the equation $x'(t) + H(t, x[g(t)]) = 0$ such that $x(t) \leq y(t)$ eventually, and satisfies $\lim_{t \to \infty} x(t) = 0$ monotonically.

2.13.2. Oscillation of Equation (3.13.1)

We are now in the position to state and prove the following result.

Theorem 3.13.1.   Let conditions (i) – (iv) hold, and $g(t) \leq t$. If for all large $T \geq T_0 \geq t_0$ so that $g(t) \geq T_0$, $t \geq T$ the first order delay equation

$$y'(t) + q(t)R^\beta[g(t), T_0]|y[g(t)]|^{\beta/\alpha} \operatorname{sgn} y[g(t)] \; = \; 0, \quad t \geq T \qquad (3.13.19)$$

is oscillatory, then equation (3.13.1) is oscillatory.

Proof.   Let $x(t)$ be a nonoscillatory solution of equation (3.13.1), say, $x(t) > 0$ for $t \geq t_0 \geq 0$. As in Lemma 3.13.3 there exists a $t_1 \geq t_0$ such that

$$x'(t) > 0 \quad \text{for} \quad t \geq t_1 \qquad (3.13.20)$$

and by Lemma 3.13.1(II), there exists a $t_2 \geq t_1$, $g(t) \geq t_1$ for $t \geq t_2$ and

$$x[g(t)] \geq a^{1/\alpha}[g(t)]R[g(t), t_1]x'[g(t)] \quad \text{for} \quad t \geq t_2. \qquad (3.13.21)$$

Using (3.13.21) in equation (3.13.1), we obtain

$$(a(t)(x'(t))^\alpha)' + q(t)R^\beta[g(t), t_1]\left(a^{1/\alpha}[g(t)]x'[g(t)]\right)^\beta$$
$$\leq (a(t)(x'(t))^\alpha)' + q(t)x^\beta[g(t)] = 0 \quad \text{for} \quad t \geq t_2.$$

Setting $w(t) = a(t)(x'(t))^\alpha$, $t \geq t_2$ in the above inequality, we get

$$w'(t) + q(t)R^\beta[g(t), t_1]w^{\beta/\alpha}[g(t)] \leq 0 \quad \text{for} \quad t \geq t_2. \qquad (3.13.22)$$

Integrating (3.13.22) from $t \geq t_2$ to u and letting $u \to \infty$, we find

$$w(t) \geq \int_t^\infty q(s)R^\beta[g(s), t_1]w^{\beta/\alpha}[g(s)]ds, \quad t \geq t_2.$$

Clearly, the function $w(t)$ is strictly decreasing for $t \geq t_2$. Hence, by Lemma 3.13.6, there exists a positive solution $y(t)$ of equation (3.13.19) with $\lim_{t\to\infty} y(t) = 0$. But this contradicts the assumption that (3.13.19) is oscillatory. This completes the proof.   ∎

Lemma 3.13.4 when applied to (3.13.19) leads to the following corollary.

Corollary 3.13.1. If for every $T \geq T_0 \geq t_0$, $t \geq g(t) \geq T_0$, $t \geq T$,

$$\liminf_{t\to\infty} \int_{g(t)}^t q(s)R^\alpha[g(s), T_0]ds > \frac{1}{e} \quad \text{when} \quad \alpha = \beta,$$

or

$$\int^\infty q(s)R^\alpha[g(s), T_0]ds = \infty \quad \text{when} \quad \alpha < \beta,$$

then equation (3.13.1) is oscillatory.

Theorem 3.13.2. Let conditions (i) – (iv) hold, $\alpha = \beta$, $g(t) > 0$ and $g'(t) > 0$ for $t \geq t_0$. If there exists a function $\rho(t) \in C^1([t_0, \infty), \mathbb{R}^+)$ such that

$$\limsup_{t\to\infty} \int_{t_0}^t \left[\rho(s)q(s) - \gamma a[g(s)]\frac{(\rho'(s))^{\alpha+1}}{(\rho(s)g'(s))^\alpha}\right]ds = \infty, \qquad (3.13.23)$$

where $\gamma = 1/(\alpha + 1)^{(\alpha+1)}$, then equation (3.13.1) is oscillatory.

Proof. Let $x(t)$ be a nonoscillatory solution of equation (3.13.1), say, $x(t) > 0$ for $t \geq t_0 \geq 0$. As in Theorem 3.13.1 there exists a $t_1 \geq t_0$ such that for $t \geq t_1$,

$$x'(t) > 0 \quad \text{and} \quad a(t)(x'(t))^\alpha \leq a[g(t)](x'[g(t)])^\alpha. \qquad (3.13.24)$$

Define $w(t) = \rho(t)a(t)(x'(t))^\alpha/x^\beta[g(t)]$ for $t \geq t_1$. Then for $t \geq t_1$,

$$w'(t) = -\rho(t)q(t) + \frac{\rho'(t)}{\rho(t)}w(t) - \beta\rho(t)g'(t)\frac{a(t)(x'(t))^\alpha x'[g(t)]}{x^{\beta+1}[g(t)]}. \quad (3.13.25)$$

Using (3.13.24) in (3.13.25) and putting $\alpha = \beta$, we get for $t \geq t_1$,

$$w'(t) \leq -\rho(t)q(t) + \frac{\rho'(t)}{\rho(t)}w(t) - \alpha\rho(t)g'(t)a(t)\left(\frac{a(t)}{a[g(t)]}\right)^{1/\alpha}\left(\frac{x'(t)}{x[g(t)]}\right)^{\alpha+1},$$

or

$$w'(t) \leq -\rho(t)q(t) + \frac{\rho'(t)}{\rho(t)}w(t) - \alpha g'(t)(a[g(t)]\rho(t))^{-1/\alpha}w^{(\alpha+1)/\alpha}(t). \quad (3.13.26)$$

Set

$$A = (\alpha g'(t))^{\alpha/(\alpha+1)}\frac{w(t)}{(a[g(t)]\rho(t))^{1/(\alpha+1)}}, \qquad \lambda = \frac{\alpha+1}{\alpha} > 1$$

and

$$B = \left(\frac{\alpha}{\alpha+1}\right)^\alpha\left[\frac{\rho'(t)}{\rho(t)}(a[g(t)]\rho(t))^{1/(\alpha+1)}(\alpha g'(t))^{-\alpha/(\alpha+1)}\right]^\alpha$$

in Lemma 3.2.1, to obtain

$$\frac{\rho'(t)}{\rho(t)}w(t) - \alpha g'(t)(a[g(t)]\rho(t))^{-1/\alpha}w^{(\alpha+1)/\alpha}(t)$$

$$\leq \left(\frac{1}{\alpha+1}\right)^{\alpha+1}\left[a[g(t)]\frac{(\rho'(t))^{\alpha+1}}{(\rho(t)g'(t))^\alpha}\right] \quad \text{for } t \geq t_1.$$

Thus, inequality (3.13.26) gives

$$w'(t) \leq -\rho(t)q(t) + \gamma a[g(t)]\frac{(\rho'(t))^{\alpha+1}}{(\rho(t)g'(t))^\alpha} \quad \text{for } t \geq t_1.$$

Integrating the above inequality from t_1 to t, we get

$$0 < w(t) \leq w(t_1) - \int_{t_1}^t\left[\rho(s)q(s) - \gamma a[g(s)]\frac{(\rho'(s))^{\alpha+1}}{(\rho(s)g'(s))^\alpha}\right]ds. \quad (3.13.27)$$

Taking $\limsup$ on both sides of (3.13.27) as $t \to \infty$, we find a contradiction to condition (3.13.23). This completes the proof. ∎

When $\alpha = \beta = 1$, Theorem 3.13.2 reduces to the following well–known result.

Corollary 3.13.2. Let conditions (i) – (iv) hold, $\alpha = \beta = 1$, $g(t) > 0$ and $g'(t) > 0$ for $t \geq t_0$. If there exists a function $\rho(t) \in C^1([t_0, \infty), \mathbb{R}^+)$ such that

$$\limsup_{t \to \infty} \int_{t_0}^t \left[\rho(s)q(s) - \frac{1}{4} a[g(s)] \frac{(\rho'(s))^2}{\rho(s)g'(s)} \right] ds = \infty,$$

then equation (3.13.1) is oscillatory.

Although, the following corollary is immediate, we shall prove it without using Lemma 3.2.1.

Corollary 3.13.3. Let condition (3.13.23) in Theorem 3.13.2 be replaced by

$$\limsup_{t \to \infty} \int_{t_0}^t \rho(s)q(s)ds = \infty \tag{3.13.28}$$

and

$$\lim_{t \to \infty} \int_{t_0}^t a[g(s)] \frac{(\rho'(s))^{\alpha+1}}{(\rho(s)g'(s))^\alpha} ds < \infty, \tag{3.13.29}$$

then the conclusion of Theorem 3.13.2 holds.

Proof. Let $x(t)$ be a nonoscillatory solution of equation (3.13.1), say, $x(t) > 0$ for $t \geq t_0 \geq 0$. We define the function $w(t)$ as in Theorem 3.13.2 and obtain (3.13.26) for $t \geq t_1$.

Integrating (3.13.26) from t_1 to t, we find

$$w(t) + \int_{t_1}^t \rho(s)q(s)ds - \int_{t_1}^t \frac{\rho'(s)}{\rho(s)} w(s)ds$$

$$+ \alpha \int_{t_1}^t g'(s)(a[g(s)]\rho(s))^{-1/\alpha} w^{(\alpha+1)/\alpha} ds \leq w(t_1) = c, \tag{3.13.30}$$

where $c > 0$ is a constant.

Suppose first that $\int_{t_1}^\infty \rho'(s)w(s)/\rho(s)ds < \infty$. Then it follows from (3.13.30) that $\int_{t_1}^t \rho(s)q(s)ds \leq c + \int_{t_1}^t \rho'(s)w(s)/\rho(s)ds$, which as $t \to \infty$ gives $\int_{t_1}^\infty \rho(s)q(s)ds < \infty$. But, this in view of (3.13.28) is impossible.

Suppose next that

$$\int_{t_1}^\infty \frac{\rho'(s)}{\rho(s)} w(s)ds = \infty. \tag{3.13.31}$$

Then by (3.13.30), we obtain

$$\int_{t_1}^t \rho(s)q(s)ds \leq c + \int_{t_1}^t \frac{\rho'(s)}{\rho(s)} w(s)ds - \alpha \int_{t_1}^t g'(s)(a[g(s)]\rho(s))^{-\frac{1}{\alpha}} w^{\frac{\alpha+1}{\alpha}}(s)ds. \tag{3.13.32}$$

Now, estimating the second integral in (3.13.32) by Hölder's inequality, we have

$$\int_{t_1}^t \frac{\rho'(s)}{\rho(s)} w(s) ds = \int_{t_1}^t \left(\frac{\rho'(s)}{\rho(s)} (a[g(s)]\rho(s))^{1/(\alpha+1)} (\alpha g'(s))^{-\alpha/(\alpha+1)} \right)$$
$$\times \left((a[g(s)]\rho(s))^{-1/(\alpha+1)} (\alpha g'(s))^{\alpha/(\alpha+1)} w(s) \right) ds$$
$$\leq \left(\int_{t_1}^t \alpha g'(s)(a[g(s)]\rho(s))^{-1/\alpha} w^{(\alpha+1)/\alpha} ds \right)^{\alpha/(\alpha+1)}$$
$$\times \left(\int_{t_1}^t \left(\frac{\rho'(s)}{\rho(s)} \right)^{\alpha+1} a[g(s)]\rho(s)(\alpha g'(s))^{-\alpha} ds \right)^{1/(\alpha+1)}.$$

$$(3.13.33)$$

Since (3.13.31) implies that $\int_{t_1}^t (\rho'(s)w(s)/\rho(s)) ds \to \infty$ as $t \to \infty$, we find from (3.13.29) and (3.13.33) that there exists a $t_2 \geq t_1$ such that for $t \geq t_2$,

$$\int_{t_1}^t \frac{\rho'(s)}{\rho(s)} w(s) ds \leq \alpha \int_{t_1}^t g'(s)(a[g(s)]\rho(s))^{-1/\alpha} w^{(\alpha+1)/\alpha}(s) ds.$$

Using this in (3.13.32), we conclude that $\limsup_{t \to \infty} \int_{t_1}^t \rho(s)q(s) ds \leq c$, which contradicts (3.3.28). This completes the proof. ∎

Example 3.13.2. Consider the half–linear differential equation

$$\left(a(t)|x'(t)|^{\alpha-1}x'(t) \right)' + \frac{k}{t^{\alpha+1}} |x[\lambda t]|^{\alpha-1} x[\lambda t] = 0, \quad t > 0 \qquad (3.13.34)$$

where α, λ, k are positive constants, $0 < \lambda \leq 1$ and $a(t) \in C([t_0, \infty), \mathbb{R}^+)$. It is easy to verify the following:

(A_1) If $a(t) = 1$, $\rho(t) = t^{\alpha+1}$ and

$$k > \frac{1}{\lambda^\alpha} \left(\frac{\alpha}{\alpha+1} \right)^{\alpha+1},$$

then all conditions of Theorem 3.13.2 are satisfied, and hence equation (3.13.34) is oscillatory.

(A_2) If $a(t) = 1/t^\beta$ where $\beta > 0$ is a constant, then for all β, λ and k the hypotheses of Corollary 3.13.2 are satisfied with $\rho(t) = t^{\alpha+1}$, and therefore, equation (3.13.34) is oscillatory.

Now, for each $t \geq t_0$, we define $\gamma(t) = \sup\{s \geq t_0 : g(s) \leq t\}$. Clearly, $\gamma(t) \geq t$ and $g \circ \gamma(t) = t$.

Theorem 3.13.3. Let conditions (i) – (iv) hold, and $\alpha = \beta$, $g(t) \leq t$ for $t \geq t_0$. Then equation (3.13.1) is oscillatory if either one of the following conditions holds

(C_1) when $a'(t) \leq 0$ then for $t \geq T \geq t_0$,

$$\limsup_{t \to \infty} R^\alpha[t,T] \int_t^\infty \left(\frac{a(s)}{a(T)}\right) \left(\frac{g(s)}{s}\right)^\alpha q(s)ds \ > \ 1, \qquad (3.3.35)$$

(C_2) when $a'(t) \geq 0$ then for $t \geq T \geq t_0$,

$$\limsup_{t \to \infty} R^\alpha[t,T] \int_t^\infty \left(\frac{g(s)}{s}\right)^\alpha q(s)ds \ > \ 1, \qquad (3.13.36)$$

(C_3) for $t \geq T \geq t_0$,

$$\limsup_{t \to \infty} R^\alpha[t,T] \int_{\gamma(t)}^\infty q(s)ds \ > \ 1. \qquad (3.13.37)$$

Proof. Let $x(t)$ be a nonoscillatory solution of equation (3.13.1), say, $x(t) > 0$ for $t \geq t_0$. As in Lemma 3.13.3, there exists a $t_1 \geq t_0$ such that $x'(t) > 0$ for $t \geq t_1$ and by Lemma 3.13.1(II) there exists a $t_2 \geq t_1$ such that

$$x(t) \ \geq \ a^{1/\alpha}(t)R[t,t_2]x'(t) \quad \text{for} \quad t \geq t_2. \qquad (3.13.38)$$

Integrating (3.13.1) from $t \geq t_2$ to u and letting $u \to \infty$, we find

$$a(t)(x'(t))^\alpha \ \geq \ \int_t^\infty q(s)x^\alpha[g(s)]ds. \qquad (3.13.39)$$

Using (3.13.38) in (3.13.39), we get for $t \geq t_2$,

$$x^\alpha(t) \ \geq \ R^\alpha[t,t_2]a(t)(x'(t))^\alpha \ \geq \ R^\alpha[t,t_2] \int_t^\infty q(s)x^\alpha[g(s)]ds. \qquad (3.13.40)$$

By Lemma 3.13.1(I), for each constant $k \in (0,1)$ there exists $T_k \geq t_1$ such that

$$x^\alpha[g(t)] \ \geq \ k^\alpha \left(\frac{a(t)}{a(t_1)}\right) \left(\frac{g(t)}{t}\right)^\alpha x^\alpha(t) \quad \text{when} \quad a'(t) \leq 0 \ \text{ and } \ t \geq T_k,$$

or

$$x^\alpha[g(t)] \ \geq \ k^\alpha \left(\frac{g(t)}{t}\right)^\alpha x^\alpha(t) \quad \text{when} \quad a'(t) \geq 0 \ \text{ and } \ t \geq T_k.$$

Thus, from (3.13.40) for $t \geq T = \max\{T_k, t_2\}$, when $a'(t) \leq 0$ and $t \geq T$, we obtain

$$1 \ \geq \ k^\alpha R^\alpha[t,t_2] \int_t^\infty \left(\frac{a(s)}{a(t_1)}\right) \left(\frac{g(s)}{s}\right)^\alpha q(s)ds, \qquad (3.13.41)$$

or when $a'(t) \geq 0$ and $t \geq T$,

$$1 \geq k^\alpha R^\alpha[t, t_2] \int_t^\infty \left(\frac{g(s)}{s}\right)^\alpha q(s)ds.$$

Hence, it follows that

$$\limsup_{t \to \infty} R^\alpha[t, t_2] \int_t^\infty \frac{a(s)}{a(t_1)} \left(\frac{g(s)}{s}\right)^\alpha q(s)ds = b_1 < \infty,$$

or

$$\limsup_{t \to \infty} R^\alpha[t, t_2] \int_t^\infty \left(\frac{g(s)}{s}\right)^\alpha q(s)ds = b_2 < \infty.$$

Suppose condition (3.13.36) holds, then there exists a sequence $\{s_n\}_{n=1}^\infty$ such that $\lim_{n \to \infty} s_n = \infty$ and for each fixed $t_2 \leq s_1$,

$$\lim_{n \to \infty} R^\alpha[s_n, t_2] \int_{s_n}^\infty \left(\frac{g(s)}{s}\right)^\alpha q(s)ds = b_2 > 1. \tag{3.13.42}$$

For $\epsilon = (b_2 - 1)/2 > 0$, there exists an integer $N > 0$ such that

$$\frac{b_2 + 1}{2} = b_2 - \epsilon < R^\alpha[s_n, t_2] \int_{s_n}^\infty \left(\frac{g(s)}{s}\right)^\alpha q(s)ds \text{ for } n \geq N. \tag{3.13.43}$$

Choose k_1 such that $(2/(b_2 + 1))^{1/\alpha} < k_1 < 1$. Then by (3.13.42) and (3.13.43), we have

$$1 \geq k_1^\alpha R^\alpha[s_n, t_2] \int_{s_n}^\infty \left(\frac{g(s)}{s}\right)^\alpha q(s)ds > \left(\frac{2}{b_2 + 1}\right)\left(\frac{b_2 + 1}{2}\right) = 1$$

for all sufficiently large n. This contradiction shows that condition (3.13.36) does not hold. The other case can be considered similarly and hence omitted.

Next, by $\gamma(t) \geq t$ and (3.13.40), we have

$$x^\alpha(t) \geq R^\alpha[t, t_2] \int_{\gamma(t)}^t q(s)x^\alpha[g(s)]ds \text{ for } t \geq t_2.$$

Since $x(t)$ is increasing and $g(s) \geq t$ for $s \geq \gamma(t)$, it follows that $1 \geq R^\alpha[t, t_2] \int_{\gamma(t)}^\infty q(s)ds$, which contradicts condition (3.13.37). ∎

Corollary 3.13.4. Let conditions (i) – (iv) hold, $\alpha = \beta$, $g(t) \leq t$ for $t \geq t_0$ and $a(t) \equiv 1$. If either

$$\limsup_{t \to \infty} t^\alpha \int_t^\infty \left(\frac{g(s)}{s}\right)^\alpha q(s)ds > 1,$$

or

$$\limsup_{t\to\infty} t^\alpha \int_{\gamma(t)}^\infty q(s)ds \; > \; 1,$$

holds, then equation (3.13.1) is oscillatory.

Example 3.13.3. Consider the half–linear differential equation

$$\left(t^\alpha |x'(t)|^{\alpha-1} x'(t)\right)' + \frac{c}{t(\ln et)^{\alpha+1}} \left|x\left[\frac{t}{2}\right]\right|^{\alpha-1} x\left[\frac{t}{2}\right] = 0, \ t \geq T = 1 \geq t_0 > 0,$$

$$(3.13.44)$$

where α and c are positive constants. Let $R[t,T] = \int_T^t a^{-1/\alpha}(s)ds = \int_1^t ds/s = \ln t$, so that

$$R^\alpha[t,T] \int_t^\infty \left(\frac{g(s)}{s}\right)^\alpha q(s)ds = (\ln t)^\alpha \int_t^\infty \left(\frac{1}{2}\right)^\alpha \frac{c}{s(\ln es)^{\alpha+1}} ds$$

$$= (\ln t)^\alpha \left(\frac{1}{2}\right)^\alpha \frac{c}{\alpha(\ln et)^\alpha} = \frac{2^{-\alpha}c}{\alpha} \left(\frac{\ln t}{\ln et}\right)^\alpha.$$

Taking $\limsup$ in the above equality as $t \to \infty$, we find in view of Theorem 3.13.3 that equation (3.13.44) is oscillatory provided $c > \alpha(2^\alpha)$.

Theorem 3.13.4. Let conditions (i) – (iv) hold, $\alpha > 1$, $\beta > 1$, $0 < g(t) \leq t$ and $g'(t) > 0$ for $t \geq t_0$, and assume that there exists a function $\rho(t) \in C^1([t_0, \infty), \mathbb{R}^+)$ such that

$$\rho'(t) \geq 0, \quad k(t) = \frac{a^{1/\alpha}[g(t)]\rho'(t)}{g'(t)} \quad \text{and} \quad k'(t) \leq 0 \quad \text{for} \quad t \geq t_0. \ (3.13.45)$$

If

$$\int^\infty \rho(s)q(s)ds = \infty, \qquad\qquad (3.13.46)$$

then equation (3.13.1) is oscillatory.

Proof. Let $x(t)$ be a nonoscillatory solution of equation (3.13.1), say, $x(t) > 0$ for $t \geq t_0 \geq 0$. As in Theorem 3.13.2, there exists a $t_1 \geq t_0$ such that (3.13.24) holds. Define the function $w(t)$ as in Theorem 3.13.2 to obtain (3.13.25) which takes the form

$$w'(t) \leq -\rho(t)q(t) + \rho'(t)\frac{a(t)(x'(t))^\alpha}{x^\beta[g(t)]} \quad \text{for} \quad t \geq t_2 \geq t_1. \ (3.13.47)$$

There exist a constant $b > 0$ and a $t_3 \geq t_2$ such that $\left(a^{1/\alpha}(t)x'(t)\right)^{\alpha-1} \leq b$ for $t \geq t_3$. Thus, (3.13.47) becomes

$$w'(t) \leq -\rho(t)q(t) + b\frac{a^{1/\alpha}(t)x'(t)}{x^\beta[g(t)]} \quad \text{for} \quad t \geq t_3. \ (3.13.48)$$

Next, there exists a $T \geq t_3$ such that $a^{1/\alpha}(t)x'(t) \leq a^{1/\alpha}[g(t)]x'[g(t)]$ for $t \geq T$ and hence

$$w'(t) \leq -\rho(t)q(t) + b\left(\frac{a^{1/\alpha}[g(t)]\rho'(t)}{g'(t)}\right)\left(\frac{x'[g(t)]g'(t)}{x^\beta[g(t)]}\right) \quad \text{for} \quad t \geq T.$$

But, by the Bonnet theorem for a fixed $t \geq T$ and for some $\xi \in [T, t]$, we have

$$\int_T^t \left(\frac{a^{1/\alpha}[g(s)]\rho'(s)}{g'(s)}\right)\left(\frac{x'[g(s)]g'(s)}{x^\beta[g(s)]}\right)ds$$

$$= \left(\frac{a^{1/\alpha}g[(T)]\rho'(T)}{g'(T)}\right)\int_T^\xi \frac{x'[g(s)]g'(s)}{x^\beta[g(s)]}ds$$

$$= k(T)\int_{x[g(T)]}^{x[g(\xi)]} u^{-\beta}du = \frac{k(T)}{\beta - 1}\left[x^{-\beta+1}[g(T)] - x^{-\beta+1}[g(\xi)]\right]$$

$$\leq \frac{k(T)}{\beta - 1}x^{1-\beta}[g(T)] \equiv K < \infty,$$

where K is a positive constant. Thus, we find

$$\int_T^t \left(\frac{a^{1/\alpha}[g(s)]\rho'(s)}{g'(s)}\right)\left(\frac{x'[g(s)]g'(s)}{x^\beta[g(s)]}\right)ds \leq K \quad \text{for} \quad t \geq T. \quad (3.13.49)$$

Therefore, in view of (3.13.49), it follows that $\int_T^t \rho(s)q(s)ds \leq -w(t) + w(T) + bK < \infty$, which contradicts (3.13.46). This completes the proof. ∎

Theorem 3.13.5. Let conditions (i) – (iv) hold, $\beta > \alpha$, $g(t) \leq t$ and $g'(t) \geq 0$ for $t \geq t_0$. If

$$\int^\infty a^{-1/\alpha}[g(s)]g'(s)\left(\int_s^\infty q(u)du\right)^{1/\alpha}ds = \infty, \quad (3.13.50)$$

then equation (3.13.1) is oscillatory.

Proof. Let $x(t)$ be a nonoscillatory solution of equation (3.13.1), say, $x(t) > 0$ for $t \geq t_0 \geq 0$. As in Theorem 3.13.4, we take $\rho(t) = 1$ to obtain $\int_t^\infty q(s)ds \leq a(t)(x'(t))^\alpha/x^\beta[g(t)]$ for $t \geq t_1 \geq t_0$, or

$$a^{-1/\alpha}[g(t)]g'(t)\left(\int_t^\infty q(s)ds\right)^{1/\alpha} \leq \frac{x'[g(t)]g'(t)}{x^{\beta/\alpha}[g(t)]} \quad \text{for} \quad t \geq T \geq t_1.$$

Integrating the above inequality from T to t, we find

$$\int_T^t a^{-1/\alpha}[g(s)]g'(s)\left(\int_s^\infty q(u)du\right)^{1/\alpha}ds \leq \int_{x[g(T)]}^{x[g(t)]} v^{-\beta/\alpha}dv$$

$$\leq \frac{\alpha}{\beta - \alpha}x^{1-(\beta/\alpha)}[g(T)] < \infty,$$

which contradicts condition (3.13.50) and completes the proof. ∎

Example 3.13.4. Consider the differential equation

$$\left(\frac{1}{t^\alpha}|x'(t)|^{\alpha-1}x'(t)\right)' + \frac{1}{t^{2\alpha+1}}\left|x\left[\frac{t}{2}\right]\right|^{\beta-1}x\left[\frac{t}{2}\right] = 0, \quad t > 0 \quad (3.13.51)$$

where $\beta > \alpha > 0$ are constants. Since,

$$\int^t a^{-1/\alpha}[g(s)]g'(s)\left(\int_s^\infty q(u)du\right)^{1/\alpha} ds = \int^t \frac{s}{4}\left(\int_s^\infty u^{-1-2\alpha}du\right)^{1/\alpha} ds$$

$$= \int^t \frac{s}{4}\left(\frac{1}{4\alpha s^2}\right) ds = \frac{1}{8\alpha}\ln t \to \infty \text{ as } t \to \infty$$

it follows from Theorem 3.13.5 that equation (3.13.51) is oscillatory. ∎

3.13.3. Some Extensions and More Oscillation Criteria

In this subsection we begin with the more general equation

$$\left(a(t)|x'(t)|^{\alpha-1}x'(t)\right)' + q(t)F(x[g(t)]) = 0, \quad (3.13.52)$$

where α, a, g, q satisfy conditions (i) – (iv), and

(v) $F \in C(\mathbb{R}, \mathbb{R})$ and $xF(x) > 0$ for $x \neq 0$.

Our interest here is to obtain oscillation criteria for (3.13.52) similar to those presented in Subsection 3.13.2 without assuming that the function F is monotonic. For this, we need the following notation and lemma. Let

$$\mathbb{R}_{t_0} = (-\infty, -t_0] \cup [t_0, \infty) \text{ if } t_0 > 0$$
$$= (-\infty, 0) \cup (0, \infty) \text{ if } t_0 = 0,$$

$$C(\mathbb{R}) = \{f \in C(\mathbb{R}, \mathbb{R}) \text{ and } xf(x) > 0 \text{ for } x \neq 0\}$$

and

$$C_B(\mathbb{R}_{t_0}) = \{f \in C(\mathbb{R}) : f \text{ is of bounded variation on every interval}$$
$$[a, b] \subseteq \mathbb{R}_{t_0}\}.$$

Lemma 3.13.7. Suppose $t_0 > 0$ and $f \in C(\mathbb{R})$. Then, $f \in C_B(\mathbb{R}_{t_0})$ if and only if $f(x) = G(x)H(x)$ for all $x \in \mathbb{R}_{t_0}$, where $G : \mathbb{R}_{t_0} \to \mathbb{R}^+$ is nondecreasing on $(-\infty, -t_0)$ and nonincreasing on (t_0, ∞), and $H : \mathbb{R}_{t_0} \to \mathbb{R}$ is nondecreasing on $\mathbb{R}_{t_0}$.

Now, we shall prove the following comparison theorem.

Theorem 3.13.6. Assume that $F \in C(\mathbb{R}_{t_0})$, $t_0 > 0$ and let G and H be a pair of continuous components of F with H being the nondecreasing one. Moreover, assume that there exists a positive constant β such that

$$H(x) \operatorname{sgn} x \geq |x|^\beta \quad \text{for} \quad x \neq 0, \tag{3.13.53}$$

conditions (i) – (v) hold, $g(t) \leq t$ for $t \geq t_0$. If for every constant $k > 0$ and all large $T \geq T_0 \geq t_0$, $g(t) \geq T_0$ for $t \geq T$, the equation

$$\left(a(t)|x'(t)|^{\alpha-1}x'(t)\right)' + q(t)G(kR[g(t), T_0])|x[g(t)]|^{\beta-1}x[g(t)] = 0 \tag{3.13.54}$$

is oscillatory, then (3.13.52) is also oscillatory.

Proof.   Let $x(t)$ be a nonoscillatory solution of equation (3.13.52). Without loss of generality assume that $x(t) > 0$ for $t \geq t_0 > 0$. As in the proofs given earlier it is easy to see that $x'(t) > 0$ and $a(t)(x'(t))^\alpha$ is nonincreasing for $t \geq t_1 \geq t_0$. Thus, there exists a $t_2 \geq t_1$ and a constant $b_1 > 0$ such that $x'(t) \leq b_1 a^{-1/\alpha}(t)$ for $t \geq t_2$. Integrating this inequality from t_2 to t, we conclude that there exist a constant $b > 0$ and a $t_3 \geq t_2$ such that

$$x[g(t)] \leq bR[g(t), t_2] \quad \text{for} \quad t \geq t_2. \tag{3.13.55}$$

Now, it follows from equation (3.13.52), (3.13.53) and (3.13.55) that

$$\begin{aligned}
&(a(t)(x'(t))^\alpha)' + q(t)G(bR[g(t), t_2])x^\beta[g(t)] \\
&\leq (a(t)(x'(t))^\alpha)' + q(t)G(x[g(t)])H(x[g(t)]) = 0.
\end{aligned} \tag{3.13.56}$$

But, in view of Lemma 3.13.3, equation (3.13.54) has an eventually positive solution, which is a contradiction. This completes the proof.   ∎

We note that Theorem 3.13.6 together with the results presented earlier can be applied to equations of the type (3.13.52) with F being any one of the following functions.

(F$_1$) $F(x) = |x|^{\beta-1}x \exp(-|x|^\gamma)$, β and γ are positive constants,

(F$_2$) $F(x) = |x|^{\beta-1}x \operatorname{sech} x$, $\beta > 0$ is a constant,

(F$_3$) $F(x) = \dfrac{|x|^{\beta-1}x}{1 + |x|^\gamma}$, β and γ are positive constants.

Next, we shall extend the results of Section 3.13.2 to equations of the neutral type of the form

$$\left(a(t)\left|(x(t) + p(t)x[\tau(t)])'\right|^{\alpha-1}(x(t) + p(t)x[\tau(t)])'\right)' + q(t)|x[g(t)]|^{\beta-1}x[g(t)]$$

$$= 0, \tag{3.13.57}$$

where α, β, a, g and q satisfy conditions (i) – (iv), and

(vi) $p(t) \in C([t_0, \infty), \mathbb{R}_0)$,

(vii) $\tau(t) \in C([t_0, \infty), \mathbb{R})$, $\tau'(t) > 0$ and $\lim_{t\to\infty} \tau(t) = \infty$.

For equation (3.13.57) we shall prove the following comparison results.

Theorem 3.13.7. Let $0 \le p(t) \le 1$, $\tau(t) < t$ and $g(t) \le t$ for $t \ge t_0$, and $p(t) \not\equiv 1$ for all large t. If the equation

$$\left(a(t)|z'(t)|^{\alpha-1}z'(t)\right)' + q(t)(1 - p[g(t)])^\beta |z[g(t)]|^{\beta-1}z[g(t)] = 0 \quad (3.13.58)$$

is oscillatory, then equation (3.13.57) is oscillatory.

Theorem 3.13.8. Let $p(t) \ge 1$, $\tau(t) > t$, $\sigma(t) = \tau^{-1} \circ g(t) \le t$ and $\sigma'(t) \ge 0$ for $t \ge t_0$, and $p(t) \not\equiv 1$ for all large t. If the equation

$$\left(a(t)|w'(t)|^{\alpha-1}w'(t)\right)' + q(t)(P[g(t)])^\beta |w[\sigma(t)]|^{\beta-1}w[\sigma(t)] = 0 \quad (3.13.59)$$

is oscillatory, where

$$P(t) = \frac{1}{p[\tau^{-1}(t)]} \left(1 - \frac{1}{p[\tau^{-1} \circ \tau^{-1}(t)]}\right)$$

and τ^{-1} is the inverse function of τ, then equation (3.13.57) is oscillatory.

Proof of Theorems 3.13.7 and 3.13.8. Let $x(t)$ be a nonoscillatory solution of equation (3.13.57), say, $x(t) > 0$ for $t \ge t_0 \ge 0$. Define $y(t) = x(t) + p(t)x[\tau(t)]$. Then, we have

$$\left(a(t)|y'(t)|^{\alpha-1}y'(t)\right)' + q(t)|x[g(t)]|^{\beta-1}x[g(t)] = 0. \quad (3.13.60)$$

It is easy to verify that $y(t) > 0$ and $y'(t) > 0$ for $t \ge t_1 \ge t_0$. Now using the hypotheses of Theorem 3.13.7, we find

$$\begin{aligned}
x(t) &= y(t) - p(t)x[\tau(t)] = y(t) - p(t)[y[\tau(t)] - p[\tau(t)]x[\tau \circ \tau(t)]] \\
&\ge y(t) - p(t)y[\tau(t)] \ge (1 - p(t))y(t) \quad \text{for} \quad t \ge T \ge t_1.
\end{aligned}$$
$$(3.13.61)$$

Using (3.13.62) in equation (3.13.60), we obtain

$$\left(a(t)(y'(t))^\alpha\right)' + q(t)(1 - p[g(t)])^\beta y^\beta[g(t)] \le 0 \quad \text{for} \quad t \ge T.$$

By Lemma 3.13.3, equation (3.13.58) has an eventually positive solution, which is a contradiction.

Next, we use the hypotheses of Theorem 3.13.8, to obtain

$$
\begin{aligned}
x(t) &= \frac{1}{p[\tau^{-1}(t)]}\left(y[\tau^{-1}(t)] - x[\tau^{-1}(t)]\right) \\
&= \frac{y[\tau^{-1}(t)]}{p[\tau^{-1}(t)]} - \frac{1}{p[\tau^{-1}(t)]}\left(\frac{y[\tau^{-1}\circ\tau^{-1}(t)]}{p[\tau^{-1}\circ\tau^{-1}(t)]} - \frac{x[\tau^{-1}\circ\tau^{-1}(t)]}{p[\tau^{-1}\circ\tau^{-1}(t)]}\right) \\
&\geq \frac{y[\tau^{-1}(t)]}{p[\tau^{-1}(t)]} - \frac{y[\tau^{-1}\circ\tau^{-1}(t)]}{p[\tau^{-1}(t)]p[\tau^{-1}\circ\tau^{-1}(t)]} \\
&\geq P(t)y[\tau^{-1}(t)] \quad \text{for} \quad t \geq T^* \quad \text{for some} \quad T^* \geq t_1.
\end{aligned}
$$

$$(3.13.62)$$

Using (3.13.62) in equation (3.13.60), we find

$$
(a(t)(y'(t))^\alpha)' + q(t)(P[g(t)])^\beta y^\beta[\sigma(t)] \leq 0 \quad \text{for} \quad t \geq T^*.
$$

Once again, by Lemma 3.13.3, we arrive at the desired contradiction. This completes the proof. ∎

Example 3.13.5. The neutral differential equations

$$
\left(t^\alpha\left|\left(x(t) + \frac{1}{2}x[ct]\right)'\right|^{\alpha-1}\left(x(t) + \frac{1}{2}x[ct]\right)'\right)' + q_1(t)|x[c_1 t]|^{\beta-1}x[c_1 t] = 0
$$

$$(3.13.63)$$

and

$$
\left(t^\alpha\left|(x(t) + 2x[c^* t])'\right|^{\alpha-1}(x(t) + 2x[c^* t])'\right)' + q_2(t)|x[c_1^* t]|^{\beta-1}x[c_1^* t] = 0,
$$

$$(3.13.64)$$

where α, β, c, c^*, c_1 and c_1^* are positive constants such that $c_1 < 1$, $c_1^* > 1$, $c_1^* \leq c^*$ and $q_1(t)$, $q_2(t) \in C([t_0, \infty), \mathbb{R}^+)$ are oscillatory if the associated differential equations

$$
\left(t^\alpha|z'(t)|^{\alpha-1}z'(t)\right)' + 2^{-\beta}q_1(t)|z[c_1 t]|^{\beta-1}z[c_1 t] = 0 \qquad (3.13.65)
$$

and

$$
\left(t^\alpha|w'(t)|^{\alpha-1}w'(t)\right)' + 4^{-\beta}q_2(t)\left|w\left[\frac{c_1^*}{c^*}t\right]\right|^{\beta-1}w\left[\frac{c_1^*}{c^*}t\right] = 0 \qquad (3.13.66)
$$

are oscillatory. Further, the oscillation of (3.13.65) and (3.13.66) can be discussed by applying the results of Section 3.13.2. The statements of such results are left to the reader.

Finally, we shall consider the damped differential equation

$$
\left(a(t)|x'(t)|^{\alpha-1}x'(t)\right)' + f(t, x[t - \tau], x'[t - \sigma]) = 0, \qquad (3.13.67)
$$

where α and a satisfy conditions (i) and (ii) respectively, τ and σ are real constants, and $F \in C([t_0, \infty) \times \mathbb{R}^2, \mathbb{R})$.

We shall assume that there exist positive constants β and μ and a function $q(t) \in C([t_0, \infty), \mathbb{R}^+)$ such that

$$f(t, x, y) \operatorname{sgn} x \geq q(t)|x|^\beta |y|^\mu \quad \text{for} \quad xy \neq 0, \ t \geq t_0. \qquad (3.13.68)$$

Theorem 3.13.9. Let condition (3.13.68) hold, $\mu \leq \alpha$, τ be any real number, and $\sigma \geq 0$. If for every positive constant θ, the delay equation

$$y'(t) + \theta a^{-\mu/\alpha}[t - \sigma]q(t)|y[t - \sigma]|^{\mu/\alpha} \operatorname{sgn} y[t - \sigma] = 0 \qquad (3.13.69)$$

is oscillatory, then (3.13.67) is also oscillatory.

Proof. Let $x(t)$ be a nonoscillatory solution of equation (3.13.67), say, $x(t) > 0$ for $t \geq t_0 \geq 0$. It is easy to verify that $x'(t) > 0$ for $t \geq t_1 \geq t_0$. There exist a positive constant c and a $t_2 \geq t_1$ such that

$$x^\beta[t - \tau] \geq c \quad \text{for} \quad t \geq t_2. \qquad (3.13.70)$$

Using condition (3.13.68) and (3.13.70) in (3.13.67), we get $(a(t)(x'(t))^\alpha)' + cq(t)(x'[t - \sigma])^\mu \leq 0$ for $t \geq t_2$. Setting $z(t) = (a^{1/\alpha}(t)x'(t))^\alpha$ for $t \geq t_2$, we find $z'(t) + cq(t)a^{-\mu/\alpha}[t - \sigma]z^{\mu/\alpha}[t - \sigma] \leq 0$ for $t \geq t_2$. Integrating this inequality from $t \geq t_2$ to u and letting $u \to \infty$, we obtain $z(t) \geq c \int_t^\infty q(s)a^{-\mu/\alpha}[s - \sigma]z^{\mu/\alpha}[s - \sigma]ds$, $t \geq t_2$. But, in view of Lemma 3.13.6, equation (3.13.69) has a positive decreasing solution $y(t)$, $\lim_{t\to\infty} y(t) = 0$, which is a contradiction. ∎

Theorem 3.13.10. Let condition (3.13.68) hold, $\beta + \mu \leq \alpha$, and $\tau \geq \sigma \geq 0$. If for all large T, $t \geq T + \tau$, the delay equation

$$v'(t) + a(t)R^\beta[t - \tau, T]a^{-\mu/\alpha}[t - \sigma]|v[t - \sigma]|^{(\beta+\mu)/\alpha} \operatorname{sgn} v[t - \sigma] = 0 \quad (3.13.71)$$

is oscillatory, then equation (3.13.67) is oscillatory.

Proof. Let $x(t)$ be a nonoscillatory solution of equation (3.13.67), say, $x(t) > 0$ for $t \geq t_0$. It is easy to verify that $x'(t) > 0$ for $t \geq t_1 \geq t_0$ and by Lemma 3.13.1(II), there exists a $T \geq t_1$ such that

$$
\begin{aligned}
x[t - \tau] &\geq R[t - \tau, T]a^{1/\alpha}[t - \tau]x'[t - \tau] \\
&\geq R[t - \tau, T]a^{1/\alpha}[t - \sigma]x'[t - \sigma] \quad \text{for} \quad t \geq T + \tau.
\end{aligned} \qquad (3.13.72)
$$

Using condition (3.13.68) and (3.13.72) in equation (3.13.67), we get

$$(a(t)(x'(t))^\alpha)' + q(t)R^\beta[t - \tau, T]a^{\beta/\alpha}[t - \sigma](x'[t - \sigma])^{\beta+\mu} \leq 0 \quad \text{for} \quad t \geq T + \tau.$$

Setting $y(t) = a(t)(x'(t))^\alpha$, $t \geq T + \tau$, we find

$$y'(t) + q(t)R^\beta[t - \tau, T]a^{-\mu/\alpha}[t - \sigma]y^{(\beta+\mu)/\alpha}[t - \sigma] \leq 0 \quad \text{for} \quad t \geq T + \tau.$$

The rest of the proof is similar to that of Theorem 3.13.9 and hence omitted. ∎

Remark 3.13.1. The results similar to that of Corollary 3.13.1 can be easily stated from Theorem 3.13.9 and 3.13.10.

Example 3.13.6. The damped differential equation

$$\left(t^2|x'(t)|x'(t)\right)' + q(t)|x'[t-\tau]|x[t-\tau] = 0, \quad t > 0 \tag{3.13.73}$$

where $\tau > 0$ is a constant, and $q(t) \in C([t_0,\infty), \mathbb{R}^+)$ is oscillatory by Theorem 3.13.10 provided for $T = 1 \geq t_0$, the equation

$$y'(t) + \frac{\ln(t-\tau)}{t-\tau} q(t)y[t-\tau] = 0, \quad t > 1+\tau \tag{3.13.74}$$

is oscillatory. We also note that (3.13.74) is oscillatory if

$$\liminf_{t\to\infty} \int_{t-\tau}^t \left(\frac{\ln(s-\tau)}{s-\tau}\right) q(s)ds > \frac{1}{e} \tag{3.13.75}$$

(cf. Lemma 3.13.4(a_3)). Therefore, (3.13.73) is oscillatory if condition (3.13.75) holds.

3.13.4. Classification of Nonoscillatory Solutions

Here, we shall consider the following second order functional differential equation

$$(a(t)\psi(x'(t)))' + F(t, x[g(t)]) = 0, \tag{3.13.76}$$

where

(i) $a(t) \in C([t_0,\infty), \mathbb{R}^+)$,

(ii) $\psi \in C(\mathbb{R}, \mathbb{R})$, ψ is strictly increasing, $\operatorname{sgn} \psi(x) = \operatorname{sgn} x$ and $\psi(\mathbb{R}) = \mathbb{R}$,

(iii) $g(t) \in C([t_0,\infty), \mathbb{R})$, $g'(t) \geq 0$, $g(t) \leq t$ and $\lim_{t\to\infty} g(t) = \infty$,

(iv) $F \in C([t_0,\infty) \times \mathbb{R}, \mathbb{R})$, $F(t, x)$ is nondecreasing in the second variable, and $\operatorname{sgn} F(t, x) = \operatorname{sgn} x$ for each fixed $t \geq t_0 \geq 0$.

In what follows, we shall assume that

$$\int_{t_0}^\infty \left|\psi^{-1}\left(\frac{k}{a(s)}\right)\right| ds = \infty \text{ for every constant } k \neq 0, \; t_0 \geq 0 \tag{3.13.77}$$

where $\psi^{-1} : \mathbb{R} \to \mathbb{R}$ denotes the inverse function of ψ, and use the notation

$$\begin{cases} \Psi_{k,T}(a;t) = \int_T^t \psi^{-1}\left(\frac{k}{a(s)}\right) ds, \quad t \geq T \\ \Psi_k(a;t) = \Psi_{k,t_0}(a;t), \quad t \geq t_0. \end{cases} \tag{3.13.78}$$

From (3.13.77) and (3.13.78) it follows that $\Psi_{k,T}(a;T) = 0$, $\lim_{t \to \infty} |\Psi_{k,T}(a;t)|$ $= \infty$ for every $k \neq 0$, $|\Psi_{k,T}(a;t)| > |\Psi_{\ell,T}(a;t)|$, $t > T$ for $|k| > |\ell|$ with $k\ell > 0$ and $\lim_{k \to 0} \Psi_{k,T}(a;t) = 0$ for all $t \geq T$.

We begin by classifying nonoscillatory solutions of (3.13.76) according to their asymptotic behavior as $t \to \infty$.

Lemma 3.13.8. Any nonoscillatory solution $x(t)$ of equation (3.13.76) is one of the following three types

(I) $\lim_{t \to \infty} a(t)\psi(x'(t)) = $ constant $\neq 0$,

(II) $\lim_{t \to \infty} a(t)\psi(x'(t)) = 0$, $\lim_{t \to \infty} |x(t)| = \infty$,

(III) $\lim_{t \to \infty} a(t)\psi(x'(t)) = 0$, $\lim_{t \to \infty} x(t) = $ constant $\neq 0$.

Proof.   Let $x(t)$ be a nonoscillatory solution of equation (3.13.76). Without loss of generality, we can suppose that $x(t) > 0$ for $t \geq t_0 > 0$. From equation (3.13.76), it follows that

$$(a(t)\psi(x'(t)))' = -F(t, x[g(t)]) < 0 \quad \text{for} \quad t \geq t_1 \geq t_0$$

and so $a(t)\psi(x'(t))$ is decreasing for $t \geq t_1$.

We claim that $a(t)\psi(x'(t)) > 0$, $t \geq t_1$ so that $\lim_{t \to \infty} a(t)\psi(x'(t))$ ≥ 0. In fact, if $a(t_2)\psi(x'(t_2)) = -k < 0$ for some $t_2 \geq t_1$, then $a(t)\psi(x'(t)) \leq -k$, $t \geq t_2$, which is equivalent to $x'(t) \leq \psi^{-1}(-k/a(t))$ for $t \geq t_2$. Integrating this inequality from t_2 to t and letting $t \to \infty$, we find in view of (3.13.77) that $x(t) \to -\infty$ as $t \to \infty$. But this contradicts the positivity of $x(t)$. Therefore, $a(t)\psi(x'(t)) > 0$ for $t \geq t_1$. A consequence of this observation is that $x'(t) > 0$ for $t \geq t_1$. Thus, the limit $\lim_{t \to \infty} a(t)\psi(x'(t))$ is either positive or zero. In the first case $x(t)$ is unbounded, since there are positive constants k_1, k_2 $(k_1 < k_2)$ and t_1 such that $\Psi_{k,t_1}(a;t) \leq x(t) - x(t_1) \leq \Psi_{k_2,t_1}(a;t)$ for $t \geq t_1$. In the second case, since $x(t)$ is increasing, $x(t)$ tends to a positive limit, finite or infinite as $t \to \infty$. This completes the proof.   ∎

Now, we shall present criteria for the existence of nonoscillatory solutions of equation (3.13.76) of the types (I), (II) and (III).

Theorem 3.13.11.   Suppose for each fixed $k \neq 0$ and $T \geq t_0 \geq 0$,

$$\lim_{\ell \to 0, \, k\ell > 0} \frac{\Psi_{\ell,T}(a;g(t))}{\Psi_{k,T}(a;g(t))} = 0 \tag{3.13.79}$$

uniformly on any interval of the form $[T^*, \infty)$ for some $T^* > T$. Then equation (3.13.76) has a nonoscillatory solution of the type (I), if and only if,

$$\int_{t_0}^{\infty} |F(s, c\Psi_k(a;g(s))|ds < \infty \tag{3.13.80}$$

for some constants $k \neq 0$ and $c > 0$,

Proof (The 'only if' part). Let $x(t)$ be a nonoscillatory solution of the type (I) of equation (3.13.76). Without loss of generality, we assume that $x(t) > 0$ eventually. There exist positive constants c_1, k_1 and t_1 such that $x[g(t)] \geq c_1 \Psi_{k_1}(a; g(t))$ for $t \geq t_1$. An integration of (3.13.76) yields $\int_t^\infty F(s, x[g(s)])ds < \infty$, which when combined with the earlier inequality leads to $\int_{t_1}^\infty F(s, c_1 \Psi_{k_1}(a; g(s)))ds < \infty$.

(The 'if' part). Suppose (3.13.80) holds for some constants $c > 0$ and $k > 0$. Because of (3.13.79) we can choose constants $\ell > 0$ and $T \geq t_0 > 0$ so that $\ell < k/2$, $T^* = \inf_{t \geq T} g(t)$ and

$$\int_T^\infty F(s, \Psi_{2\ell}(a; g(s)))ds \leq \ell. \tag{3.13.81}$$

Define the subset X of $C([T^*, \infty))$ and the mapping $W : X \to C([T^*, \infty))$ by

$$X = \{x \in C([T^*, \infty)) : \Psi_{\ell,T}(a; t) \leq x(t) \leq \Psi_{2\ell,T}(a; t),$$
$$\text{for } t \geq T, \ x(t) = 0 \text{ for } T^* \leq t \leq T\} \tag{3.13.82}$$

and

$$Wx(t) = \begin{cases} \int_T^t \psi^{-1}\left[\dfrac{1}{a(s)}\left(2\ell - \int_T^s F(u, x[g(u)])du\right)\right] ds, & t \geq T \\ 0, & T^* \leq t \leq T. \end{cases}$$
$$\tag{3.13.83}$$

(i) W maps X into itself. If $x \in X$, then since

$$0 \leq \int_T^s F(u, x[g(u)])du \leq \int_T^\infty F(u, \Psi_{2\ell}(a; g(u)))du \leq \ell, \quad s \geq T$$

we obtain from (3.13.83) that

$$\int_T^t \psi^{-1}\left(\frac{\ell}{a(s)}\right) ds \leq Wx(t) \leq \int_T^t \psi^{-1}\left(\frac{2\ell}{a(s)}\right) ds, \quad t \geq T$$

implying that $Wx \in X$.

(ii) W is continuous. Let $\{x_n\}_{n=1}^\infty$ be a sequence in X converging to $x \in X$ as $n \to \infty$ in the topology of $C([T, \infty))$. Fix $m \in \{1, 2, \cdots\}$ with $m \geq T$. The Lebesgue dominated convergence theorem implies that $\int_T^m F(s, x_n[g(s)])ds \to \int_T^m F(s, x[g(s)])ds$ as $n \to \infty$, and so $Wx_n(t) \to Wx(t)$ uniformly on $[T, m]$. Thus, $Wx_n \to Wx$ in $C([T, \infty))$.

(iii) $W(X)$ is relatively compact. This follows from the relation

$$
\begin{aligned}
0 \;\leq\; (Wx)'(t) &= \psi^{-1}\left[\frac{1}{a(t)}\left(2\ell - \int_T^t F(s, x[g(s)])ds\right)\right] \\
&\leq\; \psi^{-1}\left[\frac{1}{a(t)}\left(2\ell + \int_T^t F\left(s, \Psi_{2\ell}(a; g(s))ds\right)\right)\right], \quad t \geq T
\end{aligned}
$$

which holds for all $x \in X$.

Now, the Schauder–Tychonov fixed point theorem guarantees that there exists a $x \in X$ such that $Wx = x$. Differentiation of the integral equation $Wx(t) = x(t)$, $t \geq T$ shows that $x(t)$ is a positive solution of equation (3.13.76) for $t \geq T$. It is clear that $x(t)$ is of the type (I).

In the case when (3.13.80) holds for some constants $k < 0$ and $c > 0$, a similar argument can be used to construct a negative solution of the type (I) of equation (3.13.76). ■

Theorem 3.13.12.   Equation (3.13.76) has a nonoscillatory solution of the type (III) if and only if

$$
\int_{t_0}^{\infty}\left|\psi^{-1}\left(\frac{1}{a(s)}\int_s^{\infty} F(u, c)du\right)\right|ds \;<\; \infty, \quad t_0 \geq 0 \tag{3.13.84}
$$

for some constant $c \neq 0$.

Proof (The 'only if' part). Let $x(t)$ be a positive solution of the type (III) of equation (3.13.76). Then there is a positive constant c_1 and t_1 such that $x[g(t)] \geq c_1$ for $t \geq t_1$. Integrating (3.13.76) from $t \geq t_1$ to u and letting $u \to \infty$, we find $a(t)\psi(x'(t)) = \int_t^{\infty} F(s, x[g(s)])ds$ for $t \geq t_1$, which implies that

$$
x'(t) \;=\; \psi^{-1}\left[\frac{1}{a(t)}\int_t^{\infty} F(s, x[g(s)])ds\right] \quad \text{for} \quad t \geq t_1.
$$

Integrating this equation again from t_1 to t and using $x(t) \geq c_1$ for $t \geq t_1$ and letting $t \to \infty$, we obtain

$$
\int_{t_1}^{\infty} \psi^{-1}\left[\frac{1}{a(s)}\int_s^{\infty} F(u, c_1)du\right]ds \;<\; \infty.
$$

(The 'if' part). Suppose (3.13.84) holds. Let $c > 0$ be fixed arbitrarily and take $T \geq t_0 > 0$ so large that $T^* = \inf_{t \geq T} g(t)$, and

$$
\int_T^{\infty} \psi^{-1}\left[\frac{1}{a(s)}\int_s^{\infty} F(u, c)du\right]ds \;\leq\; \frac{c}{2}
$$

and define

$$X = \{x \in C([T^*, \infty)) : c/2 \le x(t) \le c \quad \text{for} \quad t \ge T^*\}$$

and

$$Vx(t) = \begin{cases} c - \displaystyle\int_c^\infty \psi^{-1}\left[\dfrac{1}{a(s)}\int_s^\infty F(u, x[g(u)])du\right] ds, & t \ge T \\ Vx(T) & \text{for} \quad T^* \le t \le T. \end{cases}$$

It is easy to verify that V is continuous and maps X into a compact subset of X, and hence V has a fixed point x in X, which gives the desired solution of the type (III) of equation (3.13.76).

Similarly, it can be shown that (3.13.76) possess a negative solution of the type (III) if (3.13.84) holds for some $c < 0$. This completes the proof. ∎

A characterization of the type (II) solutions of (3.13.76) is difficult. However, we have the following result.

Theorem 3.13.13. Suppose (3.13.79) holds, and $g(t) = t$. Equation (3.13.76) has a nonoscillatory solution $x(t)$ with $\lim_{t\to\infty}|x(t)| = \infty$ if (3.13.80) holds for some $k \ne 0$, $c > 0$ and

$$\int_{t_0}^\infty \left|\psi^{-1}\left[\dfrac{1}{a(s)}\int_s^\infty F(u, b)du\right]\right| ds = \infty, \quad t_0 \ge 0 \qquad (3.13.85)$$

for every nonzero constant b such that $kb > 0$.

Proof. It suffices to consider the case where $k > 0$ and $b > 0$. Let $d > 0$ be an arbitrary fixed constant and choose $\ell > 0$ small enough and $T \ge t_0 > 0$ large enough so that $d + \Psi_\ell(a; t) \le c\Psi_k(a; t)$ for $t \ge T$, and $\int_T^\infty F(s, d + \Psi_\ell(a; s))ds \le \ell$. This is possible because of (3.13.79) and the fact that $\lim_{t\to\infty}\Psi_k(a; t) = \infty$. Now, consider the set Y and the mapping U defined by

$$X = \{x \in C([T^*, \infty)) : d \le x(t) \le d + \Psi_\ell(a; t), \quad t \ge T\}$$

and

$$Ux(t) = d + \int_T^t \psi^{-1}\left[\dfrac{1}{a(s)}\left(\ell - \int_T^s F(u, x(u))du\right)\right] ds \quad \text{for} \quad t \ge T.$$

By the Schauder–Tychonov fixed point theorem it follows that U has a fixed element $x \in X$. Now, $x = x(t)$ is a solution of (3.13.76) follows from the differentiation of the integral equation $x(t) = Ux(t)$, $t \ge T$.

From (3.13.85), we have

$$
\begin{aligned}
x(t) \;&\geq\; d + \int_T^t \psi^{-1}\left[\frac{1}{a(s)}\left(\ell - \int_T^s F(u, x(u))du \right) \right] ds \\
&\geq\; d + \int_T^t \psi^{-1}\left[\frac{1}{a(s)} \int_s^\infty F(u, x(u))du \right] ds \\
&\geq\; d + \int_T^t \psi^{-1}\left[\frac{1}{a(s)} \int_s^\infty F(u, d)du \right] ds \;\to\; \infty \quad \text{as} \quad t \to \infty.
\end{aligned}
$$

We also note that $\lim_{t\to\infty} a(t)\psi(x'(t)) \geq 0.$ ∎

Example 3.13.7. Consider the equation

$$
(|x'(t)|^\alpha \,\operatorname{sgn} x'(t))' + q(t)|x[\lambda t]|^\beta \,\operatorname{sgn} x[\lambda t] \;=\; 0, \tag{3.13.86}
$$

where α, β, λ are positive constants, $0 < \lambda \leq 1$ and $q(t) \in C(\mathbb{R}^+, \mathbb{R}^+)$. Equation (3.13.86) is a special case of (3.13.76) in which $a(t) = 1$, $\psi(x) = x^\alpha$, $g(t) = \lambda t$, $F(t, x) = q(t)x^\beta$, $\psi^{-1}(x) = x^{1/\alpha}$, and $\Psi_{k,T}(a; t) = k^{1/\beta}[t - T]$, $t \geq T$. Clearly, conditions (3.13.77) and (3.13.79) are satisfied for (3.13.86).

The possible types of asymptotic behavior at infinity of nonoscillatory solutions of (3.13.86) are as follows:

(I) $\lim_{t\to\infty} x(t)/t = \text{constant} \neq 0$,

(II) $\lim_{t\to\infty} x(t)/t = 0$, $\lim_{t\to\infty} |x(t)| = \infty$,

(III) $\lim_{t\to\infty} x(t) = \text{constant} \neq 0$.

From Theorems 3.13.11 and 3.13.12 it follows that equation (3.13.86) has a solution of the type (I) if and only if

$$
\int_0^\infty s^\beta q(s)ds \;<\; \infty \tag{3.13.87}
$$

and that (3.13.86) has a solution of the type (III) if and only if

$$
\int_0^\infty \left(\int_s^\infty q(u)du \right)^{1/\alpha} ds \;<\; \infty.
$$

Theorem 3.13.13 implies that conditions (3.13.87) and

$$
\int_0^\infty \left(\int_s^\infty q(u)du \right)^{1/\alpha} ds \;=\; \infty \tag{3.13.88}
$$

are sufficient for the existence of nonoscillatory solutions $y(t)$ of (3.13.86) with $\lambda = 1$ which satisfy $\lim_{t\to\infty} |y(t)| = \infty$.

Conditions (3.13.87) and (3.13.88) are not always consistent. In fact, let $q(t) = (t+1)^m$, where m is a constant. Then, (3.13.87) holds if and only if $m < -1 - \beta$, and (3.13.88) holds if and only if $m \geq -1 - \alpha$, and hence these two conditions are inconsistent if $\alpha \leq \beta$. Thus, if $\alpha > \beta$ and $-\alpha \leq m < -1 - \beta$, then there exists a solution of the type (II) for the equation

$$(|x'(t)|^\alpha \text{ sgn } x'(t))' + (t+1)^m |x(t)|^\beta \text{ sgn } x(t) = 0.$$

Next, we shall characterize types (I) and (III) solutions of a particular case of (3.13.76), namely, the equation

$$\left(a(t)|x'(t)|^{\alpha-1}x'(t)\right)' + F(t, x[g(t)]) = 0, \quad \alpha > 0 \tag{3.13.89}$$

when the function $F(t, x)$ is not necessarily nondecreasing in x.

Theorem 3.13.14. Let conditions (i) and (iii) (for (3.13.76)) hold,

(iv)' $F \in C([t_0, \infty) \times \mathbb{R}, \mathbb{R})$ and sgn $F(t, x) = $ sgn x for $t \geq t_0$.

Suppose for positive constants ℓ and L with $\ell < L$ there exist positive constants m and M depending on ℓ and L such that for $\ell \leq |x| \leq L$,

$$mF(t, \ell) \leq |F(t, x)| \leq MF(t, L), \quad t \geq t_0. \tag{3.13.90}$$

Then equation (3.13.89) has a nonoscillatory solution $x(t)$ such that $\lim_{t\to\infty} x(t) = $ constant $\neq 0$ if and only if

$$\int^\infty a^{-1/\alpha}(s) \left(\int_s^\infty |F(u, c)| du \right)^{1/\alpha} ds < \infty \quad \text{for some constant} \quad c \neq 0. \tag{3.13.91}$$

Proof (The 'only if' part). Let $x(t)$ be a nonoscillatory solution of equation (3.13.89) such that $\lim_{t\to\infty} x(t) = $ constant $\neq 0$. Without loss of generality, we assume that $\lim_{t\to\infty} x(t) > 0$, so that there exist positive constants ℓ, L and $t_1 \geq t_0$ such that $\ell \leq x(t) \leq L$ and $\ell \leq x[g(t)] \leq L$ for $t \geq t_1$. By (3.13.90), we have

$$F(t, x[g(t)]) \geq mF(t, \ell) \tag{3.13.92}$$

for some constant $m > 0$ and $t \geq t_1$.

Integrating (3.13.89) from s to t and noting that $x'(t) > 0$ for $t \geq t_1$, we find for $t \geq s \geq t_1$,

$$\int_s^t F(u, x[g(u)]) du = a(s)(x'(s))^\alpha - a(t)(x'(t))^\alpha \leq a(s)(x'(s))^\alpha,$$

which gives

$$a^{-1/\alpha}(s) \left(\int_s^\infty F(u, x[g(u)]) du \right)^{1/\alpha} \leq x'(s), \quad s \geq t_1.$$

Thus, it follows that

$$\int_{t_1}^{t} a^{-1/\alpha}(s) \left(\int_{s}^{\infty} F(u, x[g(u)]du \right)^{1/\alpha} ds \leq x(t) - x(t_1), \quad t \geq t_1$$

which when combined with (3.13.92) yields

$$m^{1/\alpha} \int_{t_1}^{\infty} a^{-1/\alpha}(s) \left(\int_{s}^{\infty} F(u, \ell)du \right)^{1/\alpha} ds \leq L.$$

(The 'if' part). Suppose condition (3.13.91) holds for some constant $c \neq 0$. We can assume that $c > 0$. By (3.13.90) there exists a constant M such that $c/2 \leq x(t) \leq c$ implies $F(t, x[g(t)]) \leq MF(t, c)$ for $t \geq t_1 \geq t_0$. Choose $T \geq t_1$ so large that $T^* = \inf_{t \geq T} g(t)$ and

$$M^{1/\alpha} \int_{T}^{\infty} a^{-1/\alpha}(s) \left(\int_{s}^{\infty} F(u, c)du \right)^{1/\alpha} ds \leq \frac{c}{2}$$

and define the set $X \subset C([T^*, \infty))$ and the mapping $V : X \to C([T^*, \infty))$ by $X = \{x \in C([T, \infty)) : c/2 \leq x(t) \leq c, \ t \geq T^*\}$ and

$$Vx(t) = \begin{cases} c - \int_{t}^{\infty} a^{-1/\alpha}(s) \left(\int_{s}^{\infty} F(u, x[g(u)])du \right)^{1/\alpha} ds \quad \text{for} \quad t \geq T \\ Vx(T) \quad \text{for} \quad T^* \leq t \leq T. \end{cases}$$

The rest of the proof is similar to that of Theorem 3.13.12 and hence omitted. ■

Theorem 3.13.15. Let conditions (i), (iii) and (iv)' hold, and suppose that for positive constants ℓ and L there exist positive constants m and M (depending on ℓ and L respectively) such that $|z| \leq L$ implies

$$|F(t, zR[g(t), t_1])| \leq MF(t, LR[g(t), t_1]), \quad t \geq T \qquad (3.13.93a)$$

and $|z| \leq \ell$ implies

$$mF(t, \ell R[g(t), t_1]) \leq |F(t, zR[g(t), t_1])|, \quad t \geq T \qquad (3.13.93b)$$

for any $T \geq t_0$ large enough with $\inf_{t \geq T} g(t) \geq t_1$, where $R[t, T] = \int_{T}^{t} a^{-1/\alpha}(s)ds$ and $R(t) = R[t, t_0]$. Then equation (3.13.89) has a nonoscillatory solution $x(t)$ such that $x(t)/R(t) \to$ constant $\neq 0$ as $t \to \infty$ if and only if for all large $T_1 \geq T \geq t_0$, $g(t) \geq T$ for $t \geq T_1$,

$$\int^{\infty} |F(s, cR[g(s), T])|ds < \infty \quad \text{for some constant} \quad c \neq 0. \qquad (3.13.94)$$

Proof (The 'only if' part). Suppose equation (3.13.89) has a solution $x(t)$ such that $\lim_{t\to\infty} x(t)/R(t) = $ constant $\neq 0$. We can assume that $\lim_{t\to\infty} x(t)/R(t) > 0$. Then there exist positive constants ℓ, t_1 and T ($T > t_1$) such that $x(t) \geq \ell R[t, t_1]$ and $x[g(t)] \geq \ell R[g(t), t_1]$ for $t \geq T$. The rest of the proof is similar to that of Theorems 3.13.11 and 3.13.13 and hence omitted.

(The 'if' part). Let condition (3.13.94) hold with $c = 2k$ which can be assumed to be positive. By (3.13.93) there exists a constant $M > 0$ such that $0 \leq z[g(t)] \leq 2k$ implies $F(t, z[g(t)]R[g(t), t_1]) \leq MF(t, 2kR[g(t), t_1])$, $g(t) \geq t_1 \geq t_0$. Let $T \geq t_1$ be large enough and $T^* = \inf_{t\geq T} g(t) \geq t_1$ so that $M\int_T^\infty F(s, 2kR[g(s), t_1])ds \leq (2^\alpha - 1)k^\alpha$. Consider the set $X \subset C([T^*, \infty))$ and the mapping $W : X \to C([T^*, \infty))$ defined by

$$X = \{x \in C([T^*, \infty)),\ kR[t, T] \leq x(t) \leq 2kR[t, t_1]\ \text{ for }\ t \geq T$$
$$\text{and }\ x(t) = 0\ \text{ for }\ T^* \leq t \leq T\}$$

and

$$Wx(t) = \begin{cases} \int_T^t a^{-1/\alpha}(s)\left[(2k)^\alpha - \int_T^s F(u, x[g(u)])du\right]^{1/\alpha} ds & \text{for } t \geq T \\ 0 & \text{for } T^* \leq t \leq T. \end{cases}$$

The rest of the proof is similar to that of Theorem 3.13.11 and hence omitted. ∎

Example 3.13.8. Consider the differential equation

$$\left(|x'(t)|^{\alpha-1}x'(t)\right)' + \frac{t^\nu|x(t)|^{n-1}x(t)}{1+t^\mu|x(t)|^m} = 0, \tag{3.13.95}$$

where m, n, μ are positive constants, and ν is any constant. The function $F(t, x) = t^\nu|x|^{n-1}x/(1+t^\mu|x|^m)$ satisfies conditions (3.13.90) and (3.13.93), since $0 < \ell \leq x \leq L$ implies $F(t, \ell) \leq F(t, x) \leq F(t, L)$ for $n \geq m$; for $0 \leq x \leq L$ note that $F(t, xR[t, t_1]) \leq F(t, LR[t, t_1])$, $t \geq t_1$ and $n \geq m$, and for $0 < \ell \leq x \leq L$, we have

$$\left(\frac{\ell}{L}\right)^m F(t, \ell) \leq F(t, x) \leq \left(\frac{L}{\ell}\right)^m F(t, L)\ \text{ for }\ m > n.$$

It can easily be verified that (3.13.91) holds if and only if $\mu > \alpha + \nu + 1$ and (3.13.94) holds if and only if $\mu + m > \nu + n + 1$. Therefore, in view of Theorems 3.13.14 and 1.13.15 necessary and sufficient conditions for (3.13.95) to have nonoscillatory solutions $x(t)$ satisfying $\lim_{t\to\infty} x(t) = $ constant $\neq 0$ and $\lim_{t\to\infty} x(t)/R(t) = $ constant $\neq 0$ are respectively, $\mu > \alpha + \nu + 1$ and $\mu + m > \nu + n + 1$.

Remark 3.13.2. If $F(t,x)$ is nondecreasing, then conditions (3.13.90) and (3.13.93) are trivially satisfied.

Remark 3.13.3. Suppose $F(t,x) = q(t)f(x)$ where $q(t) \in ([t_0, \infty), \mathbb{R}^+)$, $f \in C(\mathbb{R}, \mathbb{R})$ and sgn $f(x) =$ sgn x. Then, $F(t,x)$ satisfies (3.13.90). If, in addition, $f(x)$ has the property that $k|f(x)f(y)| \le |f(xy)| \le K|f(x)f(y)|$, $xy \ge 0$ for some positive constants k and K, then $F(t,x)$ also satisfies condition (3.13.93).

Next, we shall consider equation (3.13.89) when $\pi(t) = \int_t^\infty a^{-1/\alpha}(s)ds$ satisfies $\pi(t_0) < \infty$, $t_0 \ge 0$, and as before, we will classify all possible nonoscillatory solutions according to their asymptotic behavior at infinity. For this, we shall need the following lemma.

Lemma 3.13.9. If $x(t)$ is a nonoscillatory solution of equation (3.13.89) which is either positive or negative on $[t_0, \infty)$, $t_0 \ge 0$ then there exist positive constants c_1 and c_2 such that

$$c_1 \pi(t) \le |x(t)| \le c_2 \quad \text{for} \quad t \ge t_0. \tag{3.13.96}$$

Proof. We can assume that $x(t) > 0$ for $t \ge t_0$. In fact, a similar argument holds if $x(t) < 0$ for $t \ge t_0$. Since $\left(a(t)|x'(t)|^{\alpha-1}x'(t)\right)' = -F(t,x[g(t)]) < 0$ for $t \ge t_0$, $a(t)|x'(t)|^{\alpha-1}x'(t)$ is decreasing on $[t_0, \infty)$, so that $x'(t)$ is eventually of constant sign, i.e., either $x'(t) > 0$ for $t \ge t_0$, or there is some $t_1 \ge t_0$ such that $x'(t) < 0$ for $t \ge t_1$.

Suppose first that $x'(t) > 0$ for $t \ge t_0$. Then, we have $a(t)(x'(t))^\alpha \le a(t_0)(x'(t_0))^\alpha$, $t \ge t_0$ which implies that $x'(t) \le a^{1/\alpha}(t_0)x'(t_0)a^{-1/\alpha}(t)$, $t \ge t_0$. Integrating this inequality from t_0 to t, we find

$$x(t) \le x(t_0) + a^{1/\alpha}(t_0)x'(t_0)\int_{t_0}^t a^{-1/\alpha}(s)ds$$
$$\le x(t_0) + a^{1/\alpha}(t_0)x'(t_0)\pi(t_0), \quad t \ge t_0.$$

Suppose next that $x'(t) < 0$ for $t \ge t_1$. Then, since $a(t)|x'(t)|^{\alpha-1}x'(t) = -a(t)(-x'(t))^\alpha$, we have $a(s)(-x'(s))^\alpha \ge a(t)(-x'(t))^\alpha$, $s \ge t \ge t_1$, or equivalently,

$$-a^{1/\alpha}(s)x'(s) \ge -a^{1/\alpha}(t)x'(t), \quad s \ge t \ge t_1. \tag{3.13.97}$$

Dividing (3.13.97) by $a^{1/\alpha}(s)$ and integrating over $[t, \tau]$, we obtain

$$x(t) > x(t) - x(\tau) \ge -a^{1/\alpha}(t)x'(t)\int_t^\tau a^{-1/\alpha}(s)ds, \quad \tau \ge t$$

which in the limit as $\tau \to \infty$ gives

$$x(t) \ge -a^{1/\alpha}(t)x'(t)\pi(t), \quad t \ge t_1. \tag{3.13.98}$$

Combining (3.13.98) with the inequality $-a^{1/\alpha}(t)x'(t) \geq -a^{1/\alpha}(t_1)x'(t_1)$, $t \geq t_1$ which follows from (3.13.97), we find $x(t) \geq -a^{1/\alpha}(t_1)x'(t_1)\pi(t)$, $t \geq t_1$. The conclusion (3.13.96) of the lemma now follows rather easily. ∎

From Lemma 3.13.9 and its proof we can conclude that one and only one of the following three possibilities occurs for the asymptotic behavior of any nonoscillatory solution $x(t)$ of (3.13.89).

(I) $\lim_{t\to\infty} |x(t)| = \text{constant} > 0$,

(II) $\lim_{t\to\infty} x(t) = 0$, $\lim_{t\to\infty} |x(t)|/\pi(t) = \infty$,

(III) $\lim_{t\to\infty} |x(t)|/\pi(t) = \text{constant} > 0$.

Now, we shall present sharp conditions which guarantee that (3.13.89) possesses nonoscillatory solutions of these three types.

Theorem 3.13.16.   A necessary and sufficient condition for equation (3.13.89) to have a nonoscillatory solution of the type (I) is that

$$\int_{t_0}^{\infty} \left(\frac{1}{a(s)} \int_{t_0}^{s} |F(u, c)|du \right)^{1/\alpha} ds < \infty \tag{3.13.99}$$

for some constant $c \neq 0$.

Proof (i). (Necessity). Let $x(t)$ be a solution of equation (3.13.89) of the type (I). We may assume that $x(t) > 0$ for $t \geq t_0$. A similar argument holds if $x(t) < 0$ for $t \geq t_0$. As in Lemma 3.13.9, $x'(t)$ is either positive for $t \geq t_0$, or negative for $t \geq t_1$ for some $t_1 \geq t_0$. Suppose $x'(t) > 0$ for $t \geq t_0$. An integration of (3.13.89) for $t \geq t_0$ yields

$$\int_{t_0}^{t} F(s, x[g(s)])ds = a(t_0)(x'(t_0))^\alpha - a(t)(x'(t))^\alpha < a(t_0)(x'(t_0))^\alpha,$$

which implies that $\int_{t_0}^{\infty} F(s, x[g(s)])ds < \infty$. Combining this inequality with the fact that $x[g(t)] \geq c_1$, $t \geq t_1 \geq t_0$ for some positive constant c_1, we obtain $\int_{t_0}^{\infty} F(s, c_1)ds < \infty$. Hence, it follows that

$$\int_{t_0}^{\infty} \left(\frac{1}{a(s)} \int_{t_0}^{s} F(u, c_1)du \right)^{1/\alpha} ds \leq \pi(t_0) \left(\int_{t_0}^{\infty} F(s, c_1)ds \right)^{1/\alpha} < \infty.$$

Next, suppose that $x'(t) < 0$ for $t \geq t_1$. Again an integration of (3.13.89) for $t \geq t_1$ gives

$$\int_{t_1}^{t} F(s, x[g(s)])ds = a(t_1)(-x'(t_1))^\alpha - a(t)(-x'(t))^\alpha \leq a(t)(-x'(t))^\alpha,$$

or

$$\left(\frac{1}{a(t)} \int_{t_1}^{t} F(s, x[g(s)]) ds \right)^{1/\alpha} \leq -x'(t) \quad \text{for} \quad t \geq t_1.$$

Integrating the above inequality over $[t_1, t_2]$ and noting that $c_1 \leq x[g(t)] \leq c_2$, $t \geq t_1$ for some positive constants c_1 and c_2, we find that

$$\int_{t_1}^{t_2} \left(\frac{1}{a(s)} \int_{t_1}^{s} F(u, c_1) du \right)^{1/\alpha} ds \leq x(t_1) - x(t_2) < c_2$$

for any $t_2 > t_1$, which implies (3.13.99).

(ii). (Sufficiency). Without loss of generality we assume that the constant c in (3.13.99) is positive. Choose $T > t_0$ so large that $T^* = \inf_{t \geq T} g(t) \geq t_0$ and

$$\int_{T}^{\infty} \left(\frac{1}{a(s)} \int_{T}^{s} F(u, c) du \right)^{1/\alpha} ds \leq \frac{c}{2} \qquad (3.13.100)$$

and define the set $X \subset C([T^*, \infty))$ and the mapping $V : X \to C([T^*, \infty))$ by $X = \{ x \in C([T^*, \infty)) : c/2 \leq x(t) \leq c, \, t \geq T^* \}$ and

$$V x(t) = \begin{cases} c - \int_{T}^{t} \left(\frac{1}{a(s)} \int_{T}^{s} F(u, x[g(u)]) du \right)^{1/\alpha} ds, & t \geq T \\ V x(T), & T^* \leq t \leq T. \end{cases}$$

In view of (3.13.100) it is clear that V maps X into itself. It can be verified that V is continuous and $V(X)$ is relatively compact in the topology of $C([T^*, \infty))$. Therefore, the Schauder–Tychonov fixed point theorem guarantees the existence of an element $x \in X$ such that $x = V(x)$, i.e.,

$$x(t) = c - \int_{T}^{t} \left(\frac{1}{a(s)} \int_{T}^{s} F(u, x[g(u)]) du \right)^{1/\alpha} ds, \quad t \geq T.$$

Differentiation of this integral equation shows that $x = x(t)$ is a positive solution of (3.13.89) on $[T, \infty)$. It is clear that $\lim_{t \to \infty} x(t) = \text{constant} \in [c/2, c]$, which means that $x(t)$ is a solution of the type (I). This completes the proof. ■

Theorem 3.13.17.   A necessary and sufficient condition for equation (3.13.89) to have a nonoscillatory solution of the type (III) is that

$$\int_{t_0}^{\infty} |F(s, c\pi[g(s)])| ds < \infty \qquad (3.13.101)$$

for some constant $c \neq 0$.

Proof (i).   (Necessity).   Let $x(t)$ be a solution of the type (III) of equation (3.13.89) which is eventually positive. Then there are positive constants c_1, c_2 and $t_1 \geq t_0$ such that $c_1\pi[g(t)] \leq x[g(t)] \leq c_2\pi[g(t)]$ for $t \geq t_1$. We can suppose that $x'(t) < 0$ for $t \geq t_1$. From equation (3.13.89), we have $a(t)(-x'(t))^\alpha \geq \int_{t_1}^t F(s, x[g(s)])ds$, $t \geq t_1$. Combining this with the inequality $a(t)(-x'(t))^\alpha \leq (x(t)/\pi(t))^\alpha$ for $t \geq t_0$, which is equivalent to (3.13.98), we obtain

$$\int_{t_1}^t F(s, c_1\pi[g(s)])ds \leq \int_{t_1}^t F(s, x[g(s)])ds \leq \left(\frac{x(t)}{\pi(t)}\right)^\alpha \leq c_2^\alpha \text{ for } t \geq t_2.$$

Thus, (3.3.101) follows in the limit as $t \to \infty$.

(ii). (Sufficiency). We only need to consider the case when the constant c in (3.13.101) is positive. Let $k > 0$ be a constant such that $2k < c^\alpha$ and take $T > t_0$ so large that $T^* = \inf_{t \geq T} g(t) \geq t_0$, and $\int_T^\infty F(s, (2k)^{1/\alpha}\pi[g(s)])ds \leq k$. Consider the set

$$Z = \{z \in C([T^*, \infty)) : \quad k^{1/\alpha}\pi(t) \leq z(t) \leq (2k)^{1/\alpha}\pi(t), \quad t \geq T,$$
$$k^{1/\alpha}\pi(T) \leq z(t) \leq (2k)^{1/\alpha}\pi(t), \quad T^* \leq t \leq T\}$$

and the mapping

$$Wz(t) = \begin{cases} \int_t^\infty \left(\frac{1}{a(s)}\left(k + \int_T^s F(u, z[g(u)]du)\right)\right)^{1/\alpha} ds, \quad t \geq T \\ Wz(T), \quad T^* \leq t \leq T. \end{cases}$$

$$(3.13.102)$$

If $z \in Z$, then for $t \geq T$, we have

$$k^{1/\alpha}\int_t^\infty a^{-1/\alpha}(s)ds \leq Wz(t)$$

$$\leq \int_t^\infty a^{-1/\alpha}(s)\left(k + \int_T^s F(u, (2k)^{1/\alpha}\pi[g(u)]du)\right)^{1/\alpha} ds$$

$$\leq (2k)^{1/\alpha}\int_t^\infty a^{-1/\alpha}(s)ds = (2k)^{1/\alpha}\pi(t),$$

which implies that $Wz \in Z$. Therefore, W maps Z into itself. Since the continuity of W and the relative compactness of $W(Z)$ can be proved without any difficulty, there exists an element $z \in Z$ such that $z = Wz$, i.e.,

$$z(t) = \int_t^\infty \left(\frac{1}{a(s)}\left(k + \int_T^s F(u, z[g(u)]du)\right)\right)^{1/\alpha} ds, \quad t \geq T.$$

From this integral equation it follows that $z(t)$ satisfies equation (3.13.89) on $[T, \infty)$, and by L'Hospital's rule $\lim_{t\to\infty} z(t)/\pi(t) = \text{constant} \in$

$[k^{1/\alpha}, (2k)^{1/\alpha}]$. Thus, $z(t)$ is a solution of the type (III) of (3.13.89). This completes the proof. ∎

Theorem 3.13.18. The equation (3.13.89) has a solution of the type (II) if

$$\int_{t_0}^{\infty} \left(\frac{1}{a(s)} \int_{t_0}^{s} |F(u,c)| du \right)^{1/\alpha} ds < \infty \qquad (3.13.103)$$

for some constant $c \neq 0$, and

$$\int_{t_0}^{\infty} |F(s, d\pi(s))| ds = \infty \qquad (3.13.104)$$

for any constant d with $cd > 0$.

Proof. We shall only consider the case where $c > 0$ in (3.13.103). Choose $k > 0$ so that $(2k)^{1/\alpha}[1 + \pi(t_0)] \leq c$, and let $T > t_0$ be large enough such that $T^* = \inf_{t \geq T} g(t) \geq t_0$ and

$$\int_{T}^{\infty} \left(\frac{1}{a(s)} \int_{T}^{s} F(u,c) du \right)^{1/\alpha} ds \leq k^{1/\alpha}.$$

Then it can be verified by the Schauder fixed point theorem that the mapping W defined by (3.13.102) has a fixed element in the set

$$Y = \{ y \in C([T^*, \infty)) : \quad k^{1/\alpha}\pi(t) \leq y(t) \leq c, \ t \geq T$$
$$k^{1/\alpha}\pi(T) \leq y(t) \leq c, \quad T^* \leq t \leq T \}.$$

In fact, if $y \in Y$, then

$$Wy(t) \leq \int_{t}^{\infty} a^{-1/\alpha}(s) \left[(2k)^{1/\alpha} + \left(2 \int_{T}^{s} F(u, y[g(u)]) du \right)^{1/\alpha} \right] ds$$

$$\leq (2k)^{1/\alpha}\pi(t) + 2^{1/\alpha} \int_{t}^{\infty} \left(\frac{1}{a(s)} \int_{T}^{s} F(u, y[g(u)]) du \right)^{1/\alpha} ds$$

$$\leq (2k)^{1/\alpha}[1 + \pi(t_0)] \leq c, \quad t \geq T$$

and so $Wy \in Y$. The verification of the continuity of W and the relative compactness of $W(Y)$ are as before. Therefore, there exists a fixed element $y \in T$ which is a solution $y(t)$ of equation (3.13.89) for $t \geq T^*$. It is clear that $y(t) \to 0$ as $t \to \infty$. Further, in view of (3.13.104), l'Hospital's rule gives

$$\lim_{t \to \infty} \frac{y(t)}{\pi(t)} = \lim_{t \to \infty} \frac{y'(t)}{\pi'(t)} = \lim_{t \to \infty} \left(k + \int_{T}^{t} F(s, y[g(s)]) ds \right)^{1/\alpha}$$

$$\geq \lim_{t \to \infty} \left(k + \int_{T}^{t} F(s, k^{1/\alpha}\pi(s)) \right) ds = \infty.$$

This shows that $y(t)$ is a type (II) solution of (3.13.89).   ∎

Example 3.13.9. Consider the equation

$$\left(e^{\lambda t}|x'(t)|^{\alpha-1}x'(t)\right)' + e^{\mu t}|x(t)|^{\beta-1}x(t) \; = \; 0, \quad t \geq 0 \qquad (3.13.105)$$

where α, β, λ are positive constants and μ is any constant. Applying Theorems 3.13.16 and 3.13.17 to (3.13.105), respectively, we find

(i)   equation (3.13.105) possesses a nonoscillatory solution of the type (I) (which is bounded from below and above by positive or negative constants) if and only if $\mu < \lambda$, and

(ii)   equation (3.13.105) possesses a nonoscillatory solution of the type (III) (which behaves like a constant and multiple of $e^{-(\lambda/\alpha)t}$ as $t \to \infty$) if and only if $\mu/\lambda < \beta/\alpha$.

Theorem 3.13.18 applies only to the case where $\alpha > \beta$. In fact, in this case the condition $\beta/\alpha \leq \mu/\lambda < 1$ ensures the existence of a nonoscillatory solution $x(t)$ of equation (3.13.105) satisfying $\lim_{t\to\infty} x(t) = 0$ and $\lim_{t\to\infty} e^{(\lambda/\alpha)t}|x(t)| = \infty$.

3.13.5. A Comparison Theorem

Most of the known oscillation results for functional differential equations are generalizations of those existing for the ordinary differential equations. Further, often the method of the proof of a generalized result is the same as that of the original result, and hence imposes a severe restriction on the deviating argument. Here, we shall reduce such a generalized problem to an ordinary differential equation so that the oscillation criteria presented in Sections 3.8 – 3.11 can be employed directly.

We need the following lemma.

Lemma 3.13.10. Suppose $g(t) \in C([t_0, \infty), \mathbb{R})$, $F \in C([t_0, \infty) \times \mathbb{R}, \mathbb{R})$, $g(t) \leq t$ for $t \geq t_0$, $\lim_{t\to\infty} g(t) = \infty$, $xF(t, x) > 0$ if $x \neq 0$, $t \geq t_0$ and $F(t, x)$ is nondecreasing in x. If the differential inequality

$$y'(t) - F(t, y[g(t)]) \; \geq \; 0 \qquad (3.13.106)$$

has a positive solution on $[t_1, \infty)$ for some $t_1 \geq t_0$, so does the equation

$$y'(t) - F(t, y[g(t)]) \; = \; 0. \qquad (3.13.107)$$

Proof.   Let $y(t)$ be a positive solution of (3.13.106) on $T \geq t_1$ so that $g(t) \geq t_1$ for $t \geq T$. Then, $y(t)$ satisfies the inequality $y(t) \geq$

$y(T) + \int_T^t F(u, y[g(u)])du.$ Let

$$
\begin{aligned}
z_1(T) &= y(t) \quad \text{for} \quad t \geq t_1, \\
z_n(t) &= y(t) \quad \text{for} \quad t \in [t_1, T] \\
&= y(T) + \int_T^t F(u, y_{n-1}[g(u)])du \quad \text{for} \quad t \geq T, \quad n = 2, 3, \cdots.
\end{aligned}
$$
(3.13.108)

It follows from the definition of z_n and (3.13.108) that the sequence $\{z_n\}$ satisfies $y(t) = z_1(t) \geq z_2(t) \geq \cdots \geq y(T)$ for all $t \geq T$. Hence, $\{z_n\}$ converges pointwise to a function $z(t)$, where $y(t) \geq z(t) \geq y(T)$ for all $t \geq T$. Let $F_n(t) = F(t, z_n[g(t)])$, $n = 1, 2, \cdots$. Then, $F_1(t) \geq F_2(t) \geq \cdots \geq 0$. Since F_1 is integrable on $[T, t]$ for any $t \geq T$, and $\lim_{n \to \infty} F_n(u) = F(u, z[g(u)])$ for any $u \in [T, t]$, by the monotone convergence theorem, we have $z(t) = y(T) + \int_T^t F(u, z[g(u)])du$ for $t \geq T$. Hence, $z(t)$ satisfies equation (3.13.107). ∎

Now, we shall prove the following main result.

Theorem 3.13.19. Let the functions g and F be as in Lemma 3.13.10 and suppose $g(t) \in C^1([t_0, \infty), \mathbb{R})$ and $g'(t) > 0$ for $t \geq t_0 \geq 0$, $a(t) \in C([t_0, \infty), \mathbb{R}^+)$ and $\int^\infty a^{-1/\alpha}(s)ds = \infty$, and $\alpha > 0$ is a constant. If (3.13.89) has a nonoscillatory solution, then the equation

$$
(a(s)|y'(s)|^{\alpha-1}y'(s))' + \frac{1}{g'[g^{-1}(s)]}F(g^{-1}(s), y(s)) = 0, \quad \left(' = \frac{d}{ds}\right)
$$
(3.13.109)

has a nonoscillatory solution.

Proof. Let $x(t)$ be a nonoscillatory solution of equation (3.13.89) and assume that $x(t) > 0$ for $t \geq t_0 \geq 0$. Clearly, there exists a $t_1 \geq t_0$ such that $x'(t) > 0$ for $t \geq t_1$. Integrating (3.13.89) from s to T, $T \geq s \geq t_1$, we obtain

$$
a(T)(x'(T))^\alpha - a(s)(x'(s))^\alpha + \int_s^T F(u, x[g(u)])du = 0
$$

and hence $x'(s) \geq a^{-1/\alpha}(s)(\int_s^\infty F(u, x[g(u)])du)^{1/\alpha}$, $s \geq t_1$. Let $v = g(u)$, then this inequality yields

$$
x'(s) \geq a^{-1/\alpha}(s)\left(\int_{g(s)}^\infty \frac{F(g^{-1}(v), x(v))}{g'(g^{-1}(v))}dv\right)^{1/\alpha}, \quad s \geq t_1.
$$

Since $g(t) \leq t$, we get

$$
x'(s) \geq a^{-1/\alpha}(s)\left(\int_s^\infty \frac{F(g^{-1}(v), x(v))}{g'(g^{-1}(v))}dv\right)^{1/\alpha} = Tx(s), \quad s \geq t_1.
$$

By Lemma 3.13.10, the equation $x'(s) = Tx(s)$ has a positive solution $y(s)$ on $[t_1, \infty)$ such that $y'(s) > 0$ for all $s \geq t_1$. Differentiating the equation $y'(s) = Ty(s)$, we find that $y(s)$ satisfies equation (3.13.109). ∎

By Theorem 3.13.19 it follows that the retarded half–linear equation

$$\left(|x'(t)|^{\alpha-1} x'(t)\right)' + q(t)|x[g(t)]|^{\alpha-1} x[g(t)] = 0, \tag{3.13.110}$$

where $q(t) \in C([t_0, \infty), \mathbb{R}^+)$, $g(t) \in C([t_0, \infty), \mathbb{R})$, $g(t) \leq t$, $g'(t) > 0$ for $t \geq t_0 \geq 0$, $\lim_{t \to \infty} g(t) = \infty$, and $\alpha > 0$ is a constant, is oscillatory if the ordinary half–linear equation

$$\left(|y'(s)|^{\alpha-1} y'(s)\right)' + \left(\frac{q[g^{-1}(s)]}{g'[g^{-1}(s)]}\right) |y(s)|^{\alpha-1} y(s) = 0, \qquad \left(' = \frac{d}{ds}\right) \tag{3.13.111}$$

is oscillatory. Further, applying Corollary 3.13.3, we find that if

$$\int^{\infty} s^\lambda \frac{q[g^{-1}(s)]}{g'[g^{-1}(s)]} ds = \infty, \quad 0 < \lambda < \alpha, \tag{3.13.112}$$

then equation (3.13.111) is oscillatory. If we let $s = g(t)$, then (3.13.112) reduces to

$$\int^{\infty} g^\lambda(t) q(t) dt = \infty. \tag{3.13.113}$$

Thus, (3.13.113) implies that (3.13.111) and hence equation (3.13.110) is oscillatory.

3.13.6. Oscillation of Equation (3.13.2)

Consider equation (3.13.2) subject to the conditions (i) – (iv), and let $x(t)$ be its nonoscillatory solution. Clearly, $x'(t)$ is eventually of constant sign, so that

$$\text{either} \quad x(t)x'(t) < 0 \quad \text{or} \quad x(t)x'(t) > 0 \quad \text{for} \quad t \geq T \tag{3.13.114}$$

provided T is sufficiently large. Thus, $x(t)$ is bounded or unbounded according to whether the first or the second inequality in (3.13.114) holds. Our first result shows that in the case when $g(t)$ is a retarded argument, equation (3.13.2) may not have any unbounded nonoscillatory solution.

Theorem 3.13.20. Suppose $g(t) < t$ for $t \geq t_0$. Then every bounded solution of equation (3.13.2) is oscillatory if either one of the following conditions holds

$(C_1) \quad \limsup_{t \to \infty} \int_{g(t)}^{t} q(s) R^\alpha[g(t), g(s)] ds > 1, \tag{3.13.115}$

(C$_2$) $\limsup\limits_{t\to\infty} \displaystyle\int_{g(t)}^{t} a^{-1/\alpha}(s)\left(\int_{s}^{t} q(u)du\right)^{1/\alpha} ds > 1,$ (3.13.116)

(C$_3$)   every solution of the delay equation

$$y'(t) + q(t)R^{\alpha}[(t + g(t))/2, g(t)]y[(t + g(t))/2] = 0 \qquad (3.13.117)$$

is oscillatory, or

(C$_4$) $\liminf\limits_{t\to\infty} \displaystyle\int_{(t+g(t))/2}^{t} q(s)R^{\alpha}[(s + g(s))/2, g(s)]ds > \frac{1}{e}.$ (3.13.118)

Proof. Let $x(t)$ be a bounded nonoscillatory solution of equation (3.13.2). Without loss of generality we can assume that $x(t)$ is eventually positive. Thus, there is a $T_0 \geq t_0$ such that $x(t) > 0$ and $x'(t) < 0$ for $t \geq T_0$.

Suppose (3.13.115) holds. Let $T > T_0$ be such that $\inf_{t \geq T} g(t) \geq T_0$. Applying Lemma 3.13.2 with $g(s)$ and $g(t)$ for σ and τ respectively, in the inequality (3.13.11), we get $x[g(s)] \geq -a^{1/\alpha}[g(t)]R[g(t), g(s)]x'[g(t)]$ for $t \geq s \geq T$, which implies

$$q(s)x^{\alpha}[g(s)] \geq q(s)R^{\alpha}[g(t), g(s)](a[g(t)](-x'[g(t)]))^{\alpha}, \quad t \geq s \geq T.$$

Replace the left–hand side of the above inequality by $(a(s)|x'(s)|^{\alpha-1}x'(s))'$ $= -(a(s)(-x'(s))^{\alpha})'$ and integrate from $g(t)$ to t, to obtain

$$a[g(t)](-x'[g(t)])^{\alpha} - a(t)(-x'(t))^{\alpha} \geq a[g(t)](-x'[g(t)])^{\alpha}$$
$$\times \int_{g(t)}^{t} q(s)R^{\alpha}[g(t), g(s)]ds, \quad t \geq T$$

and hence

$$a[g(t)](-x'[g(t)])^{\alpha}\left[\int_{g(t)}^{t} q(s)R^{\alpha}[g(t), g(s)]ds - 1\right] \leq 0 \quad \text{for} \quad t \geq T.$$

But this is inconsistent with (3.13.115).

Suppose now that (3.13.116) holds. Integration of equation (3.13.2) over $[\sigma, t]$ gives

$$a(\sigma)(-x'(\sigma))^{\alpha} = a(t)(-x'(t))^{\alpha} + \int_{\sigma}^{t} q(u)x^{\alpha}[g(u)]du$$
$$\geq \int_{\sigma}^{t} q(u)x^{\alpha}[g(u)]du \quad \text{for} \quad t \geq \sigma \geq T,$$

which implies

$$-x'(\sigma) \geq a^{-1/\alpha}(\sigma)\left(\int_{\sigma}^{t} q(u)x^{\alpha}[g(u)]du\right)^{1/\alpha}, \quad t \geq \sigma \geq T. \quad (3.13.119)$$

From (3.13.119) and the equality

$$x(s) = x(t) + \int_s^t (-x'(\sigma))d\sigma \quad \text{for} \quad t \geq s \geq T, \tag{3.13.120}$$

we find

$$x(s) \geq \int_s^t a^{-1/\alpha}(\sigma) \left(\int_\sigma^t q(u)x^\alpha [g(u)]du \right)^{1/\alpha} d\sigma, \quad t \geq s \geq T.$$

Putting $s = g(t)$ in (3.13.120) and using the fact that $x[g(t)]$ is decreasing, we get

$$x[g(t)] \left[\int_{g(t)}^t a^{-1/\alpha}(\sigma) \left(\int_\sigma^t q(u)du \right)^{1/\alpha} d\sigma - 1 \right] \leq 0 \quad \text{for} \quad t \geq T,$$

which contradicts (3.13.116).

Next, we apply Lemma 3.13.2 with $g(t)$ and $(t+g(t))/2$ for σ and τ respectively, in inequality (3.13.11), to get

$$x[g(t)] \geq a^{-1/\alpha} \left[\frac{t+g(t)}{2} \right] R \left[g(t), \frac{t+g(t)}{2} \right] \left(-x' \left[\frac{t+g(t)}{2} \right] \right), \quad t \geq T.$$

Using this inequality in (3.13.2) and letting $w(t) = a(t)(-x'(t))^\alpha$ for $t \geq T$, we obtain

$$w'(t) + q(t)R^\alpha \left[g(t), \frac{t+g(t)}{2} \right] w \left[\frac{t+g(t)}{2} \right] \leq 0 \quad \text{for} \quad t \geq t_0.$$

Integrating the above inequality from t to u and letting $u \to \infty$, we find

$$w(t) \geq \int_t^\infty q(s)R^\alpha \left[g(s), \frac{s+g(s)}{2} \right] w \left[\frac{s+g(s)}{2} \right] ds, \quad t \geq T.$$

But, then in view of Lemma 3.13.6, equation (3.13.117) has a positive solution $y(t)$ with $\lim_{t\to\infty} y(t) = 0$, which is a contradiction.

Finally, we note that condition (3.13.118) ensures the oscillation of equation (3.13.117) (cf. Lemma 3.13.4 (a_3)). This completes the proof.   ∎

Example 3.13.10. The equation

$$\left(e^{-t}|x'(t)|x'(t) \right)' = |x[t/2]|x[t/2] \tag{3.13.121}$$

has an unbounded nonoscillatory solution $x(t) = e^t$. The hypothesis of Theorem 3.13.20 (C_1) is satisfied and hence all bounded solutions of equation (3.13.121) are oscillatory.

A duality of Theorem 3.13.20 holds in the case when $g(t)$ is an advanced argument.

Theorem 3.13.21. Suppose $g(t) > t$ for $t \geq t_0$. Then every unbounded solution of equation (3.13.2) is oscillatory if either one of the following conditions holds

(A₁) $\displaystyle \limsup_{t \to \infty} \int_t^{g(t)} q(s) R^\alpha[g(s), g(t)] ds > 1,$ (3.13.122)

(A₂) $\displaystyle \limsup_{t \to \infty} \int_t^{g(t)} \left(\frac{1}{a(s)} \int_t^s q(u) du \right)^{1/\alpha} ds > 1,$ or (3.13.123)

(A₃) $\displaystyle \liminf_{t \to \infty} \int_t^{(t+g(t))/2} q(s) R^\alpha \left[g(s), \frac{s + g(s)}{2} \right] ds > \frac{1}{e}.$ (3.13.124)

Proof. Let $x(t)$ be an unbounded nonoscillatory solution of equation (3.13.2) which can be assumed to be eventually positive. There is a $T_0 \geq t_0$ such that $x(t) > 0$ and $x'(t) > 0$ for $t \geq T_0$.

Suppose (3.13.122) holds. Letting $\sigma = g(s)$ and $\tau = g(t)$ in the inequality (3.13.12), we get $x[g(s)] \geq R[g(s), g(t)] a^{1/\alpha}[g(t)] x'[g(t)]$, $s \geq t \geq T$ where T is as in Theorem 3.13.20. Thus, for $s \geq t \geq T$,

$$(a(s)(x'(s))^\alpha)' = q(s) x^\alpha[g(s)] \geq q(s) R^\alpha[g(s), g(t)] a[g(t)] (x'[g(t)])^\alpha.$$

Integration of the above inequality over $[t, g(t)]$ yields

$$a[g(t)](x'[g(t)])^\alpha \left[\int_t^{g(t)} q(s) R^\alpha[g(s), g(t)] ds - 1 \right] \leq 0, \quad t \geq T$$

which is a contradiction to (3.13.122).

Suppose (3.13.123) holds. Combining

$$x'(\sigma) \geq a^{-1/\alpha}(\sigma) \left(\int_t^\sigma q(u) x^\alpha[g(u)] du \right)^{1/\alpha}, \quad \sigma \geq t \geq T$$

with the relation $x(s) = x(t) + \int_t^s x'(\sigma) d\sigma$, $s \geq t \geq T$ we obtain

$$x(s) \geq \int_t^s \left(\frac{1}{a(\sigma)} \int_t^\sigma q(u) x^\alpha[g(u)] du \right)^{1/\alpha} d\sigma, \quad s \geq t \geq T.$$

Putting $s = g(t)$ and noting that $x[g(t)]$ is increasing, we obtain

$$x[g(t)] \left[\int_t^{g(t)} \left(\frac{1}{a(\sigma)} \int_t^\sigma q(u) du \right)^{1/\alpha} d\sigma - 1 \right] \leq 0 \quad \text{for} \quad t \geq T,$$

which contradicts (3.13.123).

Suppose (3.13.124) holds. Letting $\sigma = g(t)$ and $\tau = (t + g(t))/2$ in inequality (3.13.12), we get

$$x[g(t)] \geq a^{1/\alpha}\left[\frac{t + g(t)}{2}\right] R\left[g(t), \frac{t + g(t)}{2}\right] x'\left[\frac{t + g(t)}{2}\right], \quad t \geq T.$$

Using this inequality in (3.13.2) and letting $y(t) = a(t)(x'(t))^{\alpha}$, $t \geq T$ we obtain

$$y'(t) \geq q(t)R^{\alpha}\left[g(t), \frac{t + g(t)}{2}\right] y\left[\frac{t + g(t)}{2}\right], \quad t \geq T.$$

But, in view of Lemma 3.13.5 and condition (3.13.124), the above inequality has no eventually positive solutions, which is a contradiction. This completes the proof. ∎

Example 3.13.11. The equation

$$\left(e^{-t}|x'(t)|x'(t)\right)' = 3e^{t}|x[2t]|x[2t], \quad t \geq 0 \tag{3.13.125}$$

has a bounded nonoscillatory solution $x(t) = e^{-t}$. Clearly, the condition of Theorem 3.13.21(A_1) is satisfied and hence all unbounded solutions of (3.13.125) are oscillatory.

Remark 3.13.4. Theorems 3.3.20 and 3.3.21 also hold for the differential inequalities of the form

$$\left\{\left(a(t)|x'(t)|^{\alpha-1}x'(t)\right)' - q(t)|x[g(t)]|^{\alpha-1}x[g(t)]\right\} \operatorname{sgn} x[g(t)] \geq 0. \tag{3.3.126}$$

Remark 3.13.5. Theorems 3.13.20 and 3.13.21 fail to apply to equations of the type (3.13.2) and/or inequalities of the form (3.3.126) when $g(t) = t$. Also, oscillation of all solutions of (3.13.2) does not follow from Theorem 3.13.20 or 3.13.21, because the deviating argument under consideration is only either retarded or advanced. However, for equations of mixed type, i.e., involving both retarded and advanced arguments, it is possible to establish oscillation of all solutions. We shall show this for the mixed type equation

$$\left(a(t)|x'(t)|^{\alpha-1}x'(t)\right)' = \sum_{i=1}^{n} q_i(t)|x[g_i(t)]|^{\alpha-1}x[g_i(t)], \tag{3.13.127}$$

where

(i) $\alpha > 0$ is a constant,
(ii) $q_i(t) \in C([t_0, \infty), \mathbb{R}^{+})$ such that $\sup\{q_i(t) : t \geq T\} > 0$ for any $T \geq t_0$, $i = 1, 2, \cdots, n$,

(iii) $g_i(t) \in C^1([t_0, \infty), \mathbb{R})$, $g_i'(t) \geq 0$ for $t \geq t_0$ and $\lim_{t \to \infty} g_i(t) = \infty$, $i = 1, 2, \cdots, n$,

(iv) $a(t) \in C([t_0, \infty), \mathbb{R}^+)$ and $\int^\infty a^{-1/\alpha}(s)ds = \infty$.

The following theorem follows from the above results.

Theorem 3.13.22 (M_1). All bounded solutions of equation (3.13.127) are oscillatory if there exists an $i \in \{1, 2, \cdots, n\}$ such that $g_i(t) < t$ for $t \geq t_0$, and either one of the conditions (C_1) – (C_4) of Theorem 3.13.20 holds with g and q replaced by g_i and q_i respectively.

(M_2) All unbounded solutions of equation (3.13.127) are oscillatory if there exists a $j \in \{1, 2, \cdots, n\}$ such that $g_j(t) > t$ for $t \geq t_0$, and either one of the conditions (A_1) – (A_3) of Theorem 3.13.21 holds with g and q replaced by g_j and q_j respectively.

(M_3) All solutions of equation (3.13.127) are oscillatory if there exist i and $j \in \{1, 2, \cdots, n\}$ such that $g_i(t)$ and $g_j(t)$ satisfy the conditions (M_1) and (M_2) respectively.

Proof (M_1). Suppose to the contrary that equation (3.13.127) has a bounded nonoscillatory solution $x(t)$. Then from (3.13.127) it follows that $x(t)$ satisfies the differential inequality

$$\left\{ \left(a(t)|x'(t)|^{\alpha-1}x'(t)\right)' - q_i(t)|x[g_i(t)]|^{\alpha-1}x[g_i(t)] \right\} \text{ sgn } x[g_i(t)] \geq 0$$
$$(3.13.128)$$

for all sufficiently large t. This is however impossible, because the existence of a bounded nonoscillatory solution of (3.13.128) is excluded by Theorem 3.13.20 and Remark 3.13.4.

(M_2) An unbounded nonoscillatory solution $x(t)$ of equation (3.13.127), if exists, satisfies the inequality (3.13.128) with i replaced by j for sufficiently large t. But this is impossible because of Theorem 3.13.21 and Remark 3.13.4, and hence every unbounded solution of (3.13.127) must be oscillatory.

(M_3) This is an immediate consequence of (M_1) and (M_2). ∎

Example 3.13.12. For the equations

$$\left(|x'(t)|^{\alpha-1}x'(t)\right)' = k|x[t - \sigma]|^{\alpha-1}x[t - \sigma], \qquad (3.13.129)$$

$$\left(|x'(t)|^{\alpha-1}x'(t)\right)' = \ell|x[t + \tau]|^{\alpha-1}x[t + \tau], \qquad (3.13.130)$$

$$\left(|x'(t)|^{\alpha-1}x'(t)\right)' = k|x[t-\sigma]|^{\alpha-1}x[t-\sigma]+\ell|x[t+\tau]|^{\alpha-1}x[t+\tau], \quad (3.13.131)$$

where α, k, ℓ, σ and τ are positive constants, we find

(1) from conditions (C_1) or (C_2) of (M_1) of Theorem 3.13.22 that all bounded solutions of (3.13.129) are oscillatory if

$$k\sigma^{\alpha+1} > \alpha+1 \quad \text{or} \quad k^{1/\alpha}\sigma^{(\alpha+1)/\alpha} > \frac{\alpha+1}{\alpha}, \qquad (3.13.132)$$

(2) from conditions (A_1) or (A_2) of (M_2) of Theorem 3.13.21 that all unbounded solutions of (3.13.130) are oscillatory if

$$\ell\tau^{\alpha+1} > \alpha+1 \quad \text{or} \quad \ell^{1/\alpha}\tau^{(\alpha+1)/\alpha} > \frac{\alpha+1}{\alpha}, \qquad (3.13.133)$$

(3) from the last statement of Theorem 3.13.22 that all solutions of (3.13.131) are oscillatory provided both (3.13.132) and (3.13.133) hold.

3.13.7. Nonoscillation of Solutions of Equation (3.13.127)

We shall now study the existence and asymptotic behavior of nonoscillatory solutions of equation (3.13.127). If $x(t)$ is a nonoscillatory solution of (3.13.127), then there exists a $t_1 > t_0$ such that either

$$x(t)x'(t) > 0 \quad \text{for} \quad t \geq t_1, \qquad (3.13.134)$$

or

$$x(t)x'(t) < 0 \quad \text{for} \quad t \geq t_1. \qquad (3.13.135)$$

If (3.13.134) holds, then $x(t)$ is unbounded and the limit $a^{1/\alpha}(\infty)x'(\infty) = \lim_{t\to\infty} a^{1/\alpha}(t)x'(t)$ exists and is finite or infinite, and if (3.13.135) holds, then $x(t)$ is bounded and the finite limit $x(\infty) = \lim_{t\to\infty} x(t)$ exists.

In what follows we only need to consider eventually positive solutions of (3.13.127), since if $x(t)$ satisfies (3.13.127), then so does $-x(t)$. Let $x(t)$ be an eventually positive solution of (3.13.127) satisfying (3.13.134) and having a finite limit $a^{1/\alpha}(\infty)x'(\infty) = \lim_{t\to\infty} a^{1/\alpha}(t)x'(t) > 0$. Then twice integration of (3.13.127) for $t \geq T$ yields

$$x(t) = x(T) + \int_T^t a^{-1/\alpha}(s)\left[a(\infty)(x'(\infty))^\alpha - \int_s^\infty \sum_{i=1}^n q_i(u)x^\alpha[g_i(u)]du\right]^{1/\alpha} ds,$$

$$(3.13.136)$$

where $T > t_1$ is chosen so that $\inf_{t\geq T} g_i(t) \geq t_1$, $i = 1, 2, \cdots, n$.

Similarly, if $x(t)$ is an eventually positive solution of (3.13.127) satisfying (3.13.135), then integration of (3.13.127) twice from t to v and the letting $v \to \infty$ gives

$$x(t) = x(\infty) + \int_t^\infty a^{-1/\alpha}(s)\left[\int_s^\infty \sum_{i=1}^n q_i(u)x^\alpha[g_i(u)]du\right]^{1/\alpha}, \quad t \geq T$$

$$(3.13.137)$$

Based on these integral representations (3.13.136) and (3.13.137) of (3.13.127), we shall prove the following existence theorems.

Theorem 3.13.23. The equation (3.13.127) has a nonoscillatory solution $x(t)$ such that $\lim_{t \to \infty} x(t)/R(t) = \text{constant} \neq 0$ if and only if for all large T,

$$\int^{\infty} q_i(s) R^{\alpha}[g_i(s), T] ds < \infty, \quad i = 1, 2, \cdots, n. \tag{3.13.138}$$

Proof (The 'only if' part). Let $x(t)$ be a nonoscillatory solution of equation (3.13.127) satisfying $\lim_{t \to \infty} x(t)/R(t) = c > 0$, c is a constant. Then from (3.13.136), we find

$$\int^{\infty} \sum_{i=1}^{n} q_i(s) x^{\alpha}[g_i(s)] ds < \infty.$$

This combined with the relation $\lim_{t \to \infty} x[g_i(t)]/R[g_i(t), t_0] = c$, $i = 1, 2, \cdots, n$ where $T \geq t_0$ is large so that $\inf_{t \geq T} g_i(t) \geq t_0$ implies (3.13.138).

(The 'if' part). Suppose (3.13.138) holds. Let $k > 0$ be arbitrarily fixed, and let $T > t_0$ so large that

$$T^* = \min_i \left\{ \inf_{t \geq T} g_i(t) \right\} \geq t_0 \tag{3.13.139}$$

and

$$\sum_{i=1}^{n} \int_{T}^{\infty} q_i(s) R^{\alpha}[g_i(s), T] ds \leq \frac{2^{\alpha} - 1}{2^{\alpha}}.$$

Consider the set $X \subset C([T^*, \infty))$ and the mapping $W : X \to C([T^*, \infty))$ defined by

$$X = \{ x \in C([T^*, \infty)) : (k/2) R[t, T] \leq x(t) \leq k R[t, T], \ t \geq T,$$
$$x(t) = 0, \ T^* \leq t \leq T \}$$

and

$$W x(t) = \begin{cases} \displaystyle \int_{T}^{t} a^{-1/\alpha}(s) \left[k^{\alpha} - \int_{s}^{\infty} \sum_{i=1}^{n} q_i(u) x^{\alpha}[g_i(u)] du \right]^{1/\alpha} ds, \ t \geq T \\ 0, \ T^* \leq t \leq T. \end{cases}$$

It is clear that X is a closed convex subset of the Fréchet space $C([T, \infty))$ of continuous functions on $[T^*, \infty)$ with the usual metric topology, and

W is well–defined and continuous on X. It can be shown without difficulty that W maps X into itself and $W(X)$ is relatively compact in
$C([T^*, \infty))$. Therefore, by the Schauder–Tychonov fixed point theorem,
W has a fixed element x in X which satisfies

$$x(t) = \int_T^t a^{-1/\alpha}(s) \left[k^\alpha - \int_s^\infty \sum_{i=1}^n q_i(u)x^\alpha[g_i(u)]du \right]^{1/\alpha} ds, \quad t \geq T.$$

Finally, twice differentiation shows that $x(t)$ satisfies equation (3.13.137)
for $t \geq T$ and $\lim_{t\to\infty} x(t)/R(t) = \lim_{t\to\infty} a^{1/\alpha}(t)x'(t) = k$. ∎

Theorem 3.13.24. The equation (3.13.127) has a nonoscillatory solution
$x(t)$ such that $\lim_{t\to\infty} x(t) = \text{constant} \neq 0$ if and only if

$$\int^\infty \left[a(s) \int_s^\infty q_i(u)du \right]^{1/\alpha} ds < \infty, \quad i = 1, 2, \cdots, n. \tag{3.13.140}$$

Proof. The 'only if' part follows from (3.13.137). To prove the 'if' part,
suppose (3.13.140) is satisfied. Choose $T > t_0$ so that (3.13.39) holds and

$$\int_T^\infty \left[a^{-1/\alpha}(s) \int_s^\infty q_i(u)du \right]^{1/\alpha} ds \leq \frac{1}{2}$$

and define $X \subset C([T^*, \infty))$ and $V : X \to C([T^*, \infty))$ by

$$X = \{x \in C([T^*, \infty)) : k \leq x(t) \leq 2k, \ t \geq T^*\},$$

where $k > 0$ is a fixed constant, and

$$Vx(t) = \begin{cases} k + \int_t^\infty a^{-1/\alpha}(s) \left[\int_s^\infty \sum_{i=1}^n q_i(u)x^\alpha[g_i(u)]du \right]^{1/\alpha} ds, & t \geq T \\ Vx(T), & T^* \leq t \leq T. \end{cases}$$

As in Theorem 3.13.23 one can verify that V maps X into a relatively
compact subset of X so that there exists a $x \in X$ such that

$$x(t) = k + \int_t^\infty a^{-1/\alpha}(s) \left[\int_s^\infty \sum_{i=1}^n q_i(u)x^\alpha[g_i(u)]du \right]^{1/\alpha} ds, \quad t \geq T.$$

Finally, twice differentiation shows that $x(t)$ satisfies equation (3.13.127)
for $t \geq T$ and $x(t) \to k$ as $t \to \infty$. ∎

It remains to discuss the existence of an unbounded nonoscillatory solu-
tion $x(t)$ of (3.13.127) which has the property that $\lim_{t\to\infty} x(t)/R(t) = \infty$, and of bounded solution $x(t)$ of (3.13.127) having the property that

$\lim_{t\to\infty} x(t) = 0$. However, this is a difficult problem and there is no general criteria available for the existence of such solutions. Therefore, we confine ourself to the case where at least one of the $g_i(t)$ is retarded and present sufficient conditions which guarantee that (3.13.127) has a nonoscillatory solution which tends to zero as $t \to \infty$. Such a solution is often referred to as a *decaying nonoscillatory solution*. Our derivation is based on the following theorem.

Theorem 3.13.25. Suppose $\int^\infty a^{-1/\alpha}(s)ds = \infty$, and there exists an $i_0 \in \{1, 2, \cdots, n\}$ such that

$$g_{i_0}(t) < t, \quad q_{i_0}(t) > 0 \quad \text{for} \quad t \geq t_0. \tag{3.13.141}$$

Further, suppose there exists a positive decreasing function $y(t)$ on $[T, \infty)$ satisfying

$$y(t) \geq \int_t^\infty a^{-1/\alpha}(s) \left[\int_s^\infty q_i(u)x^\alpha[g_i(u)]du \right]^{1/\alpha} ds, \quad t \geq T \tag{3.13.142}$$

where T is chosen so that $\inf_{t \geq T} g_i(t) \geq t_0$, $i = 1, 2, \cdots, n$. Then equation (3.13.127) has a nonoscillatory solution which tends to zero as $t \to \infty$.

Proof. Let $Z = \{z \in C([T, \infty)) : 0 < z \leq y(t), \ t \geq T\}$. With each $z \in Z$ we associate the function $\tilde{z} \in C([t_0, \infty))$ defined by

$$\tilde{z}(t) = \begin{cases} z(t) & \text{for} \quad t \geq T \\ z(T) + [y(t) - y(T)] & \text{for} \quad t_0 \leq t \leq T. \end{cases} \tag{3.13.143}$$

Define the mapping $H : Z \to C([T, \infty))$ as follows

$$Hz(t) = \int_t^\infty a^{-1/\alpha}(s) \left[\int_s^\infty \sum_{i=1}^n q_i(u)(\tilde{z}[g_i(u)])^\alpha du \right]^{1/\alpha} ds, \quad t \geq T.$$

Clearly, H is continuous and maps Z into a relatively compact subset of Z. Therefore, it follows that there exists a $z \in Z$ such that $z = Hz$, i.e.,

$$z(t) = \int_t^\infty a^{-1/\alpha}(s) \left[\int_s^\infty \sum_{i=1}^n q_i(u)(\tilde{z}[g_i(u)])^\alpha du \right]^{1/\alpha} ds, \quad t \geq T. \tag{3.13.144}$$

Twice differentiation of (3.13.144) gives

$$(-a(t)(-z'(t))^\alpha)' = \sum_{i=1}^n q_i(t)(\tilde{z}[g_i(t)])^\alpha, \quad t \geq T$$

which in view of (3.13.143) implies that $z(t)$ is a solution of (3.13.127) for all sufficiently large t. It remains to show that $z(t)$ is positive. For this, we observe that $\tilde{z}(t)$ is positive on $[t_0, T]$ and $t_0 \le g_{i_0}(T) < T$ for some $i_0 \in \{1, 2, \cdots, n\}$. Therefore, it follows that $\tilde{z}[g_{i_0}(T)] > 0$ for some $i_0 \in \{1, 2, \cdots, n\}$, and hence $\sum_{i=1}^{n} q_i(T)(\tilde{z}[g_i(T)])^\alpha > 0$. Thus, $(-a(t)(-z'(t))^\alpha)'|_{t=T} \ne 0$. Furthermore, we find that z is nonnegative and decreasing on $[T, \infty)$. Hence, if $z(T) = 0$, then z is identically zero on the whole interval $[T, \infty)$, and therefore $(-a(t)(-z'(t))^\alpha)'|_{t=T} = 0$, which is a contradiction. Thus, we have shown that $z(T) > 0$.

Next, let T^* be the first zero of z in (T, ∞). Then, $\tilde{z}$ is positive on $[t_0, T^*)$, and hence $\sum_{i=1}^{n} q_i(T^*)(\tilde{z}[g_i(T)])^\alpha > 0$, since $t_0 \le g_{i_0}(T^*) < T^*$ for some $i_0 \in \{1, 2, \cdots, n\}$. Hence, $(-a(t)(-z'(t))^\alpha)'|_{t=T^*} \ne 0$. On the other hand, z is identically zero on $[T^*, \infty)$, which means that $(-a(t)(-z'(t))^\alpha)'|_{t=T^*} = 0$, a contradiction. Therefore, the solution $z(t)$ is positive. This completes the proof. ∎

To apply Theorem 3.13.25 in practice, we need to distinguish the following three cases:

$$\int^\infty \sum_{i=1}^{n} q_i(s)ds < \infty \quad \text{and} \quad \int^\infty a^{-1/\alpha}(s) \left[\int_s^\infty \sum_{i=1}^{n} q_i(u)du \right]^{1/\alpha} ds < \infty,$$
(3.13.145)

$$\int^\infty \sum_{i=1}^{n} q_i(s)ds < \infty \quad \text{and} \quad \int^\infty a^{-1/\alpha}(s) \left[\int_s^\infty \sum_{i=1}^{n} q_i(u)du \right]^{1/\alpha} ds = \infty$$
(3.13.146)

and

$$\int^\infty \sum_{i=1}^{n} q_i(s)ds = \infty.$$
(3.13.147)

The condition (3.13.145) which is the same as (3.13.140) guarantees the existence of a decaying nonoscillatory solution of equation (3.13.127).

Theorem 3.13.26. Suppose (3.13.141) holds for some $i_0 \in \{1, 2, \cdots, n\}$. If condition (3.13.140) is satisfied, then equation (3.13.127) has a nonoscillatory solution which tends to zero as $t \to \infty$.

Proof. Let T be large enough so that $\min_i \{\inf_{t \ge T} g_i(t)\} \ge \max\{1, t_0\}$ and

$$\int_T^\infty a^{-1/\alpha}(s) \left[\int_s^\infty \sum_{i=1}^{n} q_i(u)du \right]^{1/\alpha} ds \le \frac{1}{2}.$$
(3.13.148)

We shall show that $y(t) = 1 + (1/t)$ satisfies (3.13.142). Indeed, we have

$$\int_t^\infty a^{-1/\alpha}(s) \left[\int_s^\infty \sum_{i=1}^n q_i(u)(y[g_i(u)])^\alpha du \right]^{1/\alpha} ds$$

$$= \int_t^\infty a^{-1/\alpha}(s) \left[\int_s^\infty \sum_{i=1}^n q_i(u) \left(1 + \frac{1}{g_i(u)} \right)^\alpha du \right]^{1/\alpha} ds$$

$$\leq 2 \int_t^\infty a^{-1/\alpha}(s) \left[\int_s^\infty \sum_{i=1}^n q_i(u) du \right]^{1/\alpha} ds \leq 1 < y(t), \quad t \geq T.$$

The conclusion now follows from Theorem 3.13.25. ∎

Now, we shall present results which guarantee the existence of decaying nonoscillation solutions and are applicable for the cases (3.13.146) and (3.13.147).

Theorem 3.13.27. Suppose (3.13.141) holds for some $i_0 \in \{1, 2, \cdots, n\}$, and

$$\limsup_{t \to \infty} \int_{\sigma(t)}^\infty a^{-1/\alpha}(s) \left[\int_s^\infty \sum_{i=1}^n q_i(u) du \right]^{1/\alpha} ds < \frac{1}{e}, \qquad (3.13.149)$$

where $\sigma(t) = \min_i g_i(t)$. Then equation (3.13.127) has a nonoscillatory solution which tends to zero as $t \to \infty$.

Proof. Let

$$Q(t) = \left(\frac{1}{a(t)} \int_t^\infty \sum_{i=1}^n q_i(s) ds \right)^{1/\alpha}$$

and choose $T > t_0$ so that $\inf_{t \geq T} \sigma(t) \geq t_0$, and

$$Q_T = \sup_{t \geq T} \int_{\sigma(t)}^t Q(s) ds \leq \frac{1}{e}. \qquad (3.13.150)$$

Define $y(t) = \exp\left(-(1/Q_T) \int_{t_0}^t Q(s) ds \right)$. Since for $i = 1, 2, \cdots, n$,

$$y[g_i(t)] = \exp\left(\frac{1}{Q_T} \int_{g_i(t)}^t Q(s) ds \right) \exp\left(-\frac{1}{Q_T} \int_{t_0}^t Q(s) ds \right)$$

$$\leq e \exp\left(-\frac{1}{Q_T} \int_{t_0}^t Q(s) ds \right) = ey(t), \quad t \geq T$$

in view of (3.13.150), we have

$$\int_t^\infty \left(\frac{1}{a(s)} \int_s^\infty \sum_{i=1}^n q_i(u)(y[g_i(u)])^\alpha du \right)^{1/\alpha} ds$$

$$\leq e \int_t^\infty Q(s)y(s)ds \leq e \int_t^\infty Q(s) \exp\left(-\frac{1}{Q_T} \int_{t_0}^t Q(u)du \right) ds$$

$$\leq eQ_T \exp\left(-\frac{1}{Q_T} \int_{t_0}^t Q(s)ds \right) = eQ_T y(t) \leq y(t), \quad t \geq T.$$

Now, from Theorem 3.13.25 it follows that equation (3.13.127) has a de-caying nonoscillatory solution. ∎

Theorem 3.13.28. Suppose (3.13.141) holds for some $i_0 \in \{1, 2, \cdots, n\}$. Further, suppose there exists a $T > t_0$ such that $\inf_{t \geq T} \sigma(t) \geq t_0$,

$$Q_T^* = \inf_{t \geq T} Q(t) > 0 \quad \text{and} \quad \sup_{t \geq T} \int_{\sigma(t)}^t \sum_{i=1}^n q_i(u)du < \frac{\alpha + 1}{e} \left(\frac{Q_T^*}{\alpha} \right)^{\alpha/(\alpha+1)},$$

$$(3.13.151)$$

where the functions $Q(t)$ and $\sigma(t)$ are as in Theorem 3.13.27. Then equation (3.13.127) has a nonoscillatory solution which tends to zero as $t \to \infty$.

Proof. Let

$$P_T = \sup_{t \geq T} \int_{\sigma(t)}^t \sum_{i=1}^n q_i(t)ds$$

and

$$y(t) = \exp\left(-\left(\frac{\alpha + 1}{\alpha P_T} \right) \int_{t_0}^t \sum_{i=1}^n q_i(s)ds \right).$$

Then, we have $y[g_i(t)] \leq \exp\left((\alpha + 1)/\alpha \right) y(t), \ t \geq T, \ i = 1, 2, \cdots, n$ and hence

$$\int_t^\infty \sum_{i=1}^n q_i(s)(y[g_i(s)])^\alpha ds \leq e^{\alpha+1} \int_t^\infty \left(\sum_{i=1}^n q_i(s) \right) (y(s))^\alpha ds$$

$$= e^{\alpha+1} \int_t^\infty \sum_{i=1}^n q_i(s) \exp\left(-\frac{\alpha + 1}{P_T} \int_{t_0}^s q_i(u)du \right) ds$$

$$\leq \frac{P_T}{\alpha + 1} e^{\alpha+1} \exp\left(-\frac{\alpha + 1}{P_T} \int_{t_0}^t \sum_{i=1}^n q_i(s)ds \right), \quad t \geq T.$$

Consequently, we obtain

$$\int_t^\infty \left(\frac{1}{a(s)} \int_s^\infty \sum_{i=1}^n q_i(u)(y[g_i(u)])^\alpha du \right)^{1/\alpha} ds$$

$$\leq \left(\frac{P_T}{\alpha+1} \right)^{1/\alpha} e^{(\alpha+1)/\alpha} \int_t^\infty \exp \left(\frac{\alpha+1}{\alpha P_T} \int_{t_0}^s \sum_{i=1}^n q_i(u)du \right) ds$$

$$\leq \frac{1}{Q_T^*} \left(\frac{P_T}{\alpha+1} \right)^{1/\alpha} e^{(\alpha+1)/\alpha} \int_t^\infty q_i(s) \exp \left(-\frac{\alpha+1}{\alpha P_T} \int_{t_0}^s \sum_{i=1}^n q_i(u)du \right) ds$$

$$\leq \frac{\alpha P_T}{(\alpha+1)Q_T^*} \left(\frac{P_T}{\alpha+1} \right)^{1/\alpha} e^{(\alpha+1)/\alpha} \exp \left(-\frac{\alpha+1}{\alpha P_T} \int_{t_0}^t \sum_{i=1}^n q_i(s)ds \right)$$

$$\leq y(t) \quad \text{for} \quad t \geq T.$$

This establishes the existence of a strictly decreasing function $y(t) > 0$ satisfying (3.13.142). The proof now follows from Theorem 3.13.25. ∎

Example 3.13.13. Consider the equation

$$(|x'(t)|x'(t))' = t^{-\gamma}|x[\lambda t]|x[\lambda t], \quad t > 0 \tag{3.13.152}$$

where $\gamma > 1$, $0 < \lambda < 1$ are constants. This is a special case of (3.13.127) with $\alpha = 2$, $a(t) = 1$, $n = 1$, $q_1(t) = t^{-\gamma}$ and $g_1(t) = \lambda t$.

(I_1) Let $\gamma > 3$. Then both (3.13.138) and (3.13.140) hold for (3.13.152) and so by Theorems 3.13.23 and 3.13.24, equation (3.13.152) has nonoscillatory solutions $x_1(t)$ and $x_2(t)$ such that $\lim_{t\to\infty} x_1(t)/t = \text{constant} \neq 0$ and $\lim_{t\to\infty} x_2(t) = \text{constant} \neq 0$ regardless of the values of λ, $0 < \lambda < 1$.

(I_2) Let $\gamma = 3$. An easy computation shows that (3.13.149) is satisfied for (3.13.152) if $1 < \lambda < \exp(\sqrt{2}/e)$, since

$$\int_{g_1(t)}^t \left(\int_s^\infty q_1(u)du \right)^{1/\alpha} ds = \int_{\lambda t}^t \left(\int_s^\infty u^{-3}du \right)^{1/2} ds = \frac{1}{\sqrt{2}} \ln \frac{1}{\lambda}.$$

From Theorem 3.13.27 it follows that for each λ equation (3.13.152) has a nonoscillatory solution which tends to zero as $t \to \infty$.

(I_3) Let $1 < \gamma < 3$. Then, (3.13.151) is satisfied for (3.13.152) since $Q_T^* = 1$, and

$$\int_{g_1(t)}^t q_1(s)ds = \int_{\lambda t}^t s^{-\gamma}ds = \frac{1}{\gamma-1} \left(\lambda^{1-\gamma} - 1 \right) t^{1-\gamma} \to 0 \quad \text{as} \quad t \to \infty.$$

Therefore, by Theorem 3.13.28 there exists a decaying nonoscillatory solution of equation (3.13.152). ∎

Example 3.13.14.   Consider again equation (3.13.129). The condition (3.13.151) applied to (3.13.129) reduces to

$$k\sigma < \frac{\alpha+1}{e}\left(\frac{k}{\alpha}\right)^{\alpha/(\alpha+1)} \quad \text{or } \sigma < \frac{(\alpha+1)^{1/(\alpha+1)}}{e}\left(\frac{\alpha+1}{\alpha}\right)^{\alpha/(\alpha+1)} k^{-1/(\alpha+1)}.$$

$$(3.13.153)$$

It would be of interest to compare (3.13.153) with (3.13.133), which can be rewritten as

$$k > (\alpha+1)^{1/(\alpha+1)}k^{-1/(\alpha+1)} \quad \text{or} \quad k > \left(\frac{\alpha+1}{\alpha}\right)^{\alpha/(\alpha+1)} k^{-1/(\alpha+1)}.$$

This condition guarantees the nonexistence of bounded nonoscillatory solutions of (3.13.129).

3.14. Notes and General Discussions

1. The results of Sections 3.1 and 3.2 are based on the work of Li and Yeh [40].

2. The results of Section 3.3 are extracted from Agarwal et. al. [8] and generalize a well–known theorem of Levin [38].

3. Theorem 3.4.1 is due to Hartman [25], also see Swanson [58]. All other results in Section 3.4 are borrowed from Li and Yeh [45]. We note that Theorem 3.4.5 is an extension of theorem 1 of Singh [57].

4. Theorem 3.5.1 is taken from Duzurnak and Mingarilli [14], whereas Lemmas 3.5.1 and 3.5.2 and Theorem 3.5.2 are adapted from Wong and Yeh [62].

5. Section 3.6 contains the work of Chantladze et. al. [12].

6. Theorems 3.7.1 – 3.7.6 are due to Li [39]. However, Kusano and Yoshida [36] have proved Theorem 3.7.5 earlier for the particular case $q_1(t) = q(t)$ by employing the Schauder–Tychonov fixed point theorem. Theorem 3.7.6 is a generalization of theorem 3.5 in Kusano and Natito [34]. Theorems 3.7.7 – 3.7.22 are modelled after Li and Yeh [41,42,46]. Corollaries 3.7.9 and 3.7.10 are theorems 3.3 and 3.4 of Kusano and Naito [34]. Theorems 3.7.23 and 3.7.24 are taken from Jingfa [26].

7. Theorems 3.8.1 – 3.8.4 are borrowed from Elbert [15,16], Del Pino et. al. [56], Kusano et. al. [33] and Kusano and Yoshida [36]. Theorems 3.8.5 and 3.8.6, respectively are extensions of Leighton's result [37] and Wintener's criterion [60].

8. The results in Subsection 3.9.1 are extensions of some of those established by Grace [21], Manojlovic [53] and Philos [55]. The results in

Subsection 3.9.2 generalize the work of Kong [29]. The results in Subsection 3.9.3 are taken from Li and Yeh [42,43]. These results extend and improve the work of Willett [59].

9. Section 3.10 contains the work of Li and Yeh [41–44,48]. Corollary 3.10.1 is a result due to Wintner [61] (see also theorem 2.17 in Swanson [58]), whereas Theorems 3.10.3 and 3.10.4 extend and improve some results of Yan [68].

10. Theorems 3.11.1 – 3.11.3 extend oscillation criteria of Kartsatos [27], whereas Theorem 3.11.4 generalizes theorem 1 of Wong [63].

11. The results in Section 3.12 are taken from Lian et. al. [49] and extend the work of Harris and Kong [24].

12. Lemma 3.13.1 is an extension of a result of Erbe [20], whereas Lemma 3.13.3 generalizes a result of Kartsatos [28]. Lemma 3.13.4 and its duality Lemma 3.13.5 are due to Koplatadze and Chanturia [30], and Lemma 3.13.6 is extracted from Philos [54]. Most of the results in Subsection 3.13.2 are the extensions of those established by Grace and Lalli [22]. Lemma 3.13.7 is due to Mahfoud [50]. The results of Subsection 3.13.3 are new. Subsection 3.13.4 is based on the work of Elbert and Kusano [19]. Theorems 3.13.14 and 1.13.15 are due to Jingfa [26]. Lemma 3.13.9 and Theorems 3.13.16 – 3.13.18 are taken from Kusano et. al. [32]. Lemma 3.13.10 is due to Mahfoud [51] and Theorem 3.13.19 extends a result of Mahfoud [51]. For $a(t) \equiv 1$ the results of Subsections 3.13.6 and 3.13.7 reduce to those proved in Kusano and Lalli [31].

13. For several other related works to this chapter see Agarwal et. al. [1–7], Bihari [10,11], Došly [13], Elbert [17,18], Kusano et. al. [35], Li and Yeh [47], Mahfoud and Rankin [52], and Wong and Agarwal [64–67].

3.15. References

1. **R.P. Agarwal and S.R. Grace**, Oscillations of forced functional differential equations generated by advanced arguments, *Aequationes Mathematicae*, to appear.

2. **R.P. Agarwal and S.R. Grace**, Oscillation criteria for second order half–linear differential equations with deviating arguments, *Dyn. Cont. Disc. Impul. Sys.*, to appear.

3. **R.P. Agarwal and S.R. Grace**, Oscillation of certain second order differential equations, *Proc. Sixth Int. Conf. on Nonlinear Functional Analysis and Applications*, Korea 2000, *Nova Science Publishers Inc.*, New York, to appear.

4. **R.P. Agarwal and S.R. Grace**, Interval criteria for oscillation of second order half–linear ordinary differential equations, *Functional Differential Equations*, to appear.

5. **R.P. Agarwal and S.R. Grace,** Second order nonlinear forced oscillations, *Dyn. Sys. Appl.*, to appear.

6. **R.P. Agarwal, S.R. Grace and D. O'Regan,** Oscillation criteria for certain nth order differential equations with deviating arguments, *J. Math. Anal. Appl.*, to appear.

7. **R.P. Agarwal, S.R. Grace and D. O'Regan,** On the oscillation of second order functional differential equations, *Advances Mathl. Sci. Appl.*, to appear.

8. **R.P. Agarwal, W.–C. Lian and C.C. Yeh,** Levin's comparison theorems for nonlinear second order differential equations, *Applied Math. Letters* **9**(1996), 29–35.

9. **A. Besicovitch,** *Almost Periodic Functions*, Dover, New York, 1954.

10. **I. Bihari,** An oscillation theorem concerning the half–linear differential equations of second order, *Magyar Tud. Akad. Mat. Kutato Int. Kozl.* **8**(1963), 275–280.

11. **I. Bihari,** Oscillation and monotonicity theorems concerning nonlinear differential equations of the second order, *Acta Math. Sci. Hungar.* **9**(1968), 83–104.

12. **T.A. Chanturia, N. Kandelaki and A. Lomtatidze,** On zeros of solutions of a second order singular half–linear equation, *Mem. Diff. Eqns. Math. Phy.* **17**(1999), 127–154.

13. **O. Došly,** Oscillation criteria for half–linear second order differential equations, *Hiroshima Math. J.* **28**(1998), 507–521.

14. **A. Dzurnak and A.B. Mingarelli,** Sturm–Liouville equations with Besicovitch almost periodicity, *Proc. Amer. Math. Soc.* **106**(1989), 647–653.

15. **Á. Elbert,** A half–linear second order differential equation, in *Qualitative Theory of Differential equations*, Szeged–Societas János Bolyai, *Colloq. Math. Soc. János Bolyai* **30**, 1979, 153–180.

16. **Á. Elbert,** Oscillation and nonoscillation theorems for some nonlinear ordinary differential equations, *Lecture Notes in Math.*, 964, *Springer–Verlag*, New York, 1982, 187–212.

17. **Á. Elbert,** Asymptotic behavior of autonomous half–linear differential systems on the plane, *Studia Sci. Math. Hungar.* **19**(1984), 447–464.

18. **Á. Elbert,** On the half–linear second order differential equations, *Acta. Math. Hungar.* **49**(1987), 487–508.

19. **Á. Elbert and T. Kusano,** Oscillation and nonoscillation theorems for a class of second order quasilinear differential equations, *Acta. Math. Hungar.* **56**(1990), 325–336.

20. **L.H. Erbe,** Oscillation criteria for second order nonlinear delay equations, *Canad. Math. Bull.* **16**(1973), 49–56.

21. **S.R. Grace,** Oscillation theorems for nonlinear differential equations of second order, *J. Math. Anal. Appl.* **171**(1992), 220–241.

22. **S.R. Grace and B.S. Lalli,** Oscillatory behavior for nonlinear second order functional differential equations with deviating arguments, *Bull. Inst. Math. Aad. Sinica* **14**(1986), 187–196.

23. **G.H. Hardy, J.E. Littlewood and G. Polya,** *Inequalities*, 2nd ed. *Cambridge Univ. Press*, 1988.

24. **B.J. Harris and Q. Kong,** On the oscillation of differential equations with an oscillatory coefficient, *Trans. Amer. Math. Soc.* **347**(1995), 1831–1839.

25. **P. Hartman,** *Ordinary Differential Equations*, John Wiley, New York, 1964.

26. **W. Jingfa,** On second order quasilinear oscillations, *Funkcial. Ekvac.* **41**(1998), 25–54.

27. **A.G. Kartsatos,** On oscillations of nonlinear equations of second order, *J. Math. Anal. Appl.* **24**(1968), 357–361.

28. **A.G. Kartsatos,** On nth order differential inequalities, *J. Math. Anal. Appl.* **52**(1975), 1–9.

29. **Q. Kong,** Interval criteria for oscillation of second order linear ordinary differential equations, *J. Math. Anal. Appl.* **239**(1999), 258–270.

30. **R.G. Koplatadze and T.A. Chanturia,** On oscillatory and monotone solutions of first order differential equations with deviating arguments, *Diferencial'nye Uravnenija* **18**(1982), 1463–1465.

31. **T. Kusano and B.S. Lalli,** On oscillation of half–linear functional differential equations with deviating arguments, *Hiroshima Math. J.* **24**(1994), 549–563.

32. **T. Kusano, A. Ogata and H. Usami,** Oscillation theory for a class of second order quasilinear ordinary differential equations with application to partial differential equations, *Japan J. Math.* **19**(1993), 131–147.

33. **T. Kusano, A. Ogata and H. Usami,** On the oscillation of solutions of second order quasilinear ordinary differential equations, *Hiroshima Math. J.* **23**(1993), 645–667.

34. **T. Kusano and Y. Naito,** Oscillation and nonoscillation criteria for second order quasilinear differential equations, *Acta Math. Hungar.* **76**(1997), 81–99.

35. **T. Kusano, Y. Naito and A. Ogata,** Strong condition and nonoscillation of quasilinear differential equations of second order, *Diff. Eqns. Dyn. Sys.* **2**(1994), 1–10.

36. **T. Kusano and N. Yoshida,** Nonoscillation theorems for a class of quasilinear differential equations of second order, *J. Math. Anal. Appl.* **189**(1995), 115–127.

37. **W. Leighton,** The detection of the oscillation of solutions of a second order linear differential equation, *Duke J. Math.* **17**(1950), 57–61.

38. **A.Y. Levin,** A comparison principle for second order differential equations, *Sov. Math. Dokl.* **1**(1960), 1313–1316.

39. **H.J. Li,** Nonoscillation characterization of second order linear differential equations, *Math. Nachr.* **219**(2000), 147–161.

40. **H.J. Li and C.C. Yeh,** Sturmian comparison theorem for half linear second order differential equations, *Proc. Roy. Soc. Edinburgh Sect. A* **125**(1995), 1193–1204.

41. **H.J. Li and C.C. Yeh,** Nonoscillation criteria for second order half linear differential equations, *Appl. Math. Letters* **8**(1995), 63–70.

42. **H.J. Li and C.C. Yeh,** Nonoscillation theorems for second order quasi-linear differential equations, *Publ. Math.* (Debrecen) **47**(1995), 271–279.

43. **H.J. Li and C.C. Yeh,** Oscillation of half linear second order differential equations, *Hiroshima Math. J.* **25**(1995), 585–594.

44. **H.J. Li and C.C. Yeh,** Oscillation criteria for nonlinear differential equations, *Houston J. Math.* **21**(1995), 801–811.

45. **H.J. Li and C.C. Yeh,** An oscillation criterion of almost periodic Sturm–Liouville equations, *Rocky Mount. J. Math.* **25**(1995), 1417–1429.

46. **H.J. Li and C.C. Yeh,** On the nonoscillatory behavior of solutions of a second order linear differential equation, *Math. Nachr.* **182**(1996), 295–315.

47. **H.J. Li and C.C. Yeh,** Oscillation of nonlinear functional differential equations of the second order, *Appl. Math. Letters* **11**(1)(1998), 71–77.

48. **H.J. Li and C.C. Yeh,** Oscillation and nonoscillation criteria for second order linear differential equations, *Math. Nachr.* **194**(1998), 171–184.

49. **W.C. Lian, C.C. Yeh and H.J. Li,** The distance between zeros of an oscillatory solution to a half linear differential equation, *Comput. Math. Appl.* **29**(1995), 39–43.

50. **W.E. Mahfoud,** Oscillation and asymptotic behavior of solutions of nth order nonlinear delay differential equations, *J. Differential Equations* **24** (1977), 75–98.

51. **W.E. Mahfoud,** Comparison theorems for delay differential equations, *Pacific J. Math.* **83**(1979), 187–197.

52. **W.E. Mahfoud and S.M. Rankin,** Some properties of solutions of $(r(t)\Psi(x)x')' + a(t)f(x) = 0$, *SIAM J. Math. Anal.* **10**(1979), 49–54.

53. **J.V. Manojlovic,** Oscillation criteria for a second order half–linear differential equation, *Math. Comput. Modelling* **30**(5–6)(1999), 109–119.

54. **Ch.G. Philos,** On the existence of nonoscillatory solutions tending to zero at ∞ for differential equations with positive delays, *Arch. Math.* **36**(1981), 168–178.

55. **Ch.G. Philos,** Oscillation theorems for linear differential equations of second order, *Arch. Math.* **53**(1989), 482–492.

56. **M. Del Pino and R. Manasevich,** Oscillation and nonoscillation for $\left(|u'|^{p-2}u'\right)' + a(t)|u|^{p-2}u = 0$, $p > 1$, *Houston J. Math.* **14**(1988), 173–177.

57. **B. Singh,** Comparative study of asymptotic nonoscillation and quickly oscillation of second order linear differential equations, *J. Mathl. Phyl. Sci.* **4**(1974), 363–376.

58. **C.A. Swanson,** *Comparison and Oscillation Theory of Linear Differential Equations*, Academic Press, New York, 1968.

59. **D. Willett,** On the oscillatory behavior of the solutions of second order linear differential equations, *Ann. Polon. Math.* **21**(1969), 175–194.

60. **A. Wintner,** A criterion for oscillatory stability, *Quart. Appl. Math.* **7**(1949), 115–117.

61. **A. Wintner,** On the nonexistence of conjugate points, *Amer. J. Math.* **73**(1951), 368–380.

62. **F.H. Wong and C.C. Yeh,** An oscillation criterion for Sturm–Liouville equations with Besicovitch almost periodic coefficients, *Hiroshima Math. J.* **21**(1991), 521–528.

63. **J.S.W. Wong,** Oscillation criteria for a forced second order linear differential equation, *J. Math. Anal. Appl.* **231**(1999), 235–240.

64. **P.J.Y. Wong and R.P. Agarwal,** Oscillation theorems and existence criteria of asymptotically monotone solutions for second order differential equations, *Dynam. Systems Appl.* 4(1995), 477–496.

65. **P.J.Y. Wong and R.P. Agarwal,** On the oscillation and asymptotically monotone solutions of second order quasilinear differential equations, *Appl. Math. Comput.* **79**(1996), 207–237.

66. **P.J.Y. Wong and R.P. Agarwal,** Oscillatory behavior of solutions of certain second order nonlinear differential equations, *J. Math. Anal. Appl.* **198**(1996), 813–829.

67. **P.J.Y. Wong and R.P. Agarwal,** Oscillation criteria for half–linear differential equations, *Advances Math. Sci. Appl.* **9**(1999), 649–663.

68. **J. Yan,** Oscillation property for second order differential equations with an 'integral small' coefficient, *Acta Math. Sinica* **30**(1987), 206–215.

Chapter 4

Oscillation Theory for Superlinear Differential Equations

4.0. Introduction

This chapter presents oscillation and nonoscillation theory for solutions of second order nonlinear differential equations of superlinear type. Section 4.1 deals with the oscillation of superlinear equations with sign changing coefficients. Here, first we shall discuss some results which involve integrals and weighted integrals of the alternating coefficients, and then provide several criteria which use average behaviors of these integrals. More general averages such as 'weighted averages' and 'iterated averages' are also employed. In Section 4.2, we impose some additional conditions on the superlinear terms which allow us to proceed further and extend and improve some of the results established in Section 4.1. In fact, an asymptotic study has been made and interesting oscillation criteria have been proved. In Section 4.3, first we shall provide sufficient conditions which guarantee the existence of nonoscillatory solutions, and then present necessary and sufficient conditions for the oscillation of superlinear equations. Oscillation results via comparison of nonlinear equations of the same form as well as with linear ones of the same order are also established. In Section 4.4, we shall extend some of the results of the previous sections and establish several new oscillation criteria for more general superlinear equations. Necessary and sufficient conditions for such equations to be oscillatory are also given. Section 4.5 deals with the oscillation of forced–superlinear differential equations with alternating coefficients. Finally, Section 4.6 presents the oscillation and nonoscillation criteria for second order superlinear equations with nonlinear damping terms.

4.1. Superlinear Oscillation Criteria

In this section we shall present oscillation criteria for superlinear ordinary differential equations of the form

$$x''(t) + q(t)f(x(t)) = 0, \tag{4.1.1}$$

$$(a(t)x'(t))' + q(t)f(x(t)) = 0 \tag{4.1.2}$$

and the more general damped equation

$$(a(t)x'(t))' + p(t)x'(t) + q(t)f(x(t)) = 0, \tag{4.1.3}$$

where

(i) $a(t) \in C([t_0, \infty), \mathbb{R}^+)$,

(ii) $p(t), q(t) \in C([t_0, \infty), \mathbb{R})$,

(iii) $f \in C(\mathbb{R}, \mathbb{R})$, f is continuously differentiable except possibly at 0, and satisfies

$$xf(x) > 0 \quad \text{and} \quad f'(x) \geq 0 \quad \text{for} \quad x \neq 0 \tag{4.1.4}$$

also $f(x)$ is strongly superlinear in the sense that

$$\int^{\infty} \frac{du}{f(u)} < \infty \quad \text{and} \quad \int^{-\infty} \frac{du}{f(u)} < \infty. \tag{4.1.5}$$

The special case when $f(x) = |x|^\gamma \operatorname{sgn} x$, $x \in \mathbb{R}$ $(\gamma > 1)$ is of particular interest. In fact, the differential equations

$$x''(t) + q(t)|x(t)|^\gamma \operatorname{sgn} x(t) = 0, \tag{4.1.6}$$

$$(a(x)x'(t))' + q(t)|x(t)|^\gamma \operatorname{sgn} x(t) = 0 \tag{4.1.7}$$

and

$$(a(t)x'(t))' + p(t)x'(t) + q(t)|x(t)|^\gamma \operatorname{sgn} x(t) = 0 \tag{4.1.8}$$

are prototype of (4.1.1), (4.1.2) and (4.1.3) respectively, and will be discussed extensively.

We begin with the following oscillation criteria for equation (4.1.2).

Theorem 4.1.1. Suppose conditions (4.1.4) and (4.1.5) hold,

$$\int_{t_0}^{\infty} \frac{1}{a(s)} ds = \infty, \tag{4.1.9}$$

$$\int_{t_0}^{\infty} q(s)ds < \infty, \tag{4.1.10}$$

$$\liminf_{t \to \infty} \int_{T}^{t} q(s)ds \geq 0 \quad \text{for all large} \quad T \tag{4.1.11}$$

and

$$\lim_{t \to \infty} \int_{t_0}^{t} \frac{1}{a(s)} \int_{s}^{\infty} q(u) du\, ds = \infty. \tag{4.1.12}$$

Then equation (4.1.2) is oscillatory.

Proof. Let $x(t)$ be a nonoscillatory solution of (4.1.2), say, $x(t) > 0$ for $t \geq t_0 \geq 0$. Define $w(t) = a(t)x'(t)/f(x(t))$, $t \geq t_1$. Then, we have

$$w'(t) = -q(t) - \frac{f'(x(t))}{a(t)} w^2(t) \leq -q(t) \quad \text{for} \quad t \geq t_1. \tag{4.1.13}$$

For $T \geq t_1$ an integration of (4.1.13) yields

$$\frac{a(t)x'(t)}{f(x(t))} \leq \frac{a(T)x'(T)}{f(x(T))} - \int_{T}^{t} q(s) ds.$$

Now, we need to consider the following cases:

Case I. If $x'(t) \geq 0$ for all $t \geq T$, then from (4.1.10), we find

$$0 \leq \frac{a(T)x'(T)}{f(x(T))} - \int_{T}^{\infty} q(s) ds.$$

The same type of reasoning as above guarantees for $t \geq T$ that we have $\int_{t}^{\infty} q(s) ds \leq a(t)x'(t)/f(x(t))$. Dividing both sides of this inequality by $a(t)$ and integrating from T to t, we get

$$\int_{T}^{t} \frac{1}{a(s)} \int_{s}^{\infty} q(u) du\, ds \leq \int_{T}^{t} \frac{x'(s)}{f(x(s))} ds = \int_{x(T)}^{x(t)} \frac{d\xi}{f(\xi)}.$$

This in view of (4.1.12) contradicts condition (4.1.5).

Case II. If $x'(t)$ changes sign, then there exists a sequence $\{T_n\}_{n=1}^{\infty}$, $\lim_{n \to \infty} T_n = \infty$ such that $x'(T_n) < 0$. Choose N large enough so that (4.1.11) holds. Then, we have

$$\frac{a(t)x'(t)}{f(x(t))} \leq \frac{a(T_N)x'(T_N)}{f(x(T_N))} - \int_{T_N}^{t} q(s) ds,$$

and hence

$$\limsup_{t \to \infty} \frac{a(t)x'(t)}{f(x(t))} \leq \frac{a(T_N)x'(T_N)}{f(x(T_N))} + \limsup_{t \to \infty} \left(-\int_{T_N}^{t} q(s) ds \right) < 0,$$

which contradicts the fact that $x'(t)$ oscillates.

Case III. If $x'(t) < 0$ for $t \geq t_2 \geq t_1$, then condition (4.1.11) implies that for $T_0 \geq t_0$ there exists $T_1 \geq T_0$ such that $\int_{T_1}^{t} q(s) ds \geq 0$ for all $t \geq T_1$. Choosing $T_1 \geq t_2$ as indicated and then integrating (4.1.2) from

T_1 to t, we find

$$a(t)x'(t) = a(T_1)x'(T_1) - \int_{T_1}^t q(s)f(x(s))ds$$

$$= a(T_1)x'(T_1) - f(x(t)) \int_{T_1}^t q(s)ds + \int_{T_1}^t f'(x(s))x'(s) \int_{T_1}^s q(u)du ds$$

$$\leq a(T_1)x'(T_1).$$

Thus, $x(t) \leq x(T_1) + a(T_1)x'(T_1) \int_{T_1}^t ds/a(s) \to -\infty$ as $t \to \infty$, which contradicts the fact that $x(t) > 0$ for $t \geq t_1$. A similar proof holds when $x(t) < 0$ for $t \geq t_1$. ∎

Remark 4.1.1. The condition (4.1.11) on the coefficient $q(t)$ is equivalent to the following: either

$$\lim_{t \to \infty} \int_{t_0}^t q(s)ds = \infty, \tag{4.1.11$_1$}$$

or $\lim_{t \to \infty} \int_{t_0}^t q(s)ds$ exists and is finite, and

$$Q(t) = \int_t^\infty q(s)ds \geq 0 \quad \text{for all large } t. \tag{4.1.11$_2$}$$

For this, let (4.1.11) hold. Define $Q^*(t) = \int_{t_0}^t q(s)ds$. We first claim that $\liminf_{t \to \infty} Q^*(t) \neq -\infty$. Suppose this is not true. Then there exists a sequence $\{t_n\}_{n=1}^\infty$, $t_n \to \infty$ as $n \to \infty$ such that $Q^*(t_n) \leq -n$. For large T, we have $\int_T^{t_n} q(s)ds = Q^*(t_n) - Q^*(T) \leq -n - Q^*(T) \to -\infty$ as $t \to \infty$, which contradicts (4.1.11). If $\liminf_{t \to \infty} Q^*(t) = \infty$, we have (4.1.11)$_1$. Thus, we assume that $\liminf_{t \to \infty} Q^*(t) = \alpha \neq \infty$. Suppose that $\limsup_{t \to \infty} Q^*(t) = \beta > \alpha$. Then there exist two sequences $\{t_n\}_{n=1}^\infty$ and $\{T_n\}_{n=1}^\infty$, $t_n \to \infty$ and $T_n \to \infty$ as $n \to \infty$ such that $Q^*(t_n) \to \alpha$ and $Q^*(T_n) \to \beta$ as $n \to \infty$. Then, $\int_{T_n}^{t_m} q(s)ds = Q^*(t_m) - Q^*(T_n) < (\beta - \alpha)/2$ (with appropriate modification if $\beta = \infty$) for m, n sufficiently large. This contradicts (4.1.11). Thus, $\lim_{t \to \infty} Q^*(t)$ exists, and $Q(T) = \lim_{t \to \infty} \int_T^t q(s)ds = \liminf_{t \to \infty} \int_T^t q(s)ds \geq 0$ as assumed in (4.1.11). The sufficiency part is obvious.

Theorem 4.1.2. Let $a(t) \leq a_1$, $t \geq t_0$ where a_1 is a positive constant, and condition (4.1.5) holds. If for every large T there exists a $T_1 \geq T$ such that

$$\liminf_{t \to \infty} \int_{T_1}^t \int_T^s q(u)du ds > -\infty \tag{4.1.14}$$

and

$$\int_{t_0}^\infty sq(s)ds = \infty, \tag{4.1.15}$$

then equation (4.1.2) is oscillatory.

Proof. Let $x(t)$ be a nonoscillatory solution of equation (4.1.2), say, $x(t) > 0$ for $t \geq t_0$. Define $w(t)$ as in Theorem 4.1.1 and obtain (4.1.13). Next, we need to consider the following three cases:

Case I. If $x'(t)$ oscillates, choose $T \geq t_0$ so that $x'(T) = 0$ and (4.1.14) holds, then integrating inequality (4.1.13), we get $a(t)x'(t)/f(x(t)) \leq -\int_T^t q(s)ds$. It follows from (4.1.14) that there exist $T_1 \geq T$ and $A > 0$ such that

$$\int_{T_1}^t \frac{a(s)x'(s)}{f(x(s))} ds \leq -\int_{T_1}^t \int_T^s q(u)du ds < A \quad \text{for all} \quad t \geq T_1.$$

Multiplying both sides of (4.1.13) by t and integrating from T_1 to t, we obtain

$$\frac{ta(t)x'(t)}{f(x(t))} \leq \frac{T_1 a(T_1)x'(T_1)}{f(x(T_1))} + A - \int_{T_1}^t sq(s)ds,$$

which by (4.1.15) contradicts the assumption that $x'(t)$ oscillates.

Case II. If $x'(t) > 0$ for $t \geq t_1 \geq t_0$, then multiplying (4.1.13) by t and integrating from t_1 to t, we obtain

$$\frac{ta(t)x'(t)}{f(x(t))} \leq \frac{t_1 a(t_1)x'(t_1)}{f(x(t_1))} + \int_{t_1}^t \frac{a(s)x'(s)}{f(x(s))} ds - \int_{t_1}^t sq(s)ds$$

$$\leq \frac{t_1 a(t_1)x'(t_1)}{f(x(t_1))} + a_1 \int_{x(t_1)}^{x(t)} \frac{du}{f(u)} - \int_{t_1}^t sq(s)ds,$$

which in view of (4.1.5) again leads to a contradiction.

Case III. If $x'(t) < 0$ for $t \geq t_2 \geq t_1$ condition (4.1.15) implies that there exists a $t_3 \geq t_2$ such that $\int_{t_2}^{t_3} sq(s)ds = 0$ and $\int_{t_3}^t sq(s)ds \geq 0$ for $t \geq t_3$. Multiplying (4.1.2) by t and integrating by parts from t_3 to t, we obtain

$$ta(t)x'(t) \leq t_3 a(t_3)x'(t_3) - \int_{t_3}^t sq(s)f(x(s))ds$$

$$\leq t_3 a(t_3)x'(t_3) - f(x(t)) \int_{t_3}^t sq(s)ds$$

$$+ \int_{t_3}^t f'(x(s))x'(s) \int_{t_3}^s uq(u)du ds \leq t_3 a(t_3)x'(t_3),$$

and hence $x'(t) \leq t_3 a(t_3)x'(t_3)/(a_1 t)$ for $t \geq t_3$. Thus, we have

$$x(t) \leq x(t_3) + \frac{t_3 a(t_3)x'(t_3)}{a_1} \ln \frac{t}{t_3} \to -\infty \quad \text{as} \quad t \to \infty$$

which contradicts the fact that $x(t) > 0$ for $t \geq t_0$.   ∎

Remark 4.1.2. We note that the condition $a(t) \leq a_1$ for $t \geq t_0$ in Theorem 4.1.2 can be replaced by $a'(t) \leq 0$ for $t \geq t_0$. In this case we need to apply the Bonnet theorem to obtain the desired conclusion.

Theorem 4.1.3. Suppose there exists a function $\rho(t) \in C^2([t_0, \infty), \mathbb{R}^+)$ such that

$$\int^{\infty} \rho(s)q(s)ds = \infty \tag{4.1.16}$$

and

$$\int^{\infty} \frac{1}{a(s)\rho(s)}ds = \infty. \tag{4.1.17}$$

Then the following hold:

(i_1) If $\beta(t) = a(t)\rho'(t) - p(t)\rho(t) \geq 0$ and $\beta'(t) \leq 0$ for $t \geq t_0$, then all bounded solutions of (4.1.3) are oscillatory. If in addition condition (4.1.5) holds, then (4.1.3) is oscillatory.

(i_2) If

$$\int_{+0} \frac{du}{f(u)} < \infty \quad \text{and} \quad \int_{-0} \frac{du}{f(u)} < \infty \tag{4.1.18}$$

hold, and $\int^{\infty} |\beta'(s)|ds < \infty$, then all bounded solutions of (4.1.3) are oscillatory. If in addition condition (4.1.5) holds, then (4.1.3) is oscillatory.

Proof. Let $x(t)$ be a nonoscillatory solution of equation (4.1.3), say, $x(t) > 0$ for $t \geq t_0$. Define $w(t) = \rho(t)a(t)x'(t)/f(x(t))$, $t \geq t_1$ for some $t_1 \geq t_0$. Then for $t \geq t_1$, we have

$$\begin{aligned}
w'(t) &= -\rho(t)q(t) + \beta(t)\frac{x'(t)}{f(x(t))} - \frac{1}{a(t)\rho(t)}w^2(t)f'(x(t)) \\
&\leq -\rho(t)q(t) + \beta(t)\frac{x'(t)}{f(x(t))}.
\end{aligned}$$

Integrating this inequality from t_1 to t, we obtain

$$w(t) \leq w(t_1) - \int_{t_1}^{t} \rho(s)q(s)ds + \int_{t_1}^{t} \beta(s)\frac{x'(s)}{f(x(s))}ds.$$

Now, we shall show that each of the conditions (i_1) and (i_2) ensures that $\int_{t_1}^{t} \beta(s)x'(s)/f(x(s))ds$ is bounded above. Consider the following two cases:

I_1. Let (i_1) hold. By the Bonnet theorem, for a fixed $t \geq t_1$ and $\xi \in [t_1, t]$, we have

$$\int_{t_1}^{t} \beta(s)\frac{x'(s)}{f(x(s))}ds = \beta(t_1)\int_{t_1}^{\xi} \frac{x'(s)}{f(x(s))}ds = \beta(t_1)\int_{x(t_1)}^{x(\xi)} \frac{du}{f(u)}$$

and hence, since $\beta(t) \geq 0$, and

$$\int_{x(t_1)}^{x(\xi)} \frac{du}{f(u)} < \begin{cases} 0 & \text{if } x(\xi) < x(t_1) \\ \int_{x(t_1)}^{\infty} \frac{du}{f(u)} & \text{if } x(\xi) \geq x(t_1), \end{cases}$$

we find

$$\int_{t_1}^{t} \beta(s) \frac{x'(s)}{f(x(s))} ds \leq K_1 \quad \text{for all } t \geq t_1, \tag{4.1.19}$$

where $K_1 = \beta(t_1) \int_{x(t_1)}^{\infty} du/f(u)$.

From (4.1.19) it follows that

$$\int_{t_1}^{t} \rho(s)q(s)ds \leq -w(t) + w(t_1) + K_1 - \int_{t_1}^{t} \frac{1}{a(s)\rho(s)} w^2(s)f'(x(s))ds. \tag{4.1.20}$$

I_2. Let (i_2) hold. Then, we have

$$\int_{t_1}^{t} \beta(s) \frac{x'(s)}{f(x(s))} ds = \beta(t) \int_{0}^{x(t)} \frac{du}{f(u)} - \beta(t_1) \int_{0}^{x(t_1)} \frac{du}{f(u)} \\ - \int_{t_1}^{t} \beta'(s) \left(\int_{0}^{x(s)} \frac{du}{f(u)} \right) ds. \tag{4.1.21}$$

In this case, it follows from (i_2) that $0 \leq \int_{0}^{x(t)} du/f(u) \leq M$ and $|\beta(t)| \leq N$, $t \geq t_1$ where M and N are real constants. Also, the integral $\int_{t_1}^{t} \beta'(s) \left(\int_{0}^{x(s)} du/f(u) \right) ds$ converges absolutely as $t \to \infty$ and so, (4.1.21) implies

$$\int_{t_1}^{t} \beta(s) \frac{x'(s)}{f(x(s))} ds \leq M_1 + M_1 N + M \int_{t_1}^{\infty} |\beta'(s)| ds = K_2,$$

where $M_1 = \beta(t_1) \int_{0}^{x(t_1)} du/f(u)$. Thus, we obtain

$$\int_{t_1}^{t} \rho(s)q(s)ds \leq -w(t) + w(t_1) + K_2 - \int_{t_1}^{t} \frac{1}{a(s)\rho(s)} w^2(s)f'(x(s))ds. \tag{4.1.22}$$

Set $K = \max\{K_1, K_2\}$. Then, (4.1.20) and (4.1.22) can be written as

$$\int_{t_1}^{t} \rho(s)q(s)ds \leq -w(t) + w(t_1) + K - \int_{t_1}^{t} \frac{1}{a(s)\rho(s)} w^2(s)f'(x(s))ds. \tag{4.1.23}$$

It follows from condition (4.1.16) that $\lim_{t\to\infty} \rho(t)a(t)x'(t)/f(x(t)) = -\infty$, i.e. $x'(t) < 0$ for $t \geq t_2 \geq t_1$. Let $t_3 \geq t_2$ be such that $w(t_1) + K - \int_{t_1}^{t} \rho(s)q(s)ds \leq -1$ for $t \geq t_3$, then (4.1.23) implies

$$-w(t) \geq 1 + \int_{t_1}^{t} \frac{1}{a(s)\rho(s)} w^2(s) f'(x(s)) ds, \qquad (4.1.24)$$

or

$$\frac{w^2(t)f'(x(t))/a(t)\rho(t)}{1 + \int_{t_1}^{t} (w^2(s)f'(x(s))/a(s)\rho(s))\, ds} \geq -\frac{x'(t)}{f(x(t))} f'(x(t)) \quad \text{for} \quad t \geq t_3$$

and consequently for all $t \geq t_3$,

$$\ln\left\{ \frac{1}{\lambda} \left(1 + \int_{t_1}^{t} (w^2(s)f'(x(s))/a(s)\rho(s))\, ds \right) \right\} \geq \ln \frac{f(x(t_3))}{f(x(t))},$$

where $\lambda = 1 + \int_{t_1}^{t_3} (w^2(s)f'(x(s))/a(s)\rho(s))\, ds$. Hence, it follows that

$$1 + \int_{t_1}^{t} (w^2(s)f'(x(s))/a(s)\rho(s))\, ds \geq \lambda \frac{f(x(t_3))}{f(x(t))} \quad \text{for} \quad t \geq t_3.$$

Thus, (4.1.24) yields $x'(t) \leq -\lambda f(x(t_3))/a(t)\rho(t)$ for $t \geq t_3$. Therefore, we have

$$x(t) \leq x(t_3) - [\lambda\, f(x(t_3))] \int_{t_3}^{t} \frac{1}{a(s)\rho(s)}\, ds \quad \text{for all} \quad t \geq t_3,$$

which in view of (4.1.17) leads to a contradiction that $\lim_{t\to\infty} x(t) = -\infty$.

Finally, we remark that in the above proof the oscillation of all bounded solutions of (4.1.3) is trivially included. ■

The following immediate corollaries of Theorem 4.1.3 are well known basic results in the literature.

Corollary 4.1.1. If $\gamma > 1$, and

$$\int^{\infty} sq(s)ds = \infty, \qquad (4.1.25)$$

then equation (4.1.6) is oscillatory.

Corollary 4.1.2. If conditions (4.1.5) and (4.1.25) hold, then equation (4.1.1) is oscillatory.

Corollary 4.1.3. If condition (4.1.5) holds,

$$\left(\frac{t^\alpha}{a(t)} \right)' \geq 0, \quad \left(a(t) \left(\frac{t^\alpha}{a(t)} \right)' \right)' \leq 0 \quad \text{for } t \geq t_0 \quad \text{and} \quad \int^{\infty} \frac{s^\alpha q(s)}{a(s)} ds = \infty,$$

where $0 < \alpha \le 1$, then equation (4.1.2) is oscillatory.

Proof. In Theorem 4.1.3, let $p(t) \equiv 0$ and $\rho(t) = t^\alpha/a(t)$, $t \ge t_0$ and $0 < \alpha \le 1$. ∎

Example 4.1.1. Consider the differential equation

$$\left(\frac{2}{t}x'(t)\right)' + \frac{1}{t^2}x'(t) + \left[-t^{-1/2}\sin t + \frac{1}{2}t^{-3/2}(2 + \cos t)\right]$$

$$\times \; (|x(t)|^\gamma) \; \text{sgn} \; x(t) \;=\; 0, \quad t \ge t_0 = \pi/2 \tag{4.1.26}$$

where $\gamma > 1$. Let $\rho(t) = t$. Then, $\beta(t) = a(t)\rho'(t) - p(t)\rho(t) = 2/t$ and for every $t \ge t_0$, we have

$$\int_{t_0}^t sq(s)ds \;=\; \int_{\pi/2}^t \left[-s^{1/2}\sin s + \frac{1}{2}s^{-1/2}(2 + \cos s)\right]ds$$

$$=\; \int_{\pi/2}^t d\left[s^{1/2}(2 + \cos s)\right] \;=\; t^{1/2}(2 + \cos t) - 2(\pi/2)^{1/2}$$

$$\ge\; t^{1/2} - 2(\pi/2)^{1/2} \to \infty \; \text{ as } \; t \to \infty.$$

Now, Theorem 4.1.3(i_1) ensures that (4.1.26) is oscillatory.

Example 4.1.2. Consider the differential equation

$$x''(t) + \frac{1}{t^2}x'(t) + \left[-t^{1/2}\sin t + \frac{1}{2}t^{-3/2}(2 + \cos t)\right]f(x(t)) = 0, \; t \ge t_0 = \pi/2 \tag{4.1.27}$$

where $f \in C(\mathbb{R}, \mathbb{R})$, $f(x) = \sqrt{x}(1 + x)$ for $x > 0$ and $-f(-x) = f(x)$. Here, we have

$$\int_0^{x(t)} \frac{du}{\sqrt{u}(1 + u)} \;=\; 2\tan^{-1}\sqrt{x} \; \text{ and so } \; \int_{\pm 0}^{\pm \infty} \frac{du}{f(u)} \;<\; \infty.$$

Let $\rho(t) = t$. Then, $\beta(t) = 1 - (1/t)$. Hence, Theorem 4.1.3(i_2) guarantees that (4.1.27) is oscillatory. We also note that Theorem 4.1.3(i_1) is not applicable to (4.1.27).

An interesting result for (4.1.2) which follows from Theorem 4.1.3 by taking $\rho(t) = \xi_1(t)$ is the following:

Corollary 4.1.4. If condition (4.1.9) holds, and $\int^\infty \xi_1(s)q(s)ds = \infty$, where $\xi_1(t) = \int_{t_0}^t ds/a(s)$, then equation (4.1.2) is oscillatory.

Now, we shall prove the following result.

Theorem 4.1.4. Suppose condition (4.1.5) holds and there exists a real constant k such that

$$f'(x) \ge k > 0 \quad \text{for} \quad x \ne 0. \tag{4.1.28}$$

Let $\rho(t) \in C^1([t_0, \infty), \mathbb{R}^+)$ be such that condition (4.1.17) hold,

$$\beta(t) = a(t)\rho'(t) - p(t)\rho(t) \geq 0 \quad \text{and} \quad \beta'(t) \leq 0 \quad \text{for} \quad t \geq t_0 \quad (4.1.29)$$

and

$$\limsup_{t\to\infty} \frac{1}{t^2} \int_{t_0}^t a(s)\rho(s)ds < \infty. \qquad (4.1.30)$$

If

$$\liminf_{t\to\infty} \int_{t_0}^t \rho(s)q(s)ds > -\infty \qquad (4.1.31)$$

and

$$\limsup_{t\to\infty} \frac{1}{t} \int_{t_0}^t \int_{t_0}^s \rho(u)q(u)duds = \infty, \qquad (4.1.32)$$

then equation (4.1.3) is oscillatory.

Proof. Let $x(t)$ be a nonoscillatory solution of equation (4.1.3), say, $x(t) > 0$ for $t \geq t_0$. Define $w(t) = \rho(t)a(t)x'(t)/f(x(t))$, $t \geq t_0$. Then for every $t \geq t_0$, we have

$$w'(t) = -\rho(t)q(t) + \beta(t)\frac{x'(t)}{f(x(t))} - \frac{1}{a(t)\rho(t)}w^2(t)f'(x(t)).$$

Hence, for all $t \geq t_0$, it follows that

$$\int_{t_0}^t \rho(s)q(s)ds = -w(t) + w(t_0) + \int_{t_0}^t \frac{\beta(s)x'(s)}{f(x(s))}ds - \int_{t_0}^t \frac{w^2(s)f'(x(s))}{a(s)\rho(s)}ds.$$

By the Bonnet theorem, for any $t \geq t_0$ there exists a $\xi \in [t_0, t]$ such that

$$\int_{t_0}^t \beta(s)\frac{x'(s)}{f(x(s))}ds = \beta(t_0)\int_{t_0}^\xi \frac{x'(s)}{f(x(s))}ds = \beta(t_0)\int_{x(t_0)}^{x(\xi)} \frac{du}{f(u)}$$

$$\leq \beta(t_0)\int_{x(t_0)}^\infty \frac{du}{f(u)}.$$

Thus, for every $t \geq t_0$, we find

$$\int_{t_0}^t \rho(s)q(s)ds \leq -w(t) + M - \int_{t_0}^t \frac{w^2(s)f'(x(s))}{a(s)\rho(s)}ds, \qquad (4.1.33)$$

where $M = w(t_0) + \beta(t_0)\int_{x(t_0)}^\infty du/f(u)$. Therefore, from condition (4.1.28), we get

$$\int_{t_0}^t \rho(s)q(s)ds \leq -w(t) + M - k\int_{t_0}^t \frac{w^2(s)}{a(s)\rho(s)}ds \quad \text{for all} \quad t \geq t_0. \quad (4.1.34)$$

Next, we need to consider the following three cases for the behavior of $x'(t)$.

Case 1. $x'(t)$ is oscillatory. Then there exists a sequence $\{t_m\}_{m=1}^{\infty}$ in $[t_0, \infty)$ with $\lim_{m\to\infty} t_m = \infty$ such that $x'(t_m) = 0$, $m = 1, 2, \cdots$. Thus, (4.1.34) gives

$$\int_{t_0}^{t_m} \frac{w^2(s)}{a(s)\rho(s)} ds \leq \frac{M}{k} - \frac{1}{k} \int_{t_0}^{t_m} \rho(s)q(s)ds, \quad m = 1, 2, \cdots$$

and hence, in view of (4.1.31),

$$\int_{t_0}^{\infty} \frac{w^2(s)}{a(s)\rho(s)} ds < \infty. \tag{4.1.35}$$

Therefore, for some $N > 0$, $\int_{t_0}^{t} w^2(s)/(a(s)\rho(s))ds \leq N$ for every $t \geq t_0$. By the Schwartz inequality, for $t \geq t_0$, we have

$$\left| -\int_{t_0}^{t} w(s)ds \right|^2 = \left| \int_{t_0}^{t} \sqrt{a(s)\rho(s)} \left[\frac{w(s)}{\sqrt{a(s)\rho(s)}} \right] ds \right|^2$$

$$\leq \left[\int_{t_0}^{t} a(s)\rho(s)ds \right] \int_{t_0}^{t} \frac{w^2(s)}{a(s)\rho(s)} ds \leq N \int_{t_0}^{t} a(s)\rho(s)ds.$$

From condition (4.1.30) there exists a $C > 0$ such that $\int_{t_0}^{t} a(s)\rho(s)ds \leq Ct^2$ for $t \geq t_0$, and hence for all $t \geq t_0$, $-\int_{t_0}^{t} w(s)ds \leq \sqrt{NC}t$. Furthermore, (4.1.34) gives $\int_{t_0}^{t} \rho(s)q(s)ds \leq -w(t) + M$ for $t \geq t_0$, and hence for all $t \geq t_0$,

$$\int_{t_0}^{t} \int_{t_0}^{s} \rho(u)q(u)duds \leq -\int_{t_0}^{t} w(s)ds + M(t - t_0) \leq (\sqrt{NC} + M)t - Mt_0,$$

i.e.,

$$\frac{1}{t} \int_{t_0}^{t} \int_{t_0}^{s} \rho(u)q(u)duds \leq \sqrt{NC} + M - \frac{Mt_0}{t}.$$

This contradicts condition (4.1.32).

Case 2. $x'(t) > 0$ on $[T, \infty)$ for some $T \geq t_0$. In this case, from (4.1.34) it follows that for $t \geq T$, $\int_{t_0}^{t} \rho(s)q(s)ds \leq M$, and consequently

$$\frac{1}{t} \int_{T}^{t} \int_{t_0}^{s} \rho(u)q(u)duds \leq M\left(1 - \frac{T}{t}\right) \quad \text{for} \quad t \geq T,$$

which again contradicts condition (4.1.32).

Case 3. $x'(t) < 0$ on $[T, \infty)$ for some $T \geq t_0$. If

$$I = \int_{t_0}^{\infty} \frac{w^2(s)f'(x(s))}{a(s)\rho(s)} ds < \infty,$$

then condition (4.1.28) ensures that (4.1.35) holds. But, then as Case 1 we get a contradiction. Thus, we assume that $I = \infty$. Now, exactly as in the last part of the proof of Theorem 4.1.3, we arrive at the contradiction $\lim_{t\to\infty} x(t) = -\infty$. ∎

In the following result we shall employ the averaging technique and the weighted integral functions to discuss the oscillatory behavior of equation (4.1.3).

Theorem 4.1.5. Let conditions (4.1.5) and (4.1.28) hold, and assume that there exist $\rho(t) \in C^1([t_0, \infty), \mathbb{R}^+)$ and $\phi \in \Phi(t, t_0)$, where $\Phi(t, t_0)$ denotes the class of all positive and locally integrable functions on $[t_0, \infty)$ such that (4.1.29) is satisfied,

$$\int^\infty \frac{\phi(s) \left(\int_{t_0}^s \phi(u)du \right)^\lambda}{\int_{t_0}^s a(u)\rho(u)\phi^2(u))du} ds = \infty \quad \text{for some constant } \lambda, \ 0 < \lambda < 1 \tag{4.1.36}$$

and

$$\lim_{t\to\infty} \frac{1}{\int_{t_0}^t \phi(s)ds} \int_{t_0}^t \phi(s) \int_{t_0}^s \rho(u)q(u)duds = \infty. \tag{4.1.37}$$

Then equation (4.1.3) is oscillatory.

Proof. Let $x(t)$ be a nonoscillatory solution of equation (4.1.3), say, $x(t) > 0$ for $t \geq t_0$. Define $w(t) = \rho(t)a(t)x'(t)/f(x(t))$ for $t \geq t_0$ and proceed as in Theorem 4.1.4 to obtain (4.1.34). Multiplying (4.1.34) by $\phi(t)$ and integrating from t_0 to t, we get

$$\int_{t_0}^t \phi(s)w(s)ds + k \int_{t_0}^t \phi(s) \int_{t_0}^s \frac{w^2(u)}{a(u)\rho(u)} duds$$
$$\leq M \int_{t_0}^t \phi(s)ds - \int_{t_0}^t \phi(s) \int_{t_0}^s \rho(u)q(u)duds. \tag{4.1.38}$$

Using condition (4.1.37), there exists a $t_1 \geq t_0$ such that

$$M - \frac{1}{\int_{t_0}^t \phi(s)ds} \int_{t_0}^t \phi(s) \int_{t_0}^s \rho(u)q(u)duds < 0 \quad \text{for } t \geq t_1.$$

Then for all $t \geq t_1$, it follows that

$$0 \leq F(t) = k \int_{t_0}^t \phi(s) \int_{t_0}^s \frac{w^2(u)}{a(u)\rho(u)} duds < - \int_{t_0}^t \phi(s)w(s)ds$$

and

$$F^2(t) < \left| - \int_{t_0}^t \phi(s)w(s)ds \right|^2 \quad \text{for } t \geq t_1. \tag{4.1.39}$$

By the Schwartz inequality, we have

$$
\begin{aligned}
&\left[\int_{t_0}^{t} \left[\phi(s) \sqrt{a(s)\rho(s)} \right] \left(\frac{w(s)}{\sqrt{a(s)\rho(s)}} \right) ds \right]^2 \\
&\qquad \leq \left(\int_{t_0}^{t} a(s)\rho(s)\phi^2(s)ds \right) \left(\int_{t_0}^{t} \frac{w^2(s)}{a(s)\rho(s)}ds \right) \qquad (4.1.40) \\
&\qquad = \frac{1}{\phi(t)} \left(\int_{t_0}^{t} a(s)\rho(s)\phi^2(s)ds \right) F'(t) \quad \text{for} \quad t \geq t_1.
\end{aligned}
$$

We also have

$$
F(t) \geq k \int_{t_1}^{t} \phi(s) \left(\int_{t_0}^{t_1} \frac{w^2(u)}{a(u)\rho(u)}du \right) ds = c \int_{t_1}^{t} \phi(s)ds, \qquad (4.1.41)
$$

where $c = k \int_{t_0}^{t_1} w^2(s)/a(s)\rho(s)ds$. Now, from (4.1.39), (4.1.40) and (4.1.41), we obtain

$$
c^{\lambda}\phi(t) \frac{\left(\int_{t_1}^{t} \phi(s)ds \right)^{\lambda}}{\int_{t_0}^{t} a(s)\rho(s)\phi^2(s)ds} \leq F^{\lambda-2}(t)F'(t) \quad \text{for all} \quad t \geq t_1, \qquad (4.1.42)
$$

where $0 < \lambda < 1$. Integrating (4.1.42) from t_1 to t, we get

$$
c^{\lambda} \int_{t_1}^{t} \phi(s) \frac{\left(\int_{t_1}^{s} \phi(u)du \right)^{\lambda}}{\int_{t_0}^{s} a(u)\rho(u)\phi^2(u)du}ds \leq \frac{1}{1-\lambda} \frac{1}{F^{1-\lambda}(t_1)} < \infty,
$$

which contradicts condition (4.1.36). ∎

Remark 4.1.3. Condition (4.1.28) fails in the special case $f(x) = |x|^{\gamma} \operatorname{sgn} x$, $x \in \mathbb{R}$ ($\gamma > 1$). However, it is satisfied for the following functions:

(I) $f(x) = mx + |x|^{\gamma} \operatorname{sgn} x$, $x \in \mathbb{R}$ for $\gamma > 1$ and any constant $m > 0$,

(II) $f(x) = x \ln^2(\mu + |x|)$, $x \in \mathbb{R}$ for any constant $\mu > 1$,

(III) $f(x) = xe^{\lambda|x|}$, $x \in \mathbb{R}$ for any constant $\lambda > 0$,

(IV) $f(x) = \sinh x$, $x \in \mathbb{R}$.

In our next result we do not need condition (4.1.5).

Theorem 4.1.6. Let condition (4.1.28) hold, $\rho(t) \in C^1([t_0, \infty), \mathbb{R}^+)$, $\phi(t) \in \Phi(t, t_0)$ be such that $\beta(t) = a(t)\rho'(t) - p(t)\rho(t) \geq 0$ for $t \geq t_0$, and condition (4.1.36) is satisfied. If

$$
\lim_{t \to \infty} \frac{1}{\int_{t_0}^{t} \phi(s)ds} \int_{t_0}^{t} \phi(s) \int_{t_0}^{t} \left[\rho(u)q(u) - \frac{1}{4k} \frac{\beta^2(u)}{a(u)\rho(u)} \right] duds = \infty,
$$

$$(4.1.43)$$

then equation (4.1.3) is oscillatory.

Proof. Let $x(t)$ be a nonoscillatory solution of equation (4.1.3), say, $x(t) > 0$ for $t \geq t_0$. We define $w(t)$ as in Theorem 4.1.3 and obtain for $t \geq t_0$,

$$
\begin{aligned}
w'(t) &= -\rho(t)q(t) + \frac{\beta(t)}{a(t)\rho(t)}w(t) - \frac{1}{a(t)\rho(t)}w^2(t)f'(x(t)) \\
&\leq -\left[\rho(t)q(t) - \frac{1}{4k}\frac{\beta^2(t)}{a(t)\rho(t)}\right] - \frac{k}{a(t)\rho(t)}\left[w(t) - \frac{\beta(t)}{2k}\right]^2.
\end{aligned}
$$

Hence, we have

$$
w(t) + k\int_{t_0}^t \frac{1}{a(s)\rho(s)}\left[w(s) - \frac{\beta(s)}{2k}\right]^2 ds
$$

$$
\leq w(t_0) - \int_{t_0}^t \left[\rho(s)q(s) - \frac{1}{4k}\frac{\beta^2(s)}{a(s)\rho(s)}\right] ds.
$$

The rest of the proof is similar to that of Theorem 4.1.5. ∎

Corollary 4.1.5. In Theorem 4.1.6, condition (4.1.43) can be replaced by (4.1.37) and

$$
\int^\infty \frac{\beta^2(s)}{a(s)\rho(s)} ds < \infty. \tag{4.1.44}
$$

Example 4.1.3. Consider the differential equation

$$
\left(\frac{1}{t^2}x'(t)\right)' + \frac{\sin t}{t^3}x'(t) + \frac{1}{t}\left(\frac{1}{t} - \sin t\right)f(x(t)) = 0, \quad t > t_0 = 1 \tag{4.1.45}
$$

where the function f is as in Remark 4.1.2 which satisfies condition (4.1.28). Let $\rho(t) = \phi(t) = t$. Then, we have

$$
\int_{t_0}^t \frac{\beta^2(s)}{a(s)\rho(s)} ds = \int_1^t \frac{(1 - \sin s)^2}{s^3} ds < \infty,
$$

$$
\int_{t_0}^t \rho(s)q(s) ds = \int_1^t \left(\frac{1}{s} - \sin s\right) ds = \ln t + \cos t - \cos 1 \geq \ln t - 2, \quad t > 1
$$

and

$$
\frac{1}{\int_{t_0}^t \phi(s) ds}\int_{t_0}^t \phi(s)\int_{t_0}^s \rho(u)q(u) du\, ds \geq \frac{1}{t^2 - 1}\int_1^t s[\ln s - 2] ds \to \infty
$$

as $t \to \infty$. Thus, all the conditions of Corollary 4.1.5 are satisfied with $\lambda \in (0, 1)$, where λ is the constant in (4.1.36), and hence (4.1.45) is oscillatory.

Example 4.1.4. Consider the differential equation

$$(t^2 x'(t))' - 2t x'(t) + \frac{2\sqrt{t}}{2}(2 + \cos t - 2t \sin t)f(x(t)) = 0, \quad t \geq t_0 > \pi/2,$$

(4.1.46)

where $f(x)$ is as in Remark 4.1.2. We take $\rho(t) = 1/t^2$ and $\phi(t) = t$, $t \geq t_0 > \pi/2$. Then, $\beta(t) = a(t)\rho'(t) - p(t)\rho(t) = 0$ for $t > \pi/2$. Further,

$$\int_{t_0}^t \rho(s)q(s)ds = \int_{\pi/2}^t \left[-\sqrt{s}\sin s + \frac{1}{2\sqrt{s}}(2 + \cos s) \right] ds$$

$$= \int_{\pi/2}^t d\left[\sqrt{s}(2 + \cos s) \right]$$

$$= \sqrt{t}(2 + \cos t) - 2\sqrt{(\pi/2)} \geq \sqrt{t} - 2\sqrt{(\pi/2)}$$

and

$$\frac{1}{\int_{t_0}^t \phi(s)ds} \int_{t_0}^t \phi(s) \int_{t_0}^s \rho(u)q(u)duds \geq \frac{2}{t^2 - (\pi/2)^2} \int_{\pi/2}^t s\left(\sqrt{s} - 2\sqrt{\frac{\pi}{2}} \right) ds$$

$$\rightarrow \infty \quad \text{as} \quad t \rightarrow \infty.$$

Thus, all the hypotheses of Corollary 4.1.5 are satisfied with $\lambda \in [1/2, 1) \subset (0, 1)$, and hence (4.1.46) is oscillatory.

Now, we shall employ some integral inequalities to study the oscillation of some of the above equations. For this, we shall need the following lemmas.

Lemma 4.1.1. Let $k(t, s, y)$ be a real–valued function of t, s in $[b, c)$ and y in $[b_1, c_1)$ such that for fixed $t = t_0$ and $s = s_0$, $k(t_0, s_0, y)$ is a nondecreasing function of y. Let $G(t)$ be a given function on $[b, c)$; let u and v be functions on $[b, c)$ such that $u(s)$ and $v(s)$ are in $[b_1, c_1)$ for all s in $[b, c)$; let $k(t, s, v(s))$ and $k(t, s, u(s))$ be locally integrable in s for fixed t; and for all $t \in [b, c)$. Let

$$v(t) = G(t) + \int_b^t k(t, s, v(s))ds \quad \text{and} \quad u(t) \geq G(t) + \int_b^t k(t, s, u(s))ds.$$

Then, $v(t) \leq u(t)$ for all $t \in [b, c)$.

Lemma 4.1.2. Let $x(t)$ be a positive (negative) solution of equation (4.1.2) on $[b_1, c)$ for some positive number b_1 satisfying $t_0 \leq b_1 < c \leq \infty$. If there exist $b \in [b_1, c)$ and a positive constant C such that

$$-w(b_1) + \int_{b_1}^t q(s)ds + \int_{b_1}^b \frac{1}{a(s)}w^2(s)f'(x(s))ds \geq C, \qquad (4.1.47)$$

where $w(t) = a(t)x'(t)/f(x(t))$ for all $t \in [b,c)$, then $a(t)x'(t) \leq -Cf(x(b))$ $(a(t)x'(t) \geq -Cf(x(b)))$ for all $t \in [b,c)$.

Proof. Let $x(t)$ be a solution of equation (4.1.2) satisfying the hypotheses of the lemma. Since $w'(t) + w^2(t)f'(x(t))/a(t) = -q(t)$, an integration for $b_1 \leq t \leq c$ in view of (4.1.47) gives

$$-w(t) \geq C + \int_b^t \frac{1}{a(s)} w^2(s)f'(x(s))ds \qquad (4.1.48)$$

for $b \leq t \leq c$. Now, the integral in (4.1.48) is nonnegative, so from the condition $f'(x) \geq 0$ for $x \neq 0$ and the definition of w, we find that $x(t)x'(t) < 0$ on $[b,c)$. If $x(t) > 0$, we let $u(t) = -a(t)x'(t)$, then (4.1.48) becomes

$$u(t) \geq Cf(x(t)) + \int_b^t f(x(t))\frac{f'(x(s))(-x'(s))}{f^2(x(s))} u(s)ds.$$

Define

$$k(t,s,y) = f(x(t))\frac{f'(x(s))(-x'(s))}{f^2(x(s))} y$$

for t and s in $[b,c)$ and $y \in [0,\infty)$ and observe that in this domain $k(t,s,y)$ is nondecreasing in y for fixed t and s. Therefore, Lemma 4.1.1 applies with $G(t) = Cf(x(t))$, and we have $u(t) \geq v(t)$, where $v(t)$ satisfies

$$v(t) = Cf(x(t)) + \int_b^t f(x(t))\frac{f'(x(s))(-x'(s))}{f^2(x(s))} v(s)ds$$

provided $v(s) \in [0,\infty)$ for each $s \in [b,c)$. Multiplying this equality by $1/f(x(t))$ and differentiating, we obtain $v'(t)/f(x(t)) \equiv 0$, so that $v(t) \equiv v(b) = Cf(x(b)) > 0$ for all $t \in [b,c)$. Thus, by Lemma 4.1.1, $a(t)x'(t) \leq -Cf(x(b))$ for $b \leq t < c$. The proof for the case when $x(t)$ is negative follows from a similar argument by taking $u(t) = a(t)x'(t)$ and $G(t) = -Cf(x(t))$. ∎

Lemma 4.1.3. Suppose conditions (4.1.9) and (4.1.10) hold, and

$$\lim_{|x| \to \infty} |f(x)| = \infty. \qquad (4.1.49)$$

If $x(t)$ is a solution of equation (4.1.2), then

$$\int^\infty \frac{1}{a(s)} w^2(s)f'(x(s))ds < \infty, \qquad (4.1.50)$$

$$\lim_{t \to \infty} w(t) = 0, \qquad (4.1.51)$$

and

$$w(t) = \int_t^\infty \frac{1}{a(s)} w^2(s) f'(x(s)) ds + \int_t^\infty q(s) ds \qquad (4.1.52)$$

for all sufficiently large t, where $w(t)$ is as in Lemma 4.1.2.

Proof. Let $x(t)$ be a nonoscillatory solution of equation (4.1.2). If (4.1.50) does not hold, then it follows from (4.1.10) that there exist $t_2 > t_1 \geq t_0$ and a positive constant C_1 such that inequality (4.1.47) holds with $T = t_2$, $T_1 = t_1$ and $C = C_1$ for all $t \geq t_2$. For the case when $x(t) > 0$ for $t \geq t_1$ it follows from Lemma 4.1.2 and its proof that $x'(t) < 0$ and $a(t)x'(t) \leq -C_1 f(x(t_2))$ for $t \geq t_2$, or $x'(t) \leq -C_1 f(x(t_2))/a(t)$ for $t \geq t_2$. Integrating this inequality from t_2 to t, we find

$$x(t) \leq x(t_2) - C_1 f(x(t_2)) \int_{t_2}^t \frac{1}{a(s)} ds \to -\infty \quad \text{as} \quad t \to \infty,$$

which contradicts the fact that $x(t) > 0$ for $t \geq t_1$. The proof for the case when $x(t) < 0$ for $t \geq t_1$ is similar and hence omitted. This completes the proof of (4.1.50).

Now, since

$$w'(t) + \frac{1}{a(t)} w^2(t) f'(x(t)) = -q(t),$$

we have

$$w(T) + \int_t^T \frac{1}{a(s)} w^2(s) f'(x(s)) ds = w(t) + \int_t^T q(s) ds, \qquad (4.1.53)$$

which together with (4.1.10) and (4.1.50) implies that $\lim_{T\to\infty} w(T)$ exists, say, $w(T) \to C_2$ as $T \to \infty$, where C_2 is a constant. Then from (4.1.53), we find

$$w(t) = C_2 + \int_t^\infty q(s) ds + \int_t^\infty \frac{1}{a(s)} w^2(s) f'(x(s)) ds, \quad t \geq t_1. \quad (4.1.54)$$

To prove that (4.1.51) and (4.1.52) hold it suffices to show that $C_2 = 0$. First suppose that $x(t) > 0$ for $t \geq t_1$. If $C_2 < 0$, then (4.1.10) and (4.1.50) imply that there exists $T_1 > t_1$ such that

$$\left| \int_{T_1}^t q(s) ds \right| \leq -\frac{C_2}{4} \quad \text{for} \quad t \geq T_1 \quad \text{and} \quad \int_{T_1}^\infty \frac{1}{a(s)} w^2(s) f'(x(s)) ds < -\frac{C_2}{4}.$$

Now (4.1.54) implies that (4.1.47) holds on $[T_1, \infty)$ with $b = T_1$. But, then by an argument used above, Lemma 4.1.2 and its proof lead to a contradiction to $x(t) > 0$ for $t \geq t_1$. On the other hand, if $C_2 > 0$, then there exists $T_2 > t_1$ such that

$$w(t) = \frac{a(t)x'(t)}{f(x(t))} \geq \frac{C_2}{2} \quad \text{for} \quad t \geq T_2.$$

Therefore, we have

$$\int_{T_2}^{t} \frac{1}{a(s)} w^2(s) f'(x(s)) ds = \int_{T_2}^{t} a(s) \frac{f'(x(s))(x'(s))^2}{f^2(x(s))} ds$$

$$\geq \frac{C_2}{2} \int_{T_2}^{t} \frac{f'(x(s))x'(s)}{f^2(x(s))} ds = \frac{C_2}{2} \ln \frac{f(x(t))}{f(x(T_2))}.$$

The last inequality together with (4.1.49) and (4.1.50) implies that $x(t)$ is bounded from above. Since $x'(t) > 0$ for $t \geq T_2$ and $f'(x) \geq 0$ for $x \neq 0$, it follows that $f(x(t)) \geq f(x(T_2))$ for $t \geq T_2$. Therefore, we obtain

$$x'(t) \geq \frac{C_2}{2} \frac{f(x(t))}{a(t)} \geq \frac{C_2}{2} \frac{f(x(T_2))}{a(t)} \quad \text{for} \quad t \geq T_2.$$

Integrating the above inequality from T_2 to t, we get

$$x(t) \geq x(T_2) + \frac{C_2}{2} f(x(t_2)) \int_{T_2}^{t} \frac{1}{a(s)} ds,$$

which contradicts the boundedness of $x(t)$. This completes the proof that $C_2 = 0$ for the case $x(t) > 0$, $t \geq t_1$. A similar proof holds for the case when $x(t) < 0$, $t \geq t_1$. ∎

We remark that in view of Lemma 4.1.3 an alternative proof of Theorem 4.1.1 without the assumption (4.1.11) can be given as follows: Since $f(x)$ satisfies (4.1.5) it also satisfies (4.1.49), hence the integral equation (4.1.52) holds. Let $x(t)$ be a nonoscillatory solution of (4.1.2) which satisfy the integral equation (4.1.52). Then, we have

$$\frac{x'(t)}{f(x(t))} \geq \frac{1}{a(t)} \int_{t}^{\infty} q(s) ds, \quad t \geq t_1 \geq t_0.$$

Integrating the above inequality from t_1 to t and denoting $G(x) = \int_{x}^{\infty} du/f(u)$, we find

$$\int_{t_1}^{t} \frac{x'(s)}{f(x(s))} ds = \int_{x(t_1)}^{x(t)} \frac{du}{f(u)} = G(x(t_1)) - G(x(t)),$$

and hence

$$G(x(t_1)) - G(x(t)) \geq \int_{t_1}^{t} \frac{1}{a(s)} \int_{s}^{\infty} q(u) du ds \to \infty \quad \text{as} \quad t \to \infty,$$

which contradicts condition (4.1.5).

We are now in the position to prove the following result.

Theorem 4.1.7. Suppose conditions (4.1.5) and (4.1.9) hold,

$$
\left\{
\begin{array}{l}
\text{the essential infimum of } f'(x) \text{ on any} \\
\text{closed set that excludes zero is positive}
\end{array}
\right.
\qquad (4.1.55)
$$

and

$$
\int_0^1 \frac{du}{f(u)} = \infty = \int_{-1}^0 \frac{du}{f(u)}. \qquad (4.1.56)
$$

Suppose that $Q(t) = \int_t^\infty q(s)ds$ exists, and

$$
\liminf_{t\to\infty} \int_{t_0}^t \frac{1}{a(s)} Q(s)ds > -\infty, \qquad (4.1.57)
$$

then either

$$
\int_{t_0}^\infty \frac{1}{a(s)}(Q^+(s))^2 ds = \infty, \qquad (4.1.58)
$$

or

$$
\limsup_{t\to\infty} \int_{t_0}^t \frac{1}{a(s)} \left(Q(s) + \int_s^\infty \frac{1}{a(u)}(Q^+(u))^2 du \right) ds = \infty, \qquad (4.1.59)
$$

where $Q^+(t) = \max\{Q(t), 0\}$ implies that equation (4.1.2) is oscillatory.
If instead of (4.1.55) condition (4.1.28) holds, then

$$
\lim_{t\to\infty} \int_{t_0}^t \frac{1}{a(s)} \left(Q(s) + k \int_s^\infty \frac{1}{a(u)}(Q^+(u))^2 du \right) ds = \infty \qquad (4.1.60)
$$

implies that equation (4.1.2) is oscillatory even without assuming the conditions (4.1.56) and (4.1.57).

Proof. Let $x(t)$ be a nonoscillatory solution of (4.1.2), say, $x(t) > 0$ for $t \geq t_0$. Then by Lemma 4.1.3 equation (4.1.52) holds. Thus, we have

$$
\frac{x'(t)}{f(x(t))} \geq \frac{1}{a(t)} Q(t), \quad t \geq t_0. \qquad (4.1.61)
$$

This together with (4.1.57) and (4.1.56) implies that $x(t)$ is bounded below away from zero. Hence, by (4.1.55), $f'(x(t)) \geq c > 0$ for some constant c. From (4.1.61), we find

$$
\left(\frac{x'(t)}{f(x(t))} \right)^2 \geq \frac{1}{a^2(t)}(Q^+(t))^2 \quad \text{for} \quad t \geq t_1 \quad \text{for some} \quad t_1 \geq t_0.
$$

Thus, it follows that

$$\int_t^\infty a(s) \left(\frac{x'(s)}{f(x(s))} \right)^2 f'(x(s)) ds \geq c \int_t^\infty \frac{1}{a(s)} (Q^+(s))^2 ds.$$

Now, if condition (4.1.58) holds, then (4.1.52) implies that $w(t) \to \infty$ as $t \to \infty$, which is a contradiction.

Next, suppose (4.1.59) holds. It is in view of (4.1.57) equivalent to either

$$\text{(I)} \limsup_{t \to \infty} \int_{t_0}^t \frac{Q(s)}{a(s)} ds = \infty, \text{ or (II)} \int_{t_0}^\infty \frac{1}{a(s)} \int_s^\infty \frac{1}{a(u)} (Q^+(u))^2 du \, ds = \infty.$$

If (I) holds, then integrating (4.1.61) and using (4.1.5) we arrive at a contradiction.

If (II) holds, then since

$$\limsup_{t \to \infty} \int_{t_0}^t \frac{1}{a(s)} \left(Q(s) + \int_s^\infty a(u) \left(\frac{x'(u)}{f(x(u))} \right)^2 f'(x(u)) du \right) ds$$

$$\geq \limsup_{t \to \infty} \int_{t_0}^t \frac{1}{a(s)} \left(Q(s) + c \int_s^\infty \frac{1}{a(u)} (Q^+(u))^2 du \right) ds = \infty$$

$$\text{(4.1.62)}$$

it follows from (4.1.52) that

$$\frac{x'(t)}{f(x(t))} = \frac{1}{a(t)} Q(t) + \frac{1}{a(t)} \int_t^\infty a(s) \left(\frac{x'(s)}{f(x(s))} \right)^2 f'(x(s)) ds. \quad \text{(4.1.63)}$$

Integrating (4.1.63) and using (4.1.62), we obtain a contradiction to condition (4.1.5). ∎

A similar type of argument leads to the following oscillation criterion for (4.1.7) with $\gamma > 1$. It is applicable in the case when condition (4.1.57) is not satisfied. Clearly, for this equation condition (4.1.28) is also not satisfied.

Theorem 4.1.8. Suppose $Q(t) = \int_t^\infty q(s) ds$ is defined and condition (4.1.9) holds. Then either

$$\int_{t_0}^\infty \frac{1}{a(s)\xi_1(s)} (Q^+(s))^2 ds = \infty, \quad t_0 \geq 1 \quad \text{(4.1.64)}$$

where $\xi_1(t) = \int_{t_0}^t ds/a(s)$, or for some constant $\alpha > 0$ (notice that α can be very large),

$$\limsup_{t \to \infty} \int_{t_0}^t \frac{1}{a(s)} \left(Q(s) + \alpha \int_s^\infty \frac{1}{a(u)\xi_1(u)} (Q^+(u))^2 du \right) ds = \infty, \quad t \geq t_0 \geq 1$$

$$\text{(4.1.65)}$$

implies that equation (4.1.7) with $\gamma > 1$ is oscillatory.

Proof. Let $x(t)$ be an eventually positive solution of (4.1.7). Then by Lemma 4.1.3, (4.1.52) holds. This implies that $\lim_{t \to \infty} a(t)x'(t)/x^\gamma(t) = 0$. Hence, for any $\epsilon > 0$ there is a t_0 such that $-\epsilon \leq a(t)x'(t)/x^\gamma(t) \leq \epsilon$ for $t \geq t_0$. Integrating this inequality from t_0 to t, we get

$$-(\gamma - 1)\epsilon\xi_1(t) - x^{1-\gamma}(t_0) \leq -x^{1-\gamma}(t) \leq (\gamma - 1)\epsilon\xi_1(t) - x^{1-\gamma}(t_0).$$

Thus, there exist a large $t_1 \geq t_0$ and positive constants δ_1, δ_2 such that $-\delta_1\xi_1(t) \leq -x^{1-\gamma}(t) \leq \delta_2\xi_1(t)$ for $t \geq t_1$, or $\gamma x^{\gamma-1}(t) \geq \delta/\xi_1(t)$ for $t \geq t_1$, where $\delta = \gamma/\delta_1$. Now, since

$$\int_t^\infty a(s)\gamma\frac{(x'(s))^2}{x^{\gamma+1}(s)}ds \geq \int_t^\infty \frac{1}{a(s)\xi_1(s)}(Q^+(s))^2 ds, \quad t \geq t_1 \quad (4.1.66)$$

we can proceed as in Theorem 4.1.7. ∎

The following corollary is an immediate consequence of Theorem 4.1.7.

Corollary 4.1.6. Suppose conditions (4.1.5), (4.1.55) and (4.1.56) hold. If $Q(t) = \int_t^\infty q(s)ds$ exists, and

$$\liminf_{t \to \infty} \int_{t_0}^t q(s)ds > -\infty \quad (4.1.67)$$

and either

$$\int_{t_0}^\infty (Q^+(s))^2 ds = \infty, \quad (4.1.68)$$

or

$$\limsup_{t \to \infty} \int_{t_0}^t \left[Q(s) + \int_s^\infty (Q^+(u))^2 du\right] ds = \infty, \quad (4.1.69)$$

then equation (4.1.1) is oscillatory.

Remark 4.1.4.

1. Conditions (4.1.68) and (4.1.69) can be combined into the following convenient form

$$\limsup_{t \to \infty} \int_{t_0}^t \left[Q(s) + s(Q^+(s))^2\right] ds = \infty.$$

This follows by changing the order of integration in (4.1.69) and then using (4.1.67).

2. Butler's theorem [4] is in fact Corollary 4.1.6 without (4.1.56). In the proof [4, pp. 79] he attempted to show that either (4.1.56) or (4.1.28) must

hold in general, but the proof only works when f' is continuous in a neighborhood of zero, a condition which is not assumed in the hypotheses. In other words, Corollary 4.1.6 holds if (4.1.56) is replaced by the assumption that f is continuously differentiable.

3. Under further restrictions on the negative part of $Q(t)$, Butler [4] showed that (4.1.69) is also necessary for the oscillation of (4.1.1). More details of this will be given in Section 4.3.

Corollary 4.1.7. Suppose $Q(t) = \int_t^\infty q(s)ds$ is defined. If either

$$\int_{t_0}^\infty \frac{1}{s}(Q^+(s))^2 ds = \infty, \quad t_0 \geq 1 \tag{4.1.70}$$

or for some positive constant δ,

$$\limsup_{t\to\infty} \int_{t_0}^t \left[A(s) + \delta \int_s^\infty \frac{1}{u}(Q^+(u))^2 du \right] ds = \infty, \tag{4.1.71}$$

then equation (4.1.6) with $\gamma > 1$ is oscillatory.

Proof. It follows from Theorem 4.1.9 by taking $a(t) = 1$. ∎

Remark 4.1.5.

1. Either (4.1.70) or (4.1.71) is implied by the condition

$$\limsup_{t\to\infty} \int_{t_0}^t \left[Q(s) + \delta(Q^+(s))^2 \right] ds = \infty.$$

However, unlike Remark 4.1.3(1), the converse may not be true.

2. We can get a stronger condition from (4.1.71) as follows: Let $\delta_1 > 0$, and define

$$Q_1(t) = \max\left\{ Q(t) + \delta_1 \int_t^\infty \frac{1}{s}(Q^+(s))^2 ds, \ 0 \right\} \geq Q^+(t).$$

Then the condition

$$\limsup_{t\to\infty} \int_{t_0}^t \left[Q(s) + \delta_2 \int_t^\infty \frac{1}{u} Q_1^2(u) du \right] ds = \infty,$$

where $\delta_2 > 0$ is arbitrary implies oscillation. This procedure can be iterated to get successively stronger conditions. This can be proved rather easily by substituting the estimate (obtained from (4.1.52) and (4.1.66) with $a(t) = 1$ and $\xi_1(t) = t$)

$$\frac{x'(t)}{x^\gamma(t)} \geq Q(t) + \delta \int_t^\infty \frac{1}{s}(Q^+(s))^2 ds$$

into (4.1.52).

For the formulation of our next theorem, we note that if (4.1.10) holds, then the function $h_0(t) = (1/\sqrt{a(t)}) \int_t^\infty q(s)ds$ is well defined on $[t_0, \infty)$. Further, as long as the improper integrals involved converge, we can define

$$h_1(t) \quad = \quad \int_t^\infty \left([h_0(s)]^+\right)^2 ds$$

$$h_{n+1}(t) \quad = \quad \int_t^\infty \left(\left[h_0(s) + k\frac{h_n(s)}{\sqrt{a(s)}}\right]^+\right)^2 ds, \quad n = 1, 2, \cdots$$

where k is an arbitrary positive constant and $[h_0(t)]^+ = \max\{h_0(t), 0\}$.

Theorem 4.1.9. Suppose (4.1.5), (4.1.9), (4.1.10), (4.1.28) are satisfied, and there exists a positive integer N such that h_n exists for $n = 0, 1, \cdots, N$. If for every constant $k > 0$,

$$h_0(t) + k\frac{h_N(t)}{\sqrt{a(t)}} \geq 0 \tag{4.1.72}$$

for all sufficiently large t, and

$$\int^\infty \frac{1}{\sqrt{a(t)}}\left[h_0(s) + k\frac{h_N(s)}{\sqrt{a(s)}}\right] ds = \infty, \tag{4.1.73}$$

then equation (4.1.2) is oscillatory.

Proof. Suppose equation (4.1.2) has a nonoscillatory solution $x(t)$, and let $|x(t)| > 0$ for $t \geq t_1 \geq t_0$. Since the hypotheses of Lemma 4.1.3 are satisfied from (4.1.52), we find

$$w(t) \geq \sqrt{a(t)}h_0(t) + k\int_t^\infty \frac{1}{a(s)}w^2(s)ds \quad \text{for} \quad t \geq t_1. \tag{4.1.74}$$

Note that (4.1.50) implies that

$$\int_{t_1}^\infty \frac{1}{a(s)}w^2(s)ds < \infty. \tag{4.1.75}$$

From (4.1.74), we have $w(t) \geq \sqrt{a(t)}h_0(t)$, which gives

$$w^2(t) \geq a(t)\left[(h_0(t))^+\right]^2. \tag{4.1.76}$$

Thus, (4.1.74) leads to the inequality $w(t) \geq \sqrt{a(t)}h_0(t) + kh_N(t)$ for $t \geq t_1$. Now, (4.1.72) implies that there exists a $t_2 \geq t_1$ such that $x'(t)/f(x(t)) \geq 0$ for $t \geq t_2$, and hence

$$\frac{x'(t)}{f(x(t))} \geq \frac{1}{\sqrt{a(t)}}\left[h_0(t) + k\frac{h_N(t)}{\sqrt{a(t)}}\right] \quad \text{for} \quad t \geq t_2.$$

Integrating the above inequality, we obtain

$$\int_{x(t_2)}^{x(t)} \frac{du}{f(u)} \geq \int_{t_2}^{t} \frac{1}{\sqrt{a(s)}} \left[h_0(s) + k \frac{h_N(s)}{\sqrt{a(s)}} \right] ds,$$

which contradicts condition (4.1.5).   ∎

Example 4.1.5. Consider the differential equation

$$x''(t) + \frac{1}{4t^{7/4}} [6 + 3\cos t + 4t \sin t](x^3 + x) = 0, \quad t > 0. \tag{4.1.77}$$

Here $a(t) = 1$, so (4.1.9) holds. We also have

$$h_0(t) = \int_t^\infty q(s)ds = \frac{2 + \cos t}{t^{3/4}} \geq \frac{1}{t^{3/4}},$$

which implies that (4.1.72) and (4.1.73) are satisfied with $N = 1$. Observe that $f'(x) = 3x^2 + 1 > 1$ for $x > 0$, so (4.1.28) holds with $k = 1$. Thus, all the hypotheses of Theorem 4.1.9 are satisfied and hence all solutions of (4.1.77) are oscillatory.

4.2. Further Results on Superlinear Oscillations

Here, we shall establish a more general oscillation criteria for superlinear differential equations under consideration. For this, we introduce the following four conditions on the function f each of which includes as a special case the function $f(x) = |x|^\gamma \operatorname{sgn} x, \ \gamma > 1$.

(F$_1$) f is such that

$$\int^\infty \frac{\sqrt{f'(u)}}{f(u)} du < \infty \quad \text{and} \quad \int^{-\infty} \frac{\sqrt{f'(u)}}{f(u)} du < \infty \tag{4.2.1}$$

and

$$\min \left\{ \inf_{x>0} \frac{\left[\int_x^\infty \frac{\sqrt{f'(u)}}{f(u)} du \right]^2}{\int_x^\infty \frac{du}{f(u)}}, \ \inf_{x<0} \frac{\left[\int_x^{-\infty} \frac{\sqrt{f'(u)}}{f(u)} du \right]^2}{\int_x^{-\infty} \frac{du}{f(u)}} \right\} > 0.$$

(F$_2$) Condition (4.2.1) holds, and

$$\min \left\{ \inf_{x>0} \sqrt{f'(x)} \int_x^\infty \frac{\sqrt{f'(u)}}{f(u)} du, \ \inf_{x<0} \sqrt{f'(x)} \int_x^{-\infty} \frac{\sqrt{f'(u)}}{f(u)} du \right\} > 0.$$

(F$_3$) Condition (4.2.1) holds, $f'(x)$ is increasing on $\mathbb{R}^+$ and decreasing on $\mathbb{R}^-$, and

$$\min\left\{\inf_{x>0} f'(x) \int_x^\infty \frac{du}{f(u)}, \quad \inf_{x<0} f'(x) \int_x^{-\infty} \frac{du}{f(u)}\right\} > 0.$$

(F$_4$) Condition (4.2.1) holds, $f'(x) \in C^1(\mathbb{R}^- \cup \mathbb{R}^+, \mathbb{R}^+)$ with $xf''(x) \geq 0$ for $x \neq 0$ and $f(x)f''(x)/(f'(x))^2$ is bounded on $\mathbb{R} - \{0\}$.

We shall show that (F$_4$)$\Rightarrow$(F$_3$)$\Rightarrow$(F$_2$)$\Rightarrow$(F$_1$).

(i)   (F$_4$)$\Rightarrow$(F$_3$). Suppose (F$_4$) holds and consider a positive constant c so that $f(x)f''(x)/(f'(x))^2 \leq c$ for all $x \neq 0$. Then for every $x \neq 0$, we have $(f''(x)/(f'(x))^2)\,\mathrm{sgn}\,x \leq (c/f(x))\,\mathrm{sgn}\,x$, and consequently

$$\int_x^{(\mathrm{sgn}\,x)\infty} \frac{f''(u)}{(f'(u))^2}\,du \leq c \int_x^{(\mathrm{sgn}\,x)\infty} \frac{du}{f(u)}.$$

Hence, it follows that

$$\frac{1}{f'(x)} - \lim_{|u|\to\infty} \frac{1}{f'(u)} \leq c \int_x^{(\mathrm{sgn}\,x)\infty} \frac{du}{f(u)} \quad \text{for} \quad x \neq 0. \qquad (4.2.2)$$

But, since f is increasing on $\mathbb{R} - \{0\}$, $2\int_{x/2}^x du/f(u) \geq x/f(x)$ for all $x \neq 0$, which in view of the fact that f is strongly superlinear gives $\lim_{|x|\to\infty}(f(x)/x) = \infty$. This means that $\lim_{|x|\to\infty} f'(x) = \infty$. Therefore, (4.2.2) leads to

$$f'(x) \int_x^{(\mathrm{sgn}\,x)\infty} \frac{du}{f(u)} \geq \frac{1}{c} \quad \text{for every} \quad x \neq 0$$

and so (F$_3$) is satisfied.

(ii)   (F$_3$)$\Rightarrow$(F$_2$). This implication is obvious.

(iii)   (F$_2$)$\Rightarrow$(F$_1$). We assume that (F$_2$) is fulfilled. Then there exists a positive constant k such that

$$\sqrt{f'(x)} \int_x^{(\mathrm{sgn}\,x)\infty} \frac{\sqrt{f'(u)}}{f(u)}\,du \geq k \quad \text{for all} \quad x \neq 0.$$

Therefore, for $x \neq 0$, we have

$$\left(\int_x^{(\mathrm{sgn}\,x)\infty} \frac{\sqrt{f'(u)}}{f(u)}\,du\right) \frac{\sqrt{f'(x)}}{f(x)}\,\mathrm{sgn}\,x \geq \frac{k}{f(x)}\,\mathrm{sgn}\,x$$

and hence

$$\int_x^{(\mathrm{sgn}\,x)\infty} \left[\int_z^{(\mathrm{sgn}\,x)\infty} \frac{\sqrt{f'(u)}}{f(u)}\,du\right] \frac{\sqrt{f'(z)}}{f(z)}\,dz \geq k \int_x^{(\mathrm{sgn}\,x)\infty} \frac{du}{f(u)}.$$

Thus, it follows that

$$\frac{1}{2}\left[\int_x^{(\text{sgn } x)\infty} \frac{\sqrt{f'(u)}}{f(u)} du\right]^2 \geq k \int_x^{(\text{sgn } x)\infty} \frac{du}{f(u)} \quad \text{for every} \quad x \neq 0,$$

which means that (F_1) holds.

Remark 4.2.1. The case $f(x) = |x|^\gamma \text{ sgn } x, \; x \in \mathbb{R} \; (\gamma > 1)$ is a classical example of the function f which satisfies all the above conditions. Some more examples of such functions f are listed in the following:

Example 4.2.1. Consider the function

$$f(x) = |x|^\gamma [\lambda + \sin(\ln[1 + |x|])] \text{ sgn } x, \quad x \in \mathbb{R}$$

where $\gamma > 1$ and $\lambda > 1 + (1/\gamma)$. This function $f \in C(\mathbb{R}, \mathbb{R})$ has the sign property $xf(x) > 0$ for $x \neq 0$. Also, f is strongly superlinear, since $f(x) \text{ sgn } x \geq (\lambda - 1)|x|^\gamma$ for every $x \neq 0$. Moreover, f is continuously differentiable on $\mathbb{R} - \{0\}$ with

$$f'(x) = \gamma|x|^{\gamma-1}[\lambda + \sin(\ln[1 + |x|])] + \frac{|x|^\gamma}{1 + |x|}\cos(\ln[1 + |x|]), \quad x \neq 0.$$

Hence, it follows that

$$0 < \gamma\left[\lambda - \left(1 + \frac{1}{\gamma}\right)\right]|x|^{\gamma-1} \leq f'(x) \leq \gamma\left[\lambda + \left(1 + \frac{1}{\gamma}\right)\right]|x|^{\gamma-1} \quad \text{for } x \neq 0,$$

and consequently

$$\frac{f(x)}{\sqrt{f'(x)}} \text{ sgn } x \geq \frac{(\lambda - 1)|x|^\gamma}{\left(\gamma\left[\lambda + \left(1 + \frac{1}{\gamma}\right)\right]|x|^{\gamma-1}\right)^{1/2}}$$

$$= \frac{\lambda - 1}{\left(\gamma\left[\lambda + \left(1 + \frac{1}{\gamma}\right)\right]\right)^{1/2}}|x|^{(\gamma+1)/2} \quad \text{for all} \quad x \neq 0.$$

So, $f(x)/\sqrt{f'(x)}$ is strongly superlinear (i.e. (4.2.1) holds), since $(\gamma + 1)/2 > 1$. Furthermore, for any $x \neq 0$, we find

$$\int_x^{(\text{sgn } x)\infty} \frac{\sqrt{f'(u)}}{f(u)} du = \int_{|x|}^\infty \frac{\sqrt{f'(u)}}{f(u)} du$$

$$\geq \frac{\left(\gamma\left[\lambda - \left(1 + \frac{1}{\gamma}\right)\right]\right)^{1/2}}{\lambda + 1} \int_{|x|}^\infty u^{-(\gamma+1)/2} du$$

$$= \frac{2\left(\gamma\left[\lambda - \left(1 + \frac{1}{\gamma}\right)\right]\right)^{1/2}}{(\lambda + 1)(\gamma - 1)}|x|^{(1-\gamma)/2}.$$

Therefore, for all $x \neq 0$, we have

$$\sqrt{f'(x)} \int_x^{(\text{sgn } x)\infty} \frac{\sqrt{f'(u)}}{f(u)} du \geq \frac{2\gamma \left[\lambda - \left(1 + \frac{1}{\gamma}\right)\right]}{(\lambda + 1)(\gamma - 1)} > 0,$$

which means that (F$_2$) holds.

Example 4.2.2. Consider the continuous function $f(x) = x + |x|^\gamma \text{ sgn } x$, $x \in \mathbb{R}$, where $\gamma > 1$. Obviously, $xf(x) > 0$ for $x \neq 0$. Also, we observe that $f(x) \text{ sgn } x \geq |x|^\gamma$ for $x \neq 0$ and consequently f is strongly superlinear. Furthermore, $f \in C^2(\mathbb{R} - \{0\}, \mathbb{R})$ with $f'(x) = \gamma |x|^{\gamma - 1} + 1$ and $f''(x) = \gamma(\gamma - 1)|x|^{\gamma - 2} \text{ sgn } x$ for $x \neq 0$. Thus, $f'(x) > 0$ and $xf''(x) > 0$ for $x \neq 0$. Moreover, $f(x)/\sqrt{f'(x)}$ is strongly superlinear, since for $|x| \geq 1$, we have

$$\frac{f(x)}{\sqrt{f'(x)}} \text{ sgn } x = \frac{|x|^\gamma + x}{(\gamma |x|^{\gamma - 1} + 1)^{1/2}}$$

$$\geq \frac{|x|^\gamma}{(\gamma |x|^{\gamma - 1} + |x|^{\gamma - 1})^{1/2}} = \frac{1}{(\gamma + 1)^{1/2}} |x|^{(\gamma + 1)/2},$$

where $(\gamma + 1)/2 > 1$. Now, for every $x \neq 0$, we find

$$\frac{f(x)f''(x)}{(f'(x))^2} = \gamma(\gamma - 1)\frac{|x|^{\gamma - 1} \left(|x|^{\gamma - 1} - 1\right)}{(\gamma |x|^{\gamma - 1} + 1)^2}$$

$$< \gamma(\gamma - 1)\frac{|x|^{\gamma - 1}}{|x|^{\gamma - 1} + 1} < \gamma(\gamma - 1).$$

Clearly, $f(x)f''(x)/(f'(x))^2$ is bounded on $\mathbb{R} - \{0\}$, which means that (F$_4$) holds.

Example 4.2.3. Consider the function $f(x) = (|x|^{2\gamma} \text{ sgn } x)/(1 + |x|^\gamma)$, $x \in \mathbb{R}$ where $\gamma > 1$. Clearly, $f \in C(\mathbb{R}, \mathbb{R})$ and $xf(x) > 0$ for every $x \neq 0$. This function is continuously differentiable on $\mathbb{R} - \{0\}$ and

$$f'(x) = \frac{\gamma |x|^{2\gamma - 1} \left(2 + |x|^\gamma\right)}{\left(1 + |x|^\gamma\right)^2} \quad \text{for} \quad x \neq 0.$$

Moreover, f is strongly superlinear and for all $x \neq 0$,

$$\int_x^{(\text{sgn } x)\infty} \frac{du}{f(u)} = \int_{|x|}^\infty \frac{du}{f(u)} = \int_{|x|}^\infty \frac{du}{u^{2\gamma}} + \int_{|x|}^\infty \frac{du}{u^\gamma}$$

$$= \frac{|x|^{1-2\gamma}}{2\gamma - 1} + \frac{|x|^{1-\gamma}}{\gamma - 1}.$$

Furthermore, we find

$$
\begin{aligned}
\frac{\sqrt{f'(x)}}{f(x)}\,\mathrm{sgn}\,x \;&=\; \frac{\gamma^{1/2}|x|^{(2\gamma-1)/2}\,(2+|x|^{\gamma})^{1/2}}{|x|^{2\gamma}} \\
&\leq\; \frac{\gamma^{1/2}\,(|x|^{\gamma}+|x|^{\gamma})}{|x|^{(2\gamma+1)/2}} \;\leq\; \frac{(2\gamma)^{1/2}}{|x|^{(\gamma+1)/2}}
\end{aligned}
$$

for all x with $|x| \geq 2^{1/\gamma}$. This in view of $(\gamma+1)/2 > 1$, ensures the validity of (4.2.1). Now, we have

$$
\frac{\left[\int_x^{(\mathrm{sgn}\,x)\infty} \frac{\sqrt{f'(u)}}{f(u)}\,du\right]^2}{\int_x^{(\mathrm{sgn}\,x)\infty} \frac{du}{f(u)}} \;=\; \frac{\left[\gamma^{1/2}\int_{|x|}^{\infty} \frac{(2+u^{\gamma})^{1/2}}{u^{(2\gamma+1)/2}}\,du\right]^2}{\frac{|x|^{1-2\gamma}}{2\gamma-1} + \frac{|x|^{1-\gamma}}{\gamma-1}}
$$

for $x \neq 0$. It is easy to verify that this function satisfies condition (F_1).

We are now in the position to present a more general oscillation criteria for (4.1.3).

Theorem 4.2.1. Suppose condition (F_1) holds and let $\rho(t) \in C^1([t_0, \infty), \mathbb{R}^+)$ be such that

$$
\xi(t) \;=\; \int_{t_0}^t \frac{1}{a(s)\rho(s)}\,ds \quad\text{and}\quad \lim_{t\to\infty} \xi(t) \;=\; \infty. \qquad (4.2.3)
$$

Moreover, let

$$
\beta(t) \;=\; a(t)\rho'(t) - p(t)\rho(t) \;\geq\; 0 \quad\text{and}\quad \beta'(t) \leq 0 \quad\text{for}\quad t \geq t_0. \quad (4.2.4)
$$

If

$$
\liminf_{t\to\infty} \int_{t_0}^t \rho(s)q(s)ds \;>\; -\infty \qquad (4.2.5)
$$

and

$$
\limsup_{t\to\infty} \frac{1}{\xi(t)} \int_{t_0}^t \frac{1}{a(s)\rho(s)} \int_{t_0}^s \rho(u)q(u)du\,ds \;=\; \infty, \qquad (4.2.6)
$$

then equation (4.1.3) is oscillatory.

Proof. Suppose (4.1.3) possesses a nonoscillatory solution $x(t)$ on $[T, \infty)$, $T \geq t_0$. Without loss of generality, we assume that $x(t) \neq 0$ for every $t \geq T$. We observe that the substitution $y = -x$ transforms (4.1.3) into the equation

$$
(a(t)y'(t))' + p(t)y'(t) + q(t)f^*(y(t)) \;=\; 0,
$$

where $f^*(y) = -f(-y)$, $y \in \mathbb{R}$. Since the function $f^*(y)$ is subject to the same conditions as on $f(x)$, we can restrict our discussion to the case where the solution $x(t)$ is positive on $[T, \infty)$.

Let $w(t)$ be defined by $w(t) = \rho(t)a(t)x'(t)/f(x(t))$, $t \geq T$. Then for $t \geq T$, we have

$$w'(t) = \rho(t)\left[\frac{(a(t)x'(t))'}{f(x(t))} - a(t)\left(\frac{x'(t)}{f(x(t))}\right)^2 f'(x(t))\right] + a(t)\rho'(t)\frac{x'(t)}{f(x(t))}$$

and consequently for every $t \geq T$,

$$w'(t) = -\rho(t)q(t) + \beta(t)\frac{x'(t)}{f(x(t))} - \frac{1}{a(t)\rho(t)}w^2(t)f'(x(t)). \qquad (4.2.7)$$

Thus, for any $t \geq T$, we find

$$\int_T^t \rho(s)q(s)ds = -w(t) + w(T) + \int_T^t \frac{\beta(s)x'(s)}{f(x(s))}ds - \int_T^t \frac{w^2(s)f'(x(s))}{a(s)\rho(s)}ds.$$

But by the Bonnet theorem, for a fixed $t \geq T$ and for some $\eta \in [T,t]$, we have

$$\int_T^t \beta(s)\frac{x'(s)}{f(x(s))}ds = \beta(T)\int_T^\eta \frac{x'(s)}{f(x(s))}ds = \beta(T)\int_{x(T)}^{x(\eta)} \frac{du}{f(u)}$$

and hence, since $\beta(T) \geq 0$ and

$$\int_{x(T)}^{x(\eta)} \frac{du}{f(u)} < \begin{cases} 0 & \text{if} \quad x(\eta) < x(T) \\ \int_{x(T)}^{\infty} \frac{du}{f(u)} & \text{if} \quad x(\eta) \geq x(T), \end{cases}$$

we obtain

$$\int_T^t \beta(s)\frac{x'(s)}{f(x(s))}ds \leq K \quad \text{for all} \quad t \geq T, \qquad (4.2.8)$$

where $K = \beta(T)\int_{x(T)}^{\infty} du/f(u)$. Therefore, in view of (4.2.8), we conclude for $t \geq T$ that

$$\int_T^t \rho(s)q(s)ds \leq -w(t) + w(T) + K - \int_T^t \frac{w^2(s)f'(x(s))}{a(s)\rho(s)}ds. \qquad (4.2.9)$$

Now, with respect to the integral

$$I(T,t) = \int_T^t \frac{1}{a(s)\rho(s)}w^2(s)f'(x(s))ds$$

there are two cases to consider:

Case 1. $I(T,\infty)$ is finite. In this case there exists a positive constant N such that

$$I(T,t) \leq N \quad \text{for} \quad t \geq T. \qquad (4.2.10)$$

Furthermore, by using the Schwartz inequality, for $t \geq T$ we obtain

$$\left| \int_T^t \frac{x'(s)\sqrt{f'(x(s))}}{f(x(s))} ds \right|^2 = \left| \int_T^t \frac{1}{\sqrt{a(s)\rho(s)}} \left[\frac{w(s)\sqrt{f'(x(s))}}{\sqrt{a(s)\rho(s)}} \right] ds \right|^2$$

$$\leq \left[\int_T^t \frac{1}{a(s)\rho(s)} ds \right] I(T,t).$$

Thus, in view of (4.2.10), we find

$$\left| \int_T^t \frac{x'(s)}{f(x(s))} \sqrt{f'(x(s))} ds \right|^2 \leq N \int_T^t \frac{1}{a(s)\rho(s)} ds = N\xi(t) \quad \text{for all} \quad t \geq T.$$
$$(4.2.11)$$

Next, using (F$_1$), we get

$$\int_{x(t)}^\infty \frac{du}{f(u)} \leq M \left[\int_{x(t)}^\infty \frac{\sqrt{f'(u)}}{f(u)} du \right]^2, \quad t \geq T \qquad (4.2.12)$$

where M is a positive constant. Now, setting

$$K_1 = \int_{x(T)}^\infty \frac{du}{f(u)} > 0 \quad \text{and} \quad K_2 = \int_{x(T)}^\infty \frac{\sqrt{f'(u)}}{f(u)} du > 0$$

and using (4.2.12) for every $t \geq T$, we obtain

$$\left| \int_T^t \frac{w(s)}{a(s)\rho(s)} ds \right| = \left| \int_T^t \frac{x'(s)}{f(x(s))} ds \right| = \left| \int_{x(T)}^{x(t)} \frac{du}{f(u)} \right|$$

$$= \left| K_1 - \int_{x(t)}^\infty \frac{du}{f(u)} \right| \leq K_1 + \int_{x(t)}^\infty \frac{du}{f(u)}$$

$$\leq K_1 + M \left[\int_{x(t)}^\infty \frac{\sqrt{f'(u)}}{f(u)} du \right]^2 = K_1 + M \left[K_2 - \int_{x(T)}^{x(t)} \frac{\sqrt{f'(u)}}{f(u)} du \right]^2$$

$$\leq K_1 + M \left[K_2 + \left| \int_{x(T)}^{x(t)} \frac{\sqrt{f'(u)}}{f(u)} du \right| \right]^2$$

$$= K_1 + M \left[K_2 + \left| \int_T^t \frac{x'(s)}{f(x(s))} f'(x(s)) ds \right| \right]^2.$$

Thus, by (4.2.11) for every $t \geq T$, we have

$$\left| \int_T^t \frac{1}{a(s)\rho(s)} w(s) ds \right| \leq K_1 + M \left[K_2 + (N\xi(t))^{1/2} \right]^2$$

$$= (K_1 + MK_2^2) + 2MK_2\sqrt{N}\xi^{1/2}(t) + MN\xi(t).$$

By condition (4.2.3), we can choose a $T_0 > t_0$ so that

$$\int_{t_0}^{T_0} \frac{1}{a(s)\rho(s)} ds \geq 1. \tag{4.2.13}$$

and hence, in view of (4.2.13), we get

$$\left| \int_T^t \frac{1}{a(s)\rho(s)} w(s) ds \right| \leq C\xi(t) \quad \text{for every} \quad t \geq T^*, \tag{4.2.14}$$

where $T^* = \max\{T_0, T\}$ and $C = K_1 + M K_2^2 + 2 M K_2 \sqrt{N} + MN$. Now, from (4.2.9) it follows that

$$\int_T^t \rho(s)q(s) ds \leq -w(t) + w(T) + K \quad \text{for} \quad t \geq T$$

and thus, by taking into account (4.2.14), we obtain for $t \geq T^*$,

$$\int_T^t \frac{1}{a(s)\rho(s)} \int_T^s \rho(u)q(u) du\, ds$$

$$\leq -\int_T^t \frac{1}{a(s)\rho(s)} w(s) ds + [w(T) + K] \int_T^t \frac{1}{a(s)\rho(s)} ds$$

$$\leq \left| \int_T^t \frac{1}{a(s)\rho(s)} w(s) ds \right| + |w(T) + K| \int_T^t \frac{1}{a(s)\rho(s)} ds \leq C^*\xi(t),$$

where $C^* = C + |w(T) + K|$. Therefore, for every $t \geq T^*$, we have

$$\int_{t_0}^t \frac{1}{a(s)\rho(s)} \int_{t_0}^s \rho(u)q(u) du\, ds$$

$$= \int_{t_0}^T \frac{1}{a(s)\rho(s)} \int_{t_0}^s \rho(s)q(u) du\, ds + \left[\int_{t_0}^T \rho(u)q(u) du \right] \int_T^t \frac{1}{a(s)\rho(s)} ds$$

$$+ \int_T^t \frac{1}{a(s)\rho(s)} \int_T^s \rho(u)q(u) du\, ds$$

$$\leq \left| \int_{t_0}^T \frac{1}{a(s)\rho(s)} \int_{t_0}^s \rho(u)q(u) du\, ds \right| + \left| \int_{t_0}^T \rho(u)q(u) du \right| \int_T^t \frac{1}{a(s)\rho(s)} ds$$

$$+ C^*\xi(t)$$

and consequently, in view of (4.2.13) it follows that

$$\int_{t_0}^t \frac{1}{a(s)\rho(s)} \int_{t_0}^s \rho(u)q(u) du\, ds \leq C_1\xi(t) \quad \text{for all} \quad t \geq T^*,$$

where

$$C_1 = C^* + \left| \int_{t_0}^{T} \frac{1}{a(s)\rho(s)} \int_{t_0}^{s} \rho(u)q(u)duds \right| + \left| \int_{t_0}^{T} \rho(u)q(u)du \right|.$$

This contradicts condition (4.2.6).

Case 2. $I(T,\infty) = \infty$. In view of condition (4.2.5), from (4.2.9) it follows that for some constant m,

$$-w(t) \geq m + I(T,t) \quad \text{for every} \quad t \geq T. \tag{4.2.15}$$

Choose a $T_1 \geq T$ such that $M = m + I(T,T_1) > 0$. Then, (4.2.15) ensures that $w(t)$ is negative on $[T_1,\infty)$. Now, (4.2.15) gives

$$\frac{\frac{1}{a(t)\rho(t)}w^2(t)f'(x(t))}{m + I(T,t)} \geq -\frac{x'(t)f'(x(t))}{f(x(t))} \quad \text{for} \quad t \geq T_1$$

and consequently for all $t \geq T_1$,

$$\ln \frac{m + I(T,t)}{M} \geq \ln \frac{f(x(T_1))}{f(x(t))}.$$

Hence, $m + I(T,t) \geq Mf(x(T_1))/f(x(t))$ for $t \geq T_1$. Thus, (4.2.15) yields $x'(t) \leq -M_1/(a(t)\rho(t))$ for every $t \geq T_1$, where $M_1 = Mf(x(T_1)) > 0$. Thus, we have

$$x(t) \leq x(T_1) - M_1 \int_{T_1}^{t} \frac{1}{a(s)\rho(s)} ds \quad \text{for all} \quad t \geq T_1,$$

which because of condition (4.2.3) leads to the contradiction that $\lim_{t \to \infty} x(t) = -\infty$. ∎

The following corollaries, which are well known basic results in the literature, are immediate from Theorem 4.2.1.

Corollary 4.2.1. If

$$\liminf_{t \to \infty} \int_{t_0}^{t} q(s)ds > -\infty \tag{4.2.16}$$

and

$$\limsup_{t \to \infty} \frac{1}{t} \int_{t_0}^{t} \int_{t_0}^{s} q(\tau)d\tau ds = \infty, \tag{4.2.17}$$

then equation (4.1.6) is oscillatory.

The following result extends Corollary 4.2.1 to equation (4.1.1).

Corollary 4.2.2. If conditions (F_1), (4.2.16) and (4.2.17) are satisfied, then equation (4.1.1) is oscillatory.

A generalized criterion for the oscillation of (4.1.1) is the following result.

Corollary 4.2.3. Suppose condition (F_1) holds and let $\rho(t) \in C^1([t_0, \infty), \mathbb{R}^+)$ be such that $\rho'(t)$ is nonnegative and decreasing on $[t_0, \infty)$. If condition (4.2.5) holds, and

$$\limsup_{t \to \infty} \left[\int_{t_0}^t \frac{1}{\rho(s)} ds \right]^{-1} \int_{t_0}^t \frac{1}{\rho(s)} \int_{t_0}^s \rho(u)q(u)\,du\,ds \; = \; \infty, \qquad (4.2.18)$$

then equation (4.1.1) is oscillatory.

Remark 4.2.2. Since the function $\rho(t)$ in Corollary 4.2.3 is positive on $[t_0, \infty)$ and $\rho'(t)$ is nonnegative and bounded above on $[t_0, \infty)$, it follows that $\rho(t) \leq \mu t$ for all large t, where $\mu > 0$ is a constant. This ensures that

$$\int_{t_0}^{\infty} \frac{1}{\rho(s)} ds \; = \; \infty. \qquad (4.2.19)$$

Next, let Γ be the class of functions defined as follows: $\rho(t) \in \Gamma$ if and only if $\rho(t) \in C^1([t_0, \infty), \mathbb{R}^+)$ such that $\rho'(t)$ is nonnegative and decreasing on $[t_0, \infty)$. This class is large enough, in fact $\rho(t) \in \Gamma$ in each of the following cases:

(i) $\rho(t) = t^\lambda$, $t \geq t_0$ for $\lambda \in [0, 1]$.

(ii) $\rho(t) = \ln^\lambda t$, $t \geq t_0$ for $\lambda > 0$, where $t_0 > \max\{1, e^{\lambda-1}\}$.

(iii) $\rho(t) = t^\lambda \ln t$, $t \geq t_0$ for $\lambda \in (0, 1)$, where

$$t_0 \; > \; \max\left\{1, \exp\left(\frac{1}{1-\lambda} - \frac{1}{\lambda}\right)\right\}.$$

(iv) $\rho(t) = t/\ln t$, $t \geq t_0$ where $t_0 \geq e^2$,

(v) $\rho(t) = t^{1/2}[5 + \sin(\ln t)]$, $t \geq t_0$

and the simplest case

(vi) $\rho(t) = 1$ for $t \geq t_0$.

Now, if $\rho(t) = t^\lambda$, $t \geq t_0$ for $\lambda \in [0, 1]$ in Corollary 4.2.3, then we obtain the following interesting result.

Corollary 4.2.4. Suppose condition (F_1) holds. Equation (4.1.1) is oscillatory if for some $\lambda \in [0, 1]$,

$$\liminf_{t \to \infty} \int_{t_0}^t s^\lambda q(s)\,ds \; > \; -\infty$$

and

$$
\begin{cases}
\limsup_{t\to\infty} t^{\lambda-1} \int_{t_0}^{t} s^{-\lambda} \int_{t_0}^{s} u^{\lambda} q(u)\,du\,ds \;=\; \infty \quad \text{for } 0 \le \lambda < 1 \\[2mm]
\limsup_{t\to\infty} \dfrac{1}{\ln t} \int_{t_0}^{t} \dfrac{1}{s} \int_{t_0}^{s} u q(u)\,du\,ds \;=\; \infty \quad \text{for } \lambda = 1.
\end{cases}
$$

Example 4.2.4. Consider the differential equation

$$
(\sqrt{t}x'(t))' + \frac{1}{2\sqrt{t}}x'(t) + \left[-\sin t + \frac{1}{2t}(2+\cos t)\right] f(x(t)) \;=\; 0, \tag{4.2.20}
$$
$$
t \ge t_0 = \pi/2
$$

where the function f is as in Examples 4.2.1 – 4.2.3. Let $\rho(t) = \sqrt{t}$ for $t \ge t_0$. Then, we have $\beta(t) = a(t)\rho'(t) - p(t)\rho(t) = 0$ and

$$
a(t)\rho(t) \;=\; t, \quad \xi(t) \;=\; \int_{t_0}^{t} \frac{1}{a(s)\rho(s)}\,ds \;=\; \ln\frac{t}{t_0} \;\to\; \infty \quad \text{as } t \to \infty,
$$

$$
\int_{t_0}^{t} \rho(s)q(s)\,ds \;=\; \int_{\pi/2}^{t} \left[-\sqrt{s}\sin s + \frac{1}{2\sqrt{s}}(2+\cos s)\right]ds
$$
$$
\;=\; \int_{\pi/2}^{t} d[\sqrt{s}(2+\cos s)] \;=\; \sqrt{t}(2+\cos t) - 2\sqrt{\pi/2} \;\ge\; \sqrt{t} - 2\sqrt{\pi/2}
$$

and

$$
\frac{1}{\ln(2t/\pi)} \int_{t_0}^{t} \frac{1}{s} \int_{t_0}^{s} \sqrt{u}q(u)\,du\,ds \;\ge\; \frac{1}{\ln(2t/\pi)} \int_{\pi/2}^{t} \left[\frac{1}{\sqrt{s}} - \frac{2}{s}\sqrt{\frac{\pi}{2}}\right]ds
$$
$$
\;=\; \frac{2\sqrt{t}}{\ln(2t/\pi)} - \frac{2\sqrt{\pi/2}}{\ln(2t/\pi)} - 2\sqrt{\frac{\pi}{2}}
$$

and consequently,

$$
\liminf_{t\to\infty} \int_{t_0}^{t} \sqrt{s}q(s)\,ds \;>\; -\infty
$$

and

$$
\limsup_{t\to\infty} \frac{1}{\ln(2t/\pi)} \int_{t_0}^{t} \frac{1}{s} \int_{t_0}^{s} \sqrt{u}q(u)\,du\,ds \;=\; \infty.
$$

Now, Theorem 4.2.1 ensures that (4.2.20) is oscillatory.

Example 4.2.5. Consider the differential equation

$$
x''(t) + \left[-\frac{1}{\sqrt{t}}\sin t + \frac{1}{2t\sqrt{t}}(2+\cos t)\right] |x(t)|^{\gamma} \operatorname{sgn} x(t) \;=\; 0, \tag{4.2.21}
$$
$$
t \ge t_0 = \pi/2
$$

where $\gamma > 1$ is a constant. For every $t \geq t_0$, we find

$$\int_{t_0}^{t} sq(s)ds = \int_{\pi/2}^{t}\left[-\sqrt{s}\sin s + \frac{1}{2\sqrt{s}}(2 + \cos s)\right]ds \geq \sqrt{t} - 2\sqrt{\frac{\pi}{2}}$$

and

$$\frac{1}{\ln t}\int_{t_0}^{t}\frac{1}{s}\int_{t_0}^{s}uq(u)duds \geq \frac{1}{\ln t}\int_{\pi/2}^{t}\left[\frac{1}{\sqrt{s}} - \frac{2}{s}\sqrt{\frac{\pi}{2}}\right]ds$$

$$= 2\frac{\sqrt{t}}{\ln t} - 2\sqrt{\frac{\pi}{2}} + \frac{2}{\ln t}\sqrt{\frac{\pi}{2}}\left[\ln\frac{\pi}{2} - 1\right].$$

Thus, all conditions of Corollary 4.3.4 are satisfied and hence (4.2.21) is oscillatory. On the other hand, Corollary 4.2.1 is not applicable for (4.2.21). Indeed, for $t \geq t_0$, we have

$$\int_{t_0}^{t} q(s)ds = \int_{\pi/2}^{t}\left[-\frac{1}{\sqrt{s}}\sin s + \frac{1}{2s\sqrt{s}}(2 + \cos s)\right]ds$$

$$\leq \int_{\pi/2}^{t}\left(-\frac{1}{\sqrt{s}}\sin s + \frac{3}{2}\frac{1}{s\sqrt{s}}\right)ds = \frac{1}{\sqrt{t}}\cos t + \int_{\pi/2}^{t}\left(\frac{1}{2s\sqrt{s}}\cos s + \frac{3}{2s\sqrt{s}}\right)ds$$

$$\leq \frac{1}{\sqrt{t}}\cos t + 2\int_{\pi/2}^{t}\frac{1}{s\sqrt{s}}ds = \frac{1}{\sqrt{t}}\cos t - 4\left[\frac{1}{\sqrt{t}} - \sqrt{\frac{2}{\pi}}\right] \leq -\frac{3}{\sqrt{t}} + 4\sqrt{\frac{2}{\pi}}.$$

Thus, for every $t \geq t_0$,

$$\frac{1}{t}\int_{t_0}^{t}\int_{t_0}^{s}q(u)duds \leq \frac{1}{t}\int_{\pi/2}^{t}\left[-\frac{3}{\sqrt{s}} + 4\sqrt{\frac{2}{\pi}}\right]ds = -\frac{6}{\sqrt{t}} + \frac{2}{t}\sqrt{\frac{\pi}{2}} + 4\sqrt{\frac{2}{\pi}}$$

and hence

$$\limsup_{t\to\infty}\frac{1}{t}\int_{t_0}^{t}\int_{t_0}^{s}q(u)duds < \infty,$$

i.e., Corollary 4.2.1 fails to apply to equation (4.2.21).

Theorem 4.2.2. Suppose (F_2) holds and let the functions $\rho(t)$, $\beta(t)$ and $\xi(t)$ be as in Theorem 4.2.1, and conditions (4.2.3) and (4.2.4) are satisfied. If (4.2.5) holds and for every positive constant k,

$$\limsup_{t\to\infty}\frac{1}{t^n}\int_{t_0}^{t}(t - s)^{n-2}\left[(t - s)^2\rho(s)q(s) - \frac{n^2}{4}\frac{k}{v(s)}\right]ds = \infty \quad (4.2.22)$$

for some integer $n > 1$, where $v(t) = 1/(a(t)\rho(t)\xi(t))$, $t \geq t_0$ then equation (4.1.3) is oscillatory.

Proof. Let $x(t)$ be a nonoscillatory solution of equation (4.1.3). Without loss of generality, it can be assumed that $x(t) \neq 0$ for $t \geq T \geq t_0$.

Furthermore, it is enough to consider the case when $x(t) > 0$ for $t \geq T$. Define $w(t) = \rho(t)a(t)x'(t)/f(x(t))$, $t \geq T$. As in Theorem 4.2.1, we obtain (4.2.7) and conclude that (4.2.8) and (4.2.9) are satisfied with $K = \beta(T) \int_{x(T)}^{\infty} du/f(u)$. We consider the following two cases:

Case 1. $I(T, \infty) < \infty$. Then, (4.2.10) holds where N is a positive constant. Furthermore, as in Theorem 4.2.1 inequality (4.2.11) follows. Condition (F_2) ensures that

$$\sqrt{f'(x(t))} \int_{x(t)}^{\infty} \frac{\sqrt{f'(u)}}{f(u)} du \geq A \quad \text{for} \quad t \geq T, \tag{4.2.23}$$

where A is a positive constant. Now, let $B = \int_{x(T)}^{\infty} \sqrt{f'(u)}/f(u)du > 0$. Then by (4.2.23), we obtain for $t \geq T$,

$$
\begin{aligned}
f'(x(t)) &\geq A^2 \left[\int_{x(t)}^{\infty} \frac{\sqrt{f'(u)}}{f(u)} du \right]^{-2} = A^2 \left[B - \int_{x(T)}^{x(t)} \frac{\sqrt{f'(u)}}{f(u)} du \right]^{-2} \\
&= A^2 \left[B - \int_{T}^{t} \frac{x'(s)}{f(x(s))} \sqrt{f'(x(s))} ds \right]^{-2} \\
&\geq A^2 \left[B + \left| \int_{T}^{t} \frac{x'(s)}{f(x(s))} \sqrt{f'(x(s))} ds \right| \right]^{-2}.
\end{aligned}
$$

Using (4.2.11) in the above inequality, we get $f'(x(t)) \geq A^2[B + \sqrt{N\xi(t)}]^{-2}$ for $t \geq T$. As in Theorem 4.2.1 there exists a $T_0 \geq t_0$ such that (4.2.13) holds and consequently, we find

$$f'(x(t)) \geq c/\xi(t) \quad \text{for all} \quad t \geq T^* = \max\{T_0, T\}, \tag{4.2.24}$$

where $c = A^2[B + \sqrt{N}]^{-2} > 0$. In view of (4.2.24) equation (4.2.7) gives

$$\rho(t)q(t) \leq -w'(t) + \beta(t)\frac{x'(t)}{f(x(t))} - \frac{c}{a(t)\rho(t)\xi(t)} w^2(t) \quad \text{for all} \quad t \geq T^*,$$

or

$$\rho(t)q(t) \leq -w'(t) + \beta(t)\frac{x'(t)}{f(x(t))} - cv(t)w^2(t) \quad \text{for all} \quad t \geq T^*. \tag{4.2.25}$$

Thus, for $t \geq T^*$,

$$
\int_{T^*}^{t} (t-s)^n \rho(s)q(s)ds \leq -\int_{T^*}^{t} (t-s)^n w'(s)ds + \int_{T^*}^{t} (t-s)^n \beta(s)\frac{x'(s)}{f(x(s))} ds
$$
$$
- \int_{T^*}^{t} c(t-s)^n v(s)w^2(s)ds.
$$

$$\tag{4.2.26}$$

But because of (4.2.8), we have

$$\int_{T^*}^t (t-s)^n \beta(s) \frac{x'(s)}{f(x(s))} ds = n \int_{T^*}^t (t-s)^{n-1} \left[\int_{T^*}^s \beta(u) \frac{x'(u)}{f(x(u))} du \right] ds$$

$$\leq Kn \int_{T^*}^t (t-s)^{n-1} ds = K(t-T^*)^n.$$

Hence, for every $t \geq T^*$, we obtain

$$\int_{T^*}^t (t-s)^n \rho(s) q(s) ds$$

$$\leq - \int_{T^*}^t (t-s)^n w'(s) ds + K(t-T)^n - \int_{T^*}^t c(t-s)^n v(s) w^2(s) ds$$

$$= (t-T^*)^n [w(T^*) + K] - n \int_{T^*}^t (t-s)^{n-1} w(s) ds$$

$$- \int_{T^*}^t c(t-s)^n v(s) w^2(s) ds$$

$$= (t-T^*)^n [w(T^*) + K] + \frac{n^2}{4c} \int_{T^*}^t (t-s)^{n-2} \frac{1}{v(s)} ds$$

$$- \int_{T^*}^t \left[\sqrt{c(t-s)^n v(s)} w(s) + \frac{n(t-s)^{n-1}}{2\sqrt{c(t-s)^n v(s)}} \right]^2 ds$$

$$\leq (t-T^*)^n [w(T^*) + K] + \frac{n^2}{4c} \int_{T^*}^t (t-s)^{n-2} \frac{1}{v(s)} ds.$$

Now, since

$$\frac{1}{t^n} \int_{t_0}^t (t-s)^n \rho(s) q(s) ds$$

$$\leq \frac{1}{t^n} \int_{t_0}^{T^*} (t-s)^n \rho(s) |q(s)| ds + \frac{1}{t^n} \int_{T^*}^t (t-s)^n \rho(s) q(s) ds$$

$$\leq \left(1 - \frac{t_0}{t}\right)^n \int_{t_0}^{T^*} \rho(s) |q(s)| ds + \left(1 - \frac{T^*}{t}\right)^n [w(T^*) + K]$$

$$+ \frac{1}{t^n} \left(\frac{n^2}{4c}\right) \int_{T^*}^t (t-s)^{n-2} \frac{1}{v(s)} ds \quad \text{for all} \quad t \geq T^*$$

$$(4.2.27)$$

it follows that

$$\limsup_{t \to \infty} \frac{1}{t^n} \int_{t_0}^t \left[(t-s)^n \rho(s) q(s) - \frac{n^2}{4c} (t-s)^{n-2} \frac{1}{v(s)} \right] ds$$

$$\leq \int_{t_0}^{T^*} \rho(s) |q(s)| ds + w(T^*) + K < \infty,$$

which contradicts condition (4.2.22).

Case 2. $I(T, \infty) = \infty$. In this case exactly as in Theorem 4.2.1 we arrive at the contradiction $\lim_{t\to\infty} x(t) = -\infty$. ∎

The following corollaries are immediate.

Corollary 4.2.5. In Theorem 4.2.2 the condition (4.2.22) can be replaced by

$$\limsup_{t\to\infty} \frac{1}{t^2} \int_{t_0}^{t} a(s)\rho(s)\xi(s)ds < \infty \tag{4.2.28}$$

and for some integer $n > 1$

$$\limsup_{t\to\infty} \frac{1}{t^n} \int_{t_0}^{t} (t-s)^n \rho(s)q(s)ds = \infty \tag{4.2.29}$$

Corollary 4.2.6. If condition (4.2.16) holds and

$$\limsup_{t\to\infty} \frac{1}{t^n} \int_{t_0}^{t} (t-s)^n q(s)ds = \infty \tag{4.2.30}$$

for some integer $n > 1$, then equation (4.1.6) is oscillatory.

Corollary 4.2.7. If conditions (F_2), (4.2.16) and (4.2.30) are satisfied, then equation (4.1.1) is oscillatory.

Example 4.2.6. Consider the differential equation

$$\left(t^{-\frac{5}{6}}x'(t)\right)' + \frac{2}{3}t^{-\frac{11}{6}}x'(t) + \left[-t^{-\frac{1}{3}}\sin t + \frac{1}{2}t^{-\frac{4}{3}}(2+\cos t)\right]f(x(t)) = 0$$

$$\text{for} \quad t \geq t_0 = \pi/2, \tag{4.2.31}$$

where $f(x)$ is as in Examples 4.2.1 – 4.2.3. Let $\rho(t) = t^{5/6}$. Then, we have $\beta(t) = a(t)\rho'(t) - p(t)\rho(t) = 1/(6t)$, $t \geq t_0$ and $1/v(t) = a(t)\rho(t)\xi(t) = t - t_0$, $t \geq t_0$ also for all $t \geq t_0$,

$$\int_{t_0}^{t} \rho(s)q(s)ds = \int_{\pi/2}^{t} s^{5/6}\left[-s^{-1/3}\sin s + \frac{1}{2}s^{-4/3}(2+\cos s)\right]ds$$

$$= \int_{\pi/2}^{t}\left[-\sqrt{s}\sin s + \frac{1}{2\sqrt{s}}(2+\cos s)\right]ds = \int_{\pi/2}^{t} d[\sqrt{s}(2+\cos s)]$$

$$= \sqrt{t}(2+\cos t) - 2\sqrt{\frac{\pi}{2}} \geq \sqrt{t} - 2\sqrt{\frac{\pi}{2}}$$

and hence $\liminf_{t\to\infty} \int_{t_0}^{t} s^{5/6}q(s)ds > -\infty$. Furthermore, we have for all $t \geq t_0$ and every constant $c > 0$,

$$\frac{1}{t^2} \int_{t_0}^t \left[(t-s)^2 s^{5/6} q(s) - \frac{c}{v(s)} \right] ds$$

$$= \frac{2}{t^2} \left(\int_{t_0}^t (t-s) \left[\int_{t_0}^s u^{5/6} q(u) du \right] ds - c \frac{(t-t_0)^2}{2} \right)$$

$$\geq \frac{2}{t^2} \left(\int_{\pi/2}^t (t-s) \left[\sqrt{s} - 2\sqrt{\frac{\pi}{2}} \right] ds - c \frac{(t-\pi/2)^2}{2} \right)$$

$$= \frac{8}{15} \sqrt{t} - 2\sqrt{\frac{\pi}{2}} + \frac{8}{3} \left(\frac{\pi}{2} \right)^{3/2} \frac{1}{t^2} - \frac{c}{2} \left(1 - \frac{\pi}{2t} \right)^2$$

and consequently,

$$\limsup_{t \to \infty} \frac{1}{t} \int_{t_0}^t \left[(t-s)^2 s^{5/6} q(s) - \frac{c}{v(s)} \right] ds = \infty.$$

Thus, condition (4.2.22) holds for every constant $c > 0$ and $n = 2$. Now, Theorem 4.2.2 ensures that (4.2.31) is oscillatory.

We note that Corollary 4.2.5 is also applicable to equation (4.2.31).

Example 4.2.7. Consider the differential equation

$$x''(t) + \left[-t^{-1/3} \sin t + \frac{1}{2} t^{-4/3} (2 + \cos t) \right] |x(t)|^\gamma \text{ sgn } x(t) = 0 \tag{4.2.32}$$

$$\text{for} \quad t \geq t_0 = \pi/2,$$

where $\gamma > 1$ is a constant. As in Example 4.2.6 equation (4.2.32) is oscillatory by Corollary 4.2.5 for $\rho(t) = t^{5/6}$ and $n = 2$.

On the other hand, condition (4.2.30) fails and consequently Corollary 4.2.6 cannot be applied to this equation. In fact, for every $t \geq t_0$, we have

$$\int_{t_0}^t q(s) ds = \int_{\pi/2}^t \left[-s^{-1/3} \sin s + \frac{1}{2} s^{-4/3} (2 + \cos s) \right] ds$$

$$\leq \int_{\pi/2}^t \left(-s^{-1/3} \sin s + \frac{3}{2} s^{-4/3} \right) ds$$

$$= t^{-1/3} \cos t + \int_{\pi/2}^t \left[\frac{1}{3} s^{-4/3} \cos s + \frac{3}{2} s^{-4/3} \right] ds$$

$$\leq t^{-1/3} \cos t + \frac{11}{6} \int_{\pi/2}^t s^{-4/3} ds$$

$$= t^{-1/3} \cos t - \frac{11}{2} \left[t^{-1/3} - \left(\frac{\pi}{2} \right)^{-1/3} \right] \leq \frac{11}{2} \left(\frac{\pi}{2} \right)^{-1/3}.$$

Now, if n is an integer with $n > 1$, then for all $t \geq t_0$, we find

$$\frac{1}{t^n} \int_{t_0}^{t} (t-s)^n q(s) ds = \frac{n}{t^n} \int_{t_0}^{t} (t-s)^{n-1} \left[\int_{t_0}^{s} q(u) du \right] ds$$

$$\leq \frac{11}{2} \left(\frac{\pi}{2}\right)^{-1/3} \left(\frac{n}{t^n}\right) \int_{\pi/2}^{t} (t-s)^{n-1} ds = \frac{11}{2} \left(\frac{\pi}{2}\right)^{-1/3} \left(1 - \frac{\pi}{2t}\right).$$

This gives $\limsup_{t\to\infty} (1/t^n) \int_{t_0}^{t} (t-s)^n q(s) ds < \infty$, i.e., condition (4.2.30) is not satisfied.

In Theorem 4.2.2 if the condition (4.2.4) on the function $\beta(t)$ fails, then we have the following result.

Theorem 4.2.3. Suppose (F_2) holds and let the functions $\rho(t)$, $\beta(t), \xi(t)$ and $v(t)$ be as in Theorems 4.2.1 and 4.2.2 such that condition (4.2.3) is satisfied. If condition (4.2.5) holds and for every constant $c > 0$,

$$\limsup_{t\to\infty} \frac{1}{t^n} \int_{t_0}^{t} (t-s)^{n-2} \left[(t-s)^2 \rho(s)q(s) - \frac{c}{4v(s)} \left((t-s)\frac{\beta(s)}{a(s)\rho(s)} - n \right)^2 \right] ds$$

$$= \infty$$
$$(4.2.33)$$

for some integer $n > 1$, then equation (4.1.3) is oscillatory.

Proof. Let $x(t)$ be a nonoscillatory solution of equation (4.1.3), say, $x(t) > 0$ for $t \geq T \geq t_0$. Proceeding as in Theorem 4.2.2 – Case 1, we get (4.2.26) which for $t \geq T^*$ takes the form

$$\int_{T^*}^{t} (t-s)^n \rho(s)q(s) ds$$

$$\leq (t-T^*)^n w(T^*) + \int_{T^*}^{t} (t-s)^{n-1} \left[(t-s)\frac{\beta(s)}{a(s)\rho(s)} - n \right] w(s) ds$$

$$- \int_{T^*}^{t} c(t-s)^n v(s)w^2(s) ds$$

$$= (t-T^*)^n w(T^*) + \int_{T^*}^{t} (t-s)^{n-2} \left[(t-s)\frac{\beta(s)}{a(s)\rho(s)} - n \right]^2 \frac{c}{4v(s)} ds$$

$$- \int_{T^*}^{t} \left[\sqrt{c(t-s)^n v(s)} w(s) - \frac{(t-s)^{n-1} \left((t-s)\frac{\beta(s)}{a(s)\rho(s)} \right)}{2\sqrt{c(t-s)^n v(s)}} \right]^2 ds$$

$$\leq (t-T^*)^n w(T^*) + \int_{T^*}^{t} (t-s)^{n-2} \left[(t-s)\frac{\beta(s)}{a(s)\rho(s)} - n \right]^2 \frac{c}{4v(s)} ds.$$

The rest of the proof is similar to that of Theorem 4.2.2. ∎

The following corollary is immediate.

Corollary 4.2.8. In Theorem 4.2.3 condition (4.2.33) can be replaced by

$$\limsup_{t \to \infty} \frac{1}{t^n} \int_{t_0}^t (t-s)^{n-2} \left[(t-s)\frac{\beta(s)}{a(s)\rho(s)} - n \right]^2 \frac{1}{v(s)} ds < \infty \quad (4.2.34)$$

and (4.2.29) holds for some integer $n > 1$.

Example 4.2.8. Consider the differential equation

$$\left(t^{1/6} x'(t) \right)' + \frac{1}{6t} x'(t) + \left[-t^{-1/3} \sin t + \frac{1}{2} t^{-4/3} (2 + \cos t) \right] f(x(t)) = 0$$

$$\text{for} \quad t \geq t_0 = \pi/2,$$
$$(4.2.35)$$

where f is as in Example 4.2.6. Let $\rho(t) = t^{5/6}$, $t \geq \pi/2$. Then, we have $\beta(t) = a(t)\rho'(t) - p(t)\rho(t) = \left(5 - t^{-1/6} \right)/6 > 0$ for $t \geq \pi/2$ and $\beta'(t) = t^{-7/6}/36 \geq 0$ for $t \geq \pi/2$. Also, $a(t)\rho(t) = t$, $\xi(t) = \int_{\pi/2}^t ds/s = \ln(2t/\pi)$ and $1/v(t) = t \ln(2t/\pi)$, $t \geq \pi/2$. Now for all $t \geq t_0$,

$$\frac{1}{t^2} \int_{t_0}^t \left[(t-s) \frac{\beta(s)}{a(s)\rho(s)} - 2 \right]^2 \frac{1}{v(s)} ds$$

$$= \frac{1}{t^2} \int_{\pi/2}^t \left[\frac{1}{6s} (t-s) \left(5 - s^{-1/6} \right) - 2 \right]^2 s \ln \frac{2s}{\pi} ds \geq 6 \ln \frac{2t}{\pi},$$
$$(4.2.36)$$

where $b > 0$ is a constant. As in Example 4.2.7, we find

$$\frac{1}{t^2} \int_{\pi/2}^t s^{5/6} q(s) ds \geq \frac{8}{15} \sqrt{t} - 2\sqrt{\frac{\pi}{2}} + \frac{8}{3} \left(\frac{\pi}{2} \right)^{3/2} \frac{1}{t^2}. \quad (4.2.37)$$

Now, from (4.2.36) and (4.2.37) it follows that the conditions of Theorem 4.2.3 are satisfied, and hence equation (4.2.35) is oscillatory.

Next, we present the following oscillation criteria for (4.1.3) when condition (4.2.22) fails for some integer $n > 1$.

Theorem 4.2.4. Suppose condition (F_2) holds and let the functions $\rho(t)$, $\beta(t)$, $\xi(t)$ and $v(t)$ be as in Theorem 4.2.2 such that (4.2.3) – (4.2.5) and (4.2.28) are satisfied. If there exist an integer $n > 1$ and a function $\Omega(t) \in C([t_0, \infty), \mathbb{R})$ such that for any constant $c > 0$,

$$\limsup_{t \to \infty} \frac{1}{t^n} \int_T^t (t-s)^{n-2} \left[(t-s)^2 \rho(s) q(s) - \frac{n^2 c}{4} \frac{1}{v(s)} \right] ds \geq \Omega(T) \quad (4.2.38)$$

for every $T \geq t_0$, and

$$\lim_{t \to \infty} \int_{t_0}^t v(s)(\Omega^+(s))^2 ds = \infty, \quad (4.2.39)$$

where $\Omega^+(t) = \max\{\Omega(t), 0\}$, then equation (4.1.3) is oscillatory.

Proof. Let $x(t)$ be a nonoscillatory solution of equation (4.1.3), say, $x(t) > 0$ for $t \geq T \geq t_0$. We define the function $w(t)$ as in Theorem 4.2.1 and consider the two cases:

Case 1. $I(T, \infty) < \infty$. Proceeding as in Theorem 4.2.2 – Case 1, we obtain (4.2.26). The rest of the proof is exactly the same as that of Theorem 2.5.12 for $H(t, s) = (t - s)^n$, $n > 1$ and hence omitted.

Case 2. $I(T, \infty) = \infty$. The proof is similar to that of Theorem 4.2.1 – Case 2. ∎

Theorem 4.2.5. Suppose condition (F_2) holds and let the functions $\rho(t)$, $\beta(t)$, $\xi(t)$ and $v(t)$ be as in Theorem 4.2.2 such that (4.2.3) – (4.2.5) are satisfied. If there exists an integer $n > 1$ such that

$$\liminf_{t \to \infty} \frac{1}{t^n} \int_{t_0}^t (t - s)^n \rho(s) q(s) ds < \infty \qquad (4.2.40)$$

and there exists a function $\Omega(t) \in C([t_0, \infty), \mathbb{R})$ such that condition (4.2.39) holds, and for any constant $c > 0$,

$$\liminf_{t \to \infty} \frac{1}{t^n} \int_T^t (t - s)^{n-2} \left[(t - s)^2 \rho(s) q(s) - \left(\frac{n^2 c}{4} \right) \frac{1}{v(s)} \right] ds \geq \Omega(T),$$

for every $T \geq t_0$, then equation (4.1.3) is oscillatory.

Proof. The proof is similar to that of Theorem 4.2.4 except that now we employ Theorem 2.5.13 instead of Theorem 2.5.12. ∎

In Theorems 4.2.4 and 4.2.5 if condition (4.2.4) fails, then we have the following results.

Theorem 4.2.6. Suppose condition (F_2) holds and let the functions $\rho(t)$, $\beta(t)$, $\xi(t)$ and $v(t)$ be as in Theorem 4.2.2 such that conditions (4.2.3) and (4.2.5) are satisfied. If there exist an integer $n > 1$, and a function $\Omega(t) \in C([t_0, \infty), \mathbb{R})$ such that conditions (4.2.34) and (4.2.39) hold, and for every constant $c > 0$,

$$\limsup_{t \to \infty} \frac{1}{t^n} \int_T^t (t-s)^{n-2} \left[(t-s)^2 \rho(s) q(s) - \frac{c}{4v(s)} \left((t-s) \frac{\beta(s)}{a(s)\rho(s)} - n \right)^2 \right] ds$$
$$\geq \Omega(T)$$

for every $T \geq t_0$, then equation (4.1.3) is oscillatory.

Theorem 4.2.7. Suppose condition (F_2) holds and let the functions $\rho(t)$, $\beta(t)$, $\xi(t)$ and $v(t)$ be as in Theorem 4.2.2 such that conditions

(4.2.3) and (4.2.5) are satisfied. If there exist an integer $n > 1$, and a function $\Omega(t) \in C([t_0, \infty), \mathbb{R})$ such that conditions (4.2.39) and (4.2.40) hold, and for every constant $c > 0$,

$$\liminf_{t \to \infty} \frac{1}{t^n} \int_T^t (t-s)^{n-2} \left[(t-s)^2 \rho(s) q(s) - \frac{c}{4v(s)} \left((t-s) \frac{\beta(s)}{a(s)\rho(s)} - n \right)^2 \right] ds$$
$$\geq \Omega(T)$$

for every $T \geq t_0$, then equation (4.1.3) is oscillatory.

Example 4.2.9. Consider the differential equation

$$\left(e^{-t} x'(t) \right)' + (\cos t) f(x(t)) = 0, \quad t \geq t_0 > 0 \tag{4.2.41}$$

where f is one of the functions defined in Examples 4.2.1 – 4.2.3. Taking $\rho(t) = 1$ and $n = 2$, we get

$$v(t) = \left[1 - e^{-(t-t_0)} \right]^{-1} \quad \text{for} \quad t > t_0 > 0, \quad \liminf_{t \to \infty} \int_{t_0}^t \cos s\, ds > -\infty,$$

$$\liminf_{t \to \infty} \frac{1}{t^2} \int_{t_0}^t (t-s)^2 \cos s\, ds = -\sin t_0 < \infty,$$

and

$$\liminf_{t \to \infty} \frac{1}{t^2} \int_T^t \left[(t-s)^2 \cos s - c \left(1 - e^{-(t-t_0)} \right) \right] ds \geq -\sin T - K$$

for every $T \geq t_0$, where c and K are positive constants and K is sufficiently small. Let $\Omega(T) = -\sin T - K$. Next, we consider an integer N such that $(2N+1)\pi + (\pi/4) > t_0$. Then for all integers $n \geq N$ and $(2n+1)\pi + (\pi/4) \leq T \leq 2(n+1)\pi - (\pi/4)$, $\Omega(T) = -\sin T - K \geq \delta T$, where δ is a small constant. Thus, we have

$$\lim_{t \to \infty} \int_{t_0}^t v(s)(\Omega^+(s))^2 ds \geq \sum_{n=N}^{\infty} \delta^2 \int_{(2n+1)\pi + (\pi/4)}^{2(n+1) - (\pi/4)} s^2 ds = \infty.$$

Hence, all conditions of Theorem 4.2.5 are satisfied and therefore equation (4.2.41) is oscillatory.

Theorem 4.2.8. Assume that condition (4.1.5) holds,

$$\min \left\{ \inf_{x>0} f'(x) \int_x^{\infty} \frac{du}{f(u)}, \ \inf_{x<0} f'(x) \int_x^{-\infty} \frac{du}{f(u)} \right\} > 1 \tag{4.2.42}$$

and suppose there exists $\rho(t) \in C^1([t_0, \infty), \mathbb{R}^+)$ such that condition (4.2.3) holds and $a(t)\rho'(t) = p(t)\rho(t)$, i.e., $\rho(t) = \exp \left(\int^t (p(s)/a(s)) ds \right)$. If

$$\liminf_{t \to \infty} F(t) > -\infty \tag{4.2.43}$$

and

$$\lim_{t \to \infty} F(t) \quad \text{does not exist,} \tag{4.2.44}$$

where

$$F(t) = \frac{1}{\xi(t)} \int_{t_0}^t \frac{1}{a(s)\rho(s)} \int_{t_0}^s \rho(u)q(u)du\,ds,$$

then equation (4.1.3) is oscillatory.

We note that conditions (4.2.43) and (4.2.44) can be replaced by

$$-\infty \; < \; \liminf_{t \to \infty} F(t) \; < \; \limsup_{t \to \infty} F(t) \; \leq \; \infty. \tag{4.2.45}$$

Proof. Let $x(t)$ be a nonoscillatory solution of equation (4.1.3), say, $x(t) > 0$ for $t \geq t_0$. We define $w(t) = \int_{x(t)}^\infty du/f(u)$ for $t \geq t_0$. Then, we have

$$(a(t)\rho(t)w'(t))' \; = \; \rho(t)q(t) + a(t)\rho(t)(w'(t))^2 f'(x(t)) \quad \text{for} \quad t \geq t_0. \tag{4.2.46}$$

Integrating (4.2.46) twice from t_0 to t, we find

$$w'(t) \; = \; \frac{c_1}{a(t)\rho(t)} + \frac{1}{a(t)\rho(t)} \int_{t_0}^t \rho(s)q(s)ds + \frac{1}{a(s)\rho(s)} J(t_0, t),$$

where

$$J(t_0, t) \; = \; \int_{t_0}^t a(s)\rho(s)(w'(s))^2 f'(x(s))ds$$

and $c_1 = a(t_0)\rho(t_0)w'(t_0)$, and

$$\begin{aligned}
\frac{w(t)}{\xi(t)} \; &= \; \frac{w(t_0)}{\xi(t)} + c_1 + \frac{1}{\xi(t)} \int_{t_0}^t \frac{1}{a(s)\rho(s)} \int_{t_0}^s \rho(u)q(u)du\,ds \\
&\quad + \frac{1}{\xi(t)} \int_{t_0}^t \frac{1}{a(s)\rho(s)} J(t_0, s)ds.
\end{aligned} \tag{4.2.47}$$

Now, with respect to the integral $J(t_0, t)$ we need to distinguish two mutually exclusive cases:

Case 1. $J(t_0, \infty) < \infty$. First, we will show that

$$\lim_{t \to \infty} \frac{w(t)}{\xi(t)} \; = \; 0. \tag{4.2.48}$$

Let $\epsilon > 0$ be an arbitrary number. We can choose a $t_1 \geq t_0$ such that

$$J(t_1, \infty) \; \leq \; \frac{\epsilon}{4}. \tag{4.2.49}$$

Furthermore, using the Schwartz inequality for $t \geq t_1$, we have

$$w(t) - w(t_1) \leq \left| \int_{t_1}^t w'(s)ds \right| \leq [J(t_1,t)]^{1/2}[G(t_1,t)]^{1/2},$$

where

$$G(t_1,t) = \int_{t_1}^t \frac{1}{a(s)\rho(s)f'(x(s))} ds$$

and so, in view of (4.2.49), we get

$$w(t) \leq w(t_1) + \frac{\sqrt{\epsilon}}{2}[G(t_1,t)]^{1/2} \quad \text{for all} \quad t \geq t_1. \tag{4.2.50}$$

If $G(t_1,\infty) < \infty$, then from (4.2.50) it follows that the function $w(t)$ is bounded on $[t_1, \infty)$ and hence (4.2.48) is satisfied. So, we assume that $G(t_1,\infty) = \infty$. Then there exists a number $t_2 > t_1$ such that $w(t_1) \leq (\sqrt{\epsilon}/2)[G(t_1,t)]^{1/2}$ for $t \geq t_2$, and consequently (4.2.50) gives

$$w(t) \leq \sqrt{\epsilon}[G(t_1,t)]^{1/2} \quad \text{for every} \quad t \geq t_2. \tag{4.2.51}$$

But condition (4.2.42) guarantees that

$$f'(x(t))w(t) = f'(x(t)) \int_{x(t)}^\infty \frac{du}{f(u)} \geq 1 \quad \text{for} \quad t \geq t_1.$$

Thus, from (4.2.51) it follows that

$$\frac{w(t)}{a(t)\rho(t)} \leq \frac{\sqrt{\epsilon}}{a(t)\rho(t)} \left[\int_{t_1}^t \frac{w(s)}{a(s)\rho(s)} ds \right]^{1/2} \quad \text{for all} \quad t \geq t_2, \tag{4.2.52}$$

or

$$\left[\int_{t_1}^t \frac{w(s)}{a(s)\rho(s)} ds \right]^{-1/2} \frac{w(t)}{a(t)\rho(t)} \leq \sqrt{\epsilon} \quad \text{for all} \quad t \geq t_2.$$

Integrating the above inequality from t_2 to t with $t \geq t_2$, we get

$$2 \left[\int_{t_1}^t \frac{w(s)}{a(s)\rho(s)} ds \right]^{1/2} - 2 \left[\int_{t_1}^{t_2} \frac{w(s)}{a(s)\rho(s)} ds \right]^{1/2}$$

$$\leq \sqrt{\epsilon} \left[\xi(t) - \int_{t_0}^{t_2} \frac{1}{a(s)\rho(s)} ds \right] \leq \sqrt{\epsilon}\, \xi(t).$$

Therefore, on setting

$$t_3 = \max \left\{ t_2, \frac{2}{\sqrt{\epsilon}} \left[\int_{t_1}^{t_2} \frac{w(s)}{a(s)\rho(s)} ds \right]^{1/2} \right\},$$

for all $t \geq t_3$, we obtain

$$\left[\int_{t_1}^t \frac{w(s)}{a(s)\rho(s)} ds \right]^{1/2} < \sqrt{\epsilon}\, \xi(t).$$

Hence, (4.2.52) gives $w(t) < \epsilon \xi(t)$ for all $t \geq t_3$. As $\epsilon > 0$ is arbitrary, the proof of (4.2.48) is complete.

Now using (4.2.48) in (4.2.47), we find

$$\lim_{t \to \infty} \frac{1}{\xi(t)} \int_{t_0}^t \frac{1}{a(s)\rho(s)} \int_{t_0}^s \rho(u)q(u)du\,ds$$

$$= -c_1 - \lim_{t \to \infty} \frac{1}{\xi(t)} \int_{t_0}^t \frac{1}{a(s)\rho(s)} J(t_0, s)ds,$$

which contradicts condition (4.2.44).

Case 2. $J(t_0, \infty) = \infty$. Define $c = \inf_{x>0} f'(x) \int_x^\infty du/f(u)$. In view of (4.2.42), we have $c > 1$. We can choose μ such that $1 < \mu < c$. We claim that

$$\limsup_{t \to \infty} \; [(1/\mu)B(t) - w(t)] > 0, \tag{4.2.53}$$

where $B(t) = \int_{t_0}^t J(t_0, s)/(a(s)\rho(s))ds$. Assume that (4.2.53) does not hold, then there exists a $T_1 \geq t_0$ such that

$$B(t) \leq \mu w(t) \quad \text{for} \quad t \geq T_1. \tag{4.2.54}$$

It is easy to see that $B(t) \to \infty$, $a(t)\rho(t)B'(t) \to \infty$ as $t \to \infty$ and $(a(t)\rho(t)B'(t))' = a(t)\rho(t)(w'(t))^2 f'(x(t))$. Using (4.2.54) for $t \geq T_1$, we find the estimate

$$\left(\frac{(a(t)\rho(t)B'(t))'}{a(t)\rho(t)B'(t)} \right)^{1/2} \left(\frac{B'(t)}{B(t)} \right)^{1/2} = \left(\frac{(w'(t))^2 f'(x(t))}{B(t)} \right)^{1/2} \geq \mu_1 \frac{w'(t)}{w(t)} \tag{4.2.55}$$

where $\mu_1 = \sqrt{(c/\mu)} > 1$.

Note that $B(t_0) = 0$, so we can choose $B(T_2) = 1$ for some $T_2 \geq T_1$. Now, integrating (4.2.55) from T_2 to t and applying Schwartz's inequality to the left side of (4.2.55), we obtain

$$\left(\ln \left(\frac{a(t)\rho(t)B'(t)}{a(T_2)\rho(T_2)B'(T_2)} \right) \right)^{1/2} (\ln B(t))^{1/2} \geq \sqrt{\mu_1} \ln \frac{w(t)}{w(T_2)}. \tag{4.2.56}$$

Using (4.2.54) we can estimate the right side of (4.2.56) as follows

$$\sqrt{\mu_1} \ln \frac{w(t)}{w(T_2)} = \sqrt{\mu_1}\, [\ln w(t) - \ln w(T_2)]$$

$$\geq \sqrt{\mu_1}[\ln B(t) - \ln w(T_2)] \geq \sqrt{\mu_1} \ln B(t) \quad \text{for} \quad t \geq T_2.$$

Thus, $\ln(ka(t)\rho(t)B'(t)) \geq \mu_1 \ln B(t)$, where $k = 1/(a(T_2)\rho(T_2)B'(T_2)) > 0$, and hence

$$B^{-\mu_1}(t)B'(t) \geq \frac{1}{k}\frac{1}{a(t)\rho(t)}. \qquad (4.2.57)$$

Integrating (4.2.57) from T_2 to t and using condition (4.2.3), we get

$$\infty > \frac{1}{\mu_1 - 1}\left[B^{1-\mu_1}(T_2) - B^{1-\mu_1}(t)\right] \geq \frac{1}{k}\int_{T_2}^{t}\frac{1}{a(s)\rho(s)}ds \to \infty \text{ as } t \to \infty,$$

which is a contradiction. Thus, the assertion (4.2.53) holds, and there exists a sequence $\{t_n\}_{n=1}^{\infty}$ with $\lim_{n\to\infty} t_n = \infty$ and such that

$$\lim_{n\to\infty} \left[-(1/\mu)B(t_n) + w(t_n)\right] < 0. \qquad (4.2.58)$$

Using (4.2.58) in (4.2.47), we find for sufficiently large t_n,

$$\left(-1 + \frac{1}{\mu}\right)\frac{1}{\xi(t_n)}B(t_n) \geq c_1 + \frac{w(t_0)}{\xi(t_n)} + \frac{1}{\xi(t_n)}\int_{t_0}^{t_n}\frac{1}{a(s)\rho(s)}\int_{t_0}^{s}\rho(u)q(u)duds.$$
$$(4.2.59)$$

Since $1/\mu < 1$ and $B(t)/\xi(t) \to \infty$ as $t \to \infty$, (4.1.59) implies

$$\liminf_{t\to\infty}\frac{1}{\xi(t)}\int_{t_0}^{t}\frac{1}{a(s)\rho(s)}\int_{t_0}^{s}\rho(u)q(u)duds = -\infty,$$

which contradicts condition (4.2.43). This completes the proof. ∎

When $p(t) \equiv 0$ in Theorem 4.2.8 then clearly, $\rho(t) \equiv 1$. Thus, we have the following corollary.

Corollary 4.2.9. Assume that conditions (4.1.5) and (4.2.42) hold and $\xi_1(t) = \int_{t_0}^{t} ds/a(s) \to \infty$ as $t \to \infty$. If

$$\liminf_{t\to\infty}\frac{1}{\xi_1(t)}\int_{t_0}^{t}\frac{1}{a(s)}\int_{t_0}^{s}q(u)duds > -\infty$$

and

$$\lim_{t\to\infty}\frac{1}{\xi_1(t)}\int_{t_0}^{t}\frac{1}{a(s)}\int_{t_0}^{s}q(u)duds \qquad (4.2.60)$$

does not exist, then equation (4.1.2) is oscillatory.

Corollary 4.2.10. If

$$-\infty < \liminf_{t\to\infty}\frac{1}{t}\int_{t_0}^{t}\int_{t_0}^{s}q(u)duds < \limsup_{t\to\infty}\frac{1}{t}\int_{t_0}^{t}\int_{t_0}^{s}q(u)duds, \qquad (4.2.61)$$

then equation (4.1.6) is oscillatory.

Theorem 4.2.9. In Theorem 4.2.8 condition (4.2.44) can be replaced by

$$\limsup_{t\to\infty} \frac{1}{\xi(t)} \int_{t_0}^t \frac{1}{a(s)\rho(s)} \left(\int_{t_0}^s \rho(u)q(u)du \right)^2 ds = \infty. \qquad (4.2.62)$$

Proof. Let $x(t)$ be a nonoscillatory solution of equation (4.1.3), say, $x(t) > 0$ for $t \geq t_0$. Define $w(t)$ as in Theorem 4.2.8 and obtain (4.2.46). Next, we consider the two cases: 1. $J(t_0, \infty) < \infty$ and 2. $J(t_0, \infty) = \infty$. The proof of Case 2 is exactly the same as of Theorem 4.2.8 – Case 2 and hence omitted. Now, we consider

Case 1. From (4.2.46), we find

$$\int_{t_0}^t \rho(s)q(s)ds = K_1 + a(t)\rho(t)w'(t) - \int_{t_0}^t a(s)\rho(s)(w'(s))^2 f'(x(s))ds,$$

where $K_1 = -a(t_0)\rho(t_0)w'(t_0)$. Thus, if we define

$$K_2 = 3K_1^2 + 3 \left(\int_{t_0}^\infty a(s)\rho(s)(w'(s))^2 f'(x(s))ds \right)^2,$$

then, we have

$$\left(\int_{t_0}^t \rho(s)q(s)ds \right)^2 \leq 3K_1^2 + 3(a(t)\rho(t)w'(t))^2$$

$$+ 3 \left(\int_{t_0}^t a(s)\rho(s)(w'(s))^2 f'(x(s))ds \right)^2 \leq K_2 + 3(a(t)\rho(t)w'(t))^2.$$

But, condition (4.2.42) implies that $f'(x(t))w(t) = f'(x(t)) \int_{x(t)}^\infty du/f(u) > 1$ for $t \geq t_0$. Hence, for $t \geq t_0$, we have

$$\frac{1}{\xi(t)} \int_{t_0}^t \frac{1}{a(s)\rho(s)} \left(\int_{t_0}^s \rho(u)q(u)du \right)^2 ds$$

$$\leq K_2 + \frac{1}{\xi(t)} \int_{t_0}^t a(s)\rho(s)(w'(s))^2 ds$$

$$\leq K_2 + \frac{1}{\xi(t)} \int_{t_0}^t a(s)\rho(s)(w'(s))^2 f'(x(s))w(s)ds$$

$$\leq K_2 + \frac{1}{\xi(t)} \left[\max_{t_0 \leq s \leq t} w(s) \right] \int_{t_0}^t a(s)\rho(s)(w'(s))^2 f'(x(s))ds,$$

i.e.,

$$\frac{1}{\xi(t)} \int_{t_0}^t \frac{1}{a(s)\rho(s)} \left(\int_{t_0}^s \rho(u)q(u)du \right)^2 ds \leq K_2 + \frac{K_3}{\xi(t)} \max_{t_0 \leq s \leq t} w(s), \quad (4.2.63)$$

where $K_3 = 3 \int_{t_0}^{\infty} a(s)\rho(s)(w'(s))^2 f'(x(s))ds > 0$. Now, as in Theorem 4.2.8, we obtain (4.2.48), and hence we can choose a $T \geq t_0$ so that $w(t) \leq \xi(t)$ for all $t \geq T$, and

$$\max_{t_0 \leq s \leq t} w(s) \leq \max_{t_0 \leq s \leq T} w(s) + \xi(t)$$

for all $t \geq T$. Thus, for $t \geq T$, (4.2.63) gives

$$\frac{1}{\xi(t)} \int_{t_0}^{t} \frac{1}{a(s)\rho(s)} \left(\int_{t_0}^{s} \rho(u)q(u)du \right)^2 ds \leq K_2 + \frac{K_3}{\xi(t)} \left[\max_{t_0 \leq s \leq T} w(s) + \xi(t) \right].$$

Therefore, we have

$$\limsup_{t \to \infty} \frac{1}{\xi(t)} \int_{t_0}^{t} \frac{1}{a(s)\rho(s)} \left(\int_{t_0}^{s} \rho(u)q(u)du \right)^2 ds \leq K_2 + K_3 < \infty,$$

which contradicts condition (4.2.62). This completes the proof. ∎

Corollary 4.2.11. In Corollary 4.2.9 condition (4.2.60) can be replaced by

$$\limsup_{t \to \infty} \frac{1}{\xi_1(t)} \int_{t_0}^{t} \frac{1}{a(s)} \left(\int_{t_0}^{s} q(u)du \right)^2 ds = \infty.$$

Corollary 4.2.12. If

$$\liminf_{t \to \infty} \frac{1}{t} \int_{t_0}^{t} \int_{t_0}^{s} q(u)duds > -\infty$$

and

$$\limsup_{t \to \infty} \frac{1}{t} \int_{t_0}^{t} \left(\int_{t_0}^{s} q(u)du \right)^2 ds = \infty, \tag{4.2.64}$$

then equation (4.1.6) is oscillatory.

Theorem 4.2.10. In Theorem 4.2.8 condition (4.2.43) can be replaced by

$$\liminf_{t \to \infty} \frac{1}{t^n} \int_{t_0}^{t} (t-s)^n \rho(s)q(s)ds > -\infty, \tag{4.2.65}$$

where $n > 1$ is an integer and $(a(t)\rho(t))' \leq 0$ for $t \geq t_0$.

Proof. Let $x(t)$ be a nonoscillatory solution of equation (4.1.3), say, $x(t) > 0$ for $t \geq t_0$. Define the function $w(t)$ as in Theorem 4.2.8 and obtain (4.2.46). Once again, we consider the two cases when $a(t)\rho(t)(w'(t))^2 \times f'(x(t)) \in \mathcal{L}^1[t_0, \infty)$ and $a(t)\rho(t)(w'(t))^2 f'(x(t)) \notin \mathcal{L}^1[t_0, \infty)$. The proof of the first case is exactly the same as in Theorem 4.2.8 – Case 1 and hence omitted. Next, we consider the second case when $J(t_0, \infty) = \infty$. Define

$c = \inf_{x>0} f'(x) \int_x^\infty du/f(u)$. In view of condition (4.2.42), we have $c > 1$. Hence, we can choose an integer $k \geq \max\{n, 2\}$ such that

$$\frac{k}{k-1} < c. \qquad (4.2.66)$$

As $k \geq n$, it is easy to verify that condition (4.2.65) implies that

$$\liminf_{t\to\infty} \frac{1}{t^k} \int_{t_0}^t (t-s)^k \rho(s)q(s)ds > -\infty. \qquad (4.2.67)$$

Now, for all $t \geq t_0$ from equation (4.2.46), we find

$$\int_{t_0}^t (t-s)^k \rho(s)q(s)ds + \int_{t_0}^t (t-s)^k a(s)\rho(s)(w'(s))^2 f'(x(s))ds$$

$$= \int_{t_0}^t (t-s)^k (a(s)\rho(s)w'(s))'ds$$

$$= -(t-t_0)^k a(t_0)\rho(t_0)w'(t_0) + k \int_{t_0}^t (t-s)^{k-1} a(s)\rho(s)w'(s)ds.$$

Thus, for $t \geq t_0$, we have

$$\frac{1}{t^k} \int_{t_0}^t (t-s)^k \rho(s)q(s)ds$$

$$= -\left(1 - \frac{t_0}{t}\right)^k a(t_0)\rho(t_0)w'(t_0) + \frac{1}{t^k}\left[k\int_{t_0}^t (t-s)^{k-1} a(s)\rho(s)w'(s)ds\right.$$

$$\left. - \int_{t_0}^t (t-s)^k a(s)\rho(s)(w'(s))^2 f'(x(s))ds\right].$$

$$(4.2.68)$$

Next, in view of (4.2.66), we can choose a constant μ such that

$$\frac{1}{c}\left(\frac{k}{k-1}\right) < \mu < 1. \qquad (4.2.69)$$

We claim that for every $T \geq t_0$, there exists a $t \geq T$ such that

$$\int_{t_0}^t (t-s)^k a(s)\rho(s)(w'(s))^2 f'(x(s))ds > \frac{k}{\mu}\int_{t_0}^t (t-s)^{k-1} a(s)\rho(s)w'(s)ds.$$

Otherwise, there is a $T \geq t_0$ such that for $t \geq T$,

$$\int_{t_0}^t (t-s)^k a(s)\rho(s)(w'(s))^2 f'(x(s))ds \leq \frac{k}{\mu}\int_{t_0}^t (t-s)^{k-1} a(s)\rho(s)w'(s)ds.$$

$$(4.2.70)$$

By the Schwartz inequality and (4.2.70), and using the definitions of w and c for $t \geq T$, we obtain

$$0 \leq \int_{t_0}^{t} (t-s)^{k-1} a(s)\rho(s)w'(s)ds$$

$$\leq \left[\int_{t_0}^{t} (t-s)^k a(s)\rho(s)(w'(s))^2 f'(x(s))ds \right]^{1/2} \left[\int_{t_0}^{t} (t-s)^{k-2} \frac{a(s)\rho(s)}{f'(x(s))} ds \right]^{1/2}$$

$$\leq \left[\frac{k}{\mu} \int_{t_0}^{t} (t-s)^{k-1} a(s)\rho(s)w'(s)ds \right]^{1/2} \left[\frac{1}{c} \int_{t_0}^{t} (t-s)^{k-2} a(s)\rho(s)w(s)ds \right]^{1/2}$$

and hence, it follows that

$$\int_{t_0}^{t} (t-s)^{k-1} a(s)\rho(s)w'(s)ds \leq \frac{k}{\mu c} \int_{t_0}^{t} (t-s)^{k-2} a(s)\rho(s)w(s)ds. \quad (4.2.71)$$

But, for all $t \geq T$, we have

$$\int_{t_0}^{t} (t-s)^{k-2} a(s)\rho(s)w(s)ds = \frac{1}{k-1} (t-t_0)^{k-1} a(t_0)\rho(t_0)w(t_0)$$

$$+ \frac{1}{k-1} \int_{t_0}^{t} (t-s)^{k-1} [a(s)\rho(s)w'(s) + (a(s)\rho(s))'w(s)] \, ds.$$

Using the fact that $(a(t)\rho(t))' \leq 0$ for $t \geq t_0$ in the above inequality, we get

$$\int_{t_0}^{t} (t-s)^{k-2} a(s)\rho(s)w(s)ds \leq \frac{1}{k-1} (t-t_0)^{k-1} a(t_0)\rho(t_0)w(t_0)$$

$$+ \frac{1}{k-1} \int_{t_0}^{t} (t-s)^{k-1} a(s)\rho(s)w'(s)ds$$

and so (4.2.71) for all $t \geq T$, gives

$$\left[\mu c \left(\frac{k-1}{k} \right) - 1 \right] \int_{t_0}^{t} (t-s)^{k-1} a(s)\rho(s)w'(s)ds \leq (t-t_0)^{k-1} a(t_0)\rho(t_0)w(t_0).$$

Therefore, by using (4.2.70), we find

$$\left[\mu c \left(\frac{k-1}{k} \right) - 1 \right] \frac{1}{t^{k-1}} \int_{t_0}^{t} (t-s)^k a(s)\rho(s)(w'(s))^2 f'(x(s))ds$$

$$\leq \frac{k}{\mu} \left(1 - \frac{t_0}{t} \right)^{k-1} a(t_0)\rho(t_0)w(t_0) \quad \text{for} \quad t \geq T. \quad (4.2.72)$$

But, $J(t_0, \infty) = \infty$ implies

$$\lim_{t \to \infty} \frac{1}{t^{k-1}} \int_{t_0}^{t} (t-s)^k a(s)\rho(s)(w'(s))^2 f'(x(s))ds = \infty.$$

On the other hand, (4.2.69) implies that $\mu c((k-1)/k) - 1 > 0$. Thus, (4.2.72) leads to a contradiction. Hence, the claim follows and so we can consider a sequence $\{t_m\}_{m=1}^{\infty}$ with $\lim_{m\to\infty} t_m = \infty$ such that

$$\int_{t_0}^{t_m} (t_m - s)^k a(s)\rho(s)(w'(s))^2 f'(x(s))ds$$

$$> \frac{k}{\mu} \int_{t_0}^{t_m} (t_m - s)^{k-1} a(s)\rho(s)w'(s)ds \quad (m = 1, 2, \cdots).$$

Then from (4.2.68) it follows that

$$\frac{1}{t_m^k} \int_{t_0}^{t_m} (t_m - s)^k \rho(s)q(s)ds < -\left(1 - \frac{t_0}{t_m}\right)^k a(t_0)\rho(t_0)w'(t_0)$$

$$+ (\mu - 1)\frac{1}{t_m^k} \int_{t_0}^{t_m} (t_m - s)^k a(s)\rho(s)(w'(s))^2 f'(x(s))ds,$$

$$m = 1, 2, \cdots.$$

$$(4.2.73)$$

Now, since

$$\lim_{t\to\infty} \frac{1}{t^k} \int_{t_0}^{t} (t - s)^k a(s)\rho(s)(w'(s))^2 f'(x(s))ds = J(t_0, \infty) = \infty$$

and by (4.2.69) the constant $\mu < 1$, the inequality (4.2.73) ensures that

$$\lim_{m\to\infty} \frac{1}{t_m^k} \int_{t_0}^{t_m} (t_m - s)^k \rho(s)q(s)ds = -\infty.$$

This shows that

$$\liminf_{t\to\infty} \frac{1}{t^k} \int_{t_0}^{t} (t - s)^k \rho(s)q(s)ds = -\infty,$$

which contradicts condition (4.2.65). This completes the proof. ∎

Corollary 4.2.13. Assume that $a'(t) \leq 0$ for $t \geq t_0$ and condition (4.2.42) is satisfied. If condition (4.2.60) holds and

$$\liminf_{t\to\infty} \frac{1}{t^n} \int_{t_0}^{t} (t - s)^n q(s)ds > -\infty \quad \text{for some integer} \quad n > 1, \quad (4.2.74)$$

then equation (4.1.2) is oscillatory.

Corollary 4.2.14. Assume that condition (4.2.42) is satisfied. If condition (4.2.74) holds and

$$\lim_{t\to\infty} \frac{1}{t} \int_{t_0}^{t} \int_{t_0}^{s} q(u)duds \quad \text{does not exist as a real number,}$$

then equation (4.1.1) is oscillatory.

From Theorems 4.2.9 and 4.2.10 the following result is immediate.

Theorem 4.2.11. Suppose condition (4.2.42) is satisfied and there exists a function $\rho(t) \in C^1([t_0, \infty), \mathbb{R}^+)$ such that condition (4.2.3) holds, and $(a(t)\rho(t))' \leq 0$, $a(t)\rho'(t) = p(t)\rho(t)$ for $t \geq t_0$. If conditions (4.2.62) and (4.2.65) hold, then equation (4.1.3) is oscillatory.

Corollary 4.2.15. If conditions (4.2.42), (4.2.64) and (4.2.74) are satisfied, then equation (4.1.1) is oscillatory.

Next, we present a more general result.

Theorem 4.2.12. Suppose condition (4.2.42) is satisfied and assume that there exists a function $\rho(t) \in C^1([t_0, \infty), \mathbb{R}^+)$ such that $a(t)\rho'(t) = p(t)\rho(t)$, $(a(t)\rho(t))' \leq 0$ for $t \geq t_0$ and condition (4.2.3) holds. If condition (4.2.65) holds, and there exists an integer $n > 1$ such that

$$\limsup_{t \to \infty} \frac{1}{t^n} \int_{t_0}^t (t-s)^{n-2} a(s)\rho(s)\xi(s)ds < \infty \qquad (4.2.75)$$

and

$$\limsup_{t \to \infty} \frac{1}{t^n} \int_{t_0}^t (t-s)^n \rho(s)q(s)ds = \infty, \qquad (4.2.76)$$

then equation (4.1.3) is oscillatory.

Proof. Let $x(t)$ be a nonoscillatory solution of equation (4.1.3), say, $x(t) > 0$ for $t \geq t_0$. We define $w(t)$ as in Theorem 4.2.8 and obtain (4.2.46). Next, we consider the two cases: 1. $J(t_0, \infty) < \infty$ and 2. $J(t_0, \infty) = \infty$.

The proof of Case 2 is exactly the same as in Theorem 4.2.10 – Case 2. Thus, we consider Case 1. Proceeding as in Theorem 4.2.8 – Case 1, there exist a constant $c_1 > 0$ and a $T \geq t_0$ such that

$$w(t) \leq c_1\xi(t) \quad \text{for} \quad t \geq T. \qquad (4.2.77)$$

Define $y(t) = a(t)\rho(t)w'(t)$, $t \geq T$ so that equation (4.2.46) can be written as

$$y'(t) = \rho(t)q(t) + \frac{1}{a(t)\rho(t)}y^2(t)f'(x(t)), \quad t \geq T. \qquad (4.2.78)$$

Using (4.2.42) and (4.2.77) in equation (4.2.78), we get

$$y'(t) \geq \rho(t)q(t) + \frac{1}{c_1 a(t)\rho(t)\xi(t)}y^2(t) \quad \text{for} \quad t \geq T. \qquad (4.2.79)$$

Multiplying (4.2.79) by $(t-s)^n$ and integrating by parts, we find

$$\int_T^t (t-s)^n \rho(s)q(s)ds + \frac{1}{c_1}\int_T^t (t-s)^n \frac{1}{a(s)\rho(s)\xi(s)}y^2(s)ds$$
$$- n\int_T^t (t-s)^{n-1}y(s)ds \leq -(t-T)^n y(T),$$

or

$$\int_T^t \left[\left(\frac{(t-s)^n}{c_1 a(s)\rho(s)\xi(s)}\right)^{1/2} y(s) - \frac{n(t-s)^{(n-2)/2}}{2(c_1 a(s)\rho(s)\xi(s))^{-1/2}}\right]^2 ds$$
$$+ \int_T^t (t-s)^n \rho(s)q(s)ds - \frac{n^2 c_1}{4}\int_T^t (t-s)^{n-2}a(s)\rho(s)\xi(s)ds \leq -(t-T)^n y(T).$$

Thus, it follows that

$$\frac{1}{t^n}\int_T^t (t-s)^n \rho(s)q(s)ds \leq \frac{n^2 c_1}{4t^n}\int_T^t (t-s)^{n-2}a(s)\rho(s)\xi(s)ds - \left(1-\frac{T}{t}\right)^n y(T).$$
$$(4.2.80)$$

Taking $\limsup$ on both sides of (4.2.80) as $t \to \infty$ and using (4.2.75), we obtain a contradiction to condition (4.2.76). This completes the proof. ∎

Corollary 4.2.16.   Suppose condition (4.2.42) is satisfied. If condition (4.2.74) holds and

$$\limsup_{t\to\infty} \frac{1}{t^n}\int_{t_0}^t (t-s)^n q(s)ds = \infty \quad \text{for some integer } n > 1, \quad (4.2.81)$$

then equation (4.1.1) is oscillatory.

Corollary 4.2.17.   If conditions (4.2.74) and (4.2.81) are satisfied, then equation (4.1.6) is oscillatory.

Remark 4.2.3.

1. We note that (4.2.16) implies the first part of (4.2.61) and inductively condition (4.2.74).

2. The condition $\lim_{t\to\infty}\int_{t_0}^t q(s)ds = \infty$ implies (4.2.17), and condition (4.2.17) implies (4.2.61).

3. As an application of the Schwartz inequality condition (4.2.17) implies (4.2.64). We also note that condition (4.2.30) implies (4.2.17) which is equivalent to $\lim_{t\to\infty}(1/t)\int_{t_0}^t (t-s)q(s)ds = \infty$.

4. Extensions of the above theorems which involve the function $(t-s)^n$ can be obtained by replacing it with the more general function $H(t,s)$ satisfying the same conditions listed earlier (see Chapters 2 and 3).

5. It would be interesting to obtain oscillation criteria for equation (4.1.3) similar to those of Theorems 4.2.8 – 4.2.12 without the assumption that $\beta(t) = 0$, i.e., letting $\rho(t) = \exp\left(\int^t (p(s)/a(s))ds\right)$.

In the following results, we remove the assumption on $f'(x)$, i.e., the condition (4.1.28).

Theorem 4.2.13. Suppose conditions (F_2) and (4.1.9) hold and $Q(t) = \int_t^\infty q(s)ds$ exists, and $\liminf_{t\to\infty} \int_{t_0}^t Q(s)/a(s)ds > -\infty$, and either $\int_{t_0}^\infty (Q^+(s))^2/(a(s)\xi_1(s))ds = \infty$, or for every constant $c > 0$,

$$\limsup_{t\to\infty} \int_{t_0}^\infty \frac{1}{a(s)}\left[Q(s) + c\int_s^\infty \frac{1}{a_1(u)\xi_1(u)}(Q^+(u))^2 du\right] ds = \infty,$$

where $\xi_1(t) = \int_{t_0}^t ds/a(s)$, then equation (4.1.2) is oscillatory.

Proof. Let $x(t)$ be a nonoscillatory solution of equation (4.1.2), say, $x(t) > 0$ for $t \geq t_0$. It follows from Lemma 4.1.3 that (4.1.50) holds. Proceeding as in Theorem 4.2.2 – Case 1 (take $\rho(t) = 1$ and $\beta(t) = 0$), there exist a $T \geq t_0$ and a constant $c_1 > 0$ such that

$$f'(x(t)) \geq \frac{c_1}{\xi_1(t)} \quad \text{for} \quad t \geq T. \tag{4.2.82}$$

Thus, it follows that

$$\int_t^\infty a(s)\left(\frac{x'(s)}{f(x(s))}\right)^2 f'(x(s))ds \geq c_1 \int_t^\infty \frac{a(s)}{\xi_1(s)}\left(\frac{x'(s)}{f(x(s))}\right)^2 ds \text{ for } t \geq T.$$

The rest of the proof is similar to that of Theorem 4.1.7. ∎

Remark 4.2.4. We note that Theorem 4.1.8 includes Theorem 4.2.13.

Now as in Theorem 4.1.9, we define

$$h_0(t) = \frac{1}{\sqrt{a(t)\xi_1(t)}} \int_t^\infty q(s)ds, \quad \xi_1(t) = \int_{t_0}^t \frac{ds}{a(s)},$$

$$h_1(t) = \int_t^\infty \left([h_0(s)]^+\right)^2 ds$$

$$h_{n+1}(t) = \int_t^\infty \left(\left[h_0(s) + \frac{c}{\sqrt{a(s)\xi_1(s)}}h_n(s)\right]^+\right)^2 ds \quad \text{for} \quad n = 1, 2, \cdots$$

where c is a positive constant.

Theorem 4.2.14. Suppose conditions (F_2), (4.1.9) and (4.1.10) are satisfied and there exists an integer N such that $h_n(t)$ exists for $n =$

$0, 1, \cdots, N$. If for every constant $c > 0$, and for all sufficiently large t, $h_0(t) + ch_N(t)/\sqrt{a(t)\xi_1(t)} \geq 0$ and

$$\int^{\infty} \frac{1}{\sqrt{a(s)\xi_1(s)}} \left[h_0(s) + \frac{c}{\sqrt{a(s)\xi_1(s)}} h_N(s) \right] ds = \infty,$$

then equation (4.1.2) is oscillatory.

Proof. Let $x(t)$ be a nonoscillatory solution of equation (4.1.2), say, $x(t) > 0$ for $t \geq t_0$. As in Theorem 4.2.13, we obtain (4.2.82) for $t \geq T$. The rest of the proof is exactly the same as that of Theorem 4.1.9. ∎

4.3. Existence of Nonoscillatory Solutions and Comparison Results

In this section first we shall present sufficient conditions which guarantee the existence of a nontrivial nonoscillatory solution of (4.1.1), then we shall provide comparison theorems similar to those of Hille–Wintner for second order equations of the type (4.1.2). This will be followed by several oscillatory theorems involving comparison with related linear equations. This will enable us to employ the results of Chapter 2 to obtain interesting oscillation criteria for nonlinear equations.

4.3.1. A Necessary Condition for Oscillation

We shall prove the following result.

Theorem 4.3.1. Let $f(x)$ and $q(t)$ be such that

$$\begin{cases} \text{there is a nontrivial compact interval } I \text{ in which } xf(x) > 0 \text{ for} \\ x \neq 0 \text{ and } f'(x) \text{ exists at each point, is nonnegative and bounded,} \end{cases}$$
$$(4.3.1)$$

$$\begin{cases} q(t) \text{ is locally integrable and } Q(t) = \int_t^{\infty} q(s)ds \text{ exists (possibly} \\ \text{infinite) with } \liminf_{t \to \infty} \int_t^{\infty} Q(s)ds > -\infty \text{ for all } t \end{cases}$$
$$(4.3.2)$$

and

$$\int^{\infty} \left(\int_s^{\infty} (Q^-(u))^2 du \right) ds < \infty, \qquad (4.3.3)$$

where $Q^-(t) = Q^+(t) - Q(t)$. Then,

$$\int^{\infty} \left[Q(s) + \int_s^{\infty} Q^2(u)du \right] ds < \infty \qquad (4.3.4)$$

implies that there exists a nontrivial nonoscillatory solution of equation (4.1.1).

Proof. Let the hypotheses of the theorem hold, so that

$$\int^{\infty} \left[|Q(s)| + \int_s^{\infty} Q^2(u)du \right] ds < \infty$$

(Q is absolutely integrable in view of (4.3.2)).

To prove the existence of a nontrivial nonoscillatory solution of (4.1.1) we shall employ the Schauder fixed point theorem. Without loss of generality, let $I = [b, c]$ with $0 < b < c$, and choose positive constants α, ϵ with $\alpha > \epsilon$ such that $[\alpha - \epsilon, \alpha + \epsilon] \subset [b, c]$. We define the constants A, A_1, B and the function $P(t)$ as follows

$$A = \sup_{b \le u \le c} f(u), \qquad A_1 = \sup_{b \le u \le c} f'(u) \qquad (4.3.5)$$

$$B = 2AA_1, \qquad P(t) = A|Q(t)| + B \int_t^{\infty} Q^2(s)ds. \qquad (4.3.6)$$

Now, we choose $T > 0$ so large that

$$A_1 \int_T^{\infty} |Q(s)|ds \le \frac{1}{2}, \qquad A_1 P(s) \left| \int_s^{\infty} q(u)du \right| \le 1 \qquad (4.3.7)$$

for all s, $t \ge T$, and

$$\int_T^{\infty} P(s)ds \le \epsilon. \qquad (4.3.8)$$

Define X to be the space of bounded, absolutely continuous functions on $[T, \infty)$ whose derivatives are almost everywhere bounded in absolute value by some multiple of $P(t)$.

Let $\|\cdot\|_{\infty}$ denote the uniform norm. We make X into a Banach space by introducing the norm

$$\|x\| = \|x\|_{\infty} + \|P^{-1}x'\|_{\infty}.$$

For $n = 1, 2, \cdots$ define X_n to be the subset of functions x in X such that $\|x - \alpha\|_{\infty} \le \epsilon$, $\|P^{-1}x'\|_{\infty} \le 1$, and

$$\|x'(t_2) - x'(t_1)\| \le A \left| \int_{t_1}^{t_2} q(s)ds \right| + |t_2 - t_1|$$

for all t_1, $t_2 \in [T, T + n)$ and x is constant for $t \ge T + n$. Then, X_n is a convex subset of X, and by the Ascoli theorem X_n is compact.

For $x \in X_n$ define the map T_n by

$$(T_n x)(t) = \begin{cases} \alpha - \int_t^{\infty} \int_s^{\infty} q(u)f(x(u))duds & \text{for } T \le t \le T + n \\ \alpha - \int_{T+n}^{\infty} \int_s^{\infty} q(u)f(x(u))duds & \text{for } T + n \le t. \end{cases}$$

We note that since $x(t) \in X_n$ is constant on $[T+n, \infty)$, the integrability of q and $|Q|$ imply that T_n is a well–defined map of X_n at least into the space of absolutely continuous function with almost everywhere bounded derivatives.

We shall now show that T_n is a continuous map of X_n into itself. Let $x \in X_n$. Then for $T \le t < T+n$, we have

$$(T_n x)'(t) = \int_t^\infty q(s)f(x(s))ds = f(x(t))Q(t) + \int_t^\infty f'(x(s))Q(s)x'(s)ds.$$

Thus, by (4.3.5) – (4.3.7) it follows that

$$
\begin{aligned}
|(T_n x)'(t)| &\le A|Q(t)| + \int_t^\infty A_1|Q(s)| \left[A|Q(s)| + B \int_s^\infty Q^2(u)du \right] ds \\
&\le A|Q(t)| + \frac{1}{2} B \int_t^\infty Q^2(s)ds + A_1 B \left(\int_t^\infty Q^2(s)ds \right) \\
&\quad \times \left(\int_t^\infty |Q(s)|ds \right) \le P(t).
\end{aligned}
$$

Therefore, $T_n(X_n) \subset X$. Furthermore, if $T \le t_1 < t_2 < T+n$, we have

$$
\begin{aligned}
|(T_n x)'(t_2) - (T_n x)'(t_1)| &= \left| \int_{t_1}^{t_2} q(s)f(x(s))ds \right| \\
&= \left| f(x(t_1)) \int_{t_1}^{t_2} q(s)ds + \int_{t_1}^{t_2} f'(x(s))x'(s) \left(\int_s^{t_2} q(u)du \right) ds \right| \\
&\le A \left| \int_{t_1}^{t_2} q(s)ds \right| + \int_{t_1}^{t_2} A_1 P(s) \left| \int_s^\infty q(u)du \right| ds.
\end{aligned}
$$

Using (4.3.7) in the above inequality, we obtain

$$|(T_n x)'(t_2) - (T_n x)'(t_1)| \le A \left| \int_{t_1}^{t_2} q(s)ds \right| + |t_2 - t_1|.$$

For $T \le t \le T+n$, we find

$$
\begin{aligned}
|T_n x(t) - \alpha| &= \left| \int_t^\infty \int_s^\infty q(u)f(x(u))duds \right| \\
&\le \int_t^{T+n} \left[A|Q(s)| + B \int_s^\infty Q^2(u)du \right] ds + \int_{T+n}^\infty f(x[T+n]) \left| \int_s^\infty q(u)du \right| ds.
\end{aligned}
$$

Thus, by (4.3.8), we get $|T_n x(t) - \alpha| \le \int_t^\infty P(s)ds \le \epsilon$. Since $T_n x(t)$ is constant for $t \ge T+n$, this shows that $T_n(X_n) \subset X_n$.

Now, let $x, y \in X_n$ with $\|x - y\| \leq \delta$ ($\delta > 0$ is a constant). Define

$$a_0(\delta) = \sup_{\substack{b \leq u, v \leq a + \epsilon \\ \|u - v\| \leq \delta}} |f(u) - f(v)| \quad \text{and} \quad a_1(\delta) = \sup_{\substack{b \leq u, v \leq a + \epsilon \\ \|u - v\| \leq \delta}} |f'(u) - f'(v)|.$$

Then, we have

$$|(T_n x)' - (T_n y)'|(t)$$

$$= \left| \int_t^\infty [f(x(s)) - f(y(s))] q(s) ds \right|$$

$$\leq |f(x(t)) - f(y(t))||Q(t)| + \left| \int_t^\infty [f'(x(s))x'(s) - f'(y(s))y'(s)]Q(s) ds \right|$$

$$\leq |f(x(t)) - f(y(t))||Q(t)| + \int_t^\infty |f'(x(s)) - f'(y(s))||x'(s)||Q(s)| ds$$

$$+ \int_t^\infty |f'(y(s))||x'(s) - y'(s)||Q(s)| ds$$

$$\leq a_0(\delta)|Q(t)| + \int_t^\infty \left[a_1(\delta) \left(A|Q(s)| + B \int_s^\infty Q^2(u) du \right) \right.$$

$$\left. + A_1 \delta \left(A|Q(s)| + B \int_s^\infty Q^2(u) du \right) \right] |Q(s)| ds$$

$$\leq a_0(\delta)|Q(t)| + (a_1(\delta) + \delta A_1) \left[A \int_t^\infty Q^2(s) ds \right.$$

$$\left. + \left(B \int_t^\infty Q^2(s) ds \right) \left(\int_t^\infty |Q(s)| ds \right) \right]$$

$$\leq a_0(\delta)|Q(t)| + (a_1(\delta) + \delta A_1)(A + \epsilon B) \int_t^\infty Q^2(s) ds \leq b(\delta) Q(t),$$

say, where $b(\delta) \to 0$ as $\delta \to 0$. Integration of the last inequality gives

$$|T_n x - T_n y| \leq \int_t^\infty |(T_n x)' - (T_n y)'|(s) ds + |T_n x - T_n y|(T + n)$$

$$\leq b(\delta) \int_t^\infty P(s) ds + b_1(\delta) \int_t^\infty P(s) ds \leq \epsilon(b(\delta) + b_1(\delta)), \quad \text{say,}$$

where $b_1(\delta)$ has the same property as $b(\delta)$, and so $\|T_n x - T_n y\| \leq (b(\delta) + b_1(\delta))(1 + \epsilon)$. Hence, it follows that T_n is continuous. Now, we can apply the Schauder fixed point theorem to obtain a function $x_n \in X_n$ such that $T_n x_n = x_n$. Thus, for $T \leq t < T + n$, $x_n''(t) + q(t) f(x_n(t)) = 0$. The sequence $\{x_n\}$ is equicontinuous on $[T, \infty)$. It follows by a standard diagonal argument that there is a subsequence $\{x_{n_k}\}$ which converges uniformly on compact intervals to a function x which is nontrivial, bounded nonoscillatory solution on $[T, \infty)$ of equation (4.1.1). ∎

We can combine Corollary 4.1.4 and Theorem 4.3.1 to obtain the following necessary and sufficient condition for the oscillation of (4.1.1).

Corollary 4.3.1. Let $Q(t) = \int_t^\infty q(s)ds$ exist with $Q(t) \geq 0$ for $t \geq t_0$, and let f be strongly nonlinear, i.e., f satisfies the conditions

(f_1) $f \in C^1(\mathbb{R}, \mathbb{R})$, $xf(x) > 0$ for $x \neq 0$, $f'(x) > 0$ for $x \neq 0$, and

(f_2) $\liminf_{|x| \to \infty} f'(x) > 0$ and condition (4.1.5) holds.

Then equation (4.1.1) is oscillatory if and only if

$$\int^\infty \left[Q(s) + \int_s^\infty Q^2(u)du \right] ds = \infty.$$

If we remove the sign condition on Q, then a necessary condition for (4.1.1) to be oscillatory is

$$\int^\infty \left[|Q(s)| + \int_s^\infty Q^2(u)du \right] ds = \infty.$$

4.3.2. Comparison of Nonlinear Equations of the Same Form

The classical Sturm comparison theorem for the second order linear differential equations

$$(a_1(t)x'(t))' + q_1(t)x(t) = 0, \tag{4.3.9}$$

$$(a(t)x'(t))' + q(t)x(t) = 0 \tag{4.3.10}$$

is stated as follows: If a, a_1, q, q_1 are continuous functions with $0 < a(t) \leq a_1(t)$, $q_1(t) \leq q(t)$ for all t in $[t_0, \infty)$, and if equation (4.3.9) is oscillatory, then (4.3.10) is also oscillatory.

A clever application of the Sturm theorem gives the so-called Hille–Wintner comparison theorem which is in terms of the integrals of the coefficient functions q_1, q.

Theorem 4.3.2. Let $a(t) = a_1(t) = 1$, and let

$$Q_1(t) = \int_t^\infty q_1(s)ds, \quad Q(t) = \int_t^\infty q(s)ds \tag{4.3.11}$$

exist with $0 \leq Q_1(t) \leq Q(t)$ for all t in $[t_0, \infty)$. If equation (4.3.9) is oscillatory, then (4.3.10) is also oscillatory.

This result has been extended to the following form:

Theorem 4.3.3. Let $a(t)$ be bounded above on $[t_0, \infty)$. Let $Q_1(t)$, $Q(t)$ in (4.3.11) exist and let $0 < a(t) \le a_1(t)$, $0 \le |Q_1(t)| \le Q(t)$ for all t in $[t_0, \infty)$. If equation (4.3.9) is oscillatory, then (4.3.10) is also oscillatory.

We shall extend Theorems 4.3.2 and 4.3.3 to nonlinear equations of the form

$$(a_1(t)x'(t))' + q_1(t)f(x(t)) = 0, \qquad (4.3.12)$$

$$(a(t)x'(t))' + q(t)f(x(t)) = 0. \qquad (4.3.13)$$

Theorem 4.3.4. Let a, a_1, q, q_1 be continuous on $[t_0, \infty)$ such that $Q(t)$ and $Q_1(t)$ defined in (4.3.11) exist and such that $0 < a(t) \le a_1(t)$, $|Q_1(t)| \le Q(t)$ for all $t \in [t_0, \infty)$. Assume that the function f satisfies the following conditions:

(i) $f(x)$ is continuously differentiable, $xf(x) > 0$ for all $x \ne 0$, $f'(x) > 0$ for $x \ne 0$, and either

(ii) $f'(x)$ is nondecreasing on $[0, \infty)$ and is nonincreasing on $(-\infty, 0]$, or

(iii) $\liminf_{|x| \to \infty} f'(x) > 0$ and condition (4.1.5) holds.

If equation (4.3.12) is oscillatory, then (4.3.13) is also oscillatory.

Proof. It will be convenient to separate the proof into the following three cases:

(I) $\int^\infty du/a(u) = \infty$, $|f|$ convex (conditions (i) and (ii)),

(II) $\int^\infty du/a(u) = \infty$, f strongly superlinear (conditions (i) and (iii)),

(III) $\int^\infty du/a(u) < \infty$.

Case (I). Suppose (4.3.13) is not oscillatory. Then there is a solution $x(t)$ of (4.3.13) which is eventually of one sign, and without loss of generality, we assume that $x(t) > 0$ on $[t_0, \infty)$. Let $w(t) = a(t)x'(t)/f(x(t))$ to obtain the Riccati equation $w'(t) = -q(t) - w^2(t)f'(x(t))/a(t)$. Then for $t_0 \le t \le T$, we find

$$w(t) = w(T) + \int_t^T q(s)ds + \int_t^T \frac{1}{a(s)}w^2(s)f'(x(s))ds. \qquad (4.3.14)$$

Letting $T \to \infty$ in (4.3.14), the second term on the right–hand side has the nonnegative limit $Q(t)$, whereas the third term has either a nonnegative limit, or tends to ∞. Hence, it follows that $\lim_{T \to \infty} w(T) = \beta$, where $-\infty \le \beta < \infty$.

We shall show that $0 \le \beta < \infty$. Suppose on the contrary that $-\infty \le \beta < 0$. Then for sufficiently large t, say, $t \ge \bar{t}$, we have $a(t)x'(t) < 0$. If for some ϵ, $a(t)x'(t) \le -\epsilon < 0$ for $t \ge t^* \ge \bar{t}$, then it follows that $x(t) - x(t^*) \le -\epsilon \int_{t^*}^t ds/a(s) \to -\infty$ as $t \to \infty$, which contradicts

$x(t) > 0$ on $[t_0, \infty)$. Thus, we must have $\limsup_{t\to\infty} a(t)x'(t) = 0$. We obtain a contradiction as follows: Choose $t_n > \bar{t}$ with $t_n \to \infty$ as $n \to \infty$, so that for sufficiently large n and for $\bar{t} \le t \le t_n$, $-1/n = a(t_n)x'(t_n) > a(t)x'(t)$. Integrating equation (4.3.13) from t to t_n, $\bar{t} \le t < t_n$ for n large, we obtain

$$
\begin{aligned}
0 &= a(t_n)x'(t_n) - a(t)x'(t) + \int_t^{t_n} q(s)f(x(s))ds \; > \; \int_t^{t_n} q(s)f(x(s))ds \\
&= \left(\int_t^{t_n} q(s)ds \right) f(x(t)) + \int_t^{t_n} \left(\int_s^{t_n} q(u)du \right) f(x(s)) \left(\frac{x'(s)f'(x(s))}{f(x(s))} \right) ds \\
&= V(t) - \int_t^{t_n} H(s)V(s)ds,
\end{aligned}
$$

where

$$
V(t) \; = \; \left(\int_t^{t_n} q(s)ds \right) f(x(t)) \quad \text{and} \quad H(t) \; = \; \frac{-x'(t)f'(x(t))}{f(x(t))} \; > \; 0.
$$

Let $W(t) = \int_t^{t_n} H(s)V(s)ds$. Then the inequality $V(t) < \int_t^{t_n} H(s)V(s)ds$ gives $W'(t) = -H(t)V(t) > -H(t)W(t)$ from which it follows that

$$
\frac{d}{dt}\left[W(t)\exp\left(\int_{\bar{t}}^t H(s)ds \right) \right] \; > \; 0 \quad \text{for} \quad \bar{t} \le t < t_n
$$

and hence $0 = W(t_n)\exp\left(\int_t^{t_n} H(s)ds \right) > W(t)\exp\left(\int_{\bar{t}}^t H(s)ds \right)$, which implies that $W(t)$ and thus $V(t)$ is negative on $[\bar{t}, t_n)$. But this in turn implies that $\int_t^{t_n} q(s)ds < 0$ for $\bar{t} \le t < t_n$ for n sufficiently large, contradicting that $0 \le Q(t) = \int_t^\infty q(s)ds$. Hence, on taking limit in (4.3.14), we find

$$
w(t) \; = \; \beta + Q(t) + \int_t^\infty \frac{1}{a(s)}w^2(s)f'(x(s))ds, \tag{4.3.15}
$$

where $0 \le \beta < \infty$ as claimed. Now, since $x'(t)/f(x(t)) = w(t)/a(t)$, we have

$$
\int_\alpha^{x(t)} \frac{du}{f(u)} \; = \; \int_{t_0}^t \left(\frac{w(s)}{a(s)} \right) ds,
$$

where $\alpha = x(t_0) > 0$. Denoting $\int_\alpha^{x(t)} du/f(u)$ by $\Omega(x)$ for $x > 0$ and the inverse function of Ω by Γ, we get

$$
x(t) \; = \; \Gamma\left[\int_{t_0}^t \left(\frac{w(s)}{a(s)} \right) ds \right]
$$

and hence (4.3.15) can be rewritten as

$$w(t) = \beta + Q(t) + \int_t^\infty \left(\frac{w^2(s)}{a(s)}\right) f'\left(\Gamma\left[\int_{t_0}^t \left(\frac{w(u)}{a(u)}\right) du\right]\right) ds. \quad (4.3.16)$$

Conditions (i) and (ii) imply that Γ is monotone increasing from some interval $(-\infty, c)$ to $\mathbb{R}^+$ (c may be $+\infty$) and that we can write (4.3.16) as

$$w = M(w), \quad (4.3.17)$$

where M maps the set $C = \{u \in C([t_0, \infty), \mathbb{R}^+)$ with $0 \le u(t) \le w(t),\ t \in [t_0, \infty)\}$ into itself. M is monotone in the sense that $Mu < Mv$ for any $u,\ v \in C$ with $u < v$ under the natural ordering of C. Now consider the integral equation

$$z(t) = \beta + Q_1(t) + \int_t^\infty \left(\frac{z^2(s)}{a_1(s)}\right) f'\left(\Gamma\left[\int_{t_0}^s \left(\frac{z(u)}{a_1(u)}\right) du\right]\right) ds, \quad (4.3.18)$$

which we may write as

$$z = L(z). \quad (4.3.19)$$

Since $|Q_1(t)| \le Q(t)$ and $0 < a(t) \le a_1(t)$ on $[t_0, \infty)$, we find that L is a self map on $B = \{u \in C([t_0, \infty), \mathbb{R}^+)$ with $Q_1(t) \le u(t) \le z(t),\ t \in [t_0, \infty)\}$. Clearly, B is a closed convex subset of $C = C([t_0, \infty), \mathbb{R})$ the Fréchet space of continuous real–valued functions on $[t_0, \infty)$ with the compact–open topology. Let $z \in B$. Then, we have

$$|z(t)| \le w(t), \quad |Lz(t)| \le w(t), \quad t \in [t_0, \infty) \quad (4.3.20)$$

and for $t_0 \le s < t \le T$,

$$|(Lz)(t) - (Lz)(s)| \le \left|\int_s^t q_1(u) du\right| + \left|\int_s^t q(u) du\right| + |w(t) - w(s)|. \quad (4.3.21)$$

From (4.3.20) and (4.3.21) it follows that the functions in the image $L(B)$ of B are uniformly bounded and equicontinuous on compact subintervals of $[t_0, \infty)$. We claim that L is continuous on B. For this, for $z \in C$ define $\phi(z; s)$ by

$$\phi(z; s) = \left(\frac{z^2(s)}{a_1(s)}\right) f'\left(\Gamma\left[\int_{t_0}^s \left(\frac{z(u)}{a_1(u)}\right) du\right]\right)$$

so that

$$(Lz)(t) = \beta + Q_1(t) + \int_t^\infty \phi(z; s) ds. \quad (4.3.22)$$

Let $z,\ z_n \in C$ be such that $z_n \to z$ as $n \to \infty$ uniformly on compact subintervals of $[t_0, \infty)$. Let $I = [t_0, t_1]$, $\epsilon > 0$ be given and choose $T_1 \ge t_1$ so that

$$\int_{T_1}^\infty \left(\frac{w^2(s)}{a(s)}\right) f'\left(\Gamma\left[\int_{t_0}^s \left(\frac{w(u)}{a(u)}\right) du\right]\right) ds < \frac{\epsilon}{3}. \quad (4.3.23)$$

Noting that $\phi(z_n; s) \to \phi(z; s)$ as $n \to \infty$ uniformly on $[t_0, T_1]$, using (4.3.21), (4.3.22), (4.3.23) and the monotonicity of $f' \circ \Gamma$, we obtain for $t \in I$,

$$|(Lz)(t) - (Lz_n)(t)| \leq \left| \int_t^{T_1} (\phi(z; s) - \phi(z_n; s)) ds \right|$$

$$+ \int_{T_1}^{\infty} (|\phi(z; s)| + |\phi(z_n; s)|) ds < \frac{\epsilon}{3} + \frac{\epsilon}{3} < \epsilon,$$

for n sufficiently large. Thus, $Lz_n \to Lz$ as $n \to \infty$ uniformly on compact subintervals of $[t_0, \infty)$. It follows that L is continuous on B and so we can apply Schauder–Tychonov's fixed point theorem to deduce the existence of a fixed point Z of (4.3.20). Then, we find

$$X(t) = \Gamma \left[\int_{t_0}^{t} \left(\frac{Z(s)}{a_1(s)} \right) ds \right]$$

is a nonoscillatory solution of equation (4.3.12) on $[t_0, \infty)$. This completes the proof for case (I).

Case (II). A change of variables $s = \int_{t_0}^{t} d\tau/a(\tau)$, $y(s) = x(t(s))$ transforms equation (4.3.13) into

$$\frac{d^2 y}{ds^2} + a(t(s)) q(t(s)) f(y(s)) = 0. \tag{4.3.24}$$

The oscillation properties of the equation are invariant under this transformation. Let $r(s) = a(t(s)) q(l(s))$, and use the substitution $u = t(\tau)$, to obtain

$$R(s) = \int_s^{\infty} r(\tau) d\tau = \int_s^{\infty} a(t(\tau)) q(t(\tau)) d\tau = \int_{t(s)}^{\infty} q(u) du = Q(t(s)) > 0, \tag{4.3.25}$$

$$\int_T^{\infty} R(s) ds = \int_T^{\infty} Q(t(s)) ds = \int_{t(T)}^{\infty} \left(\frac{Q(u)}{a(u)} \right) du, \tag{4.3.26}$$

$$\int_T^{\infty} \left(\int_s^{\infty} R^2(\tau) d\tau \right) ds = \int_{t(T)}^{\infty} \frac{1}{a(u)} \left(\int_u^{\infty} \left(\frac{Q^2(\eta)}{a(\eta)} \right) d\eta \right) du. \tag{4.3.27}$$

It follows from Corollary 4.3.1 and (4.3.24) – (4.3.27) that equation (4.3.13) is oscillatory if and only if

$$\int^{\infty} \frac{1}{a(u)} \left[Q(u) + \int_u^{\infty} \left(\frac{Q^2(\eta)}{a(\eta)} \right) d\eta \right] du = \infty. \tag{4.3.28}$$

Similarly, we find that a necessary condition for (4.3.12) to be oscillatory is that

$$\int^{\infty} \frac{1}{a_1(u)} \left[|Q_1(u)| + \int_u^{\infty} \left(\frac{Q_1^2(\eta)}{a_1(\eta)} \right) d\eta \right] du = \infty. \tag{4.3.29}$$

Since $0 < a(t) \leq a_1(t)$ and $0 \leq |Q_1(t)| \leq Q(t)$, from (4.3.28) and (4.3.29) it follows that the theorem holds for case (II).

Case (III). We claim that condition (i) and the existence of the integral $Q_1(t)$ imply that (4.3.12) has a nonoscillatory solution. For this, we use the transformation $s = g(t) = \left(\int_t^\infty d\tau/a_1(\tau) \right)^{-1}$, $y(s) = sx(t)$. Now, if we denote $g(t_0)$ by s_0 and the inverse function of g by h, then this transformation changes (4.3.12) into

$$\frac{d^2y}{ds^2} + r_1(s)f\left(\frac{y(s)}{s}\right) = 0, \quad s \in [t_0, \infty) \qquad (4.3.30)$$

where $r_1(s) = a_1(h(s))q_1(h(s))/s^3$. Choose a constant $K \geq 1$ such that

$$0 \leq f'(x) \leq K \quad \text{whenever} \quad 0 \leq x \leq 2 \qquad (4.3.31)$$

and choose $T_0 \geq \max\{1, s_0\}$, so that

$$0 \leq Q_1^* = \max_{s \geq T_0} |Q_1(h(s))| \leq \frac{1}{24K}. \qquad (4.3.32)$$

Consider the integral equation

$$y(T) = 1 - \int_T^\infty \left(\int_s^\infty r_1(\tau)f\left(\frac{y(\tau)}{\tau}\right) d\tau \right) ds, \qquad (4.3.33)$$

which we write as $y = F(y)$. Let

$$R_1(s) = \int_s^\infty r_1(\tau)d\tau = \int_s^\infty \left(\frac{a_1(h(\tau))q_1(h(\tau))}{\tau^3} \right) d\tau$$

$$= \int_{h(s)}^\infty q_1(u) \left(\int_u^\infty \frac{d\eta}{a_1(\eta)} \right) = Q_1(h(s)) \left(\int_{h(s)}^\infty \frac{d\eta}{a_1(\eta)} \right) - \int_{h(s)}^\infty \frac{Q_1(u)}{a_1(u)} du.$$

Thus, $R_1(s)$ is well defined and satisfies

$$|R_1(s)| \leq 2Q_1^* \int_{h(s)}^\infty \frac{du}{a_1(u)} = \frac{2Q_1^*}{s} \quad \text{for} \quad s \geq T_0. \qquad (4.3.34)$$

Next, we denote $[T_0, \infty)$ by J and let X to be the Banach space of bounded continuously differentiable functions y on J such that ty' is bounded on J. In the space X we introduce the norm as $\|y\| = \|y\|_\infty + \|ty'\|_\infty$, where $\|\cdot\|_\infty$ denotes the supremum on J.

Let $\bar{B}$ be the closed ball in X with center 1 and radius $1/2$. We shall show that F maps $\bar{B}$ continuously into a compact subset of $\bar{B}$, so that the Schauder fixed point theorem can be used to guarantee the existence of $y^* \in \bar{B}$ such that $F(y^*) = y^*$. The function $y^*(s)$ will then be a nonoscillatory solution of (4.3.30), which in turn will give a nonoscillatory solution of equation (4.3.12). Now, in what follows we let $s \geq T_0 \geq 1$. If $y \in \bar{B}$, then we have $|y(s) - 1| \leq 1/2$ and $|sy'(s)| \leq 1/2$, and therefore

$$\left| \left(\frac{y(s)}{s} \right)' \right| = \left| \frac{y'(s)}{s} - \frac{y(s)}{s^2} \right| \leq \frac{2}{s^2}. \tag{4.3.35}$$

Noting that $0 \leq y(s)/s < 2/s < 2$ and using (4.3.31) (4.3.34), we get

$$|[F(y)]'(s)| = \left| \int_s^\infty r_1(\tau) f\left(\frac{y(\tau)}{\tau} \right) d\tau \right|$$

$$= \left| R_1(s) f\left(\frac{y(s)}{s} \right) + \int_s^\infty R_1(\tau) f'\left(\frac{y(\tau)}{\tau} \right) \left(\frac{y(\tau)}{\tau} \right)' d\tau \right|$$

$$\leq (2K) \left(\frac{|R_1(s)|}{s} \right) + \int_s^\infty (2K) \left(\frac{|R_1(\tau)|}{\tau^2} \right) d\tau \leq 6K \frac{Q_1^*}{s^2} \leq \frac{1}{4s}. \tag{4.3.36}$$

and so

$$|[F(y)](s) - 1| \leq 6K \frac{Q_1^*}{s} \leq \frac{1}{4}. \tag{4.3.37}$$

From (4.3.36) and (4.3.37), we find that $F(y) \in \bar{B}$ and hence F maps $\bar{B}$ into itself. Now, let $y, z \in \bar{B}$ with $\|y - z\| \leq \epsilon$. We have

$$|[F(y)]'(s) - [F(z)]'(s)| = \left| \int_s^\infty r_1(\tau) \left[f\left(\frac{y(\tau)}{\tau} \right) - f\left(\frac{z(\tau)}{\tau} \right) \right] d\tau \right|$$

$$\leq |R_1(s)| \left| f\left(\frac{y(s)}{s} \right) - f\left(\frac{z(s)}{s} \right) \right| + \int_s^\infty |R_1(\tau)||U(\tau)| d\tau, \tag{4.3.38}$$

where

$$U(\tau) = f'\left(\frac{y(\tau)}{\tau} \right) \left(\frac{y(\tau)}{\tau} \right)' - f'\left(\frac{z(\tau)}{\tau} \right) \left(\frac{z(\tau)}{\tau} \right)'.$$

Using (4.3.31) and the mean value theorem, we obtain

$$\left| f\left(\frac{y(s)}{s} \right) - f\left(\frac{z(s)}{s} \right) \right| \leq K \frac{|y(s) - z(s)|}{s} \leq \frac{K\epsilon}{s}. \tag{4.3.39}$$

Writing $U(\tau)$ as

$$\left[f'\left(\frac{y(\tau)}{\tau} \right) - f'\left(\frac{z(\tau)}{\tau} \right) \right] \left(\frac{y(\tau)}{\tau} \right)' + f'\left(\frac{z(\tau)}{\tau} \right) \left[\frac{y'(\tau) - z'(\tau)}{\tau} + \frac{z(\tau) - y(\tau)}{\tau^2} \right],$$

we can use (4.3.31) and (4.3.35), to get

$$|U(\tau)| \leq \left(\frac{2}{\tau^2} \right) \gamma\left(\frac{\epsilon}{\tau} \right) + 2K \frac{\epsilon}{\tau^2} \leq \frac{2}{\tau^2} (\gamma(\epsilon) + K\epsilon), \tag{4.3.40}$$

where γ is the modulus of continuity of f' on $[0, 2]$. Using (4.3.34), (4.3.39) and (4.3.40) in (4.3.38), we find

$$
\begin{aligned}
|[F(y)]'(s) - [F(z)]'(s)| &\leq \frac{2}{s^2} K Q_1^* \epsilon + 4 Q_1^* (\gamma(\epsilon) + K\epsilon) \left(\int_s^\infty \frac{d\tau}{\tau^3} \right) \\
&\leq \frac{1}{s^2} [Q_1^* (2K\epsilon + \gamma(\epsilon))]
\end{aligned}
$$

(4.3.41)

and therefore

$$
|[F(y)](s) - [F(z)](s)| \leq \frac{2}{s} [Q_1^* (2K\epsilon + \gamma(\epsilon))].
\tag{4.3.42}
$$

It follows from (4.3.41) and (4.3.42) that $\|F(y) - F(z)\| \leq 4Q_1^*(2K\epsilon + \gamma(\epsilon))$ and hence F is a continuous mapping of $\bar{B}$ into itself. Finally, let $y \in \bar{B}$ and $T_0 \leq \tau \leq T$. Then, we have

$$
|[F(y)]'(s) - [F(y)]'(\tau)| = \left| \int_\tau^s r_1(\mu) f\left(\frac{y(\mu)}{\mu} \right) d\tau \right| \leq 2K \int_\tau^s |r_1(\mu)| d\mu.
$$

(4.3.43)

From (4.3.43), it follows that the image $F(\bar{B})$ of $\bar{B}$ under F consists uniformly bounded functions on J, with uniformly bounded derivatives on J that are equicontinuous on compact subsets of J. Thus, we can apply the Ascoli–Arzela theorem to deduce that $F(\bar{B})$ has a compact closure in X. Now, the Schauder fixed–point theorem completes the proof of case (III). ■

Remark 4.3.1. In equations (4.3.12) and (4.3.13) the function $f(x) = x$ satisfies conditions (i) and (ii), whereas $f(x) = x^\alpha$ where $\alpha > 1$ is the ratio of odd integers satisfies conditions (i), (ii) and (iii). Hence, the following corollary is immediate.

Corollary 4.3.2. Theorem 4.3.2 for (4.3.9) and (4.3.10) holds without the restriction that $a(t)$ is bounded, and it extends to nonlinear equations of the form (4.3.12) and (4.3.13) where $f(x) \equiv x^\alpha$ and $\alpha > 1$ is the ratio of odd integers.

4.3.3. Linearization of Oscillation Theorems

In Chapter 2, we have shown via certain Riccati transformations that the oscillatory behavior of equations of type (4.1.3) is equivalent to the oscillatory behavior of linear differential equations provided conditions (i) – (iii) on the coefficients of (4.1.3) and the condition (4.1.28) are satisfied. Here, via comparison we shall relate the oscillation of the superlinear equations to that of certain linear differential equations.

Theorem 4.3.5. Suppose condition (4.1.28) holds. If the equation

$$
(a(t)y'(t))' + kq(t)y(t) = 0
\tag{4.3.44}
$$

is oscillatory, then equation (4.1.2) is oscillatory.

Proof. Let $x(t)$ be a nonoscillatory solution of (4.1.2), then $w(t) = -a(t)x'(t)/f(x(t))$ satisfies the Riccati differential equation $w'(t) = q(t) + w^2(t)f'(x(t))/a(t)$ for some $t_0 \geq 0$. Let $p(t) = a(t)/f'(x(t))$. Then the Riccati equation

$$w'(t) = q(t) + \frac{1}{p(t)}w^2(t)$$

has a solution on $[t_0, \infty)$. It is well-known that this is equivalent to the nonoscillation of the linear equation

$$(p(t)y'(t))' + q(t)y(t) = 0, \quad t \geq 0. \tag{4.3.45}$$

By assumption $p(t) \leq a(t)/k$, and hence in view of the Picone–Sturm theorem the equation $((a(t)/k)x'(t))' + q(t)x(t) = 0$, is nonoscillatory, which contradicts the hypothesis. ∎

We are now in the position to prove the following results.

Theorem 4.3.6. Suppose condition (F_2) is satisfied and let the functions $\rho(t)$, $\beta(t)$ and $\xi(t)$ be as in Theorem 4.2.1, and conditions (4.2.3) – (4.2.5) hold. If for every constant $k > 0$ the linear equation

$$\left(\frac{1}{v(t)}y'(t)\right)' + kB(t)y(t) = 0 \tag{4.3.46}$$

is oscillatory, where

$$B(t) = \rho(t)q(t) + k\left[\left(\frac{\beta(t)}{2a(t)\rho(t)v(t)}\right)' - \frac{k}{4v(t)}\left(\frac{\beta(t)}{a(t)\rho(t)}\right)^2\right]$$

and $v(t) = 1/(a(t)\rho(t)\xi(t))$, $t \geq t_0$ then equation (4.1.3) is oscillatory.

Proof. Let $x(t)$ be a nonoscillatory solution of equation (4.1.3), say, $x(t) > 0$ for $t \geq T \geq t_0$. Define $w(t) = \rho(t)a(t)x'(t)/f(x(t))$, $t \geq T$ and proceed as in Theorem 4.2.1 to obtain (4.2.7) and conclude that (4.2.8) and (4.2.9) are satisfied with $K = \beta(T)\int_{x(T)}^{\infty} du/f(u)$. Now, we consider the following two cases:

Case 1. $I(T, \infty) < \infty$. As in Theorem 4.2.2, we obtain (4.2.25) which takes the form

$$w'(t) \leq -\rho(t)q(t) + \frac{\beta(t)}{a(t)\rho(t)}w(t) - cv(t)w^2(t), \quad t \geq T.$$

Let $z(t) = w(t) - \beta(t)/(2ca(t)\rho(t)v(t))$, $t \geq T$. Then, we find that $z'(t) \leq -B(t) - cv(t)z^2(t)$ with $k = 1/c$ for $t \geq T$. Now, in view of Lemma 2.2.1 it follows that equation (4.3.46) is nonoscillatory, which is a contradiction.

Case 2. $I(T, \infty) = \infty$. Exactly as in Theorem 4.2.1, we arrive at the contradiction $\lim_{t \to \infty} x(t) = -\infty$. This completes the proof. ∎

Theorem 4.3.7. If conditions (4.2.75) and (4.2.76) in Theorem 4.3.12 are replaced by that the equation

$$\left(\frac{1}{v(t)} y'(t) \right)' + k\rho(t)q(t)y(t) = 0 \tag{4.3.47}$$

is oscillatory for every constant $k > 0$, where $v(t) = 1/(a(t)\rho(t)\xi(t))$ then equation (4.1.3) is oscillatory.

Proof. Let $x(t)$ be a nonoscillatory solution of equation (4.1.3), say, $x(t) > 0$ for $t \geq T \geq t_0$. Proceeding as in Theorem 4.2.12, we obtain (4.2.77) and the equation (4.2.78). Using condition (4.2.42), we find

$$f'(x(t)) \geq \frac{c}{\xi(t)} \quad \text{for} \quad t \geq T_1 \geq t_0, \tag{4.3.48}$$

where c is a positive constant. As in Theorem 4.3.5 since $y(t)$ satisfies (4.2.78) the corresponding linear equation

$$\left(\frac{a(t)\rho(t)}{f'(x(t))} y'(t) \right)' + \rho(t)q(t)y(t) = 0 \tag{4.3.49}$$

is nonoscillatory. Now, by (4.3.48) and as an application of the Picone–Sturm comparison theorem to (4.3.49), we conclude that the equation $(y'(t)/v(t))' + c\rho(t)q(t)y(t) = 0$ is nonoscillatory, which contradicts the given hypothesis. ∎

Remark 4.3.2.

1. We can proceed further to obtain more comparison results similar to those of Theorems 4.3.5 – 4.3.7 with different sets of conditions on the coefficients involved in the equations considered. The details are left to the reader.

2. We can use the results of Chapter 2 on linear equations to obtain interesting oscillation criteria for superlinear equations. These details are also left to the reader.

4.4. Oscillation Criteria for Certain Nonlinear Differential Equations

Here, we shall consider the nonlinear differential equation

$$\left(a(t)|x'(t)|^{\alpha-1} x'(t) \right)' + q(t)f(x(t)) = 0, \tag{4.4.1}$$

where

(i) α is a positive constant,

(ii) $a(t) \in C([t_0, \infty), \mathbb{R}^+)$ and $q(t) \in C([t_0, \infty), \mathbb{R})$,

(iii) $f \in C(\mathbb{R}, \mathbb{R})$, f is non–decreasing and $xf(x) > 0$ for $x \neq 0$.

The results we shall prove extend Lemmas 4.1.2 and 4.1.3 and Theorems 4.1.7 – 4.1.9. For this, we need the following lemma.

Lemma 4.4.1.   Let $K(t, s, x) : \mathbb{R} \times \mathbb{R} \times \mathbb{R}^+ \to \mathbb{R}$ be such that for all fixed t, s the function $K(t, s, \cdot)$ is nondecreasing. Further, let $p(t)$ be a given function and $u(t)$, $v(t)$ satisfy for $t \geq t_0$,

$$u(t) \geq (\leq) \, p(t) + \int_{t_0}^{t} K(t, s, u(s)) ds \qquad (4.4.2)$$

and

$$v(t) = p(t) + \int_{t_0}^{t} K(t, s, v(s)) ds. \qquad (4.4.3)$$

If $v^*(t)$ is the minimal (maximal) solution of the inequality (4.4.2), then $u(t) \geq (\leq) v^*(t)$ for all $t \geq t_0$.

Lemma 4.4.2.   Let $x(t)$ be a positive (negative) solution of equation (4.4.1) for $t \in [t_0, b]$, $b > t_0$. Suppose there exist $t_1 \in [t_0, b]$ and $\beta > 0$ such that for all $t \in [t_1, b]$,

$$-\frac{a(t_0)|x'(t_0)|^{\alpha-1}x'(t_0)}{f(x(t_0))} + \int_{t_0}^{t} q(s)ds + \int_{t_0}^{t_1} a(s)\frac{|x'(s)|^{\alpha+1}}{f^2(x(s))} f'(x(s))ds \geq \beta. \qquad (4.4.4)$$

Then the following holds

$$a(t)|x'(t)|^{\alpha-1}x'(t) \leq (\geq) - \beta f(x(t_1)) \quad \text{for} \quad t \in [t_1, b]. \qquad (4.4.5)$$

Proof.   Equation (4.4.1) is equivalent to

$$\frac{w'(t)}{f(x(t))} = -q(t), \qquad (4.4.6)$$

where $w(t) = a(t)|x'(t)|^{\alpha-1}x'(t)$. In view of (4.4.6) it is clear that

$$\begin{aligned}
\left(\frac{w(t)}{f(x(t))}\right)' &= \frac{w'(t)}{f(x(t))} - \frac{w(t)f'(x(t))x'(t)}{f^2(x(t))} \\
&= -q(t) - a(t)\frac{|x'(t)|^{\alpha+1}f'(x(t))}{f^2(x(t))}.
\end{aligned} \qquad (4.4.7)$$

Let $t \in [t_1, b]$. Then on integrating (4.4.7) from t_0 to t and then applying (4.4.4), we find

$$
\begin{aligned}
-\frac{w(t)}{f(x(t))} &= -\frac{w(t_0)}{f(x(t_0))} + \int_{t_0}^{t} q(s)ds + \int_{t_0}^{t} a(s)\frac{|x'(s)|^{\alpha+1}f'(x(s))}{f^2(x(s))}ds \\
&\geq \beta + \int_{t_1}^{t} a(s)\frac{|x'(s)|^{\alpha+1}f'(x(s))}{f^2(x(s))}ds > 0.
\end{aligned}
$$
(4.4.8)

Now, we consider the following two cases:

Case 1. Suppose $x(t)$ is positive. Then it follows from (4.4.8) that $-w(t) > 0$, or $x'(t) < 0$ for $t \in [t_1, t]$. Let $u(t) = -w(t) = a(t)|x'(t)|^{\beta}$. Then, (4.4.8) provides

$$
\begin{aligned}
u(t) &\geq \beta f(x(t)) + \int_{t_1}^{t} f(x(t))\frac{|x'(s)||f'(x(s))|}{f^2(x(s))}u(s)ds \\
&= \beta f(x(t)) + \int_{t_1}^{t} K(t, s, u(s))ds,
\end{aligned}
$$
(4.4.9)

where the function K is defined by

$$
K(t, s, y) = f(x(t))\frac{|x'(s)||f'(x(s))|}{f^2(x(s))}y, \quad t, s \in [t_1, b] \text{ and } y \in \mathbb{R}^+. \quad (4.4.10)
$$

It is obvious that for fixed t, s, $K(t, s, \cdot)$ is nondecreasing. With $p(t) = \beta f(x(t))$ an application of Lemma 4.4.1 to (4.4.9) yields

$$
u(t) \geq v^*(t) \quad \text{for} \quad t \in [t_1, b], \quad (4.4.11)
$$

where $v^*(t)$ is the minimal solution of the equation

$$
\begin{aligned}
v(t) &= \beta f(x(t)) + \int_{t_1}^{t} K(t, s, v(s))ds \\
&= \beta f(x(t)) + \int_{t_1}^{t} f(x(t))\frac{|x'(s)||f'(x(s))|}{f^2(x(s))}ds
\end{aligned}
$$
(4.4.12)

provided $v(t) \in \mathbb{R}^+$ for all $t \in [t_1, b]$.

It follows from (4.4.12) that

$$
\left(\frac{v(t)}{f(x(t))}\right)' = \left[\beta + \int_{t_1}^{t} \frac{|x'(s)||f'(x(s))|}{f^2(x(s))}v(s)ds\right]' = \frac{|x'(t)||f'(x(t))|}{f^2(x(t))}v(t).
$$
(4.4.13)

On the other hand a direct differentiation gives

$$
\begin{aligned}
\left(\frac{v(t)}{f(x(t))}\right)' &= \frac{v'(t)}{f(x(t))} - \frac{v(t)f'(x(t))v'(t)}{f^2(x(t))} \\
&= \frac{v'(t)}{f(x(t))} + \frac{v(t)|x'(t)||f'(x(t))|}{f^2(x(t))}.
\end{aligned}
$$
(4.4.14)

Equating (4.4.13) and (4.4.14), we get $v'(t) = 0$ for $t \in [t_1, b]$ and so, $v(t) = v(t_1) = \beta f(x(t_1))$. Hence, $v^*(t) = \beta f(x(t_1))$. The inequality (4.4.11) reduces to (4.4.5).

Case 2. Suppose $x(t)$ is negative. Then, (4.4.8) implies that $w(t) > 0$, or $x'(t) > 0$ for $t \in [t_1, b]$. With $u(t) = w(t) = a(t)|x'(t)|^\alpha$, $t \in [t_1, b]$ relation (4.4.8) yields

$$
\begin{aligned}
u(t) \ \geq \ & -\beta f(x(t)) + \int_{t_1}^t [-f(x(t))] \frac{|x'(s)||f'(x(s))|}{f^2(x(s))} u(s) ds \\
= \ & -\beta f(x(t)) + \int_{t_1}^t \bar{K}(t, s, u(s)) ds, \quad t \in [t_1, b]
\end{aligned}
\tag{4.4.15}
$$

where $\bar{K}(t, s, y) = -K(t, s, y)$.

We note that for fixed t, s, $\bar{K}(t, s, \cdot)$ is nondecreasing. Once again applying Lemma 4.4.1 to (4.4.15) with $p(t) = -\beta f(x(t))$ gives (4.4.11) where $v^*(t)$ is the minimal solution of the equation

$$
\begin{aligned}
v(t) \ = \ & -\beta f(x(t)) + \int_{t_1}^t \bar{K}(t, s, v(s)) ds \\
= \ & -\beta f(x(t)) + \int_{t_1}^t [-f(x(t))] \frac{|x'(s)||f'(x(s))|}{f^2(x(s))} v(s) ds.
\end{aligned}
\tag{4.4.16}
$$

Now, employing an argument similar to that of in Case 1, we obtain $v'(t) = 0$ and hence $v^*(t) = v(t_1) = -\beta f(x(t_1))$. The inequality (4.4.5) is now immediate from (4.4.11). ∎

Lemma 4.4.3. Suppose that

$$
\eta(t) = \int_{t_0}^t a^{-1/\alpha}(s) ds \quad \text{and} \quad \lim_{t \to \infty} \eta(t) = \infty,
\tag{4.4.17}
$$

$$
-\infty < \int_{t_0}^\infty q(s) ds < \infty
\tag{4.4.18}
$$

and

$$
\lim_{|x| \to \infty} |f(x)| = \infty.
\tag{4.4.19}
$$

If $x(t)$ is a nonoscillatory solution of equation (4.4.1), then

$$
\int_{t_0}^\infty a(s) \frac{|x'(s)|^{\alpha+1} f'(x(s))}{f^2(x(s))} ds < \infty,
\tag{4.4.20}
$$

$$
\lim_{t \to \infty} a(t) \frac{|x'(t)|^{\alpha-1} x'(t)}{f(x(t))} = 0
\tag{4.4.21}
$$

and

$$a(t)\frac{|x'(t)|^{\alpha-1}x'(t)}{f(x(t))} = \int_t^\infty a(s)\frac{|x'(s)|^{\alpha+1}f'(x(s))}{f^2(x(s))}ds + \int_t^\infty q(s)ds$$

$$(4.4.22)$$

for all sufficiently lage t.

Proof. Without loss of generality let $x(t)$ be an eventually positive solution of (4.4.1). First, we shall show (4.4.20). Suppose (4.4.20) does not hold. Then from (4.4.18) we find that (4.4.4) is satisfied for all $t \geq t_1$, where t_1 is sufficiently large. Apply Lemma 4.4.2 to conclude that $x'(t) < 0$ for $t \geq t_1$. Thus, (4.4.5) provides $a(t)|x'(t)|^\alpha \geq \beta f(x(t_1))$ for $t \geq t_1$, or

$$x'(t) \leq -[\beta f(x(t_1))]^{1/\alpha}a^{-1/\alpha}(t) \quad \text{for} \quad t \geq t_1. \tag{4.4.23}$$

Integrating (4.4.23) from t_1 to t, we get

$$x(t) \leq x(t_1) - [\beta f(x(t_1))]^{1/\alpha}\int_{t_1}^t a^{-1/\alpha}(s)ds \rightarrow -\infty \quad \text{as} \quad t \to \infty,$$

which is a contradiction to the fact that $x(t) > 0$ eventually.

Next, we shall verify (4.4.21) and (4.4.22). Integration of (4.4.7) from t_0 to t gives

$$a(t)\frac{|x'(t)|^{\alpha-1}x'(t)}{f(x(t))} = a(t_0)\frac{|x'(t_0)|^{\alpha-1}x'(t_0)}{f(x(t_0))} - \int_{t_0}^t q(s)ds$$

$$- \int_{t_0}^t a(s)\frac{|x'(s)|^{\alpha+1}f'(x(s))}{f^2(x(s))}ds. \tag{4.4.24}$$

In view of (4.4.18) and (4.4.20) it is clear from (4.4.24) that the limit

$$L = \lim_{t\to\infty} a(t)\frac{|x'(t)|^{\alpha-1}x'(t)}{f(x(t))} \tag{4.4.25}$$

exists. Now, let $t \to \infty$ in (4.4.24) and then change t_0 to t, to obtain

$$a(t)\frac{|x'(t)|^{\alpha-1}x'(t)}{f(x(t))} = L + \int_t^\infty q(s)ds + \int_t^\infty a(s)\frac{|x'(s)|^{\alpha+1}f'(x(s))}{f^2(x(s))}ds.$$

$$(4.4.26)$$

Therefore, (4.4.21) and (4.4.22) hold if $L = 0$. For this, we consider the following two cases:

Case 1. Suppose $L < 0$. Then, (4.4.18) and (4.4.20) lead to

$$\left|\int_t^\infty q(s)ds\right| \leq -\frac{L}{3} \quad \text{for} \quad t \geq t_1 \tag{4.4.27}$$

and

$$\left| \int_{t_1}^{\infty} a(s) \frac{|x'(s)|^{\alpha+1} f'(x(s))}{f^2(x(s))} ds \right| \leq -\frac{L}{3}. \tag{4.4.28}$$

Next, substituting $t = t_0$ in (4.4.26), we get

$$L = a(t_0) \frac{|x'(t_0)|^{\alpha-1} x'(t_0)}{f(x(t_0))} - \int_{t_0}^{\infty} q(s) ds - \int_{t_0}^{\infty} a(s) \frac{|x'(s)|^{\alpha+1} f'(x(s))}{f^2(x(s))} ds. \tag{4.4.29}$$

Now, using (4.4.29) and the inequalities (4.4.27) and (4.4.28), we find for $t \geq t_1$,

$$-a(t_0) \frac{|x'(t_0)|^{\alpha-1} x'(t_0)}{f(x(t_0))} + \int_{t_0}^{t} q(s) ds + \int_{t_0}^{t_1} a(s) \frac{|x'(s)|^{\alpha+1} f'(x(s))}{f^2(x(s))} ds$$

$$= -L - \int_{t}^{\infty} q(s) ds - \int_{t_1}^{\infty} a(s) \frac{|x'(s)|^{\alpha+1} f'(x(s))}{f^2(x(s))} ds$$

$$\geq -L + \frac{L}{3} + \frac{L}{3} = -\frac{L}{3} \equiv \beta > 0,$$

i.e., (4.4.4) holds. Hence, we can apply Lemma 4.4.2 to obtain a contradiction as in the proof of (4.4.20).

Case 2. Suppose $L > 0$. From the definition of L in (4.4.25), we can assume that

$$a(t) \frac{|x'(t)|^{\alpha-1} x'(t)}{f(x(t))} \geq \frac{L}{2} \quad \text{for} \quad t \geq t_1. \tag{4.4.30}$$

From (4.4.30), we find that $x'(t) > 0$ for $t \geq t_1$. Now, using (4.4.30) in (4.4.20) leads to

$$\infty > \int_{t_1}^{\infty} a(s) \frac{|x'(s)|^{\alpha+1} f'(x(s))}{f^2(x(s))} ds \geq \frac{L}{2} \int_{t_1}^{\infty} \frac{x'(s) f'(x(s))}{f(x(s))} ds$$

$$= \frac{L}{2} \lim_{t \to \infty} \ln \left[\frac{f(x(t))}{f(x(t_1))} \right],$$

from which it follows that $\lim_{t \to \infty} \ln[f(x(t))] < \infty$, or

$$\lim_{t \to \infty} f(x(t)) < \infty. \tag{4.4.31}$$

In view of (4.4.19) and (4.4.31), $x(t)$ must be bounded.

On the other hand in view of the monotonicity of f and x, relation (4.4.30) gives $a(t)|x'(t)|^{\alpha} \geq (L/2) f(x(t)) \geq (L/2) f(x(t_1))$ for $t \geq t_1$, or

$$x'(t) \geq \left[\frac{L}{2} f(x(t_1)) \right]^{1/\alpha} a^{-1/\alpha}(t) \quad \text{for} \quad t \geq t_1.$$

Integrating this inequality from t_1 to t and then using (4.4.18), we find that $x(t)$ is unbounded, which is a contradiction. This completes the proof. ∎

Theorem 4.4.1. Let condition (4.4.17) hold,

$$\int^{\infty} \frac{du}{f^{1/\alpha}(u)} < \infty \quad \text{and} \quad \int^{-\infty} \frac{du}{f^{1/\alpha}(u)} < \infty \qquad (4.4.32)$$

and suppose $Q(t) = \int_t^{\infty} q(s)ds$ exists and $Q(t) \geq 0$ for $t \geq t_0$. If either

$$\int^{\infty} \left(\frac{1}{a(s)} Q(s) \right)^{1/\alpha} ds = \infty, \qquad (4.4.33)$$

or

$$\frac{f'(x)}{f^{(\alpha-1)/\alpha}(x)} \geq k > 0 \quad \text{for} \quad x \neq 0 \quad \text{and} \quad k \quad \text{is a constant} \qquad (4.4.34)$$

and

$$\int^{\infty} \frac{1}{a^{1/\alpha}(s)} Q^{(\alpha+1)/\alpha}(s)ds = \infty, \qquad (4.4.35)$$

or condition (4.4.34) and

$$\limsup_{t \to \infty} \int_{t_0}^{t} \left(\frac{1}{a(s)} \left[Q(s) + k \int_s^{\infty} a^{-1/\alpha}(u) Q^{(\alpha+1)/\alpha}(u)du \right] \right)^{1/\alpha} ds = \infty, \qquad (4.4.36)$$

then equation (4.4.1) is oscillatory.

Proof. Suppose equation (4.4.1) has an eventually positive solution $x(t)$. Then by Lemma 4.4.3, (4.4.22) holds. Thus, we have

$$\frac{|x'(t)|^{\alpha-1} x'(t)}{f(x(t))} \geq \frac{1}{a(t)} Q(t) \geq 0 \quad \text{for} \quad t \geq t_1 \quad \text{for some} \quad t_1 \geq t_0,$$

and hence $x'(t) > 0$ for $t \geq t_1$. Now, since

$$\frac{x'(t)}{f^{1/\alpha}(x(t))} \geq \left(\frac{1}{a(t)} Q(t) \right)^{1/\alpha} \quad \text{for} \quad t \geq t_1$$

it follows that

$$\int_{x(t_1)}^{x(t)} \frac{du}{f^{1/\alpha}(u)} \geq \int_{t_1}^{t} \left(\frac{1}{a(s)} Q(s) \right)^{1/\alpha} ds \to \infty \quad \text{as} \quad t \to \infty.$$

This contradicts (4.4.32). We also find that

$$\int_t^\infty \frac{a(s)(x'(s))^{\alpha+1} f'(x(s))}{f^2(x(s))} ds = \int_t^\infty \frac{a(s)(x'(s))^{\alpha+1}}{f^{(\alpha+1)/\alpha}(x(s))} \left(\frac{f'(x(s))}{f^{(\alpha-1)/\alpha}(x(s))} \right) ds$$

$$\geq k \int_t^\infty a^{-1/\alpha}(s) Q^{(\alpha+1)/\alpha}(s) ds.$$

This contradicts (4.4.20) if (4.4.35) holds. Finally, it is easy to check from (4.4.22) that

$$\frac{a(t)(x'(t))^\alpha}{f(x(t))} \geq Q(t) + k \int_t^\infty a^{-1/\alpha}(s) Q^{(\alpha+1)/\alpha}(s) ds, \quad t \geq t_1$$

or

$$\frac{x'(t)}{f^{1/\alpha}(x(t))} \geq \left(\frac{1}{a(t)} \left[Q(t) + k \int_t^\infty a^{-1/\alpha}(s) Q^{(\alpha+1)/\alpha}(s) ds \right] \right)^{1/\alpha} \quad \text{for } t \geq t_1.$$

Integration of the above inequality from t_1 to t in view of (4.4.32) leads to a contradiction. This completes the proof. ∎

For subsequent results, we let

$$h_0(t) = \int_t^\infty q(s) ds \quad \text{exist and} \quad h_0(t) \geq 0 \quad \text{for all large } t$$

and for a positive integer n and a positive constant k define the following functions

$$h_1(t) = \int_t^\infty a^{-1/\alpha}(s) (h_0(s))^{(\alpha+1)/\alpha} ds$$

and

$$h_{n+1}(t) = \int_t^\infty a^{-1/\alpha}(s) [h_0(s) + k h_n(s)]^{(\alpha+1)/\alpha} ds, \quad n = 1, 2, \cdots.$$

Theorem 4.4.2. Let conditions (4.4.17) – (4.4.19) and (4.4.32) hold. If for every constant $k > 0$ there exists an integer N such that

$$h_n(t) \text{ exists for } n = 0, 1, \cdots, N-1 \text{ and } h_N(t) \text{ does not exist, } \quad (4.4.37)$$

then equation (4.4.1) is oscillatory.

Proof. Suppose equation (4.4.1) has an eventually positive solution $x(t)$. Then from (4.4.22), we have

$$\frac{a(t)|x'(t)|^{\alpha-1} x'(t)}{f(x(t))} = h_0(t) + \int_t^\infty \frac{a(s)|x'(s)|^{\alpha+1}}{f^{(\alpha+1)/\alpha}(x(s))} \left(\frac{f'(x(s))}{f^{(\alpha-1)/\alpha}(x(s))} \right) ds$$

$$\geq h_0(t) \geq 0 \quad \text{for} \quad t \geq t_1 \quad \text{for some} \quad t_1 \geq t_0.$$

$$(4.4.38)$$

Clearly, $x'(t) \geq 0$ for $t \geq t_1$ and $x'(t) \geq [h_0(t)f(x(t))/a(t)]^{1/\alpha}$ for $t \geq t_1$. As in Theorem 4.4.1, we find

$$\int_t^\infty a(s)\frac{(x'(s))^{\alpha+1}}{f^2(x(s))}f'(x(s))ds \geq k\int_t^\infty a^{-1/\alpha}(s)[h_0(s)]^{(\alpha+1)/\alpha}ds = kh_1(t).$$
(4.4.39)

If $N = 1$ in (4.4.37), then the right–hand side of (4.4.39) does not exist. This is a contradiction to (4.4.20).

Next, it follows from (4.4.38) and (4.4.39) that

$$\frac{a(t)|x'(t)|^{\alpha-1}x'(t)}{f(x(t))} \geq h_0(t) + kh_1(t) \quad \text{for} \quad t \geq t_1. \qquad (4.4.40)$$

Using a similar technique and relation (4.4.32), we get for $t \geq t_1$,

$$\int_t^\infty a(s)\frac{|x'(s)|^{\alpha+1}}{f^2(x(s))}f'(x(s))ds \geq \int_t^\infty ka^{-\frac{1}{\alpha}}(s)\left[h_0(s)+kh_1(s)\right]^{\frac{\alpha+1}{\alpha}}ds = kh_2(t).$$
(4.4.41)

If $N = 2$ in (4.4.37), then the right–hand side of (4.4.41) does not exist. This again contradicts (4.4.20). A similar argument yields a contradiction for any integer $N > 2$. This completes the proof. ∎

Remark 4.4.1. Lemmas 4.4.2 and 4.4.3 and Theorems 4.4.1 and 4.4.2 can be obtained for forced equations of the form

$$\left(a(t)|x'(t)|^{\alpha-1}x'(t)\right)' + q(t)f(x(t)) = e(t), \qquad (4.4.42)$$

where $e(t) \in C([t_0, \infty), \mathbb{R})$ and $\int^\infty |e(s)|ds < \infty$. In this case the conclusion of Theorems 4.4.1 and 4.4.2 need to be replaced by: every solution $x(t)$ of equation (4.4.42) is either oscillatory, or $\liminf_{t\to\infty} |x(t)| = 0$. The formulation of these results are left to the reader.

To prove our next linearization result, we need the following lemma.

Lemma 4.4.4. Let $\alpha > 0$ be a constant, $a(t) \in C([t_0, \infty), \mathbb{R}^+)$ and $q(t) \in C([t_0, \infty), \mathbb{R})$. Then the half–linear differential equation

$$\left(a(t)|x'(t)|^{\alpha-1}x'(t)\right)' + q(t)|x(t)|^{\alpha-1}x(t) = 0$$

is nonoscillatory if and only if there exist a number $T \geq t_0$ and a function $w(t) \in C^1([T, \infty), \mathbb{R})$ such that

$$w'(t) + q(t) + \alpha a^{-1/\alpha}(t)|w(t)|^{(\alpha+1)/\alpha} \leq 0 \quad \text{on} \quad [T, \infty).$$

Next, we shall consider a special case of (4.4.1), namely, the equation

$$\left(a(t)|x'(t)|^{\alpha-1}x'(t)\right)' + q(t)|x(t)|^{\beta-1}x(t) = 0, \qquad (4.4.43)$$

where $\beta > 0$ is a constant.

Theorem 4.4.3. Let $\beta > \alpha$ and condition (4.4.17) hold, and

$$\liminf_{t \to \infty} \int_{t_0}^t q(s)ds > -\infty. \tag{4.4.44}$$

If for every positive constant c the half–linear differential equation

$$\left(a(t)\eta^\alpha(t)|y'(t)|^{\alpha-1}y'(t)\right)' + cq(t)|y(t)|^{\alpha-1}y(t) = 0 \tag{4.4.45}$$

is oscillatory, then (4.4.43) is oscillatory.

Proof. Let $x(t)$ be a nonoscillatory solution of equation (4.4.43), say, $x(t) > 0$ for $t \geq t_0$. Define $w(t) = a(t)|x'(t)|^{\alpha-1}x'(t)/x^\beta(t)$ for $t \geq t_0$. Then for $t \geq t_0$, we have

$$\begin{aligned}
w'(t) &= -q(t) - \beta a(t)\frac{|x'(t)|^{\alpha+1}}{x^{\beta+1}(t)} \\
&= -q(t) - \beta a^{-1/\alpha}(t)|w(t)|^{(\alpha+1)/\alpha}\left(x^{(\beta-\alpha)/\alpha}(t)\right),
\end{aligned} \tag{4.4.46}$$

and hence

$$w(t) = w(t_0) - \int_{t_0}^t q(s)ds - \beta \int_{t_0}^t a(s)\frac{|x'(s)|^{\alpha+1}}{x^{\beta+1}(s)}ds. \tag{4.4.47}$$

Now, with respect to the integral

$$\mathcal{I}(t_0,t) = \int_{t_0}^t a(s)\frac{|x'(s)|^{\alpha+1}}{x^{\beta+1}(s)}ds$$

there are two cases to consider:

Case 1. $\mathcal{I}(t_0,\infty) < \infty$. In this case there exists a positive constant N such that $\mathcal{I}(t_0,t) \leq N$ for $t \geq t_0$. For $t \geq t_0$ we use Hölder's inequality, to obtain

$$\begin{aligned}
x^{1-\gamma}(t) - x^{1-\gamma}(t_0) &\leq (\gamma-1)\int_{t_0}^t \left|\frac{x'(s)}{x^\gamma(s)}\right| ds \\
&\leq (\gamma-1)(\mathcal{I}(t_0,t))^{\frac{1}{\alpha+1}}\left(\int_{t_0}^t a^{-\frac{1}{\alpha}}(s)ds\right)^{\frac{\alpha}{\alpha+1}} \leq (\gamma-1)N^{\frac{1}{\alpha+1}}\eta^{\frac{\alpha}{\alpha+1}}(t),
\end{aligned}$$

where $\gamma = (\beta+1)/(\alpha+1) > 1$.

By condition (4.4.17) there exist a constant $M > 0$ and $T \geq t_0$ such that $x^{1-\gamma}(t) \leq M\eta^{\alpha/(\alpha+1)}(t)$, or $x(t) \geq (\eta^{-\alpha/(\alpha+1)}(t)/M)^{1/(\gamma-1)}$ for $t \geq T$ and

$$x^{(\beta-\alpha)/\alpha}(t) \geq \frac{c_1}{\eta(t)} \quad \text{for} \quad t \geq T, \tag{4.4.48}$$

where $c_1 = (1/M)^{(\beta-\alpha)/[\alpha(\gamma-1)]}$.

Using (4.4.48) in (4.4.46), we find

$$w'(t) \leq -q(t) - \left(\frac{c_1\beta}{\alpha}\right)\alpha\left(\frac{1}{a(t)\eta^\alpha(t)}\right)^{1/\alpha}|w(t)|^{(\alpha+1)/\alpha} \quad \text{for} \quad t \geq T.$$

But, now in view of Lemma 4.4.4 equation (4.4.45) is nonoscillatory, which is a contradiction.

Case 2. $\mathcal{I}(t_0, \infty) = \infty$. From condition (4.4.44) in (4.4.47) it follows that for some constant λ,

$$-w(t) \geq \lambda + \beta\mathcal{I}(t_0, t) \quad \text{for every} \quad t \geq t_0. \tag{4.4.49}$$

Let $T \geq t_0$ be such that $\Lambda = \lambda + \beta\mathcal{I}(t_0, T) > 0$. Then, (4.4.49) implies that $w(t)$ is negative on $[T, \infty)$. Now, (4.4.49) gives

$$\frac{\beta a(t)\frac{|x'(t)|^{\alpha+1}}{x^{\beta+1}(t)}}{\lambda + \beta\mathcal{I}(t_0, t)} \geq -\beta\frac{x'(t)}{x(t)}, \quad t \geq T$$

and consequently for all $t \geq T$,

$$\ln\left[\frac{\lambda + \beta\mathcal{I}(t_0, t)}{\Lambda}\right] \geq \ln\left(\frac{x(T)}{x(t)}\right)^\beta.$$

Hence, $\lambda + \beta\mathcal{I}(t_0, t) \geq \Lambda(x(T)/x(t))^\beta$. Therefore, (4.4.49) yields $x'(t) \leq -\Lambda_1 a^{-1/\alpha}(t)$ for all $t \geq T$, where $\Lambda_1 = (\Lambda x^\beta(T))^{1/\alpha} > 0$. Thus, we have $x(t) \leq x(T) - \Lambda_1 \int_T^t a^{-1/\alpha}(s)ds$ for all $t \geq T$, which in view of (4.4.17) leads to the contradiction $\lim_{t\to\infty} x(t) = -\infty$. ∎

Now, for (4.4.1) we shall prove the following linearization result.

Theorem 4.4.4. Let conditions (4.4.17) and (4.4.34) hold. If the half–linear differential equation

$$\left(a(t)|y'(t)|^{\alpha-1}y'(t)\right)' + \left(\frac{k}{\alpha}\right)q(t)|y(t)|^{\alpha-1}y(t) = 0, \tag{4.4.50}$$

is oscillatory, then (4.4.1) is oscillatory.

Proof. Suppose equation (4.4.1) has an eventually positive solution $x(t)$, say, $x(t) > 0$ for $t \geq t_0$. Define $w(t) = a(t)|x'(t)|^{\alpha-1}x'(t)/f(x(t))$, $t \geq t_0$. Then for $t \geq t_0$, we have

$$w'(t) = -q(t) - a^{-1/\alpha}(t)|w(t)|^{(\alpha+1)/\alpha}\left(\frac{f'(x(t))}{f^{(\alpha-1)/\alpha}(x(t))}\right). \tag{4.4.51}$$

Using condition (4.4.34), we get

$$w'(t) \leq -q(t) - \frac{k}{\alpha}(\alpha)a^{-1/\alpha}(t)|w(t)|^{(\alpha+1)/\alpha} \quad \text{for} \quad t \geq t_0.$$

Now, applying Lemma 4.4.4 we find that (4.4.50) is nonoscillatory, which is a contradiction. ∎

To relax condition (4.4.34) from Theorem 4.4.4 and to extend Theorem 4.4.3 to more general equations of type (4.4.1), we require the following condition:

(F) $$\int^{\infty} \left(\frac{f'(u)}{f^2(u)}\right)^{1/(\alpha+1)} du < \infty \quad \text{and} \quad \int^{-\infty} \left(\frac{f'(u)}{f^2(u)}\right)^{1/(\alpha+1)} du < \infty$$

$$(4.4.52)$$

and

$$\min\left\{\inf_{x>0}\left(\frac{f'(x)}{f^{(\alpha-1)/\alpha}(x)}\right)^{\alpha/(\alpha+1)} \int_x^{\infty}\left(\frac{f'(u)}{f^2(u)}\right)^{1/(\alpha+1)} du,\right.$$
$$\left.\inf_{x<0}\left(\frac{f'(x)}{f^{(\alpha-1)/\alpha}(x)}\right)^{\alpha/(\alpha+1)} \int_x^{-\infty}\left(\frac{f'(u)}{f^2(u)}\right)^{1/(\alpha+1)} du\right\} > 0.$$

Remark 4.4.2. For $\alpha = 1$ conditions (F) and (F$_2$) are the same.

Theorem 4.4.5. Suppose conditions (F), (4.4.17) and (4.4.44) hold. If for every constant $c_1 > 0$ the equation

$$(a(t)\eta^{\alpha}(t)|z'(t)|^{\alpha-1}z'(t))' + c_1 q(t)|z(t)|^{\alpha-1}z(t) = 0 \qquad (4.4.53)$$

is oscillatory, then (4.4.1) is oscillatory.

Proof. Let $x(t)$ be a nonoscillatory solution of equation (4.4.1), say, $x(t) > 0$ for $t \geq t_0$. Define $w(t)$ as in Theorem 4.4.4 and obtain (4.4.51). Then, we have $w(t) = w(t_0) - \int_{t_0}^t q(s)ds - J(t_0, t)$, where

$$J(t_0, t) = \int_{t_0}^t a(s)\frac{|x'(s)|^{\alpha+1}}{f^2(x(s))}f'(x(s))ds.$$

There are two cases to consider:

Case 1. $J(t_0, \infty) < \infty$. In this case there exists a positive constant N such that $J(t_0, t) \leq N$ for $t \geq t_0$. For $t \geq t_0$, we use Hölder's inequality to obtain

$$\left|\int_{t_0}^t x'(s)\left(\frac{f'(x(s))}{f^2(x(s))}\right)^{\frac{1}{\alpha+1}} ds\right| \leq (J(t_0, t))^{\frac{1}{\alpha+1}}\left(\int_{t_0}^t a^{-1/\alpha}(s)ds\right)^{\frac{\alpha}{\alpha+1}}$$

$$\leq N^{1/(\alpha+1)}\eta^{\alpha/(\alpha+1)}(t) \quad \text{for} \quad t \geq t_0.$$

$$(4.4.54)$$

Condition (F) ensures that

$$\left(\frac{f'(x(t))}{f^{(\alpha-1)/\alpha}(x(t))}\right)^{\alpha/(\alpha+1)} \int_{x(t)}^{\infty} \left(\frac{f'(u)}{f^2(u)}\right)^{1/(\alpha+1)} du \geq S \quad \text{for} \quad t \geq t_0,$$
(4.4.55)

where S is a positive constant. Let

$$K^* = \int_{x(t_0)}^{\infty} \left(\frac{f'(u)}{f^2(u)}\right)^{1/(\alpha+1)} du > 0.$$

Then in view of (4.4.55) for $t \geq t_0$, we obtain

$$\frac{f'(x(t))}{f^{(\alpha-1)/\alpha}(x(t))}$$

$$\geq S^{(\alpha+1)/\alpha} \left[\int_{x(t)}^{\infty} \left(\frac{f'(u)}{f^2(u)}\right)^{1/(\alpha+1)} du\right]^{-(\alpha+1)/\alpha}$$

$$= S^{(\alpha+1)/\alpha} \left[K^* - \int_{x(t_0)}^{x(t)} \left(\frac{f'(u)}{f^2(u)}\right)^{1/(\alpha+1)} du\right]^{-(\alpha+1)/\alpha}$$

$$= S^{(\alpha+1)/\alpha} \left[K^* - \int_{t_0}^{t} x'(s) \left(\frac{f'(x(s))}{f^2(x(s))}\right)^{1/(\alpha+1)} ds\right]^{-(\alpha+1)/\alpha}$$

$$\geq S^{(\alpha+1)/\alpha} \left[K^* + \left|\int_{t_0}^{t} x'(s) \left(\frac{f'(x(s))}{f^2(x(s))}\right)^{1/(\alpha+1)} ds\right|\right]^{-(\alpha+1)/\alpha}$$

and hence on using (4.4.54), for every $t \geq t_0$, we get

$$\frac{f'(x(t))}{f^{(\alpha-1)/\alpha}(x(t))} \geq S^{(\alpha+1)/\alpha} \left[K^* + \left(N^{1/(\alpha+1)} \eta^{\alpha/(\alpha+1)}(t)\right)\right]^{-(\alpha+1)/\alpha}$$

and consequently, in view of (4.4.17), we find

$$\frac{f'(x(t))}{f^{(\alpha-1)/\alpha}(x(t))} \geq \frac{c}{\eta(t)} \quad \text{for all} \quad t \geq t_0, \tag{4.4.56}$$

where c is a positive constant which depends on S, K^*, N and α.

Now, from (4.4.56) equation (4.4.51) gives

$$w'(t) \leq -q(t) - \alpha \left(\frac{c}{\alpha} \frac{1}{(a(t)\eta^{\alpha}(t))^{1/\alpha}}\right) |w(t)|^{(\alpha+1)/\alpha} \quad \text{for} \quad t \geq t_0.$$

The rest of the proof is similar to that of Theorem 4.4.4 and hence omitted.

Case 2. $\mathcal{J}(t_0, \infty) = \infty$. The proof for this case is similar to that of Theorem 4.4.3 Case 2 and hence omitted.   ∎

Remark 4.4.3.

1. If $\alpha = 1$, then Theorems 4.4.4 and 4.3.5 are the same.

2. From Theorems 4.4.3 – 4.4.5 it is clear that the oscillatory results for half–linear equations presented in Chapter 3 can be used to obtain interesting oscillation criteria for nonlinear equations of type (4.4.1). The formulations of these results are left to the reader.

3. Theorems 4.4.1 and 4.4.2 hold for (4.4.1) without (4.4.34). This can be proved by using Theorem 4.4.5 and replacing the function $a(t)$ with $a(t)\eta^\alpha(t)$ in (4.4.35) and (4.4.36) of Theorem 4.4.1 and in $h_1(t)$ and $h_n(t)$ in Theorem 4.4.2. Further, in both theorems $k > 0$ is any constant.

In the following results we shall obtain oscillation criteria for (3.13.76) and (3.13.89) when $g(t) = t$.

Definition 4.4.1. We say that (3.13.76) is *strongly superlinear* if there exist a constant $c > 0$ such that $|y|^{-c}|F(t, y)|$ is nondecreasing in $|y|$ for each fixed t, and

$$\int^\infty \frac{du}{\psi^{-1}(u^c)} < \infty \quad \text{and} \quad \int^{-\infty} \frac{du}{\psi^{-1}(u^c)} < \infty. \qquad (4.4.57)$$

In view of this definition (4.4.43) is strongly superlinear if $\beta > \alpha$.

Now, we shall consider (3.13.76) with $g(t) = t$ and assume that

$$\int_{t_0}^\infty \left| \psi^{-1}\left(\frac{k}{a(s)} \right) \right| ds = \infty \quad \text{for every constant} \quad k \neq 0.$$

Theorem 4.4.6. Suppose (3.13.76) is strongly superlinear and

$$\psi^{-1}(uv) \geq \psi^{-1}(u)\psi^{-1}(v) \quad \text{for all} \quad u, \, v \quad \text{with} \quad uv > 0. \qquad (4.4.58)$$

Equation (4.4.76) is oscillatory if and only if

$$\int_{t_0}^\infty \left| \psi^{-1}\left(\frac{1}{a(s)} \int_s^\infty F(u, k)du \right) \right| ds = \infty \qquad (4.4.59)$$

for every constant $k \neq 0$.

Proof. The 'only if' part follows from Theorem 3.13.12. To prove the 'if' part assume that (3.13.76) has a nonoscillatory solution $x(t)$, say, $x(t) > 0$ for $t \geq t_0$. Integrating (3.13.76) from t to τ and letting

$\tau \to \infty$ and noting that $\lim_{t \to \infty} a(t)\psi(x'(t)) \geq 0$, we find $a(t)\psi(x'(t)) \geq \int_t^\infty F(s, x(s))ds$, which implies

$$x'(t) \geq \psi^{-1} \left(\frac{1}{a(t)} \int_t^\infty F(s, x(s))ds \right). \qquad (4.4.60)$$

We divide the above inequality by $\psi^{-1}(x^c(t))$, where $c > 0$ is a constant of strong superlinearity of equation (3.13.76) and use (4.4.58), to obtain

$$\frac{x'(t)}{\psi^{-1}(x^c(t))} \geq \psi^{-1} \left(\frac{1}{a(t)} \int_t^\infty \frac{F(s, x(s))}{x^c(s)}ds \right)$$

$$\geq \psi^{-1} \left(\frac{1}{a(t)} \int_t^\infty \frac{F(s, x(s))}{x^c(s)}ds \right) \quad \text{for} \quad t \geq t_0.$$

Since $x(t) \geq m$ for some constant $m > 0$, in view of the strong superlinearity of (3.13.76), we have $x^{-c}(t)F(t, x(t)) \geq m^{-c}F(t, m)$ for $t \geq t_0$, so that

$$\frac{x'(t)}{\psi^{-1}(x^c(t))} \geq \psi^{-1} \left((m^{-c})\frac{1}{a(t)} \int_t^\infty F(s, m)ds \right)$$

$$\geq \psi^{-1}(m^{-c})\psi^{-1} \left(\frac{1}{a(t)} \int_{t_0}^\infty F(s, m)ds \right) \quad \text{for} \quad t \geq t_0.$$

Integrating the last inequality from t_0 to t, we obtain

$$\psi^{-1}(m^{-c}) \int_{t_0}^t \psi^{-1} \left(\frac{1}{a(s)} \int_s^\infty F(u, m)du \right) ds \leq \int_{x(t_0)}^{x(t)} \frac{dw}{\psi^{-1}(w^c)},$$

which because of (4.4.57) implies

$$\int_{t_0}^\infty \psi^{-1} \left(\frac{1}{a(s)} \int_s^\infty F(u, m)du \right) ds < \infty.$$

But this contradicts (4.4.59) and completes the proof. ∎

The following result is a variant of Theorem 4.4.6.

Theorem 4.4.7. Suppose condition (4.4.58) holds and there exist functions $q(t) \in C([t_0, \infty), \mathbb{R}^+)$ and $g \in C(\mathbb{R}^+, \mathbb{R}^+)$ such that g is increasing,

$$|F(t, y)| \geq q(t)g(|y|) \quad \text{for} \quad (t, y) \in \mathbb{R}^+ \times \mathbb{R} \qquad (4.4.61)$$

and

$$\int_\epsilon^\infty \frac{du}{\psi^{-1} \circ g(u)} < \infty \quad \text{for any} \quad \epsilon > 0. \qquad (4.4.62)$$

Then equation (3.13.76) is oscillatory if and only if

$$\int_{t_0}^\infty \psi^{-1} \left(\frac{1}{a(s)} \int_s^\infty q(u)du \right) ds = \infty. \qquad (4.4.63)$$

Proof. Since the 'only if' part follows from Theorem 3.13.12, it suffices to prove the 'if' part. Assume that (3.13.76) has a nonoscillatory solution $x(t) > 0$ for $t \geq t_0 > 0$. As in Theorem 4.4.6, we obtain (4.4.60) which implies

$$x'(t) \geq \psi^{-1} \left(\frac{1}{a(t)} \int_t^\infty q(s)g(x(s))ds \right), \quad t \geq t_0.$$

Dividing the above inequality by $\psi^{-1}(g(x(t)))$ and using (4.4.58), we get

$$
\begin{aligned}
\frac{x'(t)}{\psi^{-1}(g(x(t)))} &\geq \psi^{-1} \left(\frac{1}{a(t)} \int_t^\infty q(s) \frac{g(x(s))}{g(x(t))} ds \right) \\
&\geq \psi^{-1} \left(\frac{1}{a(t)} \int_t^\infty q(s)ds \right) \quad \text{for} \quad t \geq t_0,
\end{aligned}
\tag{4.4.64}
$$

where we have used the fact that the function $g(x(s))$ is increasing and $g(x(s))/g(x(t)) \geq 1$ for $s \geq t$.

Integrating (4.4.64) from t_0 to t, we find

$$\int_{t_0}^t \psi^{-1} \left(\frac{1}{a(s)} \int_s^\infty q(u)du \right) ds \leq \int_{t_0}^t \frac{x'(s)}{\psi^{-1}(g(x(s)))} ds = \int_{x(t_0)}^{x(t)} \frac{dw}{\psi^{-1} \circ g(w)},$$

which in view of (4.4.62) implies

$$\int_{t_0}^\infty \psi^{-1} \left(\frac{1}{a(s)} \int_s^\infty q(u)du \right) ds < \infty.$$

But this contradicts condition (4.4.63) and completes the proof. ∎

Example 4.4.1. Consider (4.4.43) and assume that $\int^\infty a^{-1/\alpha}(s)ds = \infty$. From Theorem 4.4.7 it is clear that a necessary and sufficient condition for (4.4.43) to be oscillatory is that

$$\int_{t_0}^\infty \left(\frac{1}{a(s)} \int_s^\infty q(u)du \right)^{1/\alpha} ds = \infty \quad \text{if} \quad \beta > \alpha.$$

Next, we shall consider equation (3.13.89) with $g(t) = t$. Let $\pi(t) = \int_t^\infty a^{-1/\alpha}(s)ds$ and assume that $\pi(t_0) < \infty$.

Theorem 4.4.8. If for every nonzero constant k,

$$\int_{t_0}^\infty |F(s, k\pi(s))|ds = \infty \tag{4.4.65}$$

and

$$\int_{t_0}^\infty \left(\frac{1}{a(s)} \int_{t_0}^s |F(u, k\pi(u))|du \right)^{1/\alpha} ds = \infty, \tag{4.4.66}$$

then equation (3.13.89) is oscillatory.

Proof. In view of Theorem 3.13.17 condition (4.4.65) ensures the nonexistence of a nonoscillatory solution of type III of equation (3.13.89), (i.e., $\lim_{t\to\infty} |x(t)|/\,\pi(t)$ constant > 0). Since (4.4.66) implies

$$\int_{t_0}^{\infty} \left(\frac{1}{a(s)} \int_{t_0}^{s} |F(u, c_1)| du \right) ds \; = \; \infty \quad \text{for all constant} \quad c_1 \neq 0$$

the equation (3.13.89) has no nonoscillatory solution of type (I), (i.e., $\lim_{t\to\infty} |x(t)| = $ constant > 0). Thus, it suffices to show that there is no nonoscillatory solution of type II of (3.13.89), (i.e., $\lim_{t\to\infty} x(t) = 0$ and $\lim_{t\to\infty} |x(t)|/\pi(t) = \infty$). Let $x(t)$ be a type II solution of (3.13.89) which is eventually positive. Then, $x'(t)$ is eventually negative and proceeding as in Theorem 3.13.16, we obtain

$$\int_{t_0}^{\infty} \left(\frac{1}{a(s)} \int_{t_0}^{s} F(u, x(u)) du \right)^{1/\alpha} ds \; < \; \infty.$$

Since $x(t)$ satisfies $c_2 \pi(t) \leq x(t)$ for $t \geq t_0$ for some constant $c_2 > 0$, the above inequality implies that

$$\int_{t_0}^{\infty} \left(\frac{1}{a(s)} \int_{t_0}^{s} F(u, c_2 \pi(u)) du \right)^{1/\alpha} ds \; < \; \infty,$$

which contradicts the condition (4.4.66). Similarly, (3.13.89) does not admit eventually negative solutions of type II. This completes the proof. ∎

Theorem 4.4.9. Let equation (3.13.89) be strongly superlinear. Then equation (3.13.89) is oscillatory if and only if

$$\int_{t_0}^{\infty} |F(s, k\pi(s))| ds \; = \; \infty \tag{4.4.67}$$

for any nonzero constant k.

Proof. The 'only if' part follows from Theorem 3.13.17. To prove the 'if' part, assume that (3.13.89) has a nonoscillatory solution $x(t) > 0$ for $t \geq t_0 \geq 0$. Then either

(i) $x'(t) > 0, \, t \geq t_0,$ or (ii) $x'(t) < 0, \, t \geq t_1$ for some $t_1 \geq t_0$.

(i) If $x'(t) > 0$ for $t \geq t_0$, then $x(t) \geq c_1, \, t \geq t_0$ for some constant $c_1 > 0$. As in Theorem 3.13.16, we find that $\int_{t_0}^{\infty} F(s, x(s)) ds < \infty$ and hence $\int_{t_0}^{\infty} F(s, c_1) ds < \infty$. This implies that $\int_{t_0}^{\infty} F(s, c_1 \pi(s)) ds < \infty$, which contradicts condition (4.4.67).

(ii) If $x'(t) < 0$ for $t \geq t_1$, then there exists a constant $c_2 > 0$ such that $x(t) \geq c_2\pi(t)$ for $t \geq t_1$, (cf. Lemma 3.13.9), and strong superliniearity of (3.13.89) implies that

$$x^{-c}(t)F(t, x(t)) \geq (c_2\pi(t))^{-c}F(t, c_2\pi(t)) \quad \text{for} \quad t \geq t_1, \qquad (4.4.68)$$

where $c > \alpha$ is given in the Definition 4.4.1.

Next, as in Lemma 3.13.9, we obtain

$$x(t) \geq -(a^{1/\alpha}(t))\pi(t)x'(t) \quad \text{for} \quad t \geq t_1. \qquad (4.4.69)$$

Using (4.4.68) and (4.4.69), we find by noting $(x')^{\alpha^*} = |x'|^\alpha \operatorname{sgn} x'$ that

$$\left[-\left(-a(t)(x'(t))^{\alpha^*}\right)^{-(c-\alpha)/\alpha}\right]' = \frac{c-\alpha}{\alpha}\left(-a(t)(x'(t))^{\alpha^*}\right)^{-c/\alpha} F(t, x(t))$$

$$= \frac{c-\alpha}{\alpha}\left(-a(t)(x'(t))^{\alpha^*}\right)^{-c/\alpha}(x^c(t))\left(x^{-c}(t)\right)F(t, x(t))$$

$$\geq \frac{c-\alpha}{\alpha}\left(-a(t)(x'(t))^{\alpha^*}\right)^{-\frac{c}{\alpha}}\left[-a(t)(x'(t))^{\alpha^*}\right]^{\frac{c}{\alpha}}(\pi^c(t))(c_2\pi(t))^{-c}F(t, c_2\pi(t))$$

$$= \frac{c-\alpha}{\alpha}c_2^{-c}F(t, c_2\pi(t)) \quad \text{for} \quad t \geq t_1.$$

An integration of the above inequality yields

$$\frac{c-\alpha}{\alpha}c_2^{-c}\int_{t_1}^t F(s, c_2\pi(s))ds \leq \left[-a(t_1)(x'(t_1))^{\alpha^*}\right]^{-(c-\alpha)/\alpha} \quad \text{for} \quad t \geq t_1,$$

which implies that $\int_{t_1}^\infty F(s, c_2\pi(s))ds < \infty$. This again contradicts condition (4.4.67). ∎

Example 4.4.2. Once again consider (4.4.43) and assume that $a(t) = e^{\lambda t}$ and $q(t) = e^{\mu t}$, $t \geq 0$ where λ and μ are constants and $\lambda > 0$. Let $\alpha < \beta$. Then by Theorem 4.4.9 equation (4.4.43) is oscillatory if and only if $\mu/\lambda \geq \beta/\alpha$.

4.5. Superlinear Forced Oscillations

Here, we shall discuss the oscillatory behavior of the equation

$$(a(t)x'(t))' + p(t)x'(t) + q(t)|x(t)|^\gamma \operatorname{sgn} x(t) = e(t), \qquad (4.5.1)$$

where $\gamma > 0$, $a(t)$, $e(t)$, $p(t)$ and $q(t) \in C([t_0, \infty), \mathbb{R})$ and $a(t) > 0$ for $t \geq t_0$.

Theorem 4.5.1. Suppose for every $T \geq 0$ there exist $T \leq s_1 < t_1 \leq s_2 < t_2$ such that

$$e(t) \begin{cases} \leq 0 & \text{for} \quad t \in [s_1, t_1] \\ \geq 0 & \text{for} \quad t \in [s_2, t_2]. \end{cases} \tag{4.5.2}$$

If there exist $\rho(t) \in C^1([t_0, \infty), \mathbb{R}^+)$ and $u(t) \in D(s_i, t_i) = \{u(t) \in C^1([s_i, t_i], \mathbb{R}) : u(t) \neq 0, u(s_i) = u(t_i) = 0\}$, $i = 1, 2$ such that for every constant $c > 0$,

$$Q_i(u) = \int_{s_i}^{t_i} \rho(s) \left[q(s)u^2(s) - ca(s) \left(u'(s) + \frac{\beta(s)}{a(s)\rho(s)} u(s) \right)^2 \right] ds \geq 0 \tag{4.5.3}$$

for $i = 1, 2$ where $\beta(t) = a(t)\rho'(t) - p(t)\rho(t)$, then

(i) every unbounded solution of equation (4.5.1) with $\gamma > 1$ is oscillatory,

(ii) every bounded solution of equation (4.5.1) with $\gamma < 1$ is oscillatory.

Proof. Suppose $x(t)$ is a nonoscillatory solution of equation (4.5.1), say, $x(t) > 0$ for $t \geq t_0 \geq 0$. Define $w(t) = -\rho(t)a(t)x'(t)/x^\gamma(t)$ for $t \geq t_0$. Then for $t \geq t_0$, we have

$$w'(t) = \rho(t)q(t) - \frac{e(t)\rho(t)}{x^\gamma(t)} + \gamma x^{\gamma-1}(t) \frac{w^2(t)}{a(t)\rho(t)} + \frac{\beta(t)}{a(t)\rho(t)} w(t). \tag{4.5.4}$$

By the hypothesis we can choose $s_1, t_1 \geq t_0$ so that $e(t) \leq 0$ on the interval $[s_1, t_1] = I$ with $s_1 < t_1$. On the interval I, $w(t)$ satisfies (4.5.4) and the inequality

$$w'(t) \geq \rho(t)q(t) + \frac{\beta(t)}{a(t)\rho(t)} w(t) + \gamma x^{\gamma-1}(t) \frac{w^2(t)}{a(t)\rho(t)} \quad \text{on} \quad I. \tag{4.5.5}$$

Let $u(t) \in D(s_1, t_1)$ be as in the hypothesis. Multiplying (4.5.5) by $u^2(t)$ and integrating over I, we find

$$\int_I 2u(s)u'(s)w(s)ds \geq \int_I u^2(s)\rho(s)q(s)ds + \int_I \frac{\beta(s)}{a(s)\rho(s)} u^2(s)w(s)ds$$
$$+ \int_I \gamma x^{\gamma-1}(s)u^2(s) \frac{w^2(s)}{a(s)\rho(s)} ds,$$

which is equivalent to

$$0 \geq \int_I \left[\left(\frac{\gamma x^{\gamma-1}(s)}{a(s)\rho(s)} \right)^{\frac{1}{2}} u(s)w(s) + \left(u'(s) + \frac{u(s)\beta(s)}{2a(s)\rho(s)} \right) \left(\frac{a(s)\rho(s)}{\gamma x^{\gamma-1}(s)} \right)^{\frac{1}{2}} \right] ds$$
$$+ \int_I \left[\rho(s)q(s)u^2(s) - \frac{a(s)\rho(s)}{\gamma x^{\gamma-1}(s)} \left(u'(s) + \frac{u(s)\beta(s)}{2a(s)\rho(s)} \right)^2 \right] ds.$$
$$\tag{4.5.6}$$

Now, we consider the following two cases:

(i) If $x(t)$ is an unbounded nonoscillatory solution of equation (4.5.1) with $\gamma > 1$ on I, then there exists a constant $K_1 > 0$ such that $x(t) \geq K_1$ on I. Then, we have

$$\frac{1}{\gamma}x^{1-\gamma}(t) \leq \frac{K_1^{1-\gamma}}{\gamma} = c_1 \quad \text{on} \quad I, \qquad (4.5.7)$$

where c_1 is a constant. Using (4.5.7) in (4.5.6) and proceeding as in Theorem 2.7.4, we arrive at the desired contradiction.

(ii) If $x(t)$ is a bounded nonoscillatory solution of equation (4.5.1) with $\gamma < 1$ on I, then there exists a constant $K_2 > 0$ such that $x(t) \leq K_2$ on I, and hence $(1/\gamma)x^{1-\gamma}(t) \leq K_2^{1-\gamma}/\gamma = c_2$ on I, where c_2 is a constant. The rest of the proof is similar to that of the above case and hence omitted. ∎

Next, we shall present the following criterion for the oscillation of all solutions of the superlinear equation (4.5.1), i.e., when $\gamma > 1$.

Theorem 4.5.2. Suppose for every $T \geq 0$ there exist $T \leq s_1 < t_1 \leq s_2 < t_2$ such that condition (4.5.2) holds and $q(t) \geq 0$ on $[s_1, t_1] \cup [s_2, t_2]$. If there exist $\rho(t) \in C^1([t_0, \infty), \mathbb{R}^+)$ and $u(t) \in D(s_i, t_i)$, $i = 1, 2$ where $D(s_i, t_i)$ is as in Theorem 4.5.1 such that

$$Q_i(u) = \int_{s_i}^{t_i} \left[E(s)u^2(s) - a(s)\rho(s)\left(u'(s) + u(s)\frac{\beta(s)}{2a(s)\rho(s)}\right)^2 \right] ds \geq 0$$
$$(4.5.8)$$

for $i = 1, 2$ where $\beta(t) = a(t)\rho'(t) - p(t)\rho(t)$ and

$$E(t) = (\rho(t)q(t))^{1/\gamma}(|e(t)|\rho(t))^{(\gamma-1)/\gamma},$$

then equation (4.5.1) with $\gamma > 1$ is oscillatory.

Proof. Let $x(t)$ be a nonoscillatory solution of equation (4.5.1), say, $x(t) > 0$ for $t \geq t_0 \geq 0$. Define $w(t) = -\rho(t)a(t)x'(t)/x(t)$ for $t \geq t_0$. Then for $t \geq t_0$, we have

$$w'(t) = \rho(t)q(t)x^{\gamma-1}(t) + \frac{\beta(t)}{a(t)\rho(t)}w(t) - \frac{e(t)\rho(t)}{x(t)} + \frac{1}{a(t)\rho(t)}w^2(t). \quad (4.5.9)$$

By the hypothesis, we can choose s_1, $t_1 \geq t_0$ so that $e(t) \leq 0$ on the interval $[s_1, t_1] = I$. On I, $w(t)$ satisfies the equation

$$w'(t) = \rho(t)q(t)x^{\gamma-1}(t) + \frac{|e(t)|\rho(t)}{x(t)} + \frac{\beta(t)}{a(t)\rho(t)}w(t) + \frac{1}{a(t)\rho(t)}w^2(t).$$

Since, for $t \in I$,

$$\inf_{x>0} \left[\rho(t)q(t)x^{\gamma-1}(t) + \frac{|e(t)|\rho(t)}{x(t)} \right] \geq (\rho(t)q(t))^{1/\gamma}(|e(t)|\rho(t))^{(\gamma-1)/\gamma}$$

on I, we have

$$w'(t) \geq (\rho(t)q(t))^{1/\gamma}(|e(t)|\rho(t))^{(\gamma-1)/\gamma} + \frac{\beta(t)}{a(t)\rho(t)}w(t) + \frac{w^2(t)}{a(t)\rho(t)}$$

$$= E(t) + \frac{\beta(t)}{a(t)\rho(t)}w(t) + \frac{w^2(t)}{a(t)\rho(t)}.$$

(4.5.10)

Multiplying both sides of (4.5.10) by $u^2(t)$ and integrating over I, we obtain

$$0 \geq \int_I E(s)u^2(s)ds + \int_I \left[\left(2u'(s)u(s) + \frac{\beta(s)u^2(s)}{a(s)\rho(s)} \right) w(s) + \frac{u^2(s)w^2(s)}{a(s)\rho(s)} \right] ds$$

$$= Q_1(u) + \int_I \left[\frac{u(s)w(s)}{\sqrt{a(s)\rho(s)}} + \left(u'(s) + \frac{\beta(s)u(s)}{2a(s)\rho(s)} \right) \sqrt{a(s)\rho(s)} \right]^2 ds.$$

The rest of the proof is similar to that of Theorem 2.7.4. ∎

Remark 4.5.1. If $\rho(t) = \exp\left(\int^t p(s)/a(s)ds \right)$, then $\beta(t) = 0$. Thus, with this choice of the weight function $\rho(t)$ simpler forms of Theorems 4.5.1 and 4.5.2 can be reformulated. The details are left to the reader.

Example 4.5.1. Consider the superlinear equation

$$\left(\frac{1}{t}x'(t) \right)' + \frac{1}{t^2}x'(t) + \frac{m}{t^\gamma} \sin t |x(t)|^\gamma \operatorname{sgn} x(t) = \cos t, \quad t > 0 \quad (4.5.11)$$

where $\gamma > 1$ and $m > 0$. For any $T > 0$ choose n sufficiently large so that $2n\pi \geq T$, and let $s_1 = 2n\pi$ and $t_1 = 2n\pi + (\pi/2)$ in (4.5.8), take $\rho(t) = t$ and $u(t) = \sin \pi(t - s_1)/(t_1 - s_1)$ if $t \in [s_1, t_1]$. Then, we have

$$Q_1(u) = \int_{s_1}^{t_1} \left[(\rho(s)q(s))^{1/\gamma}(|e(s)|\rho(s))^{(\gamma-1)/\gamma}u^2(s) - a(s)\rho(s)(u'(s))^2 \right] ds$$

$$= 4m^{1/\gamma} \int_0^{\pi/2} \sin^{2+(1/\gamma)} s \, \cos^{3-(1/\gamma)} s \, ds - 4 \int_0^{\pi/2} \cos^2 2s ds$$

$$= 2\beta^{1/\gamma} \frac{\Gamma\left(3 + \frac{1}{\gamma}\right)\Gamma\left(4 - \frac{1}{\gamma}\right)}{\Gamma(7)} - \pi,$$

where Γ is the Gamma function. If

$$m^{1/\gamma} \geq \frac{\pi\Gamma(7)}{2\Gamma\left(3 + \frac{1}{\gamma}\right)\Gamma\left(4 - \frac{1}{\gamma}\right)}, \quad (4.5.12)$$

then $Q_1(u) \geq 0$. Similarly, for $s_2 = 2n\pi + (\pi/2)$ and $t_2 = 2n\pi + \pi$, we can show that $Q_2(u) \geq 0$. It follows from Theorem 4.5.2 that equation (4.5.11) is oscillatory if (4.5.12) holds.

4.6. Oscillation of Nonlinear Differential Equations with Nonlinear Damping Term

In this section we shall consider the differential equation

$$x''(t) + p(t)g(x'(t)) + q(t)f(x(t)) = 0, \qquad (4.6.1)$$

where f is a superlinear restoring force, i.e., $f(x) \in C^1(\mathbb{R}, \mathbb{R})$, $f'(x) \geq 0$, $xf(x) > 0$ for $x \neq 0$, $f(0) = 0$ and f satisfies condition (4.1.5), $g(x) \in C(\mathbb{R}, \mathbb{R})$ and g is locally Lipschitz. In addition $p(t)$, $q(t) \in C([t_0, \infty), \mathbb{R})$ (locally integrable would suffice).

In what follows we shall extend the well–known result, namely, if $q(t) > 0$, $t \geq t_0$ then all solutions of (4.1.6) oscillate if and only if $\int_{t_0}^{\infty} sq(s)ds = \infty$.

4.6.1. Damping with Nonlinear Growth– Oscillation Criteria

We shall need the following lemmas.

Lemma 4.6.1. Let $g(0) = 0$ and suppose $Q(t) = \int_t^{\infty} q(s)ds$ exists as a finite number and $p(t) \in \mathcal{L}^\lambda[t_0, \infty)$, $t_0 \geq 0$ where $1 \leq \lambda \leq \infty$. If $x(t)$ is a bounded, eventually monotone solution of equation (4.6.1), then $\lim_{t \to \infty} x'(t) = 0$.

Proof. Let $x(t)$ be positive and monotone increasing in $[T, \infty)$ for some $T \geq t_0$ with $\lim_{t \to \infty} x(t) = c > 0$, where c is a constant. All other possibilities can be treated similarly.

Since $x'(t) \geq 0$ on $[T, \infty)$ and $x(t)$ is bounded, we find

$$\liminf_{t \to \infty} x'(t) = 0. \qquad (4.6.2)$$

Suppose for some $\epsilon_0 > 0$,

$$\limsup_{t \to \infty} x'(t) \geq \epsilon_0. \qquad (4.6.3)$$

Define μ by $(1/\lambda) + (1/\mu) = 1$. We assume that ϵ_0 is sufficiently small so that $|g(u)| \leq 1$ whenever $|u| \leq \epsilon_0$. It follows from (4.6.2) and (4.6.3) that there are disjoint intervals $[a_n, b_n] \subset [T, \infty)$ with $\lim_{n \to \infty} a_n = \infty$

such that $\epsilon_0 = x'(a_n) > x'(t) > x'(b_n) = \epsilon_0/2$, whenever $t \in (a_n, b_n)$. Integration of equation (4.6.1) from a_n to b_n gives

$$-\frac{1}{2}\epsilon_0 = -\int_{a_n}^{b_n} p(s)g(x'(s))ds - \int_{a_n}^{b_n} q(s)f(x(s))ds. \qquad (4.6.4)$$

Now, from the integration by parts and the general mean–value theorem for integrals, we get

$$\int_{a_n}^{b_n} q(s)f(x(s))ds$$

$$= Q(b_n)f(x(b_n)) - Q(a_n)f(x(a_n)) + \int_{a_n}^{b_n} Q(s)f'(x(s))x'(s)ds$$

$$= Q(b_n)f(x(b_n)) - Q(a_n)f(x(a_n)) + Q(c_n)[f(x(b_n)) - f(x(a_n))], \qquad (4.6.5)$$

for some $c_n \in (a_n, b_n)$, since $f'(x(s))x'(s) \geq 0$ for $s \geq T$. Let C be the $\mathcal{L}^\lambda$ norm of p, Q_n be the $\sup_{t \in [a_n, b_n]} |Q(t)|$ and use (4.6.4), (4.6.5) and Hölder's inequality, to obtain the estimate $\epsilon_0/2 \leq C(b_n - a_n)^{1/\mu} + 4Q_n f(c)$. Since $Q(t) = \int_t^\infty q(s)ds < \infty$, $\lim_{n\to\infty} Q_n = 0$ and so $4Q_n f(c) \leq \epsilon_0/4$ for $n \geq N$, say. This gives $b_n - a_n \geq (\epsilon_0/4C)^\mu$, $n \geq N$. But, since $x'(t) \geq \epsilon_0/2$ on $[a_n, b_n]$, we have

$$x(b_n) - x(a_n) \geq \left(\frac{\epsilon_0}{2}\right)\left(\frac{\epsilon_0}{4C}\right)^\mu, \quad n \geq N$$

contradicting $\lim_{t\to\infty} x(t) = c$. Therefore, we must have $\limsup_{t\to\infty} x'(t) = 0$. ∎

In the remainder of this section, we shall assume that there exist constants K, b with $b > 1$, such that

$$0 < g(u) \operatorname{sgn} u \leq K|u|^b \quad \text{for all} \quad u, \qquad (4.6.6)$$

and let $\gamma(u)$ be defined by

$$\gamma(u) = \begin{cases} \dfrac{g(u)}{u} & \text{if } u \neq 0 \\ 0 & \text{if } u = 0. \end{cases}$$

It follows that γ is bounded, nonnegative function of u and $\gamma[\phi(u)]$ is locally integrable for any $\phi(u) \in C(\mathbb{R}, \mathbb{R})$. We also define by $\beta = \min\{b, 2\}$ and $Q(t) = \int_t^\infty q(s)ds$ if exists. For $1 \leq \lambda \leq \infty$, $\mathcal{L}^\lambda$ is the space of λth power integrable functions on $[t_0, \infty)$, $t_0 \geq 0$ with the usual norm denoted by $|\cdot|_\lambda$.

The next two lemmas show that certain types of nonoscillatory behavior can be excluded by appropriate sign or integral conditions on the functions $p(t)$ and $q(t)$.

Lemma 4.6.2. (i) $p(t) \geq 0$ for all sufficiently large t and the existence of $Q(t) \geq 0$ with $Q(t) \notin \mathcal{L}^1[t_0, \infty)$ imply that there is no solution $x(t)$ of equation (4.6.1) satisfying $x(t)x'(t) \geq 0$ for all large t.

(ii) $p(t) \in \mathcal{L}^{1/(2-\beta)}[t_0, \infty)$, $q(t) \geq 0$ for all sufficiently large t with $Q(t) \notin \mathcal{L}^1[t_0, \infty)$ imply that there is no bounded solution $x(t)$ of equation (4.6.1) satisfying $x(t)x'(t) \geq 0$ for all large t.

Proof. Assume that $x(t)$ is a solution of equation (4.6.1) which is positive and increasing on $[t_0, \infty)$. We also suppose that t_0 is large enough so that all sign hypotheses hold on $[t_0, \infty)$. Define $w(t) = x'(t)/f(x(t))$, $t \geq t_0$. Then for $t \geq t_0$, we have $w(t) \geq 0$ for $t \geq t_0$, and

$$w'(t) = -q(t) - p(t)\gamma(x'(t))w(t) - f'(x(t))w^2(t). \qquad (4.6.7)$$

(i)   In this case, we have

$$w'(t) \geq -q(t) \quad \text{for} \quad t \geq t_0. \qquad (4.6.8)$$

Integrating (4.6.8) from t to T with $t_0 \leq t \leq T$, we obtain $w(t) \geq w(T) + \int_t^T q(s)ds \geq \int_t^T q(s)ds$. Thus, as $T \to \infty$, we get $w(t) \geq Q(t)$ for $t \geq t_0$. Now, integrating again, we find

$$\int_{t_0}^{\infty} Q(s)ds \leq \int_{t_0}^{\infty} w(s)ds \leq \int_{x(t_0)}^{\infty} \frac{du}{f(u)} < \infty,$$

which contradicts the hypothesis that $Q(t) \notin \mathcal{L}^1[t_0, \infty)$.

(ii) For this case, we write the equation (4.6.7) in the form

$$\left(w(t) \exp\left(\int_{t_0}^t p(s)\gamma(x'(s))ds \right) \right)' = [-q(t) - f'(x(t))w^2(t)]$$

$$\times \exp\left(\int_{t_0}^t p(s)\gamma(x'(s))ds \right). \qquad (4.6.9)$$

Integrating (4.6.9) from t to T with $t_0 \leq t \leq T$, we find

$$w(t) = w(T)\phi(T) + \int_t^T [q(s) + f'(x(s))w^2(s)]\phi(s)ds,$$

where $\phi(s) = \exp\left(\int_t^s p(u)\gamma(x'(u))du \right)$. This yields the inequality

$$w(t) \geq \int_t^T q(s) \exp\left(\int_t^{\infty} p(u)\gamma(x'(u))du \right). \qquad (4.6.10)$$

Since $x(t)$ is bounded, by Lemma 4.6.1, we obtain $\lim_{t \to \infty} x'(t) = 0$. Hence, for all sufficiently large σ, we have $0 \leq \gamma(x'(\sigma)) \leq K(x'(\sigma))^{b-1} \leq$

$K(x'(\sigma))^{\beta-1}$. Thus, for s, $t \geq t_0$ sufficiently large and Hölder's inequality, we get

$$\left| \int_t^s p(\sigma)\gamma(x'(\sigma))d\sigma \right| \leq \int_t^s K|p(\sigma)|(x'(\sigma))^{\beta-1}d\sigma$$

$$\leq K \left(\int_t^s |p(\sigma)|^{\frac{1}{2-\beta}} d\sigma \right)^{2-\beta} \left(\int_t^s x'(\sigma)d\sigma \right)^{\beta-1} \leq K(|p|_m)\left(c^{\beta-1}\right) = K_0,$$

say, where $m = 1/(2-\beta)$ and $c = \lim_{t\to\infty} x(t)$. Using this estimate in (4.6.10) and the fact that $q(t) \geq 0$ for sufficiently large t, we find $w(t) \geq e^{-K_0} \int_t^T q(s)ds$. The rest of the proof is similar to that of Case (i) above.

The proof of the nonexistence of negative, decreasing solutions of equation (4.6.1) is similar. ∎

Lemma 4.6.3. If either of the following two sets of hypotheses holds, then there is no solution $x(t)$ of equation (4.6.1) satisfying $x(t)x'(t) \leq 0$ for all large t.

(I_1) $\left[1 + \int_{t_0}^t |p(s)|ds \right]^{-1} \notin \mathcal{L}^{1/(b-1)}[t_0, \infty)$, and $q(t) \geq 0$ for sufficiently large t and is not identically zero on any half–line.

(I_2) $p(t) \geq 0$ for sufficiently large t, $p(t) \in \mathcal{L}^{1/(2-\beta)}[t_0, \infty)$, and $Q(t)$ exists and nonnegative for sufficiently large t.

Proof. We shall consider only the proof of nonexistence of positive, decreasing solutions. The proof for negative, increasing solutions is similar. Assume that $x(t)$ is a positive solution of equation (4.6.1), monotone decreasing on $[t_0, \infty)$ with t_0 large enough so that sign hypotheses hold on $[t_0, \infty)$.

(I_1) $q(t) \geq 0$ implies that $x''(t) + p(t)g(x'(t)) \leq 0$ for $t \geq t_0$. Let $v(t) = -x'(t)$. Then, we have $v'(t) + p(t)g(-v(t)) \geq 0$ and so by condition (4.6.6),

$$v'(t) + K|p(t)|v^b(t) \geq 0 \quad \text{for} \quad t \geq t_0. \tag{4.6.11}$$

The hypothesis $q(t) \not\equiv 0$ on any half–line is to exclude the possibility that $x(t)$ is eventually constant. Hence, we can assume without loss of generality that $v(t_0) > 0$ and integrate (4.6.11) over an interval $[t_0, t]$ on which $v(t)$ is nonvanishing, to obtain

$$v^{1-b}(t_0) - v^{1-b}(t) \geq -K(b-1) \int_{t_0}^t |p(s)|ds. \tag{4.6.12}$$

From (4.6.12), we find that $v(t)$ must remain positive and

$$v(t) \geq \left[v^{1-b}(t_0) + K(b-1) \int_{t_0}^t |p(s)|ds \right]^{-1/(b-1)}, \quad t \geq t_0. \tag{4.6.13}$$

Now, the hypotheses imply that

$$\left[v^{1-b}(t_0) + K(b-1) \int_{t_0}^{t} |p(s)|ds\right]^{-1} \notin \mathcal{L}^{1/(b-1)}[t_0, \infty),$$

however, $0 \le \int_{t_0}^{\infty} v(s)ds \le x(t_0)$ and so (4.6.13) yields a contradiction.

(I_2) Define $u(t) = x'(t) \exp\left(\int_{t_0}^{t} p(s)\gamma(x'(s))ds\right)$, $t \ge t_0$. Then, $u(t) \le 0$ for $t \ge t_0$. Arguing as in Lemma 4.6.2(ii), we obtain

$$\int_{t_0}^{t} |p(s)|\gamma(x'(s))ds \le K_0, \quad \text{say, for all} \quad t \ge t_0. \tag{4.6.14}$$

Therefore, if $\limsup_{t\to\infty} u(t) < 0$, then $\limsup_{t\to\infty} x'(t) < 0$, which is impossible for a positive function $x(t)$. Thus, $\limsup_{t\to\infty} u(t) = 0$. Define $t_n = \inf\{t : u(t) = -1/n\}$. Clearly, t_n is well defined for n sufficiently large, and the sequence $\{t_n\}$ increases to ∞. We can write equation (4.6.1) as

$$u'(t) + q(t) \exp\left(\int_{t_0}^{t} p(s)\gamma(x'(s))ds\right) f(x(s)) = 0. \tag{4.6.15}$$

Integrating (4.6.15) from t to t_n, where $t \in [t_{n-1}, t_n)$ and using the definition of t_n, we get

$$0 > u(t) - u(t_n)$$
$$= \int_{t}^{t_n} q(s) \exp\left(\int_{t_0}^{s} p(\sigma)\gamma(x'(\sigma))d\sigma\right) f(x(s))ds$$
$$= f(x(t)) \int_{t}^{t_n} q(s)\phi(s)ds + \int_{t}^{t_n} f'(x(s))x'(s) \left(\int_{s}^{t_n} q(\sigma)\phi(\sigma)d\sigma\right) ds, \tag{4.6.16}$$

where $\phi(s) = \exp\left(\int_{t_0}^{s} p(\sigma)\gamma(x'(\sigma))d\sigma\right)$. Inequality (4.6.16) can be written as

$$F(t) - \int_{t}^{t_n} H(s)F(s)ds < 0, \tag{4.6.17}$$

where

$$F(t) = f(x(t)) \int_{t}^{t_n} q(s)\phi(s)ds \quad \text{and} \quad H(t) = -\frac{f'(x(t))x'(t)}{f(x(t))} \ge 0.$$

An application of Gronwall's lemma to (4.6.17) gives $F(t) < 0$ for $t \in [t_{n-1}, t_n)$, i.e., $\int_{t}^{t_n} q(s)\phi(s)ds < 0$. Hence, $\liminf_{T\to\infty} \int_{t}^{T} q(s)\phi(s)ds < 0$. But, since

$$\int_{t}^{T} q(s)\phi(s)ds = \phi(t)Q(t) - \phi(T)Q(T) + \int_{t}^{T} Q(s)\phi'(s)ds$$

from the hypotheses it follows that ϕ, ϕ', Q are all nonnegative on $[t_0, \infty)$ with $\lim_{T \to \infty} Q(T) = 0$, and from (4.6.14), we find that $\phi(T) \leq e^{K_0}$ for $T \geq t_0$. Thus, $\lim_{T \to \infty} \int_t^T q(s)\phi(s)ds \geq 0$. This contradiction completes the proof. ∎

Now, we are prepared to state and prove the following oscillation criteria.

Theorem 4.6.1. Let condition (4.6.6) hold. Then,

(I) $p(t)$, $q(t)$ eventually positive with $\left[1 + \int_{t_0}^t p(s)ds\right]^{-1} \notin \mathcal{L}^{1/(b-1)}[t_0, \infty)$ and $Q(t) \notin \mathcal{L}^1[t_0, \infty)$ imply that equation (4.6.1) is oscillatory.

(II) $p(t) \in \mathcal{L}^{1/(2-\beta)}[t_0, \infty)$, $q(t)$ eventually positive with $Q(t) \notin \mathcal{L}^1[t_0, \infty)$ imply that all bounded solutions of equation (4.6.1) oscillate.

(III) $p(t)$ eventually positive, $p(t) \in \mathcal{L}^{1/(2-\beta)}[t_0, \infty)$, $Q(t)$ exists and is eventually positive with $Q(t) \notin \mathcal{L}^1[t_0, \infty)$ imply that all bounded solutions of equation (4.6.1) have oscillatory derivatives.

Proof. In (I) and (II), the positivity hypothesis on $q(t)$ implies that any nonoscillatory solution $x(t)$ of equation (4.6.1) satisfies the inequality $x''(t) \operatorname{sgn} x(t) < 0$ for any sufficiently large t for which $x'(t)$ vanishes. This implies that any nonoscillatory solution must be monotonic. The result now follows from the appropriate parts of Lemmas 4.6.2 and 4.6.3.

For (III) we can use the above lemmas to conclude that any bounded nonoscillatory solution of equation (4.6.1) cannot be monotonic, which is equivalent to the statement that its derivative oscillates. ∎

Remark 4.6.1. The condition $p(t) \in \mathcal{L}^{1/(2-\beta)}[t_0, \infty)$ in (II) is stronger than the condition $\left[1 + \int_{t_0}^t p(s)ds\right]^{-1} \notin \mathcal{L}^{1/(b-1)}[t_0, \infty)$.

4.6.2. Damping with Nonlinear Growth– A Nonoscillation Theorem

We shall prove the following nonoscillation result for equation (4.6.1).

Theorem 4.6.2. Suppose the following conditions hold

(i_1) $g(0) = 0$ and there exist constants K, b with $b \geq 1$ such that for all u, v with $|u| \leq 1$, $|v| \leq 1$,

$$|g(u) - g(v)| \leq K|u - v|\left[|u|^{b-1} + |v|^{b-1}\right].$$

(i_2) $q(t) \in \mathcal{L}^1[t_0, \infty)$, $t_0 \geq 0$ and there exist $T_0 \geq t_0$, k with $k > 1$ such that

$$Q_1(t) = \int_t^\infty q^+(s)ds > k \int_t^\infty q^-(s)ds = kQ_2(t) \quad \text{for all} \quad t \geq T_0.$$

(i_3) $p(t) \in \mathcal{L}^{1/(2-\beta)}[t_0, \infty)$, $\beta = \min\{2, b\}$.

Then, $\int_t^\infty Q(s)ds < \infty$ implies that there is a nontrivial nonoscillatory solution of equation (4.6.1).

Proof. We note that the proof presented here is similar to that of Theorem 4.3.1. We observe that condition (i_2) implies that for $t \geq T_0$,

$$Q(t) = \int_t^\infty [q^+(s) - q^-(s)]\,ds \geq \left(\frac{k-1}{k}\right)Q_1(t)$$

$$\geq \left(\frac{k-1}{2k}\right)\int_t^\infty [q^+(s) + q^-(s)]\,ds = \left(\frac{k-1}{k}\right)Q_3(t),$$

where $Q_3(t) = \int_t^\infty |q(s)|ds$. Thus, $Q(t) \in \mathcal{L}^1[t_0, \infty)$ shows that $Q_3(t) \in \mathcal{L}^1[t_0, \infty)$. We also note that $Q_3(t) \to 0$ monotonically as $t \to \infty$. Let $c > 1$ and define

$$A = \max_{c-1 \leq u \leq c+1} f(u). \tag{4.6.18}$$

Choose $T \geq T_0$ so large that

$$Q_3(T) \leq \frac{1}{2A}, \tag{4.6.19}$$

$$\int_T^\infty Q_3(s)ds \leq \min\{1, 1/(2A)\} \tag{4.6.20}$$

and in view of condition (i_3),

$$\left(\int_T^\infty |p(s)|^{1/(2-\beta)}\right)^{2-\beta} \leq \frac{1}{2K}. \tag{4.6.21}$$

Let S be the space of functions $x(t)$ which are bounded and absolutely continuous on $[T, \infty)$ and whose derivatives are a.e. bounded above by some multiple of $Q_*(t) = 2AQ_3(t)$. We make S into a Banach space by introducing the norm $\|x\| = \|x\|_\infty + \|Q_*^{-1}x'\|_\infty$, where $\|\cdot\|_\infty$ is the uniform norm on $[T, \infty)$.

For $n = 1, 2, \cdots$ define $\mathcal{B}_n$ to be the set of functions $x \in S$ which are constant on $[T + n, \infty)$ and satisfy the inequalities

$$\|x - c\|_\infty \leq 1, \tag{4.6.22}$$

$$\|Q_*^{-1}x'\| \leq 1 \tag{4.6.23}$$

and

$$|x'(t_2) - x'(t_1)| \leq A \int_{t_1}^{t_2} q(s)ds + K \left[\int_{t_1}^{t_2} |p(s)|^{1/(2-\beta)} ds \right]^{2-\beta} \quad (4.6.24)$$

for all $t_1, t_2 \in [T, T+n]$.

Clearly, $\mathcal{B}_n$ is a convex subset of $\mathcal{S}$ and by the Ascoli–Arzela theorem $\mathcal{B}_n$ is compact. Define the mapping Y_n on $\mathcal{B}_n$ by

$$(Y_n x)(t) = \begin{cases} c - \displaystyle\int_t^\infty \int_s^\infty [p(\sigma)g(x'(\sigma)) + q(\sigma)f(x(\sigma))]d\sigma ds, & T \leq t \leq T+n \\ c - \displaystyle\int_{T+n}^\infty \int_s^\infty q(\sigma)f(x(\sigma))d\sigma ds, & T+n < t. \end{cases}$$

Since for any function $x \in \mathcal{B}_n$, $x(t)$ is constant on $[T+n, \infty)$, the integrability of $q(t)$ and $Q(t)$ imply that Y_n maps $\mathcal{B}_n$ into $\mathcal{S}$. We shall now verify that $Y_n(\mathcal{B}_n) \subset \mathcal{B}_n$. Let $x \in \mathcal{B}_n$. If $T \leq t \leq T+n$, we have

$$|(Y_n x)'(t)| \leq K \int_t^\infty |p(s)||x'(s)|^b ds + A \int_t^\infty |q(s)|ds. \quad (4.6.25)$$

Now, since $|x'(s)| \leq Q_*(s) = 2AQ_3(s) \leq 2AQ_3(T) \leq 1$, it follows that $|x'(s)|^b \leq |x'(s)|^\beta$. Thus, on applying Hölder's inequality and the fact that $Q_*(t)$ is a monotonic function, we obtain

$$\int_t^\infty |p(s)||x'(s)|^b ds \leq \int_t^\infty |p(s)||x'(s)|^\beta ds \leq \int_t^\infty |p(s)|Q_*^\beta(s)ds$$

$$\leq Q_*(t) \left[\int_t^\infty |p(s)|^{1/(2-\beta)} ds \right]^{2-\beta} \left(\int_t^\infty Q_*(s)ds \right)^{\beta-1}. \tag{4.6.26}$$

Combining (4.6.25) and (4.6.26) and then using the inequalities (4.6.20) and (4.6.21), we find

$$|(Y_n x)'(t)| \leq KQ_*(t) \left[\int_t^\infty |p(s)|^{1/(2-\beta)} ds \right]^{2-\beta} \left(\int_t^\infty Q_*(s)ds \right)^{\beta-1} + AQ_3(t)$$

$$\leq 2AQ_3(t) = Q_*(t).$$
$$\tag{4.6.27}$$

A similar computation for $t_1, t_2 \in [T, T+n]$ gives

$$|(Y_n x)'(t_2) - (Y_n x)'(t_1)| \leq K \left[\int_{t_1}^{t_2} |p(s)|^{1/(2-\beta)} ds \right]^{2-\beta} + AQ_3(t), \quad (4.6.28)$$

and if $t \in [T, T+n)$, we have

$$
\begin{aligned}
|(Y_n x)(t) - c| &\leq |(Y_n x)(t) - (Y_n x)(T+n)| + |(Y_n x)(T+n) - c| \\
&\leq \int_n^{T+n} 2AQ_3(s)ds + \int_{T+n}^{\infty} AQ_3(s)ds \\
&\leq \int_t^{\infty} 2AQ_3(s)ds \leq 1.
\end{aligned}
$$

$$(4.6.29)$$

We also have

$$(Y_n x)(t) = (Y_n x)(T+n) \quad \text{for all} \quad t \geq T+n. \tag{4.6.30}$$

Now $(4.6.27) - (4.6.30)$ imply that $Y_n x \in \mathcal{B}_n$.

Finally, we shall show that Y_n is a continuous map. For this, let $x, z \in \mathcal{B}_n$ with $\|x - z\| < \delta$ $(\delta > 0)$. Let

$$a(\delta) = \max_{\substack{u, v \in [c-1, c+1] \\ \|u - v\| \leq \delta}} |f(u) - f(v)|.$$

Now, by condition (i_1), the definition of $a(\delta)$, and the fact that $|x'(s)| \leq 1$ and $|z'(s)| \leq 1$ for all s, we find for $t \in [T, T+n]$,

$$
\begin{aligned}
&|(Y_n x)'(t) - (Y_n z)'(t)| \\
&\leq \int_t^{\infty} p(s)\|g(x'(s)) - g(z'(s))\|ds + \int_t^{\infty} |q(s)||f(x(s)) - f(z(s))|ds \\
&\leq \int_t^{\infty} |p(s)|K|x'(s) - z'(s)| \left[|x'(s)|^{\beta-1} + |z'(s)|^{\beta-1}\right]ds + a(\delta)\int_t^{\infty} |q(s)|ds.
\end{aligned}
$$

Thus, by Hölder's inequality, $(4.6.20)$ and $(4.6.21)$, it follows that

$$|(Y_n x)'(t) - (Y_n z)'(t)| \leq [2A\delta + a(\delta)]Q_3(t) = \left[\delta + \frac{a(\delta)}{2A}\right]Q_*(t). \tag{4.6.31}$$

Integrating $(4.6.31)$ and using $(4.6.20)$, we get

$$|(Y_n x)(t) - (Y_n z)(t)| \leq \delta + \frac{a(\delta)}{2A}. \tag{4.6.32}$$

From $(4.6.31)$ and $(4.6.32)$ it is clear that Y_n is continuous. Thus, we can apply the Schauder fixed point theorem to obtain $x_n \in \mathcal{B}_n$ which satisfies $Y_n x_n = x_n$. Finally, the equicontinuity of the sequence $\{x_n\}$ on $[T, \infty)$ completes the proof of the theorem. ■

The following result is an immediate consequence of Theorems 4.6.1 and 4.6.2.

Theorem 4.6.3. Let $\gamma > 1$, $b \geq 1$, $\beta = \min\{b, 2\}$. Assume that $p(t)$, $q(t) \in C([t_0, \infty), \mathbb{R}^+)$ and $p(t) \in \mathcal{L}^{1/(2-\beta)}[t_0, \infty)$. Then all solutions of

$$x''(t) + p(t)|x'(t)|^b \operatorname{sgn} x'(t) + q(t)|x(t)|^\gamma \operatorname{sgn} x(t) = 0$$

oscillate if and only if $\int_{t_0}^\infty sq(s)ds = \infty$.

Remark 4.6.2.

1. When the damping term is linear as in (4.1.3) and (3.1.8), in principle any oscillation result for the undamped equation yields a corresponding result for the damped equation. However, in practice, the problem is to find easily verifiable oscillation criteria for damped equations.

2. In the case of nonlinear damping it is difficult to dispense with the positivity condition on the damping. The following example illustrates this fact.

Example 4.6.1. The damped differential equation

$$x''(t) - \frac{1}{t^{2-\beta}\ln t}|x'(t)|^b \operatorname{sgn} x'(t) + \frac{1}{t^2\ln t}|x(t)|^\beta \operatorname{sgn} x(t) = 0, \quad (4.6.33)$$

for $b > 1$, $\beta = \min\{2, b\}$ and $t \geq 2$ has an unbounded nonoscillatory solution $x(t) = t$. All conditions of Theorem 4.6.1(I) are satisfied except that $p(t)$ is eventually negative. Also, all conditions of Theorem 4.6.1(II) are satisfied, and hence all bounded solutions of (4.6.33) are oscillatory.

4.7. Notes and General Discussions

1. Theorems 4.1.1 and 4.1.2 are taken from Graef et. al. [40]. Theorem 4.1.3 is an extension of a result due to Kamenev [43] and theorem I of Wong [53]. Theorem 4.1.4 includes theorems 1 and 1′ of Philos [49]. Theorems 4.1.5 and 4.1.6 are borrowed from Grace and Lalli [31]. Lemmas 4.1.1 – 4.1.3 are based on the work of Graef and Spikes [42]. Theorems 4.1.7 and 4.1.8 include theorems 3 and 4, respectively, of Kwong and Wong [45], and Theorem 4.1.7 generalizes theorem 2.1 of Butler [4]. Theorem 4.1.9 is taken from Graef and Spikes [42].

2. Conditions (F_1) – (F_4) and Examples 4.2.1 – 4.2.3 are due to Philos [48]. Theorem 4.2.1 includes as a special case the results of Grace [9,12], and theorems 1 and 1′ of Philos [50]. Corollary 4.2.2 and the main result in [48] are the same. Theorem 4.2.2 extends and improves theorems 2 and 2′ of Philos [50]. Theorems 4.2.3 – 4.2.7 are due to Grace [11] and Grace and Lalli [35]. Theorem 4.2.8 generalizes a result of Wong [54]. Theorem 4.2.9 contains theorem 2 of Philos and Purnaras [51], while Theorem 4.2.10 extends the main result of Philos and Purnaras [52]. Theorem 4.2.11 is

new whereas Theorem 4.2.12 extends the corresponding result of Wong [55]. Theorems 4.2.13 and 4.2.14 are also new.

3. Theorem 4.3.1 and theorem 2.2 in [4] are the same whereas Theorem 4.3.2 is due to Butler [6]. Theorems 4.3.5 – 4.3.7 extend some of the results of Kwong and Wong [46] and are related to the work of Garce [17], also see recent contributions of Agarwal et. al. [2].

4. Lemmas 4.4.1 – 4.4.3 and Theorem 4.4.2 are taken from Wong and Agarwal [56], while Theorem 4.4.3 is new. Lemma 4.4.5 is due to Agarwal et. al. [3]. Theorems 4.4.3 – 4.4.5 are new. Theorems 4.4.6 and 4.4.7 are borrowed from Elbert and Kusano [7], whereas Theorems 4.4.8 and 4.4.9 are due to Kusano et. al. [44].

5. Theorems 4.5.1 and 4.5.2 are taken from Agarwal and Grace [1]. Theorem 4.5.1 extends the main result of Nasr [47].

6. The results of Section 4.6 are based on Butler [5]. For several other interesting results in this direction, we refer the reader to the work of Grace et. al. [8–39], and Graef and Spikes [41].

7. We note that some of the oscillation criteria of this chapter have been extended in [8–39] to more general equations of the form

$$(a(t)\psi(x(t))x'(t))' + p(t)x'(t) + q(t)f(x(t)) = 0,$$

where $a(t)$, $p(t)$, $q(t) \in C([t_0, \infty), \mathbb{R})$, $a(t) > 0$ for $t \geq t_0$, $\psi \in C(\mathbb{R}, \mathbb{R}^+)$ and $f \in C(\mathbb{R}, \mathbb{R})$. It would be interesting to obtain oscillatory results similar to those presented in Sections 4.2 and 4.5 for equations of the type (4.4.1).

4.8. References

1. **R.P. Agarwal and S.R. Grace,** Second order nonlinear forced oscillations, *Dyn. Sys. Appl.*, to appear.

2. **R.P. Agarwal, S.R. Grace and D. O'Regan,** Linearization of second order superlinear oscillation theorems, to appear.

3. **R.P. Agarwal, S.-H. Shieh and C.-C. Yeh,** Oscillation criteria for second order retarded differential equations, *Math. Comput. Modelling* **26**(4)(1997), 1–11.

4. **G.J. Butler,** On the oscillatory behavior of a second order nonlinear differential equation, *Ann. Mat. Pura Appl.* **105**(1975), 73–92.

5. **G.J. Butler,** The oscillatory behavior of a second order nonlinear differential equation with damping, *J. Math. Anal. Appl.* **57**(1977), 273–289.

6. **G.J. Butler,** Hille–Wintner type comparison theorems for second order ordinary differential equations, *Proc. Amer. Math. Soc.* **76**(1979), 51–59.

7. **Á. Elbert and T. Kusano,** Oscillation and nonoscillation theorems for a class of second order quasilinear differential equations, *Acta. Math. Hungar.* **56**(1990), 325–336.

8. **S.R. Grace,** Oscillation criteria for second order nonlinear differential equations, *Ann. Differential Equations* 4(1988), 255–264.

9. **S.R. Grace,** An oscillation criterion for second order nonlinear differential equations, *Indian J. Pure Appl. Math.* **20**(1989), 297–306.

10. **S.R. Grace,** Oscillation theorems for second order nonlinear differential equations with damping, *Math. Nachr.* **141**(1989), 117–127.

11. **S.R. Grace,** Oscillation criteria for second order differential equations with damping, *J. Austral. Math. Soc. Ser. A* **49**(1990), 43–54.

12. **S.R. Grace,** An oscillation criterion for second order superlinear ordinary differential equations, *Utilitas Math.* **37**(1990), 251–258.

13. **S.R. Grace,** Oscillation theorems for nonlinear differential equations of second order, *J. Math. Anal. Appl.* **171**(1992), 220–241.

14. **S.R. Grace,** Oscillation theorems for damped functional differential equations, *Funkcial. Ekvac.* **35**(1992), 261–278.

15. **S.R. Grace,** Oscillation of nonlinear differential equations of second order, *Publ. Math. Debrecen* **40**(1992), 143–153.

16. **S.R. Grace,** On the oscillatory behavior of solutions of second order nonlinear differential equations, *Publ. Math. Debrecen* **43**(1993), 351–357.

17. **S.R. Grace,** Oscillation criteria of comparison type for nonlinear functional differential equations, *Math. Nachr.* **173**(1995), 177–192.

18. **S.R. Grace and B.S. Lalli,** Oscillation theorems for certain second order perturbed nonlinear differential equations, *J. Math. Anal. Appl.* **76**(1980), 205–214.

19. **S.R. Grace and B.S. Lalli,** Oscillation of solutions of damped nonlinear second order functional differential equations, *Bull. Inst. Math. Acad. Sinica* **12**(1984), 5–9.

20. **S.R. Grace and B.S. Lalli,** Oscillation and convergence to zero of solutions of damped second order nonlinear differential equations, *J. Math. Anal. Appl.* **102**(1984), 539–548.

21. **S.R. Grace and B.S. Lalli,** Oscillation theorems for nonlinear second order functional differential equations with damping, *Bull. Inst. Math. Acad. Sinica* **13**(1985), 183–192.

22. **S.R. Grace and B.S. Lalli,** Oscillatory behavior for nonlinear second order functional differential equations with deviating arguments, *Bull. Inst. Math. Aad. Sinica* **14**(1986), 187–196.

23. **S.R. Grace and B.S. Lalli,** Oscillation theorems for a second order nonlinear ordinary differential equation with damping term, *Comment. Math. Univ. Carolin.* **27**(1986), 449–453.

24. **S.R. Grace and B.S. Lalli,** Oscillatory behavior of solutions of second order differential equations with alternating coefficients, *Math. Nachr.* **127**(1986), 165–175.

25. **S.R. Grace and B.S. Lalli,** An oscillation criterion for certain second order strongly sublinear differential equations, *J. Math. Anal. Appl.* **123**(1987), 584–588.

26. **S.R. Grace and B.S. Lalli,** An oscillation criterion for second order sublinear ordinary differential equations with damping term, *Bull. Polish. Acad. Sci. Math.* **35**(1987), 181–184.

27. **S.R. Grace and B.S. Lalli,** Almost oscillation and second order functional differential equations with forcing term, *Boll. Un. Mat. Ital.* **B(7)**1(1987), 509–522.

28. **S.R. Grace and B.S. Lalli,** Oscillation theorems for second order nonlinear differential equations, *J. Math. Anal. Appl.* **124**(1987), 213–224.

29. **S.R. Grace and B.S. Lalli,** Oscillation theorems for second order nonlinear differential equations with deviating arguments, *Internat. J. Math. Math. Sci.* **10**(1987), 35–45.

30. **S.R. Grace and B.S. Lalli,** On the second order nonlinear oscillations, *Bull. Inst. Math. Acad. Sinica* **15**(1987), 297–309.

31. **S.R. Grace and B.S. Lalli,** Integral averaging and the oscillation of second order nonlinear differential equations, *Ann. Mat. Pura Appl.* **151**(1988), 149–159.

32. **S.R. Grace and B.S. Lalli,** Oscillation in second order differential equations with alternating coefficients, *Period. Math. Hungar.* **19**(1988), 69–78.

33. **S.R. Grace and B.S. Lalli,** Oscillation theorems for nonlinear second order differential equations with a damping term, *Comment. Math. Univ. Carolin.* **30**(1989), 691–697.

34. **S.R. Grace and B.S. Lalli,** Integral averaging techniques for the oscillation of second order nonlinear differential equations, *J. Math. Anal. Appl.* **149**(1990), 277–311.

35. **S.R. Grace and B.S. Lalli,** Oscillation theorems for second order superlinear differential equations with damping, *J. Austral. Math. Soc. Ser. A* **53**(1992), 156–165.

36. **S.R. Grace and B.S. Lalli,** Oscillation theorems for second order nonlinear functional differential equations with damping, *Comput. Math. Applic.* **25**(1993), 107–113.

37. **S.R. Grace and B.S. Lalli,** Oscillation criteria for damped functional differential equations, *Dynam. Stability Systems* **9**(1994), 215–222.

38. **S.R. Grace, B.S. Lalli and C.C. Yeh,** Oscillation theorems for nonlinear second order differential equations with a nonlinear damping term, *SIAM J. Math. Anal.* **15**(1984), 1082–1093.

39. **S.R. Grace, B.S. Lalli and C.C. Yeh,** Addendum: Oscillation theorems for nonlinear second order differential equations with a nonlinear damping term, *SIAM J. Math. Anal.* **19**(1988), 1252–1253.

40. **J.R. Graef, S.M. Rankin and P.W. Spikes,** Oscillation theorems for perturbed nonlinear differential equations, *J. Math. Anal. Appl.* **65**(1978), 375–390.

41. **J.R. Graef and P.W. Spikes,** Asymptotic behavior of solutions of a second order nonlinear differential equation, *J. Differential Equations* **17**(1975), 461–476.

42. **J.R. Graef and P.W. Spikes,** On the oscillatory behavior of solutions of second order nonlinear differential equations, *Czech. Math. J.* **36**(1986), 275–284.

43. **I.V. Kamenev,** Oscillation of solutions of second order nonlinear equations with sign variable coefficients, *Differencial'nye Uravnenija* **6**(1970), 1718–1721.

44. **T. Kusano, A. Ogata and H. Usami,** Oscillation theory for a class of second order quasilinear ordinary differential equations with application to partial differential equations, *Japan J. Math.* **19**(1993), 131–147.

45. **M.K. Kwong and J.S.W. Wong,** An application of integral inequality to second order nonlinear oscillation, *J. Differential Equations* **46**(1982), 63–77.

46. **M.K. Kwong and J.S.W. Wong,** Linearization of second order nonlinear oscillation theorems, *Trans. Amer. Math. Soc.* **279**(1983), 705–722.

47. **A. Nasr,** Sufficient conditions for the oscillation of forced superlinear second order differential equations with oscillatory potential, *Proc. Amer. Math. Soc.* **126**(1998), 123–125.

48. **Ch.G. Philos,** A second order superlinear oscillation criterion, *Canad. Math. Bull.* **27**(1984), 102–112.

49. **Ch.G. Philos,** Integral averages and second order superlinear oscillation, *Math. Nachr.* **120**(1985), 127–138.

50. **Ch.G. Philos,** Oscillation criteria for second order superlinear differential equations, *Canad. J. Math.* **XLI**(1989), 321–340.

51. **Ch.G. Philos and I.K. Purnaras,** On the oscillation of second order nonlinear differential equations, *Arch. Math.* **59**(1992), 260–271.

52. **Ch.G. Philos and I.K. Purnaras,** Oscillations in superlinear differential equations of second order, *J. Math. Anal. Appl.* **165**(1992), 1–11.

53. **J.S.W. Wong,** Oscillation theorems for second order nonlinear differential equations, *Bull. Inst. Math. Acad. Sinica* **3**(1975), 283–309.

54. **J.S.W. Wong,** Oscillation theorems for second order nonlinear differential equations, *Proc. Amer. Math. Soc.* **106**(1989), 1069–1077.

55. **J.S.W. Wong,** An oscillation criterion for second order nonlinear differential equations with iterated integral averages, *Differential and Integral Equations* **6**(1993), 83–91.

56. **P.J.Y. Wong and R.P. Agarwal,** Oscillation criteria for half–linear differential equations, *Advances Math. Sci. Appl.* **9**(1999), 649–663.

Chapter 5

Oscillation Theory for Sublinear Differential Equations

5.0. Introduction

In this chapter we shall present oscillation and nonoscillation criteria for all solutions of second order nonlinear differential equations of sublinear type with alternating coefficients. In Section 5.1, our sublinear oscillation results involve integrals and weighted integrals of the alternating coefficients. In some results we employ integral averaging techniques. In Section 5.2, we impose some additional conditions on the sublinear term which allow us to proceed further and extend and improve several known theorems in the literature. In fact, we make an asymptotic study which results in new oscillation criteria. Section 5.3 provides some new linearized oscillation results for second order sublinear differential equations. In Section 5.4, we shall present criteria for the nonoscillation of sublinear Emden–Fowler type equations. In Section 5.5, we compare the oscillatory behavior of certain nonlinear equations with the related half–linear differential equations. Here oscillation of general nonlinear differential equations is also discussed.

5.1. Sublinear Oscillation Criteria

We shall examine oscillatory behavior of second order sublinear ordinary differential equations of the form

$$x''(t) + q(t)f(x(t)) = 0, \qquad (5.1.1)$$

$$(a(t)x'(t))' + q(t)f(x(t)) = 0 \qquad (5.1.2)$$

and the more general damped equation

$$(a(t)x'(t))' + p(t)x'(t) + q(t)f(x(t)) = 0, \qquad (5.1.3)$$

where

(i) $a(t) \in C([t_0, \infty), \mathbb{R}^+)$,

(ii) $p(t), q(t) \in C([t_0, \infty), \mathbb{R})$,

(iii) $f \in C(\mathbb{R}, \mathbb{R})$, f is continuously differentiable except possibly at 0, and satisfies

$$xf(x) > 0 \quad \text{and} \quad f'(x) \geq 0 \quad \text{for} \quad x \neq 0 \tag{5.1.4}$$

and that f is strongly *sublinear* in the sense that

$$\int_{+0} \frac{du}{f(u)} < \infty \quad \text{and} \quad \int_{-0} \frac{du}{f(u)} < \infty. \tag{5.1.5}$$

We shall also be interested in the prototype of these equations, namely,

$$x''(t) + q(t)|x(t)|^\gamma \operatorname{sgn} x(t) = 0, \tag{5.1.6}$$

$$(a(t)x'(t))' + q(t)|x(t)|^\gamma \operatorname{sgn} x(t) = 0 \tag{5.1.7}$$

and

$$(a(t)x'(t))' + p(t)x'(t) + q(t)|x(t)|^\gamma \operatorname{sgn} x(t) = 0 \tag{5.1.8}$$

respectively, where $0 < \gamma < 1$ is a constant.

We are now in the position to prove the following results.

Theorem 5.1.1. Suppose there exists a function $\rho(t) \in C^2([t_0, \infty), \mathbb{R}^+)$ such that

$$\int^\infty \rho(s)q(s)ds = \infty \tag{5.1.9}$$

and

$$\int^\infty \frac{1}{a(s)\rho(s)} \int_{t_0}^s \rho(u)q(u)duds = \infty. \tag{5.1.10}$$

Then the following statements hold:

(i_1) If $\beta(t) = a(t)\rho'(t) - p(t)\rho(t) \leq 0$ and $\beta'(t) \geq 0$ for $t \geq t_0$, then all bounded solutions of equation (5.1.3) are oscillatory or tend monotonically to zero as $t \to \infty$. If in addition, condition (5.1.5) holds, then all solutions of (5.1.3) are oscillatory.

(i_2) If $\int^\infty |\beta'(s)|ds < \infty$, then every bounded solution of equation (5.1.3) is oscillatory.

Proof. Let $x(t)$ be a nonoscillatory solution of equation (5.1.3), say, $x(t) > 0$ for $t \geq t_0$. Define $w(t) = \rho(t)a(t)x'(t)/f(x(t))$, $t \geq t_1$ for some $t_1 \geq t_0$. Then for $t \geq t_1$, we obtain

$$\begin{aligned}
w'(t) &= -\rho(t)q(t) + \beta(t)\frac{x'(t)}{f(x(t))} - \frac{1}{a(t)\rho(t)}w^2(t)f'(x(t)) \\
&\leq -\rho(t)q(t) + \beta(t)\frac{x'(t)}{f(x(t))}.
\end{aligned}$$

Integrating this inequality from t_1 to t, we get

$$w(t) \leq w(t_1) - \int_{t_1}^t \rho(s)q(s)ds + \int_{t_1}^t \beta(s)\frac{x'(s)}{f(x(s))}ds.$$

We shall show that both conditions (i_1) and (i_2) imply that $\int_{t_1}^t (\beta(s)x'(s)$ $/f(x(s)))ds$ is bounded above. Since,

$$\begin{aligned}
\int_{t_1}^t \beta(s)\frac{x'(s)}{f(x(s))}ds &= \beta(t)\int_0^{x(t)} \frac{du}{f(u)} - \beta(t_1)\int_0^{x(t_1)} \frac{du}{f(u)} \\
&\quad - \int_{t_1}^t \beta'(s)\left(\int_0^{x(s)} \frac{du}{f(u)}\right)ds
\end{aligned} \tag{5.1.11}$$

if (i_1) holds, then (5.1.11) yields

$$\int_{t_1}^t \beta(s)\frac{x'(s)}{f(x(s))} \leq -\beta(t_1)\int_0^{x(t_1)} \frac{du}{f(u)} = k_1, \tag{5.1.12}$$

where k_1 is some constant. Also, if (i_2) holds, then $|\beta(t)| \leq b$ for some constant $b > 0$ and since $x(t)$ is bounded, we find that $\int_0^{x(t)} du/f(u)$ is also bounded. This together with the fact that $\beta'(t) \in \mathcal{L}[t_0, \infty)$ shows that

$$\int_{t_1}^t \beta(s)\frac{x'(s)}{f(x(s))}ds \leq k_2 \quad \text{for some constant} \quad k_2 > 0.$$

Let $k = w(t_1) + \max\{k_1, k_2\}$. Then, we have

$$w(t) \leq k - \int_{t_1}^t \rho(s)q(s)ds. \tag{5.1.13}$$

By condition (5.1.9) there exists a sufficiently large $T \geq t_1$ such that $\int_{t_1}^t \rho(s)q(s)ds \geq 2k$ for all $t \geq T$. Thus, it follows that

$$a(t)\rho(t)\frac{x'(t)}{f(x(t))} \leq -\frac{1}{2}\int_{t_1}^t \rho(s)q(s)ds,$$

or

$$\frac{x'(t)}{f(x(t))} \leq -\frac{1}{2}\frac{1}{a(t)\rho(t)}\int_{t_1}^t \rho(s)q(s)ds.$$

Integrating this inequality from T to t ($\geq T$), we get

$$\int_0^{x(t)} \frac{du}{f(u)} \leq \int_0^{x(T)} \frac{du}{f(u)} - \int_T^t \frac{1}{a(s)\rho(s)}\int_{t_1}^s \rho(u)q(u)duds \to -\infty \text{ as } t \to \infty,$$

which is a contradiction.

Finally, we note that the proof of the criterion for oscillation of all bounded solutions of (5.1.3) in the case (i_1) is obvious. ∎

Theorem 5.1.2. In Theorem 5.1.1 condition (5.1.10) can be replaced by

$$\int^{\infty} \frac{1}{a(s)\rho(s)} ds = \infty. \tag{5.1.14}$$

Proof. Let $x(t)$ be a nonoscillatory solution of equation (5.1.3), say, $x(t) > 0$ for $t \geq t_0$. Proceeding as in Theorem 5.1.1, we obtain (5.1.13). Now by condition (5.1.9), we find that $w(t) \to -\infty$ as $t \to \infty$. Thus, there exist a constant $c > 0$ and $T \geq t_1$ such that $a(t)\rho(t)x'(t)/f(x(t)) \leq -c$ for $t \geq T$, or

$$\frac{x'(t)}{f(x(t))} \leq -\frac{c}{a(t)\rho(t)} \quad \text{for} \quad t \geq T.$$

Integrating this inequality from t to T, we get

$$\int_0^{x(t)} \frac{du}{f(u)} \leq \int_0^{x(T)} \frac{du}{f(u)} - c \int_T^t \frac{1}{a(s)\rho(s)} ds \to -\infty \text{ as } t \to \infty,$$

which is a contradiction. ∎

The following corollary is immediate.

Corollary 5.1.1. If $0 < \gamma < 1$ and

$$\int^{\infty} s^{\gamma} q(s) ds = \infty, \tag{5.1.15}$$

then equation (5.1.6) is oscillatory.

Theorem 5.1.3. If conditions (5.1.9) and (5.1.10) in Theorem 5.1.1 are replaced by

$$\int^{\infty} \frac{1}{a(s)\rho(s)} \left[M - \int_{t_0}^s \rho(u)q(u)du \right] ds = -\infty \tag{5.1.16}$$

for every constant M, then the conclusion of Theorem 5.1.1 holds.

Proof. Let $x(t)$ be a nonoscillatory solution of equation (5.1.3), say, $x(t) > 0$ for $t \geq t_0$. Proceeding as in Theorem 5.1.1, we obtain (5.1.13), i.e.,

$$\frac{x'(t)}{f(x(t))} \leq \frac{1}{a(t)\rho(t)} \left[k - \int_{t_1}^t \rho(u)q(u)du \right].$$

Integrating this inequality from t_1 to t, we get

$$\int_{t_1}^t \frac{x'(s)}{f(x(s))} ds \leq \int_{t_1}^t \frac{1}{a(s)\rho(s)} \left[k - \int_{t_1}^t \rho(u)q(u)du \right] ds.$$

Now from condition (5.1.16), we have

$$G(t) = \int_{t_1}^{t} \frac{x'(s)}{f(x(s))} ds \to -\infty \quad \text{as} \quad t \to \infty,$$

where $G(t) = \int_{x(t_1)}^{x(t)} du/f(u)$. If $x(t) \geq x(t_1)$ for all large t, then $G(t) > 0$ which is a contradiction. Hence, for all large t, $x(t) \leq x(t_1)$, so

$$G(t) = -\int_{x(t)}^{x(t_1)} \frac{du}{f(u)} > -\int_{0}^{x(t_1)} \frac{du}{f(u)} > -\infty,$$

which is again a contradiction. ■

Example 5.1.1. Consider the differential equation

$$(t^2 x'(t))' + p(t)x'(t) + ((1/2) + \cos t)|x(t)|^{1/3} \operatorname{sgn} x(t) = 0, \quad t > 0 \quad (5.1.17)$$

where $p(t) \in C^1([t_0, \infty), \mathbb{R}_0)$ and $p'(t) \leq 0$ for $t \geq t_0$. It is easy to verify all the hypotheses of Theorem 5.1.3 with $\rho(t) = 1$, and hence all solutions of (5.1.17) are oscillatory.

Next, we present the following results.

Theorem 5.1.4. Let condition (5.1.5) hold and there exist two functions $\rho(t) \in C^2([t_0, \infty), \mathbb{R}^+)$ and $\sigma(t) \in C^1([t_0, \infty), \mathbb{R}^+)$ such that $\beta(t) = a(t)\rho'(t) - p(t)\rho(t) \leq 0$, $\beta'(t) \geq 0$, $(a(t)\rho(t)\sigma(t))' \geq 0$ for $t \geq t_0$, and

$$\limsup_{t \to \infty} a(t)\rho(t) \frac{\sigma(t)}{\int_{t_0}^{t} \sigma(s)ds} < \infty. \tag{5.1.18}$$

If

$$\lim_{t \to \infty} \frac{1}{\int_{t_0}^{t} \sigma(s)ds} \int_{t_0}^{t} \sigma(s) \int_{t_0}^{s} \rho(u)q(u)duds = \infty, \tag{5.1.19}$$

then equation (5.1.3) is oscillatory.

Proof. Let $x(t)$ be a nonoscillatory solution of equation (5.1.3), say, $x(t) > 0$ for $t \geq t_0$. Proceeding as in Theorem 5.1.1, we find that inequality (5.1.13) holds for all $t \geq t_1$. Multiplying both sides of (5.1.13) by $\sigma(t)$ and integrating from t_1 to t, we obtain

$$\int_{t_1}^{t} a(s)\rho(s)\sigma(s)\frac{x'(s)}{f(x(s))}ds + \int_{t_1}^{t} \sigma(s) \int_{t_1}^{s} \rho(u)q(u)duds \leq k \int_{t_1}^{t} \sigma(s)ds. \tag{5.1.20}$$

Integrating the first integral in (5.1.20) by parts, we have

$$
\begin{aligned}
a(t)\rho(t)\sigma(t) \int_0^{x(t)} \frac{du}{f(u)} &- \int_{t_1}^t (a(s)\rho(s)\sigma(s))' \int_0^{x(s)} \frac{du}{f(u)} ds \\
&+ \int_{t_1}^t \sigma(s) \int_{t_1}^s \rho(u)q(u) du\, ds \\
&\leq k \int_{t_1}^t \sigma(s) ds + a(t_1)\rho(t_1)\sigma(t_1) \int_0^{x(t_1)} \frac{du}{f(u)}.
\end{aligned}
\tag{5.1.21}
$$

Set $G(t) = \int_0^{x(t)} du/f(u)$ and $\Omega(t) = a(t)\rho(t)\sigma(t)$. We consider two cases: either there exists a sequence $\{T_n\}_{n=1}^\infty$, $\lim_{n\to\infty} T_n = \infty$ such that

$$
\Omega(T_n)G(T_n) - \int_{t_1}^{T_n} \Omega(s)G(s) ds \;\geq\; 0,
$$

or there exists $t_2 \geq t_1$ such that

$$
\Omega(t)G(t) - \int_{t_1}^t \Omega'(s)G(s) ds \;\leq\; 0 \quad \text{for} \quad t \geq t_2.
\tag{5.1.22}
$$

In the former case, a contradiction is easily achieved upon dividing (5.1.21) by $\int_{t_1}^t \sigma(s) ds$ and applying the hypothesis of the theorem. Otherwise, write $C(t) = \int_{t_1}^t \Omega'(s)G(s) ds$ which is nonnegative and nondecreasing. Let $t_3 \geq t_2$ be chosen so large that $C(t) > 0$ for $t \geq t_3$. From (5.1.22), we deduce that $\Omega(t)C'(t) \leq \Omega'(t)C(t)$, or $(\Omega(t)/C(t))' \geq 0$ for $t \geq t_3$. Thus,

$$
\Omega(t)G(t) \;\leq\; C(t) \;\leq\; C(t_3)\frac{\Omega(t)}{\Omega(t_3)} \quad \text{for} \quad t \geq t_3.
$$

It follows that $G(t) \leq C(t_3)/\Omega(t_3) = k_1$, where k_1 is a constant. Thus, we have

$$
\int_{t_1}^t \sigma(s) \int_{t_1}^s \rho(u)q(u) du\, ds \;\leq\; k \int_{t_1}^t \sigma(s) ds + \Omega(t)G(t_1) + k_1\Omega(t).
\tag{5.1.23}
$$

Now, we can divide (5.1.23) by $\int_{t_1}^t \sigma(s) ds$ and apply condition (5.1.19) to obtain the desired contradiction. ∎

Example 5.1.2. Consider the differential equation

$$
(tx'(t))' + x'(t) + ((1/2) + \sin t)|x(t)|^\gamma \operatorname{sgn} x(t) \;=\; 0, \quad t > 0
\tag{5.1.24}
$$

where $0 < \gamma < 1$. It is easy to verify all conditions of Theorem 5.1.4 with $\rho(t) = 1/t$ and $\sigma(t) = t$, and hence equation (5.1.24) is oscillatory.

In a similar way we can prove the following result.

Theorem 5.1.5. Let condition (5.1.5) hold and there exist two functions $\rho(t) \in C^2([t_0, \infty), \mathbb{R}^+)$ and $\sigma(t) \in C^1([t_0, \infty), \mathbb{R}^+)$ such that $\beta(t) = a(t)\rho'(t) - p(t)\rho(t) \leq 0$, $\beta'(t) \geq 0$ and $\sigma'(t) \geq 0$ for $t \geq t_0$, and

$$\limsup_{t \to \infty} \frac{\sigma(t)}{\int_{t_0}^t \frac{\sigma(s)}{a(s)\rho(s)} ds} < \infty.$$

If

$$\lim_{t \to \infty} \frac{1}{\int_{t_0}^t \frac{\sigma(s)}{a(s)\rho(s)} ds} \int_{t_0}^t \frac{\sigma(s)}{a(s)\rho(s)} \int_{t_0}^s \rho(u)q(u)du\,ds = \infty,$$

then equation (5.1.3) is oscillatory.

Proof. Let $x(t)$ be a nonoscillatory solution of equation (5.1.3), say, $x(t) > 0$ for $t \geq t_0$. Proceeding as in Theorem 5.1.1, we obtain (5.1.13) which takes the form

$$\frac{x'(t)}{f(x(t))} + \frac{1}{a(t)\rho(t)} \int_{t_1}^t \rho(u)q(u)du \leq k\frac{1}{a(t)\rho(t)}, \quad t \geq t_1.$$

Multiplying both sides of this inequality by $\sigma(t)$ and integrating from t_1 to t, we get

$$\sigma(t) \int_0^{x(t)} \frac{du}{f(u)} - \int_{t_1}^t \sigma'(s) \int_0^{x(s)} \frac{du}{f(u)} ds + \int_{t_1}^t \frac{\sigma(s)}{a(s)\rho(s)} \int_{t_1}^s \rho(u)q(u)du\,ds$$

$$\leq k \int_{t_1}^t \frac{\sigma(s)}{a(s)\rho(s)} ds + \sigma(t_1) \int_0^{x(t_1)} \frac{du}{f(u)}.$$

The rest of the proof is similar to that of Theorem 5.1.4. ∎

Next we shall prove the following theorem.

Theorem 5.1.6. Suppose $Q(t) = \int_t^\infty q(s)ds$ exists and satisfies $Q(t) \geq 0$ for $t \geq t_0$. If

$$\int_{t_0}^\infty s^{\gamma-1}Q(s)ds = \infty, \tag{5.1.25}$$

then equation (5.1.6) is oscillatory.

Proof. Let $x(t)$ be a nonoscillatory solution of equation (5.1.6), say, $x(t) > 0$ for $t \geq t_0 > 0$. Put $y(t) = x(t)/t$, $t \geq t_0$. Then, $y(t)$ satisfies the equation

$$(t^2 y'(t))' + t^{1+\gamma}q(t)y^\gamma(t) = 0. \tag{5.1.26}$$

Set $w(t) = (t^2 y'(t))/y^\gamma(t)$, $t \geq t_0$. Then, we have

$$\frac{w'(t)}{t^{1+\gamma}} = -q(t) - \gamma t^{1-\gamma} \frac{(y'(t))^2}{y^{1+\alpha}(t)}. \tag{5.1.27}$$

Integrating equation (5.1.27) from t to T where $t_0 \le t \le T$, we obtain

$$\frac{w(T)}{T^{1+\gamma}} - \frac{w(t)}{t^{1+\gamma}} + \int_t^T (1+\gamma)\frac{w(s)}{s^{2+\gamma}}ds$$
$$= -\int_t^T q(s)ds - \int_t^T \gamma \frac{s^{1-\gamma}(y'(s))^2}{y^{1+\gamma}(s)}ds. \qquad (5.1.28)$$

Let $z(t) = x'(t)/x^\alpha(t)$, $t \ge t_0$. Then, we find

$$z'(t) = -q(t) - \gamma x^{\gamma-1}(t)z^2(t). \qquad (5.1.29)$$

Integrating equation (5.1.29) between t and T, $t_0 \le t \le T$, we get

$$z(T) = z(t) - \int_t^T q(s)ds - \int_t^T \gamma x^{\gamma-1}(s)z^2(s)ds. \qquad (5.1.30)$$

Since $\int_t^\infty q(s)ds \ge 0$, we find by letting $T \to \infty$ in (5.1.30) that $\lim_{T\to\infty} z(T) = \ell$, where $-\infty \le \ell < \infty$. Now, if $\ell < 0$, then $z(t) < \ell/2$ for $t \ge t_1 \ge t_0$, and hence

$$\frac{x^{1-\gamma}(t)}{1-\gamma} - \frac{x^{1-\gamma}(t_1)}{1-\gamma} = \int_{t_1}^t z(s)ds \le \frac{1}{2}\ell(t-t_1),$$

which is impossible since $x(t) > 0$ for all $t \ge t_0$. Hence, $0 < \ell < \infty$. Thus, it follows that

$$\lim_{t\to\infty} \frac{x(t)}{t^{1/(1-\gamma)}} = c_1, \qquad \lim_{t\to\infty} \frac{x'(t)}{t^{\gamma/(1-\gamma)}} = \frac{c_1}{1-\gamma},$$
$$\lim_{t\to\infty} \frac{y(t)}{t^{\gamma/(1-\gamma)}} = c_1, \qquad \lim_{t\to\infty} \frac{y'(t)}{t^{(2\gamma-1)/(1-\gamma)}} = \frac{c_1\gamma}{1-\gamma},$$

where $c_1 = [(1-\gamma)\ell]^{1/(1-\gamma)}$. Therefore, we have

$$\lim_{T\to\infty} \frac{w(T)}{T^{1+\gamma}} = \lim_{T\to\infty} \frac{T^2 y'(T)}{T^{1+\gamma}y^\gamma(T)} = c_2, \qquad (5.1.31)$$

where $c_2 = (\gamma/(1-\gamma))c_1^{1-\gamma} \ge 0$. Letting $T \to \infty$ in (5.1.28), we find

$$\int_t^\infty \frac{w(s)}{s^{2+\gamma}}ds = c_3(t), \qquad -\infty \le c_3(t) < \infty.$$

But, since

$$\int_t^T \frac{w(s)}{s^{2+\gamma}}ds = \frac{\psi(T)}{T^\gamma} - \frac{\psi(t)}{t^\gamma} + \int_t^T \frac{\gamma}{s^{1+\gamma}}\psi(s)ds,$$

where

$$\psi(t) \;=\; \int_{t_0}^{T} \frac{w(s)}{s^2} ds \;=\; \frac{y^{1-\gamma}(t)}{1-\gamma} - \frac{y^{1-\gamma}(t_0)}{1-\gamma}$$

is bounded below on $[t_0, \infty)$, it follows that $c_3(t)$ cannot be $-\infty$, and hence

$$\int_{t}^{\infty} \frac{w(s)}{s^{2+\gamma}} ds \;=\; c_3(t), \qquad -\infty < c_3(t) < \infty. \qquad (5.1.32)$$

Using (5.1.31) and (5.1.32) and taking the limit as $T \to \infty$ in (5.1.28), we obtain

$$\int_{t}^{\infty} (1+\gamma) \frac{w(s)}{s^{2+\gamma}} ds - \frac{w(t)}{t^{1+\gamma}} \;\le\; -Q(t),$$

and hence

$$\frac{w(t)}{t^2} - \frac{1}{t^{1-\gamma}} \int_{t}^{\infty} (1+\gamma) \frac{w(s)}{s^{2+\gamma}} ds \;\ge\; t^{\gamma-1} Q(t) \quad \text{for} \quad t \ge t_0. \qquad (5.1.33)$$

Integrating (5.1.33) from t_0 to t, we find

$$\int_{t_0}^{t} \frac{w(s)}{s^2} ds - \int_{t_0}^{t} \frac{1}{s^{1-\gamma}} \int_{s}^{\infty} (1+\gamma) \frac{w(u)}{u^{2+\gamma}} du\, ds \;\ge\; \int_{t_0}^{t} s^{\gamma-1} Q(s) ds,$$

i.e.,

$$\int_{t_0}^{t} \frac{w(s)}{s^2} ds - \frac{t^\gamma}{\gamma} \int_{t}^{\infty} (1+\gamma) \frac{w(s)}{s^{2+\gamma}} ds - \frac{(1+\gamma)}{\gamma} \int_{t_0}^{t} \frac{w(s)}{s^2} ds$$

$$\ge\; c_4 + \int_{t_0}^{t} s^{\gamma-1} Q(s) ds,$$

where c_4 is some constant.

Thus, it follows that

$$-\frac{1}{\gamma} \int_{t_0}^{t} \frac{w(s)}{s^2} ds - \frac{t^\gamma}{\gamma} \int_{t}^{\infty} (1+\gamma) \frac{w(s)}{s^{2+\gamma}} ds \;\to\; \infty \quad \text{as} \quad t \to \infty. \qquad (5.1.34)$$

Now, from (5.1.31) for any $\epsilon > 0$ there is $T(\epsilon) \ge t_0$ such that

$$\int_{T(\epsilon)}^{\tau} \frac{w(s)}{s^{2+\gamma}} ds \;<\; \epsilon \quad \text{for all} \quad \tau \ge T(\epsilon).$$

Hence, we find

$$\psi(\tau) \;=\; \int_{t_0}^{\tau} \frac{w(s)}{s^2} ds \;=\; \int_{t_0}^{T(\epsilon)} \frac{w(s)}{s^2} ds + \int_{T(\epsilon)}^{\tau} \frac{w(s)}{s^{2+\gamma}} s^\gamma ds$$

$$=\; \int_{t_0}^{T(\epsilon)} \frac{w(s)}{s^2} ds + \tau^\gamma \int_{T(\epsilon)}^{\tau} \frac{w(s)}{s^{2+\gamma}} ds - \int_{T(\epsilon)}^{\tau} \gamma s^{\gamma-1} \int_{T(\epsilon)}^{s} \frac{w(u)}{u^{2+\gamma}} du\, ds$$

$$\le\; \int_{t_0}^{T(\epsilon)} \frac{w(s)}{s^2} ds + 2\epsilon \tau^\gamma \quad \text{for} \quad \tau \ge T(\epsilon).$$

Therefore, we have

$$\lim_{t \to \infty} \frac{\psi(\tau)}{\tau^\gamma} = 0. \tag{5.1.35}$$

Now, since

$$\int_t^\tau \frac{w(s)}{s^{2+\gamma}} ds = \int_t^\tau \frac{\psi'(s)}{s^\gamma} ds = \frac{\psi(\tau)}{\tau^\gamma} - \frac{\psi(t)}{t^\gamma} + \int_t^\tau \gamma \frac{\psi(s)}{s^{1+\gamma}} ds \tag{5.1.36}$$

using (5.1.35) and letting $\tau \to \infty$ in (5.1.36), we obtain

$$\int_t^\infty \frac{w(s)}{s^{2+\gamma}} ds = -\frac{\psi(t)}{t^\gamma} + \int_t^\infty \gamma \frac{\psi(s)}{s^{1+\gamma}} ds.$$

Thus, it follows that

$$-\int_{t_0}^t \frac{w(s)}{s^2} ds - t^\gamma \int_t^\infty (1+\gamma) \frac{w(s)}{s^{2+\gamma}} ds$$
$$= -\psi(t) + (1+\gamma)\psi(t) - \gamma(1+\gamma)t^\gamma \int_t^\infty \frac{\psi(s)}{s^{1+\gamma}} ds. \tag{5.1.37}$$

Clearly, (5.1.34) and (5.1.37) give

$$\psi(t) - (1+\gamma)t^\gamma \int_t^\infty \frac{\psi(s)}{s^{1+\gamma}} ds \to \infty \quad \text{as} \quad t \to \infty. \tag{5.1.38}$$

Now, since $\psi(t)$ is bounded below by $-L$, say, we find

$$-(1+\gamma)t^\gamma \int_t^\infty \frac{\psi(s)}{s^{1+\gamma}} ds \le \frac{1+\gamma}{\gamma} L.$$

Therefore, $\lim_{t \to \infty} \psi(t) = \infty$. Define t_m, $m = 1, 2, \cdots$ by $t_m = \sup\{t : \psi(t) \le m$, for m large enough$\}$. Choose $t_m \to \infty$ such that $\psi(t) \ge \psi(t_m) > 0$ for all $t \ge t_m$. Then, we have

$$\psi(t_m) - (1+\gamma)t_m^\gamma \int_{t_m}^\infty \frac{\psi(s)}{s^{1+\gamma}} ds \le \psi(t_m) - (1+\gamma)\psi(t_m)t_m^\gamma \int_{t_m}^\infty \frac{ds}{s^{1+\gamma}}$$
$$\le -\frac{1}{\gamma}\psi(t_m) < 0,$$

which contradicts (5.1.38). This completes the proof. ∎

The following result extends Corollary 5.1.1.

Theorem 5.1.7. Assume that there exists a function $\rho(t) \in C^2([t_0, \infty), \mathbb{R}^+)$ such that $\rho'(t) \ge 0$ and $\rho''(t) \le 0$ for $t \ge t_0$. If

$$\lim_{t \to \infty} \int_{t_0}^t \rho^\gamma(s)q(s)ds = \infty, \tag{5.1.39}$$

then equation (5.1.6) is oscillatory.

Proof.   Let $x(t)$ be a nonoscillatory solution of equation (5.1.6), say, $x(t) > 0$ for $t \geq t_0 \geq 0$. Define

$$w(t) = \left(\frac{x(t)}{\rho(t)}\right)^\gamma \quad \text{for} \quad t \geq t_0. \qquad (5.1.40)$$

Then, $w(t) > 0$ for $t \geq t_0$. Let $\alpha = 1/\gamma > 1$, then $x(t) = \rho(t)w^\alpha(t)$ for $t \geq t_0$. By differentiation it follows that

$$\frac{1}{w(t)}(\rho(t)w^\alpha(t))'' = \left(\frac{\alpha}{\alpha-1}\right)(\rho(t)w^{\alpha-1}(t))'' + \alpha\rho(t)w^{\alpha-3}(t)(w'(t))^2$$
$$+ \frac{1}{1-\alpha}\rho''(t)w^{\alpha-1}(t), \quad t \geq t_0.$$
$$(5.1.41)$$

Now from equation (5.1.6) and (5.1.40), we have

$$\frac{1}{w(t)}x''(t) = \frac{1}{w(t)}(\rho(t)w^\alpha(t))'' = -\rho^\gamma(t)q(t) \quad \text{for} \quad t \geq t_0. \quad (5.1.42)$$

Since $\rho''(t) \leq 0$ for $t \geq t_0$ and $\alpha > 1$, the last two terms in (5.1.41) are nonnegative, hence we may combine (5.1.41) and (5.1.42), to obtain

$$\frac{\alpha}{\alpha-1}(\rho(t)w^{\alpha-1}(t))'' \leq -\rho(t)q(t) \quad \text{for} \quad t \geq t_0. \qquad (5.1.43)$$

Integrating (5.1.43) twice from t_0 to t, we get

$$\rho(t)w^{\alpha-1}(t) \leq c_1 + c_0 t - \left(\frac{\alpha-1}{\alpha}\right)\int_{t_0}^t\int_{t_0}^s \rho^\gamma(u)q(u)du\,ds, \qquad (5.1.44)$$

where c_0 and c_1 are appropriate integration constants. Clearly, (5.1.39) implies that the right–hand side of (5.1.44) becomes eventually negative, which contradicts the assumption that $\rho(t)w^{\alpha-1}(t) > 0$ for $t \geq t_0$. This completes the proof. ∎

Remark 5.1.1.   In the above proof of Theorem 5.1.7 condition (5.1.39) can be weakened to

$$\limsup_{t\to\infty}\frac{1}{t}\int_{t_0}^t\int_{t_0}^s \rho^\gamma(u)q(u)du\,ds = \infty, \qquad (5.1.45)$$

which is, in fact, sufficient to produce the desired contradiction in (5.1.44).

Example 5.1.3.   Consider equation (5.1.6) with $0 < \gamma < 1$ and $q(t) = t^\lambda \sin t$ or $t^\lambda \cos t$, $\lambda \in \mathbb{R}$. Condition (5.1.45) with $\rho(t) = t$ shows that (5.1.6) is oscillatory if $1 - \gamma < \lambda < 1$.

5.2. Further Sublinear Oscillation Criteria

We assume that the function f satisfies condition (F), i.e.,

(F) $f \in C(\mathbb{R}, \mathbb{R})$, $xf(x) > 0$ and $f'(x) > 0$ for $x \neq 0$, and the condition of sublinearity, namely, (5.1.5) holds.

Next, we let the function $F(x)$ be defined by

$$F(x) = \int_{+0}^{x} \frac{du}{f(u)} \quad \text{for} \quad x > 0, \qquad F(x) = \int_{-0}^{x} \frac{du}{f(u)} \quad \text{for} \quad x < 0$$

and β stands for the nonnegative constant

$$\min \left\{ \frac{\inf_{x>0} F(x) f'(x)}{1 + \min_{x>0} F(x) f'(x)}, \frac{\inf_{x<0} F(x) f'(x)}{1 + \min_{x<0} F(x) f'(x)} \right\}.$$

Clearly, $0 \leq \beta < 1$ and $F(x) > 0$ for $x \neq 0$. For the special case $f(x) = |x|^{\gamma} \operatorname{sgn} x$, $0 < \gamma < 1$ it follows that $\beta = \gamma$.

The following examples produce lower bounds for β which are essential for our oscillation theory (see for example Corollary 5.2.6).

Example 5.2.1. Let $f(x) = (|x|^{\gamma} \operatorname{sgn} x)/(1 + |x|^{\gamma})$, $x \in \mathbb{R}$. We observe that $f(x)$ is continuous on $\mathbb{R}$ and has the sign property $xf(x) > 0$ for $x \neq 0$. Moreover, $f'(x) = (\gamma |x|^{\gamma - 1})/([1 + |x|^{\gamma}]^2) > 0$ for $x \neq 0$ and condition (5.1.5) holds. Furthermore, we have

$$F(x) = |x| + \frac{|x|^{1-\gamma}}{1 - \gamma} \quad \text{for} \quad x \neq 0$$

and consequently,

$$\beta = \inf_{x>0} \frac{\gamma}{(1 + x^{\gamma})^2} \left(\frac{1}{1 - \gamma} + x^{\gamma} \right) = 0.$$

Example 5.2.2. Consider the continuous function $f(x) = x + |x|^{\gamma} \operatorname{sgn} x$, $x \in \mathbb{R}$. Clearly, $xf(x) > 0$ for $x \neq 0$. Moreover, f is differentiable on $\mathbb{R} - \{0\}$ with $f'(x) = \gamma |x|^{\gamma - 1} + 1 > 0$ for all $x \neq 0$. In addition, it is easy to verify that f satisfies (5.1.5). Next, for every $x \neq 0$, we have

$$F(x) = \int_{+0}^{|x|} \frac{du}{u^{\gamma} + u} \leq \int_{+0}^{|x|} \frac{du}{u^{\gamma}} = \frac{|x|^{1-\gamma}}{1 - \gamma}$$

and consequently,

$$\inf_{x<0} F(x) f'(x) = \inf_{x>0} F(x) f'(x) \leq \inf_{x>0} \frac{x^{1-\gamma}}{1 - \gamma} [1 + \gamma x^{\gamma - 1}] = \frac{\gamma}{1 - \gamma}.$$

On the other hand, for every $x \neq 0$, we find

$$F(x) = \int_{+0}^{|x|} \frac{du}{u^\gamma + u} \geq \int_0^{|x|} \frac{du}{2u^\gamma} = \frac{|x|^{1-\gamma}}{2(1-\gamma)} \quad \text{if} \quad |x| \leq 1$$

and

$$F(x) = \int_{+0}^{|x|} \frac{du}{u^\gamma + u} \geq \int_{+0}^1 \frac{du}{u^\gamma + u} \geq \int_{+0}^1 \frac{du}{2u^\gamma} = \frac{1}{2(1-\gamma)}, \quad \text{if} \quad |x| \geq 1$$

and therefore for $x \neq 0$, we obtain

$$F(x)f'(x) \geq \frac{|x|^{1-\gamma}}{2(1-\gamma)} \left[1 + \gamma|x|^{\gamma-1}\right] = \frac{1}{2(1-\gamma)} \left[\gamma + |x|^{1-\gamma}\right]$$

$$> \frac{\gamma}{2(1-\gamma)} \quad \text{for} \quad |x| \leq 1$$

and

$$F(x)f'(x) \geq \frac{1}{2(1-\gamma)} \left[1 + \gamma|x|^{\gamma-1}\right] > \frac{1}{2(1-\gamma)} \quad \text{for} \quad |x| \geq 1.$$

Thus, $F(x)f'(x) > (\gamma/(2(1-\gamma)))$ for all $x \neq 0$, and consequently

$$\inf_{x<0} F(x)f'(x) = \inf_{x>0} F(x)f'(x) \geq \frac{\gamma}{2(1-\gamma)}.$$

Finally, we have

$$\beta \geq \frac{\gamma}{2(1-\gamma)} \bigg/ \left(1 + \frac{\gamma}{1-\gamma}\right) = \frac{\gamma}{2}.$$

Example 5.2.3. Consider the continuous function

$$f(x) = |x|^\gamma \left[k + \sin(\ln[1 + |x|])\right] \operatorname{sgn} x, \quad x \in \mathbb{R}.$$

Clearly, $xf(x) > 0$, $f'(x) > 0$ for $x \neq 0$, and condition (5.1.5) holds provided $k \geq 1 + (1/\gamma)$. Moreover, for $x \neq 0$,

$$f'(x) = \gamma|x|^{\gamma-1} \left\{k + \sin(\ln[1 + |x|]) + \frac{1}{\gamma} \frac{|x|}{1 + |x|} \cos(\ln[1 + |x|])\right\}$$

and consequently, for $x \neq 0$,

$$0 < \gamma\left[k - \left(1 + \frac{1}{\gamma}\right)\right] |x|^{\gamma-1} < f'(x) < \gamma\left[k + \left(1 + \frac{1}{\gamma}\right)\right] |x|^{\gamma-1}.$$

Furthermore, for any $x \neq 0$,

$$F(x) = \int_{+0}^{|x|} \frac{du}{u^\gamma\{k + \sin(\ln[1 + u])\}} \geq \frac{1}{k+1} \int_{+0}^{|x|} \frac{du}{u^\gamma} = \frac{1}{k+1} \frac{|x|^{1-\gamma}}{1-\gamma}$$

and

$$F(x) = \int_{+0}^{|x|} \frac{du}{u^\gamma \{k + \sin(\ln[1+u])\}} \le \frac{1}{k-1} \int_{+0}^{|x|} \frac{du}{u^\gamma} = \frac{1}{k-1} \frac{|x|^{1-\gamma}}{1-\gamma}.$$

Thus for all $x \ne 0$,

$$\frac{\gamma \left[k - \left(1 + \frac{1}{\gamma}\right)\right]}{(1-\gamma)(k+1)} < F(x)f'(x) < \frac{\gamma \left[k + \left(1 + \frac{1}{\gamma}\right)\right]}{(1-\gamma)(k-1)},$$

which gives

$$\beta \ge \frac{\gamma \left[k - \left(1 + \frac{1}{\gamma}\right)\right]}{(1-\gamma)(k+1)} \frac{(1-\gamma)(k-1)}{1 + \gamma \left[k + \left(1 + \frac{1}{\gamma}\right)\right]} = \frac{k-1}{k+1} \frac{\gamma k - \gamma - 1}{k + 2\gamma}.$$

Now, we shall prove the following results.

Theorem 5.2.1. Suppose condition (F) holds, and let $\rho(t) \in C^2([t_0, \infty), \mathbb{R}^+)$ satisfy

$$\left[\beta \rho'(t) + \left(\frac{\rho^\beta(t)}{a(t)}\right)' \left(\frac{a(t)}{\rho^{\beta-1}(t)}\right) - \rho(t) \frac{p(t)}{a(t)}\right]^2$$
$$\le \frac{4\beta^2}{1-\beta} [(1-\beta)(\rho'(t))^2 - \rho(t)\rho''(t)] \quad \text{for} \quad t \ge t_0. \tag{5.2.1}$$

If

$$\limsup_{t \to \infty} \frac{1}{t} \int_{t_0}^{t} \int_{t_0}^{s} \left(\frac{\rho^\beta(u)}{a(u)}\right) q(u) du \, ds = \infty, \tag{5.2.2}$$

then equation (5.1.3) is oscillatory.

Proof. Let $x(t)$ be a nonoscillatory solution of equation (5.1.3), say, $x(t) \ne 0$ for $t \ge t_0$. Define $w(t) = \rho^\beta(t)F(x(t))$ for $t \ge t_0$. Then for every $t \ge t_0$, we find

$$w'(t) = \beta \rho^{\beta-1}(t)\rho'(t)F(x(t)) + \left(\frac{\rho^\beta(t)}{a(t)}\right) \left(\frac{a(t)x'(t)}{f(x(t))}\right)$$

and

$$w''(t) = \beta(\beta-1)\rho^{\beta-2}(t)(\rho'(t))^2 F(x(t)) + \beta \rho^{\beta-1}(t)\rho''(t)F(x(t))$$
$$+ \beta \rho^{\beta-1}(t)\rho'(t) \left(\frac{x'(t)}{f(x(t))}\right) + \left(\frac{\rho^\beta(t)}{a(t)}\right)' \left(\frac{a(t)x'(t)}{f(x(t))}\right)$$
$$+ \frac{\rho^\beta(t)}{a(t)} \left[-q(t) - p(t) \frac{x'(t)}{f(x(t))} - a(t)f'(x(t)) \left(\frac{x'(t)}{f(x(t))}\right)^2\right]$$
$$= -\frac{\rho^\beta(t)}{a(t)} q(t) - \rho^\beta(t)f'(x(t)) \left[\left(\frac{x'(t)}{f(x(t))}\right)^2 - \left\{\beta \frac{\rho'(t)}{\rho(t)}\right.$$

$$+ \left(\frac{\rho^\beta(t)}{a(t)} \right)' \left(\frac{a(t)}{\rho^\beta(t)} \right) - \frac{p(t)}{a(t)} \right\} \frac{x'(t)}{f(x(t))} \frac{1}{f'(x(t))} \right]$$

$$+ \beta \rho^{\beta-2}(t) \left[\rho(t)\rho''(t) - (1-\beta)(\rho'(t))^2 \right] F(x(t))$$

$$= \quad - \frac{a^\beta(t)}{a(t)} q(t) - \rho^\beta(t) f'(x(t)) \left[\frac{x'(t)}{f(x(t))} - \frac{1}{2f'(x(t))} \right\{ \beta \frac{\rho'(t)}{\rho(t)}$$

$$+ \left(\frac{\rho^\beta(t)}{a(t)} \right)' \left(\frac{a(t)}{\rho^\beta(t)} \right) - \frac{p(t)}{a(t)} \right\}^2 \right]$$

$$+ \frac{\rho^{\beta-2}(t)}{4f'(x(t))} \left[\left\{ \beta\rho'(t) + \left(\frac{\rho^\beta(t)}{a(t)} \right)' \left(\frac{a(t)}{\rho^{\beta-1}(t)} \right) - \rho(t) \frac{p(t)}{a(t)} \right\}^2 \right.$$

$$\left. + 4\beta \left\{ \rho(t)\rho''(t) - (1-\beta)(\rho'(t))^2 \right\} F(x(t)) f'(x(t)) \right].$$

Using the definition of β, we have $f'(x(t))F(x(t)) \geq \beta/(1-\beta)$, and hence by condition (5.2.1), it follows that $w''(t) \leq -\rho^\beta(t)q(t)/a(t)$ for $t \geq t_0$. Thus for every $t \geq t_0$, we obtain

$$\frac{1}{t} \int_{t_0}^t \int_{t_0}^s \frac{\rho^\beta(u)}{a(u)} q(u)\,du\,ds \quad \leq \quad -\frac{w(t)}{t} + \frac{w(t_0)}{t} + w'(t_0) \left(1 - \frac{t_0}{t} \right)$$

$$\leq \quad \frac{w(t_0)}{t} + w'(t_0) \left(1 - \frac{t_0}{t} \right).$$

Therefore, it follows that

$$\limsup_{t\to\infty} \frac{1}{t} \int_{t_0}^t \int_{t_0}^s \frac{\rho^\beta(u)}{a(u)}\,du\,ds \leq w'(t_0) < \infty,$$

which contradicts condition (5.2.2).   ∎

Theorem 5.2.2. Condition (5.2.1) in Theorem 5.2.1 can be replaced by

$$\left[\beta\rho'(t) + \left(\frac{a^\beta(t)}{a(t)} \right)' \frac{a(t)}{\rho^{\beta-1}(t)} \right]^2 \leq \frac{4\beta^2}{1-\beta} \left[(1-\beta)(\rho'(t))^2 - p(t)\rho''(t) \right]$$

$$(5.2.3)$$

for $t \geq t_0$, and

$$p(t) \geq 0, \quad k(t) = \rho^\beta(t)p(t)/(a(t)) \quad \text{is a decreasing function for} \quad t \geq t_0.$$
$$(5.2.4)$$

Proof. Let $x(t)$ be a nonoscillatory solution of equation (5.1.3) and assume that $x(t) > 0$ for $t \geq t_0$. Put $w(t) = \rho^\beta(t)F(x(t))$, $t \geq t_0$. Then as in Theorem 5.2.1, for every $t \geq t_0$, we have

$$w''(t) \leq \frac{\rho^\beta(t)}{a(t)} \frac{(a(t)x'(t))'}{f(x(t))}.$$

Thus, it follows that

$$\frac{\rho^\beta(t)}{a(t)}q(t) \leq -w''(t) - \frac{\rho^\beta(t)}{a(t)}p(t)\frac{x'(t)}{f(x(t))} \quad \text{for all} \quad t \geq t_0$$

and therefore,

$$\int_{t_0}^t \left(\frac{\rho^\beta(s)}{a(s)}\right)q(s)ds \leq -w'(t) - w'(t_0) - \int_{t_0}^t k(s)\frac{x'(s)}{f(x(s))}ds \quad \text{for all} \quad t \geq t_0.$$

Now, by the Bonnet theorem, for a fixed $t \geq t_0$ and for some $\xi \in [t_0, t]$, we find

$$\int_{t_0}^t (-k(s))\frac{x'(s)}{f(x(s))}ds = -k(t_0)\int_{t_0}^\xi \frac{x'(s)}{f(x(s))}ds$$

$$= k(t_0)\int_{x(\xi)}^{x(t_0)}\frac{du}{f(u)} \leq k(t_0)F(x(t_0)),$$

since

$$\int_{x(\xi)}^{x(t_0)}\frac{du}{f(u)} < \begin{cases} 0 & \text{if } x(\xi) > x(t_0) \\ \int_{+0}^{x(t_0)}\frac{du}{f(u)} & \text{if } x(\xi) \leq x(t_0) \end{cases}$$

for $x > 0$; also, for $x < 0$,

$$\int_{x(\xi)}^{x(t_0)}\frac{du}{f(u)} < \begin{cases} 0 & \text{if } x(\xi) < x(t_0) \\ \int_{-0}^{x(t_0)}\frac{du}{f(u)} & \text{if } x(\xi) \geq x(t_0). \end{cases}$$

Hence, for every $t \geq t_0$, we obtain

$$\frac{1}{t}\int_{t_0}^t \int_{t_0}^s \left(\frac{\rho^\beta(u)}{a(u)}\right)q(u)duds \leq -\frac{w(t)}{t} + \frac{w(t_0)}{t} + M\left(1 - \frac{t_0}{t}\right)$$

$$\leq M\left(1 - \frac{t_0}{t}\right),$$

where $M = w'(t_0) + k(t_0)F(x(t_0))$. This contradicts condition (5.2.2). ∎

The following corollaries are immediate consequences of Theorems 5.2.1 and 5.2.2.

Corollary 5.2.1. Suppose condition (F) holds, and let $\rho(t) \in C^2([t_0, \infty), \mathbb{R}^+)$ satisfy

$$[2\rho'(t) - \rho(t)p(t)]^2 \leq \frac{4\beta^2}{1-\beta}[(1-\beta)(\rho'(t))^2 - \rho(t)\rho''(t)], \quad t \geq t_0.$$

If condition (5.2.2) holds, then equation (5.1.3) with $a(t) = 1$ is oscillatory.

Corollary 5.2.2. Suppose condition (F) holds, and let $\rho(t) \in C^2([t_0, \infty),$ $\mathbb{R}^+)$ satisfy

$$p(t) \geq 0, \quad \rho'(t) \geq 0, \quad \rho''(t) \leq 0 \quad \text{and} \quad [\rho^\beta(t)p(t)]' \leq 0 \quad \text{for} \quad t \geq t_0.$$

If condition (5.2.2) holds, then equation (5.1.3) with $a(t) = 1$ is oscillatory.

Example 5.2.4. Consider the differential equation

$$(\sqrt{t}x'(t))' + \frac{\sin t}{2\sqrt{t}}x'(t) + (t^{5/4}\sin t)|x(t)|^{1/2}\text{sgn } x(t) = 0, \quad t \geq t_0 > 0.$$
$$(5.2.5)$$

Since $f(x) = |x|^{1/2}\text{sgn } x$, we have $\beta = 1/2$. Next, we let $\rho(t) = t$. Then for every $t \geq t_0 > 0$, we find

$$\left[\beta\rho'(t) + \left(\frac{\rho^\beta(t)}{a(t)}\right)'\frac{a(t)}{\rho^{\beta-1}(t)} - \frac{\rho(t)}{a(t)}p(t)\right]^2$$

$$= \frac{1}{4}(1 - \sin t)^2 \leq 1 = \frac{4\beta^2}{1-\beta}[(1-\beta)(\rho'(t))^2 - \rho(t)\rho''(t)].$$

Also, it follows that

$$\frac{1}{t}\int_{t_0}^t \int_{t_0}^s \frac{\rho^\beta(u)}{a(u)}q(u)duds = \frac{1}{t}\int_{t_0}^t\int_{t_0}^s u^{5/4}\sin u\, duds \to \infty \quad \text{as} \quad t \to \infty.$$

Thus, all conditions of Theorem 5.2.1 are satisfied and hence (5.2.5) is oscillatory.

Example 5.2.5. Consider the differential equation

$$(t^2x'(t))' + \frac{t}{\sqrt{\ln t}}x'(t) + (t^{11/4}\sin t)|x(t)|^{1/2}\text{sgn } x(t) = 0, \quad t \geq t_0 > 1.$$
$$(5.2.6)$$

As in Example 5.2.4, we have $\beta = 1/2$ and letting $\rho(t) = t$, we find for all $t \geq t_0 > 1$ that

$$\left[\beta\rho'(t) + \left(\frac{\rho^\beta(t)}{a(t)}\right)'\left(\frac{a(t)}{\rho^{\beta-1}(t)}\right)\right]^2 = 1 = \frac{4\beta^2}{1-\beta}[(1-\beta)(\rho'(t))^2 - \rho(t)\rho''(t)]$$

and the function $\rho^\beta(t)p(t)/a(t) = 1/\sqrt{t\ln t} \geq 0$ is decreasing. Further,

$$\frac{1}{t}\int_{t_0}^t\int_{t_0}^s \left(\frac{\rho^\beta(u)}{a(u)}\right)q(u)duds = \frac{1}{t}\int_{t_0}^t\int_{t_0}^s u^{5/4}\sin u\, duds \to \infty \quad \text{as } t \to \infty.$$

Therefore, all conditions of Theorem 5.2.2 are satisfied and hence (5.2.6) is oscillatory.

We note that Theorem 5.2.2 is not applicable to (5.2.5) since condition (5.2.4) does not hold. Also, Theorem 5.2.1 fails to apply to (5.2.6) since condition (5.2.1) is not satisfied. We also observe that conditions of the type $\int^\infty ds/a(s) = \infty$ and $\int^\infty ds/a(s) < \infty$ are not required.

Remark 5.2.1. Condition (5.2.3) can be simplified if $a \in C^1([t_0, \infty), \mathbb{R}^+)$. In fact, it takes the form

$$\left(\frac{a'(t)}{a(t)}\right)\left(\frac{\rho'(t)}{\rho(t)}\right) \geq \frac{\beta}{1-\beta}\left(\frac{\rho''(t)}{\rho(t)}\right) + \frac{1}{4\beta}\left(\frac{a'(t)}{a(t)}\right)^2 \quad \text{for } t \geq t_0. \quad (5.2.7)$$

Next, we present an extension of Theorem 5.2.1.

Theorem 5.2.3. Suppose condition (F) holds, and let $\rho(t) \in C^2([t_0, \infty), \mathbb{R}^+)$, satisfy (5.2.1). If

$$\limsup_{t\to\infty} \frac{1}{t^n} \int_{t_0}^t (t-s)^n \left(\frac{\rho^\beta(s)}{a(s)}\right) q(s)ds = \infty, \quad (5.2.8)$$

where $n \geq 1$ is an integer, then equation (5.1.3) is oscillatory.

Proof. Let $x(t)$ be a nonoscillatory solution of equation (5.1.3), say, $x(t) > 0$ for $t \geq t_0$. Define $w(t) = \rho^\beta(t)F(x(t))$, $t \geq t_0$. Then as in Theorem 5.2.1 for $t \geq t_0$, we have $\rho^\beta(t)q(t)/a(t) \leq -w''(t)$, and consequently for $t \geq t_0$,

$$\int_{t_0}^t (t-s)^n \left(\frac{\rho^\beta(s)}{a(s)}\right) q(s)ds \leq -\int_{t_0}^t (t-s)^n w''(s)ds$$

$$= (t-t_0)^n w'(t_0) - n\int_{t_0}^t (t-s)^{n-1} w'(s)ds$$

$$= \begin{cases} (t-t_0)w'(t_0) - w(t) + w(t_0) & \text{if } n = 1 \\ (t-t_0)^n w'(t_0) - n(n-1)\int_{t_0}^t (t-s)^{n-2} w(s)ds \\ \quad + n(t-t_0)^{n-1} w(t_0) & \text{if } n > 1 \end{cases}$$

$$\leq (t-t_0)^n w'(t_0) + n(t-t_0)^{n-1} w(t_0).$$

Thus, it follows that

$$\frac{1}{t^n}\int_{t_0}^t (t-s)^n \left(\frac{\rho^\beta(s)}{a(s)}\right) q(s)ds \leq \left(1 - \frac{t_0}{t}\right)^n w'(t_0) + n\left(1 - \frac{t_0}{t}\right)^{n-1}\frac{w(t_0)}{t}$$

for $t \geq t_0$, and hence

$$\limsup_{t \to \infty} \frac{1}{t^n} \int_{t_0}^t (t-s)^n \left(\frac{\rho^\beta(s)}{a(s)} \right) q(s)ds \leq w'(t_0) < \infty,$$

which contradicts condition (5.2.8).   ■

Corollary 5.2.3. Let $n \geq 1$ be an integer and $\lambda \in [0, \beta]$. Then equation (5.1.1) is oscillatory if

$$\limsup_{t \to \infty} \frac{1}{t^n} \int_{t_0}^t (t-s)^n s^\lambda q(s)ds = \infty.$$

Proof. It suffices to apply Theorem 5.2.3 with

$$\rho(t) = \begin{cases} 1 & \text{for } t \geq t_0 \text{ if } \beta = 0 \\ t^{\lambda/\beta} & \text{for } t \geq t_0 \text{ if } \beta > 0. \end{cases} \quad ■$$

Our next result provides sufficient conditions for the oscillaion of (5.1.3) when condition (5.2.2) fails. For this, we need to define the following functions

$$g(t) = \frac{\rho''(t)}{\rho(t)} + \left(\frac{1-\beta}{4\beta^2} \right) \left(\frac{a'(t)}{a(t)} \right)^2 - \left(\frac{1-\beta}{\beta} \right) \frac{a'(t)}{a(t)} \frac{\rho'(t)}{\rho(t)}, \quad t \geq t_0 \quad (5.2.9)$$

and

$$h(t) = \frac{\rho'(t)}{\rho(t)} - \left(\frac{1-\beta}{2\beta} \right) \frac{a'(t)}{a(t)}, \quad t \geq t_0. \quad (5.2.10)$$

Theorem 5.2.4. Suppose condition (F) holds, and let $\rho(t) \in C^2([t_0, \infty), \mathbb{R}^+)$, $\Omega(t) \in C([t_0, \infty), \mathbb{R})$, $a(t) \in C^1([t_0, \infty), \mathbb{R}^+)$ and $p(t) \in C([t_0, \infty), \mathbb{R}_0)$ so that

$$\limsup_{T \to \infty} \frac{1}{T} \int_t^T \int_t^s \frac{\rho^\beta(u)}{a(u)} q(u)duds \geq \Omega(t) \quad \text{for every } t \geq t_0. \quad (5.2.11)$$

Then equation (5.1.3) is oscillatory if

$$\int^\infty \frac{[\Omega^+(s)]^2}{s} ds = \infty, \quad (5.2.12)$$

where $\Omega^+(t) = \max\{\Omega(t), 0\}$, $t \geq t_0$, and

$$\left(\rho^\beta(t) \frac{p(t)}{a(t)} \right)' \leq 0 \quad \text{for } t \geq t_0, \quad \int^\infty s \left(\frac{p(s)}{a(s)} \right)^2 ds < \infty. \quad (5.2.13)$$

$$\limsup_{t\to\infty} th(t) < \infty, \quad h^2(t) \le -cg(t) \quad \text{and} \quad h^2(t)+h'(t) \ge g(t) \quad \text{for} \quad t \ge t_0$$

$$(5.2.14)$$

for some positive constant c.

Proof. Let $x(t)$ be a nonoscillatory solution of equation (5.1.3), say $x(t) \ne 0$ for all $t \ge t_0$. Define $w(t) = \rho^\beta(t)F(x(t))$ for $t \ge t_0$. Then for every $t \ge t_0$, we have

$$w'(t) = \beta\frac{\rho'(t)}{\rho(t)}w(t) + \rho^\beta(t)\frac{x'(t)}{f(x(t))}$$

and

$$
\begin{aligned}
w''(t) &= \frac{\rho^\beta(t)}{a(t)}\frac{(a(t)x'(t))'}{f(x(t))} - \rho^\beta(t)f'(x(t))\left(\frac{x'(t)}{f(x(t))}\right)^2 \\
&\quad + \left(\frac{\rho^\beta(t)}{a(t)}\right)' a(t)\frac{x'(t)}{f(x(t))} + \beta\frac{\rho'(t)}{\rho(t)}w'(t) + \beta\left(\frac{\rho'(t)}{\rho(t)}\right)' w(t) \\
&= -\frac{\rho^\beta(t)}{a(t)}q(t) - \rho^\beta(t)\frac{p(t)}{a(t)}\frac{x'(t)}{f(x(t))} \\
&\quad - f'(x(t))F(x(t))\frac{1}{w(t)}\left[w'(t) - \beta\frac{\rho'(t)}{\rho(t)}w(t)\right]^2 \\
&\quad + \left[\beta\frac{\rho'(t)}{\rho(t)} - \frac{a'(t)}{a(t)}\right]\left[w'(t) - \beta\frac{\rho'(t)}{\rho(t)}w(t)\right] \\
&\quad + \beta\frac{\rho'(t)}{\rho(t)}w'(t) + \beta\left[\frac{\rho''(t)}{\rho(t)} - \left(\frac{\rho'(t)}{\rho(t)}\right)^2\right]w(t).
\end{aligned}
$$

Using the definition of the number β and applying the method of completing the square, we obtain for $t \ge t_0$,

$$
\begin{aligned}
w''(t) &\le -\frac{\rho^\beta(t)}{a(t)}q(t) - \rho^\beta(t)\left(\frac{p(t)}{a(t)}\right)\frac{x'(t)}{f(x(t))} \\
&\quad + \beta\left[\frac{\rho''(t)}{\rho(t)} + \left(\frac{1-\beta}{4\beta^2}\right)\left(\frac{a'(t)}{a(t)}\right)^2 - \left(\frac{1-\beta}{\beta}\right)\left(\frac{a'(t)}{a(t)}\right)\frac{\rho'(t)}{\rho(t)}\right]w(t) \\
&\quad - \left(\frac{\beta}{1-\beta}\right)\frac{1}{w(t)}\left[w'(t) - \left\{\frac{\rho'(t)}{\rho(t)} - \left(\frac{1-\beta}{2\beta}\right)\frac{a'(t)}{a(t)}\right\}w(t)\right]^2,
\end{aligned}
$$

or

$$
\begin{aligned}
w''(t) &\le -\frac{\rho^\beta(t)}{a(t)}q(t) - \rho^\beta(t)\frac{p(t)}{a(t)}\frac{x'(t)}{f(x(t))} \\
&\quad + \beta g(t)w(t) - \frac{\beta}{1-\beta}\frac{1}{w(t)}[w'(t) - h(t)w(t)]^2.
\end{aligned}
$$

$$(5.2.15)$$

Integrating (5.2.15) twice and then multiplying it by $-1/T$, we obtain for $T \geq t \geq t_0$,

$$-\frac{w(T)}{T} + \frac{w(t)}{T} + \left(1 - \frac{t}{T}\right) w'(t) \geq \frac{1}{T} \int_t^T \int_t^s \frac{\rho^\beta(u)}{a(u)} q(u) du ds$$

$$+ \frac{1}{T} \int_t^T \int_t^s \rho^\beta(u) \left(\frac{p(u)}{a(u)}\right) \frac{x'(u)}{f(x(u))} du ds + \beta \frac{1}{T} \int_t^T \int_t^s -g(u)w(u) du ds$$

$$+ \left(\frac{\beta}{1-\beta}\right) \frac{1}{T} \int_t^T \int_t^s \frac{1}{w(u)} [w'(u) - h(u)w(u)]^2 du ds.$$

$$(5.2.16)$$

Now, taking into account condition (5.2.13) and using the Bonnet theorem, we conclude that for any s, t with $s \geq t \geq t_0$, there exists a number $\xi \in [t, s]$ such that

$$-\int_t^s \left(\rho^\beta(u) \frac{p(u)}{a(u)}\right) \left(\frac{x'(u)}{f(x(u))}\right) du = \int_t^s \left(-\rho^\beta(u) \frac{p(u)}{a(u)}\right) \frac{x'(u)}{f(x(u))} du$$

$$= -\rho^\beta(t) \frac{p(t)}{a(t)} \int_t^\xi \frac{x'(u)}{f(x(u))} du = \rho^\beta(t) \frac{p(t)}{a(t)} \int_\xi^t \frac{x'(u)}{f(x(u))} du$$

$$= \rho^\beta(t) \frac{p(t)}{a(t)} \int_{x(\xi)}^{x(t)} \frac{dy}{f(y)} \leq \rho^\beta(t) \frac{p(t)}{a(t)} F(x(t)) = \frac{p(t)}{a(t)} w(t),$$

since

$$\int_{x(\xi)}^{x(t)} \frac{dy}{f(y)} < \begin{cases} 0 & \text{if } x(\xi) > x(t) \\ \int_{+0}^{x(t)} \frac{dy}{f(y)} & \text{if } x(\xi) \leq x(t) \end{cases}$$

for $x > 0$, and

$$\int_{x(\xi)}^{x(t)} \frac{dy}{f(y)} < \begin{cases} 0 & \text{if } x(\xi) < x(t) \\ \int_{-0}^{x(t)} \frac{dy}{f(y)} & \text{if } x(\xi) \geq x(t) \end{cases}$$

for $x < 0$. Hence, for $T \geq t \geq t_0$, inequality (5.2.16) becomes

$$-\frac{w(T)}{T} + \frac{w(t)}{T} + \left(1 - \frac{t}{T}\right) \left[w'(t) + \frac{p(t)}{a(t)} w(t)\right]$$

$$\geq \frac{1}{T} \int_t^T \int_t^s \frac{\rho^\beta(u)}{a(u)} q(u) du ds + \beta \frac{1}{T} \int_t^T \int_t^s (-g(u))w(u) du ds$$

$$+ \left(\frac{\beta}{1-\beta}\right) \frac{1}{T} \int_t^T \int_t^s \frac{1}{w(u)} [w'(u) - h(u)w(u)]^2 du ds.$$

$$(5.2.17)$$

Next, by (5.2.11) and (5.2.14) inequality (5.2.17) gives

$$w'(t) + \frac{p(t)}{a(t)}w(t) \geq \Omega(t) + \liminf_{T \to \infty} \frac{w(T)}{T} + \beta \int_t^\infty (-g(u))w(u)du$$
$$+ \frac{\beta}{1-\beta} \int_t^\infty \frac{1}{w(u)}[w'(u) - h(u)w(u)]^2 du$$

for all $t \geq t_0$, which proves that

$$w'(t) + \frac{p(t)}{a(t)}w(t) \geq \Omega(t) \quad \text{for every} \quad t \geq t_0, \tag{5.2.18}$$

$$\liminf_{T \to \infty} \frac{w(T)}{T} < \infty, \tag{5.2.19}$$

$$\int_{t_0}^\infty (-g(u))w(u)du < \infty, \tag{5.2.20}$$

and

$$\int_{t_0}^\infty \frac{1}{w(u)}[w'(u) - h(u)w(u)]^2 du < \infty. \tag{5.2.21}$$

By (5.2.14), for every $T \geq t_0$, we find

$$\int_{t_0}^T \frac{1}{w(u)}[w'(u) - h(u)w(u)]^2 du$$

$$= \int_{t_0}^T \frac{[w'(u)]^2}{w(u)}du - 2\int_{t_0}^T h(u)w'(u)du + \int_{t_0}^T h^2(u)w(u)du$$

$$= \int_{t_0}^T \frac{[w'(u)]^2}{w(u)}du - 2h(T)w(T) + 2h(t_0)w(t_0)$$

$$+ \int_{t_0}^T [h^2(u) + 2h'(u)]w(u)du$$

$$\geq \int_{t_0}^T \frac{[w'(u)]^2}{w(u)}du - 2Th(T)\left(\frac{w(T)}{T}\right) + 2h(t_0)w(t_0)$$

$$- (c+2)\int_{t_0}^T (-g(u))w(u)du.$$

Therefore, it follows that

$$\int_{t_0}^T \frac{1}{w(u)}[w'(u) - h(u)w(u)]^2 du$$

$$\geq \int_{t_0}^\infty \frac{[w'(u)]^2}{w(u)}du - 2\left[\limsup_{T \to \infty} Th(T)\right]\left[\liminf_{T \to \infty} \frac{w(T)}{T}\right]$$

$$+ 2h(t_0)w(t_0) - (c+2)\int_{t_0}^\infty (-g(u))w(u)du,$$

which because of (5.2.14) and (5.2.19) – (5.2.21) yields

$$\int_{t_0}^{\infty} \frac{[w'(u)]^2}{w(u)} du < \infty. \tag{5.2.22}$$

Furthermore, by the Schwartz inequality, for $t \geq t_0$, we have

$$
\begin{aligned}
w(t) &= \left[[w(t_0)]^{1/2} + \left\{ [w(t)]^{1/2} - [w(t_0)]^{1/2} \right\} \right]^2 \\
&\leq 2w(t_0) + 2 \left[[w(t)]^{1/2} - [w(t_0)]^{1/2} \right]^2 \\
&= 2w(t_0) + \frac{1}{2} \left[\int_{t_0}^{t} \frac{w'(u)}{[w(u)]^{1/2}} du \right]^2 \\
&\leq 2w(t_0) + \frac{1}{2} \left(\int_{t_0}^{t} du \right) \int_{t_0}^{t} \frac{[w'(u)]^2}{w(u)} du \\
&\leq 2w(t_0) + \frac{1}{2} t \int_{t_0}^{\infty} \frac{[w'(u)]^2}{w(u)} du
\end{aligned}
$$

and consequently,

$$w(t) \leq t \left\{ \frac{2w(t_0)}{t_0} + \frac{1}{2} \int_{t_0}^{\infty} \frac{[w'(u)]^2}{w(u)} du \right\} \quad \text{for} \quad t \geq t_0. \tag{5.2.23}$$

Finally, by using (5.2.18), (5.2.22) and (5.2.23), we get

$$
\begin{aligned}
\int_{t_0}^{\infty} \frac{[\Omega^+(s)]^2}{s} ds &\leq \int_{t_0}^{\infty} \frac{1}{s} \left[w'(s) + \left(\frac{p(s)}{a(s)} \right) w(s) \right]^2 ds \\
&\leq K \int_{t_0}^{\infty} \frac{1}{w(s)} \left[w'(s) + \left(\frac{p(s)}{a(s)} \right) w(s) \right]^2 ds \\
&\leq 2K \int_{t_0}^{\infty} \frac{[w'(s)]^2}{w(s)} ds + 2K \int_{t_0}^{\infty} \left(\frac{p(s)}{a(s)} \right)^2 w(s) ds \\
&\leq 2K \int_{t_0}^{\infty} \frac{[w'(s)]^2}{w(s)} ds + 2K^2 \int_{t_0}^{\infty} s \left(\frac{p(s)}{a(s)} \right)^2 ds,
\end{aligned}
$$

where

$$K = \frac{2w(t_0)}{t_0} + \frac{1}{2} \int_{t_0}^{\infty} \frac{[w'(s)]^2}{w(s)} ds < \infty.$$

This in view of (5.2.12) and (5.2.13) leads to a contradiction. ∎

Remark 5.2.2. When $a(t) \equiv 1$ the functions g and h defined in (5.2.9) and (5.2.10), respectively, are reduced to $g(t) = \rho''(t)/\rho(t)$ and $h(t) = \rho'(t)/\rho(t)$ for $t \geq t_0$, and hence condition (5.2.14) becomes

$$(\rho'(t))^2 \leq -c\rho(t)\rho''(t) \quad \text{for some constant } c > 0 \text{ and } t \geq t_0. \tag{5.2.24}$$

Also, by condition (5.2.24), we observe that

$$\rho(t) = \rho(t_0) + \int_{t_0}^{t} \rho'(s)ds > (t - t_0)\rho'(t), \quad t \geq t_0.$$

Therefore, $\limsup_{t \to \infty}(t\rho'(t))/\rho(t) < \infty.$

The following three corollaries are immediate.

Corollary 5.2.4. Let condition (F) hold, and let $\rho(t) \in C^2([t_0, \infty), \mathbb{R}^+)$, $\Omega(t) \in C([t_0, \infty), \mathbb{R})$ and $p(t) \in C([t_0, \infty), \mathbb{R}_0)$ be such that

$$\limsup_{T \to \infty} \frac{1}{T} \int_{t}^{T} \int_{t}^{s} \rho^\beta(u)q(u)duds \geq \Omega(t) \quad \text{for every} \quad t \geq t_0. \quad (5.2.25)$$

Then equation (5.1.3) with $a(t) = 1$ is oscillatory if conditions (5.2.12) and (5.2.24) are satisfied, and

$$(\rho^\beta(t)p(t))' \leq 0 \quad \text{for} \quad t \geq t_0 \quad \text{and} \quad \int^{\infty} sp^2(s)ds < \infty.$$

Corollary 5.2.5. Let condition (F) hold, and let $\rho(t) \in C^2([t_0, \infty), \mathbb{R}^+)$, and $\Omega(t) \in C([t_0, \infty), \mathbb{R})$ be such that condition (5.2.25) hold. Then equation (5.1.1) is oscillatory if conditions (5.2.12) and (5.2.24) are satisfied.

Corollary 5.2.6. Let condition (F) hold, and let $0 \leq \lambda < \beta$ and $\Omega \in C([t_0, \infty), \mathbb{R})$ be such that

$$\limsup_{T \to \infty} \frac{1}{T} \int_{t}^{T} \int_{t}^{s} u^\lambda q(u)duds \geq \Omega(t) \quad \text{for every} \quad t \geq t_0.$$

Then equation (5.1.1) is oscillatory if condition (5.2.12) holds.

Proof. The proof follows immediately from Theorem 5.2.4 by letting $\rho(t) = t^{\lambda/\beta}$ for $t \geq t_0$. ∎

Next, we shall prove the following oscillation result for (5.1.3).

Theorem 5.2.5. Let condition (F) hold, and let $\rho(t) \in C^2([t_0, \infty), \mathbb{R}^+)$, $\Omega(t) \in C([t_0, \infty), \mathbb{R})$, $a(t) \in C^1([t_0, \infty), \mathbb{R}^+)$ and $p(t) \in C([t_0, \infty), \mathbb{R}_0)$ be such that

$$\liminf_{T \to \infty} \frac{1}{T} \int_{t}^{T} \int_{t}^{s} \frac{\rho^\beta(u)}{a(u)}q(u)duds \geq \Omega(t) \quad \text{for every} \quad t \geq t_0. \quad (5.2.26)$$

Then equation (5.1.3) is oscillatory if condition (5.2.14) holds,

$$\limsup_{t \to \infty} \left[\int_{t_0}^{t} sh^2(s)ds\right]^{-1} \int_{t_0}^{t} \frac{[\Omega^+(s)]^2}{s}ds = \infty \quad (5.2.27)$$

and $\left(\rho^\beta(t) p(t)/a(t) \right)' \le 0$ for $t \ge t_0$, also

$$\limsup_{t \to \infty} \left[\int_{t_0}^t s h^2(s) ds \right]^{-1} \int_{t_0}^t s \left(\frac{p(s)}{a(s)} \right)^2 ds < \infty.$$

Proof. Let $x(t)$ be a nonoscillatory solution of equation (5.1.3), say, $x(t) \ne 0$ for $t \ge t_0$. Define $w(t) = \rho^\beta(t) F(x(t))$, $t \ge t_0$. Then as in Theorem 5.2.4, we obtain for $T \ge t \ge t_0$ that

$$w'(t) + \frac{p(t)}{a(t)} w(t) \ge \Omega(t) + \limsup_{T \to \infty} \frac{w(T)}{T} + \beta \int_t^\infty (-g(s)) w(s) ds$$
$$+ \frac{\beta}{1 - \beta} \int_t^\infty \frac{1}{w(s)} [w'(s) - h(s) w(s)]^2 ds.$$

Also,

$$\limsup_{T \to \infty} \frac{w(T)}{T} < \infty \qquad (5.2.28)$$

also (5.2.18), (5.2.20) and (5.2.21) hold. Furthermore, for $T \ge t_0$, we find

$$\int_{t_0}^T \frac{[w'(s)]^2}{w(s)} ds = \int_{t_0}^T \frac{1}{w(s)} [w'(s) - h(s) w(s)]^2 ds + \int_{t_0}^T 2h(s) w'(s) ds$$
$$- \int_{t_0}^T h^2(s) w^2(s) ds$$
$$= \int_{t_0}^T \frac{1}{w(s)} [w'(s) - h(s) w(s)]^2 ds + 2Th(T) \left(\frac{w(T)}{T} \right)$$
$$- 2h(t_0) w(t_0) - 2 \int_{t_0}^T h'(s) w(s) ds - \int_{t_0}^T h^2(s) w(s) ds$$
$$\le \int_{t_0}^T \frac{1}{w(s)} [w'(s) - h(s) w(s)]^2 ds + 2Th(T) \left(\frac{w(T)}{T} \right)$$
$$- 2h(t_0) w(t_0) + 2 \int_{t_0}^T -g(s) w(s) ds + \int_{t_0}^T h^2(s) w(s) ds$$
$$\le \int_{t_0}^T \frac{1}{w(s)} [w'(s) - h(s) w(s)]^2 ds + 2M (Th(T))$$
$$- 2h(t_0) w(t_0) + 2 \int_{t_0}^T -g(s) w(s) ds + M \int_{t_0}^T s h^2(s) ds,$$

where

$$M = \sup_{T \ge t_0} \frac{w(T)}{T} < \infty \quad \text{(by (5.2.28))}. \qquad (5.2.29)$$

Hence, taking (5.2.20), (5.2.21) and condition (5.2.14) into account, we conclude that there exists a positive constant N such that

$$\int_{t_0}^{t} \frac{[w'(s)]^2}{w(s)} ds \leq N \int_{t_0}^{t} sh^2(s) ds \quad \text{for} \quad t \geq t_0. \tag{5.2.30}$$

Finally, from (5.2.18) and (5.2.30) for $t \geq t_0$, we get

$$
\begin{aligned}
\int_{t_0}^{t} \frac{[\Omega^+(s)]^2}{s} ds &\leq M \int_{t_0}^{t} \frac{1}{w(s)} \left[w'(s) + \frac{p(s)}{a(s)} w(s) \right]^2 ds \\
&\leq 2M \int_{t_0}^{t} \frac{[w'(s)]^2}{w(s)} ds + 2M \int_{t_0}^{t} \left(\frac{p(s)}{a(s)} \right)^2 w(s) ds \\
&\leq 2MN \int_{t_0}^{t} sh^2(s) ds + 2M^2 \int_{t_0}^{t} s \left(\frac{p(s)}{a(s)} \right)^2 ds,
\end{aligned}
$$

where M is as in (5.2.29). Thus, it follows that

$$
\begin{aligned}
\left[\int_{t_0}^{t} sh^2(s) ds \right]^{-1} &\int_{t_0}^{t} \frac{[\Omega^+(s)]^2}{s} ds \\
&\leq 2MN + 2M^2 \left[\int_{t_0}^{t} sh^2(s) ds \right]^{-1} \int_{t_0}^{t} s \left(\frac{p(s)}{a(s)} \right)^2 ds < \infty,
\end{aligned}
$$

which contradicts condition (5.2.27). This completes the proof. ∎

Remark 5.2.3. If $a(t) \equiv 1$ in Theorem 5.2.5, then condition (5.2.14) reduces to

$$\rho'(t) \geq 0 \quad \text{and} \quad \rho''(t) \leq 0 \quad \text{for} \quad t \geq t_0. \tag{5.2.31}$$

Now, we present the following immediate corollaries.

Corollary 5.2.7. Let condition (F) hold, and let $\rho(t) \in C^2([t_0, \infty), \mathbb{R}^+)$, $\Omega(t) \in C([t_0, \infty), \mathbb{R})$ and $p(t) \in C([t_0, \infty), \mathbb{R}_0)$ be such that

$$\liminf_{T \to \infty} \frac{1}{T} \int_{t}^{T} \int_{t}^{s} \rho^\beta(u) q(u) du\, ds \geq \Omega(t) \quad \text{for every} \quad t \geq t_0. \tag{5.2.32}$$

Then equation (5.1.3) with $a(t) = 1$ is oscillatory if condition (5.2.31) holds,

$$\limsup_{t \to \infty} \left[\int_{t_0}^{t} s \left(\frac{\rho'(s)}{\rho(s)} \right)^2 ds \right]^{-1} \int_{t_0}^{t} \frac{[\Omega^+(s)]^2}{s} ds = \infty \tag{5.2.33}$$

and $(\rho^\beta(t) p(t))' \leq 0$ for $t \geq t_0$, also

$$\limsup_{t \to \infty} \left[\int_{t_0}^{t} s \left(\frac{\rho'(s)}{\rho(s)} \right)^2 ds \right]^{-1} \int_{t_0}^{t} sp^2(s) ds < \infty.$$

Corollary 5.2.8. Let condition (F) hold, and let $\rho(t) \in C^2([t_0, \infty), \mathbb{R}^+)$ and $\Omega(t) \in C([t_0, \infty), \mathbb{R})$ be such that condition (5.2.32) hold. Then equation (5.1.1) is oscillatory if conditions (5.2.31) and (5.2.33) are satisfied.

Corollary 5.2.9. Let condition (F) hold, and let $\Omega \in C([t_0, \infty), \mathbb{R})$ be such that

$$\liminf_{T \to \infty} \frac{1}{T} \int_t^T \int_t^s u^\beta q(u) du ds \geq \Omega(t) \quad \text{for every} \quad t \geq t_0. \qquad (5.2.34)$$

Then equation (5.1.1) is oscillatory if

$$\limsup_{t \to \infty} \frac{1}{\ln t} \int_{t_0}^t \frac{[\Omega^+(s)]^2}{s} ds = \infty. \qquad (5.2.35)$$

Proof. It suffices to apply Theorem 5.2.5 with $\rho(t) = t$ for $t \geq t_0$. $\blacksquare$

Remark 5.2.4. Conditions (5.2.31) and (5.2.35) imply (5.2.33). Indeed, from (5.2.31) it follows that $\limsup_{T \to \infty} T\rho'(T)/\rho(T) < \infty$ (see Remark 5.2.2), and hence there exists a positive constant b such that $t\rho'(t)/\rho(t) \leq b$ for all $t \geq t_0$. Thus, for all large t, we have

$$\int_{t_0}^t s \left(\frac{\rho'(s)}{\rho(s)} \right)^2 ds \leq b^2 \int_{t_0}^t \frac{ds}{s} = b^2[\ln t - \ln t_0] \leq 2b^2 \ln t,$$

which proves the proposed assertion.

Remark 5.2.5. As we have mentioned earlier, the function $f(x) = |x|^\gamma \operatorname{sgn} x$, $x \in \mathbb{R}$ $(0 < \gamma < 1)$ is a classical example which satisfies condition (F). Other such functions are:

(I_1) $f(x) = |x|^\gamma \operatorname{sgn} x + x$, $x \in \mathbb{R}$ $(0 < \gamma < 1)$ with $\gamma/2 \leq \beta < 1$,

(I_2) $f(x) = |x|^{1/2} \operatorname{sgn} x/(1 + |x|^{1/4})$, $x \in \mathbb{R}$ with $\beta = 1/4$,

(I_3) $f(x) = |x|^\gamma [k + \sin(\ln[1+|x|])] \operatorname{sgn} x$, $x \in \mathbb{R}$ $(0 < \gamma < 1, k \geq 1 + (1/\gamma))$ with $(k-1)(\gamma k - \gamma - 1)/(k+1)(k+2\gamma) \leq \beta < 1$.

Also, see Examples 5.2.1 – 5.2.3.

Example 5.2.6. Consider the differential equation

$$(t^\alpha x'(t))' + t^\theta x'(t) + t^\lambda (\sin t) f(x(t)) = 0, \quad t \geq t_0 > 0 \qquad (5.2.36)$$

where α, θ and λ are real constants, and the function f is any one on the above in the Remark 5.2.5. We let $-\alpha + \beta + \lambda = \delta$, $0 < \delta \leq 1$, $\theta - \alpha + \beta \leq 0$ and $\theta - \alpha < -1$. Now,

$$g(t) = \left[\frac{\alpha^2}{4\beta^2} - \alpha - \beta \right] (1-\beta) \left(\frac{1}{t^2} \right) \quad \text{and} \quad h(t) = \left[\beta - \frac{\alpha(1-\beta)}{2\beta} \right] \left(\frac{1}{t} \right).$$

We also let

$$\left[\beta - \frac{\alpha(1-\beta)}{2\beta}\right]^2 \le c\left[\alpha + \beta - \frac{\alpha^2}{4\beta^2}\right] \quad \text{for some constant} \quad c > 0$$

and

$$\left[\frac{\alpha(1-\beta)}{2\beta} - \beta\right](1-\beta)\left(1 - \frac{\alpha}{2\beta}\right) \ge \left[\frac{\alpha^2}{4\beta^2} - \alpha - \beta\right](1-\beta).$$

For all T, t with $T \ge t \ge t_0$, we obtain

$$\frac{1}{T}\int_t^T \int_t^s u^\delta \sin u\, du\, ds$$

$$= -T^{\delta-1}\sin T - 2\delta T^{\delta-2}\cos T + \delta(\delta-1)(\delta+1)T^{\delta-3}\sin T$$

$$+ \frac{1}{T}\left[t^\delta \sin t + 2\delta t^{\delta-1}\cos t - \delta(\delta-1)(\delta+1)t^{\delta-2}\sin t\right]$$

$$- \delta(\delta-1)(\delta+1)(\delta-2)\frac{1}{T}\int_t^T s^{\delta-3}\sin s\, ds$$

$$+ \left(1 - \frac{t}{T}\right)\left[t^\delta \cos t - \delta t^{\delta-1}\sin t - \delta(\delta-1)t^{\delta-2}\cos t\right]$$

$$- \delta(\delta-1)(\delta-2)\int_t^T s^{\delta-3}\cos s\, ds.$$

Thus, for $t \ge t_0$, we find

$$\limsup_{T\to\infty}\frac{1}{T}\int_t^T \int_t^s u^\beta q(u)\, du\, ds = \limsup_{T\to\infty}\frac{1}{T}\int_t^T \int_t^s u^\delta \sin u\, du\, ds$$

$$\ge t^\delta \cos t - \delta t^{\delta-1}\sin t - \delta(\delta-1)t^{\delta-2}\cos t$$

$$- \delta(\delta-1)(\delta-2)\int_t^\infty s^{\delta-3}\cos s\, ds$$

$$\ge t^\delta \cos t - \mu,$$

where μ is a positive constant. Clearly, condition (5.2.11) is satisfied with $\Omega(t) = t^\delta \cos t - \mu$, $t \ge t_0$. Next, we consider an integer N such that $2N\pi - (\pi/4) \ge \max\left\{t_0, (1+\sqrt{2}\mu)^{1/\delta}\right\}$. Then for all integers $n \ge N$, we have $\Omega(t) \ge 1/\sqrt{2}$ for every $t \in [2n\pi - (\pi/4), 2n\pi + (\pi/4)]$. Thus, we obtain

$$\int_{t_0}^\infty \frac{[\Omega^+(s)]^2}{s}\, ds \ge \frac{1}{2}\sum_{n=N}^\infty \int_{2n\pi-(\pi/4)}^{2n\pi+(\pi/4)} \frac{ds}{s}$$

$$= \frac{1}{2}\sum_{n=N}^\infty \ln\left[1 + \frac{2}{8n-1}\right] = \infty,$$

i.e., condition (5.2.12) is fulfilled. Hence, Theorem 5.2.4 can be applied to guarantee the oscillation of equation (5.2.36).

Example 5.2.7. Consider the differential equation

$$x''(t) + t[t + \ln t]^{-\gamma}(\sin t)|x(t)|^{\gamma} \operatorname{sgn} x(t) = 0, \tag{5.2.37}$$

where $0 < \gamma < 1$, $t \geq t_0 > 1$. Define $\rho(t) = t + \ln t$, $t \geq t_0$ and observe that condition (5.2.31) is satisfied. Furthermore, for every T, t with $T \geq t \geq t_0$, we have

$$\frac{1}{T} \int_t^T \int_t^s \rho^{\beta}(u)q(u)duds = \frac{1}{T} \int_t^T \int_t^s u \sin u\, du ds$$

$$= -\sin T + \left(1 - \frac{t}{T}\right)[t\cos t - \sin t] + \frac{1}{T}[-2\cos T + t\sin t + 2\sin t]$$

and consequently, for every $t \geq t_0$,

$$\liminf_{T \to \infty} \frac{1}{T} \int_t^T \int_t^s \rho^{\beta}(u)q(u)duds = t\cos t - \sin t - 1 \geq t\sin t - 2.$$

Thus, condition (5.2.34) holds with $\Omega(t) = t\cos t - 2$, $t \geq t_0$. We consider a number t_1 such that $t_1 \geq \max\{t_0, 4\sqrt{2}\}$. Next, we choose an integer N such that $2N\pi - (\pi/4) \geq t_1$. Then for every integer $n \geq N$, we have $\Omega(t) \geq t/(2\sqrt{2})$ for $t \in [2n\pi - (\pi/4), 2n\pi + (\pi/4)]$. Thus, for $n \geq N$, we get

$$\int_{t_0}^{2n\pi+(\pi/4)} \frac{[\Omega^+(s)]^2}{s}ds \geq \int_{2n\pi-(\pi/4)}^{2n\pi+(\pi/4)} \frac{[\Omega^+(s)]^2}{s}ds$$

$$\geq \frac{1}{8}\int_{2n\pi-(\pi/4)}^{2n\pi+(\pi/4)} sds = \frac{\pi^2 n}{8}$$

and therefore,

$$\limsup_{t \to \infty} \frac{1}{\ln t}\int_{t_0}^t \frac{[\Omega^+(s)]^2}{s}ds \geq \limsup_{n \to \infty} \frac{1}{\ln\left[2n\pi + \frac{\pi}{4}\right]}\int_{t_0}^{2n\pi+(\pi/4)} \frac{[\Omega^+(s)]^2}{s}ds$$

$$\geq \lim_{n \to \infty} \frac{\pi^2 n}{8\ln\left[2n\pi + \frac{\pi}{4}\right]} = \infty.$$

Hence, condition (5.2.35) is satisfied, and consequently condition (5.2.27) holds, as noted in Remark 5.2.4. Therefore, by Theorem 5.2.5 the equation (5.2.37) is oscillatory.

Example 5.2.8. Consider the differential equation

$$(tx'(t))' + \frac{1}{t}x'(t) + (t^{3/2}\sin t)|x(t)|^{1/2} \operatorname{sgn} x(t) = 0, \quad t \geq t_0 > 1. \tag{5.2.38}$$

Define $\rho(t) = t$, $t \geq t_0$. Then, $g(t) = -1/(2t^2)$ and $h(t) = 1/(2t)$, and hence condition (5.2.14) is satisfied. As in Example 5.2.7, we find that $\Omega(t) = t\cos t - 2$, $t \geq t_0$. Thus, all conditions of Theorem 5.2.5 are satisfied, and therefore equation (5.2.38) is oscillatory.

In what follows it will be convenient to use the following notation. For $\rho(t) \in C^2([t_0, \infty), \mathbb{R}^+)$ and $t \geq t_0$, we let

$$\eta(t) = \frac{a(t)}{\rho^\beta(t)}\left(\frac{a^\beta(t)}{a(t)}\right)' - \frac{p(t)}{a(t)},$$

$$h_1(t) = \frac{1}{f'(x)F(x)}\left[\frac{1}{2}\eta(t) + \frac{1}{2}\beta\frac{\rho'(t)}{\rho(t)} + \beta\frac{\rho'(t)}{\rho(t)}f'(x)F(x)\right],$$

$$g_1(t) = \beta\frac{\rho''(t)}{\rho(t)} - \beta\eta(t)\frac{\rho'(t)}{\rho(t)} - \beta[1 + \beta f'(x)F(x)]\left(\frac{\rho'(t)}{\rho(t)}\right)^2 + f'(x)F(x)h_1^2(t),$$

$$h_2(t) = \frac{1-\beta}{\beta}\left[\frac{1}{2}\eta(t) + \frac{1}{2}\beta\left(\frac{1+\beta}{1-\beta}\right)\frac{\rho'(t)}{\rho(t)}\right]$$

and

$$g_2(t) = \left[\beta\frac{\rho''(t)}{\rho(t)} - \beta\eta(t)\frac{\rho'(t)}{\rho(t)} - \beta\left[1 + \frac{\beta^2}{1-\beta}\right]\left(\frac{\rho'(t)}{\rho(t)}\right)^2 + \frac{\beta}{1-\beta}h_2^2(t)\right].$$

We are now in the position to prove the following result.

Theorem 5.2.6. Suppose condition (F) holds with $\beta > 0$. Let $n \geq 1$ be an integer and $\rho(t) \in C^2([t_0, \infty), \mathbb{R}^+)$ satisfy

$$\limsup_{t\to\infty} th_2(t) < \infty, \quad h_2^2(t) \leq -cg_2(t) \quad \text{and} \quad h_2^2(t) + h_2'(t) \geq g_2(t)$$

$$(5.2.39)$$

for some constant $c > 0$ and $t \geq t_0$. Then equation (5.1.3) is oscillatory if there exists a function $\Omega \in C([t_0, \infty), \mathbb{R})$ satisfying (5.2.12) and

$$\limsup_{t\to\infty}\frac{1}{t^n}\int_T^t (t-s)^n\frac{\rho^\beta(s)}{a(s)}q(s)ds \geq \Omega(T) \quad \text{for every} \quad T \geq t_0. \quad (5.2.40)$$

Proof. Let $x(t)$ be a nonoscillatory solution of equation (5.1.3), say, $x(t) \neq 0$ for $t \geq t_0$. Moreover, let $w(t) = \rho^\beta(t)F(x(t))$, $t \geq t_0$. Then for $t \geq t_0$, we have

$$w'(t) = \beta\frac{\rho'(t)}{\rho(t)}w(t) + \frac{\rho^\beta(t)}{a(t)}\frac{a(t)x'(t)}{f(x(t))}$$

and

$$\rho^\beta(t)\frac{x'(t)}{f(x(t))} = w'(t) - \beta\frac{\rho'(t)}{\rho(t)}w(t),$$

$$w''(t) = \beta\frac{\rho''(t)}{\rho(t)}w(t) - \beta\left(\frac{\rho'(t)}{\rho(t)}\right)^2 w(t) + \beta\frac{\rho'(t)}{\rho(t)}w(t) + \left(\frac{\rho^\beta(t)}{a(t)}\right)'\frac{a(t)x'(t)}{f(x(t))}$$

$$- \frac{\rho^\beta(t)}{a(t)}q(t) - \rho^\beta(t)\frac{p(t)}{a(t)}\frac{x'(t)}{f(x(t))} - \frac{1}{w(t)}\left[w'(t) - \beta\frac{\rho'(t)}{\rho(t)}w(t)\right]^2$$

$$\times f'(x(t))F(x(t)) \qquad (5.2.41)$$

$$= -\frac{\rho^\beta(t)}{a(t)}q(t) + \eta(t)\left[w'(t) - \beta\frac{\rho'(t)}{\rho(t)}w(t)\right] + \beta\frac{\rho''(t)}{\rho(t)}w(t)$$

$$- \beta\left(\frac{\rho'(t)}{\rho(t)}\right)^2 w(t) + \beta\frac{\rho'(t)}{\rho(t)}w'(t) - f'(x(t))F(x(t))\frac{(w'(t))^2}{w(t)}$$

$$+ 2\beta\frac{\rho'(t)}{\rho(t)}f'(x(t))F(x(t))w'(t) - \beta^2 f'(x(t))F(x(t))\left(\frac{\rho'(t)}{\rho(t)}\right)^2 w(t)$$

$$= -\frac{\rho^\beta(t)}{a(t)}q(t) - \frac{1}{w(t)}[w'(t) - h_1(t)w(t)]^2\left[\beta\frac{\rho''(t)}{\rho(t)} - \beta\eta(t)\left(\frac{\rho'(t)}{\rho(t)}\right)\right.$$

$$\left. - \beta[1 + \beta f'(x(t))F(x(t))]\left(\frac{\rho'(t)}{\rho(t)}\right)^2 + f'(x(t))F(x(t))h_1^2(t)\right]w(t)$$

$$= -\frac{\rho^\beta(t)}{a(t)}q(t) - (-g_1(t))w(t) - \frac{1}{w(t)}[w'(t) - h_1(t)w(t)]^2$$

$$\times f'(x(t))F(x(t)). \qquad (5.2.42)$$

By the definition of β and (5.2.41), we find that for $t \geq t_0$, (5.2.42) takes the form

$$w''(t) \leq -\frac{\rho^\beta(t)}{a(t)}q(t) - (-g_2(t))w(t) - \frac{\beta}{1-\beta}\frac{1}{w(t)}[w'(t) - h_2(t)w(t)]^2.$$

$$(5.2.43)$$

Thus, for any t, T with $t \geq T \geq t_0$, we obtain

$$\int_T^t (t-s)^n\frac{\rho^\beta(s)}{a(s)}q(s)ds + \int_T^t (t-s)^n(-g_2(s))w(s)ds$$

$$+ \frac{\beta}{1-\beta}\int_T^t (t-s)^n\frac{1}{w(s)}[w'(s) - h_2(s)w(s)]^2 ds$$

$$\leq -\int_T^t (t-s)^n w''(s)ds = (t-T)^n w'(T) - n\int_T^t (t-s)^{n-1}w'(s)ds$$

$$= \begin{cases} (t-T)w'(T) - w(t) + w(T) & \text{if } n = 1 \\ \\ (t-T)^n w'(T) - n(n-1)\int_T^t (t-s)^{n-2}w(s)ds + n(t-T)^{n-1}w(T) \\ \qquad\qquad\qquad\qquad\qquad\qquad\qquad\qquad\qquad\qquad\qquad \text{if } n > 1. \end{cases}$$

Therefore, for $t \geq T \geq t_0$, we have

$$\frac{1}{t^n} \int_T^t (t-s)^n \frac{\rho^\beta(s)}{a(s)} q(s)ds + \frac{1}{t^n} \int_T^t (t-s)^n (-g_2(s)) w(s)ds$$

$$+ \frac{\beta}{1-\beta} \frac{1}{t^n} \int_T^t (t-s)^n \frac{1}{w(s)} [w'(s) - h_2(s)w(s)]^2 ds$$

$$\leq \begin{cases} \left(1 - \dfrac{T}{t}\right) w'(T) - \dfrac{w(t)}{t} + \dfrac{w(T)}{t} & \text{if } n = 1 \\[2mm] \left(1 - \dfrac{T}{t}\right)^n w'(T) - n(n-1)\dfrac{1}{t^n} \int_T^t (t-s)^{n-2} w(s)ds \\[2mm] \qquad\qquad + n\left(1 - \dfrac{T}{t}\right)^{n-1} \dfrac{w(T)}{t} & \text{if } n > 1. \end{cases}$$

Thus, for every $T \geq t_0$, we get

$$\limsup_{t \to \infty} \frac{1}{t^n} \int_T^t (t-s)^n \frac{\rho^\beta(s)}{a(s)} q(s)ds + \int_T^\infty (-g_2(s)) w(s)ds$$

$$+ \frac{\beta}{1-\beta} \int_T^\infty \frac{1}{w(s)} [w'(s) - h_2(s)w(s)]^2 ds$$

$$\leq \begin{cases} w'(T) - \liminf_{t\to\infty} \dfrac{w(t)}{t} & \text{if } n = 1 \\[2mm] w'(T) - n(n-1) \liminf_{t\to\infty} \dfrac{1}{t^n} \int_T^t (t-s)^{n-2} w(s)ds & \text{if } n > 1 \end{cases}$$

and consequently, by condition (5.2.40), for all $T \geq t_0$, we obtain

$$w'(T) \geq \Omega(T) + \int_T^\infty -g_2(s)w(s)ds + \frac{\beta}{1-\beta} \int_T^\infty \frac{1}{w(s)} [w'(s) - h_2(s)w(s)]^2 ds$$

$$+ \begin{cases} \liminf_{t\to\infty} \dfrac{w(t)}{t} & \text{if } n = 2 \\[2mm] n(n-1) \liminf_{t\to\infty} \dfrac{1}{t^n} \int_T^t (t-s)^{n-2} w(s)ds & \text{if } n > 2. \end{cases}$$

This shows that $w'(T) \geq \Omega(T)$ for every $T \geq t_0$,

$$\int_{t_0}^\infty -g_2(s)w(s)ds < \infty, \qquad \int_{t_0}^\infty \frac{1}{w(s)} [w'(s) - h_2(s)w(s)]^2 ds < \infty$$

and

$$\begin{cases} \liminf_{t\to\infty} \dfrac{w(t)}{t} < \infty & \text{if } n = 1 \\[2mm] \liminf_{t\to\infty} \dfrac{1}{t^n} \int_{t_0}^t (t-s)^{n-2} w(s)ds < \infty & \text{if } n > 1. \end{cases} \tag{5.2.44}$$

Now, we shall show that (5.2.44) implies

$$\liminf_{t\to\infty} \frac{w(t)}{t} < \infty \tag{5.2.45}$$

in both cases where $n = 1$ or $n > 1$. Clearly, this is true for $n = 1$. So, we consider the case where $n > 1$. Suppose (5.2.45) fails, i.e., $\lim_{t\to\infty} w(t)/t = \infty$. Let α be a number such that $0 < \alpha < 1/(n(n-1))$, and μ be an arbitrary positive constant. Then there exists a $T_1 \geq t_0$ such that $w(t)/t \geq \mu/\alpha$ for all $t \geq T_1$. Thus, for $t \geq T_1$, we have

$$
\begin{aligned}
\frac{1}{t^n} \int_{t_0}^t (t-s)^{n-2} w(s)ds &\geq \frac{1}{t^n} \int_{T_1}^t (t-s)^{n-2} w(s)ds \\
&\geq \frac{\mu}{\alpha}\frac{1}{t^n} \int_{T_1}^t (t-s)^{n-2} s\, ds \\
&= \frac{\mu}{\alpha}\left[\frac{1}{n-1}\left(1 - \frac{T_1}{t}\right)^{n-1} - \frac{1}{n}\left(1 - \frac{T_1}{t}\right)^n \right].
\end{aligned}
$$

Since,

$$\lim_{t\to\infty}\left[\frac{1}{n-1}\left(1 - \frac{T_1}{t}\right)^{n-1} - \frac{1}{n}\left(1 - \frac{T_1}{t}\right)^n \right] = \frac{1}{n(n-1)} > \alpha,$$

we can choose a $T_2 \geq T_1$ so that

$$\frac{1}{n-1}\left(1 - \frac{T_1}{t}\right)^{n-1} - \frac{1}{n}\left(1 - \frac{T_1}{t}\right)^n \geq \alpha \quad \text{for all} \quad t \geq T_2.$$

Now, it follows that

$$\frac{1}{t^n} \int_{t_0}^t (t-s)^{n-2} w(s)ds > \mu \quad \text{for every} \quad t \geq T_2,$$

which gives

$$\lim_{t\to\infty} \frac{1}{t^n} \int_{t_0}^t (t-s)^{n-2} w(s)ds = \infty,$$

since $\mu > 0$ is arbitrary. This contradicts (5.2.44), and hence (5.2.45) follows.

The remainder of the proof proceeds exactly as in Theorem 5.2.4 with the functions $h(t)$ and $g(t)$ replaced by $h_1(t)$ and $g_2(t)$ respectively, and hence we omit the details. This completes the proof. ∎

For the special case when $p(t) = 0$ and $a(t) = 1$, we have

$$\eta(t) = \beta\frac{\rho'(t)}{\rho(t)}, \quad h_2(t) = \frac{\rho'(t)}{\rho(t)} \quad \text{and} \quad g_2(t) = \beta\frac{\rho''(t)}{\rho(t)}. \tag{5.2.46}$$

Thus, the following corollary is immediate.

Corollary 5.2.10. Suppose condition (F) holds with $\beta > 0$. Let $n \geq 1$ be an integer and the function $\rho(t) \in C^2([t_0, \infty), \mathbb{R}^+)$ satisfy condition (5.2.24). Then equation (5.1.1) is oscillatory if there exists a function $\Omega \in C([t_0, \infty), \mathbb{R})$ such that conditions (5.2.12) and (5.2.40) hold.

The following result is similar to that of Theorem 5.2.3.

Theorem 5.2.7. Suppose condition (F) holds. Let $n \geq 1$ be an integer and the function $\rho(t) \in C^2([t_0, \infty), \mathbb{R}^+)$ be such that $g_2(t) \leq 0$ for $t \geq t_0$. If condition (5.2.8) holds, then equation (5.1.3) is oscillatory.

Proof. Let $x(t)$ be a nonoscillatory solution of equation (5.1.3), say, $x(t) \neq 0$ for $t \geq t_0$. Moreover, let $w(t) = \rho^\beta(t)F(x(t))$, $t \geq t_0$. Then as in Theorem 5.2.6, (5.2.43) holds. Thus, for $t \geq t_0$, we have $w''(t) \leq -\rho^\beta(t)q(t)/a(t)$. The rest of the proof proceeds exactly as in Theorem 5.2.3, and hence we omit the details. ∎

The following corollary of Theorem 5.2.7 is immediate.

Corollary 5.2.11. Suppose condition (F) holds. Let $n \geq 1$ be an integer and the function $\rho(t) \in C^2([t_0, \infty), \mathbb{R}^+)$ be such that $\rho'(t) > 0$ and $\rho''(t) \leq 0$ for $t \geq t_0$. If condition (5.2.8) holds, then equation (5.1.1) is oscillatory.

Next, we shall prove the following oscillation result for (5.1.6).

Theorem 5.2.8. Suppose that

$$\lim_{T \to \infty} \frac{1}{T} \int_{t_0}^{T} \int_{t_0}^{s} u^\gamma q(u)\,du\,ds \quad \text{exists in} \quad \mathbb{R} \tag{5.2.47}$$

and define

$$Q(t) = \lim_{T \to \infty} \frac{1}{T} \int_{t}^{T} \int_{t}^{s} u^\gamma q(u)\,du\,ds \quad \text{for} \quad t \geq t_0. \tag{5.2.48}$$

If

$$\int_{t_0}^{\infty} \frac{1}{s}[Q(s) + k(s)]^2 ds = \infty \tag{5.2.49}$$

for every function $k(t) \in C([t_0, \infty), \mathbb{R})$ with $\lim_{t \to \infty} k(t) = 0$, then equation (5.1.6) is oscillatory.

Proof. Let $x(t)$ be a nonoscillatory solution of equation (5.1.6), say, $x(t) > 0$ for $t \geq T_0 \geq t_0$. Define $w(t) = t^\gamma x^{1-\gamma}(t)/(1-\gamma)$, $t \geq T_0$. Then for $t \geq T_0$, we have

$$w'(t) = \frac{\gamma}{1-\gamma}t^{\gamma-1}x^{1-\gamma}(t) + t^\gamma x^{-\gamma}(t)x'(t) = \gamma\frac{w(t)}{t} + t^\gamma x^{-\gamma}(t)x'(t)$$

and

$$
\begin{aligned}
w''(t) &= \gamma\frac{w'(t)}{t} - \gamma\frac{w(t)}{t^2} + \gamma t^{\gamma-1}x^{-\gamma}(t)x'(t) \\
&\quad -\gamma t^{\gamma}x^{-\gamma-1}(t)(x'(t))^2 + t^{\gamma}x^{-\gamma}(t)x''(t) \\
&= \gamma\frac{w'(t)}{t} - \gamma\frac{w(t)}{t^2} + \gamma\left[\frac{w'(t)}{t} - \gamma\frac{w(t)}{t^2}\right] \\
&\quad -\frac{\gamma}{1-\gamma}\frac{1}{w(t)}\left[w'(t) - \gamma\frac{w(t)}{t}\right]^2 + t^{\gamma}x^{-\gamma}(t)x''(t) \\
&= t^{\gamma}x^{-\gamma}(t)x''(t) - \frac{\gamma}{1-\gamma}\frac{1}{w(t)}\left[w'(t) - \frac{w(t)}{t}\right]^2.
\end{aligned}
$$

Thus, it follows that

$$
-w''(t) = t^{\gamma}q(t) + \frac{\gamma}{1-\gamma}\frac{1}{w(t)}\left[w'(t) - \frac{w(t)}{t}\right]^2 \qquad \text{for all} \quad t \geq T_0.
$$

This gives

$$
-\frac{w(T)}{T} + \frac{w(t)}{T} + \left(1 - \frac{t}{T}\right)w'(t)
$$
$$
= \frac{1}{T}\int_t^T\int_t^s u^{\gamma}q(u)duds + \frac{\gamma}{1-\gamma}\frac{1}{T}\int_t^T\int_t^s\frac{1}{w(u)}\left[w'(u) - \frac{w(u)}{u}\right]^2 duds
$$

for every T, t with $T \geq t \geq T_0$. Therefore, for every $t \geq T_0$, we obtain

$$
\begin{aligned}
w'(t) &= \lim_{T\to\infty}\frac{1}{T}\int_t^T\int_t^s u^{\gamma}q(u)duds + \lim_{T\to\infty}\frac{w(T)}{T} \\
&\quad +\frac{\gamma}{1-\gamma}\int_t^{\infty}\frac{1}{w(s)}\left[w'(s) - \frac{w(s)}{s}\right]^2 ds
\end{aligned}
$$

and consequently,

$$
\lim_{T\to\infty}\frac{w(T)}{T} < \infty \tag{5.2.50}
$$

and

$$
\int_{t_0}^{\infty}\frac{1}{w(s)}\left[w'(s) - \frac{w(s)}{s}\right]^2 ds < \infty. \tag{5.2.51}
$$

Hence, we have

$$
w'(t) = Q(t) + \lim_{T\to\infty}\frac{w(T)}{T} + G(t) \quad \text{for every} \quad t \geq T_0, \tag{5.2.52}
$$

where

$$
G(t) = \frac{\gamma}{1-\gamma}\int_t^{\infty}\frac{1}{w(s)}\left[w'(s) - \frac{w(s)}{s}\right]^2 ds, \quad t \geq T_0.
$$

Now, we set

$$k(t) = \begin{cases} \lim\limits_{T\to\infty} \dfrac{w(T)}{T} + G(t) - \dfrac{w(t)}{t} & \text{if } t > t_0 \\[2mm] \lim\limits_{T\to\infty} \dfrac{w(T)}{T} + G(t_0) - \dfrac{w(t_0)}{t_0} & \text{if } t_0 \leq t \leq T_0. \end{cases}$$

In view of (5.2.50) and (5.2.51), $k(t) \in C([t_0, \infty), \mathbb{R})$ and $\lim_{t\to\infty} k(t) = 0$. Furthermore, from (5.2.50) there exists a positive constant K such that

$$\frac{w(T)}{T} \leq K \quad \text{for all} \quad T \geq t_0. \tag{5.2.53}$$

Thus, from (5.2.52) and (5.2.53), we get

$$\int_{T_0}^{\infty} \frac{1}{w(s)} \left[w'(s) - \frac{w(s)}{s} \right]^2 ds$$

$$= \int_{T_0}^{\infty} \frac{1}{w(s)} \left[Q(s) + \lim_{T\to\infty} \frac{w(T)}{T} + G(s) - \frac{w(s)}{s} \right]^2 ds$$

$$= \int_{T_0}^{\infty} \frac{1}{w(s)} [Q(s) + k(s)]^2 ds \geq \frac{1}{K} \int_{T_0}^{\infty} \frac{1}{s} [Q(s) + k(s)]^2 ds$$

and therefore, from (5.2.51), we find

$$\int_{t_0}^{\infty} \frac{1}{s} [Q(s) + k(s)]^2 ds \leq \int_{t_0}^{T_0} \frac{1}{s} [Q(s) + k(s)]^2 ds$$

$$+ K \int_{T_0}^{\infty} \frac{1}{w(s)} \left[w'(s) - \frac{w(s)}{s} \right]^2 ds < \infty,$$

which contradicts condition (5.2.49). This completes the proof. ∎

Example 5.2.9. Consider the differential equation

$$x''(t) + (t^\lambda \sin t)|x(t)|^\gamma \operatorname{sgn} x(t) = 0, \quad t \geq t_0 > 0 \tag{5.2.54}$$

where $0 < \gamma < 1$ and $-\gamma < \lambda < 1 - \gamma$. We set $\delta = \gamma + \lambda$ so that $0 < \delta < 1$. Now, for any T, t with $T \geq t \geq t_0$, we have

$$\frac{1}{T} \int_t^T \int_t^s u^\gamma q(u)\, du\, ds = \frac{1}{T} \int_t^T \int_t^s u^\delta \sin u\, du\, ds$$

$$= -T^{\delta-1} \sin T - 2\delta T^{\delta-2} \cos T + \delta(\delta-1)(\delta+1)T^{\delta-3} \sin T$$

$$+ \frac{1}{T} \left[t^\delta \sin t + 2\delta t^{\delta-1} \cos t - \delta(\delta-1)(\delta+1)t^{\delta-2} \sin t \right]$$

$$- \delta(\delta-1)(\delta+1)(\delta-2) \frac{1}{T} \int_t^T s^{\delta-3} \sin s\, ds$$

$$+ \left(1 - \frac{t}{T}\right) [t^\delta \cos t - \delta t^{\delta-1} \sin t - \delta(\delta-1)t^{\delta-2} \cos t]$$

$$- \delta(\delta-1)(\delta-2) \int_t^T s^{\delta-3} \cos s ds$$

and consequently,

$$Q(t) = \lim_{T\to\infty} \frac{1}{T} \int_t^T \int_t^s u^\gamma q(u) du ds = t^\delta \cos t + \eta(t) \quad \text{for all} \quad t \geq t_0,$$

where for $t \geq t_0$,

$$\eta(t) = -\delta t^{\delta-1} \sin t - \delta(\delta-1)t^{\delta-2} \cos t - \delta(\delta-1)(\delta-2) \int_t^\infty s^{\delta-3} \cos s ds.$$

Obviously, $\lim_{t\to\infty} \eta(t) = 0$.

Now let us consider an arbitrary continuous function $k(t)$ on $[t_0, \infty)$ with $\lim_{t\to\infty} k(t) = 0$. Furthermore, let $c(t)$ be defined by $c(t) = \eta(t) + k(t)$. Since $\lim_{t\to\infty} c(t) = 0$, we can choose a $t_1 \geq \max\{t_0, 1\}$ so that $c(t) \geq -1/(2\sqrt{2})$ for all $t \geq t_1$. Moreover, we consider an integer N such that $2\pi N - (\pi/4) \geq t_1$. Then for any integer $n \geq N$, we have $t^\delta \cos t \geq \cos t \geq 1/\sqrt{2}$ for every $t \in [2n\pi - (\pi/4), 2n\pi + (\pi/4)]$. Thus, it follows that

$$\int_{t_0}^\infty \frac{1}{s} [Q(s) + k(s)]^2 ds$$

$$= \int_{t_0}^\infty \frac{1}{s} [s^\delta \cos s + c(s)]^2 ds \geq \sum_{n=N}^\infty \int_{2n\pi-(\pi/4)}^{2n\pi+(\pi/4)} \frac{1}{s} \left(\frac{1}{\sqrt{2}} - \frac{1}{2\sqrt{2}}\right)^2 ds$$

$$= \frac{1}{8} \sum_{n=N}^\infty \int_{2n\pi-(\pi/4)}^{2n\pi+(\pi/4)} \frac{ds}{s} = \frac{1}{8} \sum_{n=N}^\infty \ln\left[1 + \frac{2}{8n-1}\right] = \infty.$$

Hence, in view of Theorem 5.2.8, equation (5.2.54) is oscillatory.

We note that for every $t \geq t_0$, one can obtain

$$Q^*(t) = \int_{t_0}^t Q(s) ds = t^\delta \sin t + 2\delta t^{\delta-1} \cos t - t_0^\delta \sin t_0 - 2\delta t_0^{\delta-1} \cos t_0$$

$$- 3\delta(\delta-1) \int_{t_0}^t s^{\delta-2} \cos s ds - \delta(\delta-1)(\delta-2) \int_{t_0}^t \int_s^\infty u^{\delta-3} \cos u du ds$$

and hence the function $Q^*(t)$ is unbounded. Therefore, Theorem 5.2.8 can be applied to equations of type (5.1.6) without any additional hypothesis that $Q^*(t)$ is a bounded function.

The following oscillation criterion is an extension of Theorem 5.2.8.

Theorem 5.2.9. Suppose condition (F) holds, and let $\rho(t) \in C^2([t_0, \infty), \mathbb{R}^+)$ be such that for some constant $c > 0$,

$$0 \leq \frac{1}{t^2} - \frac{h_1(t)}{t} \leq -cg_1(t) \quad \text{for} \quad t \geq t_0 > 0 \tag{5.2.55}$$

and

$$\lim_{T \to \infty} \frac{1}{T} \int_{t_0}^{T} \int_{t_0}^{s} \frac{\rho^\beta(u)}{a(u)} q(u) du ds \quad \text{exists in} \quad \mathbb{R}. \tag{5.2.56}$$

Define

$$Q_1(t) = \lim_{T \to \infty} \frac{1}{T} \int_{t}^{T} \int_{t}^{s} \frac{\rho^\beta(u)}{a(u)} q(u) du ds \quad \text{for} \quad t \geq t_0$$

and, suppose that

$$\liminf_{t \to \infty} Q_1(t) = -q_0 > -\infty, \quad q_0 \quad \text{is a positive constant.} \tag{5.2.57}$$

If for every $k(t) \in C([t_0, \infty), \mathbb{R})$ with $\lim_{t \to \infty} k(t) = 0$,

$$\int_{t_0}^{\infty} \frac{1}{s} [Q_1(s) + k(s)]^2 ds = \infty, \tag{5.2.58}$$

then equation (5.1.3) is oscillatory.

Proof. Let $x(t)$ be a nonoscillatory solution of equation (5.1.3), say, $x(t) \neq 0$ for $t \geq t_0$. Define $w(t) = \rho^\beta(t) F(x(t))$, $t \geq t_0$. Then as in Theorem 5.2.6, (5.2.42) holds for every $t \geq T_0 \geq t_0$. Thus, for any T, t with $T \geq t \geq T_0$, we have

$$-\frac{w(T)}{T} + \frac{w(t)}{T} + \left(1 - \frac{t}{T}\right) w'(t) = \frac{1}{T} \int_{t}^{T} \int_{t}^{s} \frac{\rho^\beta(u)}{a(u)} q(u) du ds$$

$$+ \frac{1}{T} \int_{t}^{T} \int_{t}^{s} -g_1(u) w(u) du ds + \frac{1}{T} \int_{t}^{T} \int_{t}^{s} \frac{1}{w(u)} [w'(u) - h_1(u) w(u)]^2$$
$$\times f'(x(u)) F(x(u)) du ds$$

and hence for every $t \geq T_0$,

$$w'(t) = \lim_{T \to \infty} \frac{1}{T} \int_{t}^{T} \int_{t}^{s} \frac{\rho^\beta(u)}{a(u)} q(u) du ds + \lim_{T \to \infty} \frac{w(T)}{T} + \int_{t}^{\infty} -g_1(u) w(u) du$$

$$+ \int_{t}^{\infty} \frac{1}{w(u)} [w'(u) - h_1(u) w(u)]^2 f'(x(u)) F(x(u)) du,$$

which shows that

$$\lim_{T \to \infty} \frac{w(T)}{T} < \infty, \tag{5.2.59}$$

$$\int_{T_0}^{\infty} -g_1(u)w(u)du \ < \ \infty \tag{5.2.60}$$

and

$$\int_{T_0}^{\infty} \frac{1}{w(u)}[w'(u) - h_1(u)w(u)]^2 f'(x(u))F(x(u))du \ < \ \infty. \tag{5.2.61}$$

Now, by setting for $t \geq T_0$,

$$G_1(t) = \int_t^{\infty} -g_1(s)w(s)ds + \int_t^{\infty} \frac{1}{w(s)}[w'(s) - h_1(s)w(s)]^2 f'(x(s))F(x(s))ds,$$

we find

$$w'(t) \ = \ Q_1(t) + \lim_{T \to \infty} \frac{w(T)}{T} + G(t) \quad \text{for every} \quad t \geq T_0. \tag{5.2.62}$$

Next, we define

$$k(t) \ = \ \begin{cases} \displaystyle\lim_{T \to \infty} \frac{w(T)}{T} + G_1(t) - \frac{w(t)}{t} & \text{if} \quad t > T_0 \\[2ex] \displaystyle\lim_{T \to \infty} \frac{w(T)}{T} + G_1(T_0) - \frac{w(T_0)}{T_0} & \text{if} \quad t_0 \leq t \leq T \end{cases}$$

and observe from (5.2.59), (5.2.60) and (5.2.61) that $k(t) \in C([t_0, \infty), \mathbb{R})$ with $\lim_{t \to \infty} k(t) = 0$. Moreover, in view of (5.2.59), there is a constant $K > 0$ such that

$$\frac{w(T)}{T} \ \leq \ K \quad \text{for all} \quad T \geq T_0. \tag{5.2.63}$$

Also, since $\liminf_{t \to \infty}[Q_1(t) + k(t)] = \liminf_{t \to \infty} Q_1(t) = -q_0 > -\infty$ there exists a constant $\mu > 0$, $\mu \leq q_0$ such that

$$Q_1(t) + k(t) \ \geq \ -\mu \quad \text{for all} \quad t \geq T_0. \tag{5.2.64}$$

Now using (5.2.62), (5.2.63) and (5.2.64), we obtain for $t \geq T_0$,

$$\int_{T_0}^{t} \frac{1}{w(s)}[w'(s) - h_1(s)w(s)]^2 f'(x(s))F(x(s))ds$$

$$\geq \ \frac{\beta}{K(1-\beta)} \int_{T_0}^{t} \frac{1}{s}[w'(s) - h_1(s)w(s)]^2 ds$$

$$= \ \frac{\beta}{K(1-\beta)} \int_{T_0}^{t} \frac{1}{s}\left[Q_1(s) + \lim_{T \to \infty} \frac{w(T)}{T} + G_1(s) - h_1(s)w(s)\right]^2 ds$$

$$= \ \frac{\beta}{K(1-\beta)} \int_{T_0}^{t} \frac{1}{s}\left\{[Q_1(s) + k(s)] + \left[\frac{1}{s} - h_1(s)\right]w(s)\right\}^2 ds$$

$$= \frac{\beta}{K(1-\beta)} \left[\int_{T_0}^t \frac{1}{s} [Q_1(s) + k(s)]^2 ds + 2 \int_{T_0}^t [Q_1(s) + k(s)] \right.$$

$$\times \left. \left[\frac{1}{s^2} - \frac{h_1(s)}{s} \right] w(s) ds + \int_{T_0}^t \left[\frac{1}{s^2} - \frac{h_1(s)}{s} \right]^2 w^2(s) ds \right]$$

$$\geq \frac{\beta}{K(1-\beta)} \left[\int_{T_0}^t \frac{1}{s} [Q_1(s) + k(s)]^2 ds - 2\mu \int_{T_0}^t \left[\frac{1}{s^2} - \frac{h_1(s)}{s} \right] w(s) ds \right]$$

$$\geq \frac{\beta}{K(1-\beta)} \left[\int_{T_0}^t \frac{1}{s} [Q_1(s) + k(s)]^2 ds - 2\mu c \int_{T_0}^t -g_1(s) w(s) ds \right]$$

and consequently, because of (5.2.60) and (5.2.61), we have

$$\int_{t_0}^\infty \frac{1}{s} [Q_1(s) + k(s)]^2 ds \leq \int_{t_0}^{T_0} \frac{1}{s} [Q_1(s) + k(s)]^2 ds + 2\mu c \int_{T_0}^t -g_1(s) w(s) ds$$

$$+ \frac{K(1-\beta)}{\beta} \int_{T_0}^\infty \frac{1}{w(s)} [w'(s) - h_1(s) w(s)]^2 f'(x(s)) F(x(s)) ds < \infty,$$

which contradicts condition (5.2.58). This completes the proof. ∎

For the special case when $a(t) = 1$, $p(t) = 0$ and $\rho(t) = t$, we find

$$h_1(t) = \frac{\beta}{t} \left[\frac{1 + f'(x)F(x)}{f'(x)F(x)} \right] \quad \text{and} \quad g_1(t) = -\frac{\beta^2}{t^2} \left[\frac{1}{\beta} - 1 - \frac{1}{f'(x)F(x)} \right]$$

and hence, condition (5.2.55) is satisfied. Thus, we have the following immediate result.

Corollary 5.2.12. Suppose condition (F) holds, and

$$\lim_{T \to \infty} \frac{1}{T} \int_{t_0}^T \int_{t_0}^s u^\beta q(u) du ds \quad \text{exists in} \quad \mathbb{R}.$$

Define

$$Q_2(t) = \lim_{T \to \infty} \frac{1}{T} \int_t^T \int_t^s u^\beta q(u) du ds \quad \text{for} \quad t \geq t_0$$

and suppose that $\liminf_{t \to \infty} Q_2(t) > -\infty$. If for every $k(t) \in C([t_0, \infty), \mathbb{R})$ with $\lim_{t \to \infty} k(t) = 0$,

$$\int_{t_0}^\infty \frac{1}{s} [Q_2(s) + k(s)]^2 ds = \infty,$$

then equation (5.1.1) is oscillatory.

Example 5.2.10. Consider the differential equation

$$x''(t) + (t^\lambda \sin t) f(x(t)) = 0, \quad t \geq t_0 > 0 \tag{5.2.65}$$

where $f \in C(\mathbb{R}, \mathbb{R})$ satisfies condition (F) and λ is a real number. From Corollary 5.2.2 with $p(t) = 0$, we find that (5.2.65) is oscillatory if $\lambda > 1 - \beta$. Also, from Corollary 5.2.6, we observe that (5.2.65) is oscillatory when $-\beta < \lambda \leq 1 - \beta$. Hence, equation (5.2.65) is oscillatory if $\lambda > -\beta$. Now, Corollary 5.2.12 guarantees that (5.2.65) is also oscillatory for $\lambda = -\beta$. Indeed, in this case for $t \geq t_0$, we have

$$Q_2(t) = \lim_{T \to \infty} \int_t^T \int_t^s u^\beta q(u) du ds = \lim_{T \to \infty} \frac{1}{T} \int_t^T \int_t^s \sin u du ds = \cos t,$$

and consequently, $\liminf_{t \to \infty} Q_2(t) = -1 > -\infty$. Moreover, if $k(t) \in C([t_0, \infty), \mathbb{R})$ is an arbitrary function with $\lim_{t \to \infty} k(t) = 0$, then we can choose a number $T_1 \geq t_0$ such that $k(t) \geq -1/(2\sqrt{2})$ for all $t \geq T_1$. Furthermore, let N be an integer with $2N\pi - (\pi/4) \geq T_1$. Then, it follows that

$$\int_{t_0}^\infty \frac{1}{s} [Q_2(s) + k(s)]^2 ds \geq \sum_{n=N}^\infty \int_{2n\pi-(\pi/4)}^{2n\pi+(\pi/4)} \frac{1}{s} \left[\frac{1}{\sqrt{2}} - \frac{1}{2\sqrt{2}} \right]^2 ds$$

$$= \frac{1}{8} \sum_{n=N}^\infty \ln \left[1 + \frac{2}{8n-1} \right] = \infty.$$

Thus, all conditions of Corollary 5.2.12 are satisfied, and therefore, equation (5.2.65) is oscillatory.

The following result not only extends Theorem 5.1.6 to more general equations of the type (5.1.1), but also removes the condition $Q(t) = \int_t^\infty q(s) ds \geq 0$ for $t \geq t_0$.

Theorem 5.2.10. Let condition (F) hold, and $Q(t) = \int_t^\infty q(s) ds$ exist for $t \geq t_0$ and satisfy

$$\lim_{t \to \infty} \int_{t_0}^t s^{\beta-1} Q(s) ds = \infty, \tag{5.2.66}$$

then equation (5.1.1) is oscillatory.

Proof. First, we shall show that if $f(x)$ satisfies condition (F), then

$$\lim_{x \to \pm\infty} f(x) = \pm\infty. \tag{5.2.67}$$

In fact, from condition (F) and the definition of $F(x)$, we have $F'(x) = 1/f(x)$ and

$$\frac{f'(x)}{f(x)} \geq \frac{1}{c} \frac{F'(x)}{F(x)}, \tag{5.2.68}$$

where $1/c = \beta/(1 - \beta)$. Integrating (5.2.68) from x_0 to x, we find

$$\ln\left(\frac{f(x)}{f(x_0)}\right) \geq \frac{1}{c}\ln\left(\frac{F(x)}{F(x_0)}\right),$$

and hence

$$\left(\frac{f(x)}{f(x_0)}\right)^c \geq \frac{F(x)}{F(x_0)}. \tag{5.2.69}$$

Applying condition (F) to inequality (5.2.69), we obtain

$$cf'(x)(f(x))^c \geq \frac{1}{F(x)}(f(x))^c \geq \frac{(f(x_0))^c}{F(x_0)}. \tag{5.2.70}$$

Integrating (5.2.70) from x_0 to x, we get

$$\left(\frac{c}{c+1}\right)\left[(f(x))^{c+1} - (f(x_0))^{c+1}\right] \geq \frac{(f(x_0))^c}{F(x_0)}(x - x_0),$$

which shows that $f(x) \to \infty$ as $x \to \infty$. A similar argument applies for $x < x_0$ and results in $f(x) \to -\infty$ as $x \to -\infty$. This proves (5.2.67). So, if $x(t)$ is a nonoscillatory solution of (5.1.1), then by Lemma 4.1.3, it satisfies the integral equation

$$\frac{x'(t)}{f(x(t))} = Q(t) + \int_t^\infty \left(\frac{x'(s)}{f(x(s))}\right)^2 f'(x(s))ds, \quad t \geq t_0. \tag{5.2.71}$$

In particular, we have

$$\int_{t_0}^\infty \left(\frac{x'(s)}{f(x(s))}\right)^2 f'(x(s))ds < \infty \tag{5.2.72}$$

from which we shall show that $x(t)$ has the following asymptotic behavior

$$\lim_{t \to \infty} \frac{F(x(t))}{t} = 0. \tag{5.2.73}$$

To prove (5.2.73), we apply the Schwartz inequality to estimate $F(x)$ as follows

$$F(x(t)) - F(x(t_1)) = \left|\int_{t_1}^t \frac{x'(s)}{f(x(s))}ds\right|$$

$$\leq \left[\int_{t_1}^t \left(\frac{x'(s)}{f(x(s))}\right)^2 f'(x(s))ds\right]^{1/2}\left[\int_{t_1}^t \frac{1}{f'(x(s))}ds\right]^{1/2} \tag{5.2.74}$$

for $t \geq t_1 \geq t_0$. In view of (5.2.72), we choose t_1 sufficiently large so that for any $\epsilon > 0$,

$$\int_{t_1}^{\infty} \left(\frac{x'(s)}{f(x(s))} \right)^2 f'(x(s))ds < \frac{\epsilon}{4}.$$

Using the above inequality and condition (F) in (5.2.74), we obtain

$$F(x(t)) \leq F(x(t_1)) + \frac{\sqrt{c\epsilon}}{2} \left(\int_{t_1}^t F(x(s))ds \right)^{1/2}. \qquad (5.2.75)$$

Now suppose $F(x(t)) \in \mathcal{L}^1(t_1, \infty)$. Then, $F(x(t))$ is bounded by (5.2.75), and hence (5.2.73) is satisfied. Otherwise, we can choose a $t_2 \geq t_1$ so that

$$F(x(t_1)) \leq \frac{\sqrt{c\epsilon}}{2} \left(\int_{t_1}^t F(x(s))ds \right)^{1/2} \qquad \text{for} \quad t \geq t_2,$$

which together with (5.2.75) yields

$$F(x(t)) \leq \sqrt{c\epsilon} \left(\int_{t_1}^t F(x(s))ds \right)^{1/2}. \qquad (5.2.76)$$

Integrating (5.2.76) from t_2 to t, we get

$$\left(\int_{t_1}^t F(x(s))ds \right)^{1/2} - \left(\int_{t_1}^{t_2} F(x(s))ds \right)^{1/2} \leq \frac{\sqrt{c\epsilon}}{2}(t - t_2) \leq \frac{\sqrt{c\epsilon}}{2}t.$$
$$(5.2.77)$$

Once again, we can choose $t_3 \geq t_2$ so that

$$\int_{t_1}^{t_2} F(x(s))ds < \frac{c\epsilon}{4}t^2 \qquad \text{for all} \quad t \geq t_3.$$

Thus, from (5.2.76) and (5.2.77), (5.2.73) follows immediate. Now, we define $w(t) = t^{\beta-1}F(x(t))$, $t \geq t_0$. Then for $t \geq t_0$, we have $x'(t)/f(x(t)) = t^{1-\beta}w'(t) + (1-\beta)t^{-\beta}w(t)$ and

$$f'(x(t)) \left(\frac{x'(t)}{f(x(t))} \right)^2$$
$$= f'(x(t))F(x(t)) \left[t^{1-\beta} \frac{[w'(t)]^2}{w(t)} + 2(1-\beta)t^{-\beta}w'(t) + (1-\beta)^2 t^{-1-\beta}w(t) \right]$$
$$\geq \frac{1}{c}t^{1-\beta} \frac{[w'(t)]^2}{w(t)} + 2\lambda t^{-\beta}w'(t) + \beta(1-\beta)t^{-1-\beta}w(t).$$
$$(5.2.78)$$

Integration of the second term on the right–hand side of (5.2.78) from t to $T \geq t \geq t_0$ yields

$$2\beta \int_t^T s^{-\beta}w'(s)ds = 2\beta \left[\frac{w(T)}{T^\beta} - \frac{w(t)}{t^\beta} + \beta \int_t^T \frac{w(s)}{s^{\beta+1}}ds \right]. \qquad (5.2.79)$$

In view of (5.2.73) and the fact that $w(t)$ is nonnegative the limits of the two integrals in (5.2.79) exist and can be infinite. Noting (5.2.72) and $\beta < 1$, we can integrate both sides of (5.2.78) and deduce that the limit on the left–hand side of (5.2.79) must be finite. Hence, the integral on the right–hand side of (5.2.79) is likewise finite, and therefore, we can define an energy function

$$E(t) = w(t) - (1+\beta)t^\beta \int_t^\infty \frac{w(s)}{s^{\beta+1}} ds, \quad t \geq t_0. \tag{5.2.80}$$

Recall that $x(t)$ is a solution of (5.1.1) and satisfies the integral equation (5.2.71). Computing $E'(t)$, we find

$$E'(t) = t^{\beta-1}\frac{x'(t)}{f(x(t))} + (\beta-1)t^{1-\beta}F(x(t)) - \beta(\beta+1)t^{\beta-1}\int_t^\infty \frac{w(s)}{s^{\beta+1}} ds$$

$$+(1+\beta)t^\beta \frac{w(t)}{t^{\beta+1}} \quad \text{for} \quad t \geq t_0.$$

$$\tag{5.2.81}$$

Combining the second and the last terms in (5.2.81) and using (5.2.71), we obtain

$$E'(t) = t^{\beta-1}\left[Q(t) + \int_t^\infty \left(\frac{x'(s)}{f(x(s))}\right)^2 f'(x(s))ds\right] + 2\beta\frac{w(t)}{t}$$

$$-\beta(1+\beta)t^{\beta-1}\int_t^\infty \frac{w(s)}{s^{\beta-1}} ds, \quad t \geq t_0$$

$$\tag{5.2.82}$$

where $\beta = 1/(1+c) < 1$ and $c = (1-\beta)/\beta > 0$. Furthermore, we apply (5.2.73) and combine (5.2.78) and (5.2.79) with (5.2.82), to obtain for $t \geq t_0$,

$$E'(t) \geq t^{\beta-1}Q(t) + \frac{1}{c}t^{\beta-1}\int_t^\infty s^{1-\lambda}\frac{[w'(s)]^2}{w(s)} ds \geq t^{\beta-1}Q(t).$$

Integrating this inequality and using condition (5.2.66), we get $\lim_{t\to\infty} E(t) = \infty$. From the definition of $E(t)$ in (5.2.80), we also deduce that $\lim_{t\to\infty} w(t) = \infty$. Define $t_n = \sup\{t : w(t) \leq n\}$, which satisfies $\lim_{n\to\infty} t_n = \infty$ and also $t \geq t_n$ implies $w(t) \geq n = w(t_n)$. Replacing t in (5.2.80) by t_n and using its definition, we obtain

$$E(t_n) = w(t_n) - (1+\beta)t_n^\beta \int_{t_n}^\infty \frac{w(s)}{s^{\beta+1}} ds$$

$$\geq w(t_n) - (1+\beta)t_n^\beta w(t_n) \int_{t_n}^\infty \frac{ds}{s^{\beta+1}}$$

$$= w(t_n)\left[1 - \frac{1+\beta}{\beta}\right] = -\frac{1}{\beta}w(t_n) < 0,$$

which contradicts the fact that $\lim_{n\to\infty} E(t_n) = \infty$. This completes the proof. $\blacksquare$

Example 5.2.11. Consider the sublinear differential equation

$$x''(t) + \frac{1}{t^{\gamma+1}} \left[\gamma + \gamma\sqrt{t}\sin t + t\sqrt{t}\cos t \right] |x(t)|^{\gamma}\text{sgn } x(t) = 0, \quad (5.2.83)$$

where $0 < \gamma < 1$, $t \geq t_0 > 0$. Let $f(x) = |x|^{\gamma}\text{sgn } x$ so that $c = (1-\gamma)/\gamma$ and $\beta = \gamma$. Also,

$$Q(t) = \int_t^{\infty} q(s)ds = \frac{1}{t^{\gamma}}[1 + \sqrt{t}\sin t], \quad t \geq t_0.$$

It is easy to check that all hypotheses of Theorem 5.2.10 are satisfied and hence equation (5.2.83) is oscillatory.

We note that Theorem 5.1.6 fails to apply to equation (5.2.83) since the function $Q(t)$ assumes negative values for all large values of t and satisfies condition (5.1.25). Therefore, we conclude that Theorem 5.2.10 improves Theorem 5.1.6.

Example 5.2.12. Consider the differential equation

$$x''(t) + \frac{1}{t\sqrt{t}} \left[\frac{1}{2} + \frac{1}{2}\sqrt{t}\sin t + t\sqrt{t}\cos t \right] \left\{ |x(t)|^{-1/2} + |x(t)|^{1/2} \right\} x(t) = 0. \quad (5.2.84)$$

We let

$$f(x) = \left[|x|^{-1/2} + |x|^{1/2} \right] x \quad (5.2.85)$$

so that $F(x) = \int_0^x du/f(u) = 2\tan^{-1}\sqrt{x}$. Clearly, condition (F) is satisfied with $\beta = 1/2$, i.e., $c = (1-\beta)/\beta = 1$. All conditions of Theorem 5.2.10 are satisfied and hence equation (5.2.84) is oscillatory.

It is interesting to note that $f(x)$ in (5.2.85) satisfies both conditions of sublinearity, i.e., (5.1.5) as well as of superlinearity, i.e., (4.1.5).

Next, we shall prove the following result.

Theorem 5.2.11. Let condition (5.1.5) hold and

$$\min\left\{ \inf_{x>0} f'(x) \int_{+0}^x \frac{du}{f(u)}, \ \inf_{x<0} f'(x) \int_{-0}^x \frac{du}{f(u)} \right\} > 0. \quad (5.2.86)$$

Further, let there exist $\rho(t) \in C^1([t_0, \infty), \mathbb{R}^+)$ such that

$$\begin{cases} (a(t)\rho(t))' \leq 0, \quad a(t)\rho'(t) = p(t)\rho(t) \quad \text{for} \quad t \geq t_0, \quad \text{and} \\ \xi(t) = \int_{t_0}^t \frac{1}{a(s)\rho(s)} ds \to \infty \quad \text{as} \quad t \to \infty. \end{cases} \quad (5.2.87)$$

If

$$\limsup_{t\to\infty} \frac{1}{\xi(t)} \int_{t_0}^{t} \frac{1}{a(s)\rho(s)} \left[\int_{t_0}^{s} \rho(u)q(u)du \right]^2 ds = \infty \qquad (5.2.88)$$

and

$$\liminf_{t\to\infty} \frac{1}{t^m} \int_{t_0}^{t} (t-s)^m \rho(s)q(s)ds > -\infty \quad \text{for some integer} \quad m \geq 1,$$
$$(5.2.89)$$

then equation (5.1.3) is oscillatory.

Proof. Let $x(t)$ be a nonoscillatory solution of equation (5.1.3), say, $x(t) > 0$ for $t \geq t_0$. We define $w(t) = \int_0^{x(t)} du/f(u)$ for $t \geq t_0$. Then from (5.1.3) it follows that $w(t)$ satisfies the second order nonlinear differential equation

$$\begin{aligned}
(a(t)\rho(t)w'(t))' &= -\rho(t)q(t) + (a(t)\rho'(t) - p(t)\rho(t))\frac{x'(t)}{f(x(t))} \\
&\quad -a(t)\rho(t)(w'(t))^2 f'(x(t)) \\
&= -\rho(t)q(t) - a(t)\rho(t)(w'(t))^2 f'(x(t)), \quad t \geq t_0.
\end{aligned}$$
$$(5.2.90)$$

Now, with respect to the function $\mathcal{I}(t_0,t) = \int_{t_0}^{t} a(s)\rho(s)(w'(s))^2 f'(x(s))ds$ we shall consider two mutually exclusive cases.

Case I. $\mathcal{I}(t_0,\infty) < \infty$. In view of (5.2.86), there exists a positive constant α such that

$$f'(x(t))w(t) = f'(x(t)) \int_{+0}^{x(t)} \frac{du}{f(u)} \geq \alpha \quad \text{for all} \quad t \geq t_0. \qquad (5.2.91)$$

We shall show that

$$\lim_{t\to\infty} \frac{w(t)}{\xi(t)} = 0. \qquad (5.2.92)$$

For this, let $\epsilon > 0$ be an arbitrary number. Then we can choose a $t_1 \geq t_0$ so that

$$\mathcal{I}(t_1,\infty) \leq \frac{\epsilon\alpha}{4}. \qquad (5.2.93)$$

By Schwartz's inequality for every $t \geq t_1$, we have

$$w(t) - w(t_1) \leq \left| \int_{t_1}^{t} w'(s)ds \right| \leq (\mathcal{I}(t_1,t))^{1/2} (\mathcal{J}(t_1,t))^{1/2},$$

where $\mathcal{J}(t_1,t) = \int_{t_1}^{t} 1/(a(s)\rho(s)f'(x(s)))ds$, and consequently, by using (5.2.93), we find

$$w(t) \leq w(t_1) + \frac{\sqrt{\epsilon\alpha}}{2} (\mathcal{J}(t_1,t))^{1/2} \quad \text{for all} \quad t \geq t_1. \qquad (5.2.94)$$

If $\mathcal{J}(t_1, \infty) < \infty$, (5.2.94) ensures that $w(t)$ is bounded, and so (5.2.92) holds. Thus, we restrict our attention to the case when $\mathcal{J}(t_1, \infty) = \infty$. In this case, we can choose a $t_2 > t_1$ so that $w(t_1) \le (\sqrt{\epsilon \alpha}/2) (\mathcal{J}(t_1, t))^{1/2}$ for all $t \ge t_2$. Thus from (5.2.94) it follows that $w(t) \le \sqrt{\epsilon \alpha} (\mathcal{J}(t_1, t))^{1/2}$ for all $t \ge t_2$, and so, by using (5.2.91), we have

$$w(t) \le \sqrt{\epsilon} \left\{ \int_{t_1}^{t} \frac{w(s)}{a(s)\rho(s)} ds \right\}^{1/2} \qquad \text{for all} \quad t \ge t_2, \qquad (5.2.95)$$

or

$$\left\{ \int_{t_1}^{t} \frac{w(s)}{a(s)\rho(s)} ds \right\}^{-1/2} w(t) \le \sqrt{\epsilon} \quad \text{for all} \quad t \ge t_2.$$

Integrating this inequality from t_2 to $t \ge t_2$, we obtain

$$2 \left\{ \int_{t_1}^{t} \frac{w(s)}{a(s)\rho(s)} ds \right\}^{1/2} - 2 \left\{ \int_{t_1}^{t_2} \frac{w(s)}{a(s)\rho(s)} ds \right\}^{1/2} \le \sqrt{\epsilon} \int_{t_2}^{t} \frac{1}{a(s)\rho(s)} ds.$$

Hence, by setting

$$t_3 = \max \left\{ t_2, \frac{2}{\sqrt{\epsilon}} \left\{ \int_{t_1}^{t_2} \frac{w(s)}{a(s)\rho(s)} ds \right\}^{1/2} \right\}$$

for $t \ge t_3$, we obtain

$$\left\{ \int_{t_1}^{t} \frac{w(s)}{a(s)\rho(s)} ds \right\}^{1/2} < \frac{\sqrt{\epsilon}}{2} \xi(t) + \left\{ \int_{t_1}^{t_2} \frac{w(s)}{a(s)\rho(s)} ds \right\}^{1/2}$$

$$\le \frac{\sqrt{\epsilon}}{2} \xi(t) + \frac{\sqrt{\epsilon}}{2} t_3 \le \sqrt{\epsilon} \xi(t).$$

Thus, from (5.2.95) it follows that $w(t) < \epsilon \xi(t)$ for all $t \ge t_3$. As $\epsilon > 0$ is arbitrary, the proof of (5.2.92) is complete. Now, from equation (5.2.90), for $t \ge t_0$, we have $\int_{t_0}^{t} \rho(u)q(u)du = C_1 - a(t)\rho(t)w'(t) - \mathcal{I}(t_0, t)$, where $C_1 = a(t_0)\rho(t_0)w'(t_0)$. Hence, for every $t \ge t_0$, we find

$$\left[\int_{t_0}^{t} \rho(u)q(u)du \right]^2 = [C_1 - a(t)\rho(t)w'(t) - \mathcal{I}(t_0, t)]^2$$

$$\le 3C_1^2 + 3[a(t)\rho(t)w'(t)]^2 + 3(\mathcal{I}(t_0, t))^2$$

$$\le C_2 + 3[a(t)\rho(t)w'(t)]^2,$$

where $C_2 = 3C_1^2 + 3[\mathcal{I}(t_0, \infty)]^2$. Next, we make use of (5.2.91) to obtain for every $t \geq t_0$,

$$\frac{1}{\xi(t)} \int_{t_0}^t \frac{1}{a(s)\rho(s)} \left[\int_{t_0}^s \rho(u)q(u)du \right]^2 ds$$

$$\leq C_2 + \frac{3}{\xi(t)} \int_{t_0}^t a(s)\rho(s)(w'(s))^2 ds$$

$$\leq C_2 + \frac{(3/\alpha)}{\xi(t)} \int_{t_0}^t a(s)\rho(s)(w'(s))^2 f'(x(s))w(s)ds$$

$$\leq C_2 + \frac{(3/\alpha)}{\xi(t)} \left[\max_{t_0 \leq s \leq t} w(s) \right] \int_{t_0}^t a(s)\rho(s)(w'(s))^2 f'(x(s))ds$$

and consequently for $t \geq t_0$,

$$\frac{1}{\xi(t)} \int_{t_0}^t \frac{1}{a(s)\rho(s)} \left[\int_{t_0}^s \rho(u)q(u)du \right]^2 ds \leq C_2 + C_3 \left(\frac{1}{\xi(t)} \right) \left[\max_{t_0 \leq s \leq t} w(s) \right],$$

$$(5.2.96)$$

where $C_3 = (3/\alpha)\mathcal{I}(t_0, \infty) > 0$. But, by (5.2.92) we can choose $T \geq t_0$ so that $w(t) \leq \xi(t)$ for every $t \geq T$. Therefore, $\max_{t_0 \leq s \leq t} w(s) \leq \max_{t_0 \leq s \leq T} w(s) + \xi(t)$ for all $t \geq t_0$. Thus, (5.2.96) gives

$$\frac{1}{\xi(t)} \int_{t_0}^t \frac{1}{a(s)\rho(s)} \left[\int_{t_0}^s \rho(u)q(u)du \right]^2 ds \leq C_2 + C_3 \left(\frac{1}{\xi(t)} \right) \left[\max_{t_0 \leq s \leq T} w(s) + \xi(t) \right]$$

for all $t \geq t_0$. Hence, we have

$$\limsup_{t \to \infty} \frac{1}{\xi(t)} \int_{t_0}^t \frac{1}{a(s)\rho(s)} \left[\int_{t_0}^s \rho(u)q(u)du \right]^2 ds \leq C_2 + C_3 < \infty,$$

which contradicts condition (5.2.88).

Case II. $\mathcal{I}(t_0, \infty) = \infty$. From equation (5.2.90) for every $t \geq t_0$, we have

$$\int_{t_0}^t (t-s)^m \rho(s)q(s)ds + \int_{t_0}^t (t-s)^m a(s)\rho(s)(w'(s))^2 f'(x(s))ds$$

$$= -\int_{t_0}^t (t-s)^m (a(s)\rho(s)w'(s))'ds$$

$$= \begin{cases} (t-t_0)a(t_0)\rho(t_0)w'(t_0) - a(t)\rho(t)w(t) + a(t_0)\rho(t_0)w(t_0) \\ \quad + \int_{t_0}^t (a(s)\rho(s))'w(s)ds \quad \text{for } m = 1 \\ (t-t_0)^m a(t_0)\rho(t_0)w'(t_0) + m(t-t_0)^{m-1}a(t_0)\rho(t_0)w(t_0) \\ \quad + m \int_{t_0}^t \left[-(m-1)(t-s)^{m-2}a(s)\rho(s) + (t-s)^{m-1}(a(s)\rho(s))' \right] w(s)ds \\ \quad \text{if } m \geq 2 \end{cases}$$

$$\leq \begin{cases} (t-t_0)a(t_0)\rho(t_0)w'(t_0) + a(t_0)\rho(t_0)w(t_0) & \text{if } m=1 \\ (t-t_0)^m a(t_0)\rho(t_0)w'(t_0) + m(t-t_0)^{m-1}a(t_0)\rho(t_0)w(t_0) & \text{if } m \geq 2. \end{cases}$$

Therefore, for all $t \geq t_0$,

$$\frac{1}{t^m}\int_{t_0}^t (t-s)^m \rho(s)q(s)ds \leq -\frac{1}{t^m}\int_{t_0}^t (t-s)^m a(s)\rho(s)(w'(s))^2 f'(x(s))ds$$

$$+ \left(1 - \frac{t_0}{t}\right)^m a(t_0)\rho(t_0)w'(t_0) + \frac{m}{t}\left(1 - \frac{t_0}{t}\right)^{m-1} a(t_0)\rho(t_0)w(t_0).$$

$$(5.2.97)$$

But, since

$$\lim_{t\to\infty} \frac{1}{t^m}\int_{t_0}^t (t-s)^m a(s)\rho(s)(w'(s))^2 f'(x(s))ds = \mathcal{I}(t_0,\infty) = \infty.$$

inequality (5.2.97) gives

$$\lim_{t\to\infty} \frac{1}{t^m}\int_{t_0}^t (t-s)^m \rho(s)q(s)ds = -\infty,$$

which contradicts condition (5.2.89). This completes the proof. ■

For the special case when $p(t) = 0$ and $\rho(t) = 1$, we have the following result.

Corollary 5.2.13. Let conditions (5.1.5) and (5.2.86) hold, and $a'(t) \leq 0$ for $t \geq t_0$ and $\xi_1(t) = \int_{t_0}^t ds/a(s) \to \infty$ as $t \to \infty$. If

$$\limsup_{t\to\infty} \frac{1}{\xi_1(t)}\int_{t_0}^t \frac{1}{a(s)}\left[\int_{t_0}^s q(u)du\right]^2 ds = \infty \qquad (5.2.98)$$

and

$$\liminf_{t\to\infty} \frac{1}{t^m}\int_{t_0}^t (t-s)^m q(s)ds > -\infty \quad \text{for some integer } m \geq 1, \quad (5.2.99)$$

then equation (5.1.2) is oscillatory.

Also, when $a(t) = 1$, $p(t) = 0$ and $\rho(t) = 1$, we have the following corollary.

Corollary 5.2.14. Let conditions (5.1.5), (5.2.86) and (5.2.99) hold, and

$$\limsup_{t\to\infty} \frac{1}{t}\int_{t_0}^t \left[\int_{t_0}^s q(u)du\right]^2 ds = \infty, \qquad (5.2.100)$$

then equation (5.1.1) is oscillatory.

Remark 5.2.6.

1. The condition

$$\limsup_{t\to\infty} \frac{1}{t} \int_{t_0}^{t} \int_{t_0}^{s} q(u)\,du\,ds \; = \; \infty \qquad (5.2.101)$$

implies (5.2.100) as an application of the Schwartz inequality.

2. The condition (5.2.99) holds if

$$\liminf_{t\to\infty} \frac{1}{t} \int_{t_0}^{t} \int_{t_0}^{s} q(u)\,du\,ds \; > \; -\infty. \qquad (5.2.102)$$

Now, we can state that equation (5.1.6) with $0 < \gamma < 1$ is oscillatory if either one of the following holds:

(I_1) conditions (5.2.101) and (5.2.102),

(I_2) conditions (5.2.100) and (5.2.102),

(I_3) $-\infty < \liminf_{t\to\infty} \dfrac{1}{t} \int_{t_0}^{t} \int_{t_0}^{s} q(u)\,du\,ds < \limsup_{t\to\infty} \dfrac{1}{t} \int_{t_0}^{t} \int_{t_0}^{s} q(u)\,du\,ds \le \infty.$

$$(5.2.103)$$

Next, we shall prove the following more general result for (5.1.3). It extends the criterion (I_1) above.

Theorem 5.2.12. Let conditions (5.1.5) and (5.2.86) hold, and assume that there exists a function $\rho(t) \in C^1([t_0,\infty),\mathbb{R}^+)$ such that conditions (5.2.87) and (5.2.89) are satisfied. If, for some integer $n > 1$,

$$\limsup_{t\to\infty} \frac{1}{t^n} \int_{t_0}^{t} (t-s)^n \rho(s) q(s)\,ds \; = \; \infty \qquad (5.2.104)$$

and

$$\limsup_{t\to\infty} \frac{1}{t^n} \int_{t_0}^{t} (t-s)^{n-2} \xi(s)\,ds \; < \; \infty, \qquad (5.2.105)$$

then equation (5.1.3) is oscillatory.

Proof. Let $x(t)$ be a nonoscillatory solution of (5.1.3), say, $x(t) > 0$ for $t \ge t_0$. As in Theorem 5.2.11, we define $w(t) = \int_0^{x(t)} du/f(u)$, $t \ge t_0$ which reduces (5.1.3) to equation (5.2.90). Next, we let $z(t) = a(t)\rho(t)w'(t)$, $t \ge t_0$. Then equation (5.2.90) becomes

$$z'(t) + \frac{f'(x(t))}{a(t)\rho(t)} z^2(t) + \rho(t)q(t) \; = \; 0, \quad t \ge t_0. \qquad (5.2.106)$$

As in Theorem 5.2.11 we consider the two cases I and II. The proof of case II is exactly the same as earlier and hence omitted. Now, for the case I we proceed as in Theorem 5.2.11, to obtain (5.2.91) and (5.2.92). Using (5.2.91) in equation (5.2.106), we find

$$z'(t) + \frac{\alpha}{w(t)} z^2(t) + \rho(t)q(t) \leq 0 \quad \text{for} \quad t \geq t_0. \tag{5.2.107}$$

From (5.2.92), there exists a $t_1 \geq t_0$ such that for all $t \geq t_1$, we have

$$w(t) \leq \xi(t) \quad \text{for} \quad t \geq t_1. \tag{5.2.108}$$

Using (5.2.108) in inequality (5.2.107), we get

$$z'(t) + \frac{\alpha}{\xi(t)} z^2(t) + \rho(t)q(t) \leq 0 \quad \text{for} \quad t \geq t_1. \tag{5.2.109}$$

Multiplying (5.2.109) by $(t-s)^n$ and integrating from t_1 to t, we obtain

$$\alpha \int_{t_1}^t \frac{1}{\xi(s)}(t-s)^n z^2(s)ds + n \int_{t_1}^t (t-s)^{n-1}z(s)ds$$

$$+ \int_{t_1}^t (t-s)^n \rho(s)q(s)ds \leq (t-t_1)^n z(t_1),$$

or

$$\int_{t_1}^t \left[\sqrt{\frac{\alpha}{\xi(s)}}(t-s)^{n/2}z(s) + \frac{n}{2}\sqrt{\frac{\xi(s)}{\alpha}}(t-s)^{(n/2)-1} \right]^2 ds$$

$$+ \int_{t_1}^t (t-s)^n \rho(s)q(s)ds \leq (t-t_1)^n z(t_1) + \frac{n^2}{4\alpha}\int_{t_1}^t \xi(s)(t-s)^{n-2}ds. \tag{5.2.110}$$

Thus, it follows that

$$\frac{1}{t^n}\int_{t_1}^t (t-s)^n \rho(s)q(s)ds \leq \left(1 - \frac{t_1}{t}\right)^n z(t_1) + \left(\frac{n^2}{4\alpha}\right)\frac{1}{t^n}\int_{t_1}^t \xi(s)(t-s)^{n-2}ds.$$

Taking $\limsup$ on both sides of the above inequality as $t \to \infty$ and applying condition (5.2.105), we obtain a desired contradiction to condition (5.2.104). ∎

In Theorem 5.2.12, if (5.1.5) and (5.2.86) are replaced by condition (F), then we get the following result.

Theorem 5.2.13. Suppose condition (F) holds with $\beta > 0$ and $c = (1-\beta)/\beta$. Moreover, assume that there exists a function $\rho(t) \in C^1([t_0, \infty), \mathbb{R}^+)$ such that conditions (5.2.87) and (5.2.89) are satisfied. If for some integer $n > 1$,

$$\limsup_{t \to \infty} \frac{1}{t^n}\int_{t_0}^t (t-s)^{n-2}\left[(t-s)\rho(s)q(s) - \frac{n^2 c}{4}\xi(s)\right]ds = \infty, \tag{5.2.111}$$

then equation (5.1.3) is oscillatory.

Proof. Let $x(t)$ be a nonoscillatory solution of equation (5.1.3), say, $x(t) > 0$ for $t \geq t_0$. Proceeding as in Theorem 5.2.11, we obtain the inequality (5.2.110) which takes the form

$$\int_{t_1}^{t} (t-s)^{n-2} \left[(t-s)^2 \rho(s) q(s) - \frac{n^2}{4\alpha} \xi(s) \right] ds \leq \left(1 - \frac{t_1}{t} \right)^n z(t_1).$$

Dividing both sides of the above inequality by t^n, taking lim sup on both sides as $t \to \infty$ and noting that $\alpha = 1/c$, we obtain a desired contradiction to condition (5.2.111). ∎

The following results are immediate consequences of Theorems 5.2.11 and 5.2.12.

Corollary 5.2.15. Let conditions (5.1.5) and (5.2.86) hold, or condition (F) holds with $\beta > 0$, $a'(t) \leq 0$ for $t \geq t_0$ and $\xi_1(t) = \int_{t_0}^{t} ds/a(s) \to \infty$ as $t \to \infty$. If condition (5.2.99) holds and for some integer $n > 1$ either

$$\limsup_{t \to \infty} \frac{1}{t^n} \int_{t_0}^{t} (t-s)^n q(s) ds = \infty \tag{5.2.112}$$

and

$$\limsup_{t \to \infty} \frac{1}{t^n} \int_{t_0}^{t} \xi_1(s)(t-s)^{n-2} ds < \infty, \tag{5.2.113}$$

or

$$\limsup_{t \to \infty} \frac{1}{t^n} \int_{t_0}^{t} (t-s)^{n-2} \left[(t-s)^2 q(s) - \frac{n^2 c}{4} \xi_1(s) \right] ds = \infty, \tag{5.2.114}$$

where $c = (1-\beta)/\beta$, then equation (5.1.2) is oscillatory.

Corollary 5.2.16. Let either condition (F) hold with $\beta > 0$, or conditions (5.1.5) and (5.2.86) hold. If conditions (5.2.99) and (5.2.112) hold, then equation (5.1.1) is oscillatory.

Corollary 5.2.15 follows from Theorem 5.2.11 (Theorem 5.2.12) by letting $\rho(t) = 1$ and $p(t) = 0$, while Corollary 5.2.16 follows from Corollary 5.2.15 by letting $a(t) = 1$ and noting that for $n > 1$,

$$\lim_{t \to \infty} \frac{1}{t^n} \int_{t_1}^{t} s(t-s)^{n-2} ds = \frac{1}{n(n-1)} < \infty.$$

Remark 5.2.7. It is known that (5.2.112) alone is sufficient for the oscillation of sublinear equation (5.1.6). Also, from Theorem 5.2.12, we note that the condition

$$\lim_{t \to \infty} \frac{1}{t^n} \int_{t_0}^{t} (t-s)^n q(s) ds = \infty \quad \text{for some integer } n > 1 \tag{5.2.115}$$

alone is sufficient for the oscillation of (5.1.6) with $0 < \gamma < 1$.

We also note that conditions (5.2.112) and (5.2.115) can be improved by allowing n to be any real number > 1 and the proof remains the same.

Remark 5.2.8. In Theorem 5.2.12 the restriction of selecting the weight function $\rho(t) = \exp\left(\int_{t_0}^t (p(s)/a(s))ds\right)$ can be removed. This requires only slight modification in the proof. The details are left to the reader.

Example 5.2.13. Consider the differential equation

$$\left(\frac{1}{t}x'(t)\right)' + \frac{5}{6t^2}x'(t) + \left[-t^{-1/3}\sin t + \frac{1}{2}t^{-4/3}(2 + \cos t)\right]$$
$$\times \ \{|x(t)|^\alpha \ \text{sgn} \ x(t) + x(t)\} \ = \ 0, \quad 0 < \alpha < 1, \quad t \geq t_0 = \pi/2.$$
$$(5.2.116)$$

Here, $f(x) = |x|^\alpha \text{sgn} \ x + x$, $f'(x) = \alpha|x|^{\alpha-1} + 1 > 0$ for all $x \neq 0$, and $F(x) = \int_0^{|x|} du/(u^\alpha + u)$. To estimate $F(x)$ from below, we observe that

$$F(x) \ = \ \int_0^{|x|} \frac{du}{u^\alpha + u} \ \geq \ \int_0^{|x|} \frac{2u}{2u^\alpha} \ = \ \frac{|x|^{1-\alpha}}{2(1-\alpha)} \quad \text{if} \ |x| \leq 1,$$

and

$$F(x) \ = \ \int_0^{|x|} \frac{du}{u^\alpha + u} \ \geq \ \int_0^1 \frac{du}{2u^\alpha} \ = \ \frac{1}{2(1-\alpha)} \quad \text{if} \ |x| \geq 1.$$

Therefore, for $x \neq 0$ we have for $|x| \leq 1$,

$$f'(x)F(x) \ \geq \ \frac{|x|^{1-\alpha}}{2(1-\alpha)} \left[\alpha|x|^{\alpha-1} + 1\right] \ > \ \frac{\alpha}{2(1-\alpha)}$$

and for $|x| \geq 1$,

$$f'(x)F(x) \ \geq \ \frac{1}{2(1-\alpha)} \left[\alpha|x|^{\alpha-1} + 1\right] \ > \ \frac{1}{2(1-\alpha)} \ > \ \frac{\alpha}{2(1-\alpha)}.$$

Using condition (F), we find $\beta = \alpha/(2-\alpha)$.

Next, let $\rho(t) = t^{5/6}$. Then for $t \geq t_0 = \pi/2$, we find that (5.2.87) holds with

$$\xi(t) \ = \ \int_{(\pi/2)}^t \frac{1}{a(s)\rho(s)}ds \ = \ \int_{(\pi/2)}^t s^{1/6}ds \ = \ \frac{7}{6}\left[t^{7/6} - \left(\frac{\pi}{2}\right)^{7/6}\right].$$

Furthermore, for every $t \geq t_0$,

$$\frac{1}{t^2} \int_{t_0}^{t} (t-s)^2 s^{5/6} q(s) ds$$

$$= \frac{2}{t^2} \int_{t_0}^{t} (t-s) \left[\int_{t_0}^{s} u^{5/6} q(u) du \right] ds$$

$$= \frac{2}{t^2} \int_{t_0}^{t} (t-s) \left[\int_{t_0}^{s} u^{5/6} \left[-u^{-1/3} \sin u + \frac{1}{2} u^{-4/3}(2 + \cos u) \right] du \right] ds$$

$$= \frac{2}{t^2} \int_{(\pi/2)}^{t} (t-s) \int_{(\pi/2)}^{t} d \left[u^{1/2}(2 + \cos u) \right] ds$$

$$= \frac{2}{t^2} \int_{(\pi/2)}^{t} (t-s) \left[s^{1/2}(2 + \cos s) - 2 \left(\frac{\pi}{2} \right)^{1/2} \right] ds$$

$$\geq \frac{2}{t^2} \int_{(\pi/2)}^{t} (t-s) \left[s^{1/2} - 2 \left(\frac{\pi}{2} \right)^{1/2} \right] ds$$

$$= \frac{8}{15} t^{1/2} - 2 \left(\frac{\pi}{2} \right)^{1/2} + \frac{8}{3} \left(\frac{\pi}{2} \right)^{3/2} \frac{1}{t} - \frac{6}{5} \left(\frac{\pi}{2} \right)^{5/2} \frac{1}{t^2}$$

and consequently condition (5.2.111) is also fulfilled with $n = 2$. So, Theorem 5.2.13 guarantees that equation (5.2.116) is oscillatory.

5.3. Linearization of Sublinear Oscillation Theorems

In this section we shall relate the oscillation of sublinear differential equations to some linear second order equations. Our first result in this direction is the following.

Theorem 5.3.1. Let condition (F) hold, and suppose there exists a function $\rho(t) \in C^2([t_0, \infty), \mathbb{R}^+)$ satisfying (5.2.24). If

$$\liminf_{t \to \infty} \frac{1}{t} \int_{t_0}^{t} \int_{t_0}^{s} \rho^{\beta}(u) q(u) du ds > -\infty \tag{5.3.1}$$

and for every constant $k > 0$, the linear second order equation

$$(tv'(t))' + C(t)v(t) = 0, \tag{5.3.2}$$

is oscillatory, where

$$C(t) = k\rho^{\beta}(t)q(t) - \frac{\beta^2}{4} t \left(\frac{\rho'(t)}{\rho(t)} \right)^2 + \frac{\beta}{2} \left(t \frac{\rho'(t)}{\rho(t)} \right)' \quad \text{for all large } t,$$

then equation (5.1.1) is oscillatory.

Proof. Let $x(t)$ be a nonoscillatory solution of equation (5.1.1), say, $x(t) \neq 0$ for $t \geq t_0$, and let $w(t) = \rho^\beta(t)F(x(t))$. Then as in Theorem 5.2.5, we have

$$\limsup_{t \to \infty} \frac{w(t)}{t} < \infty. \tag{5.3.3}$$

We also have

$$w'(t) = \beta\rho^{\beta-1}(t)\rho'(t)F(x(t)) + \rho^\beta(t)\frac{x'(t)}{f(x(t))},$$

or

$$\rho^\beta(t)\frac{x'(t)}{f(x(t))} = w'(t) - \beta\frac{\rho'(t)}{\rho(t)}w(t) = z(t), \quad t \geq t_0$$

and

$$
\begin{aligned}
z'(t) &= \left(\rho^\beta(t)\frac{x'(t)}{f(x(t))}\right)' \\
&= -\rho^\beta(t)q(t) + \beta\frac{\rho'(t)}{\rho(t)}z(t) - \frac{1}{w(t)}z^2(t)f'(x(t))F(x(t)), \quad t \geq t_0.
\end{aligned}
\tag{5.3.4}
$$

From condition (F) and (5.3.3), we find that there exist a $t_1 \geq t_0$ and a constant $c_1 > 0$ such that

$$f'(x(t))F(x(t)) \geq \frac{\beta}{1-\beta} \quad \text{and} \quad w(t) \leq c_1 t \quad \text{for} \quad t \geq t_1. \tag{5.3.5}$$

Using (5.3.5) in equation (5.3.4), we obtain

$$z'(t) \leq -\rho^\beta(t)q(t) + \beta\frac{\rho'(t)}{\rho(t)}z(t) - c\frac{1}{t}z^2(t), \quad t \geq t_1$$

where $c = \beta/(c_1(1-\beta))$. Thus, for $t \geq t_1$, we have

$$
\begin{aligned}
\left(z(t) - \frac{\beta}{2c}t\frac{\rho'(t)}{\rho(t)}\right)' &\leq -\rho^\beta(t)q(t) - \frac{\beta}{2c}\left(t\frac{\rho'(t)}{\rho(t)}\right)' + \frac{\beta^2}{4c}t\left(\frac{\rho'(t)}{\rho(t)}\right)^2 \\
&\quad -\frac{c}{t}\left[z(t) - \frac{\beta}{2c}t\frac{\rho'(t)}{\rho(t)}\right]^2, \quad t \geq t_1.
\end{aligned}
$$

Now, let

$$y(t) = z(t) - \frac{\beta}{2c}t\frac{\rho'(t)}{\rho(t)}, \quad t \geq t_1$$

so that

$$y'(t) \leq -\left[\rho^\beta(t)q(t) - \frac{\beta^2}{4c}t\left(\frac{\rho'(t)}{\rho(t)}\right)^2 + \frac{\beta}{2c}\left(t\frac{\rho'(t)}{\rho(t)}\right)'\right] - \frac{c}{t}y^2(t), \quad t \geq t_1. \tag{5.3.6}$$

Finally, applying Lemma 2.2.1, we find that equation (5.3.2) is nonoscillatory, which is a contradiction. This completes the proof. ∎

The following corollary is an immediate consequence of Theorem 5.3.1.

Corollary 5.3.1. Let condition (F) hold, and suppose there exists a function $\rho(t) \in C^2([t_0, \infty), \mathbb{R}^+)$ such that $\rho'(t) \geq 0$, $\rho''(t) \leq 0$ and $(t\rho'(t)/\rho(t))' \geq 0$ for $t \geq t_0$. If condition (5.3.1) holds and for every constant $k > 0$ the second order linear equation

$$(tv'(t))' + \left[k\rho^\beta(t)q(t) - \frac{\beta^2}{t} \right] v(t) = 0$$

is oscillatory, then equation (5.1.1) is oscillatory.

Proof. Let $x(t)$ be a nonoscillatory solution of equation (5.1.1), say, $x(t) \neq 0$ for $t \geq t_0$. Proceeding as in Theorem 5.3.1, we obtain inequality (5.3.6). From the hypotheses this inequality becomes

$$y'(t) \leq -\rho^\beta(t)q(t) - \frac{\beta^2}{4c} \frac{1}{t} \left(\frac{t\rho'(t)}{\rho(t)} \right)^2 - \frac{c}{t}y^2(t) \quad \text{for} \quad t \geq t_1$$

and now by Remark 5.2.2, we find

$$y'(t) \leq -\rho^\beta(t)q(t) - \frac{\beta^2}{c} \frac{1}{t} - \frac{c}{t}y^2(t) \quad \text{for} \quad t \geq t_1.$$

The rest of the proof is similar to that of Theorem 5.3.1. ∎

Theorem 5.3.1 can be extended to more general equations of the type (5.1.3). In fact, we have the following result.

Theorem 5.3.2. Let condition (F) hold, and let $\rho(t) \in C^2([t_0, \infty), \mathbb{R}^+)$, $a(t) \in C^1([t_0, \infty), \mathbb{R}^+)$ and $p(t) \in C^1([t_0, \infty), \mathbb{R}_0)$ so that condition (5.2.14) be satisfied with the functions $g(t)$ and $h(t)$ as in (5.2.9) and (5.2.10) respectively, $(\rho^\beta(t)p(t)/a(t))' \leq 0$ for $t \geq t_0$ and

$$\liminf_{t \to \infty} \frac{1}{t} \int_{t_0}^t \int_{t_0}^s \frac{\rho^\beta(u)}{a(u)}q(u)\,du\,ds > -\infty. \tag{5.3.7}$$

If for every constant $k > 0$, the second order linear equation

$$(tv'(t))' + \left[k\frac{\rho^\beta(t)}{a(t)}q(t) + \frac{1}{2}(tP(t))' - \frac{1}{4}tP^2(t) \right] v(t) = 0 \tag{5.3.8}$$

is oscillatory, where

$$P(t) = \frac{a(t)}{\rho^\beta(t)} \left(\frac{\rho^\beta(t)}{a(t)} \right)' - \frac{p(t)}{a(t)}, \tag{5.3.9}$$

then equation (5.1.3) is oscillatory.

Proof. Let $x(t)$ be a nonoscillatory solution of equation (5.1.3), say, $x(t) \neq 0$ for $t \geq t_0$ and let $w(t) = \rho^\beta(t)F(x(t))$. Then as in Theorem 5.2.5, we obtain (5.3.3). Now, it follows that

$$\frac{\rho^\beta(t)}{a(t)}\frac{a(t)x'(t)}{f(x(t))} = w'(t) - \beta\frac{\rho'(t)}{\rho(t)}w(t) = z(t), \quad t \geq t_0$$

and

$$z'(t) = -\frac{\rho^\beta(t)}{a(t)}q(t) + \left[\frac{a(t)}{\rho^\beta(t)}\left(\frac{\rho^\beta(t)}{a(t)}\right)' - \frac{p(t)}{a(t)}\right]z(t) - \frac{1}{w(t)}z^2(t)f'(x(t))F(x(t)).$$
(5.3.10)

From condition (F) and (5.3.3) there exist a $t_1 \geq t_0$ and a constant $c_1 > 0$ such that (5.3.5) is satisfied. Using (5.3.5) in equation (5.3.10), we find

$$z'(t) \leq -\frac{\rho^\beta(t)}{a(t)}q(t) + P(t)z - c\frac{1}{t}z^2(t) \quad \text{for} \quad t \geq t_1,$$

where

$$c = \frac{\beta}{c_1(1-\beta)} \quad \text{and} \quad P(t) = \frac{a(t)}{\rho^\beta(t)}\left(\frac{\rho^\beta(t)}{a(t)}\right)' - \frac{p(t)}{a(t)}, \quad t \geq t_0.$$

Thus, for all $t \geq t_1$, we have

$$\left(z(t) - \frac{1}{2c}tP(t)\right)' \leq -\frac{\rho^\beta(t)}{a(t)}q(t) - \frac{1}{2c}(tP(t))' + \frac{1}{4c}tP^2(t)$$
$$-\frac{c}{t}\left[z(t) - \frac{1}{2c}tP(t)\right]^2, \quad t \geq t_1$$

or

$$y'(t) \leq -\left[\frac{\rho^\beta(t)}{a(t)}q(t) + \frac{1}{2c}(tP(t))' - \frac{1}{4c}tP^2(t)\right] - \frac{c}{t}y^2(t), \quad t \geq t_1$$

where $y(t) = z(t) - (1/2c)tP(t)$, $t \geq t_1$. The rest of the proof is similar to that of Theorem 5.3.1 and hence omitted. ∎

Next we present the following linearization result.

Theorem 5.3.3. Let conditions (5.1.5) and (5.2.86) hold, and let $\rho(t) \in C^1([t_0, \infty), \mathbb{R}^+)$ so that (5.2.87) and (5.2.89) be satisfied. If for every constant $k > 0$, the second order linear equation

$$(a(t)\rho(t)\xi(t)v'(t))' + k\rho(t)q(t)v(t) = 0$$
(5.3.11)

is oscillatory, then equation (5.1.3) is oscillatory.

Proof. Let $x(t)$ be a nonoscillatory solution of equation (5.1.3), say, $x(t) > 0$ for $t \geq t_0$, and let $w(t) = \int_0^{x(t)} du/f(u)$, $t \geq t_0$. Then as in Theorem 5.2.11, we find that (5.1.3) is reduced to equation (5.2.90). Next, we consider the cases I and II as in Theorem 5.2.11. The proof of case II is exactly the same and hence omitted. For case I, we proceed as in Theorem 5.2.11 to obtain (5.2.91) and (5.2.92). Now, as in Theorem 5.2.12, we let $z(t) = a(t)\rho(t)w'(t)$ so that equation (5.2.90) takes the form

$$z'(t) = -\rho(t)q(t) - \frac{1}{a(t)\rho(t)} z^2(t)f'(x(t)),$$

or

$$y'(t) = \rho(t)q(t) + \frac{1}{a(t)\rho(t)} y^2(t)f'(x(t)), \quad t \geq t_0$$

where $y(t) = -z(t)$ for $t \geq t_0$. If we let $c(t) = 1/f'(x(t))$, $t \geq t_0$ then the Riccati equation $y'(t) = \rho(t)q(t) + (c(t)y^2(t)/a(t)\rho(t))$ has a solution on $[t_0, \infty)$. It is well-known that this is equivalent to the nonoscillation of the second order linear differential equation

$$\left(\frac{a(t)\rho(t)}{c(t)} v'(t) \right)' + \rho(t)q(t)v(t) = 0. \tag{5.3.12}$$

Now, from (5.2.91) and (5.2.92) there exist a $t_1 \geq t_0$ and a constant $c_1 > 0$ such that $f'(x(t))w(t) \geq \alpha$ and $w(t) \leq c_1\xi(t)$ for $t \geq t_1$. Thus, it follows that $c(t) \geq \alpha/(c_1\xi(t)) = c_2/\xi(t)$ for $t \geq t_1$, where $c_2 = \alpha/c_1 > 0$. Finally, applying the Picone–Sturm theorem to the equation

$$\left(\frac{1}{c_2} a(t)\rho(t)\xi(t)v'(t) \right)' + \rho(t)q(t)v(t) = 0$$

we conclude that it is nonoscillatory, contradicting the fact that (5.3.11) is oscillatory. This completes the proof. ∎

The following result is immediate.

Corollary 5.3.2. Let conditions (F) and (5.2.99) hold. If for every constant $k > 0$, the second order linear equation $(ty'(t))' + kq(t)y(t) = 0$ is oscillatory, then equation (5.1.1) is oscillatory.

5.4. Nonoscillation Theorems for Sublinear Differential Equations

Here, we shall discuss the nonoscillatory behavior of the Emden–Fowler equation (5.1.6) with $0 < \gamma < 1$. It is well-known that (5.1.6) has a nonoscillatory solution if $q(t)$ satisfies

$$q(t) \geq 0 \quad \text{for} \quad t \geq t_0 \quad \text{and} \quad \int_{t_0}^{\infty} s^{\gamma} q(s) ds < \infty. \tag{5.4.1}$$

Condition (5.4.1) in particular implies that $\lim_{t \to \infty} \int_{t_0}^{t} q(s) ds$ exists and is finite when $q(t)$ is nonnegative. The following theorem extends this result to the case when $q(t) \in C([t_0, \infty), \mathbb{R})$.

Theorem 5.4.1. Suppose that

$$Q(t) = \int_{t}^{\infty} q(s) ds \quad \text{exists for all} \quad t \geq t_0. \tag{5.4.2}$$

If there exists a function $F(t) \in C^1([t_0, \infty), \mathbb{R}_0)$ such that

$$|Q(t)| \leq F(t) \quad \text{for all large} \quad t, \tag{5.4.3}$$

where $F(t) = O(t^{-\gamma})$ as $t \to \infty$, and

$$\int_{t_0}^{\infty} s^{\gamma} |F'(s)| ds < \infty, \tag{5.4.4}$$

then equation (5.1.6) has a nonoscillatory solution.

Proof. Let $x_m(t)$ be a solution of equation (5.1.6) satisfying $x_m(1) = 0$, $x_m'(1) = m$ where m is a positive number. We claim that when m is large enough, $x_m'(t) > 0$ for all $t > 1$ and so $x(t)$ is nonoscillatory. Suppose that $x_m'(t) = 0$ for some $t > 1$. Let ξ_1 be the smallest of such t. Let ξ_2 be the smallest of all those t for which $x_m'(t) = 2m$. (If no such t exists, let $\xi_2 = \infty$). Finally, let $\xi = \min\{\xi_1, \xi_2\}$. Then on $[1, \xi]$, $0 < x_m'(t) < 2m$. Thus, it follows that

$$0 < x_m(t) < 2mt, \quad t \in [1, \xi]. \tag{5.4.5}$$

At $t = \xi$, we have either

$$x_m'(\xi) = 0 \quad (\text{if } \xi = \xi_1) \quad \text{or} \quad x_m'(\xi) = 2m \quad (\text{if } \xi = \xi_2). \tag{5.4.6}$$

Integrating (5.1.6) from 1 to $t \in [1, \xi]$, we find

$$x_m'(t) = m - \int_{1}^{t} q(s) x^{\gamma}(s) ds. \tag{5.4.7}$$

We now proceed to estimate the integral in (5.4.7) as follows

$$\left| \int_{1}^{t} q(s) x_m^{\gamma}(s) ds \right| = \left| [Q(1) - Q(t)] x_m^{\gamma}(t) + \int_{1}^{t} [Q(s) - Q(1)] (x_m^{\gamma}(s))' ds \right|$$

$$\leq [2|Q(1)| + |Q(t)|] x_m^{\gamma}(t) + \int_{1}^{t} |Q(s)| (x_m^{\gamma}(s))' ds \tag{5.4.8}$$

since $x_m(t)$, $x_m'(t) > 0$ on $[1, \xi)$.

Next we integrate the last integral in (5.4.8), to obtain

$$\left| \int_1^t Q(s)(x_m^\gamma(s))'ds \right| \leq \int_1^t F(s)(x^\gamma(s))'ds$$

$$\leq F(t)x_m^\gamma(t) + \int_1^t |F'(s)|x_m^\gamma(s)ds. \qquad (5.4.9)$$

Since $Q(t) \to 0$ as $t \to \infty$, $Q(t)$ is bounded on $[1, \xi)$, i.e., $|Q(t)| \leq k_1$, $t \in [1, \xi)$ where k is a constant. By assumption there exists a constant k_2 such that $|t^\gamma F(t)| \leq k_2$, $t \in [1, \xi)$. For $t \in [1, \xi)$, we also have from (5.4.5),

$$\int_1^t |F'(s)|x_m^\gamma(s)ds \leq (2m)^\gamma \int_1^t s^\gamma |F'(s)|ds \leq k_3(2m)^\gamma, \qquad (5.4.10)$$

for some constant $k_3 > 0$. Using (5.4.9) and (5.4.10) in (5.4.8), we find

$$\left| \int_1^t q(s)x_m^\gamma(s)ds \right| \leq [3k_1 + k_2 + k_3](2m)^\gamma = K(2m)^\gamma. \qquad (5.4.11)$$

Using (5.4.11) in (5.4.7), we obtain

$$m - K(2m)^\gamma \leq x_m'(t) \leq m + K(2m)^\gamma \quad \text{for all} \quad t \in [1, \xi].$$

For $m > (2^\gamma K)^{1/(1-\gamma)}$, we have in particular $0 < x_m'(\xi) < 2m$. This contradicts (5.4.6). ∎

Example 5.4.1. Consider the equation

$$x''(t) + (t^\lambda \sin t)|x(t)|^\gamma \operatorname{sgn} x(t) = 0, \quad 0 < \gamma < 1, \quad t \geq t_0 > 0 \qquad (5.4.12)$$

where $\lambda < -\gamma$. We find that $\left| \int_t^\infty q(s)ds \right| \leq kt^\lambda$, $t \geq 1$ where k is a positive constant. Let $F(t) = ct^\lambda$, where c is a positive constant. All the hypotheses of Theorem 5.4.1 are satisfied, and hence equation (5.4.12) is nonoscillatory.

Example 5.4.2. Consider the equation

$$x''(t) + \left[\frac{1}{t^\gamma}(\ln t)^\mu \sin t \right] |x(t)|^\gamma \operatorname{sgn} x(t) = 0, \quad 0 < \gamma < 1, \quad t \geq t_0 > 0 \qquad (5.4.13)$$

where $\mu \leq -2$. We find that $F(t)$ can be taken to be a multiple of $(\ln t)^\mu/t$. If $F(t) \in C^1([t_0, \infty), \mathbb{R}_0)$ and $F'(t) \geq 0$ for $t \geq t_0$ such that

$$\int_1^\infty s^{\gamma-1}F(s)ds < \infty, \qquad (5.4.14)$$

then $F(t)$ satisfies the hypotheses of Theorem 5.4.1, i.e., $F(t) = O(t^{-\gamma})$ as $t \to \infty$, and (5.4.4) holds. To see this we apply integration by parts, to obtain

$$\frac{1}{\gamma} t^\gamma F(t) + \int_1^t \frac{1}{\gamma} s^\gamma [-F'(s)] ds = \int_1^t s^{\gamma-1} F(s) ds + \frac{1}{\gamma} F(1). \quad (5.4.15)$$

Since the right–hand side of (5.4.15) is bounded, by (5.4.14) each of the terms on the left–hand side is also bounded for all t. It can be shown by continuity argument that the theorem still holds if $F(t)$ satisfies (5.4.14) but no continuity requirement is assumed on F.

Next, we shall consider equation (5.1.6), where $q(t) \in C([t_0, \infty), \mathbb{R}^+)$ and is locally of bounded variation on $[t_0, \infty)$. Thus, $q(t)$ admits a Jordan decomposition $q(t) = q^+(t) - q^-(t)$, where $q^+(t)$ and $q^-(t)$ are continuous nondecreasing functions.

A number of nonoscillatory results for the nonlinear equation (5.1.6), $\gamma \neq 1$ require some sort of restriction on the growth of the function $q(t)$, typically that $q(t)$ be nonincreasing. Other results demand the weaker condition

$$\int_{t_0}^\infty \frac{dq^+(s)}{q(s)} < \infty. \quad (5.4.16)$$

One such result is the following.

Theorem 5.4.2.   Let condition (5.4.16) hold. Then equation (5.1.6) is nonoscillatory provided any one of the following is satisfied.

(I_1) $\displaystyle \int^\infty sq(s) ds < \infty$,

(I_2) $\displaystyle \lim_{t \to \infty} q^{(\gamma-1)/2}(t) \int_t^\infty s^\gamma q(s) ds = 0$,

(I_3) $\displaystyle \lim_{t \to \infty} q^{(\gamma-1)/(2(\gamma+1))}(t) \int_t^\infty q^{(\gamma+1)/\gamma}(s) ds = 0$,

(I_4) $\displaystyle \lim_{t \to \infty} t^2 q(t) = 0$,

(I_5) $\displaystyle \lim_{t \to \infty} t \int_t^\infty q(s) ds = 0$.

The equation

$$x''(t) + \frac{1}{t^2 (\ln t)^\gamma} |x(t)|^\gamma \operatorname{sgn} x(t) = 0, \quad 0 < \gamma < 1 \quad (5.4.17)$$

where $t > t_0 = 1$ has a nonoscillatory solution $x(t) = \ln t$. Clearly, conditions (I_2) – (I_5) are satisfied, but (I_1) fails. Motivated by this example, we expect equivalence between (I_i), $i = 2, 3, 4, 5$. To prove this we need the following lemma.

Lemma 5.4.1. If $q(t) \in C([t_0, \infty), \mathbb{R}^+)$ and is of bounded variation on $[t_0, \infty)$, then it admits the decomposition

$$q(t) = q_1(t)q_2(t) \quad \text{for} \quad t \geq t_0, \tag{5.4.18}$$

where $q_1(t)$ is a positive nonincreasing function, and $q_2(t)$ is a positive nondecreasing function. Furthermore, if $q(t)$ satisfies (5.4.16), then $0 < k_1 < q_2(t) \leq k_2 < \infty$ for all t, where k_1 and k_2 are fixed positive constants.

Proof. This follows from the identity

$$q(t) = q(t_0) \exp\left(-\int_{t_0}^t \frac{dq^-(s)}{q(s)}\right) \exp\left(\int_{t_0}^t \frac{dq^+(s)}{q(s)}\right).$$

Here, $q_2(t) = \exp\left(\int_{t_0}^t dq^+(s)/q(s)\right)$ tends to a finite limit as $t \to \infty$, if (5.4.16) holds. ∎

Remark 5.4.1. Suppose condition (5.4.16) holds. Using the fact that $q_2(t)$ is bounded from below and above by positive constants, we find that $q(t)$ satisfies any one of the conditions (I_i), $i = 2, 3, 4, 5$ if and only if $q_1(t)$ satisfies the same condition. Hence, in the proofs of some of the results of this section, we can assume without loss of generality that $q(t)$ itself is nonincreasing.

Now, we are in the position to prove the following result.

Theorem 5.4.3. Under the growth assumption (5.4.16) (or more generally (5.4.18)) on the function $q(t)$, the following chain of implications holds

$$(I_1) \Rightarrow (I_2) \Leftrightarrow (I_3) \Leftrightarrow (I_4) \Leftrightarrow (I_5). \tag{5.4.19}$$

Proof. We shall prove that $(I_4) \Leftrightarrow (I_5)$, $(I_2) \Leftrightarrow (I_5)$, $(I_5) \Rightarrow (I_3)$ and $(I_3) \Rightarrow (I_4)$. The case $(I_1) \Rightarrow (I_4)$ is left to the reader.

Case 1. $(I_4) \Leftrightarrow (I_5)$. The implication $(I_4) \Rightarrow (I_5)$ follows trivially from an application of L'Hopital's rule, i.e.,

$$0 = \lim_{t \to \infty} \frac{\int_t^\infty q(s)ds}{1/t} = \lim_{t \to \infty} \frac{-q(t)}{-1/t^2} = \lim_{t \to \infty} t^2 q(t).$$

Now, suppose that (I_5) holds, but (I_4) does not hold. Then there exists a sequence $\{t_n\}$, $\lim_{n \to \infty} t_n = \infty$ such that

$$t_n^2 q(t_n) \geq \delta > 0 \quad \text{for some constant} \quad \delta. \tag{5.4.20}$$

By (I_5), if t_n is large enough, then

$$\frac{1}{2}t_n \int_{t_n/2}^{\infty} q(s)ds \; < \; \frac{\delta}{4}. \tag{5.4.21}$$

On the other hand, since (5.4.20) holds and $q(t)$ is nonincreasing $q(t) \geq \delta t_n^{-2}$ for $t \in [t_n/2, t_n]$. Integration of this inequality over $[t_n/2, t_n]$ yields $(t_n/2)\int_{t_n/2}^{t_n} q(s)ds \geq \delta/4$, which contradicts (5.4.21).

Case 2. $(I_2) \Leftrightarrow (I_5)$. We shall first show that $(I_2) \Rightarrow (I_5)$. Given any $\epsilon > 0$, there exists a $t_1 \geq t_0$ such that

$$\left(\int_t^{\infty} s^\gamma q(s)ds \right)^{2/(1-\gamma)} \; \leq \; \epsilon^{2/(1-\gamma)} q(t) \quad \text{for} \quad t \geq t_1. \tag{5.4.22}$$

Letting $u(t) = \int_t^{\infty} s^\gamma q(s)ds$, we obtain from (5.4.22),

$$t^\gamma \; \leq \; -\epsilon^{2/(1-\gamma)} u^{-2/(1-\gamma)}(t)u'(t), \quad t \geq t_1.$$

Integrating this inequality from t_1 to $t \geq t_1$, we find

$$u(t) \; \leq \; \left[\frac{\epsilon^{-2/(1-\gamma)}}{(1-\gamma)}t^{1+\gamma} - \frac{1}{u^{(1+\gamma)/(1-\gamma)}(t_1)} \right]^{(\gamma-1)/(\gamma+1)} \tag{5.4.23}$$

and hence

$$\int_t^{\infty} s^\gamma q(s)ds \; = \; u(t) \; = \; O\left(\epsilon^{2/(1+\gamma)}t^{\gamma-1} \right) \quad \text{as} \quad t \to \infty. \tag{5.4.24}$$

Now, an integration by parts from t to $T \geq t \geq t_1$ gives

$$\int_t^T q(s)ds \; = \; \int_t^T s^\gamma q(s) \left(\frac{1}{s^\gamma} \right) ds$$

$$= \; \frac{1}{t^\gamma} \int_t^T s^\gamma q(s)ds - \gamma \int_t^T \frac{1}{s^{1+\gamma}} \int_s^T \xi^\gamma q(\xi)d\xi ds. \tag{5.4.25}$$

The fact that $\int_t^T s^\gamma q(s)ds \leq \int_t^{\infty} s^\gamma q(s)ds < \infty$, $t \geq t_1$ together with (5.4.25) shows that $\int_t^{\infty} q(s)ds$ converges. Using (5.4.24) and letting $T \to \infty$ in (5.4.25), we obtain

$$\int_t^{\infty} q(s)ds \; = \; O\left(\epsilon^{2/(1+\gamma)}t^{-1} \right) \quad \text{as} \quad t \to \infty. \tag{5.4.26}$$

Since $\epsilon > 0$ is arbitrary, (5.4.26) establishes the validity of (I_5).

Now suppose that (I_5) holds. Then for any $\epsilon > 0$ there exists $t_1 \geq t_0$ such that $\int_t^\infty q(s)ds \leq \epsilon/t$ for all $t \geq t_1$. An integration by parts from t to $T \geq t \geq t_1$ yields the following identity

$$\int_t^T s^\gamma q(s)ds = t^\gamma \int_t^T q(s)ds + \gamma \int_t^T s^{\gamma-1}\left(\int_s^T q(\xi)d\xi\right)ds. \quad (5.4.27)$$

The fact that

$$\int_t^T q(s)ds \leq \int_t^\infty q(s)ds = o\left(\frac{1}{t}\right) \quad \text{as} \quad t \to \infty$$

implies that the right–hand side of (5.4.27), and hence the left–hand side, converges as $T \to \infty$. Thus, $\int_t^\infty s^\gamma q(s)ds < \infty$. Furthermore, for $t \geq t_1$,

$$\int_t^\infty s^\gamma q(s)ds \leq \frac{\epsilon}{1-\gamma}t^{\gamma-1} = \left(\frac{\epsilon_1}{t}\right)^{1-\gamma}, \quad (5.4.28)$$

where $\epsilon_1^{1-\gamma} = \epsilon/(1-\gamma)$. Now, for any $t \geq t_1$, we need to consider the two mutually exclusive subcases (i). $q(t) \geq \epsilon_1/t^2$ and (ii). $q(t) < \epsilon_1/t^2$. In the subcase (i), estimate (5.4.27) gives

$$q^{(\gamma-1)/2}(t)\int_t^\infty s^\gamma q(s)ds \leq \left(\frac{\epsilon_1}{t^2}\right)^{(\gamma-1)/2}\left(\frac{\epsilon_1}{t}\right)^{1-\gamma} = \epsilon_1^{(1-\gamma)/2}.$$

In the subcase (ii), we set $v = (\epsilon_1/q(t))^{1/2} > t$, $t \geq t_1$ and observe that $q(t)$ is non–increasing. Then it follows from (5.4.28) that

$$\begin{aligned}
q^{(\gamma-1)/2}(t)\int_t^\infty s^\gamma q(s)ds &= q^{(\gamma-1)/2}(t)\left[\int_t^v s^\gamma q(s)ds + \int_v^\infty s^\gamma q(s)ds\right]\\
&\leq q^{(\gamma-1)/2}(t)\left[\int_{t_0}^v s^\gamma q(s)ds + \left(\frac{\epsilon_1}{v}\right)^{1-\gamma}\right]\\
&\leq \frac{\epsilon_1^{(\gamma+1)/2}}{\gamma+1} + \epsilon_1^{(1-\gamma)/2} = O(\sqrt{\epsilon}) \quad \text{as} \quad t \to \infty.
\end{aligned}$$

Since $\epsilon > 0$ is arbitrary, this establishes (I_2).

Case 3. (I_5) $\Rightarrow$ (I_3). We consider the identity for $0 < \gamma < \mu < 1$,

$$\int_t^T s^\mu q(s)ds = t^\mu \int_t^T q(s)ds + \mu \int_t^T s^{\mu-1}\int_s^T q(\xi)d\xi ds, \quad T \geq t \geq t_0. \quad (5.4.29)$$

Since $\int_t^\infty q(s)ds = o(1/t)$ as $t \to \infty$, from (5.4.29) it follows that

$$\int_t^\infty s^\mu q(s)ds = o(t^{\mu-1}). \quad (5.4.30)$$

By applying Hölder's inequality, we have

$$\int_t^\infty q^{1/(\gamma+1)}(s)ds \;\le\; \left(\int_t^\infty s^\mu q(s)ds\right)^{1/(\gamma+1)} \left(\int_t^\infty s^{-\mu/\gamma}ds\right)^{\gamma/(\gamma+1)}.$$

(5.4.31)

Using (5.4.30) in (5.4.31), we find as $t \to \infty$,

$$\int_t^\infty q^{1/(\gamma+1)}(s)ds \;\le\; o\left(t^{(\mu-1)/(\gamma+1)}\right)\left[\left(\frac{\gamma}{\mu-\gamma}\right)t^{(\gamma-\mu)/\gamma}\right]^{\gamma/(\gamma+1)}$$

$$= o\left(t^{(\gamma-1)/(\gamma+1)}\right).$$

(5.4.32)

Now, for any $\epsilon_2 > 0$ we can fix t_1 so that from (5.4.32) for all $t \ge t_1$, we have

$$\int_t^\infty q^{1/(\gamma+1)}(s)ds \;\le\; \epsilon_2^{1/(\gamma+1)}t^{(\gamma-1)/(\gamma+1)}. \tag{5.4.33}$$

For any given $t \ge t_0$, we again consider two mutually exclusive subcases (i). $q(t) \ge \epsilon_2/t^2$ and (ii). $q(t) < \epsilon_2/t^2$. In the subcase (i), from (5.4.33), we have the following estimate

$$q^{(\gamma-1)/(2(\gamma+1))}(t)\int_t^\infty q^{1/(\gamma+1)}(s)ds \;\le\; \epsilon_2^{1/2}. \tag{5.4.34}$$

In the subcase (ii), we set $v = (\epsilon_2/q(t))^{1/2} > t$. Since $q(t)$ is nonincreasing, we have

$$q^{(\gamma-1)/(2(\gamma+1))}(t)\int_t^\infty q^{1/(1+\gamma)}(s)ds \;\le\; q^{1/2}(t)(v-t) \;\le\; q^{1/2}(t)v \;=\; \epsilon_2^{1/2}.$$

(5.4.35)

On the other hand, using estimate (5.4.33), we find

$$q^{\frac{\gamma-1}{2(\gamma+1)}}(t)\int_v^\infty q^{1/(\gamma+1)}(s)ds \;\le\; q^{(\gamma-1)/(2(\gamma+1))}(t)\epsilon_2^{1/(1+\gamma)}v^{(\gamma-1)/(\gamma+1)}$$

$$= \epsilon_2^{1/(\gamma+1)}\epsilon_2^{(\gamma-1)/(2(\gamma+1))} \;=\; \epsilon_2^{1/2}.$$

(5.4.36)

Combining (5.4.35) and (5.4.36), we obtain

$$q^{(\gamma-1)/(2(\gamma+1))}(t)\int_t^\infty q^{1/(1+\gamma)}(s)ds \;\le\; 2\epsilon_2^{1/2}. \tag{5.4.37}$$

Since $\epsilon_2 > 0$ is arbitrary, (5.4.34) and (5.4.37) establish the desired assertion (I_3).

Case 4. $(I_3) \Rightarrow (I_4)$. Let $\alpha = 2/(1-\gamma) > 1$ and $1/\beta = \alpha - 1 = (\gamma+1)/(1-\gamma) > 1$. Choose $\epsilon^\alpha < 4^{-\alpha}(\alpha-1)$. By (I_3) there exists $t_1 \ge t_0$

such that for all $t \geq t_1$,

$$q^{(\gamma-1)/(2(\gamma+1))}(t) \int_t^\infty q^{1/(\gamma+1)}(s)ds \; < \; \epsilon. \tag{5.4.38}$$

Denote $f(t) = q^{\beta/2}(t)$, $t \geq t_1$, then $f^\alpha(t) = q^{1/(1+\gamma)}(t)$. Suppose (I_4) fails. Then there exist constants c and $\delta > 0$ such that $c \geq 2t_1$ and $q(c) \geq \delta/c^2$, or

$$f^\alpha(c) \; \geq \; \delta_1 c^{-2/(\gamma+1)}, \tag{5.4.39}$$

where

$$\delta_1 \; = \; \delta^{1/(\gamma+1)} \; > \; 0 \tag{5.4.40}$$

from which it follows with $d = 3c/4$ that

$$\int_\alpha^c f^\alpha(s)ds \; \geq \; f^\alpha(c)\left(\frac{c}{4}\right) \; \geq \; \frac{1}{4}\delta_1 c^{-\beta}. \tag{5.4.41}$$

Notice although the choice of c depends on ϵ, δ and δ_1 are independent of ϵ. We now rewrite (5.4.38) in terms of $f(t)$ as

$$f(t) \; > \; \frac{1}{\epsilon}\int_t^\infty f^\alpha(s)ds \; > \; \frac{1}{\epsilon}\left[\int_t^d f^\alpha(s)ds + \int_d^c f^\alpha(s)ds\right]$$

for $t_1 \leq c/2 \leq t \leq 3c/4$. Using (5.4.41) in the above estimate, we find

$$f(t) \; > \; \frac{1}{4\epsilon}\delta_1 c^{-\beta} + \frac{1}{\epsilon}\int_t^d f^\alpha(s)ds. \tag{5.4.42}$$

Now, we define $g(y) = f(t) = f(d-y)$, so that for $0 \leq y \leq c/4$, (5.4.42) becomes

$$g(y) \; > \; \frac{1}{4\epsilon}\delta_1 c^{-\beta} + \frac{1}{\epsilon}\int_0^y g^\alpha(u)du. \tag{5.4.43}$$

Comparing $g(y)$ with its minorant function $G(y)$ defined by

$$G(y) \; = \; \frac{1}{4\epsilon}\delta_1 c^{-\beta} + \frac{1}{\epsilon}\int_0^y G^\alpha(u)du \tag{5.4.44}$$

we find as an application of a standard result in the theory of integral inequalities that $g(y) \geq G(y)$ for $0 \leq y \leq c/4$. The function $G(y)$ can be found explicitly by solving (5.4.44) as

$$G(y) \; = \; \left[G^{1-\alpha}(0) + (1-\alpha)\frac{y}{\epsilon}\right]^{-\beta}, \tag{5.4.45}$$

where

$$G^{1-\alpha}(0) \; = \; \left(\frac{1}{4\epsilon}\delta_1 c^{-\beta}\right)^{-1/\beta} \; = \; \left(\frac{4\epsilon}{\delta_1}\right)^{1/\beta} c.$$

From (5.4.45) it follows that $G(y) \to \infty$ as $y \to (\epsilon/(\alpha-1))(4\epsilon/\delta)^{1/\beta}c$. Thus, unless

$$\frac{\epsilon}{\alpha-1}\left(\frac{4\epsilon}{\delta_1}\right)^{1/\beta} > \frac{1}{4}, \tag{5.4.46}$$

$G(x)$ will blow up on the finite interval $[0, c/4]$, and so will $g(y)$. But, this contradicts the fact that $f(t)$ is defined on $[c/2, 3c/4]$. Recall that δ_1 is determined only from the fact that (I_4) fails and is independent of the choice of ϵ. Hence, (5.4.46) must hold for all $\epsilon > 0$, an obvious impossibility. This completes the proof of $(I_3) \Rightarrow (I_4)$ and also that of Theorem 5.4.3. ∎

Proof of Theorem 5.4.2 (I_2). Let $x(t)$ be an oscillatory solution of equation (5.1.6) and $\{t_n\}$ be the sequence of consecutive zeros of $x(t)$. We consider the energy function

$$E(x(t)) = \frac{[x'(t)]^2}{q(t)} + \frac{2}{\gamma+1}x^{\gamma+1}(t), \quad t \geq t_0.$$

Clearly, for the solution $x(t)$ of (5.1.6), we have

$$\frac{d}{dt}E(x(t)) = -\frac{q'(t)}{q^2(t)}[x'(t)]^2 \geq 0 \quad \text{for} \quad t \geq t_0 \tag{5.4.47}$$

since $q'(t) \leq 0$ for $t \geq t_0$. From (5.4.47) it follows that $\{x'(t_n)/q^{1/2}(t_n)\}$ forms a nondecreasing sequence, and in particular

$$x'(t_n) \geq c_0 q^{1/2}(t_n), \tag{5.4.48}$$

where $c_0 = x'(t_0)/q^{1/2}(t_0)$. We let n to be sufficiently large, $t_n = \tau$, $t_{n+1} = \tau_1$ so that

$$q^{(\gamma-1)/2}(\tau)\int_\tau^\infty s^\gamma q(s)ds < c_0^{1-\gamma}. \tag{5.4.49}$$

Without loss of generality we assume that $x'(\tau) \geq 0$ so that $x(t) > 0$ in $[\tau, \tau_1]$.

Since $x(t)$ is concave in $[\tau, \tau_1]$, $x'(t) \leq x'(\tau)$ for $t \in [\tau, \tau_1]$. Thus, we have $x(t) \leq (t-\tau)x'(\tau) \leq tx'(\tau)$, $t \in [\tau, \tau_1]$. Using this inequality in equation (5.1.6), we find $-x''(t) \leq t^\gamma q(t)[x'(\tau)]^\gamma$, $t \in [\tau, \tau_1]$. Integration of this inequality from τ to $t \in [\tau, \tau_1]$ yields $x'(\tau) - x'(t) \leq [x'(\tau)]^\gamma \int_\tau^t s^\gamma q(s)ds$. In particular, when $t = \tau_1$, we obtain

$$x'(\tau) \leq [x'(\tau)]^\gamma\int_\tau^{\tau_1} s^\gamma q(s)ds \leq [x'(\tau)]^\gamma\int_\tau^\infty s^\gamma q(s)ds.$$

Thus, it follows that

$$[x'(\tau)]^{1-\gamma} \leq \int_\tau^\infty s^\gamma q(s)ds. \tag{5.4.50}$$

Using (5.4.48) in inequality (5.4.50), we get $c_0^{1-\gamma} q^{(1-\gamma)/2}(\tau) \leq \int_\tau^\infty s^\gamma q(s)ds$, which contradicts (5.4.49). This completes the proof. ∎

In case condition (5.4.16) does not hold, we have the following extension of criterion (I_2).

Theorem 5.4.4. Let $q_1(t) = \exp\left(-\int_{t_0}^t dq^-(s)/q(s)\right)$. If

$$\lim_{t\to\infty} q_1^{(\gamma-1)/2}(t) \int_t^\infty s^\gamma q(s)ds = 0, \tag{5.4.51}$$

then equation (5.1.6) is nonoscillatory.

Proof. Let $x(t)$ be an oscillatory solution of equation (5.1.6) and let $\{t_n\}$ be the sequence of consecutive zeros of $x(t)$. Consider the energy function

$$E(x(t)) = \frac{[x'(t)]^2}{q(t)} + \frac{2}{\gamma+1}x^{\gamma+1}(t), \quad t \geq t_0$$

so that

$$\frac{d}{dt}E(x(t)) = -\frac{dq(t)}{q^2(t)}[x'(t)]^2 \geq -\frac{dq^+(t)}{q(t)}E(x(t)).$$

Now, it follows that

$$E(x(t)) \geq E(x(t_0)) \exp\left(-\int_{t_0}^t \frac{dq^+(s)}{q(s)}\right) = \frac{E(x(t_0))}{q_2(t)}, \tag{5.4.52}$$

where $q_2(t)$ is as in Lemma 5.4.1.

Let n be sufficiently large, $t_n = \tau$, $t_{n+1} = \tau_1$ so that

$$q_1^{(\gamma-1)/2}(\tau) \int_\tau^\infty s^\gamma q(s)ds < [E(x(t_0))]^{(1-\gamma)/2}. \tag{5.4.53}$$

Without loss of generality we can assume that $x'(\tau) \geq 0$ so that $x(t) > 0$, $t \in [\tau, \tau_1]$. Now, in view of (5.4.52), we have $[x'(t)]^2/q(t) = E(x(\tau)) \geq E(x(t_0))/q_2(\tau)$. From this inequality and Lemma 5.4.1, we find

$$[x'(\tau)]^2 \geq E(x(t_0))\frac{q(\tau)}{q_2(\tau)} = q_1(\tau)E(x(t_0)). \tag{5.4.54}$$

Proceeding as in Theorem 5.4.2 (I_2), we obtain (5.4.50). Using (5.4.54) in (5.4.50), we get $[E(x(t_0))]^{(1-\gamma)/2}q_1^{(1-\gamma)/2}(\tau) \leq \int_\tau^\infty s^\gamma q(s)ds$, which contradicts (5.4.53). This completes the proof. ∎

5.5. Oscillation Criteria for Certain Nonlinear Differential Equations

In this section we shall consider the nonlinear differential equation

$$\left(a(t)|x'(t)|^{\alpha-1}x'(t)\right)' + q(t)f(x(t)) = 0, \qquad (5.5.1)$$

where

(i) α is a positive constant,

(ii) $a(t) \in C([t_0,\infty), \mathbb{R}^+)$, $q(t) \in C([t_0,\infty), \mathbb{R})$,

(iii) $f \in C(\mathbb{R},\mathbb{R})$, $xf(x) > 0$ and $f'(x) \geq 0$ for $x \neq 0$.

In what follows we shall denote by

$$\eta(t) = \int_{t_0}^t a^{-1/\alpha}(s)ds \quad \text{and assume that} \quad \lim_{t\to\infty} \eta(t) = \infty. \quad (5.5.2)$$

We shall relate the oscillation results for (5.5.1) to those of half–linear differential equations discussed in Chapter 3. We shall also present some sufficient conditions as well as necessary and sufficient conditions for the oscillation of (5.5.1) and/or related equations.

We begin with a special case of (5.5.1), namely, the equation

$$\left(a(t)|x'(t)|^{\alpha-1}x'(t)\right)' + q(t)|x(t)|^{\beta-1}x(t) = 0, \qquad (5.5.3)$$

where β is a positive constant.

Theorem 5.5.1. Let $\beta < \alpha$ and condition (5.5.2) hold, and

$$\liminf_{t\to\infty} \int_{t_0}^t q(s)ds > -\infty. \qquad (5.5.4)$$

If for every constant $k > 0$, the half–linear differential equation

$$\left(a(t)\eta^\alpha(t)|y'(t)|^{\alpha-1}y'(t)\right)' + kq(t)|y(t)|^{\alpha-1}y(t) = 0 \qquad (5.5.5)$$

is oscillatory, then (5.5.3) is oscillatory.

Proof. Let $x(t)$ be a nonoscillatory solution of equation (5.5.3), say, $x(t) > 0$ for $t \geq t_0$. Define $w(t) = a(t)|x'(t)|^{\alpha-1}x'(t)/x^\beta(t)$ for $t \geq t_0$.

Then for $t \geq t_0$, we have

$$
\begin{aligned}
w'(t) &= -q(t) - \beta a(t)\frac{|x'(t)|^{\alpha+1}}{x^{\beta+1}(t)} \\
&= -q(t) - \beta a^{-1/\alpha}(t)|w(t)|^{(\alpha+1)/\alpha}\left(x^{(\beta-\alpha)/\alpha}(t)\right)
\end{aligned}
\tag{5.5.6}
$$

and hence

$$
w(t) = w(t_0) - \int_{t_0}^{t} q(s)ds - \beta \int_{t_0}^{t} a(s)\frac{|x'(s)|^{\alpha+1}}{x^{\beta+1}(s)}ds.
\tag{5.5.7}
$$

We distinguish two mutually exclusive cases where the integral $\mathcal{I}(t) = \int_{t_0}^{t} a(s)|x'(s)|^{\alpha+1}/x^{\beta+1}(s)ds$ remains finite or tends to infinity as $t \to \infty$, and arrive in each case at a contradiction.

Case 1. $\mathcal{I}(\infty) < \infty$. In this case there exists a positive constant c_1 such that $\mathcal{I}(t) \leq c_1$ for $t \geq t_0$. For $t \geq t_0$, we use Hölder's inequality, to obtain

$$
\begin{aligned}
x^{1-\gamma}(t) - x^{1-\gamma}(t_0) &= (\gamma - 1)\int_{t_0}^{t}\left|\frac{x'(s)}{x^{\gamma}(s)}\right|ds \\
&\leq (\gamma - 1)[\mathcal{I}(t)]^{1/(\alpha+1)}\left[\int_{t_0}^{t} a^{-1/\alpha}(s)ds\right]^{\alpha/(\alpha+1)} \leq c_2\eta^{\alpha/(\alpha+1)}(t),
\end{aligned}
$$

where $\gamma = (\beta+1)/(\alpha+1) < 1$ and $c_2 = (\gamma-1)c_1^{1/(\alpha+1)}$. By condition (5.5.2) there exist a constant $c_3 > 0$ and a $T \geq t_0$ such that $x^{1-\gamma}(t) \leq c_3\eta^{\alpha/(\alpha+1)}(t)$, or $x(t) \leq \left(c_3\eta^{\alpha/(\alpha+1)}(t)\right)^{1/(1-\gamma)}$ for $t \geq T$ and

$$
x^{(\beta-\alpha)/\alpha}(t) \geq \frac{c_4}{\eta(t)} \quad \text{for} \quad t \geq T,
\tag{5.5.8}
$$

where $c_4 = (c_3)^{(\beta-\alpha)/(\alpha(1-\gamma))}$. Using (5.5.8) in equation (5.5.6), we get

$$
w'(t) + q(t) + \left(\frac{c_4\beta}{\alpha}\right)\alpha\left(\frac{1}{a(t)\eta^{\alpha}(t)}\right)^{1/\alpha}|w(t)|^{(\alpha+1)/\alpha} \leq 0 \quad \text{for} \quad t \geq T.
$$

Now, applying Lemma 4.4.4 we find that (5.5.5) is nonoscillatory, which is a contradiction.

Case 2. $\mathcal{I}(\infty) = \infty$. Following exactly as in Theorem 4.4.3 Case 2, we arrive at the contradiction $\lim_{t\to\infty} x(t) = -\infty$. This completes the proof. ∎

To extend Theorem 5.5.1 to general equations of type (5.5.1), we need the following conditions

$$
\int_{+0}\frac{1}{f^{1/\alpha}(u)}du < \infty \quad \text{and} \quad \int_{-0}\frac{1}{f^{1/\alpha}(u)}du < \infty
\tag{5.5.9}
$$

and

$$\min\left\{\inf_{x>0}\frac{f'(x)}{f^{\frac{\alpha-1}{\alpha}}(x)}\int_{+0}^{x}\frac{1}{f^{1/\alpha}(u)}du,\ \inf_{x<0}\frac{f'(x)}{f^{\frac{\alpha-1}{\alpha}}(x)}\int_{-0}^{x}\frac{1}{f^{1/\alpha}(u)}du\right\}>0.$$
(5.5.10)

Remark 5.5.1. Conditions (5.5.9) and (5.5.10) when $\alpha=1$ reduce to (5.1.5) and (5.2.86). Also, we can replace condition (5.5.10) by

$$\left(\frac{f'(x)}{f^{(\alpha-1)/\alpha}(x)}\right)\int_{0}^{x}\frac{1}{f^{1/\alpha}(u)}du>\frac{1}{c}>0\quad\text{for all}\quad x,\qquad(5.5.11)$$

where c is a constant.

Now we shall prove the following theorem.

Theorem 5.5.2. Let conditions (5.5.2), (5.5.4), (5.5.9) and (5.5.10) hold. If for every constant $k>0$, the half–linear differential equation (5.5.5) is oscillatory, then (5.5.1) is oscillatory.

Proof. Let $x(t)$ be a nonoscillatory solution of equation (5.5.1), say, $x(t)>0$ for $t\geq t_0$. Define $w(t)=a(t)|x'(t)|^{\alpha-1}x'(t)/f(x(t))$, $t\geq t_0$. Then for $t\geq t_0$, we have

$$w'(t)=-q(t)-a(t)\frac{|x'(t)|^{\alpha+1}}{f^2(x(t))}f'(x(t)),$$

or

$$w'(t)=-q(t)-a^{-1/\alpha}(t)|w(t)|^{(\alpha+1)/\alpha}\left(\frac{f'(x(t))}{f^{(\alpha-1)/\alpha}(x(t))}\right),\qquad(5.5.12)$$

and hence

$$w(t)=w(t_0)-\int_{t_0}^{t}q(s)ds-\int_{t_0}^{t}a(s)\frac{|x'(s)|^{\alpha+1}}{f^2(x(s))}f'(x(s))ds.$$

Now we need to consider two cases, when

$$\mathcal{I}(t_0,t)=\int_{t_0}^{t}a(s)\frac{|x'(s)|^{\alpha+1}}{f^2(x(s))}f'(x(s))ds$$

remains finite or tends to infinity as $t\to\infty$.

Case 1. $\mathcal{I}(t_0,\infty)<\infty$. In view of (5.5.10), there exist a constant $c>0$ such that

$$\left(\frac{f'(x(t))}{f^{(\alpha-1)/\alpha}(x(t))}\right)\int_{+0}^{x(t)}\frac{1}{f^{1/\alpha}(u)}du\geq c\quad\text{for}\quad t\geq t_0.\qquad(5.5.13)$$

We will show that

$$\lim_{t\to\infty} \frac{v(t)}{\eta(t)} = 0 \tag{5.5.14}$$

where $v(t) = \int_{+0}^{x(t)} du/f^{1/\alpha}(u)$. For this, let $\epsilon > 0$ be an arbitrary number. Then, we can choose a $t_1 \geq t_0$ such that

$$\mathcal{I}(t_1,\infty) \leq \left(\frac{\epsilon c}{2}\right)^{\alpha+1}. \tag{5.5.15}$$

By Hölder's inequality, for every $t \geq t_1$, we find

$$v(t) - v(t_1) \leq \left| \int_{t_1}^t \frac{x'(s)}{f^{1/\alpha}(x(s))} ds \right| \leq [\mathcal{I}(t_1,t)]^{1/(\alpha+1)} [\mathcal{J}(t_1,t)]^{\alpha/(\alpha+1)},$$

where

$$\mathcal{J}(t_1,t) = \int_{t_1}^t a^{-1/\alpha}(s) \left(\frac{f^{(\alpha-1)/\alpha} x(s))}{f'(x(s))} \right)^{1/\alpha} ds,$$

and consequently, by using (5.5.15), we get

$$v(t) \leq v(t_1) + \frac{\epsilon c}{2} [\mathcal{J}(t_1,t)]^{\alpha/(\alpha+1)}, \quad t \geq t_1. \tag{5.5.16}$$

If $\mathcal{J}(t_1,\infty) < \infty$, (5.5.16) ensures that $v(t)$ is bounded and hence (5.5.14) holds. So, we restrict our attention to the case when $\mathcal{J}(t_1,\infty) = \infty$. Let $t_2 > t_1$ be such that $v(t_1) \leq (c\epsilon/2)[\mathcal{J}(t_1,t)]^{\alpha/(\alpha+1)}$ for $t \geq t_2$. Then from (5.5.16) it follows that

$$v(t) \leq c\epsilon[\mathcal{J}(t_1,t)]^{\alpha/(\alpha+1)} \quad \text{for} \quad t \geq t_2 \tag{5.5.17}$$

and hence, from (5.5.13), we obtain

$$v(t) \leq c^{1/(\alpha+1)} \epsilon \left[\int_{t_1}^t a^{-1/\alpha}(s) v^{1/\alpha}(s) ds \right]^{\alpha/(\alpha+1)} \quad \text{for} \quad t \geq t_2,$$

or

$$\frac{a^{-1/\alpha}(t) v^{1/\alpha}(t)}{\left[\int_{t_1}^t a^{-1/\alpha}(s) v^{1/\alpha}(s) ds \right]^{1/(\alpha+1)}} \leq c^{1/(\alpha+1)} \epsilon a^{-1/\alpha}(t) \quad \text{for} \quad t \geq t_2. \tag{5.5.18}$$

Integrating inequality (5.5.18) from t_2 to t with $t \geq t_2$, we get

$$\frac{\alpha+1}{\alpha} \left[\int_{t_1}^t a^{-1/\alpha}(s) v^{1/\alpha}(s) ds \right]^{\frac{\alpha}{\alpha+1}} - \frac{\alpha+1}{\alpha} \left[\int_{t_1}^{t_2} a^{-1/\alpha}(s) v^{1/\alpha}(s) ds \right]^{\frac{\alpha}{\alpha+1}}$$

$$\leq c^{1/(\alpha+1)} \epsilon \int_{t_2}^t a^{-1/\alpha}(s) ds \leq c^{1/(\alpha+1)} \epsilon \eta(t).$$

Now, there exists a $t_3 \geq t_2$ such that

$$\left[\int_{t_1}^{t} a^{-1/\alpha}(s)v^{1/\alpha}(s)ds\right]^{\alpha/(\alpha+1)} < c_1\epsilon\eta(t) \quad \text{for all} \quad t \geq t_3,$$

where $c_1 = 2(\alpha/(\alpha+1))c^{1/(\alpha+1)}$. Thus, from (5.5.17) it follows that $v(t) < cc_1\epsilon^2\eta(t)$ for all $t \geq t_3$. As $\epsilon > 0$ is arbitrary, the proof of (5.5.14) is complete.

Now from (5.5.13) and (5.5.14) there exist a $T \geq t_3$ and a constant $c_2 > 0$ such that

$$\frac{f'(x(t))}{f^{(\alpha-1)/\alpha}(x(t))} \geq \frac{c_2}{\eta(t)} \quad \text{for} \quad t \geq T. \tag{5.5.19}$$

Using (5.5.19) in (5.5.12), we obtain

$$w'(t) + q(t) + c_2(a(t)\eta^\alpha(t))^{-1/\alpha}|w(t)|^{(\alpha+1)/\alpha} \leq 0 \quad \text{for all} \quad t \geq T.$$

Finally, applying Lemma 4.4.4 we find that (5.5.3) is nonoscillatory, which is a contradiction.

Case 2. $\mathcal{I}(t_0, \infty) = \infty$. For this case the proof can be modelled on Theorem 4.4.3 case 2. We omit the details. ∎

Remark 5.5.2. By Theorems 5.5.1 and 5.5.2 it is clear that the results established in Chapter 3 for half–linear differential equations can be employed to obtain some interesting oscillation criteria for nonlinear equations of types (5.5.1) and (5.5.3). The formulation of such results are left to the reader.

Remark 5.5.3. From the results presented in earlier chpaters it is clear that condition (5.5.4) in Theorems 5.5.1 and 5.5.2 can be weakened.

In the following results we shall present oscillation criteria for (3.13.76) and (3.13.89) with $g(t) = t$, i.e., we shall consider the equations

$$(a(t)\psi(x'(t)))' + F(t, x(t)) = 0 \tag{5.5.20}$$

and

$$\left(a(t)(x'(t))^{\alpha^*}\right)' + F(t, x(t)) = 0, \tag{5.5.21}$$

where the functions a, F, ψ and the constant α are as in equations (3.13.76) and (3.13.89), and $(x')^{\alpha^*} = |x'|^\alpha \text{ sgn } x'$.

Definition 5.5.1. We say that equation (5.5.20) is *strongly sublinear* if there exists a constant $c > 0$ such that $|y|^{-c}|F(t, y)|$ is nonincreasing in $|y|$ for each fixed t, and

$$\int_{+0} \frac{1}{[\psi^{-1}(u)]^c}du < \infty \quad \text{and} \quad \int_{-0} \frac{1}{[\psi^{-1}(u)]^c}du < \infty. \tag{5.5.22}$$

Equations (5.5.3) and (5.5.21) are strongly sublinear if $\beta < \alpha$ and $0 < c < \alpha$ respectively holds.

In what follows with respect to equation (5.5.20) we shall assume that

$$\int_{t_0}^{\infty} \left| \psi^{-1}\left(\frac{k}{a(s)}\right)\right| ds = \infty \quad \text{for every constant} \quad k \neq 0. \tag{5.5.23}$$

Theorem 5.5.3. Let equation (5.5.20) be strongly sublinear. Suppose for each fixed $k \neq 0$ and $t_0 \geq 0$,

$$\lim_{\ell \to 0, \ \ell k > 0} \frac{\Psi_{\ell,t_0}(a,t)}{\Psi_{k,t_0}(a,t)} = 0 \tag{5.5.24}$$

uniformly on any interval of the form $[t_1, \infty)$ for some $t_1 > t_0$. Suppose moreover that

$$\psi^{-1}(xy) \geq \psi^{-1}(x)\psi^{-1}(y) \quad \text{for all} \quad x, \ y \quad \text{with} \quad xy > 0. \tag{5.5.25}$$

Then equation (5.5.20) is oscillatory if and only if

$$\int_{t_0}^{\infty} |F(s, c\Psi_k(a,s))| ds = \infty \tag{5.5.26}$$

for all constants $k \neq 0$ and $c > 0$, where the function Ψ is defined in Section 3.13.4.

Proof. The 'only if' part is a consequence of Theorem 3.13.11. To prove the 'if' part, let $x(t)$ be a positive solution of equation (5.5.20) for $t \geq t_0$. First, we note that $x'(t) > 0$ for $t \geq t_0$, and

$$x'(t) \geq \psi^{-1}\left(\frac{1}{a(t)}\right)\psi^{-1}(a(t)\psi(x'(t))) \quad \text{for} \quad t \geq t_0. \tag{5.5.27}$$

Integrating (5.5.27) from t_0 to t and using the decreasing nature of $a(t)\psi(x'(t))$, we find

$$x(t) - x(t_0) \geq \psi^{-1}(a(t)\psi(x'(t)))\Psi_{1,t_0}(a,t), \quad t \geq t_0. \tag{5.5.28}$$

Now define

$$G(u) = \int_u^{u_0} \frac{dv}{[\psi^{-1}(u)]^c}, \tag{5.5.29}$$

where $c > 0$ is the strong sublinearity constant and $u_0 = a(t_0)\psi(x'(t_0))$. Then from (5.5.28) and (5.5.29) for $t \geq t_0$, we get

$$\begin{aligned}
[G(a(t)\psi(x'(t)))]' &= \frac{F(t, x(t))}{[\psi^{-1}(a(t)\psi(x'(t)))]^c} \\
&\geq [\Psi_{1,t_0}(a,t)]^c (x(t))^{-c} F(t, x(t)).
\end{aligned} \tag{5.5.30}$$

From the strong sublinearity and the inequality $x(t) \leq c_0 \Psi_{k,t_0}(a, t)$, $t \geq t_0$ where $c_0 > 0$ is a constant, it follows that

$$x^{-c}(t)F(t, x(t)) \geq c_0^{-c}[\Psi_{k,t_0}(a, t)]^{-c}F(t, c_0\Psi_{k,t_0}(a, t)) \qquad (5.5.31)$$

for $t > t_0$. Substituting (5.5.31) into (5.5.30) and using the inequality

$$\frac{\Psi_{1,t_0}(a, t)}{\Psi_{k,t_0}(a, t)} \geq \psi^{-1}\left(\frac{1}{k}\right), \qquad t > t_0$$

which follows from (5.5.25), we obtain by integration over $[t_1, t_2]$, $t_1 > t_0$ that

$$c_0^{-c}\psi^{-1}\left(\frac{1}{k}\right)\int_{t_1}^{t_2} F(s, c_0\Psi_{k,t_0}(a, s))ds \leq \int_{u_2}^{u_1} \frac{dv}{[\psi^{-1}(v)]^c},$$

where $u_i = a(t_i)\psi(x'(t_i))$, $i = 1, 2$. Letting $t_2 \to \infty$ and using (5.5.22), we obtain $\int_{t_1}^{\infty} F(s, c_0\Psi_{k,t_0}(a, s))ds < \infty$, which contradicts condition (5.5.26). This completes the proof. ∎

The following result is a variant of Theorem 5.5.3.

Theorem 5.5.4. Assume that there exist functions $q(t) \in C([t_0, \infty), \mathbb{R}^+)$ and $g(x) \in C(\mathbb{R}, \mathbb{R})$ such that $g(x)$ is increasing,

$$|F(t, y)| \geq q(t)g(|y|) \quad \text{for} \quad (t, y) \in \mathbb{R}^+ \times \mathbb{R}. \qquad (5.5.32)$$

Moreover, suppose that conditions (5.5.24) and (5.5.25) hold,

$$g(xy) \geq g(x)g(y) \quad \text{for any} \quad x, \, y > 0 \qquad (5.5.33)$$

and

$$\int_{+0} \frac{du}{g \circ |\psi^{-1}(u)|} < \infty \quad \text{and} \quad \int_{-0} \frac{du}{g \circ |\psi^{-1}(u)|} < \infty. \qquad (5.5.34)$$

Then equation (5.5.20) is oscillatory if and only if

$$\int_{t_0}^{\infty} q(s)g(\Psi_k(a, s))ds = \infty \qquad (5.5.35)$$

for every constant $k \neq 0$.

Proof. The 'only if' part is a consequence of Theorem 3.13.11. To prove the 'if' part, let $x(t)$ be a positive solution of equation (5.5.20) for $t \geq t_0$. As in Theorem 5.5.3, (5.5.28) holds, and so (5.5.33) implies

$$g(x(t)) \geq g(\psi^{-1}(a(t)\psi(x'(t)))g(\Psi_{1,t_0}(a, t)), \qquad t \geq t_0. \qquad (5.5.36)$$

Define

$$H(u) = \int_u^{u_0} \frac{dv}{g \circ \psi^{-1}(v)}, \quad u > 0 \tag{5.5.37}$$

where $u_0 = a(t_0)\psi(x'(t_0)) > 0$. Then by (5.5.36) and (5.5.37) for $t > t_0$, we have

$$
\begin{aligned}
[H(a(t)\psi(x'(t)))]' &= \frac{F(t, x(t))}{g(\psi^{-1}(a(t)\psi(x'(t))))} \\
&\geq \frac{q(t)g(x(t))}{g(\psi^{-1}(a(t)\psi(x'(t))))} \geq q(t)g(\Psi_{1,t_0}(a, t)).
\end{aligned}
\tag{5.5.38}
$$

Integrating (5.5.38) from $t_1 > t_0$ to $t_2 > t_1$, we find

$$\int_{t_1}^{t_2} q(s)g(\Psi_{1,t_0}(a, s))ds \leq \int_{u_2}^{u_1} \frac{dv}{g \circ \psi^{-1}(v)},$$

where $u_i = a(t_i)\psi(x'(t_i))$, $i = 1, 2$ which in view of (5.5.34) implies $\int_{t_1}^{\infty} q(s)g(\Psi_{1,t_0}(a, s))ds < \infty$, which contradicts condition (5.5.35). ∎

Example 5.5.1. Consider (5.5.3) and assume that $\int^{\infty} a^{-1/\alpha}(s)ds = \infty$. Clearly, from Theorem 5.5.3 if $\alpha > \beta$, a necessary and sufficient condition for equation (5.5.3) to be oscillatory is $\int_{t_0}^{\infty} s^{\beta}q(s)ds = \infty$.

Next, with respect to equation (5.5.21) we shall assume that

$$\pi(t) = \int_t^{\infty} a^{-1/\alpha}(s)ds \quad \text{and} \quad \pi(t_0) < \infty. \tag{5.5.39}$$

Theorem 5.5.5. Let equation (5.5.21) be strongly sublinear. Then it is oscillatory if and only if

$$\int_{t_0}^{\infty} \left(\frac{1}{a(s)} \int_{t_0}^{s} |F(u, k)|du \right)^{1/\alpha} ds = \infty \tag{5.5.40}$$

for every constant $k \neq 0$.

Proof. It suffices to prove the 'if' part, since the 'only if' part follows from Theorem 3.13.16. Suppose condition (5.5.40) holds and equation (5.5.21) has an eventually positive solution $x(t)$, $t \geq t_0$. We consider the two cases:

Case I. If $x'(t) > 0$ for $t \geq t_0$, then we have $\int_{t_0}^{\infty} F(s, c_1)ds < \infty$, where $c_1 > 0$ is a constant such that $x(t) \geq c_1$ for $t \geq t_0$. This together with the condition (5.5.39) shows that

$$\int_{t_0}^{\infty} \left(\frac{1}{a(s)} \int_{t_0}^{s} F(u, c_1)du \right)^{1/\alpha} ds \leq \pi(t_0) \left(\int_{t_0}^{\infty} F(s, c_1)ds \right)^{1/\alpha} < \infty,$$

which contradicts condition (5.5.40).

Case II. If $x'(t) < 0$ for $t \geq t_1 \geq t_0$, then an integration of (5.5.21) gives

$$\int_{t_1}^{t} F(s, x(s))ds \leq a(t)(-x'(t))^{\alpha} \quad \text{for} \quad t \geq t_1,$$

or equivalently,

$$-x'(t) \geq \left(\frac{1}{a(t)} \int_{t_1}^{t} F(s, x(s))ds\right)^{1/\alpha} \quad \text{for} \quad t \geq t_1. \tag{5.5.41}$$

Since, $x(t) \leq c_2$, $t \geq t_1$, for some constant $c_2 > 0$, the strong sublinearity implies

$$x^{-c}(t)F(t, x(t)) \geq c_2^{-c}F(t, c_2), \quad t \geq t_1 \tag{5.5.42}$$

for some constant $c < \alpha$ (see Definition 5.5.1). Combining (5.5.41) with (5.5.42) and using the decreasing property of $x(t)$, we find

$$-x'(t) \geq c_2^{-c/\alpha}a^{-1/\alpha}(t)\left(\int_{t_1}^{t} x^c(s)F(s, c_2)ds\right)^{1/\alpha}$$

$$\geq c_2^{-c/\alpha}a^{-1/\alpha}(t)x^{c/\alpha}(t)\left(\int_{t_1}^{t} F(s, c_2)ds\right)^{1/\alpha}, \quad t \geq t_1. \tag{5.5.43}$$

Differentiating the function $-x^{(\alpha-c)/\alpha}(t)$ and using (5.5.43), we obtain

$$\left(-x^{\frac{\alpha-c}{\alpha}}(t)\right)' = -\left(\frac{\alpha-c}{\alpha}\right)x^{-c/\alpha}(t)x'(t)$$

$$\geq \left(\frac{\alpha-c}{\alpha}\right)c_2^{-c/\alpha}x^{-\delta/\alpha}(t)a^{-1/\alpha}(t)x^{c/\alpha}(t)\left(\int_{t_1}^{t} F(s, c_2)ds\right)^{1/\alpha}$$

$$= \left(\frac{\alpha-c}{\alpha}\right)c_2^{-c/\alpha}a^{-1/\alpha}(t)\left(\int_{t_1}^{t} F(s, c_2)ds\right)^{1/\alpha}, \quad t \geq t_1$$

from which we have by integration

$$\left(\frac{\alpha-c}{\alpha}\right)c_2^{-c/\alpha}\int_{t_1}^{t_2}\left(\frac{1}{a(s)}\int_{t_1}^{s} F(u, c_2)du\right)^{1/\alpha}ds \leq x^{(\alpha-c)/\alpha}(t_1) \tag{5.5.44}$$

for any $t_2 > t_1$. Letting $t_2 \to \infty$ in (5.5.44), we conclude that

$$\int_{t_1}^{\infty}\left(\frac{1}{a(s)}\int_{t_1}^{s} F(u, c_2)du\right)^{1/\alpha}ds < \infty,$$

which again contradicts condition (5.5.40). This completes the proof.   ∎

Example 5.5.2. Consider (5.5.3) with $a(t) = e^{\lambda t}$ and $q(t) = e^{\mu t}$ for $t > 0$, where λ and μ are real constants and $\lambda > 0$. Let $\alpha > \beta$. Then by Theorem 5.5.5, equation (5.5.3) is oscillatory if and only if $\mu \geq \lambda$.

5.6. Notes and General Discussions

1. Theorems 5.1.1 and 5.1.2 extend sublinear oscillation criteria of Kamenev [28] and Kusano et. al. [30]. Theorem 5.1.3 is due to Grace and Lalli [13] and it includes theorem 8 of Graef et. al. [26]. Theorems 5.1.4 and 5.1.5 extend the corresponding results of Wong [42]. Theorem 5.1.6 is taken from Butler [5], whereas Theorem 5.1.7 is based on the work of Kwong and Wong [33] and it generalizes Belohorec criterion [3].

2. Condition (F) and Examples 5.2.1 – 5.2.3 are due to Philos [36]. Theorems 5.2.1 – 5.2.3 are extracted from Grace and Lalli [20]. Corollaries 5.2.1 and 5.2.2 are the same as theorems 2 and 3 of Philos [37]. Also Corollary 5.2.3 is the same as theorem 1 of Philos [40]. Theorems 5.2.4 and 5.2.5 are new. Corollaries 5.2.4 and 5.2.7 are the same as theorems 1' and 2' of Philos [38]. Corollary 5.2.8 is the same as theorem 1 of Kwong and Wong [31]. Theorems 5.2.6, 5.2.7 and 5.2.9 are new while Corollary 5.2.10 is the same as theorem 2 of Philos [40]. Theorem 5.2.8 is due to Philos [39]. Theorem 5.2.10 is based on the work of Wong [48]. Theorem 5.2.11 is new while Corollary 5.2.13 is due to Philos and Purnaras [41]. Theorems 5.2.12 and 5.2.13 are new while Corollary 5.2.16 is borrowed from Wong [48].

3. Theorems 5.3.1 – 5.3.3 are taken from Agarwal et. al. [1]. These results extend as well as improve theorems 4 – 6 of Kwong and Wong [32].

4. The results of Section 5.4 are taken from Agarwal et. al. [2], Goll-witzer [7], Kwong and Wong [31,34] and Wong [43,49].

5. Theorems 5.5.1 and 5.5.2 are new. Theorems 5.5.3 and 5.5.4 are taken from Elbert and Kusano [6] whereas Theorem 5.5.5 is due to Kusano et. al. [29].

6. Several other related oscillation results for sublinear differential equations are available in Butler [4], Grace [8–12], Grace and Lalli [14–19,21–24], Grace, Lalli and Yeh [25], Heidel [27], Onose [35], and Wong [44–47].

7. It will be interesting to discuss oscillation criteria for forced and damped sublinear differential equations similar to those dealt with in Sections 4.5 and 4.6.

5.7. References

1. **R.P. Agarwal, S.R. Grace and D. O'Regan**, Linearization of second order sublinear oscillation theorems, to appear.

2. **R.P. Agarwal, S.R. Grace and D. O'Regan**, *Oscillation Theory for Second Order Dynamic Equations*, to appear.

3. **Š. Belohorec**, Oscillatory solutions of certain nonlinear differential equations of second order, *Mat. Fyz. Casopis Solven. Akad. Vied.* **11**(1961),

250–255.

4. **G.J. Butler**, Integral averages and the oscillation of second order ordinary differential equations, *SIAM J. Math. Anal.* **11**(1980), 190–200.

5. **G.J. Butler**, An integral criterion for the oscillation of a second order sublinear ordinary differential equations, *Indian J. Math.* **24**(1982), 1–7.

6. **Á Elbert and T. Kusano**, Oscillation and nonoscillation theorems for a class of second order quasilinear differential equations, *Acta. Math. Hungar.* **56**(1990), 325–336.

7. **H.E. Gollwitzer**, Nonoscillation theorems for a nonlinear differential equation, *Proc. Amer. Math. Soc.* **26**(1970), 78–84.

8. **S.R. Grace**, Oscillation theorems for second order nonlinear differential equations with damping, *Math. Nachr.* **141**(1989), 117–127.

9. **S.R. Grace**, Oscillation theorems for nonlinear differential equations of second order, *J. Math. Anal. Appl.* **171**(1992), 220–241.

10. **S.R. Grace**, Oscillation theorems for damped functional differential equations, *Funkcial. Ekvac.* **35**(1992), 261–278.

11. **S.R. Grace**, Oscillation of nonlinear differential equations of second order, *Publ. Math. Debrecen* **40**(1992), 143–153.

12. **S.R. Grace**, On the oscillatory behavior of solutions of second order nonlinear differential equations, *Publ. Math. Debrecen* **43**(1993), 351–357.

13. **S.R. Grace and B.S. Lalli**, Oscillation theorems for certain second order perturbed nonlinear differential equations, *J. Math. Anal. Appl.* **77**(1980), 205–214.

14. **S.R. Grace and B.S. Lalli**, Oscillation of solutions of damped nonlinear second order functional differential equations, *Bull. Inst. Math. Acad. Sinica* **12**(1984), 5–9.

15. **S.R. Grace and B.S. Lalli**, Oscillation theorems for a second order nonlinear ordinary differential equation with damping term, *Comment. Math. Univ. Carolin.* **27**(1986), 449–453.

16. **S.R. Grace and B.S. Lalli**, Oscillatory behavior of solutions of second order differential equations with alternating coefficients, *Math. Nachr.* **127**(1986), 165–175.

17. **S.R. Grace and B.S. Lalli**, An oscillation criterion for certain second order strongly sublinear differential equations, *J. Math. Anal. Appl.* **123**(1987), 584–588.

18. **S.R. Grace and B.S. Lalli**, An oscillation criterion for second order sublinear ordinary differential equations with damping term, *Bull. Polish. Acad. Sci. Math.* **35**(1987), 181–184.

19. **S.R. Grace and B.S. Lalli**, Oscillation theorems for second order nonlinear differential equations, *J. Math. Anal. Appl.* **124**(1987), 213–224.

20. **S.R. Grace and B.S. Lalli**, On the second order nonlinear oscillations, *Bull. Inst. Math. Acad. Sinica* **15**(1987), 297–309.

21. **S.R. Grace and B.S. Lalli**, Integral averaging and the oscillation of second order nonlinear differential equations, *Ann. Mat. Pura Appl.* **151**(1988), 149–159.

22. **S.R. Grace and B.S. Lalli,** Oscillations in second order differential equations with alternating coefficients, *Period. Math. Hungar.* **19**(1988), 69–78.

23. **S.R. Grace and B.S. Lalli,** Oscillation theorems for nonlinear second order differential equations with a damping term, *Comment. Math. Univ. Carolin.* **30**(1989), 691–697.

24. **S.R. Grace and B.S. Lalli,** Integral averaging techniques for the oscillation of second order nonlinear differential equations, *J. Math. Anal. Appl.* **149**(1990), 277–311.

25. **S.R. Grace, B.S. Lalli and C.C. Yeh,** Oscillation theorems for nonlinear second order differential equations with a nonlinear damping term, *SIAM J. Math. Anal.* **15**(1984), 1082–1093.

26. **J.R. Graef, S.M. Rankin and P.W. Spikes,** Oscillation theorems for perturbed nonlinear differential equations, *J. Math. Anal. Appl.* **65**(1978), 375–390.

27. **J.W. Heidel,** A nonoscillation theorem for a nonlinear second order differential equation, *Proc. Amer. Math. Soc.* **22**(1969), 485–488.

28. **I.V. Kamenev,** Oscillation of solutions of second order nonlinear equations with sign variable coefficients, *Differencial'nye Uravnenija* **6**(1970), 1718–1721.

29. **T. Kusano, A. Ogata and H. Usami,** Oscillation theory for a class of second order quasilinear ordinary differential equations with application to partial differential equations, *Japan J. Math.* **19**(1993), 131–147.

30. **T. Kusano, H. Onose and H. Tobe,** On the oscillation of second order nonlinear ordinary differential equations, *Hiroshima Math. J.* **4**(1974), 491–499.

31. **M.K. Kwong and J.S.W. Wong,** On the oscillation and nonoscillation of second order sublinear equations, *Proc. Amer. Math. Soc.* **85**(1982), 547–551.

32. **M.K. Kwong and J.S.W. Wong,** Linearization of second order nonlinear oscillation theorems, *Trans. Amer. Math. Soc.* **279**(1983), 705–722.

33. **M.K. Kwong and J.S.W. Wong,** On an oscillation theorem of Belohorec, *SIAM J. Math. Anal.* **14**(1983), 474–476.

34. **M.K. Kwong and J.S.W. Wong,** Nonoscillation theorems for a second order sublinear ordinary differential equation, *Proc. Amer. Math. Soc.* **87**(1983), 467–474.

35. **H. Onose,** On Butler's conjecture for oscillation of an ordinary differential equation, *Quart. J. Math.* **34**(1983), 235–239.

36. **Ch.G. Philos,** Oscillation of sublinear differential equations of second order, *Nonlinear Analysis* **7**(1983), 1071–1080.

37. **Ch.G. Philos,** On second order sublinear oscillation, *Aequations Math.* **27**(1984), 242–254.

38. **Ch.G. Philos,** Integral averaging techniques for the oscillation of second order sublinear ordinary differential equations, *J. Austral. Math. Soc. Ser. A* **40**(1986), 111–130.

39. **Ch.G. Philos,** On the oscillation of second order sublinear ordinary differential equations with alternating coefficients, *Math. Nachr.* **146**(1990), 105–116.

40. **Ch.G. Philos,** Integral averages and oscillation of second order sublinear differential equations, *Differential and Integral Equations* **4**(1991), 205–213.

41. **Ch.G. Philos and I.K. Purnaras,** On the oscillation of second order nonlinear differential equations, *Arch. Math.* **59**(1992), 260–271.

42. **J.S.W. Wong,** Oscillation theorems for second order nonlinear differential equations, *Bull. Inst. Math. Acad. Sinica* **3**(1975), 283–309.

43. **J.S.W. Wong,** Remarks on nonoscillation theorems for a second order nonlinear differential equation, *Proc. Amer. Math. Soc.* **83**(1981), 541–546.

44. **J.S.W. Wong,** Oscillation theorems for second order nonlinear differential equations, *Proc. Amer. Math. Soc.* **106**(1989), 1069–1077.

45. **J.S.W. Wong,** A sublinear oscillation theorem, *J. Math. Anal. Appl.* **139**(1989), 197–215.

46. **J.S.W. Wong,** An oscillation theorem for second order sublinear differential equations, *Proc. Amer. Math. Soc.* **110**(1990), 633–637.

47. **J.S.W. Wong,** Oscillation criteria for second order nonlinear differential equations with integrable coefficients, *Proc. Amer. Math. Soc.* **115**(1992), 389–395.

48. **J.S.W. Wong,** An oscillation criterion for second order nonlinear differential equations with iterated integral averages, *Differential and integral Equations* **6**(1993), 83–91.

49. **J.S.W. Wong,** Nonoscillation theorems for second order nonlinear differential equations, *Proc. Amer. Math. Soc.* **127**(1999), 1387–1395.

Chapter 6

Further Results on the Oscillation of Differential Equations

6.0. Introduction

In this chapter we shall discuss some techniques which are different than those employed in the previous chapters to obtain oscillatory criteria for differential equations. In Section 6.1 we shall present oscillation and nonoscillation theorems for nonlinear second order differential equations by using the method of Olech, Opial and Wazewski. Section 6.2 is concerned with the oscillation of half–linear second order differential equations by employing the variational inequality given in Lemma 3.2.6. In Section 6.3 we shall begin with some preliminaries of Liapunov functions, and then apply Liapunov second method to obtain criteria for the oscillation of second order nonlinear equations.

6.1. Oscillation Criteria of Olech–Opial–Wazewski Type

Consider the second order nonlinear differential equation

$$x''(t) + q(t)f(x(t)) = 0 \qquad (6.1.1)$$

and its prototype, the so called generalized Emden–Fowler equation

$$x''(t) + q(t)|x(t)|^\gamma \operatorname{sgn} x(t) = 0, \qquad (6.1.2)$$

where $\gamma > 0$ is a constant.

In what follows we shall assume that

(i) $f \in C^2(\mathbb{R} - \{0\}, \mathbb{R})$, $xf(x) > 0$ for $x \neq 0$ and $f'(x) > 0$ for $x \neq 0$,
(ii) $q \in C(\mathbb{R}_0, \mathbb{R})$.

As a main result of this section we shall extend an oscillation result of Olech, Opial and Wazewski [8]. They showed that equation (6.1.2) with $\gamma = 1$ is oscillatory if

$$\lim_{\substack{t \to \infty}} \text{approx} \int_0^t q(s)ds = \infty, \tag{6.1.3}$$

or

$$\lim_{\substack{t \to \infty}} \text{approx inf} \int_0^t q(s)ds < \lim_{\substack{t \to \infty}} \text{approx sup} \int_0^t q(s)ds, \tag{6.1.4}$$

(see definitions in the next subsection).

As a consequence of conditions (6.1.3) and (6.1.4) it follows that if

$$\lim_{\substack{t \to \infty}} \text{approx sup} \int_0^t q(s)ds = \infty, \tag{6.1.5}$$

then equation (6.1.2) with $\gamma = 1$ is oscillatory.

6.1.1. Some Definitions and Notation

For any set $S \subset \mathbb{R}_0$, we define the density function of S by

$$\rho_S(t) = \frac{1}{t}\mu\{S \cap [0, t]\}, \tag{6.1.6}$$

where μ denotes Lebesgue measure. We denote by $g(x)$ the expression

$$g(x) = \frac{f''(x)f(x)}{(f'(x))^2}, \quad x \neq 0. \tag{6.1.7}$$

Clearly, $g(x)$ is continuous for $x \neq 0$. We note that if $f(x) = |x|^\gamma \text{ sgn } x$, $\gamma > 0$ is a constant, then

$$g(x) = \frac{\gamma - 1}{\gamma}. \tag{6.1.8}$$

With respect to the function $g(x)$, we will assume that the following condition holds: There exist numbers $M > 0$ and $m < 1$ such that

$$g(x) \leq m < 1 \quad \text{and} \quad |g(x)| \leq M \quad \text{for all} \quad x \neq 0. \tag{6.1.9}$$

Now, we recall the following definitions: If $h = h(t)$ is a real valued function defined on $[a, \infty)$, $a \geq 0$ and if $-\infty \leq \ell$, $L \leq \infty$, then we write $\lim \text{approx sup}_{t \to \infty} h(t) = L$ if $\mu\{t : h(t) > L_1\} = \infty$ for all $L_1 < L$ and $\mu\{t : h(t) > L_2\} < \infty$ for all $L_2 > L$. Similarly, we write $\lim \text{approx inf}_{t \to \infty} h(t) = \ell$ if $\mu\{t : h(t) < \ell_1\} < \infty$ for all $\ell_1 < \ell$ and $\mu\{t : h(t) < \ell_2\} = \infty$ for all $\ell_2 > \ell$. We write $\lim \text{approx}_{t \to \infty} h(t) = L$,

$-\infty \le L \le \infty$ if $\lim \operatorname{approx} \sup_{t\to\infty} h(t) = \lim \operatorname{approx} \inf_{t\to\infty} h(t) = L$. We note that $\lim \operatorname{approx}_{t\to\infty} h(t) = \infty$ implies that there exists a set $S \subset \mathbb{R}_0$ such that $\lim_{t\to\infty,\ t\in S} h(t) = \infty$ and $\lim \sup_{t\to\infty} \rho_S(t) = \lim_{t\to\infty} \rho_S(t) = 1$.

6.1.2. Oscillation Criteria

Our main result here is the following:

Theorem 6.1.1. Let condition (6.1.9) hold and assume that there exists a set $S \subset \mathbb{R}_0$ such that

$$\limsup_{t\to\infty} t[\rho_S(t) - \delta(M,m)] = \infty, \tag{6.1.10}$$

$$\delta(M,m) = \frac{M^2}{M^2 - 4m + 4} \tag{6.1.11}$$

and

$$\lim_{t\to\infty,\ t\in S} \int_0^t q(s)ds = \infty. \tag{6.1.12}$$

Then equation (6.1.1) is oscillatory.

Proof. Suppose (6.1.1) is not oscillatory, and let $x(t)$ be a nonoscillatory solution on $\mathbb{R}_0$, which we suppose satisfies $x(t) > 0$ on $[t_0, \infty)$, $t_0 \ge 0$. Define $w(t) = x'(t)/f(x(t))$, $t \ge t_0$. Then for $t \ge t_0$, we have

$$w'(t) = -q(t) - f'(x(t))w^2(t), \tag{6.1.13}$$

which on integration over $[t_0, t]$ gives

$$w(t) = w(t_0) - \int_{t_0}^t q(s)ds - \int_{t_0}^t f'(x(s))w^2(s)ds. \tag{6.1.14}$$

Define the function $\Omega(x)$, $x > 0$ by

$$\Omega(x) = \int_x^1 \frac{du}{f(u)}. \tag{6.1.15}$$

Since $\Omega(x)$ is decreasing, Ω^{-1} exists. Further, since

$$\int_{t_0}^t w(s)ds = \int_{t_0}^t \frac{x'(s)}{f(x(s))}ds = -\int_{x(t)}^{x(t_0)} \frac{du}{f(u)} = -\Omega(x(t)) + c_0,$$

where $c_0 = \Omega(x(t_0))$, if we set

$$y(t) = c_0 - \int_{t_0}^t w(s)ds = \Omega(x(t)), \tag{6.1.16}$$

then $x(t) = \Omega^{-1}(y(t))$ and $y'(t) = -w(t)$, $t \geq t_0$ and so we have from (6.1.14),

$$
\begin{aligned}
\int_{t_0}^t q(s)ds &= y'(t) - \int_{t_0}^t f'(\Omega^{-1}(y(s)))(y'(s))^2 ds - y'(t_0) \\
&= y'(t) - W(t) - y'(t_0), \quad t \geq t_0
\end{aligned}
\tag{6.1.17}
$$

where

$$
W(t) = \int_{t_0}^t f'(\Omega^{-1}(y(s)))(y'(s))^2 ds, \quad t \geq t_0.
\tag{6.1.18}
$$

Now (6.1.12) implies that $q(t)$ is not identically zero on any half–line (t_1, ∞), and so $y'(t)$ is not identically zero on any half–line (t_1, ∞) for some $t_1 \geq t_0$. It follows that $W(t) > 0$ for all t sufficiently large. Without loss of generality we may assume that $W(t) > 0$ for $t \geq t_0$. We now let the function $P(t)$ be defined by the equation

$$
y'(t) = P(t)W(t) \quad \text{for} \quad t > t_0.
\tag{6.1.19}
$$

We then have

$$
W'(t) = (y'(t))^2 f'(\Omega^{-1}(y(t))) = P^2(t)W^2(t)f'(x(t)) \quad \text{for} \quad t > t_0.
$$

If we let $V(t)$, $U(t)$ be defined by the relations

$$
V(t) = f'(x(t)) = f'(\Omega^{-1}(y(t))), \quad t \geq t_0 \quad \text{and} \quad U(t) = \frac{1}{V(t)W(t)}, \quad t > t_0,
$$

we find that $V(t) > 0$ and $U(t) > 0$ for $t > t_0$, and

$$
\begin{aligned}
V'(t) &= f''(x(t))x'(t) = f''(x(t))\frac{d}{dt}(\Omega^{-1}(y(t))) \\
&= -f''(x(t))f(x(t))y'(t) \\
&= -f''(x(t))f(x(t))P(t)W(t) \quad \text{for} \quad t > t_0.
\end{aligned}
\tag{6.1.20}
$$

We also have

$$
\begin{aligned}
U'(t) &= -\frac{V'(t)}{V^2(t)W(t)} - \frac{W'(t)}{V(t)W^2(t)} \\
&= \frac{f''(x(t))f(x(t))P(t)W(t)}{V^2(t)W(t)} - \frac{P^2(t)W^2(t)f'(x(t))}{V(t)W^2(t)} \\
&= g(x(t))P(t) - P^2(t) \quad \text{for} \quad t > t_0,
\end{aligned}
\tag{6.1.21}
$$

where $g(x)$ is defined by (6.1.7).

Now for any natural number $k > 0$ from (6.1.12) we have $\int_0^t q(s)ds \geq k$ for $t \in S \cap [t_k, \infty)$ if $t_k \geq t_0$ is sufficiently large. Define the set S_1 by

$$
S_1 = \{t > t_0 : P(t) \geq 1\}.
\tag{6.1.22}
$$

Then from (6.1.17) and (6.1.19) for all sufficiently large $k > 0$, we have

$$S \cap [t_k, \infty) \subset S_1. \tag{6.1.23}$$

Assume now that k is fixed so that $t_k > t_0$ and (6.1.23) holds, and let $\hat{S} = S_1 \cap [t_k, \infty)$. It follows that

$$\limsup_{t \to \infty} \frac{\mu(S \cap [0, t])}{t} = \limsup_{t \to \infty} \frac{\mu(S \cap [t_k, t])}{t} \leq \limsup_{t \to \infty} \frac{\mu(\hat{S} \cap [t_k, t])}{t}.$$

Now on $\hat{S}$, we have from (6.1.21) and (6.1.9) that

$$U'(t) = P(t)(g(x(t)) - P(t)) \leq P(t)(m - P(t)) \leq m - 1 < 0, \quad \text{for } t \in \hat{S} \tag{6.1.24}$$

(since the function $mP - P^2$ for $P \geq 1$ is maximized at $P = 1$). Further on $\hat{S}^c = [t_k, \infty) \backslash \hat{S}$ ($=$ the complement of $\hat{S}$ in $[t_k, \infty)$), $P(t) < 1$ so that the function $P(a - P)$ is maximized at $P = a/2$. Thus, from (6.1.21) and (6.1.9), we obtain

$$U'(t) \leq \frac{1}{2}(g(x(t)))^2 - \frac{1}{4}(g(x(t)))^2 = \frac{1}{4}(g(x(t)))^2 \leq \frac{1}{4}M^2 \quad \text{for } t \in \hat{S}^c. \tag{6.1.25}$$

If we set $\hat{S} = S \cap [t_k, t]$, $t > t_k$ then we have

$$U(t) = U(t_k) + \int_{\hat{S}(t)} U'(s)ds + \int_{\hat{S}^c(t)} U'(s)ds, \quad t \geq t_k.$$

Since $U(t) > 0$, $t > t_0$ it follows that

$$\int_{\hat{S}(t)} U'(s)ds + \int_{\hat{S}^c(t)} U'(s)ds \geq -L > -\infty \tag{6.1.26}$$

for some constant $L > 0$ and all $t \geq t_k$. From (6.1.24), we have

$$\int_{\hat{S}(t)} U'(s)ds \leq (m-1)\mu(\hat{S}(t)) \leq t(m-1)\rho_{\hat{S}}(t) \quad \text{for all large } t \tag{6.1.27}$$

and from (6.1.25), we obtain

$$\int_{\hat{S}^c(t)} U'(s)ds \leq \frac{1}{4}M^2\mu(\hat{S}^c(t)) \leq \frac{1}{4}tM^2(1 - \rho_S(t)) \quad \text{for all large } t. \tag{6.1.28}$$

Therefore, from (6.1.26) – (6.1.28), we get

$$t\left[(m-1)\rho_{\hat{S}}(t) + \frac{1}{4}M^2(1 - \rho_{\hat{S}}(t))\right] \geq -L > -\infty \quad \text{for all large } t,$$

which implies

$$t\left[\rho_{\hat{S}(t)} - \frac{M^2}{M^2 - 4m + 4}\right] \leq \frac{4L}{M^2 - 4m + 4} < \infty$$

contradicting condition (6.1.11). This proves the theorem. ∎

Corollary 6.1.1. Let condition (6.1.9) hold and

$$\lim_{t \to \infty} \text{approx} \int_0^t q(s)ds = \infty. \tag{6.1.29}$$

Then equation (6.1.1) is oscillatory.

Proof. If (6.1.29) holds, then condition (6.1.12) is satisfied for some $S \subset \mathbb{R}_0$. Since $\delta(M, m) < 1$ in (6.1.11), it follows that condition (6.1.10) holds, so the result follows from Theorem 6.1.1. ∎

For equation (6.1.2) we present the following result.

Corollary 6.1.2. Equation (6.1.2) is oscillatory if there exists a set $S \subset \mathbb{R}_0$ such that

$$\limsup_{t\to\infty} t\left[\rho_S(t) - \left(\frac{\gamma - 1}{\gamma + 1}\right)^2\right] = \infty \tag{6.1.30}$$

and

$$\lim_{t \to \infty,\, t\in S} \int_0^t q(s)ds = \infty. \tag{6.1.31}$$

Proof. Because of (6.1.8), $\delta(M, m) = ((\gamma - 1)/(\gamma + 1))^2$. So, the result follows from Theorem 6.1.1. ∎

Corollary 6.1.3. Assume that

$$\lim_{t \to \infty} \text{approx sup} \int_0^t q(s)ds = \infty. \tag{6.1.32}$$

Then equation (6.1.2) with $\gamma = 1$ is oscillatory.

Proof. If $\gamma = 1$ and (6.1.32) holds, then $\mu(S) = \infty$ for some set $S \subset \mathbb{R}_0$ for which (6.1.31) holds. Hence, the result follows from Corollary 6.1.2. ∎

Finally, we present the following result which applies in particular to functions $f(x)$ of the form $f(x) = \sum_{i=1}^{n} c_i |x|^{\gamma_i} \operatorname{sgn} x$, where the constants c_i and exponents γ_i are such that

$$xf(x) > 0 \quad \text{and} \quad f'(x) \geq c > 0 \quad \text{for some constant} \quad c > 0 \quad \text{and all} \quad x \neq 0. \tag{6.1.33}$$

If we assume $0 < \gamma_1 < \gamma_2 < \cdots < \gamma_n$, then clearly a necessary condition for (6.1.33) to hold is $c_1 > 0$, $c_n > 0$ and $\gamma_1 \leq 1 \leq \gamma_n$.

Theorem 6.1.2. Assume $f(x)$ satisfies condition (6.1.33) and assume that condition (6.1.32) holds. Then equation (6.1.1) is oscillatory.

Proof. Since (6.1.32) holds, it follows that the equation $x''(t) + cq(t)x(t) = 0$ is oscillatory by Corollary 6.1.3, where c is the same as in (6.1.33). Hence, by Theorem 4.3.5 equation (6.1.1) is oscillatory. ∎

6.1.3. Further Results

In the case of equation (6.1.2) the results of Subsection 6.1.1 are fairly sharp. In fact, first we note that if condition (6.1.31) holds, then this condition also holds with S replaced by its closure $\bar{S}$. We can therefore state the following partial converse of Corollary 6.1.2.

Theorem 6.1.3. Let S be any closed subset of $\mathbb{R}_0$ such that

$$\limsup_{t \to \infty} \rho_S(t) < \left(\frac{\gamma - 1}{\gamma + 1}\right)^2, \quad \gamma \neq 1. \tag{6.1.34}$$

Then there exists a continuous function $q = q(t)$ with

$$\lim_{t \to \infty,\ t \in S} \int_0^t q(s)ds = \infty \tag{6.1.35}$$

and such that equation (6.1.2) has a nontrivial nonoscillatory solution.

Proof. In view of (6.1.34) we can choose $\epsilon_0 > 0$ and $B_0 > 0$ such that

$$t\left[\rho_S(t) + \epsilon_0 - \left(\frac{\gamma - 1}{\gamma + 1}\right)^2\right] \leq B_0, \quad t \in \mathbb{R}_0. \tag{6.1.36}$$

Choose an open set $U \subset \mathbb{R}_0$ such that $S \subset U$ and $\mu(U \backslash S) \leq 1$ and define $Q_1 = \mathbb{R}_0 \backslash U$, $Q_2 = U \backslash S$. Thus, Q_1 is closed, $Q_1 \cap S = \emptyset$, $\mathbb{R}_0 = S \cup Q_1 \cup Q_2$. Hence, $\mu(Q_2) \leq 1$ and so, $t\rho_{Q_2}(t) \leq 1$ and $t\rho_{Q_1}(t) \geq t[1 - \rho_S(t)] - 1$. Define the function $p(t)$ by

$$p(t) = \begin{cases} 1 + \epsilon_1, & t \in S \\ \dfrac{\gamma - 1}{2\gamma}, & t \in Q_1 \end{cases}$$

where $0 < \epsilon_1 < 1$ will be chosen sufficiently small. It follows that $|p(t)| \leq 2 + |(\gamma - 1)/(2\gamma)|$ on $S \cup Q_1$, and we can extend $p(t)$ to all of $\mathbb{R}_0$ so that it is of class C^1 and satisfies $|p(t)| \leq C_0 = 2 + |(\gamma - 1)/(2\gamma)|$ for all $t \in \mathbb{R}_0$. We now let the function $w(t)$ be defined by

$$w(t) = \frac{\gamma - 1}{\gamma}p(t) - p^2(t), \quad t \in \mathbb{R}_0 \tag{6.1.37}$$

and notice that

$$
w(t) = \begin{cases} -\dfrac{1}{\gamma} - \epsilon_1 \left[\dfrac{1}{\gamma} + 1 + \epsilon_1 \right], & t \in S \\[2mm] \left(\dfrac{\gamma - 1}{2\gamma} \right)^2, & t \in Q_1 \end{cases} \tag{6.1.38}
$$

and

$$
|w(t)| \leq \left| \dfrac{\gamma - 1}{\gamma} \right| C_0 + C_0^2, \quad t \in \mathbb{R}_0. \tag{6.1.39}
$$

For convenience, we set

$$
C_1 = \left| \frac{\gamma - 1}{\gamma} \right| C_0 + C_0^2 \quad \text{and} \quad \lambda = \epsilon_1 \left(\frac{1}{\gamma} + 1 + \epsilon_1 \right). \tag{6.1.40}
$$

We next define $u(t)$ on $\mathbb{R}_0$ by

$$
u(t) = B_1 + \int_0^t w(s)\,ds, \tag{6.1.41}
$$

where B_1 will be chosen sufficiently large later.

If we write $S_t = S \cap [0, t]$, $Q_{1t} = Q_1 \cap [0, t]$, $Q_{2t} = Q_2 \cap [0, t]$, then we have

$$
u(t) = B_1 + \int_{S_t} w(s)\,ds + \int_{Q_{1t}} w(s)\,ds + \int_{Q_{2t}} w(s)\,ds. \tag{6.1.42}
$$

It follows from (6.2.38) – (6.2.40) that

$$
\int_{S_t} w(s)\,ds = -\left(\frac{1}{\gamma} + \lambda \right) t\rho_S(t), \tag{6.1.43}
$$

$$
\int_{Q_{1t}} w(s)\,ds = \left(\frac{\gamma - 1}{2\gamma} \right)^2 t\rho_{Q_1}(t) \tag{6.1.44}
$$

and

$$
\int_{Q_{2t}} w(s)\,ds \geq -C_1 t\rho_{Q_2}(t). \tag{6.1.45}
$$

Hence, we have from (6.1.42) – (6.1.45) and (6.1.36) after rearranging

$$
u(t) \geq D_0 + tD_1, \tag{6.1.46}
$$

where

$$
D_0 = B_1 - C_1 - \left(\frac{\gamma - 1}{2\gamma} \right)^2 - B_0 \left[\lambda + \left(\frac{\gamma + 1}{2\gamma} \right)^2 \right]
$$

and

$$D_1 = \left\{ \left(\frac{\gamma-1}{2\gamma}\right)^2 + \epsilon_0 \left[\left(\frac{\gamma+1}{2\gamma}\right)^2 + \lambda \right] - \left(\frac{\gamma-1}{\gamma+1}\right)^2 \left[\left(\frac{\gamma+1}{2\gamma}\right)^2 + \lambda \right] \right\}.$$

After some more rearranging, we see that D_1 can be written as

$$D_1 = \left\{ \epsilon_0 \left(\frac{\gamma+1}{2\gamma}\right)^2 + \lambda \left[\epsilon_0 - \left(\frac{\gamma-1}{\gamma+1}\right)^2 \right] \right\}.$$

Hence, since λ can be made arbitrarily small, in view of (6.1.40) it follows that $D_1 > 0$ and if we choose B_1 sufficiently large, we have $D_0 \geq 1$. Therefore, from (6.1.46) we find that $u(t) \geq 1 + D_1 t \geq 1$, $t \in \mathbb{R}_0$. Furthermore, since $u'(t) = w(t) \leq C_1$, $t \in \mathbb{R}_0$, we also have $u(t) \leq B_1 + C_1 t$, $t \in \mathbb{R}_0$. We now define $r = r(t)$ on $\mathbb{R}_0$ to be the solution of the equation

$$r'(t) = \frac{1}{u(t)} p^2(t) r(t), \quad r(0) = 1 \tag{6.1.47}$$

on $\mathbb{R}_0 \backslash Q_2 = S \cup Q_1$. We have

$$\frac{p^2(t)}{u(t)} \geq \frac{C_2^2}{B_1 + C_1 t}, \quad C_2 = \min\left\{ 1, \left|\frac{\gamma-1}{2\gamma}\right| \right\}. \tag{6.1.48}$$

Since $\mu(Q_2) \leq 1$ it follows from (6.1.47) and (6.1.48) that

$$\lim_{t \to \infty} r(t) = \infty. \tag{6.1.49}$$

We next define $v = v(t)$ by

$$v(t) = \frac{1}{r(t)u(t)} > 0, \quad t \in \mathbb{R}_0 \tag{6.1.50}$$

then in view of (6.1.49), we have $\lim_{t \to \infty} v(t) = 0$. With $f(x) = |x|^\gamma \operatorname{sgn} x$, we define the positive function $x = x(t)$ by

$$f'(x(t)) = v(t), \quad \text{i.e.,} \quad x(t) = \left(\frac{v(t)}{\gamma}\right)^{1/(\gamma-1)} \tag{6.1.51}$$

then, we have $\lim_{t \to \infty} x(t) = \begin{cases} 0 & \text{if } \gamma > 1 \\ \infty & \text{if } 0 < \gamma < 1. \end{cases}$ Now, by (6.1.41), (6.1.37), (6.1.9) and (6.1.51), we find

$$\begin{aligned} u'(t) &= w(t) = \left(\frac{\gamma-1}{\gamma}\right) p(t) - p^2(t) \\ &= \frac{f''(x(t))f(x(t))}{(f'(x(t)))^2} p(t) - p^2(t) = \frac{f''(x(t))f(x(t))}{v^2(t)} p(t) - p^2(t). \end{aligned}$$

On the other hand, we also have $u(t) = 1/(r(t)v(t))$ by (6.1.50), and therefore by (6.1.47) and (6.1.51)

$$u'(t) = -\frac{v'(t)}{r(t)v^2(t)} - \frac{r'(t)}{r^2(t)v(t)} = -\frac{f''(x(t))x'(t)}{r(t)v^2(t)} - p^2(t).$$

Hence, it follows that

$$\frac{x'(t)}{f(x(t))} = -p(t)r(t). \tag{6.1.52}$$

If we now define $q(t)$ by $q(t) = [(p(t) - 1)r(t)]'$, then in view of (6.1.50) and (6.1.52), we have

$$q(t) = (p(t)r(t))' - r'(t) = -\frac{x''(t)}{f(x(t))} + \frac{(x'(t))^2 f'(x(t))}{f^2(x(t))} - r'(t)$$

$$= -\frac{x''(t)}{f(x(t))} + p^2(t)r^2(t)v(t) - r'(t) = -\frac{x''(t)}{f(x(t))}.$$

Thus, $x(t)$ is a positive solution of equation (6.1.2). Furthermore, we have $\int_0^t q(s)ds = [p(t) - 1]r(t) - c_3$ for some constant c_3, and hence for $t \in S$ in view of the definition of $p(t)$, we have

$$\int_0^t q(s)ds = \epsilon_1 r(t) - c_3, \quad t \in S \tag{6.1.53}$$

so that (6.1.49) and (6.1.53) imply that $\lim_{t \to \infty, t \in S} \int_0^t q(s)ds = \infty$. ∎

6.2. Oscillation Criteria for Half–Linear Differential Equations

Here we shall study the oscillatory behavior of the half–linear second order differential equation

$$(\psi(x'(t)))' + q(t)\psi(x(t)) = 0, \tag{6.2.1}$$

where

(i) $q(t) \in C(\mathbb{R}_0, \mathbb{R})$,

(ii) $\psi(s) = |s|^{\alpha-1}s$ with $\alpha > 0$ is a constant.

In Chapter 3, we have shown that some of the oscillatory criteria of Chapter 2 for second order linear differential equations can be extended to equation (6.2.1). These criteria are mostly based on the Riccati transformation, since we know that if $x(t)$ is a nonzero solution of (6.2.1), then

$$w(t) = \frac{\psi(x'(t))}{\psi(x(t))} \tag{6.2.2}$$

solves the generalized Riccati equation

$$w'(t) + q(t) + \alpha |w(t)|^{(\alpha+1)/\alpha} = 0. \tag{6.2.3}$$

In these criteria, (6.2.1) is viewed as a perturbation of the nonoscillatory equation

$$(\psi(x'(t)))' = 0. \tag{6.2.4}$$

In the present section, we shall use a different approach which is based on the relationship between positivity of the '$\alpha + 1-$degree' functional

$$I(x; a, b) = \int_a^b \left[|x'(s)|^{\alpha+1} - q(s)|x(s)|^{\alpha+1} \right] ds$$

in the class of functions satisfying $x(a) = 0$, $x(b) = 0$ and the discojugacy of (6.2.1) in $[a, b]$. Moreover, we shall investigate (6.2.1) not as a perturbation of (6.2.4), but as a perturbation of the generalized Euler equation

$$(\psi(x'(t)))' + \frac{\gamma_0}{t^{\alpha+1}} \psi(x(t)) = 0, \tag{6.2.5}$$

where $\gamma_0 = (\alpha/(\alpha+1))^{\alpha+1}$ is the so called critical constant in this equation. As in the linear case, if we replace γ_0 by a constant $\gamma > \gamma_0$ $(\gamma < \gamma_0)$, then equation (6.2.5) becomes oscillatory (remains nonoscillatory).

6.2.1. Preliminary Results

We recall some results from the oscillation theory of linear equations. The well–known 'variational principle' states that the equation

$$(a(t)x'(t))' + q(t)x(t) = 0 \tag{6.2.6}$$

where $a, q : [a, b] \to \mathbb{R}$, $a(t) > 0$ is *disconjugate* in $[a, b]$, i.e., any nontrivial solution has at most one zero in $[a, b]$, if and only if

$$I(x; a, b) = \int_a^b \left[a(s)(x'(s))^2 - q(s)x^2(s) \right] ds > 0$$

for every nontrivial, piecewise $C^1[a, b]$ function for which $x(a) = 0 = x(b)$. Another important concept in oscillation theory of linear equations is the *principal solution* . A solution x_0 of (6.2.6) is said to be principal if $\lim_{t\to\infty} x_0(t)/x(t) = 0$ for any nonzero solution x of equation (6.2.6) which is linearly independent with x_0 (solution x is said to be *nonprincipal*). A principal solution of equation (6.2.6) exists (uniquely up to multiplication by a nonzero real constant) if and only if (6.2.6) is nonoscillatory.

Now consider (6.2.6) as a perturbation of the nonoscillatory equation

$$(a(t)x'(t))' + q_0(t)x(t) = 0, \tag{6.2.7}$$

where q_0 is a continuous real–valued function, and let x_0, x_1 be principal and nonprincipal solutions of (6.2.7), respectively. If

$$\lim_{t\to\infty} \frac{x_1(t)}{x_0(t)} \int_t^\infty [q(s) - q_0(s)] x_0^2(s) ds > 1, \qquad (6.2.8)$$

then equation (6.2.6) is oscillatory, and if

$$\lim_{t\to\infty} \frac{x_1(t)}{x_0(t)} \int_t^\infty [q(s) - q_0(s)]^+ x_0^2(s) ds < \frac{1}{4}, \qquad (6.2.9)$$

then (6.2.6) is nonoscillatory, here $h^+(t) = \max\{0, h(t)\}$. In particular, if $a(t) \equiv 1$, $q_0(t) = 1/(4t^2)$ then the condition

$$\lim_{t\to\infty} (\ln t) \int_t^\infty \left[q(s) - \frac{1}{4s^2} \right] s ds > 1 \qquad (6.2.10)$$

is sufficient for the equation

$$x''(t) + q(t)x(t) = 0 \qquad (6.2.11)$$

to be oscillatory, and the condition

$$\lim_{t\to\infty} (\ln t) \int_t^\infty \left[q(s) - \frac{1}{4s^2} \right] s ds < \frac{1}{4} \qquad (6.2.12)$$

is sufficient for the nonoscillation of (6.2.11).

In the following study, we shall show that condition (6.2.10) is extendable to (6.2.1), while there are some difficulties in extending condition (6.5.12) to half–linear equations. The analysis relies on Lemma 3.2.6 which we restate here for the convenience of the reader.

Lemma 6.2.1. Let x be a solution of the equation

$$(a(t)\psi(x'(t)))' + q(t)\psi(x(t)) = 0, \qquad (6.2.13)$$

on $[a, b]$ satisfying $x(t) \neq 0$ on (a, b). Denote by Y the family

$$Y = \{y(t) \in C^1[a, b] : y(a) = 0 = y(b) \quad \text{and} \quad y(t) \neq 0 \text{ on } (a, b)\}.$$

Then for every $y \in Y$,

$$I(y; a, b) = \int_a^b [a(s)|y'(s)|^{\alpha+1} - q(s)|y(s)|^{\alpha+1}] ds \geq 0, \qquad (6.2.14)$$

where equality holds if and only if y and x are proportional.

A closer examination of the proof of this lemma shows that the condition $y \in C^1[a, b]$ in definition of Y can be replaced by a weaker condition:

y is piecewise of the class C^1 in $[a,b]$, and at any discontinuous point $\bar{t} \in [a,b]$ of y' there exist finite limits $y'(\bar{t}+)$, $y'(\bar{t}-)$. This larger class of function we shall denote as Y_1.

Consequently, if we can find a nontrivial solution $y \in Y_1$ such that $I(y;a,b) \leq 0$, then (6.2.13) is *conjugate* in $[a,b]$, i.e., there exists a nontrivial solution with at least two zeros in $[a,b]$. Conversely, if $I(y;a,b) > 0$ for every nontrivial $y \in Y_1$, then (6.2.13) is disconjugate in $[a,b]$. Indeed, if y is a nontrivial solution with consecutive zeros t_1, $t_2 \in [a,b]$, then for

$$y(t) = \begin{cases} x(t), & t \in [t_1, t_2] \\ 0, & t \in [a,b] \backslash [t_1, t_2] \end{cases}$$

we have

$$I_1(y;a,b) = \int_{t_1}^{t_2} \left[a(s)|y'(s)|^{\alpha+1} - q(s)|y(s)|^{\alpha+1} \right] ds$$

$$= a(t)y(t)\psi(y'(t)) \Big|_{t_1}^{t_2} - \int_{t_1}^{t_2} y(s) \left[(a(s)\psi(y'(s)))' + q(s)\psi(y(s)) \right] ds = 0.$$

6.2.2. Oscillation Criteria

The main result of this section is the following theorem.

Theorem 6.2.1. Suppose that

$$\lim_{t\to\infty} (\ln t) \int_t^\infty \left[q(s) - \frac{\gamma_0}{s^{\alpha+1}} \right] s^\alpha ds > 2 \left(\frac{\alpha}{\alpha+1} \right)^\alpha, \qquad (6.2.15)$$

where $\gamma_0 = (\alpha/(\alpha+1))^{\alpha+1}$, then equation (6.2.1) is oscillatory.

Proof. In view of Lemma 6.2.1 it suffices to find for any $d > 0$ a piecewise differentiable function x with compact support in (d,∞), say, $[t_0,t_3]$ such that

$$I(x,t_0,t_3) = \int_{t_0}^{t_3} \left[|x'(s)|^{\alpha+1} - q(s)|x(s)|^{\alpha+1} \right] ds < 0.$$

Let $t_3 > t_2 > t_1 > t_0$ and let f, g be solutions of (6.2.5) satisfying the boundary conditions $f(t_0) = 0$, $f(t_1) = t_1^{\alpha/(\alpha+1)}$, $g(t_2) = t_2^{\alpha/(\alpha+1)}$, $g(t_3) = 0$. Define a test function x as follows

$$x(t) = \begin{cases} 0 & \text{if } t \leq t_0 \\ f(t) & \text{if } t_0 \leq t \leq t_1 \\ t^{\alpha/(\alpha+1)} & \text{if } t_1 \leq t \leq t_2 \\ g(t) & \text{if } t_2 \leq t \leq t_3 \\ 0 & \text{if } t \geq t_3, \end{cases}$$

and for convenience let

$$G(t_0, t_1) = t_1^{\alpha/(\alpha+1)} \left[\psi(f'(t_1)) - v_0 t_1^{-\alpha/(\alpha+1)} \right],$$

$$H(t_2, t_3) = t_2^{\alpha/(\alpha+1)} \left[v_0 t_2^{-\alpha/(\alpha+1)} - \psi(g'(t_2)) \right],$$

$$v_0 = \left(\frac{\alpha}{\alpha+1} \right)^\alpha, \quad x_0(t) = t^{\alpha/(\alpha+1)}.$$

Now integration by parts gives

$$I(x, t_0, t_3) = \int_{t_0}^{t_3} \left[|x'(s)|^{\alpha+1} - q(s)|x(s)|^{\alpha+1} \right] ds$$

$$= \int_{t_0}^{t_3} \left[|x'(s)|^{\alpha+1} - \frac{\gamma_0}{s^{\alpha+1}} |x(s)|^{\alpha+1} \right] ds - \int_{t_0}^{t_3} \left[q(s) - \frac{\gamma_0}{s^{\alpha+1}} \right] |x(s)|^{\alpha+1} ds$$

$$= \int_{t_0}^{t_1} \left[|f'(s)|^{\alpha+1} - \frac{\gamma_0}{s^{\alpha+1}} |f(s)|^{\alpha+1} \right] ds + \int_{t_1}^{t_2} \left[|x_0'(s)|^{\alpha+1} - \frac{\gamma_0}{s^{\alpha+1}} |x_0(s)|^{\alpha+1} \right] ds$$

$$+ \int_{t_2}^{t_3} \left[|g'(s)|^{\alpha+1} - \frac{\gamma_0}{s^{\alpha+1}} |g(s)|^{\alpha+1} \right] ds - \int_{t_0}^{t_1} \left[q(s) - \frac{\gamma_0}{s^{\alpha+1}} \right] |f(s)|^{\alpha+1} ds$$

$$- \int_{t_1}^{t_2} \left[q(s) - \frac{\gamma_0}{s^{\alpha+1}} \right] |x_0(s)|^{\alpha+1} ds - \int_{t_2}^{t_3} \left[q(s) - \frac{\gamma_0}{s^{\alpha+1}} \right] |g(s)|^{\alpha+1} ds$$

$$= f(x(s))\psi(f'(s)) \Big|_{t_0}^{t_1} + x_0(t)\psi(x_0'(s)) \Big|_{t_1}^{t_2} + g(s)\psi(g'(s)) \Big|_{t_2}^{t_3}$$

$$- \int_{t_0}^{t_1} \left[q(s) - \frac{\gamma_0}{s^{\alpha+1}} \right] |f(s)|^{\alpha+1} ds - \int_{t_1}^{t_2} \left[q(s) - \frac{\gamma_0}{s^{\alpha+1}} \right] |x_0(s)|^{\alpha+1} ds$$

$$- \int_{t_2}^{t_3} \left[q(s) - \frac{\gamma_0}{s^{\alpha+1}} \right] |g(s)|^{\alpha+1} ds$$

$$= G(t_0, t_1) + H(t_2, t_3) - \int_{t_0}^{t_1} \left[q(s) - \frac{\gamma_0}{s^{\alpha+1}} \right] |f(s)|^{\alpha+1} ds$$

$$- \int_{t_1}^{t_2} \left[q(s) - \frac{\gamma_0}{s^{\alpha+1}} \right] |x_0(s)|^{\alpha+1} ds - \int_{x_2}^{x_3} \left[q(s) - \frac{\gamma_0}{s^{\alpha+1}} \right] |g(s)|^{\alpha+1} ds.$$

Next, we claim that the functions f/x_0 and g/x_0 are strictly monotonic on (t_0, t_1) and (t_2, t_3) respectively. If $(f/x_0)'(\bar{t}) = 0$ for some $\bar{t} \in (t_0, t_1)$, then we have $(f'/f)(\bar{t}) = (x_0'/x_0)(\bar{t})$. Denote by $w_1 = \psi(f')/\psi(f)$, $w_2 = \psi(x_0')/\psi(x_0)$. Clearly, the functions w_1 and w_2 satisfy the generalized Riccati equation

$$w' + \frac{\gamma_0}{t^{\alpha+1}} + \alpha |w|^{(\alpha+1)/\alpha} = 0 \tag{6.2.16}$$

with the same initial condition at $t = \bar{t}$. Hence, these solutions coincide over the whole interval of their existence, which is the interval (t_0, t_1).

But this is an obvious contradiction (since $f(t_0) = 0$), i.e., $(f/x_0)' > 0$ on this interval. The same argument applies to prove the monotonicity of g/x_0 over (t_2, t_3).

Note that the second mean value theorem of the integral calculus implies the existence of a constant $\xi_1 \in (t_0, t_1)$ such that

$$\int_{t_0}^{t_1} \left[q(s) - \frac{\gamma_0}{s^{\alpha+1}} \right] |f(s)|^{\alpha+1} ds = \int_{t_0}^{t_1} \left[q(s) - \frac{\gamma_0}{s^{\alpha+1}} \right] \left| \frac{f(s)}{x_0(s)} \right|^{\alpha+1} |x_0(s)|^{\alpha+1} ds$$

$$= \int_{\xi_1}^{t_1} \left[q(s) - \frac{\gamma_0}{s^{\alpha+1}} \right] s^{\alpha} ds.$$

Similarly,

$$\int_{t_2}^{t_3} \left[q(s) - \frac{\gamma_0}{s^{\alpha+1}} \right] |g(s)|^{\alpha+1} ds = \int_{t_2}^{\xi_2} \left[q(s) - \frac{\gamma_0}{s^{\alpha+1}} \right] s^{\alpha} ds$$

for some $\xi_2 \in (t_2, t_3)$. Consequently,

$$\int_{t_0}^{t_3} \left[q(s) - \frac{\gamma_0}{s^{\alpha+1}} \right] |x(s)|^{\alpha+1} ds = \int_{\xi_1}^{\xi_2} \left[q(s) - \frac{\gamma_0}{s^{\alpha+1}} \right] s^{\alpha} ds.$$

Next, we shall derive the asymptotic formula for $G(t_0, t_1)$ as $t_1 \to \infty$ and for $H(t_2, t_3)$ as $t_3 \to \infty$. For this, we use the transformation $t = e^u$. Put $z(u) = x(e^u) = x(t)$, $v(u) = t^{\alpha} w(t)$, then this transformation transforms (6.2.5) and (6.2.16) into the equations (with constant coefficients)

$$(\psi(z'))' - \alpha\psi(z') + \gamma_0\psi(z) = 0 \tag{6.2.17}$$

and

$$v' = \alpha v - \gamma_0 - \alpha|v|^{(\alpha+1)/\alpha} = F(v), \tag{6.2.18}$$

respectively. Clearly, equation (6.4.18) possesses the constant solution $v = v_0 = (\alpha/(\alpha+1))^{\alpha}$, and any solution v tends to v_0 as $u \to \infty$. Indeed, $F(v_0) = 0$ and if $v = v(u)$ exists on some interval $[T, \infty)$, then

$$\int_{v(T)}^{v(u)} \frac{d\xi}{F(\xi)} = v - T. \tag{6.2.19}$$

Now, if $u \to \infty$, the integral on the left–hand side must diverge and this occurs only if $v(\infty) = v_0$.

The function $F(v)$ can be expressed in the form (since $F'(v_0) = 0$), $F(v) = (1/2)F''(v_0)(v - v_0)^2 + O((v - v_0)^3)$ as $v \to v_0$, and hence

$$\frac{1}{F(v)} = \frac{2}{F''(v_0)(v - v_0)^2} + O\left(\frac{1}{v - v_0}\right) \qquad \text{as} \quad v \to v_0,$$

this implies

$$\frac{2}{F''(v_0)(v_0 - v)} + O(\ln|v - v_0|) = u - T \quad \text{as} \quad u \to \infty.$$

Thus, we have

$$\frac{2}{F''(v_0)} + (v_0 - v)O(\ln|v - v_0|) = (v_0 - v)(u - T) \quad \text{as} \quad u \to \infty.$$

Since, $\lim_{u \to \infty}(v_0 - v(u))O(\ln|v_0 - v(u)|) = 0$, it follows that

$$\lim_{u \to \infty}(v_0 - v(u))(u - T) = \lim_{u \to \infty} u(v_0 - v(u)) = \frac{2}{F''(v_0)}.$$

Consequently, $O(\ln|v_0 - v(u)|) = O(\ln(1/u)) = O(\ln u)$ as $u \to \infty$, and thus $1/(v_0 - v) = (1/2)F''(v_0)u + O(\ln u)$ as $u \to \infty$, which means

$$
\begin{aligned}
v - v_0 &= -\frac{1}{\frac{1}{2}F''(v_0)u + O(\ln u)} = \frac{1}{\frac{1}{2}F''(v_0)u\left(1 + O\left(\frac{\ln u}{u}\right)\right)} \\
&= -\frac{2}{F''(v_0)u}\left(1 - O\left(\frac{\ln u}{u}\right)\right) = -\frac{2}{F''(v_0)u} + O\left(\frac{\ln u}{u^2}\right)
\end{aligned}
$$

as $t \to \infty$. Taking into account the relation between solutions w and v of (6.2.16) and (6.2.18), we have

$$t^\alpha w(t) - v_0 = -\frac{2}{F''(v_0)\ln t} + O\left(\frac{\ln(\ln t)}{\ln^2 t}\right) \quad \text{as} \quad t \to \infty.$$

Let $w(t, t_0) = \psi(f'(t))/\psi(f(t))$. Then the function w solves (6.2.16), and hence

$$
\begin{aligned}
G(t_0, t_1) &= t_1^{\alpha/(\alpha+1)}\left\{\psi(f'(t_1)) - \left(\frac{\alpha}{\alpha+1}\right)^\alpha t_1^{-\alpha/(\alpha+1)}\right\} = t_1^\alpha w(t_1, t_0) - v_0 \\
&= -\frac{2}{F''(v_0)\ln t_1} + O\left(\frac{\ln(\ln t_1)}{\ln^2 t_1}\right) = \frac{2v_0}{\ln t_1} + O\left(\frac{\ln(\ln t_1)}{\ln^2 t_1}\right)
\end{aligned}
$$

as $t_1 \to \infty$, since by direct computation $F''(v_0) = -1/v_0$. Moreover, observe that $G(t_0, t_1)$ is positive and decreasing. Indeed, if $u_1 = \ln t_1$ and $\bar{v}(u) = e^{\alpha u}w(e^u, t_0)$, then $G(t_0, t_1) = \bar{v}(u_1) - v_0 > 0$ for every $u_1 > u_0$ (since $f_1'(t_1) > x_0'(t_1)$) and $\bar{v}'(u) = F(\bar{v}(u)) < H(v_0) < 0$.

Concerning the asymptotic behavior of $H(t_2, t_3)$ as $t_3 \to \infty$, we proceed similarly as for $G(t_0, t_1)$, to obtain

$$H(t_2, t_3) = t_2^{\alpha/(\alpha+1)}\left[v_0 t_2^{-\alpha/(\alpha+1)} - t_2^{\alpha^2/(\alpha+1)}w(t_2, t_3)\right] = v_0 - t_2^\alpha w(t_2, t_3),$$

where $w(t, t_3) = \psi(g'(t))/\psi(g(t))$ is the solution of (6.2.3) generated by the solution g of the equation (6.2.5), i.e., the solution for which $w(t_3^-, t_3) = -\infty$. We claim that $v_0 - t_2^\alpha w(t_2, t_3) \to 0$ as $t_3 \to \infty$. For this, we use again the transformation $t = e^u$, $v(u) = t^\alpha w(t)$ which transforms (6.2.3) into equation (6.2.18). Let $\bar{u} \in \mathbb{R}$ be arbitrary and denote by $w(u, \bar{u})$ the solution of (6.2.18) determined by the solution of equation (6.2.17) satisfying $z(\bar{u}) = 0$, i.e., $\lim_{u \to \bar{u}^-} w(u, \bar{u}) = -\infty$. Similarly, as for $u \to \infty$ in the previous part of the proof, we have $\lim_{u \to \infty}[v_0 - v(u, \bar{u})] = 0$. Using the fact that (6.2.18) is autonomous, i.e., $v(u - \bar{u}, \bar{u})$ solves this equation too, we have $v_0 - v(u, \bar{u}) \to 0$ as $|u - \bar{u}| \to \infty$, regardless whether $u \to -\infty$ and $\bar{u}$ is fixed, or u is fixed and $\bar{u} \to \infty$. Consequently, if $u_2 = \ln t_2$, $u_3 = \ln t_3$ and $v(u, u_3) = t^\alpha w(t, t_3)$, $t = e^u$, we have $0 = \lim_{u_3 \to \infty}[v_0 - v(u_2, u_3)] = \lim_{t_3 \to \infty}[v_0 - t_2^\alpha w(t_2, t_3)]$, i.e., $v_0 - t_2^\alpha w(t_2, t_3) \to 0$ as $t_3 \to \infty$. Now, let $d \le t_0$ be fixed, we find

$$I(x; t_0, t_3) = G(t_0, t_1) + H(t_2, t_3) - \int_{\xi_1}^{\xi_2} \left[q(s) - \frac{\gamma_0}{s^{\alpha+1}} \right] s^\alpha ds$$

$$\le G(d, t_1) \left[\frac{G(t_0, t_1)}{G(d, t_1)} + \frac{H(t_2, t_3)}{G(d, t_1)} - \frac{1}{G(d, \xi_1)} \int_{\xi_1}^{\xi_2} \left[q(s) - \frac{\gamma_0}{s^{\alpha+1}} \right] s^\alpha ds \right].$$

(6.2.20)

Next, we let $\epsilon > 0$ be a sufficiently small number such that the limit in (6.2.15) is greater than $2(\alpha/(\alpha+1))^\alpha (1 + 6\epsilon) = 2v_0(1 + 6\epsilon)$. We have $\lim_{t \to \infty}(\ln t)G(d, t) = 2v_0$, i.e., $(\ln x)G(d, t) < 2v_0(1+\epsilon)$ if t is sufficiently large, and in view of (6.2.15), $t_0 > d$ can be chosen in such a way that

$$\frac{1}{G(d, \xi_1)} \int_{\xi_1}^\infty \left[q(s) - \frac{\gamma_0}{s^{\alpha+1}} \right] s^\alpha ds = (\ln \xi_1) \left(\int_{\xi_1}^\infty \left[q(s) - \frac{\gamma_0}{s^{\alpha+1}} \right] s^\alpha ds \right)$$

$$\times \left(\frac{1}{(\ln \xi_1)G(d, \xi_1)} \right) > \frac{2v_0(1 + 6\epsilon)}{2v_0(1 + \epsilon)} > 1 + 4\epsilon,$$

(6.2.21)

whenever $\xi_1 > t_0$ and ϵ is sufficiently small. Since

$$\lim_{t \to \infty} \frac{G(t_0, t)}{G(d, t)} = \lim_{t \to \infty} \frac{v_0 - t^\alpha w(t, t_0)}{v_0 - t^\alpha w(t, d)}$$

$$= \lim_{t \to \infty} \frac{\frac{-2}{F''(v_0) \ln t} + O\left(\frac{\ln(\ln t)}{\ln^2 t} \right)}{\frac{-2}{F''(v_0) \ln t} + O\left(\frac{\ln(\ln t)}{\ln^2 t} \right)} = 1,$$

there exists a $t_1 > t_0$ such that

$$\frac{G(t_0, t_1)}{G(d, t_1)} < 1 + \epsilon.$$

Further, (6.2.21) implies the existence of $t_2 > t_1$ such that

$$\frac{1}{G(d, \xi_1)} \int_{\xi_1}^{\xi_2} \left[q(s) - \frac{\gamma_0}{s^{\alpha+1}} \right] s^\alpha ds > 1 + 3\epsilon \quad \text{whenever} \quad \xi_2 > t_2.$$

Finally, since $H(t_2, t_3) \to 0$ as $t_3 \to \infty$, we find that $H(t_2, t_3)G^{-1}(d, t_1) < \epsilon$ if t_3 is sufficiently large. Combining the above calculations, it follows from (6.2.20) that $I(x; t_0, t_3) < -\epsilon$, and hence $I(x; t_0, t_3) < 0$, and now by Lemma 6.2.1 equation (6.2.1) is oscillatory. ∎

Remark 6.2.1.

1. As in Chapter 3, we observe that the transformation of the independent variable

$$u = \int_0^t a^{-1/\alpha}(s)ds \qquad (6.2.22)$$

transforms (6.4.13) into the equation

$$\frac{d}{du}\left(\psi\left(\frac{d}{du}x\right)\right) + a^{1/\alpha}[t(u)]q(t[u])\psi(x) = 0,$$

which is a equation of the form (6.4.1), here $t = t(u)$ is the inverse of $u = u(t)$ given by (6.2.22). Using this transformation a criterion similar to that given in Theorem 6.2.1 can also be formulated for the general equation (6.2.13).

2. In Theorem 6.2.1 and also in the earlier remarks, (6.2.1) is viewed as a perturbation of the generalized Euler equation (6.2.5). Of course, one may also consider (6.2.13) as a perturbation of the one–term nonoscillatory equation $(a(t)\psi(x'(t)))' = 0$.

Now, we shall prove the following oscillation criterion.

Theorem 6.2.2. Suppose that

$$\int^\infty a^{-1/\alpha}(s)ds = \infty \qquad (6.2.23)$$

and

$$\lim_{t\to\infty}\left(\int^t a^{-1/\alpha}(s)ds\right)^\alpha \left(\int_t^\infty q(s)ds\right) > 1,$$

then equation (6.2.13) is oscillatory.

Proof. For the exact construction of the test function $x(t)$ for which the functional I given in (6.2.14) is negative, we define

$$x(t) = \begin{cases} 0, & t \le t_0 \\ \int_{t_0}^t a^{-1/\alpha}(s)ds \Big/ \int_{t_0}^{t_1} a^{-1/\alpha}(s)ds, & t_0 \le t \le t_1 \\ 1, & t_1 \le t \le t_2 \\ \int_t^{t_3} a^{-1/\alpha}(s)ds \Big/ \int_{t_2}^{t_3} a^{-1/\alpha}(s)ds, & t_2 \le t \le t_3 \\ 0, & t \ge t_3. \end{cases}$$

Now, if $t_0 < t_1 < t_2 < t_3$ are sufficiently large, then as before we have $I(x; t_0, t_3) < 0$. ∎

Our next result is a 'nonoscillatory supplement' of Theorem 6.2.2.

Theorem 6.2.3. Suppose that condition (6.2.23) holds and

$$\lim_{t \to \infty} \left(\int^t a^{-1/\alpha}(s)ds \right)^\alpha \left(\int_t^\infty q^+(s)ds \right) < \frac{\alpha^\alpha}{(\alpha+1)^{\alpha+1}}, \qquad (6.2.24)$$

then equation (6.2.13) is nonoscillatory. Here $q^+(t) = \max\{0, q(t)\}$.

Proof. We shall show that the hypotheses of the theorem imply the existence of $N \in \mathbb{R}$ such that

$$I_N(x) = \int_N^\infty \left[a(t)|x'(t)|^{\alpha+1} - q(t)|x(t)|^{\alpha+1} \right] dt > 0$$

for any nontrivial C^1 function x with compact support in (N, ∞) (see Lemma 6.2.1). For this, first we shall establish the following inequality: Let M be a positive differentiable function for which $M'(t) \neq 0$ in $[a, b]$ and let $z \in Y$ (Y is defined as in Lemma 6.2.1). Then,

$$\int_a^b |M'(s)||z(s)|^{\alpha+1}ds \leq (\alpha+1)^{\alpha+1} \int_a^b \frac{|M(s)|^{\alpha+1}}{|M'(s)|^\alpha} |z'(s)|^{\alpha+1}ds. \quad (6.2.25)$$

Indeed, using integration by parts and the Hölder inequality, we have

$$\int_a^b |M'(s)||z(s)|^{\alpha+1}ds \leq (\alpha+1) \int_a^b M(s)|z(s)|^\alpha |z'(s)|ds$$

$$\leq (\alpha+1) \left(\int_a^b |M'(s)||z(s)|^{\alpha+1}ds \right)^{\frac{\alpha}{\alpha+1}} \left(\int_a^b \frac{M^{\alpha+1}(s)}{|M'(s)|^\alpha} |z'(s)|^{\alpha+1}ds \right)^{\frac{1}{\alpha+1}}$$

$$= \left(\int_a^b |M'(s)||z(s)|^{\alpha+1}ds \right)^{\frac{\alpha}{\alpha+1}} \left(\int_a^b \frac{M^{\alpha+1}(s)}{|M'(s)|^\alpha} |z'(s)|^{\alpha+1}ds \right)^{\frac{1}{\alpha+1}},$$

hence the required inequality follows. Now, denote

$$\nu = \frac{\alpha^\alpha}{(\alpha+1)^{\alpha+1}}, \qquad M(t) = \left(\int_0^t a^{-1/\alpha}(s)ds \right)^{-\alpha}$$

and let $N \in \mathbb{R}$ be such that the expression in (6.2.24) is less than ν for $t > N$. Using (6.2.25) and (6.2.24), we find for any differentiable x with compact support in (N, ∞),

$$\int_N^\infty q(s)|x(s)|^{\alpha+1}ds \leq \int_N^\infty q^+(s)|x(s)|^{\alpha+1}ds$$

$$= (\alpha+1)\int_N^\infty q^+(s)\left(\int_N^s x'(u)\psi(x(u))du\right)ds$$

$$\leq (\alpha+1)\int_N^\infty |x'(s)|\psi(x'(s))M(s)\left(\frac{\int_s^\infty q^+(u)du}{M(s)}\right)ds$$

$$< (\alpha+1)\nu\int_N^\infty M(s)|x'(s)|\psi(x(s))ds$$

$$\leq (\alpha+1)\nu\left(\int_N^\infty |M'(s)||x(s)|^{\alpha+1}ds\right)^{\frac{\alpha}{\alpha+1}}\left(\int_N^\infty \frac{|M(s)|^{\alpha+1}}{|M'(s)|^\alpha}|x'(s)|^{\alpha+1}ds\right)^{\frac{1}{\alpha+1}}$$

$$\leq (\alpha+1)^{\alpha+1}\nu\int_N^\infty \frac{|M(s)|^{\alpha+1}}{|M'(s)|^\alpha}|x'(s)|^{\alpha+1}ds$$

$$\leq \int_N^\infty a(s)|x'(s)|^{\alpha+1}ds$$

(since $|M(t)|^{\alpha+1}/|M'(t)|^\alpha = \alpha^{-1/\alpha}a(t)$). Hence, we have

$$\int_N^\infty \left[a(s)|x'(s)|^{\alpha+1} - q(s)|x(s)|^{\alpha+1}\right]ds > 0$$

and this completes the proof. ∎

6.3. Oscillation Criteria via Liapunov's Second Method

In this section we shall discuss the oscillatory property of solutions of second order differential equation

$$(a(t)x'(t))' + F(t, x(t), x'(t)) = 0 \qquad (6.3.1)$$

by applying Liapunov's second method. In what follows we shall assume that

(i) $a(t) \in C([t_0, \infty), \mathbb{R}^+)$, $t_0 \geq 0$, and
(ii) $F \in C([t_0, \infty) \times \mathbb{R}^2, \mathbb{R})$.

For this, we shall need to consider (6.3.1) in its equivalent system form

$$\begin{cases} x'(t) = \dfrac{y(t)}{a(t)}, \\[2mm] y'(t) = -F\left(t, x(t), \dfrac{y(t)}{a(t)}\right). \end{cases} \qquad (6.3.2)$$

6.3.1. Preliminaries

Consider the system of first order differential equations

$$x'(t) = F(t, x(t)), \qquad (6.3.3)$$

where x is an n–vector and $F(t, x)$ is an n–vector function, $F(t, x) \in C([t_0, \infty) \times \mathbb{R}^n, \mathbb{R}^n)$, and let $\|x\|$ be the Euclidean norm of x. In what follows, a *Liapunov function* will be assumed to be a scalar continuous function which satisfies locally a Lipschitz condition with respect to x, i.e., for any compact set $K \subset \mathbb{R}^n$, there exists a constant $L(K) > 0$ such that $\|F(t, x_1) - F(t, x_2)\| \leq L(K)\|x_1 - x_2\|$ for $x_1, x_2 \in K$.

Let $V(t, x)$ be a continuous scalar function defined on an open set S and is locally Lipschitzian in x. Corresponding to $V(t, x)$, we define the function

$$V'_{(6.3.3)}(t, x) = \limsup_{h \to 0^+} \frac{1}{h} \{V(t + h, x + hF(t, x)) - V(t, x)\}. \qquad (6.3.4)$$

Let $x(t)$ be a solution of (6.3.3) which stays in S and denote by $V'(t, x(t))$ the upper right–hand derivative of $V(t, x(t))$, i.e.,

$$V'(t, x(t)) = \limsup_{h \to 0^+} \frac{1}{h} \{V(t + h, x(t + h)) - V(t, x(t))\}. \qquad (6.3.5)$$

For a point $(t, x) \in S$ and small h, there exists a neighborhood U of (t, x) such that $\bar{U} \subset S$, $(t+h, x+hF(t, x)) \in U$ and $(t+h, x(t+h)) \in U$. Let L be the Lipschitz constant of $V(t, x)$ with respect to x in $\bar{U}$, and write

$$V(t+h, x(t+h)) - V(t, x(t)) = V(t + h, x + hF(t, x) + h\epsilon) - V(t, x)$$
$$\leq V(t+h, x + hF(t, x)) + Lh\|\epsilon\| - V(t, x),$$
$$(6.3.6)$$

where $\epsilon \to 0$ as $h \to 0^+$. From (6.3.6) it follows that

$$\limsup_{h \to 0^+} \frac{1}{h} \{V(t + h, x(t + h)) - V(t, x(t))\}$$
$$\leq \limsup_{h \to 0^+} \frac{1}{h} \{V(t + h, x + hF(t, x)) - V(t, x)\}. \qquad (6.3.7)$$

On the other hand, we have

$$V(t + h, x(t + h)) - V(t, x(t)) \geq V(t + h, x + hF(t, x)) - Lh\|\epsilon\| - V(t, x),$$

which implies that

$$V'_{(6.3.3)}(t, x) \leq V'(t, x(t)). \qquad (6.3.8)$$

Thus, from (6.3.8) and (6.3.7), we obtain

$$V'_{(6.3.3)}(t,x) \;=\; V'(t,x(t)). \tag{6.3.9}$$

By the same calculation, we also find the relation

$$\liminf_{h\to 0^+} \frac{1}{h}\{V(t+h,x(t+h)) - V(t,x(t))\} \tag{6.3.10}$$

$$= \liminf_{h\to 0^+} \frac{1}{h}\{V(t+h,x+hF(t,x)) - V(t,x(t))\}.$$

If $V(t,x)$ has continuous partial derivatives of the first order, it follows that

$$V'_{(6.3.3)}(t,x) \;=\; \frac{\partial V}{\partial t} + \frac{\partial V}{\partial x}\cdot F(t,x), \tag{6.3.11}$$

where '$\cdot$' denotes the scalar product.

Now if $V'_{(6.3.3)}(t,x) \le 0$ and consequently $V'(t,x(t)) \le 0$, then the function $V(t,x(t))$ is non–increasing in t, which implies that $V(t,x)$ is non–increasing along a solution of the system (6.3.3). Conversely, if $V(t,x)$ is non–increasing along a solution of the system (6.3.3), we have $V'_{(6.3.3)}(t,x) \le 0$. Further, if

$$\liminf_{h\to 0^+} \frac{1}{h}\{V(t+h,x+hF(t,x)) - V(t,x(t))\} \;\ge\; 0,$$

the function $V(t,x)$ is non–decreasing along a solution of the system (6.3.3), and conversely.

The following property of a Liapunov function $V(t,x)$ is important. Let $x(s)$, $y(s)$ be continuous and differentiable functions for $s \ge t$ such that $x(t) = y(t) = x$. Then by the definition

$$V'(t,x(t)) \;=\; \limsup_{h\to 0^+} \frac{1}{h}\{V(t+h,x(t+h)) - V(t,x(t))\}$$

and

$$V'(t,y(t)) \;=\; \limsup_{h\to 0^+} \frac{1}{h}\{V(t+h,y(t+h)) - V(t,y(t))\}.$$

Let L be a Lipschitz constant of $V(t,x)$ in a neighborhood of the point (t,x). Then for sufficiently small h,

$$V'(t,y(t)) \;\le\; \limsup_{h\to 0^+} \frac{1}{h}\{V(t+h,x(t+h)) - V(t,y(t))\}$$

$$+ \limsup_{h\to 0^+} \frac{1}{h}\{V(t+h,y(t+h)) - V(t+h,x(t+h))\}$$

$$\le\; \limsup_{h\to 0^+} \frac{1}{h}\{V(t+h,x(t+h)) - V(t,x(t))\}$$

$$+ \limsup_{h\to 0^+} \frac{1}{h}L\|y(t+h) - x(t+h)\|.$$

Thus, we have $V'(t, y(t)) \leq V'(t, x(t)) + L\|y'(t) - x'(t)\|$.

6.3.2. Oscillation Criteria

Theorem 6.3.1. Assume that there exist two continuous functions $V(t, x, y)$ and $W(t, x, y)$ which are defined on $t \geq T \geq t_0$, $x > 0$, $|y| < \infty$ and $t \geq T$, $x < 0$, $|y| < \infty$, respectively, where T can be large, and assume that $V(t, x, y)$ and $W(t, x, y)$ satisfy the following conditions:

(i) $V(t, x, y) \to \infty$ uniformly for $x > 0$, $-\infty < y < \infty$ as $t \to \infty$, and $W(t, x, y) \to \infty$ uniformly for $x < 0$ and $-\infty < y < \infty$ as $t \to \infty$,

(ii) $\dot{V}_{(6.3.2)}(t, x(t), y(t)) \leq 0$ for all sufficiently large t, where $\{x(t), y(t)\}$ is a solution of the system (6.3.2) such that $x(t) > 0$ for all large t and

$$V'_{(6.3.2)}(t, x(t), y(t))$$
$$= \limsup_{h \to 0^+} \frac{1}{h} \left\{ V(t + h, x(t + h), y(t + h)) - V(t, x(t), y(t)) \right\},$$

(iii) $W'_{(6.3.2)}(t, x(t), y(t)) \leq 0$ for all sufficiently large t, where $\{x(t), y(t)\}$ is a solution of the system (6.3.2) such that $x(t) < 0$ for all large t and

$$W'_{(6.3.2)}(t, x(t), y(t))$$
$$= \limsup_{h \to 0^+} \frac{1}{h} \left\{ W(t + h, x(t + h), y(t + h)) - W(t, x(t), y(t)) \right\}.$$

Then equation (6.3.1) is oscillatory.

Proof. Let $x(t)$ be a nonoscillatory solution of (6.3.1), say, $x(t) > 0$ for $t \geq t_0$. By condition (i), if t is sufficiently large, say, $t \geq t_1$ we find

$$V(t_0, x(t_0), y(t_0)) < V(t, x(t), y(t)) \tag{6.3.12}$$

for all $x(t) > 0$, $|y(t)| < \infty$ and $t \geq t_1$. However, by condition (ii), we have $V(t, x(t), y(t)) \leq V(t_0, x(t_0), y(t_0))$ for all $t \geq t_0$, which contradicts (6.3.12). When $x(t) < 0$ for $t \geq t_0$, by employing the function $W(t, x(t), y(t))$ we arrive again at a contradiction. This completes the proof. ∎

Now consider a Liapunov function $v(t, x, y)$, and define $v'_{(6.3.2)}(t, x, y)$ by

$$v'_{(6.3.2)}(t, x, y)$$
$$= \limsup_{h \to 0^+} \left\{ v\left(t + h, x + h\frac{y}{a(t)}, y - hF\left(t, x, \frac{y}{a(t)} \right) \right) - v(t, x, y) \right\}.$$
$$\tag{6.3.13}$$

If $v'_{(6.3.2)}(t, x, y) \leq 0$, then $v(t, x(t), y(t))$ is non–increasing in t, where $\{x(t), y(t)\}$ is a solution of the system (6.3.2).

To apply Theorem 6.3.1, we shall need the following lemma.

Lemma 6.3.1. For all large $T \geq t_0$, $t \geq T$, $x > 0$ $(x < 0)$, $-\infty < y < \infty$ assume that there exists a Liapunov function $v(t, x, y)$ which satisfies the following conditions:

(I_1) $yv(t, x, y) > 0$ for $y \neq 0$, $t \geq T$, $x > 0$ $(yv(t, x, y) < 0$ for $y \neq 0$, $t \geq T$, $x < 0)$,

(I_2) $\dot{v}_{(6.3.2)}(t, x, y) \leq -k(t)$, where $k(t) \in C([T, \infty), \mathbb{R})$, and

$$\liminf_{t \to \infty} \int_{T_1}^t k(s)ds \; \geq \; 0 \quad \text{for all large} \quad T_1 \geq T. \tag{6.3.14}$$

Moreover, assume that there exist a T_2 and a $w(t, x, y)$ for all large T such that $T_2 \geq T$ and $w(t, x, y)$ is a Liapunov function defined for $t \geq T_2$, $x > 0$, $y < 0$ $(x < 0$, $y > 0)$ which satisfies the following conditions:

(I_3) $y \leq w(t, x, y)$ $(-y \leq w(t, x, y))$ and $w(T_2, x, y) \leq m(y)$ where $m(y)$ is continuous, $m(0) = 0$ and $m(y) < 0$ $(y \neq 0)$,

(I_4) $w'_{(6.3.2)}(t, x, y) \leq -c(t)w(t, x, y)$ where $c(t) \in C([T_2, \infty), \mathbb{R}_0)$ and

$$\int_{T_2}^{\infty} \frac{1}{a(s)} \exp\left(-\int_{T_2}^s c(u)du\right) ds \; = \; \infty. \tag{6.3.15}$$

Then if $\{x(t), y(t)\}$ is a solution of the system (6.3.2) such that $x(t) > 0$ for all large t, then $y(t) \geq 0$ $(y(t) \leq 0)$ for all large t.

Proof. Let $x(t) > 0$ for $t \geq t_0$. The case when $x(t) < 0$ for $t \geq t_0$ can be treated similarly. Now assume that there exists a sequence $\{t_n\}_{n=1}^{\infty}$ such that $t_n \to \infty$ as $n \to \infty$. Let $t_n \geq T$ and t_n be sufficiently large so that

$$\liminf_{t \to \infty} \int_{t_n}^t k(s)ds \; \geq \; 0 \quad \text{for} \quad t \geq t_n. \tag{6.3.16}$$

Consider the function $v(t, x(t), y(t))$ for $t \geq t_n$. Then, we have

$$v(t, x(t), y(t)) \; \leq \; v(t_n, x(t_n), y(t_n)) - \int_{t_n}^t k(s)ds.$$

From (6.3.16) it follows that there is $T_0 > t_0$ such that for all $t \geq T_0$, $\int_{t_n}^t k(s)ds \geq (1/2)v(t_n, x(t_n), y(t_n))$, because $v(t_n, x(t_n), y(t_n)) < 0$. Therefore, for $t \geq T_0$, we obtain $v(t, x(t), y(t)) \leq (1/2)v(t_n, x(t_n), y(t_n)) < 0$, which implies that $y(t) < 0$ for all $t \geq T_0$.

For T_0 there is a $\bar{T}$ such that $\bar{T} \geq T_0$ and there is a Liapunov function $w(t, x, y)$ defined on $t \geq \bar{T}$, $x > 0$, $y < 0$. For this $w(t, x, y)$, we find

$$y(t) \leq w(t, x(t), y(t)) \leq w(\bar{T}, x(\bar{T}), y(\bar{T})) \exp\left(-\int_{\bar{T}}^t c(s)ds\right)$$

$$\leq m(y(\bar{T})) \exp\left(-\int_{\bar{T}}^t c(s)ds\right) \quad \text{for} \quad t \geq \bar{T}.$$

Since $x'(t) = y(t)/a(t)$, we have

$$x'(t) \leq m(y(\bar{T})) \frac{1}{a(t)} \exp\left(-\int_{\bar{T}}^t c(s)ds\right), \qquad (6.3.17)$$

and hence

$$x(t) \leq x(\bar{T}) + m(y(\bar{T})) \int_{\bar{T}}^t \frac{1}{a(s)} \exp\left(-\int_{\bar{T}}^s c(u)du\right) ds.$$

Since $x(t) > 0$ for $t \geq \bar{T}$ and $m(y(\bar{T})) < 0$, we find by condition (6.3.15) that $x(t) \to -\infty$ as $t \to \infty$, which is a contradiction. This completes the proof. ∎

Remark 6.3.1. When $a(t) = 1$ and $c(t) = 0$, condition (I_3) can be replaced by:

$(I_3)'$ $m_1(y) \leq w(t, x, y)$ and $w(T_2, x, y) \leq m(y)$, where $m_1(y)$ is monotone, continuous, $m_1(0) = 0$, $m_1(y) < 0$ and $m(y)$ is continuous, $m(0) = 0$ and $m(y) < 0$ $(y \neq 0)$.

If we can find Liapunov functions which satisfy the conditions of Lemma 6.3.1, then we can prove the following theorem by employing a similar argument as in Theorem 6.3.1.

Theorem 6.3.2. Let the conditions of Lemma 6.3.1 hold and assume that for each $\delta > 0$, there exist a $T(\delta) > t_0$ and Liapunov functions $V(t, x, y)$ and $W(t, x, y)$ which are defined for all $t \geq T(\delta)$, $x > \delta$, $y \geq 0$ and $t \geq T(\delta)$, $x < -\delta$, $y \leq 0$, respectively. Moreover, assume that $V(t, x, y)$ and $W(t, x, y)$ satisfy the following conditions:

(II_1) $V(t, x, y)$ and $W(t, x, y)$ tend to infinity for all x, y as $t \to \infty$,

(II_2) $V'_{(6.3.2)}(t, x, y) \leq 0$ as long as $V'_{(6.3.2)}$ is defined,

(II_3) $W'_{(6.3.2)}(t, x, y) \leq 0$ as long as $W'_{(6.3.2)}$ is defined.

Then equation (6.3.1) is oscillatory.

Since we assume the existence of Liapunov functions satisfying the conditions of Lemma 6.3.1, if $x(t) > 0$ eventually, then $x(t) > \delta$ eventually for some $\delta > 0$, because $x'(t) = y(t)/a(t) \geq 0$ eventually. A similar argument holds for $x(t) < 0$ eventually.

As applications of the above results, we consider equation (6.3.1) under the following assumptions:

(i_1) $\displaystyle\int_{t_0}^{\infty} ds/a(s) = \infty.$ (6.3.18)

(i_2) For $t \geq t_0 \geq 0$ and $x \geq 0$, there exist $q_1(t) \in C([t_0, \infty), \mathbb{R})$ and $f_1(x) \in C(\mathbb{R}, \mathbb{R})$ such that

$$\liminf_{t\to\infty} \int_T^t q_1(s)ds \geq 0 \quad \text{for all large} \quad T \geq t_0 \qquad (6.3.19)$$

and that $xf_1(x) > 0$, $f_1'(x) \geq 0$ for $x \neq 0$ and for all large t, $x \geq 0$, $|u| < \infty$,

$$q_1(t)f_1(x) \leq F(t, x, u). \qquad (6.3.20)$$

(i_3) For $t \geq t_0$ and $x \leq 0$, there exist $q_2(t) \in C([t_0, \infty), \mathbb{R})$ and $f_2(x) \in C(\mathbb{R}, \mathbb{R})$ such that

$$\liminf_{t\to\infty} \int_T^t q_2(s)ds \geq 0 \quad \text{for all large} \quad T \geq t_0 \qquad (6.3.21)$$

and that $xf_2(x) > 0$, $f_2'(x) \geq 0$ for $x \neq 0$ and for all large t, $x \leq 0$, $|u| < \infty$,

$$F(t, x, u) \leq q_2(t)f_2(x). \qquad (6.3.22)$$

Now, under the above conditions, if $\{x(t), y(t)\}$ is a solution of (6.3.2) such that $x(t) > 0$ for all large t, then $y(t) \geq 0$ for all large t. For this, we can assume that (6.3.19) – (6.3.22) hold for all $t \geq T^*$ and all $T \geq T^*$. For $t \geq T^*$, $x > 0$, $|y| < \infty$, define $v(t, x, y) = y/f_1(x)$. Then, we have

$$v'_{(6.3.2)}(t, x, y) = \frac{1}{f_1^2(x)}\left[-F\left(t, x, \frac{y}{a(t)}\right)f_1(x) - yf'(x)\frac{y}{a(t)}\right] \leq -q_1(t).$$

Hence, this $v(t, x, y)$ satisfies the hypotheses of Lemma 6.3.1 with $k(t) = q_1(t)$.

Since the condition (6.3.19) implies that for all $T \geq T^*$ there is a T_1 such that $T_1 \geq T$ and $\int_{T_1}^t q_1(s)ds \geq 0$ for all $t \geq T_1$, the function $w(t, x, y) = y + f_1(x)\int_{T_1}^t q_1(s)ds$ for $t \geq T_1$, $x > 0$, $y < 0$ satisfies the conditions of Lemma 6.3.1 with $c(t) = 0$. Thus, the conclusion follows.

If we consider functions $v(t, x, y) = y/f_2(x)$, $t \geq T^*$, $x < 0$, $|y| < \infty$ and $w(t, x, y) = -y - f_2(x)\int_{T_1}^t q_2(s)ds$, $t \geq T_1$, $x < 0$, $y > 0$ then from

Lemma 6.3.1 it follows that if $\{x(t), y(t)\}$ is a solution of (6.3.1) such that $x(t) < 0$ for all large t, then $y(t) \leq 0$ for all large t.

Now by employing Theorem 6.3.1 we shall prove the following results.

Corollary 6.3.1. Let conditions $(i_1) - (i_3)$ hold. If

$$\int_{t_0}^{\infty} q_i(s)ds = \infty, \quad i = 1, 2 \tag{6.3.23}$$

then equation (6.3.1) is oscillatory.

Proof. For $t \geq T^*$, $x > 0$ and $-\infty < y < \infty$ set

$$V(t, x, y) = \begin{cases} \dfrac{y}{f_1(x)} + \displaystyle\int_{t_0}^{t} q_1(s)ds & \text{for } y \geq 0 \\ \displaystyle\int_{t_0}^{t} q_1(s)ds & \text{for } y < 0. \end{cases}$$

Then clearly $V(t, x, y) \to \infty$ uniformly for $x > 0$ and $-\infty < y < \infty$ and we have

$$V'_{(6.3.2)}(t, x, y) = \frac{1}{f_1^2(x)} \left[-F\left(t, x, \frac{y}{a(t)}\right) f_1(x) - yf'(x)\frac{y}{a(t)} \right] + q_1(t)$$
$$\leq -q_1(t) + q_1(t) = 0 \quad \text{for } t \geq T^*, \ x > 0 \text{ and } y \geq 0.$$

Therefore, $V(t, x, y)$ satisfies the conditions of Theorem 6.3.1. Similarly,

$$W(t, x, y) = \begin{cases} \displaystyle\int_{t_0}^{t} q_2(s)ds & \text{for } y > 0 \\ \dfrac{y}{f_2(x)} + \displaystyle\int_{t_0}^{t} q_2(s)ds & \text{for } y \leq 0 \end{cases}$$

satisfies the conditions of Theorem 6.3.1. Thus, the conclusion follows. ∎

Corollary 6.3.2. Let conditions $(i_1) - (i_3)$ hold. If

$$\int_{t_0}^{\infty} q_i(s)ds < \infty, \quad \int_{t_0}^{t} \frac{1}{a(s)} \int_{s}^{\infty} q_i(u)du\,ds \to \infty \text{ as } t \to \infty, \quad i = 1, 2$$
$$\tag{6.3.24}$$

and

$$\int_{\epsilon}^{\infty} \frac{du}{f_1(u)} < \infty \quad \text{and} \quad \int_{-\epsilon}^{-\infty} \frac{du}{f_2(u)} < \infty \quad \text{for some } \epsilon > 0, \tag{6.3.25}$$

then equation (6.3.1) is oscillatory.

Proof. For $t \geq T^*$, $x > 0$, $|y| < \infty$ set

$$V(t, x, y) = \int_x^\infty \frac{du}{f_1(u)} + \int_{t_0}^t \frac{1}{a(s)} \int_s^\infty q_1(u) du ds. \qquad (6.3.26)$$

For a solution $x(t)$ which satisfies $x(t) > 0$ for all large t, we can assume that $x(t) > 0$, $y(t) \geq 0$ for $t \geq T_0 \geq t_0$, T_0 is sufficiently large, and hence

$$
\begin{aligned}
V'(t, x(t), y(t)) &= -\frac{1}{f_1(x(t))} \frac{y(t)}{a(t)} + \frac{1}{a(t)} \int_t^\infty q_1(s) ds \\
&= \frac{1}{a(t)} \left[-\frac{y(t)}{f_1(x(t))} + \int_t^\infty q_1(s) ds \right].
\end{aligned}
$$

If we set $V^*(t, x, y) = -(y/f_1(x)) + \int_t^\infty q_1(s) ds$, then $V^*(t, x(t), y(t)) \leq \int_t^\infty q_1(s) ds$, and hence $\limsup_{t \to \infty} V^*(t, x(t), y(t)) \leq 0$. On the other hand, we have

$$V'^*_{(6.3.2)}(t, x, y) = -\frac{1}{f_1^2(x)} \left[-F\left(t, x, \frac{y}{a(t)}\right) f_1(x) - y f_1'(x) \frac{y}{a(t)} \right] - q_1(t) \geq 0.$$

Therefore, $V^*(t, x(t), y(t)) \leq 0$, and consequently $V'(t, x(t), y(t)) \leq 0$ for $t \geq T_0$.

Similarly, we define $W(t, x, y)$ by

$$W(t, x, y) = \int_x^{-\infty} \frac{du}{f_2(u)} + \int_{t_0}^t \frac{1}{a(s)} \int_s^\infty q_2(u) du ds.$$

It is easy to see that $W(t, x, y)$ satisfies the conditions of Theorem 6.3.1. Thus, the conclusion follows. ∎

We can combine the conditions of Corollaries 6.3.1 and 6.3.2, to obtain

Corollary 6.3.3. Let conditions $(i_1) - (i_3)$ hold. If $\int_{t_0}^\infty q_1(s) ds = \infty$,

$$\int_{t_0}^t \frac{1}{a(s)} \int_s^\infty q_2(u) du ds \to \infty \quad \text{as} \quad t \to \infty \quad \text{and} \quad \int_{-\epsilon}^{-\infty} \frac{du}{f_2(u)} < \infty,$$

where $\epsilon > 0$, then equation (6.3.1) is oscillatory.

Corollary 6.3.4. Let conditions $(i_1) - (i_3)$ hold. If there exist a constant $b > 0$ and $\rho_i(t) \in C^1([t_0, \infty), \mathbb{R}^+)$, $i = 1, 2$ such that $f_i'(x) \geq b$ for $x \neq 0$, $i = 1, 2$ and

$$\int_{t_0}^t \rho_i(s) \left[q_i(s) - \frac{1}{4} \frac{a(s)}{b} \left(\frac{\rho_i'(s)}{\rho_i(s)} \right)^2 \right] ds \to \infty \quad \text{as} \quad t \to \infty, \quad i = 1, 2$$

then equation (6.3.1) is oscillatory.

Proof. In this case for $x > 0$, $y \geq 0$

$$V(t,x,y) = \frac{y}{f_1(x)}\rho_1(t) + \int_{t_0}^{t} \rho_1(s)\left[q_1(s) - \frac{1}{4}\frac{a(s)}{b}\left(\frac{\rho_1'(s)}{\rho_1(s)}\right)^2\right]ds$$

and for $x < 0$, $y \leq 0$

$$W(t,x,y) = \frac{y}{f_2(x)}\rho_2(t) + \int_{t_0}^{t} \rho_2(s)\left[q_2(s) - \frac{1}{4}\frac{a(s)}{b}\left(\frac{\rho_2'(s)}{\rho_2(s)}\right)^2\right]ds$$

satisfy the hypotheses of Theorem 6.3.1. ∎

Lemma 6.3.2. In addition to the assumptions of Lemma 6.3.1 assume that there exists a Liapunov function $u(t,x,y)$ defined for $t \geq T^*$, $x > 0$, $y > r$ for $r > 0$ a large constant, which satisfies

(ii$_1$) $u(t,x,y) \to \infty$ uniformly for t, x as $y \to \infty$ (as $y \to -\infty$) and $u(t,x,y) \leq \ell(y)$ $(u(t,x,y) \leq \ell(|y|))$ where $\ell(z) > 0$ is continuous,

(ii$_2$) $u'_{(6.3.2)}(t,x,y) \leq 0$.

Then if $\{x(t), y(t)\}$ is a solution of (6.3.2) such that $x(t) > 0$ $(x(t) < 0)$ for all large t, then $y(t)$ is bounded for all large t.

Proof. Let $x(t) > 0$ and $y(t) \geq 0$ for $t \geq T \geq T^*$. By Lemma 6.3.1 there is such a T. Let K be such that $y(T) < K$, $K > r$. There is a constant $c > 0$ such that $u(t,x,K) \leq c$, and there also exists an $M > 0$ for which we have $c < u(t,x,M)$ for all $t \geq T$ and $x > 0$ by condition (ii$_1$). But, in view of (ii$_2$) we arrive at a contradiction. Hence, $0 \leq y(t) \leq M$ for all $t \geq T$. ∎

Theorem 6.3.3. Let the hypotheses of Lemma 6.3.2 hold and assume that for each $\delta > 0$ and $m > 0$ there exist a $T(\delta,m) > 0$ and two Liapunov functions $V(t,x,y)$ and $W(t,x,y)$ such that $V(t,x,y)$ is defined for $t \geq T(\delta,m)$, $x > \delta$, $0 \leq y < m$ and $W(t,x,y)$ is defined for $t \geq T(\delta,m)$, $x < -\delta$, $-m < y < 0$, moreover, assume that $V(t,x,y)$ and $W(t,x,y)$ satisfy the following conditions:

(II$_1$) $V(t,x,y)$ and $W(t,x,y) \to \infty$ uniformly for x, y as $t \to \infty$,

(II$_2$) $V'_{(6.3.2)}(t,x,y) \leq 0$ as long as $V'_{(6.3.2)}$ is defined,

(II$_3$) $W'_{(6.3.2)}(t,x,y) \leq 0$ as long as $W'_{(6.3.2)}$ is defined.

Then equation (6.3.1) is oscillatory.

Proof. Let $x(t)$ be an eventually positive solution of (6.3.1). By Lemma 6.3.1 there exists a $t_1 > t_0 \geq 0$ such that $x(t) > 0$, $y(t) \geq 0$ for $t \geq t_1$.

By Lemma 6.3.2 there is a constant $m > 0$ such that $0 \leq y(t) < m$ for all $t \geq t_1$. Since $x'(t) = y(t)/a(t) \geq 0$ for $t \geq t_1$, we have $x(t) \geq x(t_1) > 0$ for $t \geq t_1$.

Consider the Liapunov function $V(t, x, y)$ defined for $t \geq T(\delta, m)$, $x > \delta$, $0 \leq y < m$ where $\delta = x(t_1)/2$, where $T \geq t_1$. The rest of the proof is similar to that of Theorem 6.3.1. When $x(t) < 0$ eventually, by employing the Liapunov function $W(t, x, y)$ and Lemma 6.3.2 we arrive at a contradiction. Thus, equation (6.3.1) is oscillatory. ∎

As applications of Theorem 6.3.3, we consider the general differential equation

$$x''(t) + a(t, x(t), x'(t))x'(t) + F(t, x(t), x'(t)) = 0 \qquad (6.3.27)$$

and its equivalent system

$$\begin{cases} x'(t) = y(t) \\ y'(t) = -a(t, x(t), y(t))y(t) - F(t, x(t), y(t)) \end{cases} \qquad (6.3.28)$$

We assume that

(α_1) $F(t, x, y) \in C([t_0, \infty) \times \mathbb{R}^2, \mathbb{R})$ and $xF(t, x, y) > 0$ for $x \neq 0$,

(α_2) $a(t, x, y) \in C([t_0, \infty) \times \mathbb{R}^2, \mathbb{R})$ and there exist two functions $g(t)$, $p(t) \in C([t_0, \infty), \mathbb{R}_0)$ such that $-g(t) \leq a(t, x, y) \leq p(t)$ for $t \geq t_0$, $x \in \mathbb{R}$, $y \in \mathbb{R}$,

(α_3) for any $\delta > 0$ and $m > 0$ there exist a $T(\delta, m)$ and a function $q(t; \delta, m) \geq 0$ defined for $t \geq T(\delta, m)$ such that $\lim_{t \to \infty} \int_{T(\delta, m)}^t q(s; \delta, m) ds = \infty$, and for $|x| \geq \delta$, $|y| \leq m$ and $xy > 0$, $|F(t, x, y)| \geq q(t; \delta, m)$,

(α_4) $\displaystyle \int_{t_0}^\infty g(s) ds < \infty$, $\displaystyle \lim_{t \to \infty} \int_{t_0}^t \exp\left(-\int_{t_0}^s p(u) du\right) ds = \infty$.

Corollary 6.3.5. If conditions $(\alpha_1) - (\alpha_4)$ hold, then equation (6.3.27) is oscillatory.

Proof. It suffices to construct Liapunov functions which satisfy the conditions of Theorem 6.3.3.

For $t \geq t_0 \geq 0$, $x > 0$, $|y| < \infty$ the function

$$v(t, x, y) = \begin{cases} \exp\left(-\displaystyle\int_{t_0}^t g(s) ds\right) & \text{if } y \geq 0 \\ \exp\left(\displaystyle\int_{t_0}^t p(s) ds\right) & \text{if } y < 0 \end{cases}$$

satisfies the conditions of Lemma 6.3.1 with $k(t) = 0$. For any $T \geq t_0$, the function $w(t, x, y) = y$ defined for $t \geq T$, $x > 0$, $y < 0$ satisfies the conditions of Lemma 6.3.1. For this, we have

$$v'_{(6.3.28)}(t, x, y) = -a(t, x, y)y - F(t, x, y) \leq -p(t)y \leq -p(t)w(t, x, y)$$

and $\int_T^\infty \exp\left(-\int_T^s p(u)du\right) ds = \infty$. Moreover, it is easily seen that $u(t, x, y) = y^2 \exp\left(-2\int_{t_0}^t g(s)ds\right)$ satisfies the conditions of Lemma 6.3.2, since $\int_{t_0}^\infty g(s)ds < \infty$ and $F(t, x, y) > 0$. Furthermore, we can see that

$$v(t, x, y) = \begin{cases} (-y)\exp\left(\int_{t_0}^t p(s)ds\right) & \text{for } t \geq t_0, \ x < 0, \ y \geq 0 \\ (-y)\exp\left(-\int_{t_0}^t g(s)ds\right) & \text{for } t \geq t_0, \ x < 0, \ y < 0 \end{cases}$$

$w(t, x, y) = -y$ for $t \geq T$, $x < 0$, $y > 0$ and $u(t, x, y) = \exp\left(-2\int_{t_0}^t g(s)ds\right)$ for $t \geq t_0$, $x < 0$, $y < 0$ satisfy the conditions of Lemma 6.3.2. Next, for each $\delta > 0$ and $m > 0$ define $V(t, x, y)$ for $t \geq T(\delta, m)$, $x > \delta$, $0 \leq y < m$ by

$$V(t, x, y) = y\exp\left(-\int_{t_0}^t g(s)ds\right) + L\int_{T(\delta,m)}^t q(s; \delta, m)ds,$$

where $L = \exp\left(-\int_{t_0}^\infty g(s)ds\right) > 0$. Then, we have

$$V'_{(6.3.28)}(t, x, y)$$
$$= \exp\left(-\int_{t_0}^t g(s)ds\right)[-g(t)y - a(t, x, y)y - F(t, x, y)] + Lq(t; \delta, m)$$
$$\leq \exp\left(-\int_{t_0}^t g(s)ds\right)[-g(t)y + g(t)y - q(t; \delta, m)] + Lq(t; \delta, m)$$
$$\leq -\left(\exp\left(-\int_{t_0}^t g(s)ds\right)\right)q(t; \delta, m) + Lq(t; \delta, m) \leq 0.$$

Thus, we see that $V(t, x, y)$ satisfies the conditions of Theorem 6.3.3. Similarly, the Liapunov function

$$W(t, x, y) = (-y)\exp\left(-\int_{t_0}^t g(s)ds\right) + L\int_{T(\delta,m)}^t q(s; \delta, m)ds$$

satisfies the conditions of Theorem 6.3.3. Thus, it follows from Theorem 6.3.3 that equation (6.3.27) is oscillatory. ∎

Corollary 6.3.6. In Corollary 6.3.5 conditions (α_3) and (α_4) can be replaced by

(α_5) for any $\delta > 0$ there exist a $T(\delta) > 0$ and a $q(t;\delta) > 0$ defined for $t \geq T(\delta)$ such that

$$\left(\exp\left(-\int_{t_0}^{t} g(s)ds\right)\right)\int_{T(\delta)}^{t} q(s;\delta)ds \to \infty \quad \text{as} \quad t \to \infty \qquad (6.3.29)$$

and for $|x| \geq \delta$, $xy \geq 0$, $|F(t,x,y)| \geq q(t;\delta)$, and

(α_6) $\displaystyle\lim_{t\to\infty}\int_{t_0}^{t}\exp\left(-\int_{t_0}^{s}p(u)du\right)ds = \infty.$ \hfill (6.3.30)

Proof. From conditions (α_1), (α_2) and (α_5) it follows that there are Liapunov functions which satisfy the conditions of Lemma 6.3.1 as was seen in Theorem 6.3.2. For $t \geq T(\delta)$, $x > \delta$, $y \geq 0$ define

$$V(t,x,y) = y\exp\left(-\int_{t_0}^{t}q(s)ds\right) + \left(\exp\int_{t_0}^{t}-g(s)ds\right)\int_{T(\delta)}^{t}q(s;\delta)ds.$$

Then, we have

$$\begin{aligned}
V'_{(6.3.28)}(t,x,y) &= \exp\left(-\int_{t_0}^{t}g(s)ds\right)[-g(t)y - a(t,x,y)y - F(t,x,y)] \\
&\quad + \exp\left(-\int_{t_0}^{t}g(s)ds\right)\left[-g(t)\int_{T(\delta)}^{t}q(s;\delta)ds + q(t;\delta)\right] \\
&\leq \exp\left(-\int_{t_0}^{t}g(s)ds\right)[-g(t)y + g(t)y - q(t;\delta)] \\
&\quad -g(t)\left(\exp\left(-\int_{t_0}^{t}g(s)ds\right)\right)\int_{T(\delta)}^{t}q(s;\delta)ds \\
&\quad +\left(\exp\left(-\int_{t_0}^{t}g(s)ds\right)\right)q(t;\delta) \leq 0.
\end{aligned}$$

For $t \geq T(\delta)$, $x < -\delta$, $y \leq 0$ if we define $W(t,x,y)$ by

$$W(t,x,y) = (-y)\exp\left(-\int_{t_0}^{t}g(s)ds\right) + \left(\exp\left(-\int_{t_0}^{t}g(s)ds\right)\right)\int_{T(\delta)}^{t}q(s;\delta)ds,$$

then we also have $W'_{(6.3.28)}(t,x,y) \leq 0$. Therefore, by Theorem 6.3.2 equation (6.3.27) is oscillatory. $\blacksquare$

From Corollaries 6.3.5 and 6.3.6 the following corollary is immediate.

Corollary 6.3.7. In Corollary 6.3.5 conditions (α_3) and (α_4) can be replaced by

(α_7) given $\delta > 0$ there exists a $T(\delta) > t_0$ and a $q(t;\delta) \geq 0$ for $t \geq T(\delta)$ with

$$\frac{1}{t}\int_{T(\delta)}^{t} q(s;\delta)ds \to \infty \quad \text{as} \quad t \to \infty$$

and such that for $|x| \geq \delta$, $xy > 0$, $|F(t,x,y)| \geq q(t;\delta)$,
(α_8) for some $T_0 \geq t_0$ and a constant M,

$$\frac{1}{t}\exp\left(\int_{T_0}^{t} g(s)ds\right) < M \quad \text{for} \quad t \geq T_0$$

and condition (6.3.30) holds.

6.4. Notes and General Discussions

1. The results of Section 6.1 are taken from Butler and Erbe [3]. Corollary 5.1.3 is a known result of Olech et. al. [8]. Theorem 6.1.1 can easily be extended to more general equations of the form

$$(a(t)x'(t))' + q(t)f(x(t)) = 0$$

and

$$(a(t)x'(t))' + q(t)|x(t)|^{\gamma}\,\text{sgn}\,x(t) = 0,$$

where $a(t) \in C([t_0,\infty),\mathbb{R}^+)$. For more details see Butler and Erbe [3].

2. The results of Section 6.2 are taken from Došlý [5]. The method used in Theorems 6.2.1 and 6.2.2 as well to establish (6.2.8) and (6.2.9) suggests the following general approach to the investigation of oscillatory properties of equation (6.2.13). In this method, this equation would be viewed as a perturbation of the general nonoscillatory equation

$$(a(t)\psi(x'(t)))' + q_0(t)\psi(x(t)) = 0. \tag{6.4.1}$$

However, except for the special case treated in Theorem 6.2.1, no 'half-linear' analogy of (6.2.8) and (6.2.9) is known. The main reason is the absence of equivalents of principal and nonprincipal solutions, and also one does not have a transformation theory similar to that for linear equation. Using the linear transformation theory, for example, criterion (6.2.8) may be proved as follows:

If x_0 is the principal solution of equation (6.2.7), then the transformation $x = x_0 u$ transforms this equation into the one term equation

$$(a(t)x_0^2(t)u'(t))' = 0, \tag{6.4.2}$$

hence this transformation converts (6.2.6) into the equation

$$(a(t)x_0^2(t)u'(t))' + [q(t) - q_0(t)]x_0^2(t)u(t) = 0.$$

Now, the last equation is treated as a perturbation of (6.4.2) in a way as suggested in Remark 6.2.1(2) (applied to the linear case) and the obtained results are then transformed 'back' into equation (6.2.6). This approach cannot be directly extended to half–linear equations because of the absence of a 'half–linear' transformation theory. In fact, this is the main difficulty in extending condition (6.2.12) to half–linear equation (6.2.1).

Therefore, it would be interesting to obtain results which can fill the gap between the theory known for linear equations and that for the half–linear equations.

3. The results of Section 6.3 are taken from Yoshizawa [9,10]. Similar results for the difference equations are available in Agarwal and Wong [1]. The obtained results offer alternative proofs for some known results. In fact, Corollaries 6.3.1 – 6.3.7 include results due to Bobisud [2], Coles [4], Macki and Wong [6] and Opial [7]. It is interesting to proceed further in this direction and establish via Liapunov second method oscillation criteria for differential and functional differential equations of arbitrary order.

6.5. References

1. **R.P. Agarwal and P.J.Y. Wong**, *Advanced Topics in Difference Equations*, *Kluwer*, Dordrecht, 1997.

2. **L.E. Bobisud**, Oscillation of solutions of damped nonlinear equations, *SIAM J. Appl. Math.* **19**(1970), 601–606.

3. **G.J. Butler and L.H. Erbe**, A generalization of Olech–Opial–Wazewski oscillation criteria to second order nonlinear equations, *Nonlinear Analysis* **11**(1987), 207–219.

4. **W.J. Coles**, Oscillation criteria for nonlinear second order equations, *Ann. Mat. Pura. Appl.* **83**(1969), 123–134.

5. **O. Došly**, Oscillation criteria for half–linear second order differential equations, *Hiroshima Math. J.* **28**(1998), 507–521.

6. **J.W. Macki and J.S.W. Wong**, Oscillation of solutions to second order nonlinear differential equations, *Pacific J. Math.* **24**(1968), 111–117.

7. **Z. Opial**, Sur les integrales oscillantes de l'equation diffrentielle $u'' + f(t)u = 0$, *Ann. Polon. Math.* **4**(1958), 308–313.

8. **C. Olech, Z. Opial and T. Wazewski**, Sur le probléme d'oscillation des intégrales de l'equation $y'' + g(t)y = 0$, *Bull. Akad. Polon. Sci.* **5**(1957), 621–626.

9. **T. Yoshizawa**, *Stability Theory by Liapunov's Second Method*, The Mathematical Society of Japan, Tokyo, 1966.

10. **T. Yoshizawa**, Oscillatory property for second order differential equations, *Tohoku Math. J.* **22**(1970), 619–634.

Chapter 7

Oscillation Results for Differential Systems

7.0. Introduction

In this chapter we are concerned with the oscillation of nonlinear two–dimensional differential systems and second order vector–matrix differential equations. In Section 7.1 we shall present criteria for the oscillation of nonlinear two–dimensional differential systems. This includes the superlinear, linear, and sublinear cases. Section 7.2 deals with the oscillation of linear second order differential systems. Here, first the system considered will be reduced to a certain scalar Riccati inequality, so that the known results from the literature can be applied to obtain oscillation criteria. Then, we shall employ the notation and definitions of Section 6.1 to present some general results. Finally, we shall use Riccati and variational techniques which involve assumptions on the behavior of the eigenvalues of the coefficient matrix (or of its integral) to present a number of sufficient conditions which guarantee the oscillation of linear second order systems. In Section 7.3 we shall discuss the oscillation of nonlinear second order differential systems with functionally commutative matrix coefficients. Here, we shall show that the oscillation theory of such systems can be effectively reduced to the study of diagonal systems of scalar second order differential equations. In Section 7.4 we shall prove some comparison theorems of Hille–Wintner type for second order operator–valued linear differential equations. In Section 7.5 some oscillation results for second order differential systems with a forcing term are given.

7.1. Oscillation of Nonlinear Two– Dimensional Differential Systems

The oscillation of second order nonlinear differential equations is of particular interest and, therefore, it is widely studied in the literature. A challenging problem is to extend known oscillation criteria for second order nonlinear differential equations to nonlinear two–dimensional differential systems. Such differential systems include in particular second order non-linear differential equations.

Consider the nonlinear two–dimensional differential system

$$\begin{aligned} x'(t) &= q_1(t)f_1(y(t)) \\ y'(t) &= -q_2(t)f_2(x(t)), \end{aligned} \tag{7.1.1}$$

where

(i) $q_i(t) \in C([t_0, \infty), \mathbb{R})$, $i = 1, 2$ and $q_1(t) \geq 0$ for $t \geq t_0$,

(ii) $f_i(x) \in C^1(\mathbb{R}, \mathbb{R})$ and satisfy $xf_i(x) > 0$ for $x \neq 0$, $i = 1, 2$, $f_1(x)$ is increasing on $\mathbb{R}$ and $f_2'(x) \geq 0$ for every $x \neq 0$.

In what follows, we shall restrict our attention only to the solutions of the differential system (7.1.1) which exist on some ray $[T_0, \infty)$, where $T_0 \geq t_0$ may depend on the particular solutions. Of course, under quite general conditions existence of such solutions can always be guaranteed. As usual, a continuous real–valued function defined on an interval $[T_0, \infty)$ is said to be oscillatory if it has arbitrarily large zeros, and otherwise, it is said to be nonoscillatory. A solution (x, y) of the system (7.1.1) is called *oscillatory* if both x and y are oscillatory, and otherwise, it is called *nonoscillatory*. The differential system (7.1.1) is said to be oscillatory if all its solutions are oscillatory. We remark that if the coefficient $q_1(t)$ is not identically zero on any interval of the form $[\tau_0, \infty)$ where $\tau_0 \geq t_0$, then from the first equation of (7.1.1) for any solution (x, y) the oscillation of x implies the same for y. So, if (x, y) is a nonoscillatory solution of (7.1.1), then x is always nonoscillatory.

The special case where $f_i(x) = |x|^{\lambda_i} \operatorname{sgn} x$, $x \in \mathbb{R}$ ($\lambda_i > 0$), $i = 1, 2$ is of particular interest. In this case the differential system (7.1.1) becomes

$$\begin{aligned} x'(t) &= q_1(t)|y(t)|^{\lambda_1} \operatorname{sgn} y(t) \\ y'(t) &= -q_2(t)|x(t)|^{\lambda_2} \operatorname{sgn} x(t), \end{aligned} \tag{7.1.2}$$

where λ_i, $i = 1, 2$ are positive constants. System (7.1.2) is the prototype of (7.1.1).

In the particular case where $q_1(t) > 0$ for $t \geq t_0$ and $f_1(x) = x$, $x \in \mathbb{R}$, (7.1.1) reduces to the equation

$$\left(\frac{1}{q_1(t)}x'(t)\right)' + q_2(t)f_2(x(t)) = 0. \tag{7.1.3}$$

Further, when $q_1(t) \equiv 1$ for $t \geq t_0$, equation (7.1.3) becomes

$$x''(t) + q_2(t)f_2(x(t)) = 0. \tag{7.1.4}$$

The prototype of equation (7.1.4) is the so called Emden–Fowler equation

$$x''(t) + q_2(t)|x(t)|^{\lambda_2}\, \operatorname{sgn} x(t) = 0, \qquad (7.1.5)$$

where $\lambda_2 > 0$. System (7.1.2) reduces to (7.1.5) when $q_1(t) \equiv 1$ for $t \geq t_0$ and $\lambda_1 = 1$.

7.1.1. Superlinear Case

It has been shown (see Theorem 4.1.3) that for equation (7.1.5) if $\lambda_2 > 1$, then a sufficient condition for oscillation is

$$\int^{\infty} s q_2(s)\,ds = \infty. \qquad (7.1.6)$$

Moreover, it is well–known that if $q_2(t) \geq 0$ for $t \geq t_0$, then condition (7.1.6) is also necessary for the oscillation of equation (7.1.5). Our first theorem generalizes this result to the system (7.1.1). For this, we define $R(t) = \int_{t_0}^{t} q_1(s)\,ds$ and assume that

$$\lim_{t \to \infty} R(t) = \infty, \qquad (7.1.7)$$

and f_2 satisfies the strong superlinearity condition, namely,

$$\int^{\infty} \frac{du}{f_2(u)} < \infty \quad \text{and} \quad \int^{-\infty} \frac{du}{f_2(u)} < \infty. \qquad (7.1.8)$$

Theorem 7.1.1. Suppose conditions (7.1.7) and (7.1.8) hold. If $q_2(t) \geq 0$ for $t \geq t_0$ and

$$\int_{t_0}^{\infty} R(s) q_2(s)\,ds = \infty, \qquad (7.1.9)$$

then the system (7.1.1) is oscillatory.

Proof. Assume that the system (7.1.1) admits a nonoscillatory solution (x, y) on an interval $[t_0, \infty)$, $t_0 \geq 0$. From condition (7.1.7) it follows that $q_1(t) \not\equiv 0$ on any interval of the form $[T_0, \infty)$, $T_0 \geq t_0$. So, as remarked above, $x(t)$ is always nonoscillatory. Without loss of generality, we assume that $x(t) \neq 0$ for all $t \geq t_0$. Furthermore, we observe that the substitution $z = -x$, $v = -y$ transforms (7.1.1) into the system

$$\begin{aligned} z'(t) &= q_1(t)\hat{f}_1(v(t)) \\ v'(t) &= -q_2(t)\hat{f}_2(z(t)), \end{aligned}$$

where $\hat{f}_i(u) = -f_i(-u)$, $u \in \mathbb{R}$, $i = 1, 2$. The functions $\hat{f}_i$, $i = 1, 2$ are subject to the conditions posed on f_i, $i = 1, 2$. Thus, we can restrict our discussion only to the case where x is positive on $[t_0, \infty)$, and assume either one of the following two cases holds:

(I_1) $y(t) \leq 0$ for $t \geq t_0$,   or   (I_2) $y(t) \geq 0$ for $t \geq t_0$.

(I_1) Let $y(t) \leq 0$ for $t \geq t_0$. The second equation of (7.1.1) implies that $y(t)$ is decreasing on $[t_0, \infty)$ and $y(t) \to$ either $-\infty$, or a finite negative number as $t \to \infty$. Similarly, it follows that $f_1(y(t)) \to$ either $-\infty$, or finite negative number as $t \to \infty$. This in view of condition (7.1.7) implies that $\int_{t_0}^{\infty} q_1(s) f_1(y(s)) ds = -\infty$. Now, integration of the first equation of (7.1.1) gives

$$x(t) = x(t_0) + \int_{t_0}^{t} q_1(s) f_1(y(s)) ds \to -\infty \quad \text{as } t \to \infty,$$

which contradicts the assumption that $x(t) > 0$ for $t \geq t_0$.

(I_2) Let $y(t) \geq 0$ for $t \geq t_0$. Define $w(t) = f_1(y(t))/f_2(x(t))$, $t \geq t_0$. Then as a consequence of the system (7.1.1) the following equation holds

$$w'(t) + q_1(t) f_2'(x(t)) w^2(t) + q_2(t) f_1'(y(t)) = 0. \tag{7.1.10}$$

Clearly, $q_1(t) f_2'(x(t)) w^2(x(t)) \geq 0$ for $t \geq t_0$, and since $x(t) > 0$ and $y(t) \geq 0$ for $t \geq t_0$, $y(t)$ decreases to a nonnegative constant. Also there is a positive constant k such that

$$f_1'(u) \geq k > 0 \quad \text{for } u \in [0, y(t_0)]. \tag{7.1.11}$$

We thus have

$$w'(t) \leq -k q_2(t) \quad \text{for } t \geq t_0. \tag{7.1.12}$$

Multiplying both sides of (7.1.12) by $R(t)$ and integrating from t_0 to t, we get

$$R(t) w(t) - \int_{t_0}^{t} q_1(s) w(s) ds \leq -k \int_{t_0}^{t} R(s) q_2(s) ds. \tag{7.1.13}$$

Since,

$$\int_{t_0}^{t} q_1(s) w(s) ds = \int_{t_0}^{t} \frac{x'(s)}{f_2(x(s))} ds = \int_{x(t_0)}^{x(t)} \frac{du}{f_2(u)} < \infty,$$

and $R(t) w(t) > 0$ for $t \geq t_0$, we find that $k \int_{t_0}^{t} R(s) q_2(s) ds < \infty$, which contradicts condition (7.1.9). This completes the proof. ∎

Next, we shall extend Theorem 4.1.1 (without condition (4.1.11)) (see alternative proof after Lemma 4.1.3) to (7.1.1). For this we shall assume that

$$\int_{t_0}^{\infty} q_2(s) ds \quad \text{exists as a finite number,} \tag{7.1.14}$$

and define $Q(t) = \int_t^\infty q_2(s)ds$, $t \geq t_0$. We shall also need the following technical lemma.

Lemma 7.1.1. Let (x,y) be a solution on $[t_0, \infty)$, $t_0 \geq 0$ of the system (7.1.1) with $x(t) > 0$ for all $t \geq t_0$. Moreover, let $T \geq t_0$ and c be a real constant. If

$$-\frac{y(t_0)}{f_2(x(t_0))} + \int_{t_0}^t q_2(s)ds + \int_{t_0}^T y(s)x'(s)\frac{f_2'(x(s))}{f_2^2(x(s))}ds \geq c$$

for every $t \geq T$, then $y(t) \leq -cf_2(x(T))$ for all $t \geq T$.

Proof. From the second equation of (7.1.1) we find for $t \geq T$,

$$\int_{t_0}^t q_2(s)ds = -\int_{t_0}^t \frac{y'(s)}{f_2(x(s))}ds$$

$$= -\frac{y(t)}{f_2(x(t))} + \frac{y(t_0)}{f_2(x(t_0))} - \int_{t_0}^t y(s)x'(s)\frac{f_2'(x(s))}{f_2^2(x(s))}ds$$

and so, we have

$$-\frac{y(t)}{f_2(x(t))} = \left[-\frac{y(t_0)}{f(x(t_0))} + \int_{t_0}^t q_2(s)ds + \int_{t_0}^T y(s)x'(s)\frac{f_2'(x(s))}{f_2^2(x(s))}ds \right]$$

$$+ \int_T^t y(s)x'(s)\frac{f_2'(x(s))}{f_2^2(x(s))}ds \quad \text{for every} \quad t \geq T.$$

This for all $t \geq T$, by the hypothesis gives

$$-\frac{y(t)}{f_2(x(t))} \geq c + \int_T^t \left[-\frac{y(s)}{f_2(x(s))} \right] \left[-x'(s)\frac{f_2'(x(s))}{f_2(x(s))} \right] ds.$$

Now by Lemma 3.1.1, we conclude that $w(t) \leq -y(t)$ for every $t \geq T$, where w satisfies

$$\frac{w(t)}{f_2(x(t))} = c + \int_T^t \frac{w(s)}{f_2(x(s))} \left[-x'(s)\frac{f_2'(x(s))}{f_2(x(s))} \right] ds \quad \text{for} \quad t \geq T.$$

Since $w'(t) = 0$ on $[T, \infty)$, we have $w(T) = cf_2(x(T))$. Thus, $w(t) = cf_2(x(T))$ for all $t \geq T$. This completes the proof. $\blacksquare$

Theorem 7.1.2. Let the function $f_1 \circ f_2(u) = f_1(f_2(u))$ satisfy

$$\int^\infty \frac{du}{f_1 \circ f_2(u)} < \infty \quad \text{and} \quad \int^{-\infty} \frac{du}{f_1 \circ f_2(u)} < \infty \qquad (7.1.15)$$

and suppose that conditions (7.1.7) and (7.1.14) hold, and for all sufficiently small u and every $v > 0$,

$$f_1(u)f_1(v) \leq f_1(uv) \leq f_1(u)(-f_1(-v)). \qquad (7.1.16)$$

If

$$\int_{t_0}^{\infty} q_1(s)f_1(Q(s))ds = \infty, \qquad (7.1.17)$$

then the system (7.1.1) is oscillatory.

Proof.   As in the proof of Theorem 7.1.1 without loss of generality, we assume that $x(t) > 0$ for $t \geq t_0$. From the first equation of (7.1.1) it follows that the function $y(t)x'(t)$ is necessarily nonnegative for all $t \geq t_0$, even though $y(t)$ is oscillatory. First, we claim that

$$\int_{t_0}^{\infty} y(s)x'(s)\frac{f_2'(x(s))}{f_2^2(x(s))}ds < \infty. \qquad (7.1.18)$$

For this, assume that (7.1.18) is not satisfied. By condition (7.1.14) there exists a constant k such that

$$-\frac{y(t_0)}{f_2(x(t_0))} + \int_{t_0}^{t} q_2(s)ds \geq k \quad \text{for} \quad t \geq t_0.$$

Furthermore, we can choose a point $t_1 \geq t_0$ such that

$$\int_{t_0}^{t_1} y(s)x'(s)\frac{f_2'(x(s))}{f_2^2(x(s))}ds \geq 1 - k.$$

So, for every $t \geq t_1$, we have

$$-\frac{y(t_0)}{f_2(x(t_0))} + \int_{t_0}^{t} q_2(s)ds + \int_{t_0}^{t_1} y(s)x'(s)\frac{f_2'(x(s))}{f_2^2(x(s))}ds \geq 1$$

and hence, by Lemma 7.1.1 with $T = t_1$ and $c = 1$, we obtain $y(t) \leq b$ for all $t \geq t_1$, where $b = -f_2(x(t_1)) \leq 0$. Next, from the first equation of (7.1.1) we have for $t \geq t_1$,

$$x(t) - x(t_1) = \int_{t_1}^{t} q_1(s)f_1(y(s))ds \leq f_1(b)\int_{t_1}^{t} q_1(s)ds,$$

which in view of condition (7.1.7) gives $\lim_{t\to\infty} x(t) = -\infty$, a contradiction. Thus (7.1.18) holds.

Now in view of (7.1.14), the definition of Q and (7.1.18), the second equation of (7.1.1) for $t \geq t_0$, gives

$$Q(t_0) - Q(t) = \int_{t_0}^{t} q_2(s)ds = -\int_{t_0}^{t} \frac{y'(s)}{f_2(x(s))}ds$$

$$= -\frac{y(t)}{f_2(x(t))} + \frac{y(t_0)}{f_2(x(t_0))} - \int_{t_0}^{t} y(s)x'(s)\frac{f_2'(x(s))}{f_2^2(x(s))}ds$$

$$+ \int_{t}^{\infty} y(s)x'(s)\frac{f_2'(x(s))}{f_2^2(x(s))}ds,$$

i.e.,

$$\frac{y(t)}{f_2(x(t))} = \theta + Q(t) + \int_t^\infty y(s)x'(s)\frac{f_2'(x(s))}{f_2^2(x(s))}\,ds \qquad (7.1.19)$$

for every $t \geq t_0$, where the real number θ is defined by

$$\theta = \frac{y(t_0)}{f_2(x(t_0))} - Q(t_0) - \int_{t_0}^\infty y(s)x'(s)\frac{f_2'(x(s))}{f_2^2(x(s))}\,ds.$$

We claim that the constant θ is nonnegative. Otherwise, from (7.1.14) and (7.1.18) it follows that there exists $T \geq t_0$ such that

$$\int_T^\infty y(s)x'(s)\frac{f_2'(x(s))}{f_2^2(x(s))}\,ds \leq -\frac{\theta}{4} \quad \text{and} \quad \int_t^\infty q_2(s)\,ds \leq -\frac{\theta}{4} \quad \text{for all } t \geq T.$$

Thus, by using (7.1.19), we find for every $t \geq T$,

$$-\frac{y(t_0)}{f_2(x(t_0))} + \int_{t_0}^t q_2(s)\,ds + \int_{t_0}^T y(s)x'(s)\frac{f_2'(x(s))}{f_2^2(x(s))}\,ds$$

$$= -\theta - Q(t_0) - \int_{t_0}^\infty y(s)x'(s)\frac{f_2'(x(s))}{f_2^2(x(s))}\,ds$$

$$= -\theta - \int_t^\infty q_2(s)\,ds - \int_T^\infty y(s)x'(s)\frac{f_2'(x(s))}{f_2^2(x(s))}\,ds \geq -\theta + \frac{\theta}{4} + \frac{\theta}{4} = -\frac{\theta}{2}$$

and so Lemma 4.3.1 ensures that $y(t) \leq b_1$ for all $t \geq T$, where $b_1 = (\theta/2)f_2(x(T)) < 0$. Hence, exactly as in establishing (7.1.18) we arrive at the contradiction $\lim_{t\to\infty} x(t) = -\infty$, which proves the claim.

Finally, (7.1.19) guarantees that $y(t) \geq Q(t)f_2(x(t))$ for every $t \geq t_0$. Hence, in view of $\lim_{t\to\infty} Q(t) = 0$ and the condition (7.1.16), from the first equation of (7.1.1), we obtain for $t \geq t_0$,

$$x'(t) = q_1(t)f_1(y(t)) \geq q_1(t)f_1(Q(t)f_2(x(t))) \geq q_1(t)f_1(Q(t))f_1 \circ f_2(x(t)),$$

and consequently

$$\int_{x(t_0)}^{x(t)} \frac{du}{f_1 \circ f_2(u)} \geq \int_{t_0}^t q_1(s)f_1(Q(s))\,ds \quad \text{for all } t \geq t_0.$$

So, because of (7.1.15), we have

$$\int_{t_0}^t q_1(s)f_1(Q(s))\,ds \leq \int_{x(t_0)}^\infty \frac{du}{f_1 \circ f_2(u)} < \infty \quad \text{for } t \geq t_0,$$

which contradicts condition (7.1.17). This completes the proof. ∎

Example 7.1.1. Consider the differential system

$$x'(t) = y^{1/3}(t)$$
$$y'(t) = -\left(\frac{\sin t}{t} + \frac{2 + \cos t}{t^2}\right) x^5(t) \tag{7.1.20}$$

for $t \geq t_0 = \pi/2$. Here, we have $q_1(t) = 1$, $q_2(t) = (\sin t/t) + (2 + \cos t)/t^2$, $f_1(y) = y^{1/3}$ and $f_2(x) = x^5$. Now,

$$Q(t) = \int_t^\infty q_2(s)ds = \int_t^\infty -d\left(\frac{2 + \cos s}{s}\right) = \frac{2 + \cos t}{t},$$

and hence $Q(t_0) = Q(\pi/2) = 4/\pi$ exists as a real number. Also, we find that $F(u) = f_1 \circ f_2(u) = u^{5/3}$ which is a superlinear function and satisfies $\int^{\pm\infty} u^{-5/3}du < \infty$. Finally,

$$\int_{t_0}^t q_1(s)f_1(Q(s))ds = \int_{\pi/2}^t \left(\frac{2 + \cos s}{s}\right)^{1/3} ds \to \infty \quad \text{as} \quad t \to \infty.$$

Thus all conditions of Theorem 7.1.2 are satisfied, and hence the system (7.1.20) is oscillatory.

7.1.2. Sublinear Case

For equation (7.1.5) it is known that if $q_2(t) > 0$ for $t \geq t_0$ and $0 < \lambda_2 < 1$, then a necessary and sufficient condition for oscillation is that $\int_{t_0}^\infty s^{\lambda_2}q_2(s)ds = \infty$. In this subsection, this result will be generalized to the system (7.1.1).

Theorem 7.1.3. Suppose $q_2(t) \geq 0$ for $t \geq t_0$ and condition (7.1.7) holds. In addition, assume that $f_i'(x) \geq 0$ for $x \neq 0$, $i = 1, 2$ and for all $v \geq 0$ and $u > 0$,

$$-f_2(-u)f_2(v) \geq f_2(uv) \geq f_2(u)f_2(v) \tag{7.1.21}$$

and the composite function $f_2 \circ f_1$ satisfies the sublinearity condition, namely,

$$\int_0^1 \frac{du}{f_2 \circ f_1(u)} < \infty \quad \text{and} \quad \int_0^{-1} \frac{du}{f_2 \circ f_1(u)} < \infty. \tag{7.1.22}$$

If

$$\int_{t_0}^\infty q_2(s)f_2(R(s))ds = \infty, \tag{7.1.23}$$

where $R(t) = \int_{t_0}^t q_1(s)ds$, then the system (7.1.1) is oscillatory.

Proof. First we shall show that for any fixed $t_1 > t_0$,

$$\int_{t_1}^{\infty} f_2(R(t) - R(t_1))q_2(t)dt = \infty. \tag{7.1.24}$$

Since $\lim_{t\to\infty} R(t) = \infty$, there is a $t_2 > t_1$ such that

$$R(t) \geq \max\{1, 2R(t_1)\} \quad \text{for} \quad t > t_2. \tag{7.1.25}$$

Thus, we have

$$R(t) - R(t_1) \geq \frac{1}{2}R(t) \quad \text{for} \quad t > t_2, \tag{7.1.26}$$

and hence by condition (7.1.21) for $t > t_2$,

$$f_2(R(t) - R(t_1)) \geq f_2(R(t)/2) \geq f_2(1/2)f_2(R(t)). \tag{7.1.27}$$

Now (7.1.24) follows from (7.1.27) and (7.1.23).

Suppose now that the system (7.1.1) is not oscillatory, so that $x(t)$ and $y(t)$ are of fixed sign for $t > t_1$. Using (7.1.24), we may use a translation to shift t_1 to t_0. Thus we may assume without loss of generality that $x(t) > 0$ and $y(t)$ is of fixed sign for all $t \geq t_0$.

As in the proof of Theorem 7.1.1 the two possibilities (I_1) and (I_2) are considered, and the case (I_1) can be disposed of in exactly the same way as in Theorem 7.1.1. Thus, we consider

(I_2) Assume that $y(t) \geq 0$ for $t \geq t_0$. From the system (7.1.1) we note that $x(t)$ is an increasing function on $[t_0, \infty)$ while $y(t)$ is a decreasing function on $[t_0, \infty)$. Integrating the first equation of (7.1.1) from t_1 to t, we obtain

$$x(t) = x(t_0) + \int_{t_0}^{t} q_1(s)f_1(y(s))ds \geq \int_{t_0}^{t} q_1(s)f_1(y(s))ds = R(t)f_1(y(t)). \tag{7.1.28}$$

Thus, for all sufficiently large t such that $R(t) \geq 1$,

$$f_2(x(t)) \geq f_2(R(t))f_1(y_1(t)) \geq f_2(R(t))f_2 \circ f_1(y(t)). \tag{7.1.29}$$

Using the second equation of the system (7.1.1), we get $y'(t) \leq -q_2(t) \times f_2(R(t))f_2 \circ f_1(y(t))$, or

$$-\frac{y'(t)}{f_2 \circ f_1(y(t))} \geq q_2(t)f_2(R(t)) \quad \text{for} \quad t \geq t_0. \tag{7.1.30}$$

Integrating (7.1.30) from t_0 to t, we find

$$\int_{y(t)}^{y(t_0)} \frac{du}{f_2 \circ f_1(u)} \geq \int_{t_0}^{t} q_2(s)f_2(R(s))ds. \tag{7.1.31}$$

The right–hand side of (7.1.31) tends to ∞ as $t \to \infty$, while the left–hand side of (7.1.31) remains bounded by (7.1.22) as $y(t) \to 0$ as $t \to \infty$. This contradiction completes the proof. ∎

Example 7.1.2. Consider the system

$$
\begin{aligned}
x'(t) &= y^{3/5}(t) \\
y'(t) &= t^{-14/9}x^{5/9}(t), \quad t \ge t_0 = 1.
\end{aligned}
\tag{7.1.32}
$$

Here we have $q_1(t) = 1$, $q_2(t) = t^{-14/9}$, $f_1(y) = y^{3/5}$ and $f_2(x) = x^{5/9}$. Clearly, $R(t) = \int_1^t ds = t - 1$ and $f_2 \circ f_1(u) = u^{1/3}$. Now

$$
\int_{t_0}^t q_2(s)f_2(R(s))ds = \int_1^t s^{-14/9}(s-1)^{5/9}ds \to \infty \quad \text{as} \quad t \to \infty.
$$

Thus all conditions of Theorem 7.1.3 are satisfied, and hence the system (7.1.32) is oscillatory.

7.1.3. Further Oscillation Criteria

The main result here is the following.

Theorem 7.1.4. Suppose f_1 satisfies

$$
\liminf_{u \to \pm\infty} |f_1(u)| > 0.
\tag{7.1.33}
$$

If

$$
\lim_{t \to \infty} \int_{t_0}^t q_i(s)ds = \infty, \quad i = 1, 2
\tag{7.1.34}
$$

then the system (7.1.1) is oscillatory.

Proof. Without loss of generality we assume that $x(t) > 0$ for $t \ge t_0$. Dividing the second equation of the system (7.1.1) by $f_2(x(t))$ and integrating over $[t_0, t]$, we obtain

$$
-\int_{t_0}^t \frac{y'(s)}{f_2(x(s))}ds = \int_{t_0}^t q_2(s)ds = Q^*(t).
\tag{7.1.35}
$$

Integrating by parts the first term of (7.1.35), we get

$$
-\frac{y(t)}{f_2(x(t))} = Q^*(t) + \left(-\frac{y(t_0)}{f_2(x(t_0))}\right) + \int_{t_0}^t y(s)x'(s)\frac{f_2'(x(s))}{f_2^2(x(s))}ds.
\tag{7.1.36}
$$

The last term of the equality (7.1.36) is nonnegative as $f_2'(x) \ge 0$ and $y(s)x'(s) = q_1(s)y(s)f_1(y) \ge 0$, $s \ge t_0$. Further, in view of (7.1.34) there exists a $t_1 \geq t_0$ such that $Q^*(t) + (-y(t_0)/f_2(x(t_0))) \geq 1$ for all $t \geq t_1$. Thus, we have

$$-\frac{y(t)}{f_2(x(t))} \geq 1 + \int_{t_1}^t \left(-x'(s)\frac{f_2'(x(s))}{f_2^2(x(s))} \right)(-y(s))ds \quad \text{for all} \quad t \geq t_1.$$

$$(7.1.37)$$

Hence, in particular $y(t) < 0$ for $t \geq t_1$. Now, Lemmas 4.1.1 and 4.4.1 allow us to compare $-y(t)$ with the solution $u(t)$ of the integral equation

$$\frac{u(t)}{f_2(x(t))} = 1 + \int_{t_1}^t \left(-x'(s)\frac{f_2'(x(s))}{f_2^2(x(s))} \right)(u(s))ds \quad \text{for all} \quad t \geq t_1 \quad (7.1.38)$$

to conclude that

$$-y(t) \geq u(t) \quad \text{for all} \quad t \geq t_1. \tag{7.1.39}$$

Differentiating (7.1.38), we find that $u'(t) = 0$, $t \geq t_1$ and hence $u(t) = 1$, $t \geq t_1$. Thus (7.1.39) reduces to $y(t) \leq -1$ for all $t \geq t_1$. From the hypothesis (7.1.33), we have $f_1(y(t)) \leq \sup_{y \leq -1} f_1(y) = k < 0$, where k is a constant. Hence, by (7.1.34) it follows that

$$\int_{t_1}^t q_1(s)f_1(y(s))ds \leq k \int_{t_1}^t q_1(s)ds \rightarrow -\infty \quad \text{as} \quad t \rightarrow \infty.$$

Finally, integration of the first equation of (7.1.1) gives

$$x(t) = x(t_1) - \int_{t_1}^t q_1(s)f_1(y(s))ds. \tag{7.1.40}$$

The fact that the right–hand side of (7.1.40) tends to $-\infty$ as $t \rightarrow \infty$ contradicts the assumption $x(t) > 0$ for $t \geq t_0$. This completes the proof. ∎

Remark 7.1.1. In all the results proved above some positivity condition has to be imposed on one of the coefficients $q_i(t)$. To see this, we will construct an example of a nonoscillatory system (7.1.1) for which condition (7.1.34) is satisfied.

We consider the linear case, namely, when $f_i(u) = u$ for $i = 1, 2$. The new variable $w(t) = -y(t)$ satisfies the Riccati differential equation $w'(t) = q_2(t) + q_1(t)w^2(t)$, $t \geq t_0 = 0$, or upon integration, the Riccati integral equation

$$w(t) = \bar{Q}(t) + \int_0^t q_1(s)w^2(s)ds, \tag{7.1.41}$$

where $\bar{Q}(t) = w(t_0) + \int_0^t q_2(s)ds$. The question whether the system (7.1.1) has a solution $x(t)$ of fixed sign (say positive) is equivalent to the question whether (7.1.41) has a continuous solution on $[0, \infty)$.

In what follows we shall construct step functions $\bar{Q}(t)$ and $q_2(t)$ so that (7.1.41) has a solution on $[0, \infty)$. We shall allow $q_i(t)$ to change sign, even though condition (7.1.34) is satisfied. Although our smoothness condition on $q_i(t)$ precludes such step functions, and the solution has jumps at each of the points $3k$, $3k + 1$, modification of our example by smoothing out the abrupt jumps easily leads to an acceptable counterexample.

Let $q_i(t) = \begin{cases} -1, & t \in [3k, 3k + 1) \\ 1, & t \in [3k + 1, 3k + 3) \end{cases}$. Then, $\lim_{t \to \infty} \int_0^t q_1(s)ds = \infty$.

Let us now construct $\bar{Q}(t)$. In each $[3k, 3k + 1)$, we choose $\bar{Q}(t)$ to be a constant $\alpha = \alpha(k)$ so large that $\alpha + \int_0^{3k} q_1(s)w^2(s)ds = k$. Later, we are going to show that the number $\int_0^{3k} q_1(s)w^2(s)ds < 0$ for all k. Hence,

$$\alpha(k) \geq k \to \infty \quad \text{as} \quad k \to \infty. \tag{7.1.42}$$

In the interval $[3k, 3k + 1)$, (7.1.41) reduces to $w(t) = k - \int_{3k}^t w^2(s)ds$, which has the solution $w(t) = [t - 3k + (1/k)]^{-1}$. Note that the right–hand side of this function is well–defined as the denominator does not vanish in $[3k, 3k + 1)$. From (7.1.41), we have

$$\int_0^{3k+1} q_1(s)w^2(s)ds = w(3k + 1) - \alpha = \frac{k}{k+1} - \alpha < 0. \tag{7.1.43}$$

In $[3k + 1, 3k + 3)$, we take $\bar{Q}(t)$ to be negative of the number in (7.1.43). Hence, we find that $\lim_{t \to \infty} \bar{Q}(t) = \infty$. In $[3k + 1, 3k + 3)$, equation (7.1.41) reduces to the simple equation $w(t) = \int_{3k+1}^t w^2(s)ds$, which has the trivial solution $w(t) = 0$ for $t \in [3k + 1, 3k + 3)$. Thus,

$$\int_0^{3(k+1)} q_1(s)w^2(s)ds = \int_0^{3k+1} q_1(s)w^2(s)ds < 0. \tag{7.1.44}$$

We assumed that the above inequality is true when the upper limit of integration is $3k$ and concluded that the same is true for $3(k + 1)$. This induction step allows us to infer that (7.1.44) holds for all $k > 0$. Continuing this process over each interval, we obtain a solution $w(t)$ defined on the whole half–line $[0, \infty)$, and so we have an example of a nonoscillatory system (7.1.1) for which condition (5.3.34) is satisfied.

It is natural to ask whether some relaxation on the positivity requirement is possible. As shown by the above example, it is not sufficient to assume that at each t one of the q_i's is nonnegative.

Remark 7.1.2. We note that Theorem 7.1.4 holds without the assumption that f_1 is increasing on $\mathbb{R}$. Clearly, the increasing character of f_1 on $\mathbb{R}$ implies condition (7.1.33).

Example 7.1.3. Consider the differential system

$$x'(t) = \frac{1}{t}\tanh y(t)$$

$$y'(t) = -\left[-\sqrt{t}\sin t + \frac{1}{2\sqrt{t}}(2+\cos t)\right]|x(t)|^\lambda \operatorname{sgn} x(t) \qquad (7.1.45)$$

for $t \geq t_0 = \pi/2$, $\lambda > 0$ is a constant. Here, we have $q_1(t) = 1/t$, $q_2(t) = -\sqrt{t}\sin t + (1/2\sqrt{t})(2+\sin t)$, $t \geq t_0$, $f_1(u) = \tanh u$, $f_2(u) = |u|^\lambda \operatorname{sgn} u$, $u \in \mathbb{R}$. For every $t \geq t_0$,

$$\int_{t_0}^t q_2(s)ds = \int_{\pi/2}^t \left[-\sqrt{s}\sin s + \frac{1}{2\sqrt{s}}(2+\sin s)\right]ds = \int_{\pi/2}^t d[\sqrt{s}(2+\cos s)]$$

$$= \sqrt{t}(2+\cos t) - 2\sqrt{\pi/2} \to \infty \quad \text{as} \quad t \to \infty.$$

All conditions of Theorem 7.1.4 are satisfied, and hence the system (7.1.45) is oscillatory. However, Theorems 7.1.1 – 7.1.3 are not applicable.

7.2. Oscillation Theorems for Linear Differential Systems

Consider the linear second order differential system

$$x''(t) + Q(t)x(t) = 0, \qquad (7.2.1)$$

where $x : [t_0, \infty) \to \mathbb{R}^n$ and $Q(t)$ is a continuous real symmetric $n \times n$ matrix function for $t \geq t_0 \geq 0$. In 1980, Hinton and Lewis conjectured that (7.2.1) is oscillatory whenever

$$\lim_{t \to \infty} \lambda_1 \left(\int_{t_0}^t Q(s)ds\right) = \infty, \qquad (7.2.2)$$

where $\lambda_1(\cdot)$ denotes the maximum eigenvalue of the matrix. In partial answer to this conjecture the following theorem is obtained.

Theorem 7.2.1. If condition (7.2.2) holds and either

$$\liminf_{t \to \infty} \frac{1}{t}\operatorname{tr}\left(\int_{t_0}^t Q(s)ds\right) > -\infty, \qquad (7.2.3)$$

or

$$\liminf_{t \to \infty} \frac{1}{t}\operatorname{tr}\left(\int_{t_0}^t \int_{t_0}^s Q(u)duds\right) > -\infty, \qquad (7.2.4)$$

where $\operatorname{tr}(\cdot)$ represents the trace of the matrix, then system (7.2.1) is oscillatory.

We recall some pertinent definitions and notation which will be subsequently used. For any $n \times n$ matrix A, the transpose will be denoted by A^*, similarly x^* denotes the transpose of the column vector x. If $t_1, t_2 \in [t_0, \infty)$, $t_1 \neq t_2$ and if there exists a nontrivial solution of (7.2.1) which vanishes at t_1 and t_2, then t_1 and t_2 are said to be (mutually) conjugate relative to (7.2.1). System (7.2.1) is said to be disconjugate on an interval $J \subset [t_0, \infty)$ if every nontrivial solution of (7.2.1) vanishes at most once in J and system (7.2.1) is said to be oscillatory if for each $t_1 > t_0$ there exists a $t_2 > t_1$ such that (7.2.1) is not disconjugate on $[t_1, t_2]$.

The matrix differential system associated to (7.2.1) is

$$X''(t) + Q(t)X(t) = 0, \quad t \in [t_0, \infty) \tag{7.2.5}$$

where $X(t)$ is an $n \times n$ matrix and $Q(t)$ is as in (7.2.1). A solution of the system (7.2.5) is said to be nontrivial if $\det X(t) \neq 0$ for at least one $t \in [t_0, \infty)$ and a nontrivial solution $X(t)$ is said to be prepared or self–conjugate if

$$X^*(t)X'(t) - (X^*(t))'X(t) = 0, \quad t \in [t_0, \infty) \tag{7.2.6}$$

(we note that for any solution $X(t)$ of system (7.2.5),

$$X^*(t)X'(t) - (X^*(t))'X(t) = C, \tag{7.2.7}$$

where C is a constant matrix). System (7.2.5) is said to be oscillatory if the determinant of every nontrivial prepared solution vanishes on $[T, \infty)$ for each $T > t_0$. This is equivalent to oscillation of (7.2.1) since any solution of (7.2.1) is of the form $x(t) = X(t)c$ for some constant vector c and some nontrivial prepared solution $X(t)$ of the system (7.2.5).

If A is a real symmetric $n \times n$ matrix, then its eigenvalues $\lambda_k(A)$, $k = 1, 2, \cdots, n$ (which are all real) will be assumed to be ordered so that

$$\lambda_1(A) \geq \lambda_2(A) \geq \cdots \geq \lambda_n(A) \tag{7.2.8}$$

and

$$\operatorname{tr} A = \sum_{k=1}^{n} \lambda_k(A). \tag{7.2.9}$$

We denote by S the linear space of all $n \times n$ real symmetric matrices. A linear functional $\phi : S \to \mathbb{R}$ is said to be positive if $\phi(A) \geq 0$ for $A \in S$ and $A \geq 0$, (i.e., A is symmetric positive semidefinite).

In the following discussion, we shall show how condition (7.2.2) leads to certain scalar Riccati inequality. We will then employ some recent results on the oscillation of scalar equations to establish a set of conditions weaker than (7.2.3) and (7.2.4) under which (7.2.2) leads to the oscillation of the system (7.2.1).

For a given solution $X(t)$ of (7.2.5) we define a matrix–valued function $W(t) = -X'(t)X^{-1}(t)$. This function satisfies the Riccati equation

$$W(t) = W(t_0) + Q_1(t) + \int_{t_0}^{t} W^2(s)ds, \qquad (7.2.10)$$

where $Q_1(t) = \int_{t_0}^{t} Q(s)ds$. Taking traces on both sides of (7.2.10), we obtain the scalar equation

$$\operatorname{tr} W(t) = \operatorname{tr} W(t_0) + \operatorname{tr} Q_1(t) + \int_{t_0}^{t} \operatorname{tr} W^2(s)ds. \qquad (7.2.11)$$

Now suppose condition (7.2.2) holds, i.e., $\lim_{t\to\infty} \lambda_1(Q_1(t)) = \infty$. Then, $\lambda_1(W(t_0) + Q_1(t)) \to \infty$ as $t \to \infty$, and given $\epsilon > 0$ we can find a large $T \geq t_0$ such that

$$\lambda_1(W(t_0) + Q_1(t)) \geq (1 - \epsilon)\lambda_1(Q_1(t)) \geq 0 \quad \text{for} \quad t \geq T. \qquad (7.2.12)$$

Hence, for all $t \geq T$,

$$\operatorname{tr} W^2(t) = \sum_{k=1}^{n} \lambda_k^2(W(t)) \geq \lambda_1^2(W(t)) \geq (1 - \epsilon)^2\lambda_1^2(Q_1(t)). \qquad (7.2.13)$$

We use this estimate in (7.2.11) to derive an inequality for the scalar quantity $\operatorname{tr} W(t)$. For this, we split the integral term in (7.2.11) as follows

$$\int_{t_0}^{t} \operatorname{tr} W^2(s)ds = \epsilon \int_{t_0}^{t} \operatorname{tr} W^2(s)ds + (1 - \epsilon) \int_{t_0}^{t} \operatorname{tr} W^2(s)ds. \qquad (7.2.14)$$

Assuming $t \geq T$ and ignoring the contribution from the interval $[t_0, T]$ to the last integral, we estimate the remainder by means of (7.2.13), to obtain

$$(1 - \epsilon) \int_{t_0}^{t} \operatorname{tr} W^2(s)ds \geq (1 - \epsilon)^3 \int_{T}^{t} \lambda_1^2(Q_1(s))ds. \qquad (7.2.15)$$

In the first integral we use the simple estimate $\operatorname{tr} W^2(t) \geq \sum_{k=1}^{n} w_{kk}^2(t) \geq (1/n)(\operatorname{tr} W(t))^2$, where $w_{kk}(t)$ is the kth diagonal entry of W, to get the inequality

$$\epsilon \int_{t_0}^{t} \operatorname{tr} W^2(s)ds \geq \frac{\epsilon}{n} \int_{t_0}^{t} (\operatorname{tr} W(s))^2 ds. \qquad (7.2.16)$$

Thus, combining equations (7.2.11), (7.2.14) with inequalities (7.2.15) and (7.2.16), we find

$$\operatorname{tr} W(t) \geq \operatorname{tr} Q_1(t) + \operatorname{tr} W(t_0) + (1-\epsilon)^3 \int_{T}^{t} \lambda_1^2(Q_1(s))ds + \frac{\epsilon}{n} \int_{t_0}^{t} (\operatorname{tr} W(s))^2 ds. \qquad (7.2.17)$$

This inequality holds for all $t \geq t_0$. Using condition (7.2.2) we get for all sufficiently large $t \geq T_1 \geq T$,

$$\operatorname{tr} W(t_0) + (1 - \epsilon)^3 \int_{t_0}^t \lambda_1^2(Q_1(s))ds \geq (1 - \epsilon)^4 \int_T^t \lambda_1^2(Q_1(s))ds.$$

Hence, if we let $\delta = (1 - \epsilon)^4$, replace ϵ by $n\epsilon$ and define $w(t) = \operatorname{tr} W(t)$, $t \geq t_0$ and

$$q(t) = \operatorname{tr} Q_1(t) + \delta \int_{t_0}^t \lambda_1^2(Q_1(s))ds, \quad t \geq t_0 \qquad (7.2.18)$$

then $w(t)$ satisfies the scalar Riccati inequality

$$w(t) \geq q(t) + \epsilon \int_{t_0}^t w^2(s)ds \quad \text{for} \quad t \geq T_1. \qquad (7.2.19)$$

Next, we state some known sufficient conditions which ensure that the inequality (7.2.19) is incompatible with the existence of a continuous function $w(t)$ on the entire half–line $[t_0, \infty)$, $t_0 \geq 0$. For $t \geq t_0$, we let $Q_1(t) = \int_{t_0}^t Q(s)ds$ and $Q_2(t) = \int_{t_0}^t Q_1(s)ds$.

(i_1) There exists a measurable subset J of $[t_0, \infty)$ with mes $J = \infty$ such that

$$\lim_{t \in J,\ t \to \infty} q(t) = \infty. \qquad (7.2.20)$$

(i_2) There exists a measurable subset J of $[t_0, \infty)$ with mes $J < \infty$ such that (7.2.20) holds, and

$$\int_J q^\gamma(s)ds = \infty \qquad (7.2.21)$$

for some constant $\gamma \in (0, 1)$.

(i_3) There exists a weight function $\rho \in \mathcal{F}_1$ such that

$$\liminf_{t \to \infty} \frac{\int_{t_0}^t \rho(s)q(s)ds}{\int_{t_0}^t \rho(s)ds} > -\infty, \quad t \geq t_0 \qquad (7.2.22)$$

where $\mathcal{F}_1$ is the family of weight functions which includes polynomial functions that are strictly positive on $[t_0, \infty)$, and

$$\lim_{t \to \infty} \int_{t_0}^t [(q(s) - \lambda)^+]^2 ds = \infty \qquad (7.2.23)$$

for every $\lambda \in \mathbb{R}$, where $(\phi(t))^+ = \max\{\phi(t), 0\}$.

(i$_4$) There exists a weight function $\rho \in \mathcal{F}_0 \subset \mathcal{F}_1$ such that (7.2.22) holds and

$$\limsup_{t \to \infty} \frac{\int_{t_0}^{t} \rho(s) q(s) ds}{\int_{t_0}^{t} \rho(s) ds} = \infty. \qquad (7.2.24)$$

Now we shall apply these criteria to present oscillation results for (7.2.1).

Theorem 7.2.2. Suppose condition (7.2.2) holds. Then system (7.2.1) is oscillatory if for some $\delta \in (0, 1)$ the function $q(t)$ defined by (7.2.18) satisfies one of the following conditions:

(I$_1$) The set $J = \{t \in [t_0, \infty) : q(t) \geq 0\}$ has infinite measure.

(I$_2$) The set $J = \{t \in [t_0, \infty) : q(t) \geq 0\}$ has finite measure and $\int_J q^\gamma(s) ds = \infty$ for some $\gamma \in (0, 1)$.

(I$_3$) Conditions (7.2.22) and (7.2.23) hold for some $\rho \in \mathcal{F}_1$.

(I$_4$) Conditions (7.2.22) and (7.2.24) hold for some $\rho \in \mathcal{F}_0$.

Proof. Suppose condition (7.2.2) holds. Then equation (7.2.1) is oscillatory or there exists a function $w(t)$ which is defined and continuous on the half–line $[t_0, \infty)$ such that inequality (7.2.19) holds for all sufficiently large t. The later possibility is excluded by each of the criteria (I$_i$), $i = 1, 2, 3, 4$. ∎

The following corollaries are immediate.

Corollary 7.2.1. Let the function $q(t)$ in Theorem 7.2.2 be replaced by $q^*(t) = \text{tr} \, Q_1(t) + ct$ for $t \geq t_0$, where c is any positive constant. Then the conclusion of Theorem 7.2.1 holds.

Corollary 7.2.2. Suppose condition (7.2.2) holds. Then equation (7.2.1) is oscillatory if for some constant c, $\text{tr} \, Q_1(t) \geq -ct$ on a set of t with infinite measure.

Corollary 7.2.3. Suppose condition (7.2.2) holds. Then equation (7.2.1) is oscillatory if

$$\liminf_{t \to \infty} \frac{1}{t^2} \text{tr} \, Q_2(t) > -\infty. \qquad (7.2.25)$$

Proof. If condition (7.2.25) holds, then there exists a constant $c > 0$ such that $(1/t) \text{tr} \, Q_2(t) + (1/4)ct \geq 0$ for $t \geq t_0$. Hence,

$$\frac{1}{t} \int_{t_0}^{t} (\text{tr} \, Q_1(s) + cs) \, ds \geq \frac{1}{4} ct.$$

The result follows from Corollary 7.2.1 (take $\rho(t) = t$ in (7.2.22) and (7.2.24)). ∎

Corollary 7.2.4. Suppose condition (7.2.2) holds. Then equation (7.2.1) is oscillatory if

$$\liminf_{t\to\infty} \frac{1}{t^{m+2}} \operatorname{tr}\left(\int_{t_0}^t s^m Q(s)ds\right) > -\infty$$

for some integer $m \geq 1$.

Proof. The proof is same as that of Corollary 7.2.3 (take $\rho(t) = t^m$). ∎

In our next result we shall use the notation and definitions of Section 6.1, and will need the following lemma.

Lemma 7.2.1. Let $V(t)$ be a continuous $n \times n$ symmetric matrix and

$$\operatorname*{lim\,approx\,sup}_{t\to\infty} \frac{1}{g(t)} \left[\operatorname{tr}\left(V(t) + \int_{t_0}^t V^2(s)ds\right)\right] = L < \infty, \quad (7.2.26)$$

where $g(t)$ is positive, absolutely continuous, and nondecreasing function on (t_0, ∞). Then,

$$\operatorname*{lim\,approx\,inf}_{t\to\infty} \frac{1}{g(t)} \int_{t_0}^t \lambda_1(V^2(s))ds < \infty. \quad (7.2.27)$$

Proof. Since $V(t) = V^*(t)$ it follows that $V^2(t) \geq 0$, and so $F(t) = \int_{t_0}^t V^2(s)ds$ satisfies $F(t_2) \geq F(t_1)$, $t_0 \leq t_1 < t_2 < \infty$, (i.e., $F(t_2) - F(t_1)$ is nonnegative definite). Therefore, since $[\lambda_1(V(t))]^2 \leq \lambda_1(V^2(t)) \leq \operatorname{tr}(V^2(t)) \leq n\lambda_1(V^2(t))$, it follows that if

$$\operatorname*{lim\,approx\,inf}_{t\to\infty} \frac{1}{g(t)} \int_{t_0}^t \lambda_1(V^2(s))ds = \infty,$$

then

$$\operatorname*{lim\,approx\,inf}_{t\to\infty} \frac{1}{g(t)} \int_{t_0}^t \operatorname{tr}(V^2(s))ds = \operatorname*{lim\,approx\,inf}_{t\to\infty} \frac{1}{g(t)}\operatorname{tr}\left(\int_{t_0}^t V^2(s)ds\right)$$

$$= \infty.$$

We define $H(t) = \operatorname{tr}\left(\int_{t_0}^t V^2(s)ds\right) - Mg(t)$, $t \geq t_0$ where $M > L$ is arbitrary. Then, $\operatorname{lim\,approx\,inf}_{t\to\infty} H(t)/g(t) = \infty$ so that the inequality $H(t) > g(t)$ holds on a set E with $\mu(E) = \infty$, where $\mu(E)$ denotes the Lebesgue measure of E. Also, if we define the set E_1 by

$$E_1 = \left\{t : \frac{1}{g(t)}[\operatorname{tr} V(t) + H(t)] \geq 0\right\},$$

then by (7.2.26) we find that $\mu(E_1) < \infty$, since $M > L$ and so the complement of E_1 relative to $[t_0, \infty)$ which we denote by E_1^c, satisfies $\mu(E_1^c) = \infty$, and $H(t) < -\text{tr } V(t)$ on E_1^c. Since $\mu(E \cap E_1^c) = \infty$ we see that the inequality $g(t) < H(t) < -\text{tr } V(t)$ holds on $E \cap E_1^c$. Thus,

$$g^2(t) \leq H^2(t) \leq (\text{tr } V(t))^2 \leq n \text{ tr } (V^2(T)) = n(H'(t) + Mg'(t))$$

holds (a.e.) on $E \cap E_1^c$ and so, we have

$$\frac{1}{n} \leq \frac{H'(t)}{H^2(t)} + M\frac{g'(t)}{H^2(t)} \leq \frac{H'(t)}{H^2(t)} + M\frac{g'(t)}{g^2(t)} \quad \text{on } E \cap E_1^c. \quad (7.2.28)$$

But, this is a contradiction since

$$\int_{t_0}^t \frac{H'(s)}{H^2(s)}ds + M\int_{t_0}^t \frac{g'(s)}{g^2(s)}ds \; < \; \int_{H(t_0)}^\infty u^{-2}du + M\int_{g(t_0)}^\infty u^{-2}du$$

$$= \frac{1}{H(t_0)} + \frac{M}{g(t_0)} < \infty$$

whereas the integral on the left–hand side is infinite since $\mu(E \cap E_1^c) = \infty$. This completes the proof. ∎

Now, we shall prove the following result:

Theorem 7.2.3. Let $g(t)$ be positive, absolutely continuous, and non-decreasing on $[t_0, \infty)$ and

$$\lim_{t \to \infty} \text{approx inf } \frac{1}{g(t)}\text{tr}\left(\int_{t_0}^t Q(s)ds\right) - l > -\infty, \; l \text{ is a constant}, \quad (7.2.29)$$

$$\lim_{t \to \infty} \text{approx inf } \frac{1}{g(t)}\int_{t_0}^t \left[\lambda_1\left(\int_{t_0}^s Q(u)du\right)\right]^2 ds = \infty \quad (7.2.30)$$

and that

$$\lim_{t \to \infty} \lambda_1\left(\int_{t_0}^t Q(s)ds\right) = \infty. \quad (7.2.31)$$

Then equation (7.2.1) is oscillatory.

Proof. If system (7.2.5) is not oscillatory, then there exists a nontrivial prepared solution $X(t)$ of system (7.2.5) with $\det X(t) \neq 0$ on $[t_1, \infty)$ for some $t_1 \geq t_0$. Let $V(t) = X'(t)X^{-1}(t)$, $t \geq t_0$, so that $V(t) = V^*(t)$. Clearly, $V(t)$ satisfies the matrix Riccati equation

$$-V'(t) = Q(t) + V^2(t), \quad t \geq t_1. \quad (7.2.32)$$

Integrating equation (7.2.32) from t_1 to t, we obtain

$$-V(t) + V(t_1) = \int_{t_1}^t Q(s)ds + \int_{t_1}^t V^2(s)ds \quad (7.2.33)$$

and hence

$$-V(t) + V(t_1) \geq \int_{t_1}^{t} Q(s)ds. \tag{7.2.34}$$

In view of the subadditivity of λ_1 and (7.2.34), we have $\lambda_1 \left(\int_{t_1}^{t} Q(s)ds \right) \leq \lambda_1(-V(t)) + \lambda_1(V(t_1))$. Since $\lambda_1 \left(\int_{t_1}^{t} V^2(s)ds \right) \geq \lambda_1 (-V(t))^2$, we find that $\lim_{t \to \infty} \lambda_1(-V(t)) = \lim_{t \to \infty} \lambda_1(V^2(t)) = \infty$. Now, by conditions (7.2.29) and (7.2.33) it follows that

$$\lim_{t \to \infty} \text{approx sup} \left[\frac{1}{g(t)} \left(\text{tr} \left(V(t) + \int_{t_1}^{t} V^2(s)ds \right) \right) \right]$$

$$= \lim_{t \to \infty} \text{approx sup} \left[\frac{1}{g(t)} \left(\text{tr} \left(V(t_1) - \int_{t_1}^{t} Q(s)ds \right) \right) \right] = k - l < \infty,$$

where $k = \lim_{t \to \infty} (1/g(t)) \, \text{tr}(V(t_1))$. Therefore, by Lemma 7.2.1, we get

$$\lim_{t \to \infty} \text{approx inf} \ \frac{1}{g(t)} \int_{t_1}^{t} \lambda_1(V^2(s))ds \ < \ \infty.$$

Again from (7.2.34), we have $\lambda_1 \left(\int_{t_1}^{t} Q(s)ds \right) \leq 2\lambda_1(-V(t))$ for all large t (since $\lambda_1(-V(t)) \to \infty$ as $t \to \infty$) and therefore,

$$\left[\lambda_1 \left(\int_{t_1}^{t} Q(s)ds \right) \right]^2 \leq 4 \left[\lambda_1(-V(t))^2 \right] \leq 4\lambda_1(V^2(t)) \tag{7.2.35}$$

for all large t. Thus, by integration, we obtain

$$\frac{1}{g(t)} \int_{t_1}^{t} \left[\lambda_1 \left(\int_{t_1}^{s} Q(u)du \right) \right]^2 ds \ \leq \ \frac{4}{g(t)} \int_{t_1}^{t} \lambda_1(V^2(s))ds \tag{7.2.36}$$

for all large t, and now taking $\lim$ approx inf on both sides of (7.2.36) as $t \to \infty$ we get a contradiction to condition (7.2.30). This completes the proof. ∎

Remark 7.2.1. If $g(t) = t$ in Theorem 7.2.2, then condition (7.2.2) implies that

$$\lim_{t \to \infty} \frac{1}{t} \int_{t_0}^{t} \left[\lambda_1 \left(\int_{t_0}^{t} Q(u)du \right) \right]^2 ds \ = \ \infty$$

so that Theorem 7.2.3 includes Theorem 7.2.1 with $\lim \inf$ replaced by $\lim$ approx inf.

The following example is illustrative.

Example 7.2.1. Let $\delta > \eta > 0$ and $\sigma = ((\eta+\sigma)/2) - 1$, $k = (\eta+\delta)/2$. Define $Q(t)$ by

$$\begin{bmatrix} 0 & -kt^\sigma \\ -kt^\sigma & -q(t) \end{bmatrix},$$

where $q(t) = \delta t^{\delta-1} - \eta t^{\eta-1}$, $t \geq t_0 = 1$. Then, $\int_1^t \text{tr}(Q(s))ds = t^\eta - t^\delta \rightarrow -\infty$ as $t \rightarrow \infty$ and $\lambda_1\left(\int_1^t Q(s)ds\right) = t^\eta \rightarrow \infty$ as $t \rightarrow \infty$. Let

$$g(t) = \left(\int_1^t \left[\lambda_1\left(\int_1^s Q(u)du\right)\right]^2 ds\right)^\gamma \quad \text{where} \quad 0 < \gamma < 1.$$

Then, $g(t) \uparrow \infty$ as $t \rightarrow \infty$ and $g(t) \sim t^{(2\eta+1)\gamma}$. We have

$$\frac{1}{g(t)} \int_1^t \text{tr}\,(Q(s))ds \sim \frac{t^\eta - t^\delta}{t^{(2\eta+1)\gamma}} \geq -t^{\delta-\gamma(2\eta+1)},$$

which is bounded below as $t \rightarrow \infty$ provided $\delta(2\eta + 1) \geq \gamma$. Now

$$\frac{1}{g(t)} \int_1^t \left[\lambda_1\left(\int_1^s Q(u)du\right)\right]^2 ds = \left[\int_1^t \left(\lambda_1\left(\int_1^t Q(u)du\right)\right)^2 ds\right]^{1-\gamma} \rightarrow \infty$$

as $t \rightarrow \infty$. By Theorem 7.2.3 oscillation is guaranteed if $0 < \eta < \delta < 2\eta+1$ (choose $\gamma = (\delta/2\eta) + 1$). If $\delta > 1$, we have $(1/t)\int_1^t \text{tr}(Q(s))ds \rightarrow -\infty$, so we cannot deduce oscillation by Theorem 7.2.1.

In our next result we will relax condition (7.2.2), and replace hypotheses (7.2.29) and (7.2.30) of Theorem 7.2.3 by a condition involving the relative rates of the growth of the largest and smallest eigenvalues of $\int_{t_0}^t Q(s)ds$ as $t \rightarrow \infty$. For this, we denote and order the eigenvalues of $\int_{t_0}^t Q(s)ds$ as $\lambda_1(t) \geq \lambda_2(t) \geq \cdots \geq \lambda_n(t)$.

We shall need the following lemma.

Lemma 7.2.2. Let $p(t)$ be locally bounded, nonnegative and measurable on $[t_0, \infty)$ and nonzero a.e. Let $q(t)$ be nonnegative and locally integrable such that $p(t) \geq q(t) \int_{t_0}^t p^2(s)ds$ for almost all $t \geq t_0$. Then for all sufficiently large $T \geq t_0$, $q \in \mathcal{L}^2[T, \infty)$.

Proof. Let $P(t) = \int_{t_0}^t p^2(s)ds$. Then, $P(t)$ is absolutely continuous and by hypothesis, $P(t) > 0$ for $t \geq t_1$, say. We have $P'(t) = p^2(t) \geq q^2(t)P^2(t)$, and hence $P'(t)/P^2(t) \geq q^2(t)$. Integrating this inequality from $T > t_1$ to t, we obtain $(1/P(T)) - (1/P(t)) = \int_T^t q^2(s)ds$, and so $\int_T^\infty q^2(s)ds \leq 1/P(T)$, proving the lemma. ∎

Theorem 7.2.4. Let one of the following set of hypotheses hold:

(I) (i) $\lim\limits_{t \to \infty} \text{approx inf } \lambda_1(t) = \infty$, and (ii) $\lim\limits_{t \to \infty} \text{approx sup } \left|\dfrac{\lambda_1(t)}{\lambda_n(t)}\right| > 0;$

(II) (i) $\lim\limits_{t \to \infty} \text{approx sup } \lambda_1(t) = \infty$, and (ii) $\lim\limits_{t \to \infty} \text{approx inf } \left|\dfrac{\lambda_1(t)}{\lambda_n(t)}\right| > 0.$

Then system (7.2.1) is oscillatory.

Proof. Assume that (I) or (II) holds and suppose that (7.2.1) has a nonoscillatory solution $X(t)$. Let $W(t) = -X'(t)X^{-1}(t)$ so that $W(t)$ satisfies

$$\Phi(t) = W(t) - \int_{t_1}^{t} W^2(s)ds = \int_{t_0}^{t} Q(s)ds + C, \quad t \geq t_1 \geq t_0 \quad (7.2.37)$$

where $C = W(t_1) - \int_{t_0}^{t_1} Q(s)ds$.

It is known that for any continuous, symmetric matrix–valued function a continuously varying orthonormal system may be selected [44]. Thus, we may choose a locally integrable vector $y(t)$ with $\|y(t)\| = 1$ such that

$$y^*(t)\left(\int_{t_0}^{t} Q(s)ds\right)y(t) = \lambda_1(t). \quad (7.2.38)$$

Let the eigenvalues of $W(t)$ be $\mu_1(t) \geq \mu_2(t) \geq \cdots \geq \mu_n(t)$. By the preceding remark, we may select a system of (orthonormal) locally integrable eigenvectors $e_i(t)$, $i = 1, 2, \cdots, n$ such that

$$W(t)e_i(t) = \mu_i(t)e_i(t), \quad e_i^*(t)e_j(t) = \delta_{ij} = \begin{cases} 1 & \text{if } i = j \\ 0 & \text{if } i \neq j. \end{cases} \quad (7.2.39)$$

Define the functions $c_i(s,t)$, $i = 1, \cdots, n$, $t_0 \leq s$, $t < \infty$ by

$$y(t) = \sum_{i=1}^{n} c_i(s,t)e_i(s). \quad (7.2.40)$$

Clearly, $c_i(s,t)$, $i = 1, \cdots, n$ are the projections of $y(t)$ on to the orthonormal system $\{e_i(s)\}_{i=1}^{n}$, and are locally integrable with respect to both s and t. We have

$$W(s)y(t) = \sum_{i=1}^{n} \mu_i(s)c_i(s,t)e_i(s). \quad (7.2.41)$$

From (7.2.37) – (7.2.41), we find

$$y^*(t)\Phi(t)y(t) = \sum_{i=1}^{n} \mu_i(t)c_i^2(t,t) - \int_{t_0}^{t}\left(\sum_{i=1}^{n} \mu_i^2(s)c_i^2(s,t)\right)ds. \quad (7.2.42)$$

At this point, in order to give a clearer presentation of the arguments, we shall concentrate on the case $n = 2$. Introduce angle functions ϕ, θ by defining

$$\begin{cases} e_1(s) & = \cos\phi(s)e_1(t_0) + \sin\phi(s)e_2(t_0), \\ e_2(s) & = -\sin\phi(s)e_1(t_0) + \cos\phi(s)e_2(t_0); \end{cases} \tag{7.2.43}$$

$$c_1(t_0, t) = \cos\theta(t), \quad c_2(t_0, t) = \sin\theta(t). \tag{7.2.44}$$

From (7.2.43) and (7.2.44), we find that $c_1(s, t) = \cos[\theta(t) - \phi(s)]$, $c_2(s, t) = \sin[\theta(t) - \phi(s)]$. If we put $\alpha(t) = \theta(t) - \phi(s)$, we may write (7.2.42) as

$$y^*(t)\Phi(t)y(t) = \mu_1(t)\cos^2\alpha(t) + \mu_2(t)\sin^2\alpha(t)$$
$$- \int_{t_1}^t \mu_1^2(s)\cos^2[\alpha(t)+\phi(t)-\phi(s)]ds - \int_{t_1}^t \mu_2^2(s)\sin^2[\alpha(t)+\phi(t)-\phi(s)]ds. \tag{7.2.45}$$

Now $y^*(t)\Phi(t)y(t) = \lambda_1(t) + y^*(t)Cy(t)$. Since C is a constant matrix and $\|y(t)\| = 1$, we have

$$\lim_{t \to \infty} \text{approx inf } y^*(t)\Phi(t)y(t) = \infty, \tag{7.2.46}$$

or

$$\lim_{t \to \infty} \text{approx sup } y^*(t)\Phi(t)y(t) = \infty \tag{7.2.47}$$

according as hypothesis (I(i)), or (II(i)) holds.

We shall demonstrate the incompatibility of (7.2.45), and (7.2.46) or (7.2.47). Part (ii) of hypotheses (I) and (II) imply that there exists a δ with $0 < \delta < 1$ such that

$$\left|\frac{\lambda_1(t)}{\lambda_2(t)}\right| > \delta \quad \text{for all} \quad t \in S_1, \tag{7.2.48}$$

where $S_1 \subset [t_0, \infty)$ satisfies

$$\mu(S_1) = \infty \quad \text{if} \quad \text{(I)} \quad \text{holds}, \tag{7.2.49}$$

and $S_1^c = [t_0, \infty) - S_1$ satisfies

$$\mu(S_1^c) < \infty \quad \text{if} \quad \text{(II)} \quad \text{holds}. \tag{7.2.50}$$

Let $\lambda(t) = \max\{|\lambda_1(t)|, |\lambda_2(t)|\}$. For $t \in S_1$ we have $\lambda(t) \le \lambda_1(t)/\delta$.

For any $x \in \mathbb{R}^n$ with $\|x\| \leq \delta/4$, we find

$$[y(t) + x]^* \Phi(t)[y(t) + x]$$
$$= y^*(t)\Phi(t)y(t) + x^*\Phi(t)x + y^*(t)\Phi(t)x + x^*\Phi(t)y(t)$$
$$\geq \lambda_1(t) - 2\lambda\|x\| - \lambda(t)\|x\|^2 - \left(1 + \frac{1}{4}\delta\right)^2 \|C\| \tag{7.2.51}$$
$$\geq \lambda_1(t) - \frac{1}{2}\lambda_1(t) - \frac{1}{16}\delta\lambda_1(t) - \left(1 + \frac{1}{4}\delta\right)^2 \|C\| \quad \text{if } t \in S_1$$
$$\geq \frac{1}{3}\lambda_1(t) - 2\|C\|.$$

Recall $y(t) = \cos\theta(t)e_1(t_0) + \sin\theta(t)e_2(t_0)$. It is easy to see that we may choose $x = \hat{x}(t)$ with $\|\hat{x}(t)\| \leq \delta/4$ such that $y(t) + \hat{x}(t) = \cos\hat{\theta}(t)e_1(t_0) + \sin\hat{\theta}(t)e_2(t_0)$, where $\hat{\theta}(t) = \theta(t) + (\delta/2)$. Now define $x(t)$ by

$$x(t) = \begin{cases} 0 & \text{if } |\alpha(t)| \geq \delta/8 \pmod{\pi} \text{ and } |\alpha(t) - (\pi/2)| \geq \delta/8 \pmod{\pi}, \\ x(t) & \text{otherwise}, \end{cases}$$

and let $\hat{y}(t) = y(t) + x(t)$. If $\hat{a}(t)$ is defined by $\hat{y}(t) = \cos\hat{\theta}(t)e_1(t_0) + \sin\hat{\theta}(t)e_2(t_0)$, $\hat{a}(t) = \hat{\theta}(t) - \phi(t)$, then we have

$$|\hat{a}(t)| \geq \frac{\delta}{8} \pmod{\pi}, \qquad \left|\hat{a}(t) - \frac{\pi}{2}\right| \geq \frac{\delta}{8} \pmod{\pi} \tag{7.2.52}$$

for all $t \geq t_0$. Now (7.2.45) and (7.2.51) give

$$\hat{y}^*(t)\Phi(t)\hat{y}(t) = \mu_1(t)\cos^2\hat{a}(t) + \mu_2(t)\sin^2\hat{a}(t)$$
$$- \int_{t_0}^t \mu_2^2(s)\cos^2[\hat{a}(t) + \phi(t) - \phi(s)]\,ds$$
$$- \int_{t_0}^t \mu_2^2(s)\sin^2[\alpha(t) + \phi(t) - \phi(s)]\,ds \tag{7.2.53}$$
$$\geq \frac{1}{3}\lambda_1(t) - 2\|C\|, \quad t \in S_1.$$

Part (i) of the hypotheses of the theorem yields the existence of a subset S_2 of $[t_0, \infty)$ such that

$$\frac{1}{3}\lambda_1(t) - 2\|C\| > 0 \quad \text{for } t \in S_2 \tag{7.2.54}$$

and

$$\mu(S_2^c) < \infty \quad \text{if (I) holds}, \tag{7.2.55}$$
$$\mu(S_2) = \infty \quad \text{if (II) holds}, \tag{7.2.56}$$

where $S_2^c = [t_0, \infty) - S_2$. Now (7.2.50), (7.2.52) – (7.2.56) show that there exists a subset $S = S_1 \cap S_2$ of $[t_0, \infty)$ with $\mu(S) = \infty$ such that

$$\mu_1(t) \cos^2 \hat{\alpha}(t) + \mu_2(t) \sin^2 \hat{\alpha}(t) - \int_{t_0}^t \mu_1^2(s) \cos^2[\hat{\alpha}(t) + \phi(t) - \phi(s)] ds$$

$$- \int_{t_0}^t \mu_2^2(s) \sin^2[\hat{\alpha}(t) + \phi(t) - \phi(s)] ds > 0 \quad \text{for} \quad t \in S,$$

(7.2.57)

where $\alpha(t)$ is bounded away from 0 and $\pi/2$ (mod π). Let

$$U_1(t) = \mu_1(t) \cos^2 \hat{\alpha}(t) - \int_{t_0}^t \mu_1^2(s) \cos^2[\hat{\alpha}(t) + \phi(t) - \phi(s)] ds \quad (7.2.58)$$

and let $J_1 = \{t : U_1(t) > 0\}$. By (7.2.52) there exist $\delta_1 > 0$ and a positive integer m such that

$$\cos^2[\hat{\alpha}(t) + w] \geq \delta_1 \quad \text{for} \quad t \geq t_0, \quad |w| \leq \frac{1}{m} \pmod{\pi}. \quad (7.2.59)$$

For $i = 1, 2, \cdots, m$ define

$$A_i = \left\{ t \geq t_0 : \frac{i-1}{m} \leq \phi(t) < \frac{i}{m} \pmod{\pi} \right\} \quad \text{and} \quad \hat{\mu}_i(t) = \begin{cases} |\mu_1(t)|, & t \in A_i \\ 0, & \text{otherwise.} \end{cases}$$

If $t \in A_i$, we have from (7.2.59) that

$$U_1(t) \leq \hat{\mu}_i(t) - \delta_1 \int_{t_0}^t (\hat{\mu}_i(s))^2 ds. \quad (7.2.60)$$

By Lemma 7.2.2 either $\mu_i(t) = 0$ a.e. or if $q_i(t)$ is defined by $\hat{\mu}_i(t) = \hat{q}_i(t) \int_{t_0}^t (\hat{\mu}_i(s))^2 ds$, we have $\hat{q}_i \in \mathcal{L}^2[T, \infty)$ for $T(\geq t_0)$ sufficiently large. In the latter case the set $\{t \geq t_0 : \hat{q}_i(t) \geq \delta_1\}$ has a finite measure, and in either case, the set $\{t \in A_i : U_1(t)\}$ has finite measure. Since $[t_0, \infty) = \cup_{i=1}^m A_i$, we find $\mu(S_1) < \infty$. Similarly, if

$$U_2(t) = \mu_2(t) \sin^2 \hat{\alpha}(t) - \int_{t_0}^t \mu_2^2(s) \sin^2[\hat{\alpha}(t) + \phi(t) - \phi(s)] ds$$

and $J_2 = \{t : U_2(t) > 0\}$, then $\mu(S_2) < \infty$. But this contradicts (7.2.56), and this completes the proof of the theorem for the case $n = 2$.

For the general case $n > 2$, the basic idea of the proof is the same. We introduce the orthogonal matrix $U(s)$ whose rows are the eigenvectors $e_i(s)$ of $W(s)$ (see (7.2.39)). Denote the vector $(c_1(s,t), \cdots, c_n(s,t))^*$ by $c(s,t)$. Assuming without loss of generality that $U(t_0) = I$, where I is the identity matrix, then (7.2.40) gives

$$c(s,t) = U^{-1}(s)c(t_0, t). \quad (7.2.61)$$

Now denote $U^{-1}(s)$ by $V(s)$, the components of $V(t)c(t_0, t)$ by $v_i(t)$ and the components of $[V(s) - V(t)]c(t_0, t)$ by $w_i(s, t)$, $i = 1, 2, \cdots, n$. From (7.2.42) we may write $y^*(t)\Phi(t)y(t)$ as

$$\sum_{i=1}^{n} \mu_i(t)v_i^2(t) = \int_{t_0}^{t} \sum_{i=1}^{n} \mu_i^2(s)[v_i(t) + w_i(s, t)]^2 ds. \qquad (7.2.62)$$

Replacing $y(t)$ by an appropriate $\hat{y}(t)$ and obtaining the corresponding functions $\hat{v}_i(t)$, $\hat{w}_i(s, t)$, we find

$$\hat{y}^*(t)\Phi(t)\hat{y}(t) = \sum_{i=1}^{n} \mu_i(t)\hat{v}_i^2(t) - \int_{t_0}^{t} \sum_{i=1}^{n} \mu_i^2(s)[\hat{v}_i(t) + \hat{w}_i(s, t)]^2 ds > 0, \quad t \in S$$

$$(7.2.63)$$

where $|v_i(t)| \geq \delta_0 > 0$ on S and S is a subset of $[t_0, \infty)$ of infinite measure, $\hat{w}_i(s, t)$ is the ith component of $[V(s) - V(t)]\hat{c}(t_0, t)$ and $\|\hat{c}(t_0, t)\| = 1$. Note that (7.2.63) is the analog of (7.2.53) and (7.2.56). Since the $n \times n$ orthogonal matrices (identified with Z^{n-1}) form a compact set, we may find a finite decomposition of Z^{n-1}, $\{E_k\}_{k=1}^m$, say, such that $G, H \in E_k$ imply $\|G - H\| < \delta_0/2$, $k = 1, 2, \cdots, m$. Now define $A_k = \{t \geq t_0 : V(t) \in E_k\}$. Then we shall have

$$[\hat{v}_j(t) + \hat{w}_j(s, t)]^2 \geq \frac{1}{4}\delta_0^2, \qquad (7.2.64)$$

whenever $s, t \in A_k$, $k = 1, 2, \cdots, m$, $j = 1, 2, \cdots, n$.

With (7.2.63), (7.2.64) and Lemma 7.2.2, we may complete the proof as in the case $n = 2$. ∎

Example 7.2.2. We define the 3×3 matrix $Q(t)$ by specifying its integral from 0 to t as follows

$$\int_0^t Q(s)ds$$

$$= \begin{bmatrix} t^{1/2}\cos^2 t - t^n \sin^2 t(1 + \cos t) & \frac{1}{2}t^{1/2}\sin 2t[1 + t^{n-1}(1 + \cos t)] & 0 \\ \frac{1}{2}t^{1/2}\sin 2t[1 + t^{n-1}(1 + \cos t)] & t^{1/2}\sin^2 t - t^n \cos^2 t(1 + \cos t) & 0 \\ 0 & 0 & 0 \end{bmatrix}$$

where $n \geq 1$. A straightforward computation shows that the eigenvalues of $\int_0^t Q(s)ds$ are $\lambda_1(t) = t^{1/2}$, $\lambda_2(t) = 0$, $\lambda_3(t) = -t^n[1 + \cos t]$. Let $t_m = (2m + 1)\pi$ and $t = t_m + s$. Then it is easily verified that for t near to t_m, $|\lambda_3(t)| \leq c_1 s^2 m^n$ for some constant c_1. Since $|\lambda_1(t)| \geq c_2 m^{1/2}$ for some constant c_2, for t near t_m, we have

$$\left|\frac{\lambda_1(t)}{\lambda_3(t)}\right| \geq 1 \quad \text{if} \quad |s| \leq c_3 m^{(1/4)-(\pi/2)}, \quad c_3 = \left(\frac{c_2}{c_1}\right)^{1/2}.$$

Thus, $\{t : |\lambda_1(t)|/|\lambda_3(t)| \geq 1\}$ has measure at least $c_4 \sum_{m=1}^{\infty} m^{(1/4)-(\pi/2)}$ for some constant $c_4 > 0$, i.e., has infinite measure if $n \leq \delta/2$. It follows that hypothesis (I) of Theorem 7.2.4 holds, and hence equation (7.2.1) is oscillatory. It is easily verified that Theorem 7.2.3 is not applicable for $2 \leq n \leq 5/2$.

Now we shall present some sufficient conditions so that all prepared solutions $X(t)$ of (7.2.5) are oscillatory. Once again the results obtained are based on Riccati techniques. We shall need the following lemma.

Lemma 7.2.3. Assume equation (7.2.5) is nonoscillatory on $[t_0, \infty)$. Then a necessary and sufficient condition that

$$\lim_{T \to \infty} \int_t^T W^2(s)ds \qquad (7.2.65)$$

exists for any solution $W(t) = -X'(t)X^{-1}(t)$ of (7.2.10), where $X(t)$ is a prepared solution of (7.2.5) is that

$$L(Q) = \liminf_{t \to \infty} \frac{1}{t} \int_{t_0}^t \int_{t_0}^s \text{tr } Q(u)du\,ds > -\infty. \qquad (7.2.66)$$

Proof. Suppose first that the limit in (7.2.65) exists for some solution $W(t) = -X'(t)X^{-1}(t)$ of (7.2.10) where $\det X(t) \neq 0$ for $t \geq t_0$. Then from (7.2.10), we have

$$\text{tr } W(t) + \text{tr } \int_t^{\infty} W^2(s)ds = \int_{t_0}^t \text{tr } Q(s)ds - C, \qquad (7.2.67)$$

where $C = -\text{tr } W(t_0) - \text{tr } \int_{t_0}^{\infty} W^2(s)ds$. Since

$$\frac{1}{t} \int_{t_0}^t \left[\text{tr } W(s) - \int_s^{\infty} \text{tr } W^2(u)du \right]^2 ds$$

$$\leq \frac{2}{t} \left[\int_{t_0}^t \left[(\text{tr } W(s))^2 + \left(\int_s^{\infty} \text{tr } W^2(u)du \right)^2 \right] ds \right] \qquad (7.2.68)$$

and since $(\text{tr } W(t))^2 \leq n \text{ tr } W^2(t)$, we find $(1/t) \int_{t_0}^t \text{tr } W^2(s)ds \to 0$ as $t \to \infty$. Similarly,

$$\frac{1}{t} \int_{t_0}^t \left(\int_s^{\infty} \text{tr } W^2(u)du \right)^2 ds \to 0 \quad \text{as} \quad t \to \infty$$

so that (7.2.67) and (7.2.68) imply

$$\frac{1}{t} \int_{t_0}^t \left[\int_{t_0}^s \text{tr } Q(u)du - C \right]^2 ds \to 0 \quad \text{as} \quad t \to \infty. \qquad (7.2.69)$$

Now since in view of Cauchy–Schwarz inequality

$$\left| \frac{1}{t} \int_{t_0}^{t} \left[\int_{t_0}^{s} \operatorname{tr} Q(u)du - C \right] ds \right| \leq \left[\frac{1}{t} \int_{t_0}^{t} \left[\int_{t_0}^{s} \operatorname{tr} Q(u)du - C \right]^2 ds \right]^{1/2}$$

it follows that

$$\lim_{t \to \infty} \frac{1}{t} \int_{t_0}^{t} \left[\int_{t_0}^{s} \operatorname{tr} Q(u)du \right] ds = C \quad \text{exists} \qquad (7.2.70)$$

so that (7.2.66) holds.

Conversely, suppose (7.2.66) holds and let $W(t) = -X'(t)X^{-1}(t)$, where $X(t)$ is a prepared solution of (7.2.5) with $\det X(t) \neq 0$, $t \geq t_0$. From (7.2.10), we find

$$-\frac{1}{t} \int_{t_0}^{t} W(s)ds + \frac{1}{t} \int_{t_0}^{t} \int_{t_0}^{s} W^2(s)duds - \frac{1}{t}(t-t_0)W(t_0) - \frac{1}{t} \int_{t_0}^{t} \int_{t_0}^{s} Q(u)duds$$
$$(7.2.71)$$

and hence in view of (7.2.66) it follows that

$$-\frac{1}{t} \int_{t_0}^{t} \operatorname{tr} W(s)ds + \frac{1}{t} \int_{t_0}^{t} \int_{t_0}^{s} \operatorname{tr} W^2(u)duds \leq M \qquad (7.2.72)$$

for some $M > 0$. Since $\operatorname{tr} W^2(t) \geq 0$ for $t \geq t_0$, $\lim_{t \to \infty} \int_{t_0}^{t} \operatorname{tr} W^2(s)ds$ exists and is finite or infinite. Suppose that $\int_{t_0}^{t} \operatorname{tr} W^2(s)ds \to \infty$ as $t \to \infty$. Then,

$$\frac{1}{t} \int_{t_0}^{t} \int_{t_0}^{s} \operatorname{tr} W^2(u)duds \to \infty \quad \text{as} \quad t \to \infty.$$

From (7.2.72) it follows that $(1/t) \int_{t_0}^{t} \operatorname{tr} W(s)ds \to \infty$ as $t \to \infty$ and so for large t again from (7.2.72), we find

$$\frac{1}{t} \int_{t_0}^{t} \int_{t_0}^{s} \operatorname{tr} W^2(u)duds \leq \frac{1}{t} \int_{t_0}^{t} \operatorname{tr} W(s)ds + M \leq \frac{2}{t} \int_{t_0}^{t} \operatorname{tr} W(s)ds.$$
$$(7.2.73)$$

Now by the Cauchy–Schwarz inequality, we have

$$\left| \frac{1}{t} \int_{t_0}^{t} \operatorname{tr} W(s)ds \right| \leq \left[\frac{1}{t} \int_{t_0}^{t} (\operatorname{tr} W(s))^2 ds \right]^{1/2} \leq \left[\frac{n}{t} \int_{t_0}^{t} \operatorname{tr} W^2(s)ds \right]^{1/2}$$

so (7.2.73) gives

$$\left[\frac{1}{t} \int_{t_0}^{t} \int_{t_0}^{s} \operatorname{tr} W^2(u)duds \right]^2 \leq \frac{4n}{t} \int_{t_0}^{t} \operatorname{tr} W^2(s)ds. \qquad (7.2.74)$$

If we set $H(t) = \int_{t_0}^{t} \int_{t_0}^{s} \operatorname{tr} W^2(u) du\, ds$, then from (7.2.74), we obtain

$$H^2(t) \leq 4nt H'(t), \quad t \geq t_1 \quad \text{for some large} \quad t_1 \geq t_0 \tag{7.2.75}$$

and so

$$\frac{1}{4nt} \leq \frac{H'(t)}{H^2(t)}, \quad t \geq t_1. \tag{7.2.76}$$

Now an integration of (7.2.76) over $[t_1, \infty)$ gives $\infty = (1/4n) \ln t|_{t=t_1}^{\infty} \leq 1/H(t_1) < \infty$, which is a contradiction. Thus, $\lim_{t\to\infty} \int_{t_0}^{t} \operatorname{tr} W^2(s) ds$ exists as a finite limit. This implies the existence of $\lim_{t\to\infty} \int_{t_0}^{t} W^2(s) ds$, as the following argument shows: Let the (operator) norm of a matrix A be denoted by $|A|$. For $t_0 \leq s \leq t$, define $A[s,t]$ by $A[s,t] = \int_{s}^{t} W^2(u) du$. Then, $A[s,t]$ is a nonnegative definite matrix and

$$|A[s,t]| = \lambda_1(A[s,t]) \leq \operatorname{tr} A[s,t] = \int_{s}^{t} \operatorname{tr} W^2(u) du \to 0 \quad \text{as} \quad s, t \to \infty.$$

Hence, we have $|A[s,t]| \to 0$ as $s, t \to \infty$, i.e., $\int_{s}^{t} W^2(u) du \to 0$ as $s, t \to \infty$, which implies the existence of $\lim_{t\to\infty} \int_{t_0}^{t} W^2(s) ds$. This completes the proof of the lemma. $\blacksquare$

Now we are ready to prove the following two results.

Theorem 7.2.5. Assume that condition (7.2.66) holds. Then equation (7.2.5) is oscillatory in case any one of the following conditions is satisfied

(I_1) $\quad \limsup\limits_{t\to\infty} \dfrac{1}{t} \int_{t_0}^{t} \lambda_1 \left(\int_{t_0}^{s} Q(u) du \right) ds = \infty,$ $\hfill (7.2.77)$

(I_2) $\quad \limsup\limits_{t\to\infty} \dfrac{1}{t} \int_{t_0}^{t} \left[\lambda_1 \left(\int_{t_0}^{s} Q(u) du \right) \right]^2 ds = \infty,$ $\hfill (7.2.78)$

(I_3) $\quad \lim\limits_{t\to\infty} \operatorname{approx\,sup} \lambda_1 \left(\int_{t_0}^{t} Q(s) ds \right) = \infty,$ $\hfill (7.2.79)$

(I_4) $\quad \lim\limits_{t\to\infty} \operatorname{approx\,inf} \lambda_1 \left(\int_{t_0}^{t} Q(s) ds \right) = -\infty.$ $\hfill (7.2.80)$

Proof. (I_1) Assume conditions (7.2.66) and (7.2.77) hold. Suppose there exists a prepared solution $X(t)$ of equation (7.2.5) which is not oscillatory. Without loss of generality, we may suppose that $\det X(t) \neq 0$, $t \geq t_0$ so that from equation (7.2.10), we have

$$\lambda_1(W(t) - W(t_0)) = \lambda_1 \left(\int_{t_0}^{t} Q(s) ds + \int_{t_0}^{t} W^2(s) ds \right). \tag{7.2.81}$$

By the convexity of λ_1 and the fact that $\int_{t_0}^{t} W^2(s)ds \geq 0$, $t \geq t_0$, (i.e., nonnegative definite for $t \geq t_0$), we find from (7.2.81) that $\lambda_1(W(t)) + \lambda_1(-W(t_0)) \geq \lambda_1\left(\int_{t_0}^{t} Q(s)ds\right)$, and hence

$$\frac{1}{t}\int_{t_0}^{t} \lambda_1(W(s))ds + \frac{t-t_0}{t}\lambda_1(-W(t_0)) \geq \frac{1}{t}\int_{t_0}^{t} \lambda_1\left(\int_{t_0}^{s} Q(u)du\right)ds$$

(7.2.82)

so that from condition (7.2.77) there exists a sequence $\{T_n\}_{n=1}^{\infty}$, $T_n \to \infty$ as $n \to \infty$ with

$$\frac{1}{T_n}\int_{t_0}^{T_n} \lambda_1(W(s))ds \to \infty \quad \text{as} \quad T_n \to \infty. \tag{7.2.83}$$

Since $\lambda_1(W(s))^2 \leq \lambda_1(W^2(s))$ by the Cauchy–Schwarz inequality, we have

$$\left|\frac{1}{T_n}\int_{t_0}^{T_n} \lambda_1(W(s))ds\right| \leq \left[\frac{1}{T_n}\int_{t_0}^{T_n} (\lambda_1(W(s)))^2 ds\right]^{1/2}$$

$$\leq \left[\frac{1}{T_n}\int_{t_0}^{T_n} \lambda_1(W^2(s))ds\right]^{1/2} \to \infty \quad \text{as} \quad T_n \to \infty.$$

But since $L(Q) > -\infty$, Lemma 6.2.3 implies that $\int_{t_0}^{\infty} \lambda_1(W^2(s))ds \leq \int_{t_0}^{\infty} \text{tr } W^2(s)ds < \infty$. This contradiction proves part (I_1).

(I_2) Assume conditions (7.2.66) and (7.2.78) hold. As in part (I_1) above, we may assume $W(t) = -X'(t)X^{-1}(t)$ is a solution of equation (7.2.10) for $t \geq t_0$, and so by Lemma 7.2.3, we obtain

$$W(t) + \int_{t}^{\infty} W^2(s)ds + C = \int_{t_0}^{t} Q(s)ds, \tag{7.2.84}$$

where $C = -W(t_0) - \int_{t_0}^{\infty} W^2(s)ds$. Hence, we have

$$\left(\lambda_1\left(\int_{t_0}^{t} Q(s)ds\right)\right)^2 = \left(\lambda_1\left[w(t) + \int_{t}^{\infty} W^2(s)ds + C\right]\right)^2$$

$$\leq \lambda_1\left(W(t) + \int_{t}^{\infty} W^2(s)ds + C\right)^2$$

$$\leq \text{tr}\left(W(t) + \int_{t}^{\infty} W^2(s)ds + C\right)^2$$

$$\leq 2\text{tr}\left(W(t) + \int_{t}^{\infty} W^2(s)ds\right)^2 + 2\text{tr } C^2$$

$$\leq 4\text{tr } W^2(t) + 4\text{tr}\left(\int_{t}^{\infty} W^2(s)ds\right)^2 + 2\text{tr } C^2,$$

(since tr $(A+B)^2 \leq 2(\text{tr } A^2 + \text{tr } B^2)$). Therefore,

$$\frac{1}{t}\int_{t_0}^{t}\left[\lambda_1\left(\int_{t_0}^{s}Q(u)du\right)\right]^2 ds \leq \frac{4}{t}\int_{t_0}^{t}\text{tr}\left(\int_{s}^{\infty}W^2(u)du\right)^2 ds$$

$$+\frac{4}{t}\int_{t_0}^{t}\text{tr } W^2(s)ds + \frac{2(t-t_0)}{t}\text{tr } C^2.$$

$$(7.2.85)$$

If we set $A(t) = \int_{t}^{\infty}W^2(s)ds$, then tr $A(t) \to 0$ as $t \to \infty$ and $A(t) \geq 0$ so that $\lambda_1(A(t)) \to 0$, and hence $\lambda_1(A^2(t)) \to 0$ as $t \to \infty$. Therefore, tr $A^2(t) \to 0$ as $t \to \infty$. Thus, the integrals on the right–hand side of (7.2.85) tend to zero as $t \to \infty$, and the last term is bounded. However, condition (7.2.78) implies that the left–hand side of (7.2.85) is not bounded. This contradiction completes the proof of part (I_2).

(I_3) Assume conditions (7.2.66) and (7.2.79) hold. As in part (I_2), we obtain equation (7.2.84) so that

$$\lambda_1(W(t)) + \int_{t}^{\infty}\lambda_1(W^2(s))ds \geq \lambda_1\left(\int_{t_0}^{t}Q(s)ds\right) + \lambda_n(-C). \quad (7.2.86)$$

Since $\int_{t}^{\infty}\lambda_1(W^2(s))ds \to 0$ as $t \to \infty$, it follows that $\int_{t}^{\infty}[\lambda_1(W(s))]^2 ds \to 0$ as $t \to \infty$ and $\lambda_1\left(\int_{t}^{\infty}W^2(s)ds\right) \to 0$ as $t \to \infty$. Now for any $k \geq 1$, $\mu\left\{t : \lambda_1\left(\int_{t_0}^{t}Q(s)ds\right) \geq k\right\} = \infty$ so that if $k \geq |\lambda_n(-C)| + 1$, then from (7.2.86), we have

$$\mu\left\{t : \lambda_1(W(t)) + \lambda_1\left(\int_{t}^{\infty}W^2(s)ds\right) \geq 1\right\} = \infty. \quad (7.2.87)$$

Since $\lambda_1\left(\int_{t}^{\infty}W^2(s)ds\right) < 1/2$ if $t \geq t_1$, $t_1 \geq t_0$, we find

$$\mu\{t : \lambda_1(w(t)) \geq 1/2\} = \infty, \quad (7.2.88)$$

i.e., $\int_{E_k}(\lambda_1(W(s)))^2 ds = \infty$, where $E_k = \{t : \lambda_1(W(t)) \geq 1/2\}$. This contradiction proves part (I_3).

(I_4) Assume conditions (7.2.66) and (7.2.80) hold. The proof is similar to part (I_3). Since for any $M > 0$,

$$\mu\left\{t : \lambda_1\left(\int_{t_0}^{t}W(s)ds\right) < -M\right\} = \infty \quad (7.2.89)$$

and since

$$\lambda_1(W(t)) \leq \lambda_1\left[W(t) + \int_{t}^{\infty}W^2(s)ds\right] \leq \lambda_1\left(\int_{t_0}^{t}Q(s)ds\right) + \lambda_1(-C)$$

it follows that if $M > 1 + |\lambda_1(-C)|$, we have $\mu\{t : \lambda_1(W(t)) \leq -1\} = \infty$ so that $\int_{t_0}^{\infty} (\lambda_1(W(s)))^2 ds = \infty$, which is a contradiction. This completes the proof of the theorem. ■

Theorem 7.2.6. Assume condition (7.2.66) holds. Then equation (7.2.5) is oscillatory if

$$\lim_{t \to \infty} \text{approx sup} \; \lambda_n \left(\int_{t_0}^{t} Q(s)ds \right) > -\infty. \qquad (7.2.90)$$

Proof. Assume $L(Q) = -\infty$, and

$$\lim_{t \to \infty} \text{approx sup} \; \lambda_n \left(\int_{t_0}^{t} Q(s)ds \right) = m > -\infty.$$

If equation (7.2.5) is nonoscillatory, then without loss of generality there is a solution $X(t)$ of (7.2.5) for which $\det X(t) \neq 0$ on $[t_0, \infty)$. Then with $W(t) = -X'(t)X^{-1}(t)$, $t \geq t_0$, we get equation (7.2.10). By Lemma 7.2.3, since $L(Q) = -\infty$, it follows that $\text{tr} \int_{t_0}^{t} W^2(s)ds \to \infty$ as $t \to \infty$, and hence $\lambda_1 \left(\int_{t_0}^{t} W^2(s)ds \right) \to \infty$ as $t \to \infty$. Since $\lambda_1 \left(-\int_{t_0}^{t} Q(s)ds \right) = -\lambda_n \left(\int_{t_0}^{t} Q(s)ds \right)$, from equation (7.2.10), we have

$$\lambda_1 \left(W(t) - \int_{t_0}^{t} Q(s)ds \right) \leq \lambda_1(W(t)) + \lambda_1 \left(-\int_{t_0}^{t} Q(s)ds \right)$$
$$= \lambda_1(W(t)) - \lambda_n \left(\int_{t_0}^{t} Q(s)ds \right). \qquad (7.2.91)$$

Thus for any $\epsilon > 0$, $\mu\left\{t : \lambda_n \left(\int_{t_0}^{t} Q(s)ds \right) \geq m - \epsilon\right\} = \infty$. Now again from equation (7.2.10), we have

$$\frac{1}{n}\text{tr} \int_{t_0}^{t} W^2(s)ds = \frac{1}{n}\text{tr} \left(W(t) - \int_{t_0}^{t} Q(s)ds - W(t_0) \right)$$
$$= \frac{1}{n}\text{tr} \left(W(t) - \int_{t_0}^{t} Q(s)ds \right) - \frac{1}{n}\text{tr}\, W(t_0)$$
$$\leq \lambda_1 \left(W(t) - \int_{t_0}^{t} Q(s)ds \right) - \frac{1}{n}\text{tr}\, W(t_0)$$
$$\leq \lambda_1(W(t)) + \lambda_1 \left(-\int_{t_0}^{t} Q(s)ds \right) - \frac{1}{n}\text{tr}\, W(t_0)$$
$$= \lambda_1(W(t)) - \lambda_n \left(\int_{t_0}^{t} Q(s)ds \right) - \frac{1}{n}\text{tr}\, W(t_0)$$
$$\qquad (7.2.92)$$

and since $(1/n)\mathrm{tr}\int_{t_0}^t W^2(s)ds \geq (1/n)\int_{t_0}^t \lambda_1(W^2(s))ds$ from (7.2.92) it follows that

$$\frac{1}{n}\mathrm{tr}\, W(t_0) + \frac{1}{n}\int_{t_0}^t \lambda_1(W^2(s))ds \leq \lambda_1(W(t)) - \lambda_n\left(\int_{t_0}^t Q(s)ds\right), \quad t \geq t_0.$$
(7.2.93)

Hence, for every $\epsilon > 0$,

$$\mu\left\{t : \frac{1}{n}\mathrm{tr}\, W(t_0) + \frac{1}{n}\int_{t_0}^t \lambda_1(W^2(s))ds \leq \lambda_1(W(t)) - m + \epsilon\right\} = \infty$$
(7.2.94)

and since $\int_{t_0}^t \lambda_1(W^2(s))ds \geq \lambda_1\left(\int_{t_0}^t W^2(s)ds\right) \to \infty$ as $t \to \infty$, we find that if E is defined by

$$E = \left\{t : \frac{1}{2n}\int_{t_0}^t \lambda_1(W^2(s))ds \leq \lambda_1(W(t))\right\} \cap [t_0 + 1, \infty),$$

then $\mu(E) = \infty$. But now with $P(t) = \int_{t_0}^t \lambda_1(W^2(s))ds$, we have $P'(t) = \lambda_1(W^2(t)) \geq \lambda_1(W(t))^2$, and so $P'(t) \geq (1/4n^2)P^2(t)$ for $t \in E$, and hence it follows that

$$\int_E \frac{P'(s)}{P^2(s)}ds \geq \frac{1}{4n^2}\mu(E) = \infty,$$

which is a contradiction since

$$\int_E \frac{P'(s)}{P^2(s)}ds \leq \frac{1}{P(t_0 + 1)}.$$

This completes the proof. ∎

To prove our next result, we need the following lemma.

Lemma 7.2.4. Let $p(t)$ be locally bounded, nonnegative and measurable on $[t_0, \infty)$ with $p(t)$ not almost everywhere zero. Let $q(t)$ be nonnegative and locally integrable such that

$$\int_{t_0}^t p(s)ds \geq q(t)\int_{t_0}^t \int_{t_0}^s p^2(u)du\,ds \quad \text{for almost all } t \geq t_0. \quad (7.2.95)$$

Then for all sufficiently large $T \geq t_0$, $q(t)/\sqrt{t} \in \mathcal{L}^2[T, \infty)$.

Proof. Let $P(t) = \int_{t_0}^t \int_{t_0}^s p^2(u)du\,ds$. Then, $P(t)$ is continuously differentiable and $P(t) > 0$ for $t > t_1$, say. Clearly, $P'(t) = \int_{t_0}^t p^2(s)ds$ and by the Cauchy–Schwarz inequality

$$P'(t) \geq \left(\frac{1}{t - t_0}\right)\left(\int_{t_0}^t p(s)ds\right)^{1/2} \geq \left(\frac{1}{t - t_0}\right)q^2(t)P^2(t). \quad (7.2.96)$$

Now, since $\int_T^\infty (P'(s)/P^2(s))ds < \infty$ for $T > t_1$, the result follows. ∎

Theorem 7.2.7. Assume that

$$\lambda_1 \left(\int_{t_0}^t Q(s)ds \right) > 0 \quad \text{for sufficiently large} \quad t \tag{7.2.97}$$

and

$$\liminf_{t\to\infty} \left| \frac{\lambda_1 \left(\int_{t_0}^t Q(s)ds \right)}{\lambda_n \left(\int_{t_0}^t Q(s)ds \right)} \right| > 0. \tag{7.2.98}$$

Then equation (7.2.5) is oscillatory if

$$\lim_{t\to\infty} \frac{1}{t} \int_{t_0}^t \lambda_1 \left(\int_{t_0}^s Q(u)du \right) ds = \infty. \tag{7.2.99}$$

Proof. The argument is similar to that of used in Theorem 7.2.4, so we content ourself here with sketching the proof. If (7.2.5) is nonoscillatory, we may without loss of generality find a prepared solution $X(t)$ with $\det X(t) \neq 0$ on $[t_0, \infty)$. If $W(t) = -X'(t)X^{-1}(t)$, then $W(t)$ is symmetric and we have

$$\Phi(t) = W(t) - \int_{t_0}^t W^2(s)ds = \int_{t_0}^t Q(s)ds + C, \quad t \geq t_0 \tag{7.2.100}$$

for some constant matrix C.

As in the proof of Theorem 7.2.4, we obtain (7.2.39). Let $U(t)$ be the orthogonal matrix whose rows are the $e_i(t)$ and let $c_i(s,t)$ be the projection of $y(t)$ onto $e_i(s)$. Let the components of $U^{-1}(t)c(t_0,t)$ be $v_i(t)$ and let the components of $[U^{-1}(s)-U^{-1}(t)]c(t_0,\infty)$ be $w_i(s,t)$, $i = 1,2,\cdots,n$. Then, we have

$$\lambda_1 \left(\int_{t_0}^t Q(s)ds \right) + y^*(t)Cy(t) = y^*(t)\Phi(t)y(t)$$
$$= \sum_{i=1}^n \mu_i(t)v_i^2(t) - \int_{t_0}^t \sum_{i=1}^n \mu_i^2(s)[v_i(t) + w_i(s,t)]^2 ds.$$

Again as in the proof of Theorem 7.2.4 the given hypotheses of the theorem allow us to find a unit vector $\hat{y}(t)$, functions $\hat{\mu}_{ij}(t)$, $i = 1,\cdots,n$, $j = 1,\cdots,m$ and $\delta > 0$ such that

$$\frac{1}{3}\lambda_1 \left(\int_{t_0}^t Q(s)ds \right) - 2\|C\| \leq \hat{y}^*(t)\Phi(t)\hat{y}(t)$$
$$\leq \sum_{i,j} \left[\hat{\mu}_{ij}(t) - \delta \int_{t_0}^t \hat{\mu}_{ij}(s)ds \right]. \tag{7.2.101}$$

Hypothesis (7.2.99) and (7.2.101) imply that there exists a $t_1 \geq t_0$ such that

$$\int_{t_0}^t \sum_{i,j} \left[\hat{\mu}_{ij}(s) - \delta \int_{t_0}^s \mu_{ij}^2(u) du \right] ds > 0 \quad \text{for} \quad t \geq t_1. \qquad (7.2.102)$$

If $\hat{\mu}_{ij} \equiv 0$ a.e. for $t \geq t_0$, let $\hat{q}_{ij}(t) = 0$, otherwise define $\hat{q}_{ij}$ by

$$\int_{t_0}^t |\hat{\mu}_{ij}(s)| ds = \hat{q}_{ij}(t) \int_{t_0}^t \int_{t_0}^s |\mu_{ij}(u)|^2 du ds. \qquad (7.2.103)$$

By Lemma 7.2.4 each $\hat{q}_{ij}$ has the property that $\hat{q}_{ij}(t)/\sqrt{t} \in \mathcal{L}^2[T, \infty)$ if $T > t_1$ is sufficiently large, and so $\hat{q}(t) = \sum_{i,j} \hat{q}_{ij}(t) \in \mathcal{L}^2[T, \infty)$ if T is sufficiently large. However, (7.2.102) implies that

$$[\hat{q}(t) - \delta] \sum_{i,j} \int_{t_0}^t \int_{t_0}^s \hat{\mu}_{ij}(u) du ds > 0 \quad \text{for} \quad t > T$$

and so $\hat{q}(t) > \delta$ for $t > T$, contradicting the fact that $\hat{q}(t)/\sqrt{t} \in \mathcal{L}^2[T, \infty)$. This proves the theorem. ∎

To illustrate Theorems 7.2.5 and 7.2.7, we present the following counterexamples.

Example 7.2.3. In Theorem 7.2.5 parts (I_1) and (I_2), one cannot replace the average of $\lambda_1 \left(\int_{t_0}^t Q(s) ds \right)$, or $\left[\lambda_1 \left(\int_{t_0}^t Q(s) ds \right) \right]^2$ by these expressions alone, and that one cannot replace 'lim approx sup' by 'lim sup' in parts (I_3) and (I_4). For this, it suffices to consider a scalar counterexample. We choose $v(t) \in C^1([t_0, \infty), \mathbb{R}) \cap \mathcal{L}^2[t_0, \infty)$ and define $q(t) = v'(t) - v^2(t)$. Then, $x(t) = \exp\left(-\int_{t_0}^t v(s) ds \right)$ is a nonoscillatory solution of the equation $x''(t) + q(t)x(t) = 0$, and we have $\int_{t_0}^t q(s) ds = v(t) - v(t_0) - \int_{t_0}^t v^2(s) ds$. It is clear that we can choose $v(t)$ such that $\limsup_{t \to \infty} v(t)$ takes on any value M with $-\infty < M \leq \infty$, $\liminf_{t \to \infty} v(t)$ takes on any value m, $-\infty \leq m \leq M \leq \infty$ and such that $\liminf_{t \to \infty} (1/t) \int_{t_0}^t v(s) ds > -\infty$.

Example 7.2.4. The condition (7.2.99) is not by itself an oscillation criterion for (7.2.5) when $n > 1$. To see this consider the case when $Q(t)$ is a 2×2 diagonal matrix for all t, say $Q(t) = \text{diag}\{q(t), q_2(t)\}$. Then (7.2.5) is uncoupled as two scalar equations

$$x_i''(t) + q_i(t)x_i(t) = 0, \quad i = 1, 2. \qquad (7.2.104)_i$$

If we take $t_0 = 0$ and set $p_i(t) = \int_0^t q_i(s) ds$, we will have the required counterexample if we can find $q_1(t), q_2(t)$ so that equations $(7.2.104)_1$ and

$(7.2.104)_2$ are both nonoscillatory with

$$p_2(t) \geq 0 \quad \text{for} \quad t \geq 0 \tag{7.2.105}$$

and

$$\lim_{t \to \infty} \frac{1}{t} \int_0^t p_1^+(s)ds = \infty, \quad \text{where} \quad p_1^+(t) = \max\{0, p_1(t)\}. \tag{7.2.106}$$

For this, on choosing

$$c_1(t) = \begin{cases} 1, & t \in \text{support } p_1^+(t) \\ 0, & \text{otherwise,} \end{cases} \qquad c_2(t) = \begin{cases} 0, & t \in \text{support } p_1^+(t) \\ 1, & \text{otherwise} \end{cases}$$

and defining $y(t) = (c_1(t), c_2(t))^*$, we have

$$\begin{aligned}
\lambda_1 \left(\int_0^t Q(s)ds \right) &\geq y^*(t) \left(\int_0^t Q(s)ds \right) y(t) \\
&= c_1^2(t)p_1(t) + c_2^2(t)p_2(t) \geq p_1^+(t).
\end{aligned}$$

Thus condition (7.2.99) holds, but clearly equation (7.2.5) is nonoscillatory. The choice of $q_2(t)$ is straightforward, e.g., $q_2(t) = 1/(4(1+t)^2)$.

To construct a suitable $q_1(t)$, we first give an inductive definition of a suitable Riccati scalar variable $w_1(t)$. Define $w_1(t)$ on $[0,1]$ by $w_1(t) = \begin{cases} 0, & \text{if } t = 0 \\ 1, & \text{if } 1/2 \leq t \leq 1, \end{cases}$ so that $w_1(t)$ is continuously differentiable and monotone nondecreasing on $[0,1]$. Set $a_1 = \int_0^1 w_1^2(s)ds$, $b_1 = 4a_1$ and $\epsilon_1 = 1/(8b_1)$. Then, $a_1 \geq 1/2$, $b_1 \geq 2$ and $2\epsilon_1 \leq 1/8 < 1/2$. Let $s_m = \sum_{i=1}^m 1/i$. Extend $w_1(t)$ to $[s_1, s_2]$ by taking $w_1(t) = b_1$ on $[1 + \epsilon_1, s_2]$ and making $w_1(t)$ continuously differentiable and monotone nondecreasing on $[0, s_2]$. Inductively, suppose that a_i, b_i, ϵ_i have been defined for $1 \leq i \leq m$ with $b_i = 4a_i$, $\epsilon_i = 1/(8b_i) < 1/(2(i+1))$, and $w_1(t)$ has been defined as a continuously differentiable monotone nondecreasing function on $[0, s_{m+1}]$ such that $w_1(t) = b_1$ on $[s_i + \epsilon_i, s_{i+1}]$ and $\int_0^{s_i} w_1^2(s)ds = a_i$, $1 \leq i \leq m$. Then,

$$a_{m+1} \geq a_m + \int_{s_m+\epsilon_m}^{m+2\epsilon_m} w_1^2(s)ds = a_m + b_m^2 \epsilon_m \geq \frac{3}{2}a_m.$$

By the inductive hypothesis, we find that

$$a_{m+1} \geq \left(\frac{3}{2} \right)^m a_1 \geq \frac{1}{2} \left(\frac{3}{2} \right)^m.$$

If $b_{m+1} = 4a_{m+1}$, $\epsilon_{m+1} = 1/(8b_{m+1})$, then in view of

$$\left(\frac{3}{2} \right)^m = \left(1 + \frac{1}{2} \right)^m \geq 1 + \frac{m}{2} \geq \frac{2(m+2)}{16},$$

we have

$$\epsilon_{m+1} \leq \frac{1}{16}\left(\frac{3}{2}\right)^{-m} \leq \frac{1}{2(m+2)}.$$

Now we extend $w_1(t)$ to $[s_{m+1}, s_{m+2}]$ so that $w_1(t) = b_{m+1}$ on $[s_{m+1} + \epsilon_{m+1}, s_{m+2}]$ and is continuously differentiable and monotone nondecreasing on $[0, s_{m+2}]$. With this inductive definition of $w_1(t)$ on $[0, \infty)$, note that if $t \in [s_m + \epsilon_m, s_m + 2\epsilon_m]$, we have

$$w_1(t) - \int_0^t w_1^2(s)ds \geq b_m - a_m - 2\epsilon_m b_m^2 \geq \frac{1}{2}b_m.$$

Define $q_1(t)$ by $q_1(t) = w_1'(t) - w_1^2(t)$. Then for $t \in [s_k + \epsilon_k, s_k + 2\epsilon_k]$, we find $p_1(t) = \int_0^t q_1(s)ds \geq (1/2)b_k$. Thus, if $s_m \leq t \leq s_{m+1}$, we obtain

$$\frac{1}{t}\int_0^t p_1^+(s)ds \geq \frac{1}{t}\sum_{k=1}^m \frac{1}{2}b_k\epsilon_k = \frac{1}{16}\left(\frac{m}{t}\right).$$

Since $s_m \sim \ln m$ for $s_m \leq t \leq s_{m+1}$, we have $m \sim e^t$ so that

$$\frac{1}{t}\int_0^t p_1^+(s)ds \sim \frac{1}{16}\left(\frac{e^t}{t}\right).$$

Thus (7.2.106) holds, yet (7.2.104)$_1$ is nonoscillatory.

Finally, we turn our attention to variational principles for obtaining oscillatory criteria for (7.2.5). We assume that $Q(t)$ is symmetric and locally integrable on $[t_0, \infty)$. On any subinterval $[\alpha, \beta]$ of $[t_0, \infty)$, we define

$$\begin{aligned}A_1(\alpha, \beta) = \ \ \{&\eta : [\alpha, \beta] \to \mathbb{R}^n : \eta(\alpha) = \eta(\beta) = 0, \\ &\eta(t) \in AC[\alpha, \beta], \ \eta'(t) \in \mathcal{L}^2(\alpha, \beta)\}.\end{aligned}$$
$$(7.2.107)$$

The basic result we need here is that system (7.2.5) is oscillatory if and only if there is a sequence of intervals $\{[a_n, b_n]\}$ with $\lim_{n\to\infty} a_n = \infty$ and a sequence of functions $\{\phi_n\}$, $\phi_n \in A_1(a_n, b_n)$ such that

$$\int_{a_n}^{b_n}\left[|\phi_n'(s)|^2 - \phi_n^*(s)Q(s)\phi_n(s)\right]ds < 0. \qquad (7.2.108)$$

The idea now is to utilize conditions on the behavior of $Q(t)$, or its integral to construct sequences fulfilling (7.2.108).

Our first result in this direction is the following.

Theorem 7.2.8. Suppose that

$$\limsup_{t\to\infty} \lambda_1\left(\int_{t_0}^t Q(s)ds\right) = \infty. \qquad (7.2.109)$$

Then equation (7.2.5) is oscillatory if either

(i_1) $\lambda_1(Q(t))$ is bounded above on $[t_0, \infty)$, or

(i_2) $\lambda_n(Q(t))$ is bounded below on $[t_0, \infty)$.

Proof. Choose any sequence $a_k > t_0$ with $\lim_{k\to\infty} a_k = \infty$. Since

$$\lambda_1\left(\int_{a_k}^t Q(s)ds\right) \geq \lambda_1\left(\int_{t_0}^t Q(s)ds\right) - \lambda_1\left(\int_{t_0}^{a_k} Q(s)ds\right)$$

by condition (7.2.109) it follows that $\limsup_{t\to\infty} \lambda_1\left(\int_{a_k}^t Q(s)ds\right) = \infty$. Therefore, we can choose $b_k > a_{k+2}$ and $y_k \in \mathbb{R}^n$ with $\|x_k\| = 1$, so that

$$y_k^*\left(\int_{a_k}^{b_k} Q(s)ds\right) y_k \geq k, \quad k = 1, 2, \cdots. \tag{7.2.110}$$

Suppose that condition (i_1) holds. Then there exists M such that $\lambda_1(Q(t)) \leq M$ for $t \geq t_0$. Define $\phi_k \in A(a_k, b_k)$ as follows:

$$\phi_k(t) = \begin{cases} (t - a_k)y_k, & a_k \leq t \leq a_k + 1 \\ y_k, & a_k + 1 \leq t \leq b_k - 1 \\ (-t + b_k)y_k, & b_k - 1 \leq t \leq b_k. \end{cases} \tag{7.2.111}$$

Clearly, $\int_{a_k}^{b_k} |\phi_k'(s)|^2 ds = 2$, and

$$\int_{a_k}^{b_k} \phi_k^*(s)Q(s)\phi_k(s)ds$$

$$= \int_{a_k}^{b_k} y_k^* Q(s)y_k ds - \int_{a_k}^{a_k+1} [1 - (s - a_k)^2]y_k^* Q(s)y_k ds \tag{7.2.112}$$

$$- \int_{b_k-1}^{b_k} [1 - (b_k - s)^2]y_k^* Q(s)y_k ds \geq k - \frac{4}{3}M.$$

Thus, for sufficiently large k,

$$\int_{a_k}^{b_k} \left[|\phi_k'(s)|^2 - \phi_k^*(s)Q(s)\phi_k(s)\right] ds \leq 2 + \frac{4}{3}M - k < 0.$$

It follows that system (7.2.5) is oscillatory.

If condition (i_2) holds, we have $\lambda_n(Q(t)) \geq -M$ for $t \geq t_0$. Define $\phi_k \in A_1(a_k - 1, b_k + 1)$ by

$$\phi_k(t) = \begin{cases} (t + 1 - a_k)y_k, & a_k - 1 \leq t \leq a_k \\ y_k, & a_k \leq t \leq b_k \\ (-t + b_k + 1)y_k, & b_k \leq t \leq b_k + 1. \end{cases} \tag{7.2.113}$$

Again $\int_{a_k-1}^{b_k+1} |\phi'_k(s)|^2 ds = 2$, and we have

$$\int_{a_k-1}^{b_k+1} \phi_k^*(s)Q(s)\phi_k(s)ds$$

$$= \int_{a_k-1}^{a_k} \phi_k^*(s)Q(s)\phi_k(s)ds + \int_{b_k}^{b_k+1} \phi_k^*(s)Q(s)\phi_k(s)ds$$

$$+ \int_{a_k}^{b_k} \phi_k^*(s)Q(s)\phi_k(s)ds$$

$$= \int_{a_k}^{b_k} y_k^* Q(s)y_k ds + \int_{a_k-1}^{a_k} [s+1-a_k]^2 y_k^* Q(s)y_k ds \qquad (7.2.114)$$

$$+ \int_{b_k}^{b_k+1} [b_k+1-t]^2 y_k^* Q(s)y_k ds$$

$$\geq k - M \left[\int_{a_k-1}^{a_k} [s+1-a_k]^2 ds + \int_{b_k}^{b_k+1} [b_k+1-s]^2 ds \right]$$

$$= k - \frac{2}{3}M,$$

and so for sufficiently large k,

$$\int_{a_k-1}^{b_k+1} \left[|\phi'_k(s)|^2 - \phi_k^*(s)Q(s)\phi_k(s) \right] ds \leq 2 + \frac{2}{3}M - k < 0,$$

and again we conclude that equation (7.2.5) is oscillatory. ∎

Example 7.2.5. In Theorem 7.2.8 condition (7.2.109) alone is sufficient has been shown by Byers et. al. [12]. However, this condition cannot be replaced by $\lim_{t\to\infty} \int_{t_0}^t \lambda_1(Q(s))ds = \infty$. In fact, Example 7.2.4 can be used to provide a counterexample. For this, we define $q_1(t)$, $q_2(t)$, $w_1(t)$, $\cdots$ etc. as in Example 7.2.4, then

$$\int_0^t \lambda_1(Q(s))ds \geq \int_0^t q^+(s)ds = \int_0^t [w'_1(s) - w_1^2(s)]^+ ds.$$

Now on $[s_m, s_m + \epsilon_m]$, we have $w_1(s) \leq b_m$, and so

$$\int_{s_m}^{s_m+\epsilon_m} [w'_1(s) - w_1^2(s)]^+ ds \geq \int_{s_m}^{s_m+\epsilon_m} [w'_1(s) - w_1^2(s)]ds$$

$$= w_1(s_m + \epsilon_m) - w_1(s_m) - \int_{s_m}^{s_m+\epsilon_m} w_1^2(s)ds$$

$$\geq b_m - b_{m-1} - b_m^2 \epsilon_m$$

$$\geq 4[a_m - a_{m-1}] - \frac{1}{8}b_m = \frac{4}{3}a_m - \frac{1}{8}b_m$$

$$= \frac{5}{28}b_m,$$

and hence,

$$\int_0^{s_{m+1}} \lambda_1(Q(s))ds \geq \frac{5}{24} \sum_{k=1}^{m} b_k \rightarrow \infty \quad \text{as} \quad m \rightarrow \infty.$$

Theorem 7.2.9. Let $Q(t) = (q_{ij}(t))$, $i, j = 1, \cdots, n$ and each $q_{ij}(t)$ is bounded on $[t_0, \infty)$, and $q_{ii}(t) = 0$, $i = 1, \cdots, n$. Then system (7.2.5) is oscillatory if there exist i, j such that

$$\limsup_{t \to \infty} \int_{t_0}^{t} q_{ij}(s)ds = \infty. \tag{7.2.115}$$

Proof. Applying the technique of Theorem 7.2.8 to the scalar equation

$$x''(t) + q_{ij}(t)x(t) = 0 \tag{7.2.116}$$

there exist intervals $[a_m, b_m]$ with $\lim_{m \to \infty} a_m = \infty$ and a scalar function $\phi_m \in A_1(a_m, b_m)$ (with $n = 1$) such that

$$\int_{a_m}^{b_m} \left[|\phi'_m(s)|^2 - q_{ij}(s)|\phi_m(s)|^2 \right] ds < 0. \tag{7.2.117}$$

Define $\psi_m \in A_1(a_m, b_m)$ by $\psi_m(t) = \text{col}\,(0, \cdots, \phi_m(t), 0, \cdots, \phi_m(t), \cdots, 0)$, where the only nonzero entries are the ith and jth. Then,

$$\int_{a_m}^{b_m} \left[|\psi'(s)|^2 - \psi_m^*(s)Q(s)\psi(s) \right] ds = 2 \int_{a_m}^{b_m} \left[|\phi'_m(s)|^2 - q_{ij}(s)||\phi_m(s)|^2 \right] ds < 0$$

implies oscillation of (7.2.5). ∎

Theorem 7.2.10. Suppose for each positive integer $m \geq t_0$ there exists a positive number ϵ_m, and for each positive integer k there exists a unit vector $y_{mk} \in \mathbb{R}^n$ such that the set

$$S_{mk} = \left\{ t \geq m : y_{mk}^* \left(\int_m^t Q(s)ds \right) y_{mk} \geq k \right\}$$

has measure at least ϵ_m. Then equation (7.2.5) is oscillatory.

Proof. Let $m \geq t_0 + 1$ be a positive integer. Without loss of generality we can assume that $0 < \epsilon_m < 1$. Let

$$\max_{m-(\epsilon/2) \leq t \leq m} \|Q(t)\| = \alpha_m \tag{7.2.118}$$

and choose the positive integer $k = k_m$ so that

$$k > 2 + \frac{1}{12}\alpha_m \epsilon_m^2. \tag{7.2.119}$$

Let y_{mk} be the unit vector given by the hypothesis of the theorem. Since $\mu(S_{mk}) \geq \epsilon_m$, we can find a finite collection of closed intervals $I_j = [a_j, b_j]$, $1 \leq j \leq r$ where $m < a_1 < b_1 < \cdots < a_r < b_r$ are such that $\cup_{j=1}^r I_j \subset S_{mk}$ and $\mu\left(\cup_{j=1}^r I_j\right) = \sum_{j=1}^r \delta_j \geq \epsilon_m/2$, where $\delta_j = b_j - a_j$. Since $\int_m^t Q(s)ds$ is a continuous function of t, we can by maximizing the length of the intervals, assume that

$$y_{mk}^*\left(\int_m^{a_j} Q(s)ds\right) y_{mk} = k, \quad 1 \leq j \leq r \tag{7.2.120}$$

and

$$y_{mk}^*\left(\int_m^{b_j} Q(s)ds\right) y_{mk} = k, \quad 1 \leq j \leq r-1. \tag{7.2.121}$$

Furthermore, by discarding intervals if necessary and shrinking the final interval, we can suppose that

$$\sum_{j=1}^r \delta_j = \frac{1}{2}\epsilon_m. \tag{7.2.122}$$

Let a_0 and b_0 be defined to m, δ_0 to be 0, and let $\Delta_j = \sum_{i=0}^j \delta_j$, $j = 0, 1, \cdots, r$.

We define $\phi_m \in A_1(m - (1/2)\epsilon_m, b_r)$ as follows

$$\phi_m(t) = \rho_m(t)y_{mk}, \tag{7.2.123}$$

where

$$\rho_m(t) = \begin{cases} t + (1/2)\epsilon_m - m, & m - (1/2)\epsilon_m \leq t \leq m \\ (1/2)\epsilon_m - \Delta_j, & b_j \leq t \leq a_{j+1}, \ 0 \leq j \leq r-1 \\ (1/2)\epsilon_m - \Delta_{j-1} + a_j - t, & a_j \leq t \leq b_{j+1}, \ 0 \leq j \leq r. \end{cases} \tag{7.2.124}$$

Clearly,

$$\int_{m-(1/2)\epsilon_m}^{b_r} |\phi_m'(s)|^2 ds = \frac{1}{2}\epsilon_m + \Delta_r = \epsilon_m \tag{7.2.125}$$

and

$$\int_{m-(1/2)\epsilon_m}^{b_r} \phi_m^*(s)Q(s)\phi_m(s)ds = \int_{m-(1/2)\epsilon_m}^{b_r} \phi_m^*(s)Q(s)\phi_m(s)ds$$
$$+ \sum_{j=0}^{r-1} \int_{b_j}^{a_{j+1}} \phi_m^*(s)Q(s)\phi(s)ds + \sum_{j=1}^{r-1} \int_{a_j}^{b_j} \phi_m^*(s)Q(s)\phi_m(s)ds.$$
$$\tag{7.2.126}$$

By (7.2.120) and (7.2.121), it follows that

$$\int_{m-(1/2)\epsilon_m}^{m} \phi_m^*(s)Q(s)\phi_m(s)ds \geq -\alpha_m \int_{m-(1/2)\epsilon_m}^{m} [s+(1/2)\epsilon_m - m]^2 ds$$

$$= -\frac{1}{24}\alpha_m \epsilon_m^3$$

(7.2.127)

and by (7.2.120),

$$\int_{b_0}^{a_1} \phi_m^*(s)Q(s)\phi_m(s)ds = \frac{1}{2}\epsilon_m \int_{m}^{a_1} y_{mn}^* Q(s)y_{mk}ds$$

$$\geq \frac{1}{2}\epsilon_m y_{mk}^* \left(\int_{m}^{a_1} Q(s)ds \right) y_{mk} = \frac{1}{2}\epsilon_m k.$$

(7.2.128)

For $1 \leq j \leq r-1$, by (7.2.120) and (7.2.121) we also have

$$\int_{b_j}^{a_{j+1}} \phi_m^*(s)Q(s)\phi_m(s)ds = ((1/2)\epsilon_m - \Delta_j) \int_{b_j}^{a_{j+1}} y_{mk}^* Q(s)y_{mk}ds$$

$$= ((1/2)\epsilon_m - \Delta_j) \int_{m}^{a_j} y_{mk}^* Q(s)y_{mk}ds - \int_{m}^{b_j} y_{mk}^* Q(s)y_{mk}ds = 0.$$

(7.2.129)

Noting that for $1 \leq j \leq r$, $\rho_m(t) \geq 0$ and $\rho_m'(t) = -1$ on $[a_j, b_j]$, and hence on integration by parts, we find

$$\int_{a_j}^{b_j} \phi_m^*(s)Q(s)\phi_m(s)ds = \int_{a_j}^{b_j} \rho_m^2(s)y_{mk}^* Q(s)y_{mk}ds$$

$$= \rho_m^2(b_j) \int_{a_j}^{b_j} y_{mk}^* Q(s)y_{mk}ds + \int_{a_j}^{b_j} 2\rho_m(s) \left(\int_{a_j}^{s} y_{mk}^* Q(u)y_{mk}du \right) ds.$$

(7.2.130)

By (7.2.120) and (7.2.121), we further have

$$\int_{a_j}^{b_j} y_{mk}^* Q(s)y_{mk}ds = \int_{m}^{b_j} y_{mk}^* Q(s)y_{mk}ds - \int_{m}^{a_j} y_{mk}^* Q(s)y_{mk}ds \geq 0,$$

(7.2.131)

(with equality if $1 \leq j \leq r-1$). Since $t \in [a_j, b_j]$ implies $t \in S_{mk}$, using (7.2.120) we find that for $t \in [a_j, b_j]$,

$$\int_{a_j}^{t} y_{mk}^* Q(s)y_{mk}ds = \int_{m}^{t} y_{mk}^* Q(s)y_{mk}ds - \int_{m}^{a_j} y_{mk}^* Q(s)y_{mk}ds \geq 0.$$

(7.2.132)

From (7.2.130) – (7.2.132), we get

$$\int_{a_j}^{b_j} \phi_m^*(s)Q(s)\phi_m(s)ds \geq 0, \quad j=1,\cdots,r.$$

(7.2.133)

Finally, from (7.2.125) – (7.2.129), (7.2.133) and (7.2.119), we find

$$\int_{m-(\epsilon/2)}^{b_r} \left[|\phi'(s)|^2 - \phi_m^*(s)Q(s)\phi_m(s) \right] ds \leq \epsilon_m + \frac{1}{24}\alpha_m \epsilon_m^3 - \frac{1}{2}\epsilon_m k < 0.$$

Since $m - (1/2)\epsilon_m > m - (1/2)$ is an arbitrary positive integer exceeding t_0, the theorem follows. ∎

7.3. Oscillation Theorems for Differential Systems with Functionally Commutative Matrix Coefficients

Consider the second order nonlinear differential system

$$X''(t) + G(X^*(t))Q(t)H(X(t)) = 0 \qquad (7.3.1)$$

where

(i) $G, H \in \mathcal{F}$, and the set $\mathcal{F}$ will be defined later,

(ii) $Q(t)$ is a real–valued, continuous, symmetric $n \times n$ matrix valued function on $[t_0, \infty)$, $t_0 > 0$,

(iii) $X(t)$ is a twice continuously differentiable $n \times n$ matrix valued function on $[t_0, \infty)$, and X^* denotes the transpose of X.

A solution of the system (7.3.1) is a real–valued $n \times n$ matrix $X(t)$ satisfying (7.3.1) on $[T, \infty)$, $T \geq t_0$ (T may depend on X) and not identically singular on a neighborhood of infinity. Any solution $X(t)$ of (7.3.1) satisfies (7.2.7), and if it satisfies (7.2.6), $X(t)$ is said to be prepared. $X(t)$ oscillates if del $X(t)$ has arbitrarily large zeros.

The continuous $n \times n$ matrix–valued function $Q(t)$ is said to be functionally commutative on an interval J if

$$Q(s)Q(t) = Q(t)Q(s) \quad \text{for all} \quad s, t \in J. \qquad (7.3.2)$$

We shall consider the general system (7.3.1) with symmetric, functionally commutative coefficients and show that the oscillation theory for such systems can be effectively reduced to the study of a diagonal system of scalar second order equations. In the case of equations like (7.3.1), this system of scalar equations is uncoupled, allowing one to directly employ the existing theory to obtain oscillation and nonoscillation criteria.

The following notation and preliminaries are required: Let $\mathcal{M}_n$ be the Banach space of real $n \times n$ matrices with any of the usual norms. Let $\mathcal{D}_n$ and $\mathcal{N}_n$ be the subsets of diagonal and nonsingular matrices, respectively.

We denote by $\mathcal{F}$ the set of continuous functions $F : \mathcal{M}_n \to \mathcal{M}_n$ which are invariant on $\mathcal{D}_n$ and $\mathcal{D}_n \cap \mathcal{N}_n$, and for which $F(P^{-1}DP) = P^{-1}F(D)P$ for all $D \in \mathcal{D}_n$ and $P \in \mathcal{N}_n$. Let $\mathcal{S}_n$ denote the symmetric $n \times n$ matrices. Evidently the values of F on $\mathcal{D}_n$ completely determine it on $\mathcal{S}_n$. Also the following properties hold:

(a_1) F is invariant of $\mathcal{N}_n$,

(a_2) $F(A^*) = (F(A))^*$ for all $A \in \mathcal{M}_n$, in particular F is invariant on $\mathcal{S}_n$.

Let $F(\mathrm{diag}\{x_1, \cdots, x_n\}) = \mathrm{diag}\{f_1(x_1, \cdots, x_n), \cdots, f_n(x_1, \cdots, x_n)\}$. Denote by $\mathcal{F}_0$ the subset of $\mathcal{F}$ for which $f_i(x_1, \cdots, x_n) = f(x_i)$, $i = 1, 2, \cdots, n$.

As examples of $F \in \mathcal{F}_0$, we can take $F(X) = X^k$ where k is a nonzero integer, or $F(X) = X^{1/q}$ where q is an odd integer.

It is easy to check that both $\mathcal{F}$ and $\mathcal{F}_0$ are closed under multiplication.

In what follows, we shall consider the linear system (7.2.5) and the general system (7.3.1), where G, $H \in \mathcal{F}$, and Q is continuous, symmetric and functionally commutative. We shall denote the set of such Q by $\mathcal{C}$.

The following lemma gives a characterization of functionally commutative matrices.

Lemma 7.3.1. Let the $n \times n$ matrix $A(t)$ be diagonalizable, and functionally commutative on an interval J. Then, $A(t)$ has a constant set of eigenvalues, and therefore there exists a constant matrix P such that

$$A(t) = P^{-1}D(t)P \quad \text{where} \quad D(t) \in \mathcal{D}_n, \quad t \in J.$$

In the case of linear system (7.2.5), we can readily reduce our discussion to that of an associated diagonal system.

Theorem 7.3.1. Let $Q \in \mathcal{C}$, and assume that there exists a constant matrix P so that $Q(t) = P^{-1}D(t)P$, where $D(t) = \mathrm{diag}\{q_1(t), \cdots, q_n(t)\}$, where $q_i(t) \in C([t_0, \infty), \mathbb{R})$, $i = 1, 2, \cdots, n$. Then all prepared solutions of (7.2.5) oscillate if and only if at least one of an associated system of n scalar equations

$$y_i''(t) + q_i(t)y_i(t) = 0, \quad i = 1, 2, \cdots, n \qquad (7.3.3)$$

is oscillatory.

Proof. Since $Q(t)$ is symmetric, the matrix P is actually orthogonal. Let $Y(t) = PX(t)$, then the system (7.2.5) is transformed into

$$Y''(t) + D(t)Y(t) = 0. \qquad (7.3.4)$$

Now $X(t)$ is a prepared solution of (7.2.5) if and only if $Y(t)$ is a prepared solution of (7.3.4) and evidently both systems have the same oscillatory and nonoscillatory character. To establish the oscillatory character of (7.3.4), it is enough to consider the solution $Y_0(t)$ satisfying

$$Y_0(t_0) = 0, \quad Y_0'(t_0) = I, \quad I \text{ is the identify matrix.} \tag{7.3.5}$$

In fact, in view of the Morse separation theorem [42] the oscillation of this solution will guarantee oscillation of all solutions. Clearly, the solution of (7.3.3) satisfying the initial condition (7.3.5) is diagonal for all $t \geq t_0$, and so for this solution, system (7.3.4) may be written in the form (7.3.3), where $Y_0(t) = \text{diag}\{y_1(t), \cdots, y_n(t)\}$, and $D(t) = \{q_1(t), \cdots, q_n(t)\}$. Now since $\det Y_0(t) = \prod_{i=1}^{n} y_i(t)$, $Y_0(t)$ oscillates if and only if one of the functions $y_i(t)$ oscillates, i.e., if and only if one of the scalar equations (7.3.3) is oscillatory. This completes the proof. ∎

Now we shall study the general system (7.3.1). For this, first we note the following simple consequence of a solution which is symmetric and prepared.

Theorem 7.3.2. Any symmetric and prepared solution $X(t)$ of (7.3.1) which is nonsingular on an interval J, commutes with $Q(t)$ for all $t \in J$.

Proof. Since $X(t)$ is symmetric and prepared, (7.3.1) shows that $X(t)$ commutes with $X'(t)$ for all $t \geq t_0$. Differentiation of the identity

$$X'(t)X(t) - X(t)X'(t) = 0$$

leads to the fact that $X(t)$ commutes with $X''(t)$ for all $t \geq t_0$. If $X(t)$ is nonsingular on J, system (7.3.1) gives $X(t)G(X(t))Q(t)H(X(t)) = G(X(t))Q(t)H(X(t))X(t)$. Since $H(X(t))$ is nonsingular on J, we can write this as

$$X(t)G(X(t))Q(t) = G(X(t))Q(t)H(X(t))X(t)H^{-1}(X(t)) \quad \text{for all } t \in J. \tag{7.3.6}$$

Now, let $X(t) = P^{-1}(t)Y(t)P(t)$, where $P(t) \in \mathcal{N}_n$ and $X(t) \in \mathcal{D}_n$. Then, since $Y(t)$ and $H(Y(t)) \in \mathcal{D}_n$, we have

$$H(X(t))X(t) = P^{-1}(t)H(Y(t))Y(t)P(t) = P^{-1}(t)Y(t)H(Y(t))P(t)$$
$$= X(t)H(x(t)).$$

Similarly, $X(t)G(X(t)) = G(X(t))X(t)$. Thus, (7.3.6) becomes $G(X(t)) \times X(t)Q(t) = G(X(t))Q(t)X(t)$ for all $t \in J$. The result now follows from the fact that $G(X(t))$ is nonsingular on J. ∎

Next, we state the following two well known lemmas.

Lemma 7.3.2. Let A be an $n \times n$ diagonalizable matrix. Then the matrix S diagonalizes A, i.e., $SAS^{-1} \in \mathcal{D}_n$ if and only if S is a matrix whose columns are linearly independent eigenvectors of A.

Lemma 7.3.3. Let $A_1, \cdots, A_N$ be a given set of $n \times n$ matrices. Then the following statements are equivalent:

(I) A_i is diagonalizable for each i, and the A_i commute pairwise.

(II) $A_1, \cdots, A_N$ can be simultaneously diagonalized.

Theorem 7.3.3. Let $Q \in \mathcal{C}$ be as in Theorem 7.3.1. Moreover, assume that $G, H \in \mathcal{F}$ and let $F = GH$ with $F(\text{diag} \{x_1, \cdots, x_n\}) = \text{diag} \{f_1(x_1, \cdots, x_n), \cdots, f_n(x_1, \cdots, x_n)\}$. Then all symmetric, prepared solutions of (7.3.1) oscillate if and only if the diagonal system

$$y_i''(t) + q_i(t) f_i(y_1(t), \cdots, y_n(t)) = 0, \quad i = 1, 2, \cdots, n \qquad (7.3.7)$$

is oscillatory, i.e., if and only if in each solution $(y_1(t), \cdots, y_n(t))$ of system (7.3.7) at least one of the scalar functions $y_i(t)$ oscillates.

Proof. Assume that (7.3.7) is oscillatory and (7.3.1) has a nonoscillatory solution $X(t)$. Then $X(t)$ is nonsingular on $[t_0, \infty)$, (say). By Theorem 7.3.2, $X(t)$ commutes with $Q(t)$ for all $t \geq t_0$. $X(t)$ is symmetric and therefore diagonalizable, and so by Lemma 7.3.3 it can be simultaneously diagonalized with $Q(t)$ for $t \geq t_0$. By Lemma 7.3.2 it follows that for each $t \geq t_0$, $X(t)$ and $Q(t)$ have the same set of linearly independent eigenvectors. But $Q(t)$ has a constant set of eigenvectors for $t \geq t_0$, so $X(t)$ and $Q(t)$ can be diagonalized by the same constant matrix P for all $t \geq t_0$. Now observe that

$$
\begin{aligned}
G(X(t))Q(t) &= G(P^{-1}Y(t)P)P^{-1}D(t)P, \quad \text{where} \quad Y(t) \in \mathcal{D}_n \\
&= P^{-1}G(Y(t))PP^{-1}D(t)P = P^{-1}G(Y(t))D(t)P \\
&= P^{-1}D(t)G(Y(t))P = Q(t)G(X(t)),
\end{aligned}
$$

$$(7.3.8)$$

since $G(Y(t))$, $D(t) \in \mathcal{D}_n$.

Therefore, equation (7.3.1) can be written as

$$X''(t) + Q(t)F(X(t)) = 0, \quad t \geq t_0. \qquad (7.3.9)$$

Thus, we have

$$PY''(t)P^{-1} + PQ(t)F(P^{-1}Y(t)P)P^{-1} = 0,$$

which reduces to

$$Y''(t) + D(t)F(Y(t)) = 0. \qquad (7.3.10)$$

Let $Y(t) = \text{diag} \{y_1(t), \cdots, y_n(t)\}$. Then, $Y(t)$ is a nonsingular solution of (7.3.10) for $t \geq t_0$. But (7.3.10) is equivalent to (7.3.7), and we have a contradiction to the assumption that (7.3.7) is oscillatory. Therefore, the system (7.3.1) is oscillatory.

Conversely, suppose that (7.3.7) has a nonoscillatory solution $(y_1(t), \cdots, y_n(t))$. Then, $X(t) = P^{-1}Y(t)P = \text{diag}\{y_1(t), \cdots, y_n(t)\}$ is a nonoscillatory solution of the system (7.3.1). This completes the proof. ∎

The following corollary is immediate.

Corollary 7.3.1. Let $Q \in \mathcal{C}$ be as in Theorem 7.3.1 and $G,\ H \in \mathcal{F}_0$. Let $F = GH$ with $F(\text{diag}\{x_1, \cdots, x_n\}) = \text{diag}\{f(x_1), \cdots, f(x_n)\}$. Then all symmetric prepared solutions of the system (7.3.1) oscillate if and only if all solutions (for some i), of

$$y_i''(t) + q_i(t)f(y_i(t)) = 0 \qquad (7.3.11)$$

oscillate.

Next we shall present oscillation and nonoscillation theorems for the systems (7.2.5) and (7.3.1). First we shall consider the linear system (7.2.5). Since in view of Theorem 7.3.1 oscillation of (7.2.5) is equivalent to oscillation of the uncoupled system of scalar equations (7.3.3), explicit knowledge of the functions $q_i(t)$ (which exists if $Q(t)$ is known explicitly) will allow us to use any of the known oscillation or nonoscillation criteria for the second order linear differential equations of the form

$$y''(t) + q(t)y(t) = 0. \qquad (7.3.12)$$

As an application of Corollary 7.3.1 we present the following result.

Theorem 7.3.4. Let $Q \in \mathcal{C}$ be as in Theorem 7.3.1 and suppose that

$$\lim_{t \to \infty} \lambda_1 \left(\int_{t_0}^t Q(s)ds \right) = \infty.$$

Then all prepared solutions of the system (7.2.5) oscillate.

Proof. Since $P^{-1}Q(t)P = \text{diag}\{q_1(t), \cdots, q_n(t)\}$, we find

$$P^{-1} \left(\int_{t_0}^t Q(s)ds \right) P = \text{diag} \left\{ \int_{t_0}^t q_1(s)ds, \cdots, \int_{t_0}^t q_n(s)ds \right\},$$

so

$$\lambda_1 \left(\int_{t_0}^t Q(s)ds \right) = \max_{1 \leq i \leq n} \int_{t_0}^t q_i(s)ds.$$

Now, $\lim_{t \to \infty} \lambda_1 \left(\int_{t_0}^t Q(s)ds \right) = \infty$ implies $\lim \text{approx sup}_{t \to \infty} \int_{t_0}^t q_i(s)ds = \infty$ for at least one index i. By Corollary 7.3.1 it follows that for such indices i the equation $y_i''(t) + q_i(t)y_i(t) = 0$ is oscillatory. The oscillation of (7.2.5) now follows from Theorem 7.3.1. ∎

Remark 7.3.1. It is evident that weaker hypothesis

$$\lim_{t \to \infty} \text{approx sup} \; \lambda_1 \left(\int_{t_0}^t Q(s)ds \right) \; = \; \infty$$

would suffice. Other oscillation criteria similar to those in Chapter 2 and Section 6.1 can also be applied to equation (7.2.5) when $Q \in \mathcal{C}$.

We conclude our discussion of the linear system (7.2.5) with the following nonoscillation result.

Theorem 7.3.5. Let $Q \in \mathcal{C}$ be as in Theorem 7.3.1 and suppose that

$$\lim_{t \to \infty} \int_{t_0}^t s\lambda_1(Q(s))ds \quad \text{exists as a finite limit.}$$

Then the system (7.2.5) is nonoscillatory.

Proof. Let $q(t) = \lambda_1(Q(t))$. Then, $\lim_{t \to \infty} \int_{t_0}^t sq(s)ds$ exists as a finite limit, which implies that equation (7.3.12) is nonoscillatory. Since $q_i(t) \leq q(t)$ for all $t \geq t_0$, $i = 1, 2, \cdots, n$ it follows from the Sturm comparison theorem that each of the equations (7.3.5) is nonoscillatory. Hence the system (7.2.5) is nonoscillatory. ∎

Next, we shall consider the nonlinear system (7.3.1) when $Q(t)$ is nonnegative definite for all $t \geq t_0$ and $G, H \in \mathcal{F}_0$ with $F = GH$ and $F(\text{diag}\{x_1, \cdots, x_n\}) = \text{diag}\{f(x_1), \cdots, f(x_n)\}$. We recall the following known oscillation criteria for the equation

$$x''(t) + q(t)f(x(t)) \; = \; 0, \tag{7.3.13}$$

where $q(t) \in C([t_0, \infty), \mathbb{R}^+)$ and $f \in C(\mathbb{R}, \mathbb{R})$, $xf(x) > 0$, $f'(x) \geq 0$ for $x \neq 0$.

(I_1) Let $\int^{\pm \infty} du/f(u) < \infty$. Then equation (7.3.13) is oscillatory if and only if $\int^\infty sq(s)ds = \infty$.

(I_2) Let $f(x) = x^{\alpha/\beta}$, where α, β are odd integers with $0 < \gamma = \alpha/\beta < 1$. Then equation (7.3.13) is oscillatory if and only if $\int^\infty s^\gamma q(s)ds = \infty$.

We shall extend the above criteria to the system (7.3.1).

Theorem 7.3.6. Let the function f be strictly increasing with $f(0) = 0$, $\lim_{x \to \pm \infty} f(x) = \pm \infty$ and suppose that

$$\int^{\pm \infty} \frac{du}{f(u)} \; < \; \infty. \tag{7.3.14}$$

Let $Q \in \mathcal{C}$ be as in Theorem 7.3.1 and $Q(t)$ nonnegative definite for all $t \geq t_0$. Then all symmetric prepared solutions of the system (7.3.1) oscillate if and only if

$$\lim_{t \to \infty} \lambda_1 \left(\int_{t_0}^t sQ(s)ds \right) = \infty. \qquad (7.3.15)$$

Proof. $\lim_{t \to \infty} \lambda_1 \left(\int_{t_0}^t sQ(s)ds \right) = \infty$ if and only if $\lim_{t \to \infty} \max_{1 \leq i \leq n} \int_{t_0}^t sq_i(s)ds = \infty$, if and only if $\sum_{i=1}^n \int_{t_0}^t sq_i(s)ds = \infty$, and since the q_i's are all nonnegative if and only if $\lim_{t \to \infty} \int_{t_0}^t sq_i(s)ds = \infty$, for at least one index i. Theorem 7.3.6 now follows from the above criterion (I_1) and Corollary 7.3.1. ∎

Theorem 7.3.7. Let $f(x) = x^{\alpha/\beta}$, where α, β are odd integers with $0 < \gamma = \alpha/\beta < 1$. Let $Q \in \mathcal{C}$ be as in Theorem 7.3.1 and $Q(t)$ is nonnegative definite for all $t \geq t_0$. Then all symmetric prepared solutions of the system (7.3.1) oscillate if and only if

$$\lim_{t \to \infty} \lambda_1 \left(\int_{t_0}^t s^\gamma Q(s)ds \right) = \infty.$$

Proof. The proof is the same as that of Theorem 7.3.6 except now we use criterion (I_2). ∎

Theorem 7.3.6 when applied to system

$$X''(t) + (X^*(t))^k \, Q(t) X^{k+1}(t) = 0, \qquad (7.3.16)$$

where k is a positive integer, and Q and X satisfy (ii) and (iii) of the system (7.3.1) gives the following corollary.

Corollary 7.3.2. Let $Q \in \mathcal{C}$ be as in Theorem 7.3.1 and $Q(t)$ is nonnegative definite for all $t \geq t_0$. Then condition (7.3.15) is necessary and sufficient for all symmetric prepared solutions of the system (7.3.16) to be oscillatory.

Finally, we present the following nonoscillation result. The arguments are similar to those of earlier theorems and hence the proof is omitted.

Theorem 7.3.8. Let $f(x) = x^{2k+1}$, where k is a positive integer. Let the eigenvalues of $Q(t)$ be $\mu_1(t) \geq \mu_2(t) \geq \cdots \geq \mu_n(t) \geq 0$, and $Q \in \mathcal{C}$ be as in Theorem 7.3.1. Further, let $\mu_j'(t) \leq 0$ for each j, and

$$\lim_{t \to \infty} \int_{t_0}^t s^{2k+1} \mu_1(s)ds < \infty.$$

Then all symmetric prepared solutions of the system (7.3.1) are nonoscillatory.

7.4. Comparison Theorems for Operator–Valued Linear Differential Equations

The Hille–Wintner theorem asserts that if $q(t)$, $q_1(t) \in C([t_0, \infty), \mathbb{R})$, $p(t) = \int_t^\infty q(s)ds$, $p_1(t) = \int_t^\infty q_1(s)ds$, $t \geq t_0$ exist and satisfy $|p_1(t)| \leq p(t)$ for all $t \in [t_0, \infty)$, then the existence of a nonoscillatory solution of the equation $x''(t) + q(t)x(t) = 0$ on $[t_0, \infty)$ implies the existence of a nonoscillatory solution of $x''(t) + q_1(t)x(t) = 0$ on $[t_0, \infty)$.

We shall consider the systems

$$X''(t) + Q(t)X(t) = 0 \qquad\qquad (7.4.1)$$

and

$$X''(t) + Q_1(t)X(t) = 0 \qquad\qquad (7.4.2)$$

and prove several generalizations of the Hille–Wintner result in the Banach lattice case. To obtain such results we need the following notation, definitions and preliminaries. We denote by B, a Banach lattice. We recall that a Banach lattice B is a Banach space with a vector lattice structure such that $|x| \leq |y|$ implies $\|x\| \leq \|y\|$, $x, y \in B$, where $|x| \equiv x_v(-x)$. Let $B^+ = \{x \in B : x \geq 0\}$. A subset $S \subset B$ is order bounded if there exists $z \in B^+$ with $|x| \leq z$ for all $x \in S$. The Banach lattice B is said to be order complete if for every nonempty majorized (with respect to ordering) subset A of B, $\sup A$ exists in B. Also, B is said to have order continuous norm if each downward directed family A of B such that $\inf A = 0$ converges in norm to 0. A Banach lattice with order continuous norm is automatically order complete ([46], Theorem II.5.10, p.89). Familiar examples of such Banach lattice are ℓ_p, $\mathcal{L}_p$, $1 \leq p < \infty$, c_0 and any reflexive Banach lattice.

Let $\mathcal{L}^+(B)$ denote the set of positive bounded linear operators on B, i.e., $T \in \mathcal{L}^+(B)$ if and only if $T(B^+) \subset B^+$. $\mathcal{L}(B)$ represents the Banach algebra of bounded linear operators $T : B \to B$. An operator $T \in \mathcal{L}(B)$ is said to be regular if $T = T_1 - T_2$ for some T_1, $T_2 \in \mathcal{L}^+(B)$. $\mathcal{L}^r(B)$ denotes the vector space of regular operators on B. It can be shown (cf. [46], p. 229) that if B is order complete, then T is regular if and only if T maps order bounded sets onto order bounded sets if and only if $|T| \equiv T_v(-T)$ exists in the lattice structure induced on $\mathcal{L}(B)$. Furthermore, if B is order complete, then $\mathcal{L}^r(B)$ is an order complete vector lattice ([46], p.229). We refer to [45] and [46] for further discussion of Banach lattices and their properties.

Remark 7.4.1. If B is a Hilbert space, then $\mathcal{L}(B)$ is a B^*-algebra. In this case comparison and oscillation theorems have been obtained which are generalizations of the known theorems in the classical scalar case, (e.g., the Sturm comparison theorem and the Hille–Wintner comparison theorem). In these results the notion of comparison of operator in $\mathcal{L}(B)$ is used which is induced by the inner product, i.e., if T_1, $T_2 \in \mathcal{L}(B)$, then $T_1 \geq T_2$ means $\prec (T_1 - T_2)x, x \succ \geq 0$ for all x in the Hilbert space B.

In this section we shall be interested in employing an alternative notion of positivity of $\mathcal{L}(B)$ which is induced by a vector lattice structure on B. This has the advantage that the class of positive operators is not only a positive cone but is also closed under multiplication, a fact which is not true in the B^*-algebra case. A further advantage of the Banach lattice is that it enables us to obtain certain results for nonself–adjoint equations.

Now, we consider systems (7.4.1) and (7.4.2), where Q, $Q_1 : [t_0, \infty) \to \mathcal{L}(B)$. By a solution $X(t)$ of (7.4.1) (or (7.4.2)), we mean $X : [t_0, \infty) \to \mathcal{L}(B)$ which is twice continuously differentiable in the uniform operator topology and satisfy the system (7.4.1) (or (7.4.2)) for $t \in [t_0, \infty)$. A solution $X = X(t)$ is said to be nonsingular at a point $t_1 \in [t_0, \infty)$ if it has a bounded inverse $X^{-1}(t_1) \in \mathcal{L}(B)$. If $X(t)$ is nonsingular for all $t \in [t_1, \infty)$, $t_1 \geq t_0$ then $X = X(t)$ is said to be a nonoscillatory solution of (7.4.1) on $[t_1, \infty)$. Otherwise, $X = X(t)$ is said to be oscillatory on $[t_0, \infty)$. Note that the inverse $X^{-1}(t)$ of a nonoscillatory solution $X(t)$ of (7.4.1) is continuously differentiable.

We are now prepared to prove the following result.

Theorem 7.4.1. Let B be a Banach lattice with order continuous norm. Suppose the limits $P(t) = \lim_{T \to \infty} \int_t^T Q(s)ds$, $P_1(t) = \lim_{T \to \infty} \int_t^T Q_1(s)ds$ exist (in the uniform operator topology of $\mathcal{L}(B)$), and

(i) $P(t)$, $P_1(t)$, $P(t) - P_1(t) \in \mathcal{L}^+(B)$, $t \in [t_0, \infty)$.

If there exists a nonoscillatory solution $X(t)$ of (7.4.1) such that

(ii) $W(t) = X'(t)X^{-1}(t) \in \mathcal{L}^+(B)$ for all $t \in [t_0, \infty)$,

then the system (7.4.2) has a nonoscillatory solution on $[t_0, \infty)$.

Proof. In what follows inequality sign will refer to an appropriate positive cone. The Riccati transformation $W(t) = X'(t)X^{-1}(t)$ in (7.4.1) yields the system

$$W(t) = W(T) + \int_t^T Q(s)ds + \int_t^T W^2(s)ds, \quad t_0 \leq t < T < \infty. \quad (7.4.3)$$

We assert that (7.4.3) implies that the set $\left\{ \int_t^T W^2(s)ds : t_0 \leq T < \infty \right\}$ is an upward directed set in $\mathcal{L}^r(B)$ and is order bounded above by $W(t)$. For this, note that the monotonicity of $\left\{ \int_t^T W^2(s)ds : t_0 \leq T < \infty \right\}$ follows from the hypothesis that $W(s)$ (and therefore $W^2(s)$) is nonnegative for all $s \geq t_0$. Denote by $\Phi(t,T) = \int_t^T W^2(s)ds$, so that (7.4.3) can be written as

$$\Phi(t,T) \;=\; W(t) - W(T) - P(t) + P(T). \tag{7.4.4}$$

If $t_0 \leq T \leq \tau$, then in view of (7.4.4), we have $\Phi(t,T) \leq \Phi(t,\tau) \leq W(t) + P(\tau)$, and hence

$$W(t) + P(\tau) - \Phi(t,T) \;\geq\; 0 \quad \text{for} \quad t_0 \leq T \leq \tau. \tag{7.4.5}$$

For each $\tau \geq T$ the left–hand side of (7.4.5) is in $\mathcal{L}^r(B)$, which is a normed vector lattice and therefore has norm–closed positive cone $\mathcal{L}^+(B)$. Letting $\tau \to \infty$ in (7.4.5), we obtain $W(t) - \Phi(t,T) \geq 0$, which completes the proof of the assertion. Since $\mathcal{L}^r(B)$ is order complete, it follows that $\Phi(t) = \sup_T \Phi(t,T)$ exists. Let $y \in B^+$. Then, (7.4.3) implies that

$$W(T)y \;=\; W(t)y - \left(\int_t^T Q(s)ds \right) y - \Phi(t,T)y. \tag{7.4.6}$$

Since B has order continuous norm, $\Phi(t,T)y \to \Phi(t)y$ in norm, so $\Phi(t)$ is the strong operator limit of $\Phi(t,T)$ as $T \to \infty$. Thus the limit in norm as $T \to \infty$ exists for the right–hand side of (7.4.6). We conclude that the limit in norm of $W(T)y$ as $T \to \infty$ exists (in B^+ by norm–closedness of B^+), we call it $C(y)$. Thus,

$$C(y) \;=\; W(t)y - P(t)y - \Phi(t)y. \tag{7.4.7}$$

It follows that $C(y)$ in fact defines a positive linear operator on B^+ and extends uniquely to a linear operator on B (cf. [46], p. 58). Thus (7.4.7) holds for all y and we may write $C(y) = Cy$. Hence,

$$W(t) \;=\; C + P(t) + \Phi(t), \quad C \in \mathcal{L}^+(B), \quad t \in [t_0, \infty). \tag{7.4.8}$$

We now define a sequence of operator–valued functions $\{Z_n(t)\}$ as follows:

$$\begin{cases} Z_0(t) = C, \quad t \geq t_0 \\[2mm] Z_{n+1}(t) = C + P_1(t) + s\text{–}\lim_{T \to \infty} \int_t^T Z_n^2(s)ds, \quad n = 0,1,\cdots, \quad t \geq t_0 \end{cases}$$
$$\tag{7.4.9; n}$$

where s–$\lim$ denotes the strong operator limit. To justify this, inductively we shall show

$$\begin{cases} 0 \leq Z_n(t) \leq W(t) \quad \text{and} \quad Z_n(t) \text{ is continuous in} \\ \text{the uniform operator topology } \mathcal{L}(B), \quad t \geq t_0. \end{cases} \tag{7.4.10; n}$$

Clearly (7.4.10;0) is true. Now assume that (7.4.10;n) holds. Then, $0 \leq Z_n(s) \leq W(s)$ implies $0 \leq Z_n^2(s) \leq W^2(s)$. This is a consequence of the cone $\mathcal{L}^+(B)$ being closed under multiplication. (It does not, in general, hold in the B^*-algebra case). Thus, $\left\{ \int_t^T Z_n^2(s)ds \right\}$ is monotone increasing in T and is order bounded by $W(t)$ since $\int_t^T Z_n^2(s)ds \leq \int_t^T W^2(s)ds \leq W(t)$, and hence $s-\lim_{T \to \infty} \int_t^T Z_n^2(s)ds$ exists. Therefore, $Z_{n+1}(t)$ is well-defined by (7.4.9;n). Now it follows from the hypothesis that $0 \leq P_1(t) \leq P(t)$ and $0 \leq Z_{n+1}(t) \leq W(t)$. We also have

$$Z_{n+1}(t) - Z_{n+1}(s) = P_1(t) - P_1(s) - \int_s^t Z_n^2(\tau)d\tau, \qquad (7.4.11)$$

so it follows that $Z_{n+1}(t)$ is continuous, hence (7.4.10;n+1) is true and thus (7.4.10;n) holds for all $n \geq 0$. Let $\Psi_n(t,T) = \int_t^T Z_n^2(s)ds$ and denote $s-\lim_{T \to \infty} \Psi_n(t,T) = \Psi_n(t)$. Since for fixed t, $\{Z_n(t)\}_{n=1}^\infty$ is a monotone increasing sequence in $\mathcal{L}^+(B)$, bounded above by $W(t)$, it follows that $s-\lim_{n \to \infty} Z_n(t)$ exists. Let $Z(t) = s-\lim_{n \to \infty} Z_n(t)$. By (7.4.9;n), we have

$$Z(t) = C + P_1(t) + \Psi(t), \qquad (7.4.12)$$

where $\Psi(t) = s-\lim_{n \to \infty} \Psi_n(t)$. Note that $Z(t)$ is continuous, since

$$Z(t) - Z(s) = P_1(t) - P_1(s) - s-\lim_{n \to \infty} \int_s^t Z_n^2(\tau)d\tau \qquad (7.4.13)$$

and $0 \leq Z_n(\tau) \leq W(\tau)$ implies

$$0 \leq \left\| \int_s^t Z_n^2(u)du \right\| \leq \int_s^t \|W^2(u)\|du \to 0 \quad \text{as} \quad t - s \to 0$$

by continuity of W. For each τ and each $y \in B^+$, $Z_n^2(\tau)y \to Z^2(\tau)y$ in norm and $Z^2(\tau)y$ is continuous in τ. Now it follows from Dini's theorem that this convergence is uniform on any compact subinterval $[s,t]$ of $[t_0, \infty)$. Thus, $\lim_{n \to \infty} \left(\int_s^t Z_n^2(\tau)d\tau \right) y = \left(\int_s^t Z^2(\tau)d\tau \right) y$, $y \in B^+$, so that $s-\lim_{n \to \infty} \int_s^t Z_n^2(\tau)d\tau = \int_s^t Z^2(\tau)d\tau$. We may therefore write (7.4.13) as

$$Z(t) - Z(s) = P_1(t) - P_1(s) - \int_s^t Z^2(\tau)d\tau. \qquad (7.4.14)$$

Dividing both sides of (7.4.14) by $t - s$ $(t \neq s)$ and taking limits in the uniform operator topology as $s \to t$, we get

$$Z'(t) = -Q_1(t) - Z^2(t). \qquad (7.4.15)$$

Now define $\hat{X}(t)$ to be the solution of $\hat{X}'(t) = Z(t)\hat{X}(t)$, $\hat{X}(t_0) = I$ where I is the identity operator in $\mathcal{L}(B)$. Then, $\hat{X}(t)$ is a nonoscillatory solution of (7.4.2), and this proves the theorem. ∎

Example 7.4.1. Let B be a Banach lattice with order continuous norm and let B_1 be its dual space. (B_1 is also a Banach lattice, under the dual norm, which is order complete). Let $c \in B^+$ and $\phi \in B_1^+$. Let $q(t) \in C^2([t_0, \infty), \mathbb{R}^+)$ such that $q'(t) > 0$ and $q''(t) < 0$ for all $t \in [t_0, \infty)$. Define $Q : [t_0, \infty) \to \mathcal{L}(B)$ by

$$Q(t)y = -\frac{q''(t)\phi(y)c}{1 + q(t)\phi(c)}, \quad y \in B.$$

Then,

$$Q(t) \in \mathcal{L}^+(B) \quad \text{for all} \quad t \in [t_0, \infty), \quad \left(\|Q(t)\| = \frac{|q''(t)|\|\phi\|\|c\|}{1 + q(t)\|\phi(c)\|}\right).$$

Note that

$$\left\|\int_t^\infty Q(s)ds\right\| = \left(\int_t^\infty -\frac{q''(s)}{1 + q(s)(c)}ds\right)\|\phi\|\|c\| \le q'(t)\|\phi\|\|c\|.$$

It may be verified that $X''(t) + Q(t)X(t) = 0$ has a solution $X(t)$ defined by $X(t)y = y + q(t)\phi(y)c$, $y \in B$. Now it follows that $X(t)$ is a nonoscillatory solution and furthermore, the Riccati variable $W(t) = X'(t)X^{-1}(t)$ is given by

$$W(t)y = \frac{q'(t)\phi(y)c}{1 + q(t)\phi(c)}, \quad y \in B$$

so $W(t) \in \mathcal{L}^+(B)$ on $[t_0, \infty)$. Therefore, if $c_1 \in B^+$, $c_1 \le c$, $\phi_1 \in B_1^+$, $\phi_1 \le \phi$ and if $q_1(t)$ is a real–valued function such that

$$0 \le \int_t^\infty -\frac{q_1''(s)}{1 + q_1(s)\phi_1(c_1)}ds \le \int_t^\infty -\frac{q''(s)}{1 + q(s)\phi(c)}ds, \quad t \in [t_0, \infty)$$

then equation $X''(t) + Q_1(t)X(t) = 0$ has a nonoscillatory solution by Theorem 7.4.1, where $Q_1(t)$ is given by

$$Q_1(t)y = -\frac{q_1''(t)\phi_1(y)c_1}{1 + q_1(t)\phi_1(c)}, \quad y \in B.$$

We note that in general $Q_1(t) \notin \mathcal{L}^+(B)$.

One would like to weaken the hypothesis (i) in Theorem 7.4.1 to

(i)′ $P_1(t) \in \mathcal{L}^r(B)$, $P(t)$, $P(t) - |P_1(t)| \in \mathcal{L}^+(B)$, $t \in [t_0, \infty)$

and to dispense with hypothesis (ii) (which holds automatically in the scalar case for a nonoscillatory solution $X(t)$ of (7.4.1) when $P(t) \ge 0$).

As far as (ii) is concerned, we may make the following remarks:

Assume $P(t) \in \mathcal{L}^+(B)$, $t \in [t_0, \infty)$, and define

$$
\begin{cases}
P_0(t) = P(t) \\
P_n(t) = P(t) + s\text{-}\lim_{T \to \infty} \int_t^T P_{n-1}^2(s)ds, \quad n = 1, 2, \cdots.
\end{cases}
$$

Then the existence of nonoscillatory solution $X(t)$ of (7.4.1) satisfying (ii) is equivalent to the following:

(iii) $P_n(t)$ is defined for all $t \in [t_0, \infty)$, $n = 0, 1, 2, \cdots$ and $s\text{-}\lim_{n \to \infty} P_n(t) = \hat{P}(t)$ exists.

The equivalence of (ii) and (iii) can be established by using essentially the same arguments as were used in the proof of Theorem 7.4.1.

When B is finite dimensional we can replace (i) by (i)'.

Theorem 7.4.2. Let $B = \mathbb{R}^n$ and suppose that

$$
P(t) = \lim_{T \to \infty} \int_t^T Q(s)ds, \quad P_1(t) = \lim_{T \to \infty} \int_t^T Q_1(s)ds \qquad (7.4.16)
$$

exist such that (i)' holds. If there exists a nonoscillatory solution $X(t)$ of (7.4.1) such that (ii) holds, then the system (7.4.2) has a nonoscillatory solution on $[t_0, \infty)$.

Proof. As in the proof of Theorem 7.4.1, we put $W(t) = X'(t)X^{-1}(t)$ in (7.4.1) and obtain (7.4.3) which leads to (7.4.8). Let $\mathcal{Z}$ be the set of continuous functions $Z : [t_0, \infty) \to \mathcal{L}(\mathbb{R}^n)$ such that $|Z(t)| \leq W(t)$ for all $t \in [t_0, \infty)$. Then, $\mathcal{Z}$ is closed, convex subset of $\mathcal{M}$, the Fréchet space of continuous operator–valued functions on $[t_0, \infty)$ with the compact–open topology.

Now if $Z \in \mathcal{Z}$, then for each $t \in [t_0, \infty)$, $Z(t)$ may be represented by an $n \times n$ matrix, say, $Z(t) = [(Z(t))_{ij}]$. Similarly, we may represent $W(t)$ by the $n \times n$ matrix $W(t) = [(W(t))_{ij}]$, and we have $|(Z(t))_{ij}| \leq (W(t))_{ij}$, $i, j = 1, \cdots, n$. It follows that $|Z^2(t)| \leq W^2(t)$. Furthermore, we have $\|Z^2(t)\| \leq \|W^2(t)\|$, where $\|\cdot\|$ denotes the uniform operator norm on $\mathcal{L}(B)$. Since $W^2(t)$ is integrable on $[t_0, \infty)$, it follows that $Z^2(t)$ is integrable on $[t_0, \infty)$, and

$$
\left\| \int_t^\infty Z^2(s)ds \right\| \leq \int_t^\infty W^2(s)ds. \qquad (7.4.17)
$$

It follows from (7.4.8), (7.4.17) and (i)' that F on $\mathcal{Z}$ defined by

$$
(FZ)(t) = C + P_1(t) + \int_t^\infty Z^2(s)ds \qquad (7.4.18)
$$

is a map from $\mathcal{Z}$ into itself. Let $\{Z_n\}_{n=1}^{\infty}$, $Z \in \mathcal{Z}$ with $\lim_{n \to \infty} Z_n = Z$. For any $T \geq t > t_0$, we find

$$\|(FZ_n)(t) - (FZ)(t)\| = \left\| \int_t^{\infty} Z_n^2(s)ds - \int_t^{\infty} Z^2(s)ds \right\|$$

$$\leq \int_t^T \|Z_n^2(s) - Z^2(s)\| ds + \left\| \int_T^{\infty} Z_n^2(s)ds \right\| + \left\| \int_T^{\infty} Z^2(s)ds \right\|.$$

$$(7.4.19)$$

Now since $\left\| \int_T^{\infty} Z_n^2(s)ds \right\| \leq \left\| \int_T^{\infty} W^2(s)ds \right\|$ and $\left\| \int_T^{\infty} Z^2(s)ds \right\| \leq \left\| \int_T^{\infty} W^2(s)ds \right\|$, we can for any given $\epsilon > 0$ choose $T = T(\epsilon)$ so that $\left\| \int_T^{\infty} Z_n^2(s)ds \right\| < \epsilon/3$, $n = 1, 2, \cdots$ and $\left\| \int_T^{\infty} Z^2(s)ds \right\| < \epsilon/3$. By the uniform convergence of Z_n to Z on $[t_0, \infty)$, we have

$$\int_t^T \|Z_n^2(s) - Z^2(s)\| ds < \frac{\epsilon}{3} \quad \text{for all} \quad n \geq N(\epsilon), \quad \text{say}$$

and all $t \in [t_0, T]$. Thus, (7.4.19) gives $\|(FZ_n)(t) - (FZ)(t)\| < \epsilon$ for all $n \geq N(\epsilon)$, $t \in [t_0, \infty)$. It follows that $\lim_{n \to \infty} FZ_n = FZ$, i.e., $F : \mathcal{Z} \to \mathcal{Z}$ is continuous.

Now for s, $t \geq t_0$, we have $\|FZ(t) - FZ(s)\| \leq \int_s^t \|Q_1(\tau)\| d\tau + \int_s^t \|W^2(\tau)\| d\tau$. Hence, $F(\mathcal{Z})$ consists of equicontinuous, uniformly bounded operator–valued functions. Thus, $F(\mathcal{Z})$ is precompact. By Tychonov's theorem, F has a fixed point in Z, i.e., there exists $Z(t)$ such that

$$Z(t) = C + P_1(t) + \int_t^{\infty} Z^2(s)ds. \tag{7.4.20}$$

It follows that the system (7.4.2) has a nonoscillatory solution. ∎

Next, we compare the system (7.4.2) in $\mathcal{L}(\mathbb{R}^n)$ with an appropriate scalar equation

$$x''(t) + q(t)x(t) = 0, \tag{7.4.21}$$

where $q(t) \in C([t_0, \infty), \mathbb{R})$.

Theorem 7.4.3. Let $Q_1 : [t_0, \infty) \to \mathcal{L}(\mathbb{R}^n)$ be continuous and let $[(Q_1(t))_{ij}]$ be the usual matrix representation of $Q_1(t)$. Assume that $(P_1(t))_{ij} = \int_t^{\infty} (Q_1(s))_{ij} ds$ exists for all i, j. Let $q(t) \in C([t_0, \infty), \mathbb{R})$ and suppose $p(t) = \int_t^{\infty} q(s)ds$ exists with

(iv) $p(\sqrt{n}t) \geq \sqrt{n} \max_{ij}(P_1(t))_{ij}$, $t \in [t_0, \infty)$.

If equation (7.4.21) has a nonoscillatory solution on $[t_0, \infty)$, then so does (7.4.2).

Proof. Let $x(t)$ be a nonoscillatory solution of equation (7.4.21) and let $u(t) = x'(t)x^{-1}(t)$ to obtain the Riccati equation $u'(t) = -q(t) - u^2(t)$.

Put $v(t) = (1/\sqrt{n})u(\sqrt{n}t)$, to get

$$v'(t) = -q(\sqrt{n}t) - nv^2(t). \tag{7.4.22}$$

Integrate (7.4.22), to find

$$v(t) = v(T) + \int_t^T q(\sqrt{n}s)ds + n \int_t^T v^2(s)ds. \tag{7.4.23}$$

Letting $T \to \infty$ in (7.4.23), it follows that $\lim_{T \to \infty} v(T) = b$, where $-\infty \le b < \infty$. It is easy to see that $0 \le b < \infty$, and we have

$$v(t) = b + \frac{1}{\sqrt{n}}p(\sqrt{n}t) + n \int_t^\infty v^2(s)ds. \tag{7.4.24}$$

Let $\mathcal{Z}_0$ be the set of continuous functions $Z(t) = [(Z(t))_{ij}]$ from $[t_0, \infty)$ to $\mathcal{L}(\mathbb{R}^n)$ such that $|(Z(t))_{ij}| \le v(t)$, $t \in [t_0, \infty)$ for each i, j. Define the map $F_0 : \mathcal{Z}_0 \to \mathcal{Z}_0$ by

$$((F_0 Z)(t))_{ij} = (P_1(t))_{ij} + \int_t^\infty \left(\sum_{k=1}^n (Z(s))_{ik}(Z(s))_{kj} \right) ds, \tag{7.4.25}$$

for $t \in [t_0, \infty)$, i, $j = 1, 2 \cdots, n$.

It follows from the hypothesis (iv) and (7.4.24) that F_0 is well defined, $\mathcal{Z}_0$ is closed, convex subset of $\mathcal{M}$, the Fréchet space of continuous operator–valued functions on $[t_0, \infty)$ with the compact–open topology. Arguments similar to those used in the proof of Theorem 7.4.2 show that F_0 is continuous on $\mathcal{Z}_0$ and that $F_0(\mathcal{Z}_0)$ is precompact subset of $\mathcal{Z}_0$. Hence Tychonov's theorem yields the existence of a fixed point of F_0, i.e., there exists $Z(t) = [(Z(t))_{ij}]$ such that

$$(Z(t))_{ij} = (P_1(t))_{ij} + \int_t^\infty \left(\sum_{k=1}^n (Z(s))_{ik}(Z(s))_{kj} \right) ds \tag{7.4.26}$$

for $t \in [t_0, \infty)$, i, $j = 1, 2, \cdots, n$.

From the matrix Riccati equation of (7.4.2), it is clear that (7.4.26) enables us to define a nonoscillatory solution of the system (7.4.2). ∎

Example 7.4.2. As an application of Theorem 7.4.2 consider the case $n = 2$. Let

$$Q(t) = \frac{1}{t^2} \begin{bmatrix} a & b \\ c & d \end{bmatrix},$$

where a, b, c, d are nonnegative constants and $t \in [1, \infty)$. Invoking condition (iii), there is a nonoscillatory solution $X(t)$ of (7.4.1) satisfying

(ii) if and only if the sequence $\{P_n(t)\}$ converges as $n \to \infty$, where

$$P_0(t) = P(t) = \frac{1}{t}\begin{bmatrix} a & b \\ c & d \end{bmatrix},$$

$$P_n(t) = P(t) + \int_t^\infty P_{n-1}^2(s)ds, \quad n = 1, 2, \cdots.$$

If

$$P_n(t) = \frac{1}{t}\begin{bmatrix} a_n & b_n \\ c_n & d_n \end{bmatrix},$$

we obtain the relations

$$\begin{cases} a_{n+1} = a + a_n^2 + b_n c_n \\ b_{b+1} = b + b_n(a_n + d_n) \\ c_{n+1} = c + c_n(a_n + d_n) \\ d_{n+1} = d + d_n^2 + b_n c_n, \quad n = 0, 1, 2, \cdots. \end{cases} \tag{7.4.27}$$

Since a, b, c, $d \geq 0$, the sequences a_n, b_n, c_n, d_n are increasing with n, and hence have finite limits α, β, γ, δ if and only if the system of equations

$$\begin{cases} \alpha = a + \alpha^2 + \beta\gamma \\ \beta = b + \beta(\alpha + \delta) \\ \gamma = c + \gamma(\alpha + \delta) \\ \delta = d + \delta^2 + \beta\gamma \end{cases} \tag{7.4.28}$$

has a solution.

It can be shown that the solvability of (7.4.28) is equivalent to the existence of positive solution m of the equation

$$m^2[1 - 2(d + a) - m^2] = (d - a)^2 + 4bc, \tag{7.4.29}$$

and α, β, γ and δ are given by

$$\alpha = \frac{1}{2}\left[1 - m - \frac{d-a}{m}\right], \quad \beta = \frac{b}{m}$$

$$\gamma = \frac{c}{m}, \quad \delta = \frac{1}{2}\left[1 - m + \frac{d-a}{m}\right].$$

This in turn is easily seen to be equivalent to the condition

$$\begin{cases} \alpha < 1/4 \quad \text{if} \quad d = a \quad \text{and} \quad bc = 0 \\ d + a + \sqrt{(d-a)^2 + 4bc} \leq 1/2 \quad \text{if} \quad d \neq a \quad \text{or} \quad bc \neq 0. \end{cases} \tag{7.4.30}$$

Let a, b, c, d be arbitrary nonnegative constants satisfying (7.4.30) with equality. Further, let $Q_1(t) = [q_{ij}(t)]$ be chosen so that

$$\begin{cases} \limsup_{t \to \infty} t \int_t^\infty q_{11}(s)ds < a, & \limsup_{t \to \infty} t \int_t^\infty q_{12}(s)ds < b \\ \limsup_{t \to \infty} t \int_t^\infty q_{21}(s)ds < c, & \limsup_{t \to \infty} t \int_t^\infty q_{22}(s)ds < d. \end{cases} \tag{7.4.31}$$

Then by Theorem 7.4.2 the system (7.4.2) has a nonoscillatory solution.

This example may be thought of as an extension to the 2×2 matrix case of Hille's nonoscillation criterion

$$\limsup_{t \to \infty} tq_1(t) < \frac{1}{4} \tag{7.4.32}$$

for the scalar equation $x''(t) + q_1(t)x(t) = 0$. In fact, in the case $b = c = d = 0$, we have $a = 1/4$ and (7.4.31) reduces to (7.4.32).

7.5. Oscillation Results for Differential Systems with Forcing Terms

Consider the system

$$X''(t) + Q(t, X(t)) = F(t). \tag{7.5.1}$$

Let $\mathcal{M}_n$ denote the space of all $n \times n$ matrices and assume that

(i) $Q : [t_0, \infty) \times \mathcal{M}_n \to \mathcal{M}_n$ is continuous and

$$\|Q(t, X)\| \le q(t, \|X\|), \quad (t, X) \in [t_0, \infty) \times \mathcal{M}_n, \quad t_0 \ge 0$$

where $q \in C([t_0, \infty) \times \mathbb{R}^+, \mathbb{R}^+)$ and $q(t, z)$ is increasing in z,

(ii) $F : [t_0, \infty) \to \mathcal{M}_n$ is continuous and there exists a continuous $V : [t_0, \infty) \to \mathcal{M}_n$ such that $F(t) = V''(t)$, $t \in [t_0, \infty)$.

The following theorem guarantees the existence of solutions of (7.5.1) with a certain asymptotic behavior.

Theorem 7.5.1. Let conditions (i) and (ii) hold, and assume that for some constant $\lambda > 0$,

$$\int_{t_0}^\infty sq(s, \lambda + \|V(s)\|)ds < \infty. \tag{7.5.2}$$

Then if $K \in \mathcal{M}_n$ with $\|K\| < \lambda$, there exists a solution $X(t)$ of the system (7.5.1) which satisfies $\lim_{t \to \infty}[X(t) - V(t)] = K$.

Proof. Let $X(t)$ be a solution of (7.5.1) and set $Y(t) = X(t) - V(t)$. Then it follows from the system (7.5.1) that

$$Y''(t) + Q(t, Y(t) + V(t)) = 0. \tag{7.5.3}$$

We shall show that the integral equation

$$Y(t) = K + \int_t^\infty (t - s)Q(s, Y(s) + V(s))ds \tag{7.5.4}$$

has a solution on $[t_1, \infty)$ for some $t_1 > t_0$. For this, it suffices show that the operator

$$(TY)(t) = K + \int_t^\infty (t - s)Q(s, Y(s) + V(s))ds, \quad t \geq t_1 \tag{7.5.5}$$

has a fixed point in a suitable Banach space of matrix–valued functions. Let $\bar{t} \in [t_0, \infty)$ and consider

$$B_{\bar{t}} = \left\{ U : [\bar{t}, \infty) \to \mathcal{M}_n; \ U \text{ is continuous and} \right.$$
$$\left. \lim_{t \to \infty} U(t) \text{ exists as a finite matrix} \right\}.$$

Clearly, $B_{\bar{t}}$ is a Banach space under the sup norm. From (7.5.2), there exists a $t_1 > t_0$ such that for every $t \geq t_1$,

$$\|K\| + \int_t^\infty (s - t)q(s, \lambda + \|V(s)\|)ds \leq \lambda. \tag{7.5.6}$$

Denoting $S = \{U \in B_{t_1} : \|U\| \leq \lambda\}$, we get from (7.5.6) that $T(S) \subset S$. Now, it is easy to see that the set $T(S)$ is equicontinuous. In fact, this is a consequence of the fact that every function in $T(S)$ has derivative bounded above, for all $t \geq t_1$, by the integral

$$\int_t^\infty q(s, \lambda + \|V(s)\|)ds < \infty. \tag{7.5.7}$$

It is also easy to see that T is continuous and if $Y \in S$, then

$$\|TY - K\| \leq \int_t^\infty (s - t)q(s, \lambda + \|V(s)\|)ds. \tag{7.5.8}$$

Since the integral on the right–hand side of (7.5.8) tends to zero independently of Y, it follows that the set $T(S)$ is equiconvergent. Consequently, $T(S)$ is relatively compact in B_{t_1}. Thus, by Tychonov's theorem, the operator T has a fixed point in the ball S. This fixed point $Y(t)$ satisfies equation (7.5.1). The conclusion of the theorem now follows by letting $X(t) = Y(t) + V(t)$. ∎

Next, we employ Theorem 7.5.1 to obtain an infinity of oscillatory solutions of the system (7.5.1).

Theorem 7.5.2. Let conditions (i) and (ii) hold, and assume that for some $\lambda > 0$ condition (7.5.2) is satisfied, and

(iii) $\|V(t)\|$ is bounded on $[t_1, \infty)$, $t_1 > t_0 \geq 0$,

(iv) there exist $L \in \mathcal{M}_n$ and $\delta > 0$ such that $\|L\| < \lambda$ and

$$\limsup_{t \to \infty} \det[L + V(t)] > \delta, \tag{7.5.9}$$

and

$$\liminf_{t \to \infty} \det[L + V(t)] < -\delta. \tag{7.5.10}$$

Then there exists an oscillatory solution of the system (7.5.1).

Proof. Theorem 7.5.1 implies the existence of a solution $X(t)$ of (7.5.1) such that $X(t) - V(t) \to L$ as $t \to \infty$. We can assume that this solution $X(t)$ is defined for all $t > t_1$. Let $\{t_j\}$, $j = 1, 2, \cdots$ be such that $\lim_{j \to \infty} t_j = \infty$, and $\det[L + V(t_j)] = \delta$, $j = 1, 2, \cdots$. Since $V(t)$ is bounded, $X(t)$ is also bounded on $[t_1, \infty)$ and this implies the existence of a compact set Z in $\mathcal{M}_n$ such that $V(t)$, $X(t) \in Z$, $t \in [t_1, \infty)$. By the uniform continuity of the determinant function on Z, there exist $\rho(\delta) > 0$ and a positive integer j_0 such that

$$\|X(t_j) - [L + V(t_j)]\| < \rho(\delta) \tag{7.5.11}$$

and

$$|\det X(t_j) - \det[L + V(t_j)]| < \frac{\delta}{2} \tag{7.5.12}$$

for every $j \geq j_0$. Consequently, from (7.5.12) we deduce that $\det(X(t_j)) > \delta/2$ for all $j \geq j_0$. Similarly, we can show that $\det(X(s_i)) < -\delta/2$ for all large i, where $\{s_i\}$, $i = 1, 2, \cdots$ is a sequence in $[t_1, \infty)$ with $\lim_{i \to \infty} s_i = \infty$. Obviously, $\det(X(t))$ has an unbounded set of zeros on $[t_1, \infty)$ and this completes the proof. ∎

Example 7.5.1. In the system (7.5.1) let

$$Q(t, X) = \frac{1}{(t+1)^3} X^2, \quad q(t, \|X\|) = \frac{1}{(t+1)^3} \|X\|^2$$

and

$$F(t) = \begin{bmatrix} \sin t & 1/(t+1) \\ 1/(t+1) & \cos t \end{bmatrix}''.$$

It is easy to see that the hypotheses of Theorem 7.5.2 are satisfied, and hence (7.5.1) has at least one oscillatory solution.

Remark 7.5.1. It is worth noting that if the hypotheses of Theorem 7.5.2 are satisfied, there is an infinity of oscillatory solutions of (7.5.1). This follows easily from the fact that the integral equation (7.5.4) has solutions for all 'small' matrices K.

7.6. Notes and General Discussions

1. Theorems 7.1.1, 7.1.3 and 7.1.4 are taken from Kwong and Wong [33]. These theorems extend the sufficiency part of the results due to Atkinson [5], Belohorec [7] and Waltman [49]. Lemma 7.1.1 and Theorem 7.1.2 are due to Kordonis and Philos [27]. For some other related work we refer to Mirzov [39–41].

2. The results presented in Section 7.1 can be extended to systems of type (7.1.1) (and its prototype the system (7.1.2)) with deviating arguments, namely,

$$x'(t) = q_1(t)f_1(y[g_1(t)])$$
$$y'(t) = -q_2(t)f_2(x[g_2(t)]),$$

where q_i and f_i, $i = 1, 2$ are as in (7.1.1) with $q_i(t) \geq 0$ for $t \geq t_0$, $i = 1, 2$ and $g_i(t) \in C([t_0, \infty), \mathbb{R})$, $\lim_{t \to \infty} g_i(t) = \infty$, $i = 1, 2$. The formulation of these results are left to the reader.

3. Another possible extension is the study of the system (7.1.2) when $\lambda_1 \lambda_2 = 1$. This is so called *half–linear* case. With the introduction of the variable $w(t) = -(y_2(t)/x^{\lambda_2}(t))\operatorname{sgn} x(t)$, we get the Riccati equation $w'(t) = q_2(t) + \lambda_2 q_1(t)w^\lambda(t)$, where $\lambda = (\lambda_2 + 1)/\lambda_2$. All Sturmian comparison–type theorems continue to hold and oscillation criteria can easily be obtained via the traditional methods (see Chapter 3).

4. Theorem 7.2.1 is due to Mingarelli [37,38], whereas Theorem 7.2.2 and Corollaries 7.2.1 – 7.2.4 are from Kwong et. al. [29]. Criteria (i_1) – (i_4) are extracted from the work of Kwong and Zettl [34,35]. Lemmas 7.2.1, 7.2.2 and Theorems 7.2.3 and 7.2.4 are taken from Butler and Erbe [10]. Lemmas 7.2.3, 7.2.4 and Theorems 7.2.5 – 7.2.10 are borrowed from Butler et. al. [11].

5. The scalar Riccati inequality (7.2.19) remains valid if instead of the trace, we use any positive continuous linear functional on the space of real symmetric $n \times n$ matrices to define the function $w(t)$. Applying such a functional ϕ, say, to both sides of the equation (7.2.10), we obtain, instead of (7.2.11),

$$\phi(W(t)) = \phi(W(t_0)) + \phi(Q_1(t)) + \int_{t_0}^t \phi(W^2(s))ds. \qquad (7.6.1)$$

It has been shown by Akiyama [3], that any continuous positive linear functional on the cone of nonnegative real symmetric matrices is equivalent with the trace functional. Hence, there exists a positive constant c, which depends only on ϕ, such that $\phi(W^2(t)) \geq c\,\mathrm{tr}(W^2(t))$. Thus, instead of (7.2.13) we have the inequality

$$\phi(W^2(t)) \geq c(1-\epsilon)^2\lambda_1^2(Q_1(t)) \quad \text{for all} \quad t \geq T. \tag{7.6.2}$$

The analogue of (7.2.14) is

$$\int_{t_0}^t \phi(W^2(s))ds = \epsilon \int_{t_0}^t \phi(W^2(s))ds + (1-\epsilon)\int_{t_0}^t \phi(W^2(s))ds. \tag{7.6.3}$$

Here we use (7.6.2) to establish the last integral, whereas in the first integral we use the elementary inequality $\phi(W^2(t)) \geq (\phi(I))^{-1}[\phi(W(t))]^2$, where I is the $n \times n$ identity matrix. Thus, we obtain instead of (7.2.17),

$$\begin{aligned}\phi(W(t)) \geq{}& \phi(Q_1(t)) + \phi(W(t_0)) + c(1-\epsilon)^3\int_T^t \lambda_1^2(Q_1(s))ds \\ & + \frac{\epsilon}{\phi(I)}\int_{t_0}^t [\phi(W(s))]^2 ds.\end{aligned} \tag{7.6.4}$$

Now, it follows that

$$\phi(W(t_0)) + c(1-\epsilon)^3\int_T^t \lambda_1^2(Q_1(s))ds \geq c(1-\epsilon)^4\int_{t_0}^t \lambda_1^2(Q_1(s))ds \tag{7.6.5}$$

for $t \geq T_1 \geq T$. Thus if we let $\delta = c(1-\epsilon)^4$, replace ϵ by $\phi(I)\epsilon$, and define $w(t) = \phi(W(t))$, $t \geq t_0$ then

$$q(t) = \phi(Q_1(t)) + \delta\int_{t_0}^t \lambda_1^2(Q_1(s))ds.$$

Hence, inequality (7.2.19) holds again.

Therefore, at least in some results, we can replace the functional trace by any positive linear functional ϕ in the space of real symmetric $n \times n$ matrices. For a more general discussion of this we refer the reader to Akiyama [3] and Etgen and Pawlowski [18,19].

6. Theorem 7.2.3 can be generalized by considering the principal submatrices of $\int_{t_0}^t Q(s)ds$. We recall (cf. 6], p. 113): For any $n \times n$ symmetric matrix A, the sequence of symmetric matrices $A_k = (a_{ij})$, $i,j = 1,2,\cdots,k$ for $k = 1,2,\cdots,n$ satisfies $\lambda_{j+1}(A_{k+1}) \leq \lambda_j(A_k) \leq \lambda_j(A_{k+1})$, where $\lambda_j(A_k)$ denotes the j th characteristic roots of A_k.

Theorem 7.2.3'. Let $g = g(t)$ be positive, continuous and nondecreasing in $[t_0,\infty)$ and assume there exists k, $1 \leq k \leq n$ such that

$$\lim_{t \to \infty} \operatorname{approx\,inf} \frac{1}{g(t)}\,\mathrm{tr}\left(\int_{t_0}^t Q_k(s)ds\right) > -\infty, \tag{7.6.6; k}$$

$$\lim_{\substack{t \to \infty}} \operatorname{approx\,inf} \frac{1}{g(t)} \int_{t_0}^{t} \left(\lambda_1 \left(\int_{t_0}^{s} Q_k(\tau) d\tau \right) \right)^2 ds \; = \; \infty, \qquad (7.6.7;k)$$

and

$$\lim_{t \to \infty} \lambda_1 \left(\int_{t_0}^{t} Q_k(s) ds \right) \; = \; \infty. \qquad (7.6.8;k)$$

Then equation (7.2.1) is oscillatory.

Proof. The proof proceeds as in Theorem 7.2.3 upto (7.2.33). Then, we have

$$-V_k(t) + V_k(t_1) \; = \; \int_{t_1}^{t} Q_k(s) ds + \int_{t_1}^{t} [V^2(s)]_k ds \; \geq \; \int_{t_1}^{t} Q_k(s) ds$$

and therefore $\lambda_1(-V_k(t)) \to \infty$ as $t \to \infty$ by (7.6.8;k). A straight–forward modification of the proof of Theorem 7.2.3 now yields a contradiction to condition (7.6.7;k). ∎

7. Lemma 7.3.1 is the Theorem 8 in [20], whereas Lemmas 7.3.2 and 7.3.3 can be found in [16]. The rest of the results in Section 7.3 are taken from Butler and Erbe [9], except Corollary 7.3.2 and the two criteria (I_1) and (I_2) which are due to Kura [28], Macki and Wong [36], and Belohorec [7], respectively. We also remark that Theorems 7.3.6, 7.3.7 and 7.3.8 are the extensions of the results of Macki and Wong [36], Belohorec [7], and Atkinson [5], respectively to second order matrix equations.

8. The results of Section 7.4 are taken from Butler and Erbe [8]. We note that Theorem 7.4.1 compares two operator–valued equations, whereas the results of Etgen and Lewis [17] compare an operator–valued equation with a scalar equation. We also refer the reader to the papers [1,2] for some comparison results of nonself–adjoint equations.

9. The results of Section 7.5 are taken from Kartsatos and Walters [26].

7.7. References

1. **S. Ahmad and A.C. Lazer**, A n–dimensional extension of the Sturm separation and comparison theory to a class of nonself–adjoint systems, *SIAM J. Math. Anal.* 9(1978), 1137–1150.

2. **S. Ahmad and A.C. Lazer**, On an extension of Sturm's comparison theorem to a class of nonself–adjoint second order systems, *Nonlinear Analysis* 4(1980), 497–501.

3. **K. Akiyama**, On the maximum eigenvalue conjecture for the oscillation of second order differential systems, *M.Sc. Thesis, University of Ottawa,* 1983.

4. **W. Allegretto and L. Erbe,** Oscillation criteria for matrix differential inequalities, *Canad. Math. Bull.* **16**(1973), 5–10.

5. **F.V. Atkinson,** On second order nonlinear oscillation, *Pacific J. Math.* **5**(1955), 643–647.

6. **R. Bellman,** *Introduction to Matrix Analysis,* 2nd ed. *McGraw Hill,* New York, 1970.

7. **Š. Belohorec,** Oscillatory solutions of certain nonlinear differential equations of second order, *Mat. Fyz. Casopis Solven. Akad. Vied.* **11**(1961), 250–255.

8. **G.J. Butler and L.H. Erbe,** Comparison theorems for second order operator–valued linear differential equations, *Pacific J. Math.* **112**(1984), 21–34.

9. **G.J. Butler and L.H. Erbe,** Oscillation theory for second order differential systems with functionally commutative matrix coefficients, *Funkcial. Ekvac.* **28**(1985), 47–55.

10. **G.J. Butler and L.H. Erbe,** Oscillation results for second order differential systems, *SIAM J. Math. Anal.* **17**(1986), 19-29.

11. **G.J. Butler, L.H. Erbe and A.B. Mingarelli,** Riccati techniques and variational principles in oscillation theory for linear systems, *Trans. Amer. Math. Soc.* **303**(1987), 263–282.

12. **R. Byers, J. Harris and M.K. Kwong,** Weighted means and oscillation conditions for second order matrix differential equations, *J. Differential Equations* **61**(1986), 164–177.

13. **W.J. Coles,** Oscillation criteria for nonlinear second order equations, *Ann. Mat. Pura. Appl.* **82**(1969), 123–134.

14. **W.J. Coles and D. Willet,** Summability criteria for oscillation of second order linear differential equations, *Ann. Mat. Pura. Appl.* **79**(1968), 391–398.

15. **W.A. Coppel,** *Disconjugacy, Lecture Notes in Math.* **220**, *Springer–Verlag,* New York, 1971.

16. **M.P. Drazin, J.W. Dungey and K.W. Greuenberg,** Some theorems on commutative matrices, *J. London Math. Soc.* **25**(1950), 221–228.

17. **G.J. Etgen and R.T. Lewis,** A Hille–Wintner comparison theorem for second order differential systems, *Czech. Math. J.* **30**(1980), 98–107.

18. **G.J. Etgen and J.F. Pawlowski,** Oscillation criteria for second order selfadjoint differential systems, *Pacific J. Math.* **66**(1976), 99–110.

19. **G.J. Etgen and J.F. Pawlowski,** A comparison theorem and oscillation criteria for second order differential systems, *Pacific J. Math.* **72**(1977), 59–69.

20. **H.I. Freedman,** Functionally commutative matrices and matrices with constant eigenvectors, *Linear and Multilinear Algebra* **4**(1976), 197–213.

21. **P. Hartman,** *Ordinary Differential Equations,* John Wiley, New York, 1964.

22. **T.L. Hayden and H.C. Howard,** Oscillation of differential equations in Banach spaces, *Ann. Mat. Pura Appl.* **85**(1970), 383–394.

23. **E. Hille,** Nonoscillation theorems, *Trans. Amer. Math. Soc.* **64**(1948), 234–252.

24. **E. Hille,** *Lectures on Ordinary Differential Equations,* Addison–Wesley, Reading, Mass. 1969.

25. **D. Hinton and R.T. Lewis,** Oscillation theory for generalized second order differential equations, *Rocky Mountain J. Math.* **10**(1980), 751–766.

26. **A.G. Kartsatos and T. Walters,** Some oscillation results for matrix and vector differential equations with forcing term, *J. Math. Anal. Appl.* **73**(1980), 506–513.

27. **I.-G.E. Kordonis and Ch.G. Philos,** On the oscillation of nonlinear two–dimensional differential systems, *Proc. Amer. Math. Soc.* **126**(1998), 1661–1667.

28. **T. Kura,** A matrix analog of Atkinson's oscillation theorem, *Funkcial. Ekvac.* **28**(1985), 47–55.

29. **M.K. Kwong, H. Kaper, K. Akiyama and A. Mingarelli,** Oscillation of linear second order differential systems, *Proc. Amer. Math. Soc.* **91**(1984), 85–91.

30. **M.K. Kwong and J.S.W. Wong,** An application of integral inequality to second order nonlinear oscillation, *J. Differential Equations* **46**(1982), 63–77.

31. **M.K. Kwong and J.S.W. Wong,** Linearization of second order nonlinear oscillation theorems, *Trans. Amer. Math. Soc.* **279**(1983), 705–722.

32. **M.K. Kwong and J.S.W. Wong,** On the oscillation theorem of Belohorec, *SIAM J. Math. Anal.* **14**(1983), 474–476.

33. **M.K. Kwong and J.S.W.Wong,** oscillation of Emden–Fowler systems, *Differential and Integral Equations* **1**(1988), 133–141.

34. **M.K. Kwong and A. Zettl,** Integral inequalities and second order linear oscillation, *J. Differential Equations* **45**(1982), 16–33.

35. **M.K. Kwong and A. Zettl,** Asymptotically constant functions and second order linear oscillation, *J. Math. Anal. Appl.* **93**(1983), 475–494.

36. **J.W. Macki and J.S.W. Wong,** Oscillation of solutions to second order nonlinear differential equations, *Pacific J. Math.* **24**(1968), 111–117.

37. **A.B. Mingarelli,** An oscillation criterion for second order self adjoint differential systems, *C.R. Math. Rep. Acad. Sci. Canada* **2**(1980), 287–290.

38. **A.B. Mingarelli,** On a conjecture for oscillation of second order ordinary systems, *Proc. Amer. Math. Soc.* **82**(1981), 593–598.

39. **D.D. Mirzov,** On oscillatoriness of the solutions of a system of nonlinear differential equations, *Differencial'nye Uravnenija* **9**(1973), 581–583.

40. **D.D. Mirzov,** The oscillation of solutions of a system of nonlinear differential equations, *Math. Zametki* **16**(1974), 571–576.

41. **D.D. Mirzov,** Oscillation properties of solutions of a nonlinear Emden–Fowler differential system, *Differencial'nye Uravnenija* **16**(1980), 1980–1984.

42. **M. Morse,** *The Calculus of Variations in the Large,* Amer. Math. Soc. Colloq. Publ. **18**, New York, 1934.

43. **C. Olech, Z. Opial and T. Wazewski,** Sur le probléme d'oscillation des intégrales de l'equation $y'' + q(t)y = 0$, *Bull. Akad. Polon. Sci.* **5**(1957), 621–626.

44. **B.N. Parlett,** *The Symmetric Eigenvalue Problem*, Prentice–Hall, Englewood Cliffs, NJ, 1980.

45. **A.P. Robertson and W. Robertson,** *Topological Vector Spaces*, 2nd ed. *Cambridge University Press*, Cambridge, 1973.

46. **H.H. Schaefer,** *Banach Lattices and Positive Operators*, Springer–Verlag, New York, 1974.

47. **E.C. Tomastik,** Oscillation of systems of second order differential equations, *J. Differential Equations* **9**(1971), 436–442.

48. **T. Walters,** A characterization of positive linear functionals and oscillation criteria for matrix differential equations, *Proc. Amer. Math. Soc.* **78**(1980), 198–202.

49. **P. Waltman,** An oscillation criterion for a nonlinear second order equation, *J. Math. Anal. Appl.* **10**(1965), 439–441.

50. **D. Willet,** Classification of second order linear differential equations with respect to oscillation, *Advances Math.* **3**(1969), 594–693.

51. **D. Willet,** A necessary and sufficient condition for the oscillation of some linear second order differential equations, *Rocky Mountain J. Math.* **1** (1971), 357–365.

52. **A. Wintner,** On the comparison theorem of Kneser–Hille, *Math. Scand.* **5**(1957), 255–260.

53. **J.S.W. Wong,** On two theorems of Waltman, *SIAM J. Appl. Math.* **14**(1966), 724–728.

54. **J.S.W. Wong,** Oscillation theorems for second order nonlinear differential equations, *Proc. Amer. Math. Soc.* **106**(1989), 1069–1077.

55. **J.S.W. Wong,** Oscillation criteria for second order nonlinear differential equations with integrable coefficients, *Proc. Amer. Math. Soc.* **115**(1992), 389–395.

Chapter 8

Asymptotic Behavior of Solutions of Certain Differential Equations

8.0. Introduction

The study of behavioral properties of solutions of differential equations near infinity is of immense importance and hence it continues to attract many researchers. Therefore, in this chapter we shall present some recent contributions on the asymptotic behavior of solutions of second order differential equations as well as the behavioral properties of positive solutions of singular Emden–Fowler–type equations. In Section 8.1 it is shown that for a large class of differential equations, not only can the existence of nonoscillatory solutions be proved, but also an explicit asymptotic form of the nonoscillatory solutions may be provided. Then, we shall impose more restrictions on the sign of the integrable coefficient of the equation, and get necessary and sufficient conditions so that the solutions have the specified asymptotic behavior as $t \to \infty$, i.e., solutions which behave asymptotically like a nonzero constant and also those which behave asymptotically like ct, $c \neq 0$. For this, various averaging techniques of the types employed in the previous chapters to study the oscillatory behavior of such equations have been used. Section 8.2 is devoted to the study of existence, uniqueness and asymptotic behavior of positive solutions of singular Emden–Fowler–type equations. The cases when the coefficient of the equation under consideration is of constant sign, or of an alternating sign are systematically discussed. Then the existence as well as nonexistence results for the positive solutions of Emden–Fowler–type systems are proved.

8.1. Asymptotic Behavior of Solutions of Nonlinear Differential Equations

Here we shall study the asymptotic behavior of solutions of the second order nonlinear differential equation

$$x''(t) + q(t)f(x(t)) = 0, \tag{8.1.1}$$

where $q(t) \in C([t_0, \infty), \mathbb{R})$, $f(x) \in C^1(\mathbb{R}, \mathbb{R})$, $t_0 > 0$ and f satisfies $xf(x) > 0$ and $f'(x) \geq 0$ for all $x \neq 0$. When $f(x) = |x|^\gamma \mathrm{sgn}\, x$, $x \in \mathbb{R}$ $(\gamma > 0)$, equation (8.1.1) reduces to

$$x''(t) + q(t)|x(t)|^\gamma \mathrm{sgn}\, x(t) = 0. \tag{8.1.2}$$

The differential equation (8.1.2) is the prototype of (8.1.1) and is known as the generalized Emden–Fowler equation.

8.1.1. Asymptotic Behavior of Nonoscillatory Solutions

In the study of the asymptotic behavior of solutions of differential equations of the form (8.1.1) an interesting problem is to establish necessary and/or sufficient conditions for the existence of solutions which behave like nontrivial linear functions $c_1 + c_2 t$ as $t \to \infty$. Such a solution satisfies the asymptotic condition

$$x(t) = c + o(1) \quad \text{as} \quad t \to \infty, \tag{8.1.3}$$

$$x(t) = o(t) \quad \text{and} \quad \lim x(t) = \pm\infty \quad \text{as} \quad t \to \infty, \tag{8.1.4}$$

or, the condition

$$x(t) = ct + o(t) \quad \text{as} \quad t \to \infty, \tag{8.1.5}$$

where c is a nonzero constant.

In this subsection, we shall discuss the asymptotic properties of nonoscillatory solutions of (8.1.2) when

$$\lim_{t \to \infty} \int_{t_0}^t q(s)ds \quad \text{exists and is finite.} \tag{8.1.6}$$

If condition (8.1.6) holds, then we can define the function Q by

$$Q(t) = \int_t^\infty q(s)ds, \quad t \geq t_0. \tag{8.1.7}$$

If $q(t) \geq 0$ for $t \geq t_0$, then it is easy to show that a nonoscillatory solution $x(t)$ of (8.1.1) satisfies exactly one of the three asymptotic conditions (8.1.3) – (8.1.5). In the following result, we shall prove that this fact remains valid even when $Q(t) \geq 0$ for $t \geq t_0$.

Theorem 8.1.1. Let $\gamma > 0$. Suppose in addition to condition (8.1.6), $Q(t) \geq 0$ for $t \geq t_0$. Then for each nonoscillatory solution $x(t)$ of equation (8.1.2), exactly one of the three asymptotic conditions (8.1.3) – (8.1.5) is satisfied.

Proof. Let $x(t)$ be a nonoscillatory solution of equation (8.1.2), say, $x(t) > 0$ for $t \geq t_0 > 0$. It is known (cf. Lemma 4.1.3) that $x(t)$ satisfies the equality

$$x'(t) \; = \; Q(t)x^\gamma(t) + \gamma x^\gamma(t) \int_t^\infty x^{-\gamma-1}(s)(x'(s))^2 ds \quad \text{for} \quad t \geq t_0,$$

where $Q(t)$ is defined in (8.1.7). Therefore, we have

$$x'(t) \; \geq \; Q(t)x^\gamma(t) \quad \text{for} \quad t \geq t_0. \tag{8.1.8}$$

Since $Q(t) \geq 0$ for $t \geq t_0$, it follows that $x'(t) \geq 0$ for $t \geq t_0$. An integration by parts of equation (8.1.2) gives

$$x'(u) - Q(u)x^\gamma(u) + \gamma \int_t^u Q(s)x^{\gamma-1}(s)x'(s)ds \; = \; x'(t) - Q(t)x^\gamma(t), \tag{8.1.9}$$

where $u \geq t \geq t_0$. Let t be fixed. Since $Q(s)x^{\gamma-1}(s)x'(s)$ is nonnegative, the integral term in (8.1.9) has a finite limit, or diverges to ∞ as $t \to \infty$. If the latter case occurs, then $x'(u) - Q(u)x^\gamma(u) \to -\infty$ as $u \to \infty$, which is a contradiction to (8.1.8). Thus, the former case occurs, i.e.,

$$\int^\infty Q(s)x^{\gamma-1}(s)x'(s)ds \; < \; \infty. \tag{8.1.10}$$

Define the function $k_1(t)$ as

$$k_1(t) \; = \; \int_t^\infty Q(s)x^{\gamma-1}(s)x'(s)ds \tag{8.1.11}$$

and the finite constant $\alpha = \lim_{u \to \infty}[x'(u) - Q(u)x^\gamma(u)]$. Then equality (8.1.9) yields

$$x'(t) \; = \; \alpha + Q(t)x^\gamma(t) + \gamma k_1(t) \quad \text{for} \quad t \geq t_0. \tag{8.1.12}$$

Observe by (8.1.8) that $\alpha \geq 0$. From (8.1.8) and (8.1.10) it follows that

$$\int^\infty Q^2(s)x^{2\gamma-1}(s)ds \; < \; \infty. \tag{8.1.13}$$

Next, we define the function $k_2(t)$ by

$$k_2(t) \; = \; \int_t^\infty Q^2(s)x^{2\gamma-1}(s)ds \quad \text{for} \quad t \geq t_0. \tag{8.1.14}$$

Integrating the equation (8.1.12) from t_0 to t, we get

$$x(t) = x(t_0) + \alpha(t - t_0) + \int_{t_0}^{t} Q(s)x^{\gamma}(s)ds + \gamma \int_{t_0}^{t} k_1(s)ds. \qquad (8.1.15)$$

By Schwartz's inequality and the fact that $x'(t) \geq 0$ for $t \geq t_0$, the first integral term in (8.1.15) can be estimated as follows

$$\int_{t_0}^{t} Q(s)x^{\gamma}(s)ds \leq \left(\int_{t_0}^{t} Q^2(s)x^{2\gamma-1}(s)ds \right)^{1/2} \left(\int_{t_0}^{t} x(s)ds \right)^{1/2}$$
$$\leq [k_2(t_0)(t - t_0)x(t)]^{1/2} \quad \text{for} \quad t \geq t_0. \qquad (8.1.16)$$

Thus, for $t \geq t_0$, we find

$$x(t) \leq x(t_0) + \alpha(t - t_0) + [k_2(t_0)(t - t_0)x(t)]^{1/2} + \gamma k_1(t_0)(t - t_0).$$

The above inequality may be regarded as a quadratic inequality in $x^{1/2}(t)$. Then, we have

$$x^{1/2}(t) \leq \frac{1}{2} \left[(k_2(t)(t - t_0))^{1/2} + D^{1/2}(t) \right] \quad \text{for} \quad t \geq t_0,$$

where $D(t) = k_2(t_0)(t - t_0) + 4[x(t_0) + \alpha(t - t_0) + \gamma k(t_0)(t - t_0)]$. It is obvious that $D(t) = O(t)$ as $t \to \infty$, and consequently, there exists a positive constant m such that

$$x(t) \leq mt \quad \text{for} \quad t \geq t_0. \qquad (8.1.17)$$

Let $T \geq t_0$ be an arbitrary number. It is clear that

$$0 \leq \frac{1}{t} \int_{t_0}^{t} Q(s)x^{\gamma}(s)ds = \frac{1}{t} \int_{t_0}^{T} Q(s)x^{\gamma}(s)ds + \frac{1}{t} \int_{T}^{t} Q(s)x^{\gamma}(s)ds$$

for $t \geq T$. Arguing as in (8.1.16), we find $\int_{T}^{t} Q(s)x^{\gamma}(s)ds \leq [k_2(T)(t - T)x(t)]^{1/2}$ for $t \geq T$, which when combined with (8.1.17) yields

$$\int_{T}^{t} Q(s)x^{\gamma}(s)ds \leq [mk_2(T)t(t - T)]^{1/2} \quad \text{for} \quad t \geq T. \qquad (8.1.18)$$

Taking the upper limit as $t \to \infty$ in the above equality and using (8.1.18), we obtain

$$0 \leq \limsup_{t \to \infty} \frac{1}{t} \int_{t_0}^{t} Q(s)x^{\gamma}(s)ds \leq [mk_2(T)]^{1/2}. \qquad (8.1.19)$$

Since T is arbitrary and $k_2(T) \to 0$ as $T \to \infty$, letting $T \to \infty$ in (8.1.19), we get

$$\lim_{t \to \infty} \frac{1}{t} \int_{t_0}^{t} Q(s)x^{\gamma}(s)ds = 0. \qquad (8.1.20)$$

In view of (8.1.15), (8.1.20) and the fact that $k_1(t) \to 0$ as $t \to \infty$, we find $\lim_{t\to\infty} x(t)/t = \alpha$.

Recall that $x(t)$ is nondecreasing for $t \geq t_0$. Now, there are three cases to consider:

(i) $\alpha = 0$ and $x(t)$ is bounded above,

(ii) $\alpha = 0$ and $x(t)$ is unbounded,

(iii) $\alpha > 0$ (and hence $x(t)$ is unbounded).

Case (i) implies (8.1.3) with $c = \lim_{t\to\infty} x(t) > 0$, while case (iii) implies (8.1.5) with $c = \alpha > 0$. It is also clear that case (ii) implies (8.1.4). This completes the proof. ∎

8.1.2. Bounded Asymptotically Linear Solutions

In this subsection, we shall present necessary and/or sufficient conditions for (8.1.1) to have solutions which behave asymptotically like nonzero constants.

Theorem 8.1.2. Suppose

$$\int^{\infty} |Q(s)|\,ds \; < \; \infty \tag{8.1.21}$$

and

$$\int^{\infty} sQ^2(s)\,ds \; < \; \infty. \tag{8.1.22}$$

Then for any constant $c \neq 0$, equation (8.1.1) has a solution $x(t)$ such that

$$x(t) \; = \; c + O\left(\int_t^{\infty} [|Q(s)| + P(s)]\,ds \right) \tag{8.1.23}$$

and

$$x'(t) \; = \; O(|Q(t)| + P(t)), \tag{8.1.24}$$

as $t \to \infty$, where $P(t) = \int_t^{\infty} Q^2(s)\,ds$.

Proof. Condition (8.1.22) implies that $P(t)$ is nonincreasing and integrable on $[t_0, \infty)$. We may assume that $c > 0$. Let

$$m \; = \; \max\left\{ f(x) : x \in \left[\frac{c}{2}, \frac{3c}{2}\right] \right\}, \quad m' \; = \; \max\left\{ |f'(x)| : x \in \left[\frac{c}{2}, \frac{3c}{2}\right] \right\}$$

and choose b, $T \geq t_0$ so that

$$mm' + bm' \int_T^{\infty} |Q(s)|\,ds \; \leq \; b \tag{8.1.25}$$

and

$$m \int_T^\infty |Q(s)|ds + b \int_T^\infty P(s)ds \le \frac{c}{2}. \qquad (8.1.26)$$

Let F be the Fréchet space of continuously differentiable functions on $[T,\infty)$ with the family of seminorms $\{\|\cdot\|_\ell,\ \ell = 1, 2, \cdots\}$ defined by $\|x\|_\ell = \sup\{|x(t)| + |x'(t)| : T \le t \le T + \ell\}$. We have the convergence $x_k \to x\ (k \to \infty)$ in the topology of F if and only if $x_k(t) \to x(t)\ (k \to \infty)$ and $x'_k(t) \to x'(t)\ (k \to \infty)$ uniformly on every compact subinterval of $[T,\infty)$. Let X denote the set

$$X = \left\{x \in F : |x(t) - c| \le \frac{c}{2},\ |x'(t)| \le mQ(t) + bP(t)\ \text{ for }\ t \ge T\right\}.$$

Note that X is a nonempty closed convex subset of F. For $t \ge T$, we define the operator Φ on X as follows

$$(\Phi x)(t) = c - \int_t^\infty Q(s)f(x(s))ds - \int_t^\infty \left(\int_s^\infty Q(u)f'(x(u))x'(u)du\right)ds. \qquad (8.1.27)$$

We shall find a fixed point of Φ in X by employing the Schauder–Tychonov theorem.

(I_1) Φ is well–defined on X and maps X into itself. For this, let $x \in X$. For $r \ge s \ge T$, we have

$$\left|\int_s^r Q(u)f'(x(u))x'(u)du\right| \le \int_s^r |Q(u)|m'\left[m|Q(u)| + bP(u)\right]du$$

$$\le mm' \int_s^r Q^2(u)du + m'bP(s) \int_s^r |Q(u)|du.$$

Therefore, letting $r \to \infty$ and using (8.1.25), we find that $\int_s^r Q(u)f'(x(u)) \times x'(u)du$ converges and satisfies $|\int_s^\infty Q(u)f'(x(u))x'(u)du| \le bP(t)$ for $s \ge T$. This implies

$$|(\Phi x)'(t)| \le |Q(t)f(x(t))| + \left|\int_t^\infty Q(u)f'(x(u))x'(u)du\right|$$

$$\le m|Q(t)| + bP(t)\ \text{ for }\ t \ge T,$$

and in view of (8.1.26), for $t \ge T$, we have

$$|(\Phi x)(t) - c| \le \int_T^\infty |Q(s)f(x(s))|ds + \int_T^\infty \left|\int_s^\infty Q(u)f'(x(u))x'(u)du\right|ds \le \frac{c}{2}.$$

Thus, Φ is well–defined on X and maps X into X.

(I_2) Φ is continuous on X. For this, let x, $x_k\ (k = 1, 2, \cdots)$ be functions in X such that $x_k(t) \to x(t)$, $x'_k(t) \to x'(t)$ as $k \to \infty$, uniformly on every compact subinterval of $[T, \infty)$. Then, we have

$$|(\Phi x_k)'(t) - (\Phi x)'(t)| \leq |Q(t)||f(x_k(t)) - f(x(t))|$$
$$+ \int_t^\infty |Q(u)||f'(x_k(u))x_k'(u) - f'(x(u))x'(u)|du$$

for $t \geq T$. Observe that $f(x_k(t)) \to f(x(k))$ as $t \to \infty$, uniformly on compact subintervals of $[T, \infty)$ and $|Q(u)||f'(x_k(u))x_k'(u) - f'(x(u))x'(u)|$ is bounded above by $\int_T^\infty 2m'|Q(u)|[m|Q(u)| + bP(u)]du$ and $|Q(u)| \times |f'(x_k(u))x_k'(u) - f'(x(u))x'(u)| \to 0$ as $k \to \infty$. Applying the Lebesgue dominated convergence theorem, we find that $(\Phi x_k)'(t) \to (\Phi x)'(t)$ as $k \to \infty$ uniformly on every compact subinterval of $[T, \infty)$. Moreover, since

$$|(\Phi x_k)(t) - (\Phi x)(t)| \leq \int_t^\infty |(\Phi x_k)'(s) - (\Phi x)'(s)|ds$$

and $|(\Phi x_k)'(s) - (\Phi x)'(s)|$ is bounded above by $\int_T^\infty 2[m|Q(s)| + bP(s)]ds$. Once again by applying the Lebesgue dominated convergence theorem, we conclude that $(\Phi x_k)(t) \to (\Phi x)(t)$ as $k \to \infty$ uniformly on every finite subinterval of $[T, \infty)$. Thus, Φ is continuous on X.

(I_3) $\overline{\Phi X}$ is compact. This can be proved by using Arzela–Ascoli's theorem. Since the proof is standard (see the proofs of Theorems 4.3.1 and 4.3.4). The details are left to the reader.

Now the Schauder–Tychonov fixed point theorem guarantees the existence of a fixed point $x \in X$ of Φ. It can be easily verified that $x = x(t)$ is a solution of equation (8.1.1) for $t \geq T$, and satisfies the properties (8.1.23) and (8.1.24). This completes the proof. ∎

Remark 8.1.1. In Theorem 8.1.2 the assumption that $f'(x) \geq 0$ for $x \neq 0$ is unnecessary. In fact, this sign condition of f' is not used in the proof.

It is known that if conditions (8.1.21) and (8.1.22) are satisfied, then (8.1.1) has a nonoscillatory solution (see Theorem 4.3.1). Thus, Theorem 8.1.2 asserts the existence of a nonoscillatory solution of (8.1.1) with the asymptotic behavior (8.1.23) and (8.1.24).

Example 8.1.1. Consider the differential equation

$$x''(t) + (kt^\lambda \sin t)|x(t)|^\gamma \operatorname{sgn} x(t) = 0, \tag{8.1.28}$$

where k, λ and $\gamma > 0$ are constants. Applying Theorem 8.1.2 to the case $f(x) = |x|^\gamma \operatorname{sgn} x$, $|Q(t)| = \left|\int_t^\infty ks^\lambda \sin s \, ds\right| \leq 2|k|t^\lambda$ ($\lambda < 0$), we find that if $\lambda < -1$, then for any nonzero constant c, equation (8.1.28) has a solution $x(t)$ such that $x(t) = c + O(t^{\lambda+1})$ and $x'(t) = O(t^\lambda)$ as $t \to \infty$. Note that (8.1.28) has a nonoscillatory solution if and only if

(i) $\lambda < -1$ for $\gamma > 1$ (see [1–3]),

(ii) $\begin{cases} \lambda < -1, \ k \ \text{arbitrary} \\ \lambda = -1, \ |k| \leq 1/\sqrt{2} \end{cases}$ for $\gamma = 1$ (see [4,15]),

(iii) $\lambda < -\gamma$ for $0 < \gamma < 1$ (see [4,5]).

The superlinear case $(\gamma > 1)$ shows that Theorem 8.1.2 is 'sharp' in the sense that (8.1.28) never has a nonoscillatory solution unless $\lambda < -1$.

The following corollaries of Theorem 8.1.2 are immediate.

Corollary 8.1.1. Suppose conditions (8.1.21) and (8.1.22) are satisfied. Then equation (8.1.1) has a nonoscillatory solution $x(t)$ satisfying (8.1.3).

Corollary 8.1.2. Suppose condition (8.1.21) and

$$\lim_{t\to\infty} tQ(t) = 0 \tag{8.1.29}$$

are satisfied. Then equation (8.1.1) has a nonoscillatory solution $x(t)$ such that for any constant $c \neq 0$,

$$x(t) = c + o(1) \quad \text{and} \quad x'(t) = o(t^{-1}) \quad \text{as} \quad t \to \infty. \tag{8.1.30}$$

Proof.   It suffices to note that conditions (8.1.21) and (8.1.29) imply (8.1.22) and $t \int_t^\infty Q^2(s)ds \leq \int_t^\infty sQ^2(s)ds \to 0$ as $t \to \infty$. ∎

In what follows it will be shown that the converse of Corollaries 8.1.1 and 8.1.2 in some sense is possible provided $Q(t)$ is of fixed sign.

Theorem 8.1.3. Suppose $Q(t) \geq 0$ for all large t and $f'(x) > 0$ for $x \neq 0$. Then the following statements are equivalent:

(i)   for any constant $c \neq 0$, there exists a solution $x(t)$ of equation (8.1.1) satisfying (8.1.3),

(ii)   for some constant $c \neq 0$, there exists a solution $x(t)$ of equation (8.1.1) satisfying (8.1.3),

(iii) the integral conditions (8.1.21) and (8.1.22) are satisfied.

Proof. (i) implies (ii) trivially, and (iii) implies (i) by Corollary 8.1.1. We claim that (ii) implies (iii). Let $x(t)$ be a solution of equation (8.1.1) for which (8.1.3) holds for some constant $c \neq 0$. We may assume that $c > 0$. There is a number T such that $c/2 \leq x(t) \leq 2c$ for $t \geq T$. Recall Lemma 4.1.3, if we remove condition (4.1.49) and let $a(t) = 1$, then the equality (4.1.52) takes the form

$$\frac{x'(t)}{f(x(t))} = \alpha + Q(t) + \int_t^\infty \left(\frac{x'(s)}{f(x(s))}\right)^2 f'(x(s))ds, \tag{8.1.31}$$

where α is a nonnegative constant. Since $Q(t) \geq 0$, (8.1.31) gives

$$\frac{x'(t)}{f(x(t))} \geq Q(t) + \int_t^\infty m_1 Q^2(s)ds \geq 0 \quad \text{for} \quad t \geq T,$$

where $m_1 = \min\{f'(x) : c/2 \leq x \leq 2c\} > 0$. An integration of the above inequality over $[T, t]$ gives

$$\int_{x(T)}^{x(t)} \frac{du}{f(u)} \geq \int_T^t \left[Q(s) + m_1 \int_s^\infty Q^2(u)du \right] ds \quad \text{for} \quad t \geq T.$$

Since the left–hand side of the above inequality remains bounded as $t \to \infty$, we conclude that (8.1.21) and (8.1.22) are satisfied. This completes the proof. ∎

For the superlinear case, i.e., equation (8.1.2) with $\gamma > 1$, the next result is now clear.

Theorem 8.1.4. Let $\gamma > 1$. Suppose condition (8.1.6) holds and that $Q(t) \geq 0$ for $t \geq t_0$. Then the following three statements are equivalent:

(i)   equation (8.1.2) has a nonoscillatory solution $x(t)$ satisfying (8.1.3),

(ii) equation (8.1.2) has a nonoscillatory solution,

(iii) the integral conditions (8.1.21) and (8.1.22) are satisfied.

In fact, the equivalence of (ii) and (iii) can be obtained from the contrapositive form of a special case of Corollary 4.3.1, and the equivalence of (i) and (iii) is given in Theorem 8.1.3.

Theorem 8.1.5. Suppose either $Q(t) \geq 0$ or $Q(t) \leq 0$ for all large t. Then the following three statements are equivalent:

(i)   for any constant $c \neq 0$, there exists a solution $x(t)$ of equation (8.1.1) satisfying (8.1.30),

(ii) for some constant $c \neq 0$, there exists a solution $x(t)$ of equation (8.1.1) satisfying (8.1.30),

(iii) the integral conditions (8.1.21) and (8.1.22) are satisfied.

Proof. As (i) implies (ii) trivially, and (iii) implies (i) by Corollary 8.1.2, it suffices to show that (ii) implies (iii). Suppose $x(t)$ is a solution of (8.1.1) such that (8.1.30) holds for some constant $c \neq 0$. As in the proof of Theorem 8.1.3, we employ Lemma 4.1.3 to obtain (8.1.30) which is satisfied for all large t. Since $\alpha = \lim_{t \to \infty} x'(t)/f(x(t)) = 0$, it follows that

$$\frac{x'(t)}{f(x(t))} = Q(t) + \int_t^\infty \left(\frac{x'(s)}{f(x(s))} \right)^2 f'(x(s))ds,$$

or equivalently,

$$tQ(t) = \frac{tx'(t)}{f(x(t))} - t \int_t^\infty \left(\frac{x'(s)}{f(x(s))} \right)^2 f'(x(s))ds \qquad (8.1.32)$$

for all large t. The term $tx'(t)/f(x(t)) \to 0$ as $t \to \infty$ since $x(t)$ satisfies (8.1.30). Using l'Hospital's rule and (8.1.30), we find

$$\lim_{t\to\infty} t \int_t^\infty \left(\frac{x'(s)}{f(x(s))} \right)^2 f'(x(s))ds = \lim_{t\to\infty} \frac{\frac{d}{dt} \int_t^\infty \left(\frac{x'(s)}{f(x(s))} \right)^2 f'(x(s))ds}{\frac{d}{dt}\left(\frac{1}{t}\right)}$$

$$= \lim_{t\to\infty} \frac{f'(x(t))}{f^2(x(t))}[tx'(t)]^2 = 0.$$

Therefore, we have (8.1.29).

Next, we shall show that (8.1.21) holds. In view of (8.1.1), we have

$$(x(t) - tx'(t))' = tq(t)f(x(t))$$
$$= -(tQ(t)f(x(t)))' + Q(t)f(x(t)) + tQ(t)f'(x(t))x'(t).$$

An integration of the above equality from T to t gives

$$x(t) - tx'(t) = \alpha_1 - tQ(t)f(x(t))$$
$$+ \int_T^t Q(s)[f(x(s)) + sf'(x(s))x'(s)]ds, \qquad (8.1.33)$$

where $\alpha_1 = x(T) - Tx'(T) + TQ(T)f(x(T))$. Since $tQ(t) \to 0$ as $t \to \infty$ and $x(t)$ satisfies (8.1.30), we have $x(t) - tx'(t) \to c$, $tQ(t)f(x(t)) \to 0$ and $f(x(t)) + tx'(t)f'(x(t)) \to f(c) \neq 0$ as $t \to \infty$. These facts together with (8.1.33) and the sign property of $Q(t)$ easily lead to condition (8.1.21). This completes the proof.   ∎

Remark 8.1.2. From Theorems 8.1.3 and 8.1.5, we note that even for solutions which have the same limit as $t \to \infty$, there is an essential difference between restricting and not restricting the asymptotic behavior of their derivatives.

8.1.3. Unbounded Asymptotically Linear Solutions

In this subsection we shall present necessary and/or sufficient conditions for equation (8.1.1) to have solutions which behave asymptotically like ct (c is a nonzero constant). In the above results, no growth condition on f was required in proving the existence of a solution asymptotic to a nonzero constant as $t \to \infty$. The situation now becomes different and we shall need to impose one of the following growth conditions on f:

$f'(x)$ is nondecreasing for $x > 0$ and nonincreasing for $x < 0$,   (8.1.34)

$f'(x)$ is nonincreasing for $x > 0$ and nondecreasing for $x < 0$.   (8.1.35)

Theorem 8.1.6. Suppose either condition (8.1.34) or (8.1.35) is satisfied. Also, suppose for all constants $k \neq 0$, $k_1 \neq 0$ and $k_2 \neq 0$

$$\frac{1}{t} \int^t |f(ks)||Q(s)|ds \to 0 \quad \text{as} \quad t \to \infty, \tag{8.1.36}$$

$$\int^t f'(ks)|Q(s)|ds < \infty, \tag{8.1.37}$$

and

$$\int^\infty |f(k_1 s)||f'(k_2 s)Q^2(s)ds < \infty. \tag{8.1.38}$$

Then for any nonzero constant c equation (8.1.1) has a solution $x(t)$ such that

$$x(t) = ct + O\left(\int^t [|f(c,s)||Q(s)| + P_c(s)]ds\right), \tag{8.1.39}$$

and

$$x'(t) = c + O\left(|f(c,t)||Q(t)| + P_c(t)\right), \tag{8.1.40}$$

as $t \to \infty$, where

$$P_c(t) = \sup_{r \geq t} \max \left\{ \int_r^\infty f'(c_2 s)|Q(s)|ds, \int_r^\infty |f(c_1 s)|f'(c_2 s)Q^2(s)ds \right\},$$

$$c_1 = 3c/2, \quad \text{and} \quad c_2 = \begin{cases} 3c/2 & \text{if condition (8.1.34) holds} \\ c/2 & \text{if condition (8.1.35) holds} \end{cases}$$

Proof. The proof can be modelled on that of Theorem 8.1.2. So, we shall only give a few details. Let c be a given nonzero number. Without loss of generality, we may assume that $c > 0$. By conditions (8.1.36) and (8.1.37) and the fact that $P_c(t) \to 0$ as $t \to \infty$, there is a sufficiently large T such that for $t \geq T$ the following three conditions are satisfied $\int_T^t f(c_1 s)|Q(s)|ds \leq ct/4$, $(c+2)\int_t^\infty f'(c_2 u)|Q(u)|du \leq 1$ and $(c+2)$ $\times \int_T^t P_c(s)ds \leq c/4$. Let F be the Fréchet space of all continuously differentiable functions on $[T, \infty)$ with the same topology as in the proof of Theorem 8.1.2, and let X be defined by

$$X = \left\{ x \in F : |x(t) - ct| \leq (c/2)t, \ |x'(t)| \leq c + f(c_1 t)Q(t) \right.$$
$$\left. + (c+2)P_c(t) \text{ for } t \geq T \right\}.$$

Clearly, X is a nonempty closed convex subset of F. Now, for $t \geq T$ consider a mapping $\Phi : X \to F$ defined by

$$(\Phi x)(t) = ct + \int_T^t Q(s)f(x(s))ds + \int_T^t \left(\int_s^\infty Q(u)f'(x(u))x'(u)du\right) ds.$$

As in the proof of Theorem 8.1.2 it can be shown that (I_1) Φ is well–defined on X and maps X into itself, (I_2) Φ is continuous on X, and (I_3) $\overline{\Phi X}$ is compact. By the Schauder–Tychonov fixed point theorem the operator Φ has a fixed point $x \in X$. This fixed point $x = x(t)$ is a solution of equation (8.1.1) satisfying (8.1.39) and (8.1.40). ∎

Example 8.1.2. Consider the equation (8.1.28). Applying Theorem 8.1.6 to the case $f(x) = |x|^\gamma \operatorname{sgn} x$, $|Q(t)| \leq 2|k|t^\lambda$, $\lambda < 0$, we see that if $\lambda < -\gamma$, then (8.1.28) has a solution $x(t)$ such that $x(t) = ct + O(t^\delta)$ as $t \to \infty$ if $\lambda + \gamma + 1 \neq 0$ where $\delta = \max\{\lambda + \gamma + 1, 0\}$ and $x(t) = ct + O(\ln t)$ as $t \to \infty$ if $\lambda + \gamma + 1 = 0$.

We note that Theorem 8.1.6 is also 'sharp' in the sense that equation (8.1.28) in the sublinear case $(0 < \gamma < 1)$ never has a nonoscillatory solution unless $\lambda < -\gamma$.

The following corollaries of Theorem 8.1.6 are immediate.

Corollary 8.1.3.   Suppose either condition (8.1.34) or (8.1.35) holds. Also, suppose conditions (8.1.36) – (8.1.38) hold. Then for any nonzero constant c equation (8.1.1) has a solution $x(t)$ such that the asymptotic condition (8.1.5) holds.

Corollary 8.1.4. Suppose either condition (8.1.34) or (8.1.35) holds. Also suppose for every $k \neq 0$,

$$f(kt)Q(t) \to 0 \quad \text{as} \quad t \to \infty \tag{8.1.41}$$

and

$$\int^\infty f'(ks)|Q(s)|ds < \infty. \tag{8.1.42}$$

Then for any nonzero constant c equation (8.1.1) has a solution $x(t)$ such that

$$x(t) = t[c + o(1)], \quad x'(t) = c + o(1) \quad \text{as} \quad t \to \infty. \tag{8.1.43}$$

Next, we shall prove the converse of Corollaries 8.1.3 and 8.1.4 when $Q(t)$ is of constant sign.

Theorem 8.1.7. Suppose $Q(t) \geq 0$ for all large t. Moreover, suppose either condition (8.1.34) or (8.1.35) holds, $f'(x) > 0$ for $x \neq 0$ and for all nonzero constants k_1, k_2,

$$\limsup_{t \to \infty} \frac{1}{t^2} \int^t |f(k_1 s)|[f'(k_2 s)]^{-1} ds < \infty. \tag{8.1.44}$$

Then equation (8.1.1) has a solution $x(t)$ satisfying (8.1.5) for some nonzero constant c provided

$$\lim_{t \to \infty} \frac{1}{t} \int^t |f(ks)|Q(s)ds = 0 \quad \text{for some nonzero constant } k, \quad (8.1.45)$$

$$\int^\infty f'(ks)Q(s)ds < \infty \quad \text{for some nonzero constant } k, \quad (8.1.46)$$

$$\int^\infty |f(k_1 s)||f'(k_2 s)|Q^2(s)ds < \infty \quad \text{for some nonzero constants } k_1 \text{ and } k_2. \quad (8.1.47)$$

Proof. Let $x(t)$ be a solution of equation (8.1.1) which satisfies (8.1.5). We may assume that $c > 0$. There is a number T such that $(c/2)t \leq x(t) \leq 2ct$ for $t \geq T$. As in the proof of Theorem 8.1.3, by employing Lemma 4.1.3, it follows for $t \geq T$ that

$$x'(t) = \alpha f(x(t)) + Q(t)f(x(t)) + f(x(t)) \int_t^\infty \left(\frac{x'(s)}{f(x(s))}\right)^2 f'(x(s))ds, \quad (8.1.48)$$

where α is a nonnegative constant. On the other hand, an integration by parts of (8.1.1) from T to t gives

$$x'(t) = \beta + Q(t)f(x(t)) - \int_T^t Q(s)f'(x(s))x'(s)ds, \quad (8.1.49)$$

where $\beta = x'(T) - Q(T)f(x(T))$. Combining (8.1.48) and (8.1.49), we get

$$\alpha f(x(t)) + f(x(t)) \int_t^\infty \left(\frac{x'(s)}{f(x(s))}\right)^2 f'(x(s))ds = \beta - \int_T^t Q(s)f'(x(s))x'(s)ds. \quad (8.1.50)$$

Since $x'(t) \geq 0$ by (8.1.48), (8.1.50) implies

$$\int_T^t Q(s)f'(x(s))x'(s)ds < \infty. \quad (8.1.51)$$

From (8.1.48), we get $x'(t) \geq Q(t)f(x(t))$ for $t \geq T$, and hence we conclude that $\int_T^t f(x(s))f'(x(s))Q(s)ds < \infty$, which in view of (8.1.34) or (8.1.35) implies (8.1.47). By (8.1.51) we find that the right–hand side of (8.1.50) has a finite limit as $t \to \infty$. Let θ denote that limit, i.e.,

$$\theta = \lim_{t \to \infty} \left[\alpha f(x(t)) + f(x(t)) \int_t^\infty \left(\frac{x'(s)}{f(x(s))}\right)^2 f'(x(s))ds\right]. \quad (8.1.52)$$

Integrating (8.1.48) from $\tau \geq T$ to t and dividing by t, we get

$$
\begin{aligned}
\frac{x(t) - x(\tau)}{t} &= \frac{1}{t} \int_\tau^t Q(s) f(x(s)) ds + \frac{\alpha}{t} \int_\tau^t f(x(s)) ds \\
&\quad + \frac{1}{t} \int_\tau^t f(x(s)) \left(\int_s^\infty \left(\frac{x'(u)}{f(x(u))} \right)^2 f'(x(u)) du \right) ds.
\end{aligned}
$$
(8.1.53)

Since $x(t)$ satisfies (8.1.5), from (8.1.52) and (8.1.53) it follows that

$$
\lim_{t \to \infty} \frac{1}{t} \int_\tau^t Q(s) f(x(s)) ds = c - \theta.
$$
(8.1.54)

Now, it is easy to see that

$$
\begin{aligned}
0 &\leq \frac{1}{t} \int_\tau^t Q(s) f(x(s)) ds \\
&\leq \frac{1}{t} \left(\int_\tau^t \frac{f(cs)}{f'(\bar{c}s)} ds \right)^{1/2} \left(\int_\tau^t f(x(s)) f'(x(s)) Q^2(s) ds \right)^{1/2}
\end{aligned}
$$
(8.1.55)

for $t \geq \tau$, where $\bar{c} = \begin{cases} c/2 & \text{if condition (8.1.34) holds} \\ 2c & \text{if condition (8.1.35) holds.} \end{cases}$ Note that

(8.1.44) implies that there exists a positive constant m independent of τ such that

$$
\frac{1}{t^2} \int_\tau^t \frac{f(2cs)}{f'(\bar{c}s)} ds \leq m^2 \quad \text{for} \quad t \geq \tau.
$$

Taking the limit as $t \to \infty$ in (8.1.55), we have from (8.1.54),

$$
0 \leq c - \theta \leq m \left(\int_\tau^\infty f(x(s)) f'(x(s)) Q^2(s) ds \right)^{1/2}.
$$
(8.1.56)

Since τ is arbitrary, letting $\tau \to \infty$ in (8.1.56), we find that $c = \theta$. Therefore, we have from (8.1.52) and (8.1.54) that

$$
\lim_{t \to \infty} \left[\alpha f(x(t)) + f(x(t)) \int_t^\infty \left(\frac{x'(s)}{f(x(s))} \right)^2 f'(x(s)) ds \right] = c
$$
(8.1.57)

and

$$
\lim_{t \to \infty} \frac{1}{t} \int^t Q(s) f(x(s)) ds = 0.
$$
(8.1.58)

From (8.1.58), (8.1.45) is immediate. By (8.1.48) and (8.1.57), we obtain $x'(t) \geq c/2$ for all large t. Combining this with (8.1.51), we see that (8.1.46) is satisfied. This completes the proof. ∎

For equation (8.1.2), Corollary 8.1.3 and Theorem 8.1.7 yield the following result.

Theorem 8.1.8. Suppose $Q(t) \geq 0$ for all large t. Then the following statements are equivalent:

(i)   for any $c \neq 0$ there exists a solution $x(t)$ of equation (8.1.2) satisfying (8.1.5),

(ii)   for some $c \neq 0$ there exists a solution $x(t)$ of equation (8.1.2) satisfying (8.1.5),

(iii)   the following integral conditions are satisfied

$$\int^{\infty} s^{\gamma-1}Q(s)ds < \infty \qquad\qquad (8.1.59)$$

and

$$\int^{\infty} s^{2\gamma-1}Q^2(s)ds < \infty. \qquad\qquad (8.1.60)$$

Now we are ready to prove the following result.

Theorem 8.1.9. Let $0 < \gamma < 1$. Suppose condition (8.1.6) holds and that $Q(t) \geq 0$ for $t \geq t_0$. Then the following three statements are equivalent:

(i)   equation (8.1.2) has a nonoscillatory solution $x(t)$ satisfying (8.1.5),

(ii)   equation (8.1.2) has a nonoscillatory solution,

(iii)   the integral conditions (8.1.59) and (8.1.60) are satisfied.

Proof. It is trivial that (i) implies (ii). The equivalence of (i) and (iii) follows from Theorem 8.1.8. Here, we claim that (ii) implies (iii). Theorem 5.1.6 shows that if $\int^{\infty} s^{\gamma-1}Q(s)ds = \infty$, then equation (8.1.2) is oscillatory. This means that if equation (8.1.2) has a nonoscillatory solution, then the condition (8.1.59) must be satisfied. Therefore, it is enough to show that if equation (8.1.2) has a nonoscillatory solution, then the condition (8.1.60) is satisfied.

Suppose that equation (8.1.2) has a solution $x(t)$, say, $x(t) > 0$ for $t \geq t_0$. Then the equalities and inequalities in the proof of Theorem 8.1.1 remain valid. Now, we consider the following two possibilities:

(I_1) $2\gamma - 1 \leq 0$. Then the desired condition (8.1.60) follows from (8.1.13) and (8.1.17).

(I_2) $2\gamma - 1 > 0$. In (8.1.15), the first three terms of the right–hand side are nonnegative and $k_1(t)$ is nonincreasing on $[t_0, \infty)$. Thus, we have $x(t) \geq \gamma(t - t_0)k_1(t)$ for $t \geq t_0$. Hence, in view of (8.1.8), (8.1.11) and (8.1.14), we get $x(t) \geq \gamma(t-t_0)k_2(t)$ for $t \geq t_0$, and so by the assumption

$2\gamma - 1 > 0$,

$$Q^2(t)x^{2\gamma-1}(t)k_2^{1-2\gamma}(t) \geq \gamma^{2\gamma-1}(t-t_0)^{2\gamma-1}Q^2(t) \quad \text{for} \quad t \geq t_0. \quad (8.1.61)$$

Since $k_2'(t) = -Q^2(t)x^{2\gamma-1}(t)$ for $t \geq t_0$, an integration of (8.1.61) for $t \geq t_0$, gives

$$-\frac{1}{2-2\gamma}k_2^{2-2\gamma}(t) + \frac{1}{2-2\gamma}k_2^{2-2\gamma}(t_0) \geq \gamma^{2\gamma-1}\int_{t_0}^{t}(s-t_0)^{2\gamma-1}Q^2(s)ds.$$

Note that the left–hand side of the above inequality is bounded on $[t_0, \infty)$. This implies that the desired condition (8.1.60) is satisfied. This completes the proof.   ∎

The equivalence of (ii) and (iii) in Theorem 8.1.9 can be restated as follows.

Corollary 8.1.5. Let $0 < \gamma < 1$. Suppose condition (8.1.6) holds and that $Q(t) \geq 0$ for $t \geq t_0$. Then equation (8.1.2) is oscillatory if and only if

$$\int^{\infty}\left(s^{\gamma-1}Q(s) + \int_s^{\infty}\left[u^{\gamma-1}Q(u)\right]^2 du\right)ds = \infty.$$

Theorem 8.1.10. Suppose either $Q(t) \geq 0$ or $Q(t) \leq 0$ for all large t. Also suppose either condition (8.1.34) or (8.1.35) is satisfied. Then in order for equation (8.1.1) to have a solution $x(t)$ satisfying (8.1.43) for some nonzero constant c it is necessary that

$$\lim_{t\to\infty} f(kt)Q(t) = 0 \quad \text{for some nonzero constant } k \quad (8.1.62)$$

and

$$\int^{\infty} f'(ks)|Q(s)|ds < \infty \quad \text{for some nonzero constant } k. \quad (8.1.63)$$

Proof. First, we shall prove (8.1.62). If $f(x)$ is bounded as $x \to \infty$ or $-\infty$, then (8.1.62) is trivially satisfied since $\lim_{t\to\infty} Q(t) = 0$. Thus, we assume $\lim_{x\to\infty} f(x) = \pm\infty$. Let $x(t)$ be a solution of equation (8.1.1) satisfying (8.1.43). By Lemma 4.1.3, we have

$$Q(t)f(x(t)) = x'(t) - f(x(t))\int_t^{\infty}\left(\frac{x'(s)}{f(x(s))}\right)^2 f'(x(s))ds \quad \text{for} \quad t \geq T.$$
$$(8.1.64)$$

Using l'Hospital's rule we find that the second term of the right–hand side of (8.1.64) tends to c as $t \to \infty$, i.e.,

$$\left(\frac{d}{dt}\int_t^{\infty}\left(\frac{x'(s)}{f(x(s))}\right)^2 f'(x(s))ds\right)\left(\frac{d}{dt}\frac{1}{f(x(t))}\right)^{-1} = x'(t) \to c \quad \text{as} \quad t \to \infty.$$

It follows from (8.1.64) and (8.1.43) that $\lim_{t\to\infty} Q(t)f(x(t)) = 0$, which implies (8.1.62). Integrating (8.1.1) by parts, we have (8.1.49). Since $\lim_{t\to\infty} x'(t) = c$ and $\lim_{t\to\infty} Q(t)f(x(t)) = 0$, we find that $\int^\infty |Q(s)$ $\times |f'(x(s))||x'(s)|ds < \infty$ and from this (8.1.63) follows. This completes the proof. ■

Combining Corollary 8.1.3 with Theorem 8.1.10, we get the following result.

Theorem 8.1.11. Suppose either $Q(t) \geq 0$ or $Q(t) \leq 0$ for all large t. Then the following three statements are equivalent:

(i)   for any constant $c \neq 0$, there exists a solution $x(t)$ of equation (8.1.2) satisfying (8.1.43),

(ii)   for some constant $c \neq 0$, there exists a solution $x(t)$ of equation (8.1.2) satisfying (8.1.43),

(iii) the following integral conditions are satisfied

$$\lim_{t\to\infty} t^\gamma Q(t) = 0 \quad \text{and} \quad \int^\infty s^{\gamma-1}|Q(s)|ds < \infty.$$

It is to be noted that the condition (iii) of Theorem 8.1.11 can be replaced by $\lim_{t\to\infty} \int^t s^\gamma q(s)ds$ exists and is finite.

8.1.4. Further Extensions and Improvements

The purpose of this subsection is to show that if solutions of (8.1.1) satisfy (8.1.3) or (8.1.5), then (8.1.6) can be relaxed to the rather weak hypothesis

$$\limsup_{t\to\infty} \frac{1}{t} \int_{t_0}^t \int_{t_0}^s q(u)duds > -\infty. \tag{8.1.65}$$

Further, it will be shown that under the hypothesis (8.1.65) if equation (8.1.1) has a solution which satisfies either (8.1.3) or (8.1.5), then

$$\lim_{t\to\infty} \frac{1}{t} \int_{t_0}^t \int_{t_0}^s q(u)duds \quad \text{exists and is finite.} \tag{8.1.66}$$

Therefore, we may introduce the function $Q_2(t)$ by

$$Q_2(t) = \lim_{T\to\infty} \frac{1}{T} \int_t^T \int_t^s q(u)duds, \quad t \geq t_0. \tag{8.1.67}$$

The role of $Q(t)$ in the above results is thus taken by the function $Q_2(t)$. In fact, it can be shown that if $Q(t)$ is replaced by $Q_2(t)$ in the results of Subsections 8.1.1 – 8.1.3, then the corresponding results remain valid.

Clearly, condition (8.1.6) implies (8.1.65) and the function $Q_2(t)$ is equal to $Q(t)$. Thus, the results we shall present in this subsection generalize those given above.

We begin this subsection with a result on the existence of solutions x satisfying the asymptotic conditions (8.1.43) and

$$x(t) = c + o(1), \quad x'(t) = o(t^{-1}) \quad \text{as} \quad t \to \infty, \tag{8.1.68}$$

where c is a nonzero constant.

Theorem 8.1.12. If equation (8.1.1) has a solution $x(t)$ satisfying either (8.1.43) or (8.1.68) for some constant $c \neq 0$, then (8.1.6) holds.

Proof. Suppose equation (8.1.1) has a solution $x(t)$ satisfying either (8.1.43) or (8.1.68) for some constant $c \neq 0$. There is a number $T \geq t_0$ such that $x(t) \neq 0$ for $t \geq T$. Dividing both sides of equation (8.1.1) by $f(x(t))$ and integrating over $[T, t]$, we find

$$\frac{x'(t)}{f(x(t))} - \frac{x'(T)}{f(x(T))} + \int_T^t \left(\frac{x'(s)}{f(x(s))} \right)^2 f'(x(s)) ds + \int_T^t q(s) ds = 0 \tag{8.1.69}$$

for $t \geq T$. To complete the proof, it is enough to show that

$$\lim_{t \to \infty} \frac{x'(t)}{f(x(t))} \quad \text{exists and is finite} \tag{8.1.70}$$

and

$$\int_T^\infty \left(\frac{x'(s)}{f(x(s))} \right)^2 f'(x(s)) ds < \infty. \tag{8.1.71}$$

When $x(t)$ satisfies (8.1.68) with $c \neq 0$, we have

$$\frac{x'(t)}{f(x(t))} = o(t^{-1}) \quad \text{and} \quad \left(\frac{x'(t)}{f(x(t))} \right)^2 f'(x(t)) = o(t^{-2}) \quad \text{as} \quad t \to \infty$$

and hence (8.1.70) and (8.1.71) are satisfied.

When $x(t)$ satisfies (8.1.43) with $c \neq 0$, we have $x(t) \to \infty$ or $-\infty$ as $t \to \infty$, so by the hypothesis on the function f, $\lim_{t \to \infty} f(x(t))$ exists in $[-\infty, \infty]$, (i.e., the extended real line). Then, (8.1.70) is trivially satisfied. Next, note the equality

$$\int_T^t \frac{f'(x(s))x'(s)}{f^2(x(s))} ds = \frac{1}{f(x(T))} - \frac{1}{f(x(t))}. \tag{8.1.72}$$

Since the right–hand side of (8.1.72) has a finite limit as $t \to \infty$, so does the left–hand side. Therefore, in view of the fact that $x'(t) \to c \neq 0$

as $t \to \infty$, we conclude that (8.1.71) is satisfied. This completes the proof. ∎

Remark 8.1.3. Theorem 8.1.12 implies that when condition (8.1.6) does not hold, equation (8.1.1) does not have any asymptotically linear solution x of the form (8.1.43) or (8.1.68). Therefore, the assumption of the condition (8.1.6) in Subsections 8.1.1 – 8.1.3 is justified.

We now return to the problem of existence of solutions x satisfying the asymptotic conditions (8.1.3) and (8.1.5). Note that the asymptotic behavior of the derivatives of solutions x is not restricted in (8.1.3) and (8.1.5).

Theorem 8.1.13. Assume that the condition (8.1.65) is satisfied. If equation (8.1.1) has a solution $x(t)$ which satisfies either (8.1.3) or (8.1.5) for some constant $c \neq 0$, then in addition to the condition (8.1.66) the equality

$$\frac{x'(t)}{f(x(t))} = \alpha + Q_2(t) + \int_t^\infty \left(\frac{x'(s)}{f(x(s))} \right)^2 f'(x(s)) ds \qquad (8.1.73)$$

holds for all large t, where $\alpha \geq 0$ is a constant.

Proof. Let $x(t)$ be a solution of equation (8.1.1) which satisfies either (8.1.3) or (8.1.5) for some constant $c \neq 0$, and let $T \geq t_0$ be a number such that $x(t) \neq 0$ for $t \geq T$. Then as in the proof of Theorem 8.1.12 we find that (8.1.69) is satisfied for $t \geq T$. Integrating (8.1.69) over $[T, t]$ and dividing by t, we obtain

$$\frac{1}{t} \int_{x(T)}^{x(t)} \frac{du}{f(u)} - \frac{x'(T)}{f(x(T))} \left(\frac{t-T}{t} \right) + \frac{1}{t} \int_T^t \int_T^s \left(\frac{x'(\tau)}{f(x(\tau))} \right)^2 f'(x(\tau)) d\tau ds$$

$$+ \frac{1}{t} \int_T^t \int_T^s q(\tau) d\tau ds = 0 \quad \text{for} \quad t \geq T. \quad (8.1.74)$$

The first term of the left–hand side of (8.1.74) has a finite limit as $t \to \infty$. In fact, for the case when $x(t)$ satisfies (8.1.3) with $c \neq 0$, it is clear that this term tends to zero as $t \to \infty$. For the case when $x(t)$ satisfies (8.1.5) with $c \neq 0$, we assume that $c > 0$ (the same will be done for the ‘case $c < 0$), then it follows that

$$\lim_{t \to \infty} \frac{1}{t} \int_{x(T)}^{x(t)} \left[\frac{1}{f(u)} - \frac{1}{f(\infty)} \right] du = 0$$

and consequently, the same term in (8.1.74) tends to $c/f(\infty)$, where $f(\infty) = \lim f(x)$ as $x \to \infty$. Denote by α the nonnegative limit of the first term of the left–hand side of (8.1.74).

The limit as $t \to \infty$ of the third term of the left–hand side of (8.1.74) is

$$\int_T^\infty \left(\frac{x'(\tau)}{f(x(\tau))} \right)^2 f'(x(\tau))d\tau, \qquad (8.1.75)$$

which is finite or infinite. Taking the upper limit as $t \to \infty$ in (8.1.74), we find

$$\alpha - \frac{x'(T)}{f(x(T))} + \int_T^\infty \left(\frac{x'(\tau)}{f(x(\tau))} \right)^2 f'(x(\tau))d\tau + \limsup_{t\to\infty} \frac{1}{t} \int_T^t \int_T^s q(\tau)d\tau ds = 0.$$

From this equality and condition (8.1.65) it follows that the integral (8.1.75) is finite. Then in the limit as $t \to \infty$ in (8.1.74), we find that (8.1.66) holds, and moreover, we have

$$\alpha - \frac{x'(T)}{f(x(T))} + \int_T^\infty \left(\frac{x'(\tau)}{f(x(\tau))} \right)^2 f'(x(\tau))d\tau + Q_2(T) = 0, \qquad (8.1.76)$$

where $Q_2(t)$ is defined in (8.1.67). Since equality (8.1.76) holds as long as the number T is chosen such that $x(t) \neq 0$ for $t \geq T$, we may regard T as arbitrary provided T is chosen sufficiently large. Thus, we see that (8.1.73) is satisfied for all sufficiently large t. This completes the proof. ∎

A further improvement of Theorem 8.1.13 can be obtained by relaxing (8.1.66) to the weaker condition: there exists an integer $n \geq 1$ such that

$$\lim_{t\to\infty} \frac{1}{t^{n-1}} \int_{t_0}^t (t-s)^{n-1} q(s)ds \quad \text{exists and is finite.} \qquad (8.1.77)$$

Then, we can define the function Q_n as

$$Q_n(t) = \lim_{T\to\infty} \frac{1}{T^{n-1}} \int_t^T (T-s)^{n-1} q(s)ds, \quad t \geq t_0 \qquad (8.1.78)$$

which coincides with the function Q when $n = 1$ and with Q_2 for $n = 2$. Obviously, each one of the hypotheses (8.1.6) and (8.1.66) implies (8.1.77).

Now we shall prove that if equation (8.1.1) has a solution $x(t)$ satisfying (8.1.3) or (8.1.5) for some constant $c \neq 0$ and there exists an integer $n \geq 1$ such that

$$\limsup_{t\to\infty} \frac{1}{t^{n-1}} \int_{t_0}^t (t-s)^{n-1} q(s)ds > -\infty, \qquad (8.1.79)$$

then condition (8.1.77) holds.

Theorem 8.1.14. Let $n \geq 2$ and let condition (8.1.79) be satisfied. If equation (8.1.1) has a solution $x(t)$ which satisfies either condition (8.1.3) or (8.1.5) for some constant $c \neq 0$, then (8.1.77) holds, and

$$\frac{x'(t)}{f(x(t))} = \alpha + Q_n(t) + \int_t^\infty \left(\frac{x'(s)}{f(x(s))} \right)^2 f'(x(s))ds \qquad (8.1.80)$$

for all large t, where α is a nonnegative constant.

Proof. The proof for the case $n = 2$ is given in Theorem 8.1.13. So, we shall assume that $n \geq 3$. Let $x(t)$ be a solution of equation (8.1.1) defined for $t \geq T_0$ for some $T_0 \geq t_0$ and satisfies (8.1.3) or (8.1.5), where c is a nonzero constant. From (8.1.3) or (8.1.5) it follows that $x(t) \neq 0$ for all large t. Clearly, there is no loss of generality in assuming that $x(t) \neq 0$ for all $t \geq T_0$.

From equation (8.1.1), for every T, t with $T \geq t \geq T_0$, we have

$$\int_t^T (T-s)^{n-1}q(s)ds = -\int_t^T (T-s)^{n-1} \left(\frac{x''(s)}{f(x(s))} \right) ds$$

$$= (T-t)^{n-1} \left(\frac{x'(t)}{f(x(t))} \right) - (n-1)(n-2) \int_t^T (T-s)^{n-3} \left[\int_t^s \frac{x'(\tau)}{f(x(\tau))}d\tau \right] ds$$

$$- \int_t^T (T-s)^{n-1} \left(\frac{x'(s)}{f(x(s))} \right)^2 f'(x(s))ds.$$

So, for $T \geq t \geq T_0$ it follows that

$$\frac{1}{T^{n-1}} \int_t^T (T-s)^{n-1}q(s)ds = \left(1 - \frac{t}{T} \right)^{n-1} \left(\frac{x'(t)}{f(x(t))} \right)$$

$$- (n-1)(n-2)\frac{1}{T^{n-1}} \int_t^T (T-s)^{n-3} \left[\int_t^s \frac{x'(\tau)}{f(x(\tau))}d\tau \right] ds \qquad (8.1.81)$$

$$- \frac{1}{T^{n-1}} \int_t^T (T-s)^{n-1} \left(\frac{x'(s)}{f(x(s))} \right)^2 f'(x(s))ds.$$

Next, we shall show that the

$$\lim_{T\to\infty} \frac{1}{T^{n-1}} \int_t^T (T-s)^{n-3} \left[\int_t^s \frac{x'(\tau)}{f(x(\tau))}d\tau \right] ds \qquad (8.1.82)$$

exists as a nonnegative real number, and it is independent of t ($\geq T_0$). Denote this limit by δ. For this, we assume that the point $t \geq T_0$ is fixed. We first observe that the limit in (8.1.82) can be written as

$$\lim_{T\to\infty} \frac{1}{T^{n-1}} \int_t^T (T-s)^{n-3} \left[\int_{x(t)}^{x(s)} \frac{du}{f(u)} \right] ds. \qquad (8.1.83)$$

If the solution $x(t)$ satisfies condition (8.1.3) for some constant $c \neq 0$, then $x(t)$ is bounded and hence $\int_{x(t)}^{x(s)} du/f(u)$ is also bounded for all $s \geq t$. In this case it is clear that the above limit equals zero. It remains to consider the case when $x(t)$ satisfies condition (8.1.5) for some constant $c \neq 0$. Assume that $c > 0$. (When $c < 0$, the proof is similar and hence omitted). This means that $x(t)$ is positive on $[T_0, \infty)$. Consider an arbitrary number $\epsilon > 0$ with $0 < \epsilon < c$. The limit $f(\infty) = \lim_{u \to \infty} f(u)$ exists as a positive number or infinity. So, we can choose a $u_0 > 0$, so that $(1/f(u)) - (1/f(\infty)) < \epsilon$ for all $u \geq u_0$. Furthermore, by taking into account (8.1.5), we can conclude that there exists a $t_1 \geq t$ with $(c - \epsilon)t_1 \geq u_0$ and such that

$$(c - \epsilon)r \leq x(r) \leq (c + \epsilon)r \quad \text{for every} \quad r \geq t_1. \tag{8.1.84}$$

Then for any T, s with $T \geq s \geq t_1$, we obtain

$$\left| \int_{x(t)}^{x(s)} \left[\frac{1}{f(u)} - \frac{1}{f(\infty)} \right] du \right|$$

$$\leq \left| \int_{x(t)}^{x(t_1)} \left[\frac{1}{f(u)} - \frac{1}{f(\infty)} \right] du \right| + \left| \int_{x(t_1)}^{x(s)} \left[\frac{1}{f(u)} - \frac{1}{f(\infty)} \right] du \right|$$

$$\leq \left| \int_{x(t)}^{x(t_1)} \left[\frac{1}{f(u)} - \frac{1}{f(\infty)} \right] du \right| + \int_{(c-\epsilon)t_1}^{(c+\epsilon)s} \epsilon \, du$$

$$\leq \left| \int_{x(t)}^{x(t_1)} \left[\frac{1}{f(u)} - \frac{1}{f(\infty)} \right] du \right| + \epsilon[(c + \epsilon)T - (c - \epsilon)t_1].$$

Thus, for every $T \geq t_1$ it follows that

$$\frac{1}{T^{n-1}} \left| \int_t^T (T - s)^{n-3} \left\{ \int_{x(t)}^{x(s)} \left[\frac{1}{f(u)} - \frac{1}{f(\infty)} \right] du \right\} ds \right|$$

$$\leq \frac{1}{T^{n-1}} \int_t^T (T - s)^{n-3} \left| \int_{x(t)}^{x(s)} \left[\frac{1}{f(u)} - \frac{1}{f(\infty)} \right] du \right| ds$$

$$\leq \left\{ \left| \int_{x(t)}^{x(t_1)} \left[\frac{1}{f(u)} - \frac{1}{f(\infty)} \right] du \right| + \epsilon[(c + \epsilon)T - (c - \epsilon)t_1] \right\}$$

$$\times \frac{1}{(n - 2)T} \left(1 - \frac{t}{T} \right)^{n-2},$$

and consequently,

$$\limsup_{T \to \infty} \frac{1}{T^{n-1}} \left| \int_t^T (T - s)^{n-3} \int_{x(t)}^{x(s)} \left[\frac{1}{f(u)} - \frac{1}{f(\infty)} \right] du \, ds \right| \leq \frac{1}{(n - 2)} \epsilon(c + \epsilon).$$

Since ϵ is arbitrary, we find

$$\lim_{T \to \infty} \int_t^T (T-s)^{n-3} \left\{ \int_{x(t)}^{x(s)} \left[\frac{1}{f(u)} - \frac{1}{f(\infty)} \right] du \right\} ds = 0. \qquad (8.1.85)$$

By using (8.1.84), we obtain for $T \geq t_1$,

$$\frac{1}{T^{n-1}} \int_t^T (T-s)^{n-3} \left[\int_{x(t)}^{x(s)} du \right] ds$$

$$= \frac{1}{T^{n-1}} \int_t^{t_1} (T-s)^{n-3} x(s) ds + \frac{1}{T^{n-1}} \int_{t_1}^T (T-s)^{n-3} x(s) ds$$

$$\quad - \frac{1}{(n-2)} \frac{1}{T} \left(1 - \frac{t}{T} \right)^{n-2} x(t)$$

$$\leq \frac{1}{T^2} \int_T^{t_1} x(s) ds + (c+\epsilon) \frac{1}{T^{n-1}} \int_{t_1}^T (T-s)^{n-3} s\, ds$$

$$\quad - \frac{1}{(n-2)} \frac{1}{T} \left(1 - \frac{t}{T} \right)^{n-2} x(t)$$

$$= \frac{1}{T^2} \int_t^{t_1} x(s) ds + (c+\epsilon) \left[\frac{1}{(n-2)} \frac{t_1}{T} \left(1 - \frac{t_1}{T} \right)^{n-2} \right.$$

$$\quad \left. - \frac{1}{(n-2)(n-1)} \left(1 - \frac{t_1}{T} \right)^{n-1} \right] - \frac{1}{(n-2)} \frac{1}{T} \left(1 - \frac{t}{T} \right)^{n-2} x(t)$$

$$\to \frac{c+\epsilon}{(n-2)(n-1)} \quad \text{as } T \to \infty$$

and

$$\frac{1}{T^{n-1}} \int_t^T (T-s)^{n-3} \left[\int_{x(t)}^{x(s)} du \right] ds$$

$$\geq \frac{1}{T^2} \left(1 - \frac{t_1}{T} \right)^{n-3} \int_t^{t_1} x(s) ds + (c-\epsilon) \frac{1}{T^{n-1}} \int_{t_1}^T (T-s)^{n-3} s\, ds$$

$$\quad - \frac{1}{(n-2)} \frac{1}{T} \left(1 - \frac{t}{T} \right)^{n-2} x(t)$$

$$= \frac{1}{T^2} \left(1 - \frac{t_1}{T} \right)^{n-3} \int_t^{t_1} x(s) ds + (c-\epsilon) \left[\frac{1}{(n-2)} \frac{t_1}{T} \left(1 - \frac{t_1}{T} \right)^{n-2} \right.$$

$$\quad \left. + \frac{1}{(n-2)(n-1)} \left(1 - \frac{t_1}{T} \right)^{n-1} \right] - \frac{1}{(n-2)} \frac{1}{T} \left(1 - \frac{t}{T} \right)^{n-2} x(t)$$

$$\to \frac{c-\epsilon}{(n-2)(n-1)} \quad \text{as } T \to \infty.$$

Therefore, since ϵ is arbitrary, we conclude that

$$\lim_{T \to \infty} \frac{1}{T^{n-1}} \int_t^T (T - s)^{n-3} \left[\int_{x(t)}^{x(s)} du \right] ds = \frac{c}{(n-2)(n-1)}. \tag{8.1.86}$$

From (8.1.85) and (8.1.86) it follows that the limit (8.1.83) is equal to the real number

$$\delta = \frac{c}{(n-2)(n-1)} \left(\frac{1}{f(\infty)} \right)$$

and this proves our assertion.

Now, from (8.1.81), we have

$$\limsup_{T \to \infty} \frac{1}{T^{n-1}} \int_{T_0}^T (T - s)^{n-1} q(s) ds$$

$$= \frac{x'(T_0)}{f(x(T_0))} - \alpha - \lim_{T \to \infty} \frac{1}{T^{n-1}} \int_{T_0}^T (T - s)^{n-1} \left(\frac{x'(s)}{f(x(s))} \right)^2 f'(x(s)) ds,$$

where $\alpha = (n-1)(n-2)\delta$. Thus, condition (8.1.79) guarantees that

$$\lim_{T \to \infty} \frac{1}{T^{n-1}} \int_{T_0}^T (T - s)^{n-1} \left(\frac{x'(s)}{f(x(s))} \right)^2 f'(x(s)) ds < \infty.$$

Hence, for all $t \geq T_0$,

$$\lim_{T \to \infty} \frac{1}{T^{n-1}} \int_t^T (T - s)^{n-1} \left(\frac{x'(s)}{f(x(s))} \right)^2 f'(x(s)) ds$$

$$= \int_t^\infty \left(\frac{x'(s)}{f(x(s))} \right)^2 f'(x(s)) ds < \infty.$$

Finally, from (8.1.81) we conclude that condition (8.1.77) is satisfied, and

$$Q_n(t) = \lim_{T \to \infty} \frac{1}{T^{n-1}} \int_t^T (T - s)^{n-1} q(s) ds$$

$$= \frac{x'(t)}{f(x(t))} - \alpha - \int_t^\infty \left(\frac{x'(s)}{f(x(s))} \right)^2 f'(x(s)) ds \quad \text{for} \quad t \geq T_0.$$

So, the solution $x(t)$ satisfies (8.1.80). This completes the proof. ∎

In view of Theorem 8.1.14 it is reasonable to restrict our attention to the case when condition (8.1.77) holds in discussing the existence of asymptotically linear solutions $x(t)$ of the form (8.1.3) and (8.1.5). Now, it is important to note the following lemma.

Lemma 8.1.1. Let $n \geq 2$ be an integer. Then

(i) if (8.1.77) holds, then the function Q_n is continuously differentiable on $[t_0, \infty)$ and $Q'_n(t) = -q(t)$, moreover

$$\lim_{t \to \infty} \frac{1}{t^{n-1}} \int_{t_0}^t (t-s)^{n-2} Q_n(s) ds = 0, \qquad (8.1.87)$$

(ii) if the function $h(t)$ is continuously differentiable on $[t_0, \infty)$ such that $h'(t) = -q(t)$ and

$$\lim_{t \to \infty} \frac{1}{t^{n-1}} \int_{t_0}^t (t-s)^{n-2} h(s) ds = 0, \qquad (8.1.88)$$

then condition (8.1.77) holds and $Q_n(t) = h(t)$.

Proof. (i) For any T, t with $T \geq t \geq t_0$, we have

$$\frac{1}{T^{n-1}} \int_t^T (T-s)^{n-1} q(s) ds$$

$$= \frac{1}{T^{n-1}} \int_{t_0}^T (T-s)^{n-1} q(s) ds - \frac{1}{T^{n-1}} \int_{t_0}^t (T-s)^{n-1} q(s) ds$$

$$= \frac{1}{T^{n-1}} \int_{t_0}^T (T-s)^{n-1} q(s) ds - \left(1 - \frac{t}{T}\right)^{n-1} \int_{t_0}^t q(\tau) d\tau$$

$$- (n-1) \frac{1}{T^{n-1}} \int_{t_0}^t (T-s)^{n-2} \left[\int_{t_0}^s q(\tau) d\tau\right] ds.$$

However, since for $T \geq t \geq t_0$,

$$\left| \frac{1}{T^{n-1}} \int_{t_0}^t (T-s)^{n-2} \left[\int_{t_0}^s q(\tau) d\tau\right] ds \right|$$

$$\leq \frac{1}{T^{n-1}} \int_{t_0}^t (T-s)^{n-2} \left[\int_{t_0}^s |q(\tau)| d\tau\right] ds \leq \frac{1}{T} \int_{t_0}^t \left[\int_{t_0}^s |q(\tau)| d\tau\right] ds$$

it follows that

$$\lim_{T \to \infty} \frac{1}{T^{n-1}} \int_{t_0}^t (T-s)^{n-2} \left[\int_{t_0}^s q(\tau) d\tau\right] ds = 0 \quad \text{for} \quad t \geq t_0.$$

Thus, for all $t \geq t_0$,

$$\lim_{T \to \infty} \frac{1}{T^{n-1}} \int_t^T (T-s)^{n-1} q(s) ds = \lim_{T \to \infty} \frac{1}{T^{n-1}} \int_{t_0}^T (T-s)^{n-1} q(s) ds - \int_{t_0}^t q(\tau) d\tau,$$

i.e.,

$$Q_n(t) = Q_n(t_0) - \int_{t_0}^t q(\tau) d\tau \quad \text{for} \quad t \geq t_0. \qquad (8.1.89)$$

This ensures that Q_n is continuously differentiable on $[t_0, \infty)$ and satisfies $Q'_n(t) = -q(t)$. Furthermore, for every $t \geq t_0$,

$$\frac{1}{t^{n-1}} \int_{t_0}^t (t-s)^{n-2} Q_n(s) ds$$

$$= \frac{1}{t^{n-1}} \int_{t_0}^t (t-s)^{n-2} \left[Q_n(t_0) - \int_{t_0}^s q(\tau) d\tau \right] ds$$

$$= \frac{1}{(n-1)} \left(1 - \frac{t_0}{t}\right)^{n-1} Q_n(t_0) - \frac{1}{t^{n-1}} \int_{t_0}^t (t-s)^{n-2} \left[\int_{t_0}^s q(\tau) d\tau \right] ds$$

$$= \frac{1}{(n-1)} \left(1 - \frac{t_0}{t}\right)^{n-1} Q_n(t_0) - \frac{1}{(n-1)} \frac{1}{t^{n-1}} \int_{t_0}^t (t-s)^{n-1} q(s) ds$$

from which (8.1.87) follows immediately in view of the definition of Q_n.

(ii)   From $h'(t) = -q(t)$, it follows that

$$h(t) = h(t_0) - \int_{t_0}^t q(s) ds \quad \text{for} \quad t \geq t_0. \tag{8.1.90}$$

So, we can derive for every $t \geq t_0$ (by using the same arguments as above for Q_n)

$$\frac{1}{t^{n-1}} \int_{t_0}^t (t-s)^{n-2} h(s) ds = \frac{1}{(n-1)} \left(1 - \frac{t_0}{t}\right)^{n-1} h(t_0)$$

$$- \frac{1}{(n-1)} \frac{1}{t^{n-1}} \int_{t_0}^t (t-s)^{n-1} q(s) ds.$$

Thus, in view of (8.1.88) we conclude that (8.1.77) holds, and $Q_n(t_0) = h(t_0)$. Finally, from (8.1.89) and (8.1.90) we find that $Q_n(t) = h(t)$. This completes the proof of the lemma. ∎

Remark 8.1.4.   It can be easily verified that the results presented in Subsections 8.1.1 – 8.1.3 remain valid with the function $Q_n(t)$ (where $n \geq 1$ is an integer) in place of $Q(t)$, and condition (8.1.77) instead of (8.1.6).

To illustrate the above results, we discuss the following example.

Example 8.1.3.   Consider equation (8.1.2) with

$$q(t) = \begin{cases} n^\lambda \sin\left(n^\lambda (t - 2n\pi)\right) & \text{for } t \in I_n, \ n = 1, 2, \cdots \\ 0 & \text{otherwise,} \end{cases} \tag{8.1.91}$$

where $\lambda > 0$ is a constant and

$$I_n = \left\{ t : (2n - n^{-\lambda})\pi \leq t \leq (2n + n^{-\lambda})\pi \right\}, \quad n = 1, 2, \cdots.$$

Here $q(t)$ is regarded as a function on the interval $\mathbb{R}_0 = [0, \infty)$. Let

$$
h(t) = \begin{cases} 1 + \cos(n^\lambda(t - 2n\pi)) & \text{for } t \in I_n, \ n = 1, 2, \cdots \\ 0 & \text{otherwise.} \end{cases} \tag{8.1.92}
$$

Then the function h is continuously differentiable on $\mathbb{R}_0$ and $h'(t) = -q(t)$ for $t \geq 0$. Since $h(t)$ has no limit as $t \to \infty$, Theorem 8.1.12 implies that equation (8.1.2) with (8.1.91) has no solutions x of the form (8.1.43) and (8.1.68). Noting that $0 \leq h(t) \leq 2$ for $t \in I_n$, $n = 1, 2, \cdots$ we find that if $t \in I_n$ for some $n = 1, 2, \cdots$, then

$$
0 \leq \frac{1}{t} \int_0^t h(s)\,ds \leq \frac{2}{(2n - n^{-\lambda})\pi} \sum_{i=1}^n \int_{I_i} ds = \frac{4}{2n - n^{-\lambda}} \sum_{i=1}^n \frac{1}{i^\lambda}.
$$

Thus, $\lim_{t \to \infty}(1/t) \int_0^t h(s)\,ds = 0$ and so, by Lemma 8.1.1, (8.1.88) with $n = 2$ is satisfied, and $Q_2(t) = h(t)$. Noting that $Q_2(t) \ (= h(t))$ satisfies $Q_2(t) \leq 2$ for $t \in I_n$, $n = 1, 2, \cdots$ and $Q_2(t) \geq 1$ for $t \in J_n$, $n = 1, 2, \cdots$, where

$$
J_n = \left\{ t : \left(2n - \frac{n^{-\lambda}}{2}\right)\pi \leq t \leq \left(2n + \frac{n^{-\lambda}}{2}\right)\pi \right\},
$$

we can easily show that integral conditions (8.1.21), (8.1.22), (8.1.59) and (8.1.60) with $Q_2(t)$ inplace of $Q(t)$, are satisfied if and only if $\lambda > 1$, $\lambda > 2$, $\lambda > \gamma$ and $\lambda > 2\gamma$, respectively. Therefore, we can conclude by Theorem 8.1.2 with $Q_2(t)$ inplace of $Q(t)$ and condition (8.1.66) instead of (8.1.6) that equation (8.1.2) with (8.1.92) has a solution $x(t)$ of the form (8.1.3) $(c \neq 0)$ if and only if $\lambda > 2$. Similarly, we can conclude by Theorem 8.1.8 with $Q_2(t)$ inplace of $Q(t)$ and condition (8.1.66) instead of (8.1.6) that equation (8.1.2) with (8.1.91) has a solution $x(t)$ of the form (8.1.5) $(c \neq 0)$ if and only if $\lambda > 2\gamma$.

8.2. Asymptotic Behavior of Positive Solutions of Singular Emden–Fowler–Type Equations

In this section we shall discuss the positive solutions of the singular Emden–Fowler–type equations

$$
x''(t) + q(t)x^{-\lambda}(t) = 0, \tag{8.2.1}
$$

$$
x''(t) = q(t)x^{-\lambda}(t) \tag{8.2.2}
$$

and the system

$$\begin{cases} x''(t) = p(t)y^{-\lambda}(t) \\ y''(t) = q(t)x^{-\mu}(t), \end{cases}$$

$$(8.2.3)$$

where $p(t)$, $q(t) \in C([t_0, \infty), \mathbb{R})$ and λ, μ are positive constants.

In Subsections 8.2.1 and 8.2.2 we shall investigate the problem of existence and asymptotic behavior of positive solutions of equations (8.2.1) and (8.2.2). Subsection 8.2.3 is concerned with the positive solutions of the system (8.2.3).

8.2.1. Positive Solutions of Equation (8.2.1)

If $x(t)$ is a solution of (8.2.1) which exists and is positive for all sufficiently large t, then we call $x(t)$ a *proper solution* of equation (8.2.1). If $x(t)$ is a proper solution of (8.2.1), then $x''(t) \leq 0$ and since $x(t) > 0$ we find that $x'(t)$ decreases to nonnegative limit as $t \to \infty$. Therefore, if $x(t)$ is a positive solution of (8.2.1) then either: (I) $x(t)$ is a proper solution in which case either (I_1) $0 < \lim_{t\to\infty} x'(t) < \infty$ or (I_2) $\lim_{t\to\infty} x'(t) = 0$, or (II) $x(t)$ is not a proper solution and there exists $b < \infty$ such that $\lim_{t\to b} x(t) = 0$ and $x(t)$ cannot be continued after b as a twice continuously differentiable solution of (8.2.1).

In what follows we are concerned with the proper solutions of equation (8.2.1) and discuss existence, uniqueness and asymptotic behavior of solutions satisfying (I_1) or (I_2).

Existence and Uniqueness of Positive Solutions

Let $\alpha > 0$ and c be fixed real numbers. In what follows by $x(t, \beta)$ we shall denote the solution of

$$x''(t) + q(t)x^{-\lambda}(t) = 0, \quad x(c) = \alpha, \quad x'(c) = \beta.$$

Theorem 8.2.1. If

$$\int_{t_0}^{\infty} s^{-\lambda}q(s)ds = \infty, \quad t_0 \geq 1$$

$$(8.2.4)$$

then for all β, $x(t, \beta)$ falls into case (II). If

$$\int_{t_0}^{\infty} s^{-\lambda}q(s)ds < \infty, \quad t_0 \geq 1$$

$$(8.2.5)$$

then there exists β_0 such that

(a_1) if $\beta < \beta_0$, then $x(t, \beta)$ falls into case (II),

(a_2) $x(t, \beta_0)$ satisfies (I_2),

(a_3) if $\beta > \beta_0$, then $x(t, \beta)$ satisfies (I_1).

Furthermore, $\lim_{t \to \infty} x(t, \beta_0) < \infty$ if and only if $\int_{t_0}^{\infty} sq(s)ds < \infty$.

Theorem 8.2.1 is a direct consequence of the following lemma and two propositions.

Lemma 8.2.1. Suppose $a < b$, $q(t)$, $q_1(t) \in C([a, b], \mathbb{R}_0)$, $q(t) < q_1(t)$ and $x(t)$ and $y(t)$ are positive solutions on $[a, b]$ of equations (8.2.1) and

$$y''(t) + q_1(t)y^{-\lambda}(t) \ = \ 0$$

respectively. If $x(a) \geq y(a)$ and $x'(a) > y'(a)$, then $x(t) > y(t)$ and $x'(t) > y'(t)$ for $t \in (a, b]$.

Proof. For $t \in [a, b]$, we have

$$x(t) \ = \ x(a) + (t - a)x'(a) - \int_a^t (t - s)q(s)x^{-\lambda}(s)ds$$

and

$$y(t) \ = \ y(a) + (t - a)y'(a) - \int_a^t (t - s)q_1(s)y^{-\lambda}(s)ds.$$

Subtracting these equations, we get

$$\begin{aligned} x(t) - y(t) \ = \ & [x(a) - y(a)] + [x'(a) - y'(a)](t - a) \\ & + \int_a^t (t - s)\left[q_1(s)y^{-\lambda}(s) - q(s)x^{-\lambda}(s)\right]ds. \end{aligned} \qquad (8.2.6)$$

Since $x'(a) > y'(a)$, there exists an $\epsilon > 0$ such that $x(t) > y(t)$ for $a < t \leq a + \epsilon$. Suppose $x(t) \leq y(t)$ for some $t \in (a, b]$. Then there exists $c \in (a, b]$ such that $x(c) = y(c)$ and $x(t) > y(t)$ for $a < t < c$. Letting $t = c$ in (8.2.6), then the left–hand side is zero and the right–hand side is positive, which is a contradiction. Hence, $x(t) > y(t)$ for $a < t \leq b$, and since

$$x'(t) \ = \ x'(a) - \int_a^t q(s)x^{-\lambda}(s)ds \quad \text{and} \quad y'(t) \ = \ y'(a) - \int_a^t q_1(s)y^{-\lambda}(s)ds,$$

we have $x'(t) > y'(t)$ for $a \leq t \leq b$. ∎

Proposition 8.2.1. Let c be a real number. Then there exist solutions of (8.2.1) which are positive for $t \geq c$ if and only if condition (8.2.5) holds. Furthermore, if $\alpha > 0$ and condition (8.2.5) holds, then there exist solutions of

$$x''(t) + q(t)x^{-\lambda}(t) \ = \ 0, \quad x(c) = \alpha \qquad (8.2.7)$$

which are positive for $t \geq c$ and $\lim_{t \to \infty} x'(t) > 0$.

Proof. Let $x(t)$ be a solution of equation (8.2.1) which is positive for $t \geq c$. Let $z(t)$ be the solution of

$$-z''(t) = q(t)[x(c) + x'(c)(t - c)]^{-\lambda}, \quad z(c) = x(c), \quad z'(c) = x'(c).$$

Since $0 \leq x(t) \leq x(c) + x'(c)(t - c)$ for $t \geq c$, we have $-z''(t) \leq -y''(t)$ for $t \geq c$. Hence, $z'(t) - x'(t) = \int_c^t [z''(s) - x''(s)]ds \geq 0$ and since $x'(t) \geq 0$ for $t \geq c$, we have $z'(t) \geq 0$ for $t \geq c$. But,

$$z'(t) = x'(c) - \int_c^t q(s)[x(c) + x'(c)(s - c)]^{-\lambda}ds$$

and therefore,

$$\int_c^\infty q(s)[x(c) + x'(c)(s - c)]^{-\lambda}ds < \infty.$$

Hence, $\int_1^\infty s^{-\lambda}q(s)ds < \infty$. Conversely, suppose that condition (8.2.5) holds and $\alpha > 0$. Choose $\beta > 1 + \int_c^\infty q(s)[\alpha + s - c]^{-\lambda}ds$. Let $x(t)$ be the solution of equation (8.2.7) with $x'(c) = \beta$. If $x'(t) > 1$ for all $t \geq c$, then clearly $x(t)$ is positive and $\lim_{t \to \infty} x'(t) > 0$. If $x'(t) \leq 1$ for some $t > c$, let $a = \inf\{t > c : x'(t) = 1\}$. Then, $y'(t) \geq 1$ for $c \leq t \leq a$ and hence $x(t) \geq \alpha + (t - c)$ for $c \leq t \leq a$. So,

$$\beta = x'(c) = x'(a) + \int_c^a q(s)x^{-\lambda}(s)ds \leq 1 + \int_c^a q(s)[\alpha + s - c]^{-\lambda}ds$$

$$\leq 1 + \int_c^\infty q(s)[\alpha + s - c]^{-\lambda}ds$$

and this contradicts the choice of β. ∎

Proposition 8.2.2. Let $\alpha > 0$ and c be real numbers. If condition (8.2.5) holds, then there exists a unique solution $x(t)$ of

$$x''(t) + q(t)x^{-\lambda}(t) = 0, \quad x(c) = \alpha, \quad x'(\infty) = 0. \tag{8.2.8}$$

Furthermore, $x(\infty) < \infty$ if and only if

$$\int_0^\infty sq(s)ds < \infty. \tag{8.2.9}$$

Proof. Without any loss of generality we can assume that $c = 0$. Suppose $x(t)$ and $y(t)$ are two positive solutions of (8.2.8) with $y'(0) > x'(0)$. Then by Lemma 8.2.1, $y(t) > x(t)$ for $t > 0$. Hence,

$$(y(t) - x(t))'' = q(t)\left[x^{-\lambda}(t) - y^{-\lambda}(t)\right] > 0.$$

So, $y'(t) - x'(t) \geq y'(0) - x'(0)$ for $t \geq 0$, and hence $y'(\infty) - x'(\infty) \geq y'(0) - x'(0)$. This contradicts $y'(\infty) = x'(\infty) = 0$, and hence the uniqueness follows.

Let $x_\beta(t)$ be the solution of

$$x''(t) + q(t)x^{-\lambda}(t) = 0, \quad x(0) = \alpha, \quad x'(0) = \beta.$$

Let $T = \{\beta : x'_\beta(t) \geq 0 \text{ for } t \geq 0\}$ and $S = \{\beta : x'_\beta(t) < 0 \text{ for some } t \geq 0\}$. By Lemma 8.2.1 if $\beta_1 \in S$ and $\beta_2 \in T$, then $\beta_1 < \beta_2$. By Proposition 8.2.1, $T \neq \emptyset$ and every negative number belongs to S. Thus, $\sup S = \inf T$ and if $b = \sup S$, then $0 \leq b < \infty$. By continuous dependence on initial conditions $b \notin S$. So, $b \in T$ and we need only to show that $x'_b(\infty) = 0$.

Let $x(t) = x_b(t)$ and suppose $x'(\infty) > 0$. Then, $x(t)/(t+1)$ is continuous and positive for $t \geq 0$ and $\lim_{t\to\infty} x(t)/(t+1) = x'(\infty) > 0$. Therefore, if $M = \inf\{x(t)/(t+1) : t \geq 0\}$, then $M > 0$. Choose $\delta > 0$ such that

$$\delta \exp\left(\lambda \left[\frac{2}{M}\right]^{\lambda+1} \int_0^\infty q(s)(s+1)^{-\lambda} ds\right) < \frac{M}{4}$$

and $b - \delta > M/2$ (since $b \geq x'(\infty) \geq M$ this is possible). Let $y(t) = x_{b-\delta}(t)$. If $y(t)/(t+1) > M/2$ for all $t > 0$, then clearly $b - \delta \in T$ contradicting $b = \inf T$. Suppose $y(t)/(t+1) \leq M/2$ for some $t > 0$. Since $y(0) = x(0) > M/2$, $y(t)/(t+1) \geq M/2$ for $0 \leq t \leq a$, where $a = \inf\{t > 0 : y(t)/(t+1) \leq M/2\}$. Also, $y(a)/(a+1) = M/2$. Now, for $0 \leq t \leq a$, we have

$$0 < \frac{x(t)}{t+1} - \frac{y(t)}{t+1}$$

$$= \frac{\delta t}{t+1} + \int_0^t \left(1 - \frac{s+1}{t+1}\right)\left(\frac{q(s)}{(s+1)^\lambda}\right)\left[\left(\frac{s+1}{y(s)}\right)^\lambda - \left(\frac{s+1}{x(s)}\right)^\lambda\right] ds$$

$$\leq \delta + \int_0^t \frac{q(s)}{(s+1)^\lambda} \int_{y(s)/(s+1)}^{x(s)/(s+1)} \lambda u^{-\lambda-1} du\, ds$$

$$\leq \delta + \int_0^t \frac{q(s)}{(s+1)^\lambda} \lambda \left(\frac{2}{M}\right)^{\lambda+1} \left[\frac{x(s)}{s+1} - \frac{y(s)}{s+1}\right] ds.$$

Now, by Gronwall's inequality, we have for $0 \leq t \leq a$,

$$\frac{x(t)}{t+1} - \frac{y(t)}{t+1} \leq \delta \exp\left[\lambda\left(\frac{2}{M}\right)^{\lambda+1} \int_0^t \frac{q(s)}{(s+1)^\lambda} ds\right] < \frac{M}{4},$$

and hence

$$\frac{x(a)}{a+1} < \frac{y(a)}{a+1} + \frac{M}{4} = \frac{3M}{4}$$

contradicting the choice of M. Thus, $x'(\infty) = 0$ and this proves the existence.

Let $x(t)$ be the solution of (8.2.8). Then integrating (8.2.8) from t to u and letting $u \to \infty$, we obtain

$$x'(t) = \int_t^\infty q(s)x^{-\lambda}(s)ds. \tag{8.2.10}$$

Integration of (8.2.10) from 0 to ∞ now gives

$$x(\infty) = \alpha + \int_0^\infty sq(s)x^{-\lambda}(s)ds \tag{8.2.11}$$

and since $\alpha < x(t) < x(\infty)$ it follows from (8.2.11) that $x(\infty) < \infty$ if and only if condition (8.2.9) holds.

Asymptotic Behavior

The following theorem deals with the asymptotic behavior of proper solutions of equation (8.2.1) which satisfy case (I_1).

Theorem 8.2.2. Let $x(t)$ be a proper solution of equation (8.2.1) satisfying (I_1). Then there exists a constant $a > 0$ such that

$$x(t) = at + O\left(\int_{t_0}^t \int_s^\infty u^{-\lambda}q(u)duds\right) \quad \text{as } t \to \infty. \tag{8.2.12}$$

Furthermore, if

$$\int_{t_0}^\infty s^{-\lambda+1}q(s)ds < \infty, \quad t_0 \geq 1 \tag{8.2.13}$$

then there exists a constant $b > 0$ such that

$$x(t) = at + b - a^{-\lambda}[1 + o(1)]\int_t^\infty (s-t)s^{-\lambda}q(s)ds \quad \text{as } t \to \infty. \tag{8.2.14}$$

Proof. Let $a = \lim_{t\to\infty} x'(t)$. Then, $a > 0$ and $x(t) = at[1 + o(1)]$ as $t \to \infty$. Integrating equation (8.2.1) from t to v and letting $v \to \infty$, we obtain

$$x'(t) - a = \int_t^\infty q(s)x^{-\lambda}(s)ds = a^{-\lambda}[1 + o(1)]\int_t^\infty s^{-\lambda}q(s)ds. \tag{8.2.15}$$

Integrating (8.2.15) from $t_0 \geq 1$ to t, we find (8.2.12).

Suppose (8.2.13) holds. Then the right–hand side of (8.2.15) is integrable on $[1, \infty)$ and hence $\lim_{t\to\infty}[x(t) - at]$ exists. Denote this limit by b. Then integrating (8.2.15) from t to ∞ gives

$$x(t) = at + b - a^{-\lambda}[1 + o(1)] \int_t^\infty \int_s^\infty u^{-\lambda} q(u) du ds. \qquad (8.2.16)$$

Interchanging the order of integration in (8.2.16), we get (8.2.14). ∎

Next, we present the asymptotic behavior of the proper solutions of equation (8.2.1) satisfying (I₂) and condition (8.2.9).

Theorem 8.2.3. Let $x(t)$ be a proper solution of equation (8.2.1) satisfying (I₂). If condition (8.2.9) holds, then there exists a constant $a > 0$ such that

$$x(t) = a - a^{-\lambda}[1 + o(1)] \int_t^\infty (s - t) q(s) ds \quad \text{as } t \to \infty. \qquad (8.2.17)$$

Proof. Since condition (8.2.9) holds, we have by Theorem 8.2.1 that $\lim_{t\to\infty} x(t) = a < \infty$ for some constant $a > 0$. Integrating equation (8.2.1) from t to u and letting $u \to \infty$, we obtain

$$x'(t) = \int_t^\infty q(s) x^{-\lambda}(s) ds = a^{-\lambda}[1 + o(1)] \int_t^\infty q(s) ds. \qquad (8.2.18)$$

Integrating (8.2.18) from t to ∞ gives

$$x(t) = a - a^{-\lambda}[1 + o(1)] \int_t^\infty \int_s^\infty q(u) du ds. \qquad (8.2.19)$$

Interchanging the order of integration in (8.2.19), we get (8.2.17). ∎

To study the asymptotic behavior of proper solutions of equation (8.2.1) satisfying (I₂) and the condition

$$\int_0^\infty sq(s) ds = \infty, \qquad (8.2.20)$$

we need the following three lemmas.

Lemma 8.2.2. Let c be a positive constant. Then, $x(t)$ is a solution of equation (8.2.1) if and only if $c^{1/(\lambda+1)} x(t)$ is a solution of the equation $y''(t) + cq(t) y^{-\lambda}(t) = 0$.

Proof. The verification is obvious, and hence left to the reader. ∎

Lemma 8.2.3. Let $y(t)$ and $w(t)$ be two positive proper solutions of equation (8.2.1) satisfying (I₂). If condition (8.2.20) holds, then $y(t) \sim z(t)$ as $t \to \infty$. (The notation $y(t) \sim z(t)$ as $t \to \infty$ means that $\lim_{t\to\infty} y(t)/z(t) = 1$).

Proof. Without any loss of generality, we can assume that $y(t)$ and $z(t)$ are defined for $t \geq 0$ and $y(0) > z(0)$. By Proposition 8.2.1, condition (8.2.5) holds. And, by Proposition 8.2.2, $y(t)$ cannot intersect with $z(t)$. Thus, $y(t) > z(t)$ for $t \geq 0$. Also, $y'(t) \leq z'(t)$ for $t \geq 0$, if for $y'(a) > z'(a)$ for some constant $a \geq 0$, then for $t \geq a$,

$$y'(t) - z'(t) = y'(a) - z'(a) + \int_0^t q(s) \left[z^{-\lambda}(s) - y^{-\lambda}(s) \right] ds \geq y'(a) - z'(a) > 0,$$

contradicting $\lim_{t \to \infty} y'(t) = \lim_{t \to \infty} z'(t) = 0$. Hence, $(y(t) - z(t))$ decreases to a nonnegative limit as $t \to \infty$. Therefore, for $t \geq 0$,

$$1 \leq \frac{y(t)}{z(t)} = 1 + \frac{y(t) - z(t)}{z(t)} \leq 1 + \frac{y(0) - z(0)}{z(t)}. \tag{8.2.21}$$

Since condition (8.2.20) holds, we have by Proposition 8.2.2 that $z(t) \to \infty$ as $t \to \infty$. Thus, by (8.2.21), $y(t) \sim z(t)$ as $t \to \infty$. ∎

Lemma 8.2.4. Suppose $q(t), q_1(t) \in C([a, \infty), \mathbb{R}_0)$, $a \geq 0$, $q(t) \leq q_1(t)$ and $\gamma > 0$ is a constant. Let $x(t)$ and $y(t)$ be positive solutions of equation (8.2.1) with $x(a) = \gamma$, $x'(\infty) = 0$ and the equation

$$y''(t) + q_1(t) y^{-\lambda}(t) = 0, \quad y(a) = \gamma, \quad y'(\infty) = 0, \tag{8.2.22}$$

respectively. Then, $y(t) \geq x(t)$ for $t \geq a$.

Proof. Suppose there exists a constant $b > a$ such that $y(b) < x(b)$. Let $c = \max\{t < b : y(t) \geq x(t)\}$. Then, $y(c) = x(c)$ and $y(t) < x(t)$ for $c < t \leq b$. Thus, for some $d \in (c, b)$, $y'(d) < x'(d)$ and $y(d) < x(d)$. Hence, by Lemma 8.2.1, $y(t) < x(t)$ for $t \geq d$. Therefore,

$$x''(t) - y''(t) = q_1(t) y^{-\lambda}(t) - q(t) x^{-\lambda}(t) > 0 \quad \text{for} \quad t \geq d.$$

Hence, $x'(t) - y'(t) \geq x'(d) - y'(d) > 0$ for $t \geq d$, contradicting $\lim_{t \to \infty} x'(t) = \lim_{t \to \infty} y'(t) = 0$. Thus, $y(t) \geq x(t)$ for $t \geq a$. ∎

Theorem 8.2.4. Suppose $q(t), q_1(t) \in C(\mathbb{R}_0, \mathbb{R}^+)$, $\lim_{t \to \infty} q_1(t)/q(t) = r > 0$, where r is a constant, and $x(t)$ and $y(t)$ are positive solutions of equation (8.2.1) with $x'(\infty) = 0$, and

$$y''(t) + q_1(t) y^{-\lambda}(t) = 0, \quad y'(\infty) = 0,$$

respectively. If (8.2.20) holds, then $y(t) \sim r^{1/(\lambda+1)} x(t)$ as $t \to \infty$.

Proof. Let $p(t) = q_1(t)/q(t)$, $t \geq 0$ and the numbers $\epsilon, \eta \in (0, 1)$. Choose a constant $a > 0$ such that $(1 - \epsilon)r < p(t) < (1 + \epsilon)r$ for $t \geq a$. We can assume that $x(t)$ and $y(t)$ are defined for $t \geq 0$. By Proposition

8.2.1, condition (8.2.5) holds. And, by Proposition 8.2.2 there exist unique solutions $y_1(t)$ and $y_2(t)$ of

$$y_1''(t) + (1 - \epsilon)rq(t)y_1^{-\lambda}(t) = 0, \quad y_1(a) = y(a), \quad y_1'(\infty) = 0$$

and

$$y_2''(t) + (1 + \epsilon)rq(t)y_2^{-\lambda}(t) = 0, \quad y_2(a) = y(a), \quad y_2'(\infty) = 0,$$

respectively. Since $(1 - \epsilon)rq(t) \leq q_1(t) \leq (1 + \epsilon)rq(t)$ for $t \geq a$, we have by Lemma 8.2.4 that $y_1(t) \leq y(t) \leq y_2(t)$ for $t \geq a$. By Lemma 8.2.2, $[(1 - \epsilon)r]^{1/(\lambda+1)}x(t)$ and $[(1 + \epsilon)r]^{1/(\lambda+1)}x(t)$ are solutions of

$$y_1''(t) + (1 - \epsilon)rq(t)y_1^{-\lambda}(t) = 0, \quad y_1'(\infty) = 0$$

and

$$y_1''(t) + (1 + \epsilon)rq(t)y_1^{-\lambda}(t) = 0, \quad y_1'(\infty) = 0,$$

respectively. By Lemma 8.2.3, $[(1 - \epsilon)r]^{1/(\lambda+1)}x(t) \sim y_1(t)$ and $[(1 + \epsilon)r]^{1/(\lambda+1)}x(t) \sim y_2(t)$. Hence, there exists $b \geq a$ such that for $t \geq b$,

$$(1-\eta)[(1-\epsilon)r]^{1/(\lambda+1)}x(t) \leq y_1(t) \leq y(t) \leq y_2(t) \leq (1+\eta)[(1+\epsilon)r]^{1/(\lambda+1)}x(t).$$

Since ϵ and η are arbitrary numbers in $(0,1)$, $y(t) \sim r^{1/(\lambda+1)}x(t)$ as $t \to \infty$. ∎

Theorem 8.2.5. Suppose $q(t) \in C(\mathbb{R}_0, \mathbb{R}^+)$, $q_1(t) \in C^1(\mathbb{R}_0, \mathbb{R}^+)$, $q(t) \sim q_1(t)$ as $t \to \infty$,

$$\int_0^\infty sq_1(s)ds = \infty \tag{8.2.23}$$

and $x(t)$ is a positive solution of equation (8.2.1) with $x'(\infty) = 0$. Let $f(t) = \left[\int_t^\infty s^{-\lambda}q_1(s)ds\right]^{1/(\lambda+1)}$. If $\lim_{t \to \infty} tf''(t)/f'(t) = r$ where $r > -2$ is a constant, then $x(t) \sim ((\lambda + 1)/(2 + r))^{1/(\lambda+1)}(tf(t))$ as $t \to \infty$.

Proof. By Proposition 8.2.1 condition (8.2.5) holds. Since $q(t) \sim q_1(t)$ as $t \to \infty$ condition (8.2.23) is also satisfied. Hence, $f(t)$ is well–defined and by Proposition 8.2.2 there exists a solution $y(t)$ of $y''(t)+q_1(t)y^{-\lambda}(t) = 0$, $y'(\infty) = 0$. Let $z(t) = tf(t)$, $t \geq 0$. Then,

$$z''(t) + \theta(t)z^{-\lambda}(t) = 0, \tag{8.2.24}$$

where

$$\theta(t) = \frac{1}{\lambda + 1}\left[2 + \frac{tf''(t)}{f'(t)}\right]q_1(t).$$

Since $z(t) > 0$ and $\theta(t) > 0$ for all sufficiently large t, we have by (8.2.24) that $\lim_{t\to\infty} z'(t) \geq 0$. Let $\alpha = \lim_{t\to\infty} z'(t)$. If $\alpha > 0$, then $z(t) \sim \alpha(t)$ as $t \to \infty$, and hence $f(t) \sim \alpha$ as $t \to \infty$, contradicting $f(t) \to 0$ as $t \to \infty$. Hence, $z'(\infty) = 0$. Also, $\lim_{t\to\infty} \theta(t)/q_1(t) = (2+r)/(1+\lambda) > 0$. So, by Theorem 8.2.4,

$$z(t) \sim \left(\frac{2+r}{1+\lambda}\right)^{1/(\lambda+1)} \qquad y(t) \sim \left(\frac{2+r}{1+\lambda}\right)^{1/(\lambda+1)} x(t) \quad \text{as} \quad t \to \infty.$$

Therefore,

$$x(t) \sim \left(\frac{1+\lambda}{2+r}\right)^{1/(\lambda+1)} tf(t) \quad \text{as} \quad t \to \infty.$$

Remark 8.2.1.

1. In Theorem 8.2.5, the number r can be calculated by the following equation

$$\frac{tf''(t)}{f'(t)} = \frac{tq_1'(t)}{q(t)} + \frac{\lambda q_1(t)}{(1+\lambda)f^{\lambda-1}(t)} - \lambda. \qquad (8.2.25)$$

2. In Theorem 8.2.4, the asymptotic behavior of proper solutions of (8.2.1) depends on the asymptotic behavior of $q(t)$. Using Theorem 8.2.4, we can find in Theorem 8.2.5 the asymptotic behavior of proper solutions of equation (8.2.1) under rather mild conditions on the function $q(t)$.

As an application of Theorems 8.2.1 – 8.2.5, we shall determine the asymptotic behavior of all proper solutions of equation (8.2.1) when $q(t) \sim q_1(t)$ as $t \to \infty$, where $q_1(t) = t^\alpha (\ln t)^\beta$, and α and β are constants. One of the following possibilities must hold:

(i_1) $0 < \alpha + 2 < \lambda + 1$,

(i_2) $0 < \alpha + 2 = \lambda + 1$ and $\beta > -1$,

(i_3) $0 = \alpha + 2$ and $\beta > -1$,

(i_4) $0 = \alpha + 2$ and $\beta = -1$,

(i_5) $\alpha + 2 < 0$ or ($\alpha + 2 = 0$ and $\beta < -1$),

(i_6) $\alpha + 2 > \lambda + 1$ or ($\alpha + 2 = \lambda + 1$ and $\beta \geq -1$).

Each case (i_1) – (i_4) holds if and only if $\int_{t_0}^\infty s^{-\lambda} q(s)ds < \infty$ and $\int_{t_0}^\infty sq(s)ds = \infty$, $t_0 \geq 1$. Case (i_5) holds if and only if $\int_{t_0}^\infty sq(s)ds < \infty$, $t_0 \geq 1$, and case (i_6) holds if and only if $\int_{t_0}^\infty s^{-\lambda} q(s)ds = \infty$, $t_0 \geq 1$.

By Proposition 8.2.1, there is no proper solution of equation (8.2.1) if (i_6) holds.

If (i_5) holds and $x(t)$ is a proper solution of equation (8.2.1), then by Theorems 8.2.1 and 8.2.2 there exist constants a and b, $a > 0$ such that

$$x(t) = at + b - a^{-\lambda}[1 + o(1)] \int_t^\infty (s - t)s^{-\lambda}s^\alpha (\ln s)^\beta ds \qquad (8.2.26)$$

as $t \to \infty$, or

$$x(t) = a - a^{-\lambda}[1 + o(1)] \int_t^\infty (s - t)s^\alpha (\ln s)^\beta ds. \qquad (8.2.27)$$

By using the fact that

$$\int_t^\infty s^m (\ln s)^n ds \sim \begin{cases} \dfrac{t^{m+1}}{-m-1}(\ln t)^n & \text{if } m < -1 \\[2mm] \dfrac{(\ln t)^{n+1}}{-n-1} & \text{if } m = -1 \text{ and } n < -1 \end{cases} \qquad (8.2.28)$$

the integrals in (8.2.26) and (8.2.27) can be simplified.

Suppose (i_1) or (i_2) holds. Let $f(t) = \left(\int_t^\infty s^{-\lambda} q_1(s) ds \right)^{1/(\lambda+1)}$. Then using (8.2.28) and (8.2.25), we get

$$r = \lim_{t \to \infty} \frac{tf''(t)}{f'(t)} = \begin{cases} \dfrac{\alpha + 2}{\lambda + 1} - 2 & \text{if } (i_1) \text{ holds} \\[2mm] -1 & \text{if } (i_2) \text{ holds.} \end{cases}$$

Since $r > -2$, we have by Theorem 8.2.5 that if $x(t)$ is a proper solution of equation (8.2.1) satisfying (I_2), then

$$x(t) \sim \left(\frac{\lambda + 1}{r + 2} \right)^{1/(\lambda+1)} tf(t)$$

$$\sim \begin{cases} \left[\dfrac{(\lambda + 1)^2}{(\alpha+2)(\lambda-1-\alpha)} \right]^{1/(\lambda+1)} (\ln t)^{\beta/(\lambda+1)} t^{(2+\alpha)/(\lambda+1)}, & \text{if } (i_1) \text{ holds} \\[4mm] \left[\dfrac{\lambda + 1}{-\beta - 1} \right]^{1/(\lambda+1)} t(\ln t)^{(\beta+1)/(\lambda+1)} & \text{if } (i_2) \text{ holds.} \end{cases}$$

If $x(t)$ is a proper solution satisfying (i_1), then by Theorem 8.2.1 there exists $a > 0$ such that

$$x(t) = at + O \left(\int_{t_0}^t \int_s^\infty u^{\alpha-\lambda}(\ln u)^\beta du ds \right) \qquad \text{as } t \to \infty.$$

If (i_3) or (i_4) holds, we cannot use Theorem 8.2.5 because $r = -2$. If (i_3) holds, let

$$y(t) = \left(\frac{\lambda + 1}{\beta + 1} \right)^{1/(\lambda+1)} (\ln t)^{(\beta+1)/(\lambda+1)}.$$

Then, $-y''(t)y^\lambda(t) \sim q_1(t) \sim q(t)$ as $t \to \infty$. Hence, by Theorem 8.2.4, if $x(t)$ is a proper solution of equation (8.2.1) satisfying (i$_3$) then

$$x(t) = at + b - \frac{a^{-\lambda}}{\lambda(\lambda+1)}[1 + o(1)]t^{-\lambda}(\ln t)^\beta \qquad (8.2.29)$$

for some constants a and b, $a > 0$.

If (i$_4$) holds, we let $y(t) = (\lambda+1)^{1/(\lambda+1)}(\ln \ln t)^{1/(\lambda+1)}$. Then, $-y''(t)$ $\times y^\lambda(t) \sim q_1(t) \sim q(t)$ as $t \to \infty$. Hence, by Theorem 8.2.4, if $x(t)$ is a proper solution of equation (8.2.1) satisfying (i$_4$), then $x(t) \sim (\lambda + 1)^{1/(\lambda+1)}(\ln \ln t)^{1/(\lambda+1)}$.

8.2.2. Positive Decaying Solutions of Equation (8.2.2)

By a positive decaying solution of equation (8.2.2) we mean a function $x(t) \in C^2([t_0, \infty), \mathbb{R}^+)$, $t_0 \geq 0$ which solves equation (8.2.2) and satisfies the asymptotic condition

$$\lim_{t\to\infty} x(t) = 0 \quad \text{and} \quad \lim_{t\to\infty} x'(t) = 0. \qquad (8.2.30)$$

In what follows we shall assume that

(i)   the support of $q(t)$ is not compact, and

(ii) the improper integral

$$Q(t) = \int_t^\infty q(s)ds \qquad (8.2.31)$$

converges for all $t \geq t_0$ and $Q(t) \geq 0$ on $[t_0, \infty)$.

Existence and Uniqueness of Positive Decaying Solutions

We shall prove two existence theorems by means of a variant of the shooting method; one applies to the case when $q(t) \in C([t_0, \infty), \mathbb{R})$, and the other when $q(t) \in C([t_0, \infty), \mathbb{R}_0)$. For this, we shall need the following basic lemma.

Lemma 8.2.5. Suppose the function $Q(t)$ satisfies

$$\int_{t_0}^\infty Q(s)ds < \infty \qquad (8.2.32)$$

and

$$\int_{t_0}^\infty sQ^2(s)ds < \infty. \qquad (8.2.33)$$

Then for any constant $\ell > 0$ equation (8.2.2) has a positive solution $x(t)$ satisfying

$$\lim_{t\to\infty} x(t) = \ell, \quad \lim_{t\to\infty} x'(t) = 0 \qquad (8.2.34)$$

and $x'(t) < 0$ on $[t_0, \infty)$.

Proof. Clearly, $x(t) \in C^2([t_0, \infty), \mathbb{R}^+)$ is a positive solution of equation (8.2.2) satisfying (8.2.34) if and only if $x(t)$ solves the integral equation

$$x(t) = \ell + \int_t^\infty \left(\int_s^\infty q(u)x^{-\lambda}(u)du \right) ds \quad \text{for} \ t \geq t_0.$$

In view of (ii) this equation is equivalent to the integro–differential equation

$$x(t) = \ell + \int_t^\infty Q(s)x^{-\lambda}(s)ds - \lambda \int_t^\infty \left(\int_s^\infty Q(u)x^{-\lambda-1}(u)x'(u)du \right) ds, \ t \geq t_0.$$
$$(8.2.35)$$

First, we shall solve equation (8.2.35) in some neighborhood of infinity, say, for $t \geq T$, T being sufficiently large. As in Theorem 8.1.1, choose $k, m > 0$ and $T \geq t_0$ so that

$$\ell^{-\lambda} \int_T^\infty Q(s)ds + \lambda\ell^{-\lambda-1}m \int_T^\infty sQ^2(s)ds$$

$$+ \lambda\ell^{-\lambda-1}k \left(\int_T^\infty Q(s)ds \right) \left(\int_T^\infty sQ^2(s)ds \right) \leq \ell,$$

$$\ell^{-\lambda} \leq k \quad \text{and} \quad \lambda\ell^{-\lambda-1} \left(m + k\int_T^\infty Q(s)ds \right) \leq k.$$

With this choice of k, m and T consider the closed convex subset X of the Fréchet space $C^1([T, \infty), \mathbb{R})$ (equipped with the usual topology) consisting of all functions $x(t)$ satisfying $\ell \leq x(t) \leq 2\ell$, and

$$0 \leq -x(t) \leq mQ(t) + k\int_t^\infty Q^2(s)ds \quad \text{for} \ t \geq T.$$

For $t \geq T$ and each $x \in X$, we define the mapping $F : X \to X$ by

$$(Fx)(t) = \ell + \int_t^\infty Q(s)x^{-\lambda}(s)ds - \lambda \int_t^\infty \left(\int_s^\infty Q(u)x^{-\lambda-1}(u)x'(u)du \right) ds.$$

Then it is easy to verify that F maps X continuously into a compact subset of X. Therefore, the Schauder–Tychonov theorem implies that F has a fixed point x in X, i.e., $x(t) = (Fx)(t)$ for $t \geq T$. Clearly, this $x(t)$ gives a positive solution of the equation (8.2.35) on $[T, \infty)$.

We claim that $x'(t) < 0$ for $t \geq T$. To this end suppose that $x'(\tau) = 0$ for some $\tau \in [T, \infty)$. Since

$$x'(t) = -Q(t)x^{-\lambda}(t) + \lambda \int_t^\infty Q(s)x^{-\lambda-1}(s)x'(s)ds \quad \text{for} \quad t \geq T \quad (8.2.36)$$

from the nonpositivity of each term on the right–hand side of (8.2.36), we find that $\int_\tau^\infty Q(s)x^{-\lambda-1}(s)x'(s)ds = \infty$, which implies that $Q(t)x'(t) \equiv 0$ on $[\tau, \infty)$. Multiplying (8.2.36) by $Q(t)$, we get $Q(t) \equiv 0$ on $[\tau, \infty)$, i.e., $q(t) \equiv 0$ on $[\tau, \infty)$. This contradicts the assumption (i). Thus, we have $x'(t) < 0$ for $t \geq T$.

Now, we can continue $x(t)$ to the left as a solution of equation (8.2.2). Let $J \subset [t_0, \infty)$ be the maximal interval of existence for $x(t)$. Clearly, $x(t)$ satisfies (8.2.35) and (8.2.36) for $t \in J$. We claim that $x'(t) < 0$ throughout J. In fact, if this is not true, we can find $\tau \in J$, $\tau < T$ such that $x'(\tau) = 0$, $x'(t) < 0$ in (τ, ∞). Then by putting $t = \tau$ in (8.2.35) we arrive at a contradiction

$$0 = -Q(\tau)x^{-\lambda}(\tau) + \lambda \int_\tau^\infty Q(s)x^{-\lambda-1}(s)x'(s)ds < 0.$$

Thus, $x'(t) < 0$ for $t \in J$.

Finally, we shall show that $J = [t_0, \infty)$. Suppose to the contrary that $J \neq [t_0, \infty)$. Since $x(t) > 0$ and $x'(t) < 0$ on J, there exists a $t_1 \in [t_0, \infty)$ such that

$$x(t) \to \infty \quad \text{as} \quad t \to t_1^+. \quad (8.2.37)$$

Then, $x^{-\lambda}(t)$ is bounded on (t_1, T), and this fact combined with the equation

$$x(t) = x(T) + x'(T)(t - T) + \int_T^t \left(\int_T^s q(u)x^{-\lambda}(u)du \right) ds, \quad t \in (t_1, T]$$

which follows from (8.2.2) shows that $x(t)$ tends to a finite limit as $t \to t_1^+$. This contradicts (8.2.37), and so, we must have $J = [t_0, \infty)$, i.e., $x(t)$ exists on the whole interval $[t_0, \infty)$. This completes the proof. ∎

Theorem 8.2.6. Suppose condition (8.2.32) holds. If

$$\int_{t_0}^\infty s|q(s)| \left(\int_s^\infty Q(u)du \right)^{-\lambda/(\lambda+1)} ds < \infty, \quad (8.2.38)$$

then equation (8.2.2) has a positive decaying solution.

Proof. We note that (8.2.38) implies $\int_{t_0}^\infty s|q(s)|ds < \infty$ so that condition (8.2.33) is satisfied. For each $n \in \mathbb{N}$, let $x_n(t)$ be a positive solution of equation (8.2.2) satisfying $x_n'(t) < 0$ for $t \geq t_0$, and $x_n(t) \to 1/n$, $x_n'(t) \to 0$ as $t \to \infty$. The existence of such x_n (for each n) is guaranteed by Lemma 8.2.5. Also, $x_n(t)$ for $t \geq t_0$ satisfies

$$
\begin{aligned}
x_n(t) &= \frac{1}{n} + \int_t^\infty \left(\int_s^\infty q(u) x_n^{-\lambda}(u) du \right) ds \\
&= \frac{1}{n} + \int_t^\infty Q(s) x_n^{-\lambda}(s) ds - \lambda \int_t^\infty \left(\int_s^\infty Q(u) x_n^{-\lambda}(u) x'(u) du \right) ds.
\end{aligned}
$$
(8.2.39)

Since $x_n'(t) < 0$ for $t \geq t_0$, from (8.2.39), we obtain

$$
x_n(t) \geq x_n^{-\lambda}(t) \int_t^\infty Q(s) ds, \quad t \geq t_0
$$

which implies

$$
x_n(t) \geq \left(\int_t^\infty Q(s) ds \right)^{1/(\lambda+1)}, \quad t \geq t_0.
$$
(8.2.40)

From (8.2.40) and the first equality of (8.2.39) it follows that

$$
\begin{aligned}
x_n(t) &\leq \frac{1}{n} + \int_t^\infty (s-t)|q(s)| \left(\int_s^\infty Q(u) du \right)^{-\lambda/(\lambda+1)} ds \\
&\leq 1 + \int_t^\infty s|q(s)| \left(\int_s^\infty Q(u) du \right)^{-\lambda/(\lambda+1)} ds, \quad t \geq t_0.
\end{aligned}
$$
(8.2.41)

Similarly, $x_n'(t)$ can be estimated as follows

$$
|x_n'(t)| \leq \int_t^\infty |q(s)| \left(\int_s^\infty Q(u) du \right)^{-\lambda/(\lambda+1)} ds, \quad t \geq t_0.
$$
(8.2.42)

Inequalities (8.2.41) and (8.2.42) assert that the sequence $\{x_n(t)\}$ is uniformly bounded and euicontinuous on each compact subset of $[t_0, \infty)$. Therefore, Ascoli–Arzela's theorem implies that $\{x_n\}$ has a subsequence $\{x_{n_i}\}$ converging to some $\bar{x} \in C([t_0, \infty), \mathbb{R})$ uniformly on each compact subset of $[t_0, \infty)$. From (8.2.40) it follows that $\bar{x}(t) > 0$ for $t \geq t_0$. Letting $n_i \to \infty$ in the equation

$$
x_{n_i}(t) = \frac{1}{n_i} + \int_t^\infty \left(\int_s^\infty q(u) x_{n_i}^{-\lambda}(u) du \right) ds, \quad t \geq t_0
$$

we find by Lebesgue dominated convergence theorem that $\bar{x}(t)$ satisfies

$$
\bar{x}(t) = \int_t^\infty \left(\int_s^\infty q(u)(\bar{x}(u))^{-\lambda} du \right) ds, \quad t \geq t_0.
$$

Hence, $\bar{x}(t)$ is a positive decaying solution of equation (8.2.2). This completes the proof. ∎

Example 8.2.1. Consider equation (8.2.2) with

$$q(t) = -\left(\frac{1+\sin t}{t^{2+\epsilon}}\right)', \quad t \geq 1, \quad \epsilon > \lambda.$$

Since $q(t) = O\left(t^{-2-\epsilon}\right)$ as $t \to \infty$, and

$$\int_t^\infty Q(s)ds = \frac{t^{-1-\epsilon}}{1+\epsilon} + O\left(t^{-2-\epsilon}\right) \geq Ct^{-1-\epsilon} \quad \text{as} \quad t \to \infty,$$

for some constant $C > 0$, all conditions of Theorem 8.2.6 are satisfied and hence there exists a positive decaying solution of this equation.

Theorem 8.2.7. Let $q(t) \geq 0$ for $t \geq t_0$ and $\lambda \in (0,1)$. Suppose there exists a nonincreasing function $p(t) \in C([t_0, \infty), \mathbb{R}^+)$ such that

$$q(t) \leq p(t) \quad \text{for} \quad t \geq t_0 \tag{8.2.43}$$

and

$$\int_{t_0}^\infty p^{1/2}(s)ds < \infty. \tag{8.2.44}$$

Then equation (8.2.2) has a positive decaying solution.

Proof. We notice that (8.2.44) implies $\int_{t_0}^\infty sp(s)ds < \infty$, which in turn implies conditions (8.2.32) and (8.2.33). Let $\{x_n(t)\} \subset C([t_0, \infty), \mathbb{R}^+)$ denote the same sequence as in the proof of Theorem 8.2.6. Now, we shall show that the following estimates hold

$$x_n^{(\lambda+1)/2}(t) \leq \left(\frac{1}{n}\right)^{(\lambda+1)/2} + \left(\frac{1+\lambda}{2}\right)\left(\frac{2}{1-\lambda}\right)^{1/2} \int_t^\infty p^{1/2}(s)ds, \quad t \geq t_0 \tag{8.2.45}$$

and

$$(x_n'(t))^2 \leq \left(\frac{2}{1-\lambda}\right)p(t)x_n^{1-\lambda}(t), \quad t \geq t_0. \tag{8.2.46}$$

Integrating equation (8.2.2) (with $x = x_n$), we find

$$\left(\frac{1}{2}[x_n'(t)]^2\right)' = q(t)x_n^{-\lambda}(t)x_n'(t) \quad \text{for} \quad t \geq t_0$$

and noting that $x_n'(t) < 0$ on $[t_0, \infty)$ and $\lim_{t\to\infty} x_n'(t) = 0$, we obtain for $t \geq t_0$,

$$\frac{1}{2}[x'(t)]^2 = \int_t^\infty q(s)x_n^{-\lambda}(s)(-x_n'(s))ds \leq p(t)\int_t^\infty x_n^{-\lambda}(s)(-x_n'(s))ds$$

$$= -\left(\frac{p(t)}{1-\lambda}\right)\left[\left(\frac{1}{n}\right)^{1-\lambda} - x_n^{1-\lambda}(t)\right].$$

This establishes (8.2.46). Since (8.2.46) can be rewritten as

$$- \left(x_n^{(1+\lambda)/2}(t) \right)' \leq \left(\frac{1+\lambda}{2} \right) \left(\frac{2}{1-\lambda} \right)^{1/2} p^{1/2}(t) \quad \text{for} \quad t \geq t_0$$

an integration of this yields (8.2.45).

By (8.2.45), $\{x_n\}$ is uniformly bounded on each compact subset of $[t_0, \infty)$ and so is $\{x_n'\}$ by (8.2.46). Moreover, equation (8.2.2) with $x = x_n$ together with (8.2.40) shows the uniform boundedness and the equicontinuity of $\{x_n''\}$ on each compact subset of $[t_0, \infty)$. Hence, it follows by Ascoli–Arzela's theorem that there exists a subsequence $\{x_{n_i}\}$ of $\{x_n\}$ which converges to a function $\bar{x} \in C^2([t_0, \infty), \mathbb{R})$ in the C^2-topology on each compact subset of $[t_0, \infty)$. It is clear that $\bar{x}(t) > 0$ on $[t_0, \infty)$. Letting $n_i \to \infty$ in the equation $x_{n_i}''(t) = q(t) x_{n_i}^{-\lambda}(t)$, $t \geq t_0$ and the inequality

$$x_{n_i}^{(\lambda+1)/2}(t) \leq \left(\frac{1}{n_i} \right)^{(1+\lambda)/2} + \left(\frac{1+\lambda}{2} \right) \left(\frac{2}{1-\lambda} \right)^{1/2} \int_t^\infty p^{1/2}(s) ds, \quad t \geq t_0$$

we see that $\bar{x}(t)$ is the desired positive decaying solution of equation (8.2.2). This completes the proof. ∎

Remark 8.2.2.

1. When $\int_{t_0}^\infty Q(s) ds = \infty$, equation (8.2.2) has no positive decaying solution regardless of the sign of the function $q(t)$. To show this, let $x(t)$ be a positive decaying solution of (8.2.2). Multiplying (8.2.2) by $x^\lambda(t)$ and integrating from t to u and letting $u \to \infty$, we get

$$-x'(t) x^\lambda(t) - \lambda \int_t^\infty x^{\lambda-1}(s)(x'(s))^2 ds = Q(t), \quad t \geq t_0$$

which implies $- \left(x^{\lambda+1}(t)/(\lambda+1) \right)' \geq Q(t)$ for $t \geq t_0$. Integrating this inequality from $t \geq t_0$ to $\tau \geq t$, we obtain

$$\frac{1}{\lambda+1} \left[x^{\lambda+1}(t) - x^{\lambda+1}(\tau) \right] \geq \int_t^\tau Q(s) ds.$$

Letting $\tau \to \infty$, we find that

$$\frac{1}{\lambda+1} x^{\lambda+1}(t) \geq \int_t^\infty Q(s) ds, \quad t \geq t_0. \tag{8.2.47}$$

Therefore, the integrability condition on $Q(t)$ is natural for the equation (8.2.2) to have a positive decaying solution.

2. The above results can be extended easily to more general equations of the form

$$(a(t) x'(t))' + \delta q(t) x^{-\lambda}(t) = 0,$$

where $\delta = \pm 1$, $a(t) \in C([t_0, \infty), \mathbb{R}^+)$ and $\int^\infty ds/a(s) = \infty$, by changing the independent variable $s = \int_{t_0}^t du/a(u)$. The details are left to the reader (this transformation has been already used in the previous chapter).

Next, we present the following uniqueness theorem for positive decaying solutions of equation (8.2.2).

Theorem 8.2.8. Let $q(t) \geq 0$ for $t \geq t_0$. Suppose condition (8.2.32) holds and that the equation

$$w'(t) + \frac{\lambda}{\lambda+1} q(t) \left(\int_t^\infty Q(s)ds \right)^{-1} w(t) = 0, \quad t \geq t_0 \qquad (8.2.48)$$

is nonoscillatory. Then the positive decaying solution of (8.2.2) is unique.

Proof. Let $x_1(t)$ and $x_2(t)$ be positive decaying solutions of equation (8.2.2). Suppose $x_1(t) \geq x_2(t)$ in some neighborhood of infinity, say, for $t \geq t_1 \geq t_0$. Since $x_i(t) \geq \int_t^\infty \left(\int_s^\infty q(u)x_i^{-\lambda}(u)du \right) ds$, $t \geq t_0$, $i = 1, 2$ it follows that $x_1(t) \leq x_2(t)$ for $t \geq t_1$ and consequently that $x_1(t) \equiv x_2(t)$ for $t \geq t_1$. Since the uniqueness of solutions of the initial value problem for equation (8.2.2) ensures that $x_1(t) \equiv x_2(t)$ on $[t_0, t_1]$, we conclude that $x_1(t) \equiv x_2(t)$ on $[t_0, \infty)$.

From the above observation it suffices to show that the difference $w(t) = x_2(t) - x_1(t)$ cannot change sign infinitely often on $[t_0, \infty)$. Suppose the contrary. Then, $w(t)$ is an oscillatory solution of the linear differential equation

$$w'(t) + c(t)w(t) = 0, \quad t \geq t_0 \qquad (8.2.49)$$

where

$$c(t) = \begin{cases} q(t)\dfrac{x_2^{-\lambda}(t) - x_1^{-\lambda}(t)}{x_1(t) - x_2(t)} & \text{if } x_1(t) \neq x_2(t) \\ \lambda q(t) x_1^{-\lambda-1}(t) & \text{if } x_1(t) = x_2(t). \end{cases}$$

Since each $x = x_i$ satisfies (8.2.47), it follows from the mean value theorem that

$$c(t) \leq \frac{\lambda}{\lambda+1} q(t) \left(\int_t^\infty Q(s)ds \right)^{-1} \quad \text{for } t \geq t_0.$$

In view of the nonoscillatory character of (8.2.48), Sturm's comparison theorem implies that equation (8.2.49) must be nonoscillatory, which is a contradiction. This completes the proof. ∎

The following corollary is an immediate consequence of Theorem 8.2.8 and the well-known Hille's theorem.

Corollary 8.2.1. Let $q(t) \geq 0$ for $t \geq t_0$ and condition (8.2.32) hold. If

$$\frac{\lambda}{\lambda+1} t^2 q(t) \left(\int_t^\infty Q(s)ds \right)^{-1} \leq \frac{1}{4} \quad \text{for all large} \quad t,$$

then the positive decaying solution of equation (8.2.2) is unique.

Example 8.2.2. Consider (8.2.2) under the condition that $c_1 t^{-2-\alpha} \leq q(t) \leq c_2 t^{-2-\alpha}$, $t \geq 1$ for some constants $c_i > 0$, $i = 1, 2$ and $\alpha > 0$. By Theorem 8.2.6 and Corollary 8.2.1, if

$$\left(\frac{c_2}{c_1} \right) \left(\frac{\lambda}{\lambda+1} \right) \alpha(\alpha+1) \leq \frac{1}{4},$$

then equation (8.2.2) has exactly one positive decaying solution. In fact, the singular Emden–Fowler equation $x''(t) = t^{-2-\alpha} x^{-\lambda}(t)$, $t \geq 1$, $\alpha > 0$ has exactly one positive decaying solution

$$x(t) = \left[\frac{(\lambda+1)^2}{\alpha(\alpha+\lambda+1)} \right]^{1/(\lambda+1)} t^{-\alpha/(\lambda+1)} \quad \text{for} \quad t \geq 1$$

provided $\lambda\alpha(\alpha+1)/(\lambda+1) \leq 1/4$.

Asymptotic Behavior of Positive Decaying Solutions

Consider (8.2.2) and the equation

$$y''(t) = p(t)y^{-\lambda}(t), \tag{8.2.50}$$

where $p(t)$, $q(t) \in C^1([t_0, \infty), \mathbb{R}^+)$. In the following theorem, we assert that if $p(t)$ and $q(t)$ have the same asymptotic behavior in some sense, then so do the positive decaying solutions of these equations as $t \to \infty$. For this, we notice that when equation (8.2.50) admits a positive decaying solution, then the integral $P(t) = \int_t^\infty (s-t)p(s)ds$ converges for $t \geq t_0$. To prove our result we shall need the following elementary lemma.

Lemma 8.2.6. If $x(t) \in C^1([t_0, \infty), \mathbb{R})$ satisfies $\int^\infty x^2(s)ds < \infty$ and $|x'(t)| \leq c$, $t \geq t_0$ for some constant $c > 0$, then $\lim_{t \to \infty} x(t) = 0$.

Theorem 8.2.9. Let $x(t)$ and $y(t)$ be positive decaying solutions of equations (8.2.2) and (8.2.50), respectively. Suppose that

$$\lim_{t \to \infty} \frac{q(t)}{p(t)} = 1, \tag{8.2.51}$$

$$\limsup_{t \to \infty} P^{-1/(\lambda+1)}(t) \int_t^\infty (s-t)p(s)P^{-\lambda/(\lambda+1)}ds < \infty, \tag{8.2.52}$$

$$0 < \liminf_{t\to\infty} Y(t) \le \limsup_{t\to\infty} Y(t) < \infty, \tag{8.2.53}$$

where $Y(t) = \left[(p(t)y^{3-\lambda}(t))^{-1/2}\right]'$, and either

$$\int^\infty p^{1/2}(s)y^{-(1+\lambda)/2}(s)\left|\frac{q(s)}{p(s)} - 1\right| ds < \infty, \tag{8.2.54}$$

or

$$\int^\infty \left|\left(\frac{q(s)}{p(s)}\right)'\right| ds < \infty \tag{8.2.55}$$

is satisfied. Then, $x(t) \sim y(t)$ as $t \to \infty$.

Proof. The hypotheses of the theorem imply that for $t \ge t_0$,

$$c_1 P^{1/(\lambda+1)}(t) \le x(t), \quad y(t) \le c_2 \int_t^\infty (s-t)p(s)P^{-\lambda/(\lambda+1)}(s)ds \tag{8.2.56}$$

for some constants $c_i = c_i(\lambda) > 0$, $i = 1,2$. In fact, the first inequality in (8.2.56) is obtained from (8.2.47), while the second inequality in (8.2.56) follows from $y(t) = \int_t^\infty \left(\int_s^\infty p(u)y^{-\lambda}(u)du\right) ds$, $t \ge t_0$. Define the new function v by $v(t) = x(t)/y(t)$ for $t \ge t_0$. Then, $v(t)$ satisfies the equation

$$v''(t) + 2\left(\frac{y'(t)}{y(t)}\right)v'(t) + p(t)y^{-\lambda}(t)v(t) = q(t)y^{-\lambda-1}(t)v^{-\lambda}(t) \tag{8.2.57}$$

for $t \ge t_0$. Introducing the new independent variable $\tau = \int_{t_0}^t y^{-2}(u)du$, equation (8.2.57) transforms into

$$\frac{d^2v}{d\tau^2} + p(t)y^{3-\lambda}(t)v = q(t)y^{3-\lambda}(t)v^{-\lambda}, \quad \tau \ge 0. \tag{8.2.58}$$

Moreover, since $\int_0^\infty (p[t(\tau)]y^{3-\lambda}[t(\tau)])^{1/2} d\tau = \infty$ by (8.2.53), the change of variable $s = \int_0^\tau (p[t(\xi)]y^{3-\lambda}[t(\xi)])^{1/2} d\xi$ transforms equation (8.2.58) into

$$\ddot{v} - f(s)\dot{v} + v = g(s)v^{-\lambda}, \quad s \ge 0, \quad \left(\cdot = \frac{d}{ds}\right) \tag{8.2.59}$$

where

$$f(s) = \left[(p(s)y^{3-\lambda}(s))^{-1/2}\right]' y^2(s) \quad \text{and} \quad g(s) = q(s)/p(s) \quad \text{for} \quad s \ge 0.$$

Notice that (8.2.51), (8.2.53) – (8.2.55) are equivalent to the conditions

$$\lim_{s\to\infty} g(s) = 1, \tag{8.2.60}$$

$$0 < \liminf_{s\to\infty} f(s) \leq \limsup_{s\to\infty} f(s) < \infty, \qquad (8.2.61)$$

$$\int^\infty |g(s) - 1| ds < \infty \qquad (8.2.62)$$

and

$$\int^\infty |\dot{g}(s)| ds < \infty, \qquad (8.2.63)$$

respectively. It follows from (8.2.52) and (8.2.56) that

$$m \leq v(s) \leq M, \quad s \geq 0 \qquad (8.2.64)$$

for some constants m, $M > 0$. Condition (8.2.61) implies that

$$k_1 \leq f(s) \leq k_2, \quad s \geq s_0 \qquad (8.2.65)$$

for s_0 sufficiently large and some constants k_1, $k_2 > 0$. For simplicity, we suppose that $s_0 = 0$ and $\lambda \neq 1$. The proof will be complete if we can show that $\lim_{s\to\infty} v(s) = 1$. From (8.2.60) and (8.2.64) there is a constant $K > 0$ such that $|g(s)v^{-\lambda}(s) - v(s)| \leq K$ for $s \geq s_0$. Let L be a constant satisfying $L > K/k_1$. We claim that $\dot{v}(s)$ is bounded in $\mathbb{R}_0$, i.e.,

$$|\dot{v}(s)| \leq L \quad \text{for all large} \quad s. \qquad (8.2.66)$$

For this, suppose that (8.2.66) fails to hold. Since clearly (8.2.64) implies $\liminf_{s\to\infty} |\dot{v}(s)| = 0$, we can find a sequence of intervals $\{[a_n, b_n]\}$, $n \in \mathbb{N}$ such that $b_n < a_{n+1}$, $\lim_{n\to\infty} a_n = \infty$, $|\dot{v}(a_n)| = |\dot{v}(b_n)| = L$ $|\dot{v}(s)| > L$, $a_n < s < b_n$. Then Rolle's theorem shows that there exists $c_n \in (a_n, b_n)$ such that $\ddot{v}(c_n) = 0$, $|\dot{v}(c_n)| > L$. However, by setting $s = c_n$ in (8.2.59) we arrive at a contradiction for all large n, namely, $k_1 L \leq f(c_n)|\dot{v}(c_n)| = |g(c_n)v^{-\lambda}(c_n)| \leq K$. Hence, (8.2.66) holds.

Note that by the hypotheses of the theorem and boundedness of $\dot{v}(s)$ it follows that $\ddot{v}(s)$ is also bounded on $\mathbb{R}_0$.

First suppose that condition (8.2.54) holds. By rewriting equation (8.2.59) as

$$\left(\frac{1}{2}\dot{v}^2\right)^{\cdot} - f(s)\dot{v}^2 + \left(\frac{1}{2}v^2\right)^{\cdot} = g(s)\left(\frac{1}{1-\lambda}v^{1-\lambda}\right)^{\cdot}, \quad s \geq 0$$

and integrating over $[0, s]$, we find

$$\frac{1}{2}\dot{v}^2(s) - \int_0^s f(u)\dot{v}^2(u)du + \left(\frac{1}{2}v^2(s) - \frac{1}{(1-\lambda)}v^{1-\lambda}(s)\right)$$
$$= \int_0^s [g(u) - 1]v^{-\lambda}(u)\dot{v}(u)du + c_3, \quad s \geq 0 \qquad (8.2.67)$$

where c_3 is a constant. Since (8.2.62), (8.2.64) and (8.2.66) give

$$\int_0^s \left|[g(u) - 1]v^{-\lambda}(u)\dot{v}(u)\right| du \leq c_4 \int_0^\infty |g(s) - 1|ds < \infty,$$

for some constant $c_4 > 0$, (8.2.66) together with (8.2.64) implies that $\int^\infty f(s)\dot{v}^2(s)ds < \infty$, and hence by (8.2.65), we obtain

$$\int^\infty \dot{v}^2(s)ds < \infty. \tag{8.2.68}$$

From the boundedness of $\ddot{v}(s)$, (8.2.68) and Lemma 8.2.6, we find that $\lim_{t\to\infty} \dot{v}(s) = 0$. Accordingly, it follows from (8.2.67) that the limit

$$\lim_{s\to\infty} \left(\frac{1}{2}v^2(s) - \frac{1}{(1-\lambda)}v^{1-\lambda}(s)\right)$$

must exist as a finite value, i.e., $\lim_{s\to\infty} v(s) = \ell$ (finite). Letting $s \to \infty$ in equation (8.2.59), we get $\lim_{s\to\infty} \ddot{v}(s) = \ell^{-\lambda} - \ell$. If $\ell^{-\lambda} - \ell \neq 0$, then the boundedness of $\dot{v}(s)$ is violated. Hence, $\ell^{-\lambda} = \ell$, i.e., $\ell = 1$ as desired.

Next, let condition (8.2.55) hold. We notice that equation (8.2.67) is equivalent to

$$\frac{1}{2}\dot{v}^2(s) - \int_0^s f(u)\dot{v}^2(u)du + \left(\frac{1}{2}v^2(s) - \frac{1}{(1-\lambda)}g(s)v^{1-\lambda}(s)\right)$$
$$= -\frac{1}{(1-\lambda)}\int_0^s \dot{g}(u)v^{1-\lambda}(u)du + c_5 \tag{8.2.69}$$

for some real constant c_5. Condition (8.2.63) ensures the convergence of the right–hand side of (8.2.59). Hence, we have (8.2.68) and $\lim_{s\to\infty} \dot{v}(s) = 0$, which in turn imply that the limit

$$\lim_{s\to\infty} \left(\frac{1}{2}v^2(s) - \frac{1}{(1-\lambda)}g(s)v^{1-\lambda}(s)\right)$$

exists in $\mathbb{R}$. The rest of the proof is similar to that of presented above and hence omitted. This completes the proof. ∎

Remark 8.2.3. As a simple consequence of the proof of Theorem 8.2.7, we observe that if $0 < \lambda < 1$ and there exists a non–increasing function $p^*(t) \in C([t_0,\infty), \mathbb{R}^+)$ such that $p(t) \leq p^*(t)$ for $t \geq t_0$, then (8.2.52) can be replaced by the condition

$$\limsup_{t\to\infty} P^{-1/2}(t) \int_t^\infty (p^*(s))^{1/2}ds < \infty.$$

Consider the case when the function $p(t)$ behaves like ct^α where c and α are constants, $c > 0$ and $\alpha < -2$, i.e., assume that

$$0 < \liminf_{t\to\infty} \frac{p(t)}{t^\alpha} \leq \limsup_{t\to\infty} \frac{p(t)}{t^\alpha} < \infty \tag{8.2.70}$$

and

$$0 < \liminf_{t\to\infty} \frac{-p'(t)}{t^{\alpha-1}} \leq \limsup_{t\to\infty} \frac{-p'(t)}{t^{\alpha-1}} < \infty. \tag{8.2.71}$$

Then for any positive decaying solution $y(t)$ of equation (8.2.50), (8.2.56) shows that $C_1 t^{(\alpha+2)/(\lambda+1)} \leq y(t) \leq C_2 t^{(\alpha+2)/(\lambda+1)}$, $t \geq t_0$ for some positive constants C_1 and C_2. Moreover, the equation $-y'(t) = \int_t^\infty p(s) y^{-\lambda}(s) ds$, $t \geq t_0$ gives the estimates $C_3 t^{(\alpha+2)/(\lambda+1)-1} \leq -y'(t) \leq C_4 t^{(\alpha+2)/(\lambda+1)-1}$, $t \geq t_0$.

Now, the following corollary is immediate.

Corollary 8.2.2. Let $\lambda \leq 3$, $\alpha < -2$ and let $x(t)$ and $y(t)$ be positive decaying solutions of equations (8.2.2) and (8.2.50), respectively. Suppose that conditions (8.2.51), (8.2.70), (8.2.71) and either

$$\int^\infty \frac{1}{s} \left| \frac{q(s)}{p(s)} - 1 \right| ds < \infty, \quad \text{or} \quad \int^\infty \left| \left(\frac{q(s)}{p(s)} \right)' \right| ds < \infty$$

are satisfied. Then, $x(t) \sim y(t)$ as $t \to \infty$.

Next, we present the following result.

Corollary 8.2.3. Let $\alpha < -2$ and let $q(t) = ct^\alpha[1 + \epsilon(t)]$, where c is a positive constant and $\epsilon(t) \in C([t_0, \infty), \mathbb{R})$ satisfies $\lim_{t\to\infty} \epsilon(t) = 0$, and either

$$\int^\infty \frac{|\epsilon(s)|}{s} ds < \infty, \tag{8.2.72}$$

or

$$\int^\infty |\epsilon'(s)| ds < \infty \tag{8.2.73}$$

holds. Then any positive decaying solution $x(t)$ of equation (8.2.2) satisfies

$$x(t) \sim \left[\frac{c(\lambda+1)^2}{(\alpha+2)(\alpha-\lambda+1)} \right]^{1/(1+\lambda)} t^{(\alpha+2)/(\lambda+1)} \quad \text{as} \quad t \to \infty.$$

Proof.   It suffices to note that the singular Emden–Fowler equation $y''(t) = ct^\alpha y^{-\lambda}(t)$, $t \geq t_0$ with $c > 0$ and $\alpha < -2$ has a positive decaying solution $y(t)$ given by

$$y(t) = \left[\frac{c(\lambda+1)^2}{(\alpha+2)(\alpha-\lambda+1)} \right]^{1/(\lambda+1)} t^{(\alpha+2)/(\lambda+1)} \quad \text{for} \quad t \geq t_0$$

for which (8.2.53) (with $p(t) = ct^\alpha$) is satisfied.   ∎

Next, we shall employ the transformation given in the proof of Theorem 8.2.9 to establish the following uniqueness theorem of positive decaying solutions of equation (8.2.2).

Theorem 8.2.10. Let $\lambda \le 3$ and $\alpha < -2$. Suppose the function $q(t)$ satisfies

$$0 < \liminf_{t\to\infty} \frac{q(t)}{t^\alpha} \le \limsup_{t\to\infty} \frac{q(t)}{t^\alpha} < \infty$$

and

$$0 < \liminf_{t\to\infty} \frac{-q'(t)}{t^{\alpha-1}} \le \limsup_{t\to\infty} \frac{-q'(t)}{t^{\alpha-1}} < \infty.$$

Then equation (8.2.2) has a unique positive decaying solution.

Proof. Suppose to the contrary that we have two distinct positive decaying solutions $x(t)$ and $y(t)$ of equation (8.2.2). If the difference $x(t) - y(t)$ is of one sign for all sufficiently large t, then the simple observation we had in Theorem 8.2.8 implies that $x(t) \equiv y(t)$. Therefore, we need to treat the case when $x(t) - y(t)$ changes sign infinitely many times. As in the proof of Theorem 8.2.9, we define the new functions $v(t) = x(t)/y(t)$ and $\tau(t) = \int_{t_0}^t y^{-2}(u)du$, $t \ge t_0$. Then calculations give

$$\frac{d^2v}{d\tau^2} + q(t)y^{3-\lambda}(t)[v - v^{-\lambda}] = 0, \quad \tau \ge 0. \tag{8.2.74}$$

Moreover, the change of variable $s = \int_0^\tau \left(q[t(\xi)]y^{3-\lambda}[t(\xi)] \right)^{1/2} d\xi$ transforms equation (8.2.74) into

$$\ddot{v} - f(s)\dot{v} + v = v^{-\lambda}, \quad s \ge 0, \quad \left(\cdot = \frac{d}{ds} \right) \tag{8.2.75}$$

where $f(s) = \left((p(s)y^{3-\lambda}(s))^{-1/2} \right)' y^2(s)$, $s \ge 0$. Now, in view of the proofs of Theorem 8.2.9 and Corollary 8.2.1, we have $\lim_{s\to\infty} \dot{v}(s) = 0$. But, this leads to a contradiction. To see this, let s_n, $n \in \mathbb{N}$ be the point at which $v(s)$ cuts the horizontal line $v = 1$ and $s_n \uparrow \infty$ as $n \to \infty$. The existence of such points is guaranteed by the fact that $v(s)$ oscillates around $v = 1$. Then an integration of equation (8.2.75) multiplied by $\dot{v}(s)$ over $[s_n, s_{n+1}]$ yields

$$(\dot{v}(s_{n+1}))^2 - (\dot{v}(s_n))^2 = 2 \int_{s_n}^{s_{n+1}} f(s)(\dot{v}(s))^2 ds,$$

which contradicts the fact that $\lim_{t\to\infty} \dot{v}(s) = 0$ unless $v(s) \equiv 1$. This completes the proof.   ∎

In the following uniqueness theorem, we shall employ Corollary 8.2.3.

Theorem 8.2.11. Let $\alpha < -2$. Suppose the function $q(t)$ is of the form $q(t) = ct^\alpha[1 + \epsilon(t)]$, where c is a positive constant, $\epsilon(t) \in C([t_0, \infty), \mathbb{R})$, $\lim_{t\to\infty} t\epsilon'(t) = 0$ and either condition (8.2.72) or (8.2.73) holds. Then equation (8.2.2) has a unique positive decaying solution.

Proof. Let $x(t)$ and $y(t)$ be two positive decaying solutions of equation (8.2.2). We may suppose that $x(t) - y(t)$ changes sign infinitely many times as before. We notice from Corollary 8.2.3 and l'Hospital's rule that $x(t)$ and $y(t)$ satisfy

$$x(t),\ y(t) \sim Kt^\sigma \quad \text{and} \quad x'(t),\ y'(t) \sim \sigma Kt^{\sigma-1} \quad \text{as} \quad t \to \infty, \quad (8.2.76)$$

where

$$K = \left[\frac{c(\lambda+1)^2}{(\alpha+2)(\alpha-\lambda+1)}\right]^{1/(\lambda+1)} \quad \text{and} \quad \sigma = \frac{\alpha+2}{\lambda+1}.$$

Then as in the proof of Theorem 8.2.10 we obtain equation (8.2.75). By taking into account (8.2.76) it is easily seen that

$$\lim_{s\to\infty} f(s) = c^{-1/2} K^{(\lambda+1)/2}[\lambda - 2\alpha + 3](\lambda+1)^{-1} > 0$$

and now the proof can be completed as that of Theorem 8.2.10. ∎

Asymptotic Behavior of Positive Increasing Solutions

This subsection address the asymptotic behavior of positive increasing solutions of equation (8.2.2) when $q(t) \in C([t_0, \infty), \mathbb{R}^+)$. If $x(t)$ is a positive solution of equation (8.2.2), then $x''(t) > 0$ eventually, and hence $x'(t)$ is eventually of one sign. Therefore, we have either (i) $x'(t) < 0$ eventually, or (ii) $x'(t) > 0$ eventually. The case (i) when $\lim_{t\to\infty} x(t) = 0$, i.e., decaying solutions of equation (8.2.2) is already studied in the previous subsection. The other case, i.e., case (ii) is our objective here. A positive increasing solution of equation (8.2.2) is defined to be $x(t) \in C^2([t_0, \infty), \mathbb{R}^+)$ having the asymptotic behavior

$$\lim_{t\to\infty} x'(t) = \lim_{t\to\infty} \frac{x(t)}{t} = c, \quad c \in \mathbb{R}^* = \mathbb{R}^+ \cup \{\infty\}. \quad (8.2.77)$$

Remark 8.2.4. Suppose $q(t)$ satisfies the condition

$$\int^\infty s^{-\lambda} q(s)ds < \infty. \quad (8.2.78)$$

Then any positive increasing solution of (8.2.2) satisfies $\lim_{t\to\infty} x'(t) = \lim_{t\to\infty} x(t)/t = c_1 < \infty$, where c_1 is a constant. For this, since $x'(t) >$

$0, t \geq t_0$ it follows that $x(t) \geq kt, t \geq t_0$ for some constant $k > 0$. Integrating (8.2.2) from t_0 to t and using $x(t) \geq kt, t \geq t_0$ we obtain

$$x'(t) = x'(t_0) + \int_{t_0}^{t} q(s)x^{-\lambda}(s)ds \leq x'(t_0) + k^{-\lambda}\int_{t_0}^{t} s^{-\lambda}q(s)ds < \infty$$

for $t \geq t_0$. Since $x'(t)$ is an increasing function for $t \geq t_0$ it must have a finite limit as $t \to \infty$, and this completes the proof of the assertion.

From this observation, we restrict our attention to the case when condition (8.2.78) fails to hold. We shall use the following notation: For $t \geq t_0$, $Q_1(t) = \int_{t_0}^{t} s^{-\lambda}q(s)ds$, $Q_2(t) = \int_{t_0}^{t} Q_1^{1/(\lambda+1)}(s)ds$ and $Q_3(t) = \int_{t}^{\infty} Q_2^{-2}(s)ds$. Now, we shall prove the following lemma.

Lemma 8.2.7. Suppose

$$\int^{\infty} s^{-\lambda}q(s)ds = \infty. \tag{8.2.79}$$

Then any positive increasing solution $x(t)$ of equation (8.2.2) satisfies

$$\liminf_{t\to\infty} \frac{x(t)}{Q_2(t)} \geq (\lambda+1)^{1/(\lambda+1)}.$$

Proof. Let $x(t)$ be a positive increasing solution of equation (8.2.2). Since $x''(t) > 0$ for $t \geq t_0$, it follows that

$$x(t) = x(t_0) + \int_{t_0}^{t} x'(s)ds \leq x(t_0) + (t - t_0)x'(t) \quad \text{for} \quad t \geq t_0.$$

Hence, $(x'(t))^{\lambda} \geq t^{-\lambda}x^{\lambda}(t)[1 + o(1)]$ as $t \to \infty$, which is equivalent to $(x'(t))^{\lambda}x''(t) \geq t^{-\lambda}q(t)[1 + o(1)]$ as $t \to \infty$, or

$$\left(\frac{(x'(t))^{\lambda+1}}{\lambda+1}\right)' \geq t^{-\lambda}q(t)[1 + o(1)] \quad \text{as} \quad t \to \infty.$$

Integrating the above inequality twice from t_0 to t, we get

$$x'(t) \geq (\lambda+1)^{1/(\lambda+1)}Q_1^{1/(\lambda+1)}(t)[1 + o(1)]$$

and

$$x(t) \geq (\lambda+1)^{1/(\lambda+1)}Q_2(t)[1 + o(1)] \quad \text{as} \quad t \to \infty.$$

This completes the proof. ∎

Lemma 8.2.8. Let the functions $f(t), g(t) \in C^1([t_0, \infty), \mathbb{R})$ be defined near infinity such that $g'(t) \neq 0$ and $\lim_{t\to\infty} g(t) = \infty$. Then,

$$\liminf_{t\to\infty} \frac{f'(t)}{g'(t)} \leq \liminf_{t\to\infty} \frac{f(t)}{g(t)} \leq \limsup_{t\to\infty} \frac{f(t)}{g(t)} \leq \limsup_{t\to\infty} \frac{f'(t)}{g'(t)}.$$

Remark 8.2.5. If condition (8.2.79) holds, then the function $y(t) = CQ_2(t)$, $t \geq t_0$ is a positive increasing solution of the equation

$$y''(t) = \frac{C^{\lambda+1}}{\lambda+1} q(t) \left[\frac{Q_2(t)}{tQ_1^{1/(\lambda+1)}(t)} \right]^{\lambda} y^{-\lambda}(t), \quad t \geq t_1 > t_0 \geq 0, \quad (8.2.80)$$

where C is a positive constant. Hence, it is natural to expect that the positive increasing solutions of (8.2.2) behave like $y(t)$ when the coefficient of (8.2.80) is asymptotic to the function $q(t)$. This observation leads to the following results.

Theorem 8.2.12. Let $0 < \lambda < 1$. Suppose (8.2.79) and

$$\lim_{t \to \infty} \frac{Q_2(t)}{tQ_1^{1/(\lambda+1)}(t)} = a \in \mathbb{R}^+ \qquad (8.2.81)$$

hold. Then any positive increasing solution $x(t)$ of equation (8.2.2) satisfies

$$x(t) \sim (\lambda+1)^{1/(\lambda+1)} a^{-\lambda/(\lambda+1)} Q_2(t) \quad \text{as} \quad t \to \infty. \qquad (8.2.82)$$

Theorem 8.2.13. Suppose conditions (8.2.79) and (8.2.81) hold. Further, assume that

$$\lim_{t \to \infty} q(t)Q_2^{3-\lambda}(t)Q_3^2(t) = b \in \mathbb{R}^+ \qquad (8.2.83)$$

and either

$$\int^{\infty} Q_3^{-1}(s)Q_2^{-2}(s) \left| \frac{Q_2(s)}{sQ_1^{1/(\lambda+1)}(s)} - a \right| ds < \infty, \qquad (8.2.84)$$

or

$$\int^{\infty} \left| \left[\frac{Q_2(s)}{sQ_1^{1/(\lambda+1)}(s)} \right]' \right| ds < \infty$$

hold. Then any positive increasing solution $x(t)$ of equation (8.2.2) has the asymptotic form (8.2.82).

Theorem 8.2.14. Suppose in addition to (8.2.79), (8.2.81), (8.2.83), (8.2.84) the following condition holds

$$\int^{\infty} Q_3^{-1}(s)Q_2^{-2}(s) \left| q(s)Q_2^{3-\lambda}(s)Q_3^2(s) - b \right| ds < \infty.$$

Then any positive increasing solution $x(t)$ of equation (8.2.2) has the asymptotic form (8.2.82).

Also, from Lemmas 8.2.7 and 8.2.8 the following result is immediate.

Theorem 8.2.15. Let $0 < \lambda < 1$, and $x(t)$ and $y(t)$ be positive increasing solutions of equations (8.2.2) and (8.2.50), respectively, such that

$$x(t) \geq my(t) \quad \text{for} \quad t \geq t_0, \tag{8.2.85}$$

for some constant $m > 0$. Suppose further that condition (8.2.79) and

$$\lim_{t \to \infty} \frac{q(t)}{p(t)} = 1 \tag{8.2.86}$$

hold. Then, $x(t) \sim y(t)$ as $t \to \infty$.

Since the function $z(t) = (\lambda + 1)^{1/(\lambda+1)} Q_2(t)$ for $t \geq t_0$ solves the equation

$$z''(t) = q(t) \left[\frac{Q_2(t)}{t Q_1^{1/(\lambda+1)}(t)} \right]^\lambda z^{-\lambda}(t) \quad \text{for} \quad t \geq t_1 > t_0 \geq 0$$

Lemma 8.2.7 and Theorem 8.2.15 show that the assumption $0 < \lambda < 1$ in Theorem 8.2.12 is superfluous if $a = 1$ in condition (8.2.81). We state this result in the following.

Corollary 8.2.4. If conditions (8.2.79) and (8.2.81) with $a = 1$ hold, then any positive increasing solution $x(t)$ of equation (8.2.2) satisfies $x(t) \sim (\lambda + 1)^{1/(\lambda+1)} Q_2(t)$ as $t \to \infty$.

Theorem 8.2.16. Let $x(t)$ and $y(t)$ be positive increasing solutions of equations (8.2.2) and (8.2.50) respectively, satisfying (8.2.85). Suppose conditions (8.2.79), (8.2.86),

$$\lim_{t \to \infty} p(t) y^{3-\lambda}(t) \left(\int_t^\infty y^{-2}(s) ds \right)^2 = \ell \in \mathbb{R}^+, \tag{8.2.87}$$

$$\int^\infty \left| \left[p(s) y^{3-\lambda}(s) \left(\int_s^\infty y^{-2}(u) du \right)^\lambda \right]' \right| ds < \infty \tag{8.2.88}$$

and either

$$\int^\infty \left(\int_s^\infty y^{-2}(u) du \right)^{-1} y^{-2}(s) \left| \frac{q(s)}{p(s)} - 1 \right| ds < \infty, \tag{8.2.89}$$

or

$$\int^\infty \left| \left(\frac{q(s)}{p(s)} \right)' \right| ds < \infty \tag{8.2.90}$$

hold. Then, $x(t) \sim y(t)$ as $t \to \infty$.

Proof. From Theorem 8.2.15 it follows that

$$my(t) \leq x(t) \leq My(t) \quad \text{for} \quad t \geq t_0 \tag{8.2.91}$$

for some constant $M > 0$. Set $v(t) = x(t)/y(t)$ for $t \geq t_0$. Since $v(t)$ satisfies equation (8.2.57), the change of variable $\tau = \left(\int_t^\infty y^{-2}(u)du \right)^{-1}$ implies that $v(\tau)$ satisfies

$$\frac{d^2v}{d\tau^2} + \frac{2}{\tau}\frac{dv}{d\tau} + p(t)y^{3-\lambda}(t)\tau^{-4}v = q(t)y^{3-\lambda}(t)\tau^{-4}v^{-\lambda}, \quad \tau \geq \tau_0$$

for some $\tau_0 > 0$. Furthermore, introducing the new independent variable $s = \ln \tau$, this equation is reduced to

$$\ddot{v} + \dot{v} + \bar{p}(s)v = \bar{q}(s)v^{-\lambda} \quad \text{for} \quad s \geq s_0 = \ln \tau_0, \quad \left(\cdot = \frac{d}{ds} \right) \tag{8.2.92}$$

where $\bar{p}(s) = p(t)y^{3-\lambda}(t)\tau^{-2}$ and $\bar{q}(s) = q(t)y^{3-\lambda}(t)\tau^{-2}$ for $s \geq s_0$. Now, conditions (8.2.86) – (8.2.90) become

$$\lim_{s \to \infty} \frac{\bar{q}(s)}{\bar{p}(s)} = 1, \tag{8.2.93}$$

$$\lim_{s \to \infty} \bar{q}(s) = \ell, \tag{8.2.94}$$

$$\int^\infty |(\bar{q}(s))\dot{}| ds < \infty \tag{8.2.95}$$

$$\int^\infty \left| \frac{\bar{q}(s)}{\bar{p}(s)} - 1 \right| ds < \infty \tag{8.2.96}$$

and

$$\int^\infty \left| \left(\frac{\bar{q}(s)}{\bar{p}(s)} \right)\dot{} \right| ds < \infty \tag{8.2.97}$$

respectively. By (8.2.91), we have

$$m \leq v(s) \leq M, \quad s \geq s_0. \tag{8.2.98}$$

Since (8.2.93) and (8.2.94) imply that $\lim_{s \to \infty} \bar{q}(s) = \ell$, the argument used in the proof of Theorem 8.2.9 shows that $\dot{v}(s)$ is bounded on $[s_0, \infty)$, and from equation (8.2.92), we see that $\ddot{v}(s)$ is also bounded. Multiplying equation (8.2.92) by $\dot{v}(s)$, we get

$$\left(\frac{\dot{v}^2}{2} \right)\dot{} + \dot{v}^2 + \bar{p}(s)\left(\frac{v^2}{2} - \frac{v^{1-\lambda}}{1-\lambda} \right)\dot{} = \bar{p}(s)\left(\frac{\bar{q}(s)}{\bar{p}(s)} - 1 \right)v^{-\lambda}\dot{v}, \quad s \geq s_0. \tag{8.2.99}$$

We shall prove that $\lim_{s\to\infty} v(s) = 1$ when condition (8.2.89) holds. The other case when (8.2.90) holds can be proved similarly, and will not be included. Integrating equation (8.2.99) over $[s_0, s]$, we obtain

$$
\begin{aligned}
\frac{\dot{v}^2(s)}{2} &+ \int_{s_0}^s \dot{v}^2(\eta)d\eta + \bar{p}(s)\left[\frac{v^2(s)}{2} - \frac{v^{1-\lambda}(s)}{1-\lambda}\right] \\
&- \int_{s_0}^s (\bar{p}(\eta))^\cdot \left[\frac{v^2(\eta)}{2} - \frac{v^{1-\lambda}(\eta)}{1-\lambda}\right] d\eta \\
&= r + \int_{s_0}^s \bar{p}(\eta)\left[\frac{\bar{q}(\eta)}{\bar{p}(\eta)} - 1\right]v^{-\lambda}(\eta)\dot{v}(\eta)d\eta \quad \text{for} \quad s \geq s_0,
\end{aligned}
\tag{8.2.100}
$$

where r is a real constant. Conditions (8.2.95) and (8.2.98) ensure the convergence of the second integral on the left–hand side of (8.2.100) as $s \to \infty$. Similarly, conditions (8.2.94) and (8.2.96) and the boundedness of $\dot{v}(s)$ imply the convergence of the integral on the right–hand side, from which we have $\int^\infty \dot{v}^2(s)ds < \infty$. Hence, $\lim_{s\to\infty} \dot{v}(s) = 0$ by Lemma 8.2.6. The rest of the proof is similar to that of Theorem 8.2.9 and hence omitted. This completes the proof. ∎

Consider the case when $q(t)$ behaves like a positive constant multiple of t^α, $\alpha - \lambda + 1 > 0$. Lemma 8.2.7 shows that for every positive increasing solution $x(t)$ of equation (8.2.2), $x(t) \geq mt^{(\alpha+2)/(\lambda+1)}$, $t \geq t_0$ for some constant $m > 0$. On the other hand, it is easily seen that the equation $y''(t) = ct^\alpha y^{-\lambda}(t)$, $t \geq t_0$ where $c > 0$ is a constant, admits a positive increasing solution $y(t)$ given by

$$
y(t) = \left[\frac{c(\lambda+1)^2}{(\alpha+2)(\alpha-\lambda+1)}\right]^{1/(\lambda+1)} t^{(\alpha+2)/(\lambda+1)} \quad \text{for} \quad t \geq t_0.
$$

Clearly, this $y(t)$ and $p(t) = ct^\alpha$, $t \geq t_0$ satisfy (8.2.87). Therefore, Theorems 8.2.15 and 8.2.16 give the next result, which can be regarded as an analogue to Corollary 8.2.3.

Corollary 8.2.5. Let $\alpha > \lambda - 1$. Suppose $q(t) = ct^\alpha[1+\epsilon(t)]$, where c is a positive constant, $\epsilon(t) \in C([t_0, \infty), \mathbb{R})$, $\lim_{t\to\infty} \epsilon(t) = 0$. If $0 < \lambda < 1$, then

$$
x(t) \sim \left[\frac{c(\lambda+1)^2}{(\alpha+2)(\alpha-\lambda+1)}\right]^{1/(\lambda+1)} t^{(\alpha+2)/(\lambda+1)} \quad \text{as} \quad t \to \infty.
$$

Finally, we present the following result.

Theorem 8.2.17. Let $x(t)$ and $y(t)$ be positive increasing solutions of equations (8.2.2) and (8.2.50), respectively, satisfying (8.2.85). Suppose conditions (8.2.79), (8.2.86), (8.2.87),

$$
\int^\infty \left(\int_s^\infty y^{-2}(u)du\right)^{-1} y^{-2}(s) \left|p(s)y^{3-\lambda}(s)\left(\int_s^\infty y^{-2}(u)du\right)^2 - \ell\right| ds < \infty
$$

and either (8.2.89) or (8.2.90) hold. Then, $x(t) \sim y(t)$ as $t \to \infty$.

Proof. The proof is similar to that of Theorem 8.2.16. ∎

8.2.3. Positive Solutions of Singular Emden–Fowler–Type Systems

Here we shall discuss positive solutions of the singular Emden–Fowler-type system (8.2.3), where

(i) $\lambda > 0$, $\mu > 0$ are constants,

(ii) $p(t)$, $q(t) \in C([t_0, \infty), \mathbb{R})$,

(iii) p, q have unbounded support,

(iv) the improper integrals $P(t) = \int_t^\infty p(s)ds$ and $Q(t) = \int_t^\infty q(s)ds$ converge for $t \geq t_0$, $P(t) \geq 0$ and $Q(t) \geq 0$ for $t \geq t_0$, and

(v) PQ has unbounded support.

A vector function $(x, y) \in C^2([t_0, \infty), \mathbb{R}) \times C^2([t_0, \infty), \mathbb{R})$ is called a positive solution of the system (8.2.3) when it satisfies (8.2.3) and $x(t) > 0$, $y(t) > 0$ for $t \geq t_0$. In what follows, we shall provide sufficient conditions which ensure the existence of positive solutions (x, y) of the system (8.2.3) satisfying

$$\begin{cases} \lim_{t \to \infty} x(t) = 0, & \lim_{t \to \infty} x'(t) = 0 \\ \lim_{t \to \infty} y(t) = 0, & \lim_{t \to \infty} y'(t) = 0 \end{cases} \tag{8.2.101}$$

and

$$\begin{cases} \lim_{t \to \infty} x(t) = \ell, & \lim_{t \to \infty} x'(t) = 0 \\ \lim_{t \to \infty} y(t) = 0, & \lim_{t \to \infty} y'(t) = 0, \end{cases} \tag{8.2.102}$$

where $\ell > 0$ is a constant. For this, we need the following basic lemma.

Lemma 8.2.9. Suppose

$$\int_{t_0}^\infty P(s)ds < \infty, \tag{8.2.103}$$

$$\int_{t_0}^\infty Q(s)ds < \infty \tag{8.2.104}$$

and

$$\int_{t_0}^\infty sP(s)Q(s)ds < \infty. \tag{8.2.105}$$

Then for any constants $\ell > 0$, $m > 0$ the system (8.2.3) admits a positive solution (x, y) satisfying

$$\begin{cases} \lim_{t \to \infty} x(t) = \ell, & \lim_{t \to \infty} x'(t) = 0 \\ \lim_{t \to \infty} y(t) = m, & \lim_{t \to \infty} y'(t) = 0 \end{cases} \tag{8.2.106}$$

and $x'(t) < 0$ and $y'(t) < 0$ for $t \geq t_0$.

Proof. It is easy to verify from condition (iv) that (x, y) is a positive solution of (8.2.3) satisfying (8.2.106) if and only if it solves the system

$$x(t) = \ell + \int_t^\infty P(s) y^{-\lambda}(s) ds - \lambda \int_t^\infty \left(\int_s^\infty P(u) y^{-\lambda-1}(u) y'(u) du \right) ds \tag{8.2.107}$$

$$y(t) = m + \int_t^\infty Q(s) x^{-\mu}(s) ds - \mu \int_t^\infty \left(\int_s^\infty Q(u) x^{-\mu-1}(u) x'(u) du \right) ds \tag{8.2.108}$$

for $t \geq t_0$. First, we solve this system in some neighborhood of infinity, say, $t \geq T \geq t_0$. Choose positive constants c, k and $T \geq t_0$ so that

$$\lambda m^{-\lambda-1} \left(\ell^{-\mu} + k \int_T^\infty P(s) ds \right) \leq c, \quad \mu \ell^{-\mu-1} \left(m^{-\lambda} + c \int_T^\infty Q(s) ds \right) \leq k,$$

$$m^{-\lambda} \int_T^\infty P(s) ds + \lambda m^{-\lambda-1} \left(\ell^{-\mu} + k \int_T^\infty P(s) ds \right) \left(\int_T^\infty sP(s)Q(s) ds \right) \leq \ell,$$

$$\ell^{-\mu} \int_T^\infty Q(s) ds + \mu \ell^{-\mu-1} \left(m^{-\lambda} + c \int_T^\infty Q(s) ds \right) \left(\int_T^\infty sP(s)Q(s) ds \right) \leq m.$$

Consider the set X of all functions $(x, y) \in C^1([T, \infty), \mathbb{R}^+) \times C^1([T, \infty), \mathbb{R}^+) = Y$, where components satisfy the inequalities $\ell \leq x(t) \leq 2\ell$, $m \leq y(t) \leq 2m$,

$$0 \leq -x'(t) \leq m^{-\lambda} P(t) + c \int_t^\infty P(s)Q(s) ds$$

$$0 \leq -y'(t) \leq \ell^{-\mu} Q(t) + k \int_t^\infty P(s)Q(s) ds$$

for $t \geq T$. Clearly, X is a nonempty closed convex subset of the Fréchet space Y. Define the mapping $F : X \to Y$ by $F(x, y) = (x_1, y_1)$, where

$$x_1(t) = \ell + \int_t^\infty P(s) y^{-\lambda}(s) ds - \lambda \int_t^\infty \left(\int_s^\infty P(u) y^{-\lambda-1}(u) y'(u) du \right) ds$$

$$x_1'(t) = -P(t) y^{-\lambda}(t) + \lambda \int_t^\infty P(s) y^{-\lambda-1}(s) y'(s) ds$$

and

$$y_1(t) = m + \int_t^\infty Q(s)x^{-\mu}(s)ds - \mu \int_t^\infty \left(\int_s^\infty Q(u)x^{-\mu-1}(u)x'(u)du \right) ds$$

$$y_1'(t) = -Q(t)x^{-\mu}(t) + \mu \int_t^\infty Q(s)x^{-\mu-1}(s)x'(s)ds$$

for $t \geq T$. We shall show that F maps X continuously into a relatively compact set of itself.

To prove $FX \subset X$, let $(x, y) \in X$. Since

$$\int_s^\infty P(u)y^{-\lambda-1}(u)y'(u)du$$

$$\leq m^{-\lambda-1} \left[\ell^{-\mu} \int_s^\infty P(u)Q(u)du + k \int_s^\infty P(u) \left(\int_u^\infty P(\xi)Q(\xi)d\xi \right) du \right]$$

$$\leq m^{-\lambda-1} \left[\ell^{-\mu} \int_s^\infty P(u)Q(u)du + k \left(\int_s^\infty P(u)du \right) \left(\int_s^\infty P(u)Q(u)du \right) \right]$$

for $s \geq T$, we obtain

$$|x_1'(t)| \leq m^{-\lambda}P(t) + \lambda m^{-\lambda-1} \left(\ell^{-\mu} + k \int_T^\infty P(s)ds \right) \left(\int_t^\infty P(s)Q(s)ds \right)$$

$$\leq m^{-\lambda}P(t) + c \int_t^\infty P(s)Q(s)ds \quad \text{for} \quad t \geq T$$

and

$$0 \leq x(t) - \ell$$

$$\leq m^{-\lambda} \int_T^\infty P(s)ds + \lambda \int_t^\infty \left(\int_s^\infty P(u)y^{-\lambda-1}(u)|y'(u)|du \right) ds$$

$$\leq m^{-\lambda} \int_T^\infty P(s)ds + \lambda m^{-\lambda-1} \left[\ell^{-\mu} \int_t^\infty \left(\int_s^\infty P(u)Q(u)du \right) ds \right.$$

$$\left. +k \int_t^\infty \left(\int_s^\infty P(u)du \right) \left(\int_s^\infty P(u)Q(u)du \right) ds \right]$$

$$\leq m^{-\lambda} \int_T^\infty P(s)ds + \lambda m^{-\lambda-1} \left(\ell^{-\mu} + k \int_T^\infty P(s)ds \right) \left(\int_T^\infty sP(s)Q(s)ds \right)$$

$$\leq \ell \quad \text{for} \quad t \geq T.$$

The estimates for $y_1(t)$ and $y_1'(t)$ can be obtained similarly. Thus, we find that $FX \subset X$. Furthermore, it is easy to show (see previous arguments) that the mapping F is continuous and FX is compact. Therefore, the Schauder–Tychonov theorem implies that F has a fixed element $(x, y) \in X$. Hence, the system (8.2.107), (8.2.108) admits a positive solution (x, y) in $[T, \infty)$.

Note that the derivative (x', y') is given by

$$x'(t) = -P(t)y^{-\lambda}(t) + \lambda \int_t^\infty P(s)y^{-\lambda-1}(s)y'(s)ds \tag{8.2.109}$$

$$y'(t) = -Q(t)x^{-\mu}(t) + \mu \int_t^\infty Q(s)x^{-\mu-1}(s)x'(s)ds \tag{8.2.110}$$

for $t \geq T$. Now, we claim that $x'(t) < 0$, $y'(t) < 0$ in $[T, \infty)$. Suppose not, then $x'(\tau) = 0$ for some $\tau \geq T$. It follows from (8.2.109) that $\lambda \int_\tau^\infty P(s)y^{-\lambda-1}(s)y'(s)ds = 0$, i.e., $P(t)y'(t) \equiv 0$ in $[\tau, \infty)$. Multiplying (8.2.110) by $P(t)$, we find that $P(t)Q(t) \equiv 0$ in $[\tau, \infty)$. This contradicts condition (v). Hence, $x'(t) < 0$ in $[T, \infty)$. Similarly, we can show that $y'(t) < 0$ in $[T, \infty)$.

Next, we extend (x, y) to the left as a solution of the system (8.2.3). Let $I \subset [t_0, \infty)$ be the maximal interval of existence of (x, y). It is clear that (8.2.107) – (8.2.110) are still valid for $t \in I$. We claim again that $x'(t) < 0$, $y'(t) < 0$, $t \in I$. In fact, if this is not true, we can find $\tau \in I$, $\tau < T$ such that $x'(t) < 0$, $y'(t) < 0$ for $t > \tau$ and either $x'(t)$ or $y'(t)$ vanishes at τ. Suppose $x'(\tau) = 0$, then putting $t = \tau$ in (8.2.109), we find $P(t)y'(t) \equiv 0$ in $[\tau, \infty)$, and therefore, multiplying (8.2.110) by $P(t)$ we arrive at a contradiction as before. Hence, $x'(t) < 0$, $y'(t) < 0$ for $t \in I$.

From the above observation it is clear that I coincides with the whole interval $[t_0, \infty)$, and therefore $x'(t) < 0$, $y'(t) < 0$ for $t \in I = [t_0, \infty)$. This completes the proof. ∎

Now, for the case $\lambda\mu < 1$ we shall prove the following results which guarantee the existence of positive solutions of (8.2.3) satisfying (8.2.101) or (8.2.102) with $\ell > 0$.

Theorem 8.2.18. Let $\lambda\mu < 1$ and conditions (8.2.103) and (8.2.104) hold. Further, suppose that

$$\int_{t_0}^\infty s|p(s)| \left(\int_s^\infty Q(u)du \right)^{-\lambda} ds < \infty \tag{8.2.111}$$

and

$$\int_{t_0}^\infty s|q(s)| \left(\int_s^\infty P(u)du \right)^{-\mu} ds < \infty. \tag{8.2.112}$$

Then the system (8.2.3) has a positive solution (x, y) satisfying (8.2.101).

Proof. Clearly, the assumptions of Lemma 8.2.9 are fulfilled. Thus, for $n \in \mathbb{N}$ there exists a positive solution $(x_n, y_n) \in C^2([t_0, \infty), \mathbb{R}^+) \times C^2([t_0, \infty), \mathbb{R}^+) = Y$ of the system (8.2.3) satisfying

$$x_n'(t) < 0, \quad y_n'(t) < 0 \quad \text{for } t \geq t_0 \tag{8.2.113}$$

and $\lim_{t\to\infty} x_n(t) = 1/n$, $\lim_{t\to\infty} y_n(t) = 1/n$, $\lim_{t\to\infty} x'_n(t) = 0$, and $\lim_{t\to\infty} y'_n(t) = 0$. Moreover, for $t \geq t_0$,

$$
\begin{aligned}
x_n(t) &= \frac{1}{n} + \int_t^\infty \left(\int_s^\infty p(u)y_n^{-\lambda}(u)du \right) ds \\
&= \frac{1}{n} + \int_t^\infty P(s)y_n^{-\lambda}(s)ds - \lambda \int_t^\infty \left(\int_s^\infty P(u)y_n^{-\lambda-1}(u)y'_n(u)du \right) ds
\end{aligned}
$$
$$(8.2.114)$$

$$
\begin{aligned}
y_n(t) &= \frac{1}{n} + \int_t^\infty \left(\int_s^\infty q(u)x_n^{-\mu}(u)du \right) ds \\
&= \frac{1}{n} + \int_t^\infty Q(s)x_n^{-\mu}(s)ds - \mu \int_t^\infty \left(\int_s^\infty Q(u)x_n^{-\mu-1}(u)x'_n(u)du \right) ds.
\end{aligned}
$$
$$(8.2.115)$$

Hence, (8.2.113) – (8.2.115) give

$$x_n(t) \geq y_n^{-\lambda}(t) \int_t^\infty P(s)ds, \quad t \geq t_0 \qquad (8.2.116)$$

$$y_n(t) \geq x_n^{-\mu}(t) \int_t^\infty Q(s)ds, \quad t \geq t_0. \qquad (8.2.117)$$

Thus, it follows from (8.2.114) and (8.2.117) and the decreasing nature of $y_n(t)$ that for $t \geq t_0$,

$$
\begin{aligned}
-x'_n(t) = \int_t^\infty p(s)y_n^{-\lambda}(s)ds &\leq \int_t^\infty |p(s)| \left(\int_s^\infty Q(u)du \right)^{-\lambda} x_n^{\lambda\mu}(s)ds \\
&\leq x_n^{\lambda\mu}(t) \int_t^\infty |p(s)| \left(\int_s^\infty Q(u)du \right)^{-\lambda} ds,
\end{aligned}
$$

or

$$-\left(\frac{x_n^{1-\lambda\mu}(t)}{1-\lambda\mu} \right)' \leq \int_t^\infty |p(s)| \left(\int_s^\infty Q(u)du \right)^{-\lambda} ds. \qquad (8.2.118)$$

Integrating (8.2.118) from t to ξ and letting $\xi \to \infty$, we get

$$\frac{x_n^{1-\lambda\mu}(t)}{1-\lambda\mu} \leq \frac{(1/n)^{1-\lambda\mu}}{1-\lambda\mu} + \int_t^\infty s|p(s)| \left(\int_s^\infty Q(u)du \right)^{-\lambda} ds, \quad t \geq t_0.$$

Thus, the sequence $\{x_n\}$ is uniformly bounded on each compact subset of $[t_0, \infty)$. Moreover, by the same computation the sequences $\{x'_n\}$, $\{y_n\}$ and $\{y'_n\}$ are also uniformly bounded on each compact subset of $[t_0, \infty)$. Hence, by Ascoli–Arzela's theorem we can find a subsequence $\{(x_{n_i}, y_{n_i})\}$ of $\{(x_n, y_n)\}$ and a function $(\bar{x}, \bar{y}) \in Y$ to which $\{(x_{n_i}, y_{n_i})\}$ converges uniformly on each compact subset of $[t_0, \infty)$. Inequalities (8.2.116) and

(8.2.117) show that $\bar{x}(t) > 0$, $\bar{y}(t) > 0$ in $[t_0, \infty)$. Let $n_i \to \infty$ in the equations

$$x_{n_i}(t) = \frac{1}{n_i} + \int_t^\infty \left(\int_s^\infty p(u) y_{n_i}^{-\lambda}(u) du \right) ds, \quad t \geq t_0$$

$$y_{n_i}(t) = \frac{1}{n_i} + \int_t^\infty \left(\int_s^\infty q(u) x_{n_i}^{-\mu}(u) du \right) ds, \quad t \geq t_0.$$

Now, the Lebesgue dominated convergence theorem asserts that $(\bar{x}, \bar{y})$ is a positive solution of the system (8.2.3) satisfying (8.2.101). This completes the proof. ∎

Theorem 8.2.19. Let $\lambda\mu < 1$, (8.2.111) and $\int_{t_0}^\infty s|q(s)|ds < \infty$ hold. Then for any constant $\ell > 0$, the system (8.2.3) admits a positive solution (x, y) satisfying (8.2.102).

Proof. As in Theorem 8.2.18, we consider the positive solutions (x_n, y_n), $n \in \mathbb{N}$ of the system (8.2.3) such that $x_n'(t) < 0$, $y_n'(t) < 0$ for $t \geq t_0$ and $\lim_{t\to\infty} x_n(t) = \ell$, $\lim_{t\to\infty} y_n(t) = 1/n$, $\lim_{t\to\infty} x_n'(t) = 0$, $\lim_{t\to\infty} y_n'(t) = 0$. Note that inequality (8.2.117) remains valid. Thus, we have

$$\frac{x_n^{1-\lambda\mu}(t)}{1 - \lambda\mu} \leq \frac{\ell^{1-\lambda\mu}}{1 - \lambda\mu} + \int_t^\infty s|p(s)| \left(\int_s^\infty Q(u)du \right)^{-\lambda} ds, \quad t \geq t_0.$$

Now, it is easy to see that $x_n(t) \geq \ell$ for $t \geq t_0$, and hence by the above inequality, we get the following estimate for $y_n(t)$

$$y_n(t) \leq \frac{1}{n} + \ell^{-\mu} \int_t^\infty s|q(s)|ds, \quad t \geq t_0.$$

Finally, using similar arguments as in Theorem 8.2.18, we arrive at the desired conclusion. This completes the proof. ∎

When $\lambda\mu \geq 1$, it is unknown whether or not the system (8.2.3) has such positive solutions. However, similar manipulations as in Theorem 8.2.18 give the following nonexistence criterion.

Theorem 8.2.20. Let $\lambda\mu \geq 1$.

(I_1) If conditions (8.2.104) and (8.2.111) hold, then the system (8.2.3) admits no positive solution (x, y) satisfying

$$\begin{cases} \lim_{t\to\infty} x(t) = 0, \quad \lim_{t\to\infty} x'(t) = 0, \quad \lim_{t\to\infty} y'(t) = 0 \\ \lim_{t\to\infty} y(t) = c_1 \in \mathbb{R}_0, \quad c_1 \text{ is a constant} \\ x'(t) < 0, \quad y'(t) < 0 \quad \text{for all large } t. \end{cases} \quad (8.2.119)$$

(I_2)   If conditions (8.2.103) and (8.2.112) hold, then the system (8.2.3) admits no positive solution (x, y) satisfying

$$
\begin{cases}
\lim\limits_{t\to\infty} y(t) = 0, \quad \lim\limits_{t\to\infty} y'(t) = 0, \quad \lim\limits_{t\to\infty} x'(t) = 0 \\[2mm]
\lim\limits_{t\to\infty} x(t) = c_2 \in \mathbb{R}_0, \quad c_2 \text{ is a constant} \\[2mm]
x'(t) < 0, \quad y'(t) < 0 \quad \text{for all large } t.
\end{cases}
\tag{8.2.120}
$$

Proof.   We only consider (I_1), because (I_2) can be treated similarly. Let (x, y) be a positive solution of the system (8.2.3) with the required property (8.2.119). As in Lemma 8.2.9, we find that it satisfies (8.2.107) and (8.2.108) with $\ell = 0$ and $m = y(\infty) \in \mathbb{R}_0$. Now choose $T \geq t_0$ so large that $x'(t) < 0$, $y'(t) < 0$ for $t \geq T$. Then as in Theorem 8.2.18, we get

$$
\begin{aligned}
x(t) &= \int_t^\infty \left(\int_s^\infty p(u) y^{-\lambda}(u) du \right) ds \\
&\leq \int_t^\infty \left[\int_s^\infty |p(u)| \left(\int_u^\infty Q(v) dv \right)^{-\lambda} x^{-\lambda\mu}(u) du \right] ds \\
&\leq x^{-\lambda\mu}(t) \int_t^\infty (s-t)|p(s)| \left(\int_s^\infty Q(u) du \right)^{-\lambda} ds,
\end{aligned}
$$

i.e.,

$$
x^{1-\lambda\mu}(t) \leq \int_t^\infty s|p(s)| \left(\int_s^\infty Q(u) du \right)^{-\lambda} ds, \quad t \geq T.
$$

Letting $t \to \infty$, we get a contradiction. This completes the proof.   ∎

The following examples are illustrative.

Example 8.2.3.   Consider the system (8.2.3) with the functions p and q satisfying

$$
a_1 t^{-2-\alpha} \leq p(t) \leq a_2 t^{-2-\alpha}, \quad b_1 t^{-2-\beta} \leq q(t) \leq b_2 t^{-2-\beta} \tag{8.2.121}
$$

for $t \geq 1$, where α, β, a_i, b_i, $i = 1, 2$ are positive constants.

(I)   Let $\lambda\mu < 1$. From Theorem 8.2.18, we find that if

$$
\alpha - \lambda\beta > 0 \quad \text{and} \quad \beta - \mu\alpha > 0, \tag{8.2.122}
$$

then the system (8.2.3) has a positive solution (x, y) satisfying (8.2.101). In particular, the singular Emden–Fowler system

$$
\begin{cases}
x''(t) = t^{-2-\alpha} y^{-\lambda}(t) \\
y''(t) = t^{-2-\beta} x^{-\mu}(t), \quad t \geq 1
\end{cases}
\tag{8.2.123}
$$

where α and β are positive constants, has a positive solution (x, y) of the form

$$\begin{cases} x(t) = at^{-\gamma_1} \\ y(t) = bt^{-\gamma_2}, \end{cases} \tag{8.2.124}$$

where $a,\ b,\ \gamma_i,\ i = 1, 2$ are positive constants, if and only if condition (8.2.122) holds. In fact, (x, y) of the form (8.2.124) becomes a positive solution of (8.2.123) if and only if the system

$$a\gamma_1(\gamma_1 + 1) = b^{-\lambda}, \quad -\gamma_1 - 2 = \lambda\gamma_2 - 2 - \alpha$$
$$b\gamma_2(\gamma_2 + 1) = a^{-\mu}, \quad -\gamma_2 - 2 = \mu\gamma_1 - 2 - \beta$$

has a solution $(a, b, \gamma_1, \gamma_2)$.

(II) Let $\lambda\mu > 1$. It is easy to see that if

$$\alpha - \lambda\beta > 0 \quad \text{or} \quad \beta - \mu\alpha > 0, \tag{8.2.125}$$

then the system (8.2.123) never has a positive solution (x, y) of the form (8.2.124). On the other hand, Theorem 8.2.20 concludes that if condition (8.2.125) holds, then the system (8.2.3) with (8.2.121) does not have any positive solution (x, y) satisfying (8.2.101).

Example 8.2.4. Consider the system (8.2.3) with the functions p and q satisfying

$$a_1 t^{-\alpha_1 - 2} \le p(t) \le a_2 t^{-\alpha_2 - 2}, \quad b_1 t^{-\beta_1 - 2} \le q(t) \le b_2 t^{-\beta_2 - 2}$$

for $t \ge 1$, where $a_i,\ b_i,\ i = 1, 2$ are positive constants, $\alpha_1 > \alpha_2 > 0$ and $\beta_1 > \beta_2 > 0$ are parameters. Let $\lambda\mu < 1$. Then it follows from Theorem 8.2.18 that there exist $\alpha_i = \alpha_i(\lambda, \mu) > 0$ and $\beta_i = \beta_i(\lambda, \mu)$ such that the system (8.2.3), with this $\alpha_i(\lambda, \mu)$ and $\beta_i(\lambda, \mu)$ has a positive solution (x, y) satisfying (8.2.101). To see this it suffices to notice the fact that for any given $\lambda,\ \mu$ satisfying $\lambda\mu < 1$ inequalities

$$\alpha_2 - \lambda\beta_1 > 0, \quad \alpha_1 - \alpha_2 > 0$$
$$\beta_2 - \mu\alpha_1 > 0, \quad \beta_1 - \beta_2 > 0$$

admit a positive solution $\alpha_i,\ \beta_i,\ i = 1, 2$. Here, we may adopt the following known results from convex analysis. For each $n \times n$ matrix M exactly one of the following two cases holds:

(M_1) There exists n–vector ξ satisfying $M\xi > 0$ and $\xi > 0$.

(M_2) There exists n–vector $\eta \ne 0$ satisfying $M\eta < 0$ and $\eta \ge 0$. (The order relation $v \ge w$ $[v > w]$ for vectors $v = (v_i),\ w = (w_i)$ is defined as $v_i \ge w_i$ $[v_i > w_i]$ for all i).

Example 8.2.5. Consider the system (8.2.3) with

$$p(t) = -\left(\frac{1+\sin t}{t^{2+\epsilon}}\right)', \quad q(t) = -\left(\frac{1+\sin t}{t^{2+\delta}}\right)'$$

for $t \geq 1$, where ϵ, δ are positive constants. Let $\lambda\mu < 1$. Then, we find

$$p(t) = O\left(t^{-2-\epsilon}\right), \quad \int_t^\infty P(s)ds \geq c_1 t^{-1-\epsilon},$$

$$q(t) = O\left(t^{-2-\delta}\right), \quad \int_t^\infty Q(s)ds \geq c_2 t^{-1-\delta}$$

as $t \to \infty$ for some positive constants c_1, c_2. Therefore, Theorems 8.2.18 and 8.2.19 assert that system (8.2.3) has a positive solution (x, y) satisfying (8.2.101) if $\lambda(1+\delta) < \epsilon$ and $\mu(1+\epsilon) < \delta$ and that the system (8.2.3) has a positive solution (x, y) satisfying (8.2.102), $\ell > 0$ if $\lambda(1+\delta) < \epsilon$.

8.3. Notes and General Discussions

1. The results of Section 8.1 except Theorem 8.1.14 and Lemma 8.1.1 are taken from Naito [6–8], while Theorem 8.1.14 and Lemma 8.1.1 are due to Philos et. al. [9].

2. It is known [3, theorem 1] that there is a class X of $f(x)$ which contains the function $f(x) = |x|^\gamma \operatorname{sgn} x$ $(\gamma \geq 1)$, such that if equation (8.1.1) with $f \in X$ has a nonoscillatory solution, then either (8.1.66) holds, or $\liminf_{t\to\infty}(1/t)\int_{t_0}^t \int_{t_0}^s q(u)duds = -\infty$. Thus, for the special case when $f(x)$ belongs to X and $q(t)$ satisfies

$$\liminf_{t\to\infty} \frac{1}{t}\int_{t_0}^t \int_{t_0}^s q(u)duds > -\infty, \tag{8.3.1}$$

if equation (8.1.1) has a nonoscillatory solution, then (8.1.66) holds. In Theorem 8.1.13, however, no condition on f is needed except that

$$xf(x) > 0 \quad \text{and} \quad f'(x) \geq 0 \quad \text{for} \quad x \neq 0 \tag{8.3.2}$$

and (8.1.65) is weaker than (8.3.1).

3. It is also known [3, theorem 2] that there is a class Y of $f(x)$ which contains the function $f(x) = |x|^\gamma \operatorname{sgn} x$ $(0 < \gamma < 1)$, such that if equation (8.1.1) with $f \in Y$ has a nonoscillatory solution, then either (8.1.66) holds, or $\lim_{t\to\infty}(1/t)\int_{t_0}^t \int_{t_0}^s q(u)duds = -\infty$. This means that when $f(x)$ belongs to Y and $q(t)$ satisfies (8.1.65), if equation (8.1.1) has a nonoscillatory solution, then (8.1.66) holds. Note that no condition

on f is needed in Theorem 8.1.13 except (8.3.2). However, instead of the asymptotic condition as $t \to \infty$, the nonoscillatory solution is required.

4. The results of Section 8.1 can be extended to equations of the form

$$(a(t)x'(t))' + q(t)f(x(t)) = 0,$$

where $a(t) \in C([t_0, \infty), \mathbb{R}^+)$. Also, to equations with deviating arguments of the form

$$x''(t) + q(t)f(x[g(t)]) = 0,$$

where $g(t) \in C([t_0, \infty), \mathbb{R})$ and $\lim_{t \to \infty} g(t) = \infty$. The details are left to the reader.

5. It would be interesting to obtain results similar to those presented in Section 8.1 for equations of the form

$$\left(|x'(t)|^{\alpha-1}x'(t)\right)' + q(t)|x(t)|^{\beta-1}x(t) = 0,$$

where $\alpha > 0$, $\beta > 0$ are constants and $q(t) \in C([t_0, \infty), \mathbb{R})$, and also for the more general equations

$$\left(|x'(t)|^{\alpha-1}x'(t)\right)' + q(t)f(x(t)) = 0$$

and

$$\left(|x'(t)|^{\alpha-1}x'(t)\right)' + q(t)f(x[g(t)]) = 0,$$

where $g(t), q(t) \in C([t_0, \infty), \mathbb{R})$, $f \in C(\mathbb{R}, \mathbb{R})$ and $\lim_{t \to \infty} g(t) = \infty$.

6. The results on the positive solutions of equation (8.2.1) are taken from Tallaferro [10], while the results on the positive decaying solutions and asymptotic behavior of solutions of singular Emden–Fowler equation (8.2.2) are due to Usami [12,14]. The results concerning the positive solutions of singular Edmen–Fowler-type system (8.2.3) are from Usami [13]. For several other related works, see Tallaferro [11].

7. The results of Section 8.2 can be extended rather easily to more general equations of the form

$$(a(t)x'(t))' + \delta q(t)x^{-\lambda}(t) = 0$$

and equations with deviating arguments of the type

$$(a(t)x'(t))' + \delta q(t)x^{-\lambda}[g(t)] = 0,$$

where λ and q are as in equations (8.2.1) and (8.2.2), $\delta = \pm 1$ and $a(t) \in C([t_0, \infty), \mathbb{R}^+)$.

8. It is desirable to extend some of the results of Section 8.2 to equations of the form

$$\left(|x'(t)|^{\alpha-1}x'(t)\right)' + \delta q(t)x^{-\lambda}(t) = 0,$$

where $\delta = \pm 1$, α and λ are positive constants, and $q(t) \in C([t_0, \infty), \mathbb{R})$.

8.4. References

1. **G.J. Butler**, On the oscillatory behavior of a second order nonlinear differential equation, *Ann. Mat. Pura Appl.* **105**(1975), 73–92.

2. **G.J. Butler**, Oscillation theorems for a nonlinear analogue of Hill's equation, *Quart. J. Math.* **27**(1976), 159–171.

3. **G.J. Butler**, Integral averages and the oscillation of second order ordinary differential equations, *SIAM J. Math. Anal.* **11**(1980), 190–200.

4. **T. Kura**, Oscillation theorems for a second order sublinear ordinary differential equation, *Proc. Amer. Math. Soc.* **84**(1982), 535–538.

5. **M.K. Kwong and J.S.W. Wong**, On the oscillation and nonoscillation of second order sublinear equations, *Proc. Amer. Math. Soc.* **85**(1982), 547–551.

6. **M. Naito**, Asymptotic behavior of solutions of second order differential equations with integrable coefficients, *Trans. Amer. Math. Soc.* **282**(1984), 577–588.

7. **M. Naito**, Nonoscillatory solutions of second order differential equations with integrable coefficients, *Proc. Amer. Math. Soc.* **109**(1990), 769–774.

8. **M. Naito**, Integral averages and the asymptotic behavior of solutions of second order ordinary differential equations, *J. Math. Anal. Appl.* **164**(1992), 370–380.

9. **Ch.G. Philos and I.K. Purnaras**, Asymptotic behavior of solutions of second order nonlinear ordinary differential equations, *Nonlinear Analysis* **24**(1995), 81–90.

10. **S.D. Taliaferro**, On the positive solutions of $y'' + \phi(t)y^{-\lambda} = 0$, *Nonlinear Analysis* **2**(1978), 437–446.

11. **S.D. Taliaferro**, Asymptotic behavior of solutions of $y'' = \phi(t)f(y)$, *SIAM J. Math. Anal.* **17**(1981), 853–865.

12. **H. Usami**, On positive decaying solutions of singular Emden–Fowler–type equations, *Nonlinear Analysis* **18**(1991), 795–803.

13. **H. Usami**, Positive solutions of singular Emden–Fowler–type systems, *Hiroshima Math. J.* **22**(1992), 421–431.

14. **H. Usami**, Asymptotic behavior of positive solutions of singular Emden–Fowler–type equations, *J. Math. Soc. Japan* **40**(1994), 195–211.

15. **D. Willet**, On the oscillatory behavior of the solutions of second order linear differential equations, *Ann. Polon. Math.* **21**(1969), 175–194.

Chapter 9

Miscellaneous Topics

9.0. Introduction

This chapter contains some special results. In Section 9.1, we shall extend the Sturm–Picone theorem, obtain nonoscillation theorems for perturbed second order nonlinear differential equations, and present a nonlinear Picone type identity to enable us to prove some Sturm–Picone type comparison theorems for nonlinear equations. Section 9.2 is devoted to the study of nonoscillatory solutions of forced differential equations of second order. In Section 9.3 we shall present some limit cycle criteria and discuss its related properties for nonlinear second order differential equations. Finally, in Section 9.4 we shall present some properties of solutions of very general second order differential equations.

9.1. Comparison and Nonoscillation Results

The classical Sturm–Picone theorem compares equations

$$(a(t)x'(t))' + q(t)x(t) = 0 \tag{9.1.1}$$

and

$$(a_1(t)y'(t))' + q_1(t)y(t) = 0, \tag{9.1.2}$$

where $a(t),\ a_1(t) \in C^1([t_0, \infty), \mathbb{R}^+),\ q(t),\ q_1(t) \in C([t_0, \infty), \mathbb{R})$. It states that if

$$a(t) \le a_1(t) \quad \text{and} \quad q(t) \ge q_1(t) \quad \text{on} \quad [t_1, t_2] \subset [t_0, \infty), \quad t_0 > 0$$

then if any solution $y(t)$ of (9.1.2) has two (or more) zeros on $[t_1, t_2]$, every solution of (9.1.1) also has at least one zero on $[t_1, t_2]$.

We shall compare the oscillatory behavior of solutions of (9.1.1) with the solutions of the equation

$$(a(t)\rho(t)y'(t))' + q(t)\rho(t)y(t) = 0, \tag{9.1.3}$$

where $\rho(t) \in C^1([t_0, \infty), \mathbb{R})$. For this, we shall need the following lemma.

Lemma 9.1.1. If there exists a function $y(t) \in C^1([t_1, t_2], \mathbb{R})$ with y not identically zero, and $y(t_1) = y(t_2) = 0$ such that

$$\int_{t_1}^{t_2} \left[a(s)(y'(s))^2 - q(s)y^2(s) \right] ds \le 0,$$

then every solution of equation (9.1.1) except a constant multiple of $y(t)$ vanishes at some point of (t_1, t_2).

In our first result we shall prove a Sturm–type theorem concerning the existence of zeros on the interval $[t_1, t_2]$.

Theorem 9.1.1. Suppose there exists a solution $y(t)$ of equation (9.1.3) which has no zeros on $[t_1, t_2]$. If $a(t)\rho(t) \ge 0$ and $(a(t)\rho'(t))' \le 0$ on $[t_1, t_2]$, then each solution $x(t)$ of equation (9.1.1) can have at most one zero on $[t_1, t_2]$.

Proof. Suppose there exists a solution $x(t)$ of equation (9.1.1) which has zeros at $t = \tau_1$ and $t = \tau_2$ on the interval $[t_1, t_2]$. Assume without loss of generality that $x(t) > 0$ on (τ_1, τ_2). Then,

$$
\begin{aligned}
0 &= \int_{\tau_1}^{\tau_2} (a(s)\rho(s)x(s)x'(s))' ds \\
&= \int_{\tau_1}^{\tau_2} \left[\frac{1}{2} a(s)\rho'(s)\frac{d}{ds}(x^2(s)) + a(s)\rho(s)(x'(s))^2 + \rho(s)x(s)(a(s)x'(s))' \right] ds \\
&= -\frac{1}{2} \int_{\tau_1}^{\tau_2} (a(s)\rho'(s))' x^2(s) ds + \int_{\tau_1}^{\tau_2} \left[a(s)\rho(s)(x'(s))^2 - q(s)\rho(s)x^2(s) \right] ds.
\end{aligned}
$$

Thus, we have

$$\int_{\tau_1}^{\tau_2} \left[a(s)\rho(s)(x'(s))^2 - q(s)\rho(s)x^2(s) \right] ds \le 0. \qquad (9.1.4)$$

Hence there exists a function $x(t) \in C^1([\tau_1, \tau_2], \mathbb{R})$ vanishing at $t = \tau_1$ and $t = \tau_2$ and satisfying inequality (9.1.4). An application of Lemma 9.1.1 shows that every solution $y(t)$ of equation (9.1.3) has a zero on $[\tau_1, \tau_2]$. This contradiction proves the theorem. ∎

Remark 9.1.1. An obvious consequence of Theorem 9.1.1 states that if (9.1.3) is nonoscillatory, then so is equation (9.1.1).

Theorem 9.1.2. Subject to the same hypotheses of Theorem 9.1.1 any solution of (9.1.3) has a zero on any interval containing two zeros of any solution of equation (9.1.1).

Proof. Suppose to the contrary that a solution of equation (9.1.3) has no zeros on such an interval. Then an application of Theorem 9.1.1 yields a contradiction. ∎

Example 9.1.1. Consider the differential equation

$$((\sin t)y'(t))' + (k^2 \sin t)y(t) = 0, \quad k > 2. \tag{9.1.5}$$

We shall show that any solution of (9.1.5) has a zero on the interval $[0, \pi]$. For this, we compare (9.1.5) with the equation $x''(t) + k^2 x(t) = 0$, which has solutions with two zeros on the interval $[0 + \epsilon, \pi - \epsilon]$ for a sufficiently small $\epsilon > 0$. Here $a(t) = 1$, $q(t) = k^2$ and $\rho(t) = \sin t$. Thus, $a(t)\rho(t) = \sin t > 0$ and $(a(t)\rho'(t))' = -\sin t < 0$ on $[\epsilon, \pi - \epsilon]$. Now by Theorem 9.1.2 every solution of (9.1.5) has a zero on $[\epsilon, \pi - \epsilon]$, and hence on $[0, \pi]$.

Next, we shall consider the equation

$$(a(t)x'(t))' + q(t)x(t) = g(t)f(x(t)), \tag{9.1.6}$$

where the functions $a(t)$ and $q(t)$ are as in (9.1.1), $g(t) \in \mathcal{L}^1[t_0, \infty)$ locally, $f(x) \in C(\mathbb{R}, \mathbb{R})$ and $xf(x) > 0$ for $x \neq 0$.

Theorem 9.1.3. Suppose $x(t)$ is a solution of (9.1.6) such that $x(t_1) = 0 = x(t_2)$, $t_2 > t_1 > t_0$. If there exists a function $\rho(t) \in C^1([t_1, t_2], \mathbb{R})$ such that $a(t)\rho(t) \geq 0$, $(a(t)\rho'(t))' \leq 0$, $g(t)\rho(t) \geq 0$ on $[t_1, t_2]$ and $g(t)\rho(t) \in \mathcal{L}^1[t_0, \infty)$ locally, and if $a_1(t) \leq a(t)\rho(t)$ and $q_1(t) \geq q(t)\rho(t)$ on $[t_1, t_2]$, then any solution of equation (9.1.2) will have a zero on $[t_1, t_2]$.

Proof. Proceeding as in Theorem 9.1.1, we have

$$
\begin{aligned}
0 &= \int_{t_1}^{t_2} (a(s)\rho(s)x(s)x'(s))' ds \\
&= -\frac{1}{2} \int_{t_1}^{t_2} \left[(a(s)\rho'(s))'x^2(s) - 2g(s)\rho(s)x(s)f(x(s)) \right] ds \\
&\quad + \int_{t_1}^{t_2} \left[a(s)\rho(s)(x'(s))^2 - q(s)\rho(s)x^2(s) \right] ds.
\end{aligned}
$$

Thus, it follows that

$$\int_{t_1}^{t_2} \left[a(s)\rho(s)(x'(s))^2 - q(s)\rho(s)x^2(s) \right] ds \leq 0.$$

The rest of the proof is similar to that of Theorem 9.1.1 and hence omitted. ∎

The following corollary is immediate:

Corollary 9.1.1. Subject to the hypotheses of Theorem 9.1.3 if any solution of (9.1.2) has no zero on the interval $[t_1, t_2]$, then any solution of equation (9.1.6) will have at most one zero on $[t_1, t_2]$.

Example 9.1.2. Consider the differential equations

$$y''(t) + y(t) = 0 \tag{9.1.7}$$

and

$$(2x'(t))' + x(t) = (\sin t)x^3(t). \tag{9.1.8}$$

Here $a(t) = 2$, $q(t) = 1$, $g(t) = \sin t$ and $f(x) = x^3$. We claim that any solution of (9.1.8) has at most one zero on the interval $[\pi/3, \pi/2]$. For this, we choose $\rho(t) = \sin t$. Then, $a(t)\rho(t) = 2\sin t \geq 1 = a_1(t)$, $q(t)\rho(t) = \sin t \leq 1 = q_1(t)$, $(a(t)\rho'(t))' = -2\sin t < 0$ and $g(t)\rho(t) = \sin^2 t > 0$. Hence, Corollary 9.1.1 is applicable, since there exists a solution of (9.1.7) having no zero on $[\pi/3, \pi/2]$.

Remark 9.1.2. If the condition $xf(x) > 0$ for $x \neq 0$ in Theorem 9.1.3 is replaced by $xf(x) < 0$ for $x \neq 0$, then Theorem 9.1.3 remains true provided $g(t)\rho(t) \geq 0$ is replaced by $g(t)\rho(t) \leq 0$.

Now consider the differential equations

$$(a_1(t)x'(t))' + Q_1(t, x(t), x'(t)) = P(t, x(t), x'(t)) \tag{9.1.9}$$

and

$$(a_2(t)y'(t))' + Q_2(t, y(t), y'(t)) = 0, \tag{9.1.10}$$

where $a_i(t) \in C([t_0, \infty), \mathbb{R}^+)$, P, $Q_i \in C([t_0, \infty) \times \mathbb{R}^2, \mathbb{R})$, $i = 1, 2$. We assume that for $t \geq t_0$, (t, x, x'), $(t, y, y') \in [t_0, \infty) \times \mathbb{R}^2$,

$$a_1(t) \geq a_2(t), \tag{9.1.11}$$

$$\frac{Q_1(t, x, x')}{x} \leq \frac{Q_2(t, y, y')}{y} \tag{9.1.12}$$

and

$$P(t, x, x') \geq 0. \tag{9.1.13}$$

The following result extends the Sturm–Picone theorem.

Theorem 9.1.4. Let conditions (9.1.11) – (9.1.13) hold. If $x(t)$ is a solution of (9.1.9) with $x(t_1) = 0 = x(t_2)$ and $x(t) > 0$ on $(t_1, t_2) \subset (t_0, \infty)$, then every solution of equation (9.1.10) has a zero on $[t_1, t_2]$.

Proof. Let $x(t)$ be such a solution of (9.1.9) and assume that $y(t)$ is a solution of (9.1.10) such that $y(t) \not\equiv 0$ on $[t_1, t_2]$. From the well-known

Picone identity (see Section 1.1), we have

$$\left[\frac{x(t)}{y(t)} (a_1(t)x'(t)y(t) - a_2(t)y'(t)x(t)) \right]'$$

$$= \left[\frac{Q_2(t, y(t), y'(t))}{y(t)} - \frac{Q_1(t, x(t), x'(t))}{x(t)} \right] + [a_1(t) - a_2(t)](x'(t))^2$$

$$+ a_2(t) \left[\frac{1}{y(t)} (x'(t)y(t) - x(t)y'(t)) \right]^2 + x(t)P(t, x(t), x'(t)).$$

An integration of this identity from t_1 to t_2 yields the desired contradiction. ∎

The following corollary is an immediate consequence of Theorem 9.1.4.

Corollary 9.1.2. Suppose conditions (9.1.11) – (9.1.13) hold.

(i) If (9.1.9) has an oscillatory solution, then every solution of (9.1.10) is oscillatory.

(ii) If there is a solution of (9.1.10) with no zeros on $[t_1, t_2]$, then no solution of (9.1.9) which is nonnegative on $[t_1, t_2]$ can vanish more than once there.

Remark 9.1.3. We can obtain analogous results when

$$P(t, x, x') \leq 0 \quad \text{for} \quad (t, x, x') \in [t_0, \infty) \times \mathbb{R}^2. \tag{9.1.14}$$

In this case, however, we would need to take a solution $x(t)$ of (9.1.9) satisfying $x(t_1) = 0 = x(t_2)$ and $x(t) < 0$ for $t_1 < t < t_2$. Then every solution of (9.1.10) must vanish on $[t_1, t_2]$.

Next, we state the following result.

Theorem 9.1.5. If in addition to conditions (9.1.11) – (9.1.13),

$$Q_1(t, x, x') \leq 0 \quad \text{if} \quad x \leq 0 \tag{9.1.15}$$

and equation (9.1.10) has a nonoscillatory solution, then every solution of (9.1.9) is nonoscillatory.

Remark 9.1.4. If in Theorem 9.1.5 we replace (9.1.13) by (9.1.14), then the conclusion holds provided we replace (9.1.15) by

$$Q_1(t, x, x') \geq 0 \quad \text{if} \quad x \geq 0. \tag{9.1.16}$$

Example 9.1.3. Consider the equation

$$y''(t) + q(t)y(t) = 0, \tag{9.1.17}$$

where $q(t) \in C([t_0, \infty), \mathbb{R}_0)$ is such that it is nonoscillatory. By applying Theorem 9.1.5, we find that the equation

$$x''(t) + q(t) \frac{x(t)}{x^2(t) + 1} = x^2(t)$$

is also nonoscillatory.

Now we shall consider the special case of equations (9.1.9) and (9.1.10), namely,

$$(a_1(t)x'(t))' + q_1(t)f_1(x(t)) = P(t, x(t), x'(t)) \qquad (9.1.18)$$

and

$$(a_2(t)y'(t))' + q_2(t)f_2(y(t)) = 0, \qquad (9.1.19)$$

where a_1, a_2 and P are as in equations (9.1.9) and (9.1.10), $q_1(t)$, $q_2(t) \in C([t_0, \infty), \mathbb{R})$ and f_1, $f_2 \in C(\mathbb{R}, \mathbb{R})$.

Theorem 9.1.6. Assume that condition (9.1.13) holds,

$$q_2(t) \geq q_1(t), \quad t \geq t_0 \qquad (9.1.20)$$

$$f_1(x) \geq 0 \quad \text{if} \quad x \geq 0 \quad \text{and} \quad f_1(0) = 0, \quad f_2(y) \neq 0 \quad \text{if} \quad y \neq 0, \quad (9.1.21)$$

$$f_2'(y) > 0 \qquad (9.1.22)$$

and there is a positive constant k such that

$$0 \leq \frac{f_1'(x)}{f_2'(y)} \leq k \qquad (9.1.23)$$

and

$$a_1(t) \geq k a_2(t), \quad t \geq t_0. \qquad (9.1.24)$$

If $x(t)$ is a solution of (9.1.18) with $x(t_1) = 0 = x(t_2)$ and $x(t) > 0$ for $t_1 < t < t_2$, then every solution of (9.1.19) has a zero on $[t_1, t_2]$.

Proof. Let $x(t)$ be a solution of (9.1.18) with the above properties and suppose that $y(t)$ is a solution of (9.1.19) with $y(t) \neq 0$ on $[t_1, t_2]$. Since

$$\left[\frac{f_1(x(t))}{f_2(y(t))} \left(a_1(t)x'(t)f_2(y(t)) - a_2(t)y'(t)f_1(x(t)) \right) \right]'$$

$$= (a_1(t)x'(t))' f_1(x(t)) + a_1(t)f_1'(x(t))(x'(t))^2 - (a_2(t)y'(t))' \frac{f_1^2(x(t))}{f_2(y(t))}$$

$$- 2a_2(t) \frac{f_1(x(t))}{f_2(y(t))} f_1'(x(t))x'(t)y'(t) + a_2(t) \frac{f_1^2(x(t))}{f_2^2(y(t))} f_2'(y(t))(y'(t))^2$$

$$= [q_2(t) - q_1(t)]f_1^2(x(t)) + f_1(x(t))P(t, x(t), x'(t)) + a_2(t)f_2'(y(t)) \times$$

$$\left[\left(\frac{f_1(x(t))}{f_2(y(t))} y'(t) \right)^2 - 2\frac{f_1(x(t))}{f_2(y(t))} \frac{f_1'(x(t))}{f_2'(y(t))} x'(t)y'(t) + \left(\frac{f_1'(x(t))}{f_2'(y(t))} x'(t) \right)^2 \right]$$

$$- a_2(t) \frac{(f_1'(x(t)))^2}{f_2'(y(t))} (x'(t))^2 + a_1(t)f_1'(x(t))(x'(t))^2$$

$$= [q_2(t) - q_1(t)]f_1^2(x(t)) + f_1(x(t))P(t, x(t), x'(t))$$

$$+ a_2(t)f_2'(y(t)) \left[\frac{f_1(x(t))}{f_2(y(t))} y'(t) - \frac{f_1'(x(t))}{f_2'(y(t))} x'(t) \right]^2$$

$$+ f_1'(x(t))(x'(t))^2 \left[a_1(t) - a_2(t)\frac{f_1'(x(t))}{f_2'(y(t))} \right]$$

an integration of the above identity from t_1 to t_2 gives the desired contradiction. ■

Next, we state the following nonoscillation theorem.

Theorem 9.1.7. In addition to conditions (9.1.13) and (9.1.20) – (9.1.24) assume that

$$q_1(t) \geq 0 \quad \text{for} \quad t \geq t_0 \tag{9.1.25}$$

and

$$f_1(x) \leq 0 \quad \text{if} \quad x \leq 0. \tag{9.1.26}$$

If (9.1.19) has a nonoscillatory solution, then (9.1.18) is nonoscillatory.

Of particular interest in Theorems 9.1.6 and 9.1.7 is the fact that if $k < 1$, then we may have $a_2(t) > a_1(t)$, contrary to what is needed in the Sturm–Picone theorem for linear equations. The following example illustrates how Theorem 9.1.7 can be applied in practice.

Example 9.1.4. Consider the differential equations

$$((1+t)x'(t))' + q(t)\tanh(x(t)/2) = 0, \quad t \geq 2 \tag{9.1.27}$$

and

$$((2t + \sin t)y'(t))' + q(t)y(t) = 0, \quad t \geq 2 \tag{9.1.28}$$

where $q(t) \in C([t_0, \infty), \mathbb{R}_0)$, and assume that (9.1.28) is nonoscillatory. Since $f_1'(x) = (1/2)\text{sech}^2(x/2) \leq 1/2$ and $a_1(t) = 1+t \geq (2t+\sin t)/2 = a_2(t)/2$ for $t \geq t_2$, in view of Theorem 9.1.7 equation (9.1.27) is nonoscillatory.

In the following result, we shall replace condition (9.1.20) by

$$q_2(t) \geq kq_1(t) \quad \text{for} \quad t \geq t_0. \tag{9.1.29}$$

Theorem 9.1.8. Suppose conditions (9.1.11), (9.1.13), (9.1.21) – (9.1.23) and (9.1.29) hold. If $x(t)$ is a solution of (9.1.18) with $x(t_1) = 0 = x(t_2)$ and $x(t) > 0$ on (t_1, t_2), then every solution of (9.1.19) has a zero on $[t_1, t_2]$. If in addition (9.1.25) and (9.1.26) hold and (9.1.19) has a nonoscillatory solution, then all solutions of (9.1.18) are nonoscillatory.

Proof. Proceeding as in Theorem 9.1.6, we have

$$
\left[\frac{f_1(x(t))}{f_2(y(t))} \left(k^2 a_1(t) x'(t) f_2(y(t)) - k a_2(t) y'(t) f_1(x(t)) \right) \right]'
$$
$$
= -k^2 q_1(t) f_1^2(x(t) + k^2 f_1(x(t)) P(t, x(t), x'(t)) + k^2 a_1(t) f_1'(x(t))(x'(t))^2
$$
$$
+ k q_2(t) f_1^2(x(t)) - 2 k a_2(t) y'(t) \frac{f_1(x(t)) f_1'(x(t))}{f_2(y(t))} x'(t)
$$
$$
+ k a_2(t)(y'(t))^2 \frac{f_1^2(x(t)) f_2'(y(t))}{f_2^2(y(t))}
$$
$$
\geq k f_1^2(x(t))[q_2(t) - k q_1(t)] + k^2 f_1(x(t)) P(t, x(t), x'(t))
$$
$$
+ k^2 a_1(t) f_1'(x(t))(x'(t))^2 + a_2(t) \frac{f_1'(x(t))}{f_2^2(y(t))} \left[f_1^2(x(t)) - 2 k f_1(x(t)) f_2(y(t)) \right.
$$
$$
\times x'(t) y'(t) + k^2 f_2^2(y(t))(x'(t))^2 \big] - k^2 a_2(t) f_1'(x(t))(x'(t))^2
$$
$$
= k f_1^2(x(t))[q_2(t) - k q_1(t)] + k^2 f_1(x(t)) P(t, x(t), x'(t))
$$
$$
+ k^2 f_1'(x(t))(x'(t))^2 [a_1(t) - a_2(t)]
$$
$$
+ a_2(t) f_1'(x(t)) \left[f_1(x(t)) y'(t) - k f_2(y(t)) x'(t) \right]^2 .
$$

The remainder of the proof follows as before. ∎

Remark 9.1.5.
1. It is easy to obtain many corollaries and theorems similar to those of Corollary 9.1.2 and Theorems 9.1.5 and 9.1.7. The formulation of such results are left to the reader.

2. It would be interesting to extend the above results to higher order equations, to equations with deviating arguments, and to compare two perturbed equations with some conditions on the relative sizes of the perturbations.

9.2. Nonoscillatory Solutions of Forced Differential Equations

Here we shall discuss the existence of nonoscillatory solutions of the forced differential equation

$$
(a(t)(b(t)x(t))')' + F(t, x(t)) + e(t) = 0. \tag{9.2.1}
$$

In particular, we shall show that if the forcing term $e(t)$ is small in certain sense, then equation (9.2.1) has nonoscillatory solutions provided the associated unforced equation

$$(a(t)(b(t)x(t))')' + F(t, x(t)) = 0 \qquad (9.2.2)$$

has the same property.

We begin by comparing the nonoscillatory behavior of the equations

$$(a(t)(b(t)x(t))')' + F_1(t, x(t)) + e_1(t) = 0 \qquad (9.2.3)$$

and

$$(a(t)(b(t)y(t))')' + F_2(t, y(t)) + e_2(t) = 0, \qquad (9.2.4)$$

where $a(t)$, $b(t) \in C([t_0, \infty), \mathbb{R}^+)$, $e_i(t) \in C([t_0, \infty), \mathbb{R})$ and $F_i \in C([t_0, \infty) \times \mathbb{R}^+, \mathbb{R}^+)$, $i = 1, 2$.

In what follows it will be convenient to distinguish the following two cases:

$$R[t, T] = \int_T^t \frac{ds}{a(s)}, \quad T \geq t_0 \quad \text{and} \quad \lim_{t \to \infty} R[t, T] = \infty \qquad (9.2.5)$$

and

$$\pi(t) = \int_t^\infty \frac{ds}{a(s)} \quad \text{and} \quad \pi(t_0) < \infty. \qquad (9.2.6)$$

In the following lemma, we let t_0 and T be such that $T \geq t_0$, $u \in C([t_0, \infty), \mathbb{R}^+)$, $w \in C([T, \infty), \mathbb{R}^+)$, $H \in C([T, \infty) \times \mathbb{R}_0, \mathbb{R}_0)$, X, $Y \in C(\Omega, \mathbb{R}_0)$, where $\Omega = \{(t, s) : t \geq s \geq T\}$, and H is nondecreasing in the second variable.

Lemma 9.2.1. Suppose the functions u, w, H, X and Y satisfy $\int_T^\infty Y^*(s)H(s, u(s))ds < \infty$, and

$$u(t) \geq w(t) + \int_T^t X(t, s) \int_s^\infty Y(v, s)H(v, u(v))dvds \quad \text{for} \quad t \geq T,$$

where $Y^*(s) = \max\{Y(t, s) : s \in [T, t]\}$. Then the integral equation

$$z(t) = w(t) + \int_T^t X(t, s) \int_s^\infty Y(v, s)H(v, z(v))dvds$$

has a solution $z \in C([t_0, \infty), \mathbb{R}^+)$ such that $w(t) \leq z(t) \leq u(t)$ for $t \geq T$.

Theorem 9.2.1. Suppose that

$$F_1(t, x) \geq F_2(t, x) \quad \text{for} \quad x > 0 \qquad (9.2.7)$$

and $F_2(t, x)$ is nondecreasing in x. Further, suppose that there exist two functions $\eta_1,\ \eta_2 \in C([t_0, \infty), \mathbb{R})$ such that $\eta_1(t)$ is oscillatory,

$$(a(t)(b(t)\eta_i(t))')' + e_i(t) = 0, \quad t \geq t_0, \quad i = 1, 2 \tag{9.2.8}_i$$

and

$$\lim_{t \to \infty} b(t)\eta_i(t) = 0, \quad i = 1, 2 \tag{9.2.9}$$

when condition (9.2.5) holds, also

$$\lim_{t \to \infty} \frac{b(t)}{\pi(t)} \eta_i(t) = 0, \quad i = 1, 2 \tag{9.2.10}$$

when condition (9.2.6) holds.

If (9.2.3) has an eventually positive solution $x(t)$, then (9.2.4) has an eventually positive solution $y(t)$ such that $y(t) \leq x(t)$ for all large t.

Proof. Let $x(t)$ be a positive solution of (9.2.3) defined on $[t_0, \infty)$. Set $w(t) = x(t) - \eta_1(t)$. Then for $t \geq t_0$, we find

$$(a(t)(b(t)w(t))')' = -F_1(t, x(t)) < 0. \tag{9.2.11}$$

Thus, $w(t)$ is eventually of constant sign. If $w(t) < 0$ for $t \geq t_1$ for some $t_1 \geq t_0$, then $0 < x(t) < \eta_1(t)$, $t \geq t_1$ which contradicts the fact that $\eta_1(t)$ is oscillatory and hence we must have $w(t) > 0$ for $t \geq t_1$. Now we need to consider two cases:

(i) Suppose condition (9.2.5) holds. It is easy to see that $a(t)(b(t)w(t))' > 0$ for $t \geq t_1$ and the finite limit $c_2 = \lim_{t \to \infty} a(t)(b(t)w(t))' \geq 0$ exists. Integrating equation (9.2.11) from $t \geq t_1$ to $u \geq t$ and letting $u \to \infty$, we obtain

$$a(t)(b(t)w(t))' = c_2 + \int_t^\infty F_1(s, x(s))ds, \quad t \geq t_1. \tag{9.2.12}$$

Dividing equation (9.2.12) by $a(t)$ and integrating from t_1 to t, we get

$$b(t)w(t) = c_1 + c_2 R[t, t_1] + \int_{t_1}^t \frac{1}{a(s)} \int_s^\infty F_1(u, x(u))du\,ds,$$

or

$$b(t)x(t) = c_1 + b(t)\eta_1(t) + c_2 R[t, t_1] + \int_{t_1}^t \frac{1}{a(s)} \int_s^\infty F_1(u, x(u))du\,ds \tag{9.2.13}$$

for $t \geq t_1$, where $c_1 = b(t_1)w(t_1) > 0$. In view of (9.2.9) there is a $t_2 \geq t_1$ such that

$$c_1 + b(t)\eta_1(t) \geq \frac{1}{2}c_1 + b(t)\eta_2(t) > 0, \quad t \geq t_2. \tag{9.2.14}$$

From (9.2.13), (9.2.14) and (9.2.7), we find

$$b(t)x(t) \geq \frac{1}{2}c_1 + b(t)\eta_2(t) + c_2 R[t, t_1] + \int_{t_2}^{t} \frac{1}{a(s)} \int_{s}^{\infty} F_2(u, x(u)) du ds,$$
(9.2.15)

$t \geq t_2$. Now by Lemma 9.2.1 it follows that there exists a continuous solution $y(t)$ of the integral equation

$$b(t)y(t) = \frac{1}{2}c_1 + b(t)\eta_2(t) + c_2 R[t, t_1] + \int_{t_2}^{t} \frac{1}{a(s)} \int_{s}^{\infty} F_2(u, y(u)) du ds, \quad t \geq t_2$$
(9.2.16)

satisfying

$$b(t)x(t) \geq b(t)y(t) \geq \frac{1}{2}c_1 + b(t)\eta_2(t) + c_2 R[t, t_1] \quad \text{for} \quad t \geq t_2. \quad (9.2.17)$$

Differentiating equation (9.2.16) twice, it follows that $y(t)$ is a solution of (9.2.4) on $[t_1, \infty)$. Finally, $x(t) \geq y(t) > 0$ on $[t_2, \infty)$ follows from (9.2.14) and (9.2.17).

(ii) Suppose condition (9.2.6) holds. Then we rewrite equations (9.2.3) and (9.2.4), respectively, as

$$\left(a(t)\pi^2(t) \left(\frac{b(t)}{\pi(t)} x(t) \right)' \right)' + \pi(t)F_1(t, x(t)) + \pi(t)e_1(t) = 0, \quad (9.2.18)$$

and

$$\left(a(t)\pi^2(t) \left(\frac{b(t)}{\pi(t)} y(t) \right)' \right)' + \pi(t)F(t, y(t)) + \pi(t)e_2(t) = 0. \quad (9.2.19)$$

Since

$$\int_{t_0}^{t} \frac{1}{a(s)\pi^2(s)} ds = \frac{1}{\pi(t)} - \frac{1}{\pi(t_0)} \to \infty \quad \text{as} \quad t \to \infty,$$

the above arguments in (i) apply to equations (9.2.18) and (9.2.19), and the desired conclusion follows immediately. This completes the proof. ∎

Remark 9.2.1. Suppose $\eta_1(t)$ is oscillatory, $(9.2.8)_1$ and either (9.2.9) or (9.2.10) are satisfied. Let $x(t)$ be an eventually positive solution of equation (9.2.3). Then it is easy to verify that there are constants $k_i > 0$, $i = 1, 2$ and $T \geq t_0$ such that $k_1 \leq b(t)x(t) \leq k_2 R[t, t_0]$ for $t \geq T$ if condition (9.2.5) holds, and $k_1 \pi(t) \leq b(t)x(t) \leq k_2$ for $t \geq T$ if condition (9.2.6) holds. This observation shows that:

(I_1) If condition (9.2.5) holds, a solution $x(t)$ of (9.2.3) with the property $\lim_{t \to \infty} b(t)x(t)/R[t, t_0] = \text{constant} > 0$ (respectively, $\lim_{t \to \infty} b(t)x(t) =$

constant > 0) can be regarded as a 'maximal' (respectively 'minimal') positive solution of (9.2.3).

(I_2) If condition (9.2.6) holds, a solution $x(t)$ of equation (9.2.3) with the property $\lim_{t\to\infty} b(t)x(t) = $ constant > 0 (respectively $\lim_{t\to\infty} b(t)$ $\times x(t)/\pi(t) = $ constant > 0) can be regarded as 'maximal' (respectively 'minimal') positive solution of (9.2.3).

The following result establishes the existence of minimal positive solutions of equations (9.2.3) and (9.2.4).

Theorem 9.2.2. Suppose conditions (9.2.7), (9.2.8)$_i$ – (9.2.10), $i = 1, 2$ are satisfied.

(i) Let condition (9.2.5) hold. If equation (9.2.3) has a solution $x(t)$ such that $\lim_{t\to\infty} b(t)x(t) = $ constant > 0, then (9.2.4) has a solution $y(t)$ such that $y(t) \leq x(t)$ for all large t and $\lim_{t\to\infty} b(t)y(t) = $ constant > 0.

(ii) Let condition (9.2.6) hold. If equation (9.2.3) has a solution $x(t)$ such that $\lim_{t\to\infty} b(t)x(t)/\pi(t) = $ constant > 0, then (9.2.4) has a solution $y(t)$ such that $y(t) \leq x(t)$ for all large t and $\lim_{t\to\infty} b(t)y(t)/\pi(t) = $ constant > 0.

Proof. It suffices to prove the statement (i). Let $x(t)$ be a positive solution on $[t_0, \infty)$ of (9.2.3) such that $\lim_{t\to\infty} b(t)x(t) = c > 0$, where c is a constant. Put $w(t) = x(t) - \eta_1(t)$. As in the proof of Theorem 9.2.1, we see that $w(t)$ is eventually of constant sign, say, for $t \geq t_1 \geq t_0$. If $w(t) < 0$, $t \geq t_1$ then $x(t) < \eta_1(t)$, $t \geq t_1$ which implies that $\liminf_{t\to\infty} b(t)\eta_1(t) \geq c > 0$, a contradiction to condition (9.2.9). Thus, we have $w(t) > 0$ for $t \geq t_1$. Proceeding as in the proof of Theorem 9.2.1(i) and noting that $\lim_{t\to\infty} b(t)w(t) = c$, $c_2 = \lim_{t\to\infty} a(t)(b(t)w(t))' = 0$, we conclude that there exists a continuous solution $y(t)$ of the integral equation

$$b(t)y(t) = \frac{1}{2}c_1 + b(t)\eta_2(t) + \int_{t_2}^{t} \frac{1}{a(s)} \int_{s}^{\infty} F_2(u, y(u))du\,ds$$

satisfying $b(t)x(t) \geq b(t)y(t) \geq (c_1/2) + b(t)\eta_2(t)$ for $t \geq t_2$, where $c_1 > 0$ is a constant and $t_2 > t_1$ is sufficiently large. Now it is easy to see that $y(t)$ is a minimal positive solution of (9.2.4). This completes the proof. ∎

Next, for the existence of maximal positive solutions of equations (9.2.3) and (9.2.4), we only state the following result.

Theorem 9.2.3. Let conditions (9.2.7) and (9.2.8)$_i$ hold,

$$\lim_{t\to\infty} \frac{b(t)}{R[t, t_0]}\eta_1(t) = 0, \qquad \lim_{t\to\infty} \frac{b(t)}{R[t, t_0]}\eta_2(t) = 0 \qquad (9.2.20)$$

when condition (9.2.5) holds, and

$$\lim_{t\to\infty} b(t)\eta_1(t) = 0, \qquad \lim_{t\to\infty} b(t)\eta_2(t) = 0 \qquad (9.2.21)$$

when condition (9.2.6) holds.

(i) Let condition (9.2.5) hold. If equation (9.2.3) has a solution $x(t)$ such that $\lim_{t\to\infty} b(t)x(t)/R[t,t_0] = $ constant > 0, then (9.2.4) has a solution $y(t)$ such that $y(t) \le x(t)$ for all large t and $\lim_{t\to\infty} b(t)y(t)/R[t,t_0] = $ constant > 0.

(ii) Let condition (9.2.6) hold. If equation (9.2.3) has a solution $x(t)$ such that $\lim_{t\to\infty} b(t)x(t) = $ constant > 0, then (9.2.4) has a solution $y(t)$ such that $y(t) \le x(t)$ for all large t and $\lim_{t\to\infty} b(t)y(t) = $ constant > 0.

Now we shall consider equations (9.2.1) and (9.2.2) where $a(t)$, $b(t) \in C([t_0,\infty),\mathbb{R}^+)$, $e(t) \in C([t_0,\infty),\mathbb{R})$, $F \in C([t_0,\infty) \times \mathbb{R}^+, \mathbb{R}^+)$ and $F(t,x)$ is nondecreasing in x.

First we state necessary and sufficient conditions for equation (9.2.2) so that it possess maximal and minimal positive solutions.

Theorem 9.2.4. (i) Suppose condition (9.2.5) holds. Then, (9.2.2) has a maximal (respectively minimal) positive solution if and only if

$$\int^\infty F\left(s, \frac{c}{b(s)}R[s,t_0]\right) ds < \infty \quad \left(\int^\infty R[s,t_0]F\left(s, \frac{c}{b(s)}\right) ds < \infty\right)$$

for some constant $c > 0$.

(ii) Suppose condition (9.2.6) holds. Then, (9.2.2) has a maximal (respectively minimal) positive solution if and only if

$$\int^\infty \pi(s)F\left(s, \frac{c}{b(s)}\right) ds < \infty \quad \left(\int^\infty F\left(s, \frac{c}{b(s)}\pi(s)\right) ds < \infty\right)$$

for some constant $c > 0$.

In particular, for equations (9.2.1) and (9.2.2), Theorems 9.2.1 – 9.2.3 yield the following results, which roughly conclude that the nonoscillatory character of equation (9.2.2) is not affected by adding a small forcing term $e(t)$.

Corollary 9.2.1. Suppose there exists a function $\eta(t) \in C([t_0,\infty),\mathbb{R})$ with the properties

$$(a(t)(b(t)\eta(t))')' + e(t) = 0, \quad t \ge t_0 \qquad (9.2.22)$$

$$\lim_{t\to\infty} b(t)\eta(t) = 0 \quad \text{when condition (9.2.5) holds,} \qquad (9.2.23)$$

$$\lim_{t\to\infty} \frac{b(t)}{\pi(t)}\eta(t) \ = \ 0 \quad \text{when condition (9.2.6) holds.} \qquad (9.2.24)$$

If equation (9.2.2) has an eventually positive solution $x(t)$, then (9.2.1) has an eventually positive solution $y(t)$ such that $y(t) \le x(t)$ for all large t.

Corollary 9.2.2. Suppose there exists a function $\eta(t) \in C([t_0,\infty),\mathbb{R})$ satisfying condition (9.2.22) and either condition (9.2.23) or (9.2.24).

(i) Let condition (9.2.5) hold. If equation (9.2.2) has a solution $x(t)$ such that $\lim_{t\to\infty} b(t)x(t) = \text{constant} > 0$, then (9.2.1) has a solution $y(t)$ such that $y(t) \le x(t)$ for all large t and $\lim_{t\to\infty} b(t)y(t) = \text{constant} > 0$.

(ii) Let condition (9.2.6) hold. If equation (9.2.2) has a solution $x(t)$ such that $\lim_{t\to\infty} b(t)x(t)/\pi(t) = \text{constant} > 0$, then (9.2.1) has a solution $y(t)$ such that $y(t) \le x(t)$ for all large t and $\lim_{t\to\infty} b(t)y(t)/\pi(t) = \text{constant} > 0$.

Corollary 9.2.3. Suppose there exists a function $\eta(t) \in C([t_0,\infty),\mathbb{R})$ satisfying condition (9.2.22) and

$$\lim_{t\to\infty} \frac{b(t)}{R[t,t_0]}\eta(t) \ = \ 0 \quad \text{when condition (9.2.5) holds,}$$

$$\lim_{t\to\infty} b(t)\eta(t) \ = \ 0 \quad \text{when condition (9.2.6) holds.}$$

(i) Let condition (9.2.5) hold. If equation (9.2.2) has a solution $x(t)$ such that $\lim_{t\to\infty} b(t)x(t)/R[t,t_0] = \text{constant} > 0$, then (9.2.1) has a solution $y(t)$ such that $y(t) \le x(t)$ for all large t and $\lim_{t\to\infty} b(t)y(t)/ R[t,t_0] = \text{constant} > 0$.

(ii) Let condition (9.2.6) hold. If equation (9.2.2) has a solution $x(t)$ such that $\lim_{t\to\infty} b(t)x(t) = \text{constant} > 0$, then (9.2.1) has a solution $y(t)$ such that $y(t) \le x(t)$ for all large t and $\lim_{t\to\infty} b(t)y(t) = \text{constant} > 0$.

Remark 9.2.2. In the above results we were concerned only with eventually positive solutions of the equations under consideration. It is clear that these results have counterparts for eventually negative solutions provided $F_i,\ F \in C([t_0,\infty) \times \mathbb{R}^-,\mathbb{R}^-),\ i = 1,2$.

9.3. Limit Circle Criteria and Related Properties For Nonlinear Equations

In this section we shall present some sufficient conditions for the integrability of solutions of the forced second order nonlinear differential equation

$$(a(t)x'(t))' + q(t)f(x(t)) \ = \ e(t). \qquad (9.3.1)$$

We will say that equation (9.3.1) is of *limit circle type* if all its solutions satisfy

$$\int^{\infty} x(u)f(x(u))du < \infty. \tag{9.3.2}$$

Otherwise equation (9.3.1) is said to be of *limit point type*.

Now we shall consider equation (9.3.1) where $a(t)$, $q(t) \in C([t_0, \infty), \mathbb{R}^+)$, $e(t) \in C([t_0, \infty), \mathbb{R})$, $f \in C(\mathbb{R}, \mathbb{R})$, $a'(t)$, $q'(t) \in AC_{loc}[t_0, \infty)$, $a''(t)$, $q''(t) \in L^2_{loc}[t_0, \infty)$ and $xf(x) > 0$ for $x \neq 0$.

As earlier for any $h(t) \in C([t_0, \infty), \mathbb{R})$ we let $h^+(t) = \max\{h(t), 0\}$ and $h^-(t) = \max\{-h(t), 0\}$. Also, we define $F(x) = \int_0^x f(u)du$. We shall assume that there exist positive constants k and n and nonnegative constants A, B and K such that

$$k \geq 2(n+1), \tag{9.3.3}$$

$$0 \leq \frac{1}{k}xf(x) - \frac{n}{n+1}F(x) \leq BF(x) \tag{9.3.4}$$

and

$$\frac{1}{2}x^2 \leq AF(x) + K. \tag{9.3.5}$$

To simplify the notation in what follows, we let $\alpha = 1/(2(n+1))$ and $\beta = (2n+1)/(2(n+1))$. We use the transformation

$$s = \int_{t_0}^{t} a^{-\beta}(u)q^{\alpha}(u)du \quad \text{and} \quad y(s) = x(t) \tag{9.3.6}$$

so that equation (9.3.1) becomes

$$\ddot{y} + \alpha p(t)\dot{y} + Q(t)f(y) = E(t), \quad (\cdot = d/ds) \tag{9.3.7}$$

where $p(t) = (a(t)q(t))'/(a^{\alpha}(t)q^{\alpha+1}(t))$, $Q(t) = (a(t)q(t))^{\beta-\alpha}$ and $E(t) = a^{\beta-\alpha}(t)e(t)q^{-2\alpha}(t)$. Note also that $\beta - \alpha = 2\beta - 1 = n/(n+1)$.

In system form equation (9.3.7) can be written as

$$\begin{cases} \dot{y} = z - \dfrac{1}{k}p(t)y \\[2mm] \dot{z} = \left(\dfrac{1}{k} - \alpha\right)p(t)z - Q(t)f(y) + \left[\dot{p}(t) - \left(\dfrac{1}{k} - \alpha\right)p^2(t)\right]\left(\dfrac{1}{k}y\right) + E(t). \end{cases} \tag{9.3.8}$$

Theorem 9.3.1. In addition to conditions (9.3.3) – (9.3.5) assume that

$$\int_{t_0}^{\infty} \frac{[(a(u)q(u))']^-}{a(u)q(u)}du < \infty, \tag{9.3.9}$$

$$\int_{t_0}^{\infty} \left| \left[\frac{(a(u)q(u))'}{a^{\alpha}(u)q^{\alpha+1}(u)} \right]' - \left(\frac{1}{k} - \alpha \right) \frac{[(a(u)q(u))']^2}{a^{\alpha+1}(u)q^{\alpha+2}(u)} \right| du \; < \; \infty, \quad (9.3.10)$$

$$\int_{t_0}^{\infty} \frac{|e(u)|}{(a(u)q(u))^{\alpha}} du \; < \; \infty \quad\quad\quad (9.3.11)$$

and

$$\int_{t_0}^{\infty} (a(u)q(u))^{\alpha-\beta} du \; < \infty. \quad\quad\quad (9.3.12)$$

Then any solution $x(t)$ of (9.3.1) satisfies (9.3.2) and $\int_{t_0}^{\infty} F(x(u))du < \infty$.

Proof. Define $V(y, z, s) = (1/2)z^2 + Q(t)F(y)$, then

$$\dot{V}(s) = \left(\frac{1}{k} - \alpha \right) p(t)z^2 + \left[p(t) - \left(\frac{1}{k} - \alpha \right) p^2(t) \right] \left(\frac{1}{k} yz \right) + E(t)z$$

$$-Q(t)p(t) \left(\frac{1}{k} yf(y) \right) + \dot{Q}(t)F(y)$$

$$= \left(\frac{1}{k} - \alpha \right) p(t)z^2 + \left[p(t) - \left(\frac{1}{k} - \alpha \right) p^2(t) \right] \left(\frac{1}{k} yz \right) + E(t)z$$

$$+Q(t) \left[(\beta - \alpha)F(y) - \frac{1}{k} yf(y) \right] \left[\frac{(a(t)q(t))'}{a(t)q(t)} \right] \left(\frac{a^{\beta}(t)}{q^{\alpha}(t)} \right).$$

Applying conditions (9.3.3) and (9.3.4), we get

$$\dot{V}(s) \leq \left[B - 2 \left(\frac{1}{k} - \alpha \right) \right] \left(\frac{[(a(t)q(t))']^-}{a(t)q(t)} \right) \left(\frac{a^{\beta}(t)}{q^{\alpha}(t)} \right) V$$

$$+E(t)z + \left[p(t) - \left(\frac{1}{k} - \alpha \right) p^2(t) \right] \left(\frac{1}{k} yz \right).$$

Now from (9.3.5), we find

$$|yz| \leq \frac{1}{2} y^2 + \frac{1}{2} z^2 \leq \frac{1}{2} z^2 + AF(y) + K,$$

so

$$\dot{V}(s) \leq \left[B - 2 \left(\frac{1}{k} - \alpha \right) \right] \left(\frac{[(a(t)q(t))']^-}{a(t)q(t)} \right) \left(\frac{a^{\beta}(t)}{q^{\alpha}(t)} \right) V$$

$$+|E(t)| \left(\frac{1}{2} + V \right) + \left| p(t) - \left(\frac{1}{k} - \alpha \right) p^2(t) \right| \frac{1}{k} \left[V + K + A \frac{V}{Q(t)} \right].$$

Since $p(t) = \dot{p}(t)a^{\beta}(t)q^{-\alpha}(t)$, we have

$$I(t) = \left| \dot{p}(t) - \left(\frac{1}{k} - \alpha \right) p^2(t) \right|$$

$$= \left| p'(t) - \left(\frac{1}{k} - \alpha \right) p^2(t)q^{\alpha}(t)a^{-\beta}(t) \right| a^{\beta}(t)q^{-\alpha}(t).$$

Let $\tau(s)$ denote the inverse function of $s(t)$, we obtain by (9.3.10) that

$$\int_{s_0}^{s} I[\tau(\xi)]d\xi = \int_{t_0}^{t} I(u)q^\alpha(u)a^{-\beta}(u)du$$

converges. Similarly,

$$\int_{s_0}^{s} |E[\tau(\xi)]|d\xi = \int_{t_0}^{t} |E(u)|q^\alpha(u)a^{-\beta}(u)du = \int_{t_0}^{t} \frac{|e(u)|}{(a(u)q(u))^\alpha}du$$

converges by (9.3.11). Next observe that condition (9.3.7) implies that $Q(t)$ is bounded from below. Hence if we integrate $\dot{V}$ from s_0 to s, use the bound indicated above, apply Gronwall's inequality and then transform the integrals from s back to t, we find that conditions (9.3.9) – (9.3.11) are precisely those needed to insure that $V(s)$ is bounded. Thus,

$$(a(t)q(t))^{\beta-\alpha}F(x(t)) = (a(t)q(t))^{\beta-\alpha}F(y(s)) \leq K_1$$

for some constant $K_1 \geq 0$, and now the conclusion of the theorem follows from conditions (9.3.4) and (9.3.12). ■

In particular, for the Emden–Fowler equation

$$x''(t) + q(t)x^\gamma(t) = 0, \tag{9.3.13}$$

where γ is an odd positive integer and $q(t)$ is as in equation (9.3.1), Theorem 9.3.1 yields the following result.

Corollary 9.3.1. Assume that

$$\int_{t_0}^{\infty} \left| \frac{q''(u)}{q^\theta(u)} - \theta\frac{[q'(u)]^2}{q^\lambda(u)} \right| du < \infty \tag{9.3.14}$$

and

$$\int_{t_0}^{\infty} \left| \frac{q''(u)}{q^\theta(u)} - \theta\frac{[q'(u)]^2}{q^\lambda(u)} \right| q^{-n/(n+1)}(u)du < \infty, \tag{9.3.15}$$

where $\theta = (2n+3)/(2(n+1))$ and $\lambda = (4n+5)/(2(n+1))$. If

$$\int_{t_0}^{\infty} q^{-n/(n+1)}(u)du < \infty, \tag{9.3.16}$$

then every solution $x(t)$ of equation (9.3.13) satisfies $\int_{t_0}^{\infty} x^{2n}(u)du < \infty$.

Actually, the corollary stated above follows by using the transformation (9.3.6) and the technique of the proof of Theorem 9.3.1 directly to equation (9.3.13) (the n in (9.3.6) is chosen so that $\gamma = 2n-1$). Conditions (9.3.3) and (9.3.4) are satisfied if we take $k = 2(n + 1)$ and $B = 0$. Clearly, (9.3.5) holds, (e.g., let $A = n$ and $k = 1$).

Next, we shall discuss some relationship between the nonlinear limit circle property and the boundedness, oscillation and convergence to zero of solutions of equation (9.3.1).

Theorem 9.3.2. Under the hypotheses of Theorem 9.3.1 all solutions of equation (9.3.1) are bounded. If in addition $F(x) > 0$, $x \neq 0$ and $a(t)b(t) \to \infty$ as $t \to \infty$, then all solutions of (9.3.1) converge to zero as $t \to \infty$.

Proof. From the proof of Theorem 9.3.1, we have $(a(t)q(t))^{\beta - \alpha} F(x(t)) \leq K_1$. Condition (9.3.9) implies that $a(t)b(t)$ is bounded below away from zero, so $F(x(t))$ is bounded. The boundedness of $x(t)$ now follows from condition (9.3.5). If $a(t)b(t) \to \infty$ as $t \to \infty$, then the above inequality shows that $F(x(t)) \to 0$ as $t \to \infty$. Since $F(x) > 0$ for $x \neq 0$, we find that $x(t) \to 0$ as $t \to \infty$. ∎

Next, we present the following result.

Theorem 9.3.3. Suppose there exist constants $M > 0$, $N \geq 0$, $m > 0$ and $c > 0$ such that

$$x^2 \leq Mxf(x) + N, \qquad \left| a^{1/2}(t)q'(t)q^{-3/2}(t) \right| \leq m, \qquad (9.3.17)$$

$$\left| \frac{e(t)}{q(t)} \right| \leq c, \qquad \int_{t_0}^{\infty} \left| \frac{e(u)}{q(u)} \right| du < \infty \qquad (9.3.18)$$

and

$$\int_{t_0}^{\infty} a(u)[q'(u)]^2 q^{-3}(u) du < \infty. \qquad (9.3.19)$$

If $x(t)$ is a limit circle solution of equation (9.3.1), i.e., (9.3.2) holds, then

$$\int_{t_0}^{\infty} \frac{a(u)}{q(u)} [x'(u)]^2 du < \infty. \qquad (9.3.20)$$

Proof. Multiply equation (9.3.1) by $x(t)/q(t)$ and note that $(a(t)x')'x = (a(t)xx')' - a(t)[x']^2$, and integrate by parts from $t_1 \geq t_0$ to t, to find

$$\frac{a(t)}{q(t)} x'(t)x(t) - \frac{a(t_1)}{q(t_1)} x'(t_1)x(t_1) + \int_{t_1}^{t} \frac{a(u)}{q^2(u)} q'(u)x'(u)x(u) du$$

$$+ \int_{t_1}^{t} x(u)f(x(u)) du - \int_{t_1}^{t} \frac{a(u)}{q(u)} [x'(u)]^2 du = \int_{t_1}^{t} \frac{e(u)}{q(u)} x(u) du.$$

$$(9.3.21)$$

By the Schwarz inequality, we have

$$\left| \int_{t_1}^{t} \frac{a(u)}{q^2(u)} q'(u) x'(u) x(u) du \right| \leq \left[\int_{t_1}^{t} \frac{a(u)}{q(u)} [x'(u)]^2 du \right]^{1/2}$$

$$\times \left[\int_{t_1}^{t} \frac{a(u)}{q^3(u)} [q'(u)]^2 x^2(u) du \right]^{1/2}.$$

Now from condition (9.3.17), we find

$$\frac{a(t)}{q^3(t)} [q'(t)]^2 x^2(t) \leq m^2 M x(t) f(x(t)) + N \frac{a(t)}{q^3(t)} [q'(t)]^2,$$

which on integrating and then applying (9.3.2) and (9.3.19), gives

$$\int_{t_1}^{\infty} \frac{a(u)}{q^3(u)} [q'(u)]^2 x^2(u) du \leq K_1 < \infty, \quad K_1 \text{ is a constant.}$$

Since

$$|x(t)| \frac{e(t)}{q(t)} \leq \frac{e(t)}{2q(t)} x^2(t) + \frac{e(t)}{2q(t)} \leq \frac{1}{2} M x(t) f(x(t)) + \left(\frac{N+1}{2} \right) \frac{e(t)}{q(t)},$$

the integral of the right–hand side of (9.3.21) converges in view of (9.3.18) and (9.3.2). If $x(t)$ is not eventually monotonic, let $\{t_j\}$, $\lim_{j \to \infty} t_j = \infty$ be an increasing sequence of zeros of $x'(t)$. Then from (9.3.21), we have $K_1 H^{1/2}(t_1) + K_2 \geq H(t_1)$, where K_2 is a constant, and $H(t) = \int_{t_1}^{t} (a(u)(x'(u))^2/q(u)) du$. It follows that $H(t_j) \leq K_3 < \infty$ for all j, where K_3 is a constant, and so (9.3.20) holds. If $x(t)$ is eventually monotonic, then $x(t) x'(t) \leq 0$, $t \geq t_1$ for all sufficiently large $t_1 \geq t_0$, since otherwise condition (9.3.2) will be violated. Using this fact in (9.3.21) we can repeat the type of arguments used above to again obtain that (9.3.20) holds. ∎

It is known that equation (9.3.1) with $f(x) = x$ and $e(t) = 0$ is oscillatory if it is limit circle. But, this is not true in general for forced equations. For example, Theorem 9.3.1 guarantees that all solutions of

$$(tx'(t))' + t^2 x(t) = \frac{9}{t^2} + \frac{1}{t} \quad \text{for} \quad t \geq 1$$

belong to $\mathcal{L}^2[1, \infty)$, but $x(t) = 1/t^3$ is a nonoscillatory solution of this equation. However, the following result holds for the forced equation (9.3.1).

Theorem 9.3.4. Assume that $f(x)$ is bounded away from zero if x is bounded away from zero,

$$\int_{t_0}^{\infty} \left(\int_{t_0}^{u} \frac{1}{a(v)} dv \right) |e(u)| du < \infty \tag{9.3.22}$$

and equation (9.3.1) is limit circle. Then every solution of equation (9.3.1) either oscillates or converges monotonically to zero as $t \to \infty$.

Proof. Let $x(t)$ be a nonoscillatory solution of equation (9.3.1), say, $x(t) > 0$ for $t \geq t_1 \geq t_0$. Clearly, $\liminf_{t \to \infty} x(t) = 0$ for otherwise condition (9.3.2) will be violated. If $x(t)$ is eventually monotonic, we are done, so we assume that $x(t)$ is not eventually monotonic. If $x(t)$ does not converge to zero as $t \to \infty$, then there exists a constant $K_1 > 0$ such that for any $t_2 > t_1$ there exists $t_3 > t_2$ with $x(t_3) \geq K_1$. Choose $t_2 > t_1$ so that

$$\int_{t_2}^{\infty} \left(\int_{t_2}^{u} \frac{1}{a(v)} dv \right) |e(u)| du \; < \; \frac{1}{2} K_1$$

and choose $t_3 > t_2$ such that $x'(t_3) = 0$, and $x(t_3) \geq K_1$. Integrating equation (9.3.1) from t_3 to t, we find

$$x'(t) \;=\; \frac{1}{a(t)} \int_{t_3}^{t} [e(u) - q(u)f(x(u))] du. \qquad (9.3.23)$$

Another integration yields

$$x(t) \;=\; x(t_3) + \left(\int_{t_3}^{t} \frac{1}{a(u)} du \right) \int_{t_3}^{t} [e(u) - q(u)f(x(u))] du$$
$$- \int_{t_3}^{t} \left(\int_{t_3}^{u} \frac{1}{a(v)} dv \right) [e(u) - q(u)f(x(u))] du.$$

If $t_4 > t_3$ is any zero of $x'(t)$, then (9.3.23) shows that the first integral above vanishes at $t = t_4$, and then we have

$$x(t_4) \;\geq\; x(t_3) - \int_{t_3}^{t_4} \left(\int_{t_3}^{u} \frac{1}{a(v)} dv \right) |e(u)| du \;\geq\; \frac{1}{2} K_1,$$

i.e., $x(t)$ is bounded from below by $K_1/2$ at every zero of $x'(t)$ for $t \geq t_3$, which contradicts $\liminf_{t \to \infty} x(t) = 0$. The proof for the case $x(t) < 0$, $t \geq t_1$ is similar. ∎

Next, we state two results for the oscillation of unforced and forced equations, respectively.

Theorem 9.3.5. If in addition to the hypotheses of Theorem 9.3.4, $e(t) = 0$ and $\int^{\infty} ds/a(s) = \infty$, then all solutions of equation (9.3.1) are oscillatory.

Theorem 9.3.6. Suppose the hypotheses of Theorem 9.3.4 hold except possibly condition (9.3.22). If for every constant $c > 0$ and all large $T \geq t_0$,

$$\liminf_{t \to \infty} \left[\int_{T}^{t} \frac{1}{a(u)} \left(\int_{T}^{u} e(v) dv - c \right) du \right] \;=\; -\infty$$

and

$$\limsup_{t\to\infty} \left[\int_T^t \frac{1}{a(u)} \left(\int_T^u e(v)dv + c \right) du \right] = \infty,$$

then all solutions of equation (9.3.1) are oscillatory.

The proofs of Theorems 9.3.5 and 9.3.6 are easy and hence omitted.

Next, we are concerned with the limit point criteria for second order nonlinear equations of type (9.3.1).

To simplify the notation used in our next result, we define the functions h, $H : [t_0, \infty) \to \mathbb{R}$ by

$$h(t) = \frac{[(a(t)q(t))']^-}{a(t)q(t)} + |e(t)| \quad \text{and} \quad H(t) = \frac{[(a(t)q(t))']^+}{a(t)q(t)} + \frac{|e(t)|}{(a(t)q(t))^{1/2}}.$$

Theorem 9.3.7. Suppose conditions (9.3.17) – (9.3.19) hold, there exists a constant $M_1 > 0$ such that

$$F(x) \leq M_1 x f(x) \tag{9.3.24}$$

and either

(I_1)

$$\int_{t_0}^\infty \frac{q(u)}{a(u)} \exp\left(-\int_{t_0}^u h(v)dv \right) du = \infty,$$

$$\int_{t_0}^\infty |e(u)| \left(\exp \int_{t_0}^u h(v)dv \right) du < \infty, \tag{9.3.25}$$

or
(I_2)

$$\int_{t_0}^\infty \exp\left(-\int_{t_0}^u H(v)dv \right) du = \infty,$$

$$\int_{t_0}^\infty \frac{|e(u)|}{(a(u)q(u))^{1/2}} \left(\exp \int_{t_0}^u H(v)dv \right) du < \infty. \tag{9.3.26}$$

Then equation (9.3.1) is of limit point type, i.e., there is a solution of (9.3.1) which does not satisfy condition (9.3.2).

Proof. We will write equation (9.3.1) as the system

$$\begin{cases} x' = y \\ y' = [-a'(t)y - q(t)f(x) + e(t)](1/a(t)). \end{cases} \tag{9.3.27}$$

If (I_1) holds, we define $V(x, y, t) = (1/2)a^2(t)y^2(t) + a(t)q(t)F(x)$. Then,

$$V'(t) = a(t)e(t)y + (a(t)q(t))'F(x) \geq a(t)e(t)y - \frac{[(a(t)q(t))']^-}{a(t)q(t)}V.$$

Now, we have

$$|a(t)e(t)y| \leq |e(t)| \left[\frac{1}{2} + \frac{1}{2}a^2(t)y^2 \right],$$

so it follows that

$$V'(s) \geq -\left(\frac{[(a(t)q(t))']^-}{a(t)q(t)} + |e(t)| \right) V - \frac{1}{2}|e(t)|. \qquad (9.3.28)$$

From condition (9.3.25), we find

$$\frac{1}{2} \int_{t_0}^{\infty} |e(u)| \left(\exp \left(\int_{t_0}^{u} h(v)dv \right) \right) du \leq K_1 < \infty.$$

Let $(x(t), y(t))$ be a solution of system (9.3.27) such that $(x(t_0), y(t_0)) = (x_0, y_0)$ and $V(x_0, y_0, t_0) = V(t_0) > K_1 + 1$. Then from (9.3.28), we get

$$\left(V(t) \exp \left(\int_{t_0}^{t} h(u)du \right) \right)' \geq -\frac{1}{2}|e(t)| \exp \left(\int_{t_0}^{t} h(u)du \right).$$

Integrating this inequality from t_0 to t, we obtain

$$V(t) \exp \left(\int_{t_0}^{t} h(u)du \right) \geq V(t_0) - K_1 > 1.$$

Hence,

$$\int_{t_0}^{t} \frac{V(u)}{a(u)q(u)} du \geq \int_{t_0}^{t} \frac{1}{a(u)q(u)} \exp \left(-\int_{t_0}^{u} h(v)dv \right) du \to \infty \text{ as } t \to \infty.$$

In view of Theorem 9.3.3 this shows that $x(t)$ cannot be a limit circle solution of equation (9.3.1).

If (I_2) holds, we define

$$V_1(x, y, t) = a(t)\frac{y^2}{2q(t)} + F(x).$$

Then, we have

$$V_1'(t) \geq \frac{e(t)}{q(t)}y - \frac{[a(t)q(t)]^+}{a(t)q(t)}V_1.$$

However, since

$$\left| \frac{e(t)}{q(t)}y \right| \left[\frac{|e(t)|}{(a(t)q(t))^{1/2}} \right] \left(V_1 + \frac{1}{2} \right),$$

we find

$$V_1'(t) + H(t)V_1(t) \geq -\frac{1}{2}\frac{|e(t)|}{(a(t)q(t))^{1/2}}.$$

The rest of the proof is similar to the case (I_1) and hence omitted. ∎

In our next limit point theorem we shall need the following lemma.

Lemma 9.3.1. Suppose there exist constants $M > 0$, $N \geq 0$ and $m_1 > 0$ such that

$$x^2 \leq Mxf(x) + N, \tag{9.3.29}$$

$$\left| \frac{(a(t)q(t))'}{a^{1/2}(t)q^{3/2}(t)} \right| \leq m_1 \quad \text{and} \quad \int_{t_0}^{\infty} \frac{[(a(u)q(u))']^2}{a(u)q^3(u)} du < \infty. \tag{9.3.30}$$

If $x(t)$ is a limit circle solution of equation (9.3.1), (i.e., condition (9.3.2) holds), then

$$\int_{t_0}^{\infty} \frac{[(a(u)q(u))']^2}{a(u)q^3(u)} x^2(u) du < \infty.$$

Proof. Clearly, by condition (8.3.30)

$$\int_{t_0}^{\infty} \frac{[(a(u)q(u))']^2}{a(u)q^3(u)} x^2(u) du \leq m_1^2 M \int_{t_0}^{\infty} x(u) f(x(u)) du$$

$$+ \int_{t_0}^{\infty} \frac{[(a(u)q(u))']^2}{a(u)q^3(u)} du < \infty. \quad ∎$$

Theorem 9.3.8. Suppose there exist constants $k_1 > 0$, $n > 0$ and $B_1 \geq 0$ such that

$$k_1 \leq 2(n+1) \quad \text{and} \quad 0 \leq (\beta - \alpha)F(x) - \frac{1}{k_1}xf(x) \leq B_1 F(x). \tag{9.3.31}$$

In addition assume that conditions (9.3.5), (9.3.9) – (9.3.11), (9.3.18), (9.3.19), (9.3.24) and (9.3.30) hold with the k in condition (9.3.10) replaced by k_1, and

$$\left| a^{1/2}(t)q'(t)q^{-3/2}(t) \right| \leq m. \tag{9.3.32}$$

If

$$\int_{t_0}^{\infty} (a(u)q(u))^{\alpha - \beta} du = \infty, \tag{9.3.33}$$

then equation (9.3.1) is of limit point type.

Proof. Apply the transformation (9.3.6) to (9.3.1) to obtain equation (9.3.7), and then write (9.3.7) in the form of system (9.3.8) with k now replaced by k_1. As in Theorem 9.3.1, we define $V(y, z, s) = (1/2)z^2 +$

$(a(t)q(t))^{\beta-\alpha}F(y)$ and differentiate, to obtain

$$\dot{V}(s) \geq -\left(\frac{1}{k_1} - \alpha\right)\frac{[(a(t)q(t))']^-}{a^\alpha(t)q^{\alpha+1}(t)}z^2 - |E(t)|\left(V + \frac{1}{2}\right)$$

$$-\left|\dot{p}(t) - \left(\frac{1}{k_1} - \alpha\right)p^2(t)\right|\frac{1}{k_1}\left[V + K + A\frac{V}{Q(t)}\right]$$

$$-Q(t)\left[(\beta - \alpha)F(y) - \frac{1}{k_1}yf(y)\right]\frac{[a(t)q(t))']^-}{a(t)q(t)}\frac{a^\beta(t)}{q^\alpha(t)}$$

$$\geq -\left[B_1 + 2\left(\frac{1}{k_1} - \alpha\right)\right]\left(\frac{[(a(t)q(t))']^-}{a(t)q(t)}\right)\left(\frac{a^\beta(t)}{q^\alpha(t)}\right)V$$

$$-\left|\dot{p}(t) - \left(\frac{1}{k_1} - \alpha\right)p^2(t)\right|\frac{1}{k_1}\left[V + K + A\frac{V}{Q(t)}\right] - |E(t)|\left(V + \frac{1}{2}\right).$$

Now define the functions $G, g : [t_0, \infty) \to \mathbb{R}$ by

$$G(t) = \left\{\left[B_1 + 2\left(\frac{1}{k_1} - \alpha\right)\right]\frac{[(a(t)q(t))']^-}{a(t)q(t)} + \frac{|e(t)|}{(a(t)q(t))^\alpha}\right.$$

$$\left. + \left|p'(t) - \left(\frac{1}{k_1} - \alpha\right)p^2(t)\frac{q^\alpha(t)}{a^\beta(t)}\right|\left[1 + A\frac{1}{Q(t)}\right]\frac{1}{k_1}\right\}\frac{a^\beta(t)}{q^\alpha(t)}$$

and

$$g(t) = \frac{K}{k_1}\left|p'(t) - \left(\frac{1}{k_1} - \alpha\right)p^2(t)\frac{q^\alpha(t)}{a^\beta(t)}\right| + \frac{|e(t)|}{2(a(t)q(t))^\alpha}\frac{a^\beta(t)}{q^\alpha(t)},$$

so that $\dot{V}(s) + G(t)V(s) \geq -g(t)$, and hence

$$\left(V\exp\left(\int_{s_0}^s G[\tau(\xi)]d\xi\right)\right)^{\cdot} \geq -g(t)\exp\left(\int_{s_0}^s G[\tau(\xi)]d\xi\right) \geq -K_1g(t)$$

$$(9.3.34)$$

since conditions (9.3.9) – (9.3.11) guarantee that $\exp\left(\int_{s_0}^s G[\tau(\xi)]d\xi\right) \leq K_1 < \infty$ for some constant $K_1 > 0$. In view of (9.3.10) and (9.3.11), we have $K_1\int_{s_0}^\infty g[\tau(\xi)]d\xi \leq K_2 < \infty$ for some constant $K_2 > 0$. Now let $x(t)$ be any solution of (9.3.1) such that $V(y(s_0), z(s_0), s_0) > K_2 + 1$. Integrating (9.3.34) from s_0 to s, we find

$$V(s)\exp\left(\int_{s_0}^s G[\tau(\xi)]d\xi\right) \geq V(s_0) - K_2 > 1$$

and thus $V(s) \geq 1/K_1$ for $s \geq s_0$. Now dividing this inequality by $(a(t)q(t))^{\beta-\alpha}$ and rewriting the left–hand side in terms of t, we get

$$\frac{a(t)}{2q(t)}(x'(t))^2 + \frac{(a(t)q(t))'}{kq^2(t)}x(t)x'(t) + \frac{[(a(t)q(t))']^2}{2k^2a(t)q^3(t)}x^2(t) + F(x(t))$$

$$\geq \frac{1}{K_1}(a(t)q(t))^{\alpha-\beta}.$$

$$(9.3.35)$$

If $x(t)$ was a limit circle solution of equation (9.3.1), then since conditions (9.3.5) and (9.3.24) imply (9.3.29), in view of Theorem 9.3.3, we have $\int_{t_0}^{\infty}(a(u)/q(u))(x'(u))^2 du < \infty$. Also, by Lemma 8.3.1

$$\int_{t_0}^{\infty}\frac{[(a(u)q(u))']^2}{a(u)q^3(u)}x^2(u)du \; < \; \infty$$

and by Lemma 8.3.1 and Theorem 8.3.3

$$\left|\int_{t_0}^{\infty}\frac{(a(u)q(u))'}{q^2(u)}x(u)x'(u)du\right|$$

$$\leq \; \left[\int_{t_0}^{\infty}\frac{[(a(u)q(u))']^2}{a(u)q^3(u)}x^2(u)du\right]^{1/2}\left[\int_{t_0}^{\infty}\frac{a(u)}{q(u)}[x'(u)]^2 du\right]^{1/2} \; < \; \infty$$

and by condition (9.3.24) and the assumption that $x(t)$ was a limit circle solution $\int_{t_0}^{\infty}F(x(u))du < \infty$. Now integrating (9.3.35) we get the desired contradiction. ∎

By combining Theorem 9.3.8 with Theorems 9.3.1 and 9.3.3 we can obtain necessary and sufficient conditions for the equation

$$(a(t)x'(t))' + q(t)x^{2n-1}(t) \; = \; e(t), \tag{9.3.36}$$

where n is a positive integer, to be of limit circle type, i.e., all solutions belong to $\mathcal{L}^{2n}[t_0,\infty)$. For this, first note that conditions (9.3.3), (9.3.4) and (9.3.31) are satisfied with $k = k_1 = 2(n+1)$ and $B = B_1 = 0$. In addition, conditions (9.3.5) and (9.3.24) are automatically satisfied.

Theorem 9.3.9. Suppose conditions (9.3.9), (9.3.11), (9.3.18), (9.3.19), (9.3.30) and (9.3.32) hold, and

$$\int_{t_0}^{\infty}\left|\left[\frac{(a(u)q(u))'}{a^{\alpha}(u)q^{\alpha+1}(u)}\right]'\right|du \; < \; \infty.$$

Then (9.3.36) is of limit circle type if and only if $\int_{t_0}^{\infty}[a(u)q(u)]^{-n/(n+1)}du < \infty$.

In particular, Theorem 9.3.9 for

$$x''(t) + q(t)x^{2n-1}(t) \; = \; 0 \tag{9.3.37}$$

immediately gives the following corollary.

Corollary 9.3.2. If

$$\int_{t_0}^{\infty}\frac{[q'(u)]^-}{q(u)}du \; < \; \infty, \qquad \left|\frac{q'(t)}{q^{3/2}(t)}\right| \; \leq \; m_2, \qquad \int_{t_0}^{\infty}\frac{[q'(u)]^2}{q^3(u)}du \; < \; \infty,$$

where $m_2 \geq 0$ is a constant and condition (9.3.14) holds, then equation (9.3.37) is of limit circle type if and only if

$$\int_{t_0}^{\infty} q^{-n/(n+1)}(u)du < \infty. \tag{9.3.38}$$

Example 9.3.1. Consider the equation

$$x''(t) + t^{\sigma} x^{2n-1}(t) = 0, \tag{9.3.39}$$

where σ is any constant and n is a positive integer. Now condition (9.3.38) implies that $\sigma n/(n+1) > 1$, or

$$\sigma > 1 + (1/n). \tag{9.3.40}$$

It is easy to check that all the hypotheses of Corollary 9.3.2 are satisfied, and hence condition (9.3.40) is a necessary and sufficient condition for all solutions of equation (9.3.39) to belong to $\mathcal{L}^{2n}[t_0, \infty)$.

9.4. Properties of Certain Differential Equations

Consider the second order differential equation

$$(a(t)\psi(x(t))x'(t))' + q(t)f(x(t)) = e(t), \tag{9.4.1}$$

where $a(t) \in C^1([t_0, \infty), \mathbb{R}^+)$, $e(t)$, $q(t) \in C([t_0, \infty), \mathbb{R})$, $t_0 \geq 0$, $\psi(x) \in C^1(\mathbb{R}, \mathbb{R})$, $f(x) \in C(\mathbb{R}, \mathbb{R})$ and $xf(x) > 0$ for $x \neq 0$.

We shall show that under appropriate conditions on $a(t)$ and $\psi(x)$, (9.4.1) can be reduced to the equation

$$x''(t) + q(t)f(x(t)) = e(t). \tag{9.4.2}$$

We define $h : \mathbb{R} \to \mathbb{R}$ by

$$h(x) = \int_0^x \psi(u)du. \tag{9.4.3}$$

If ψ is assumed to satisfy $\psi(x) > 0$ for $x \neq 0$, then clearly h is increasing, continuously differentiable and $xh(x) > 0$ for $x \neq 0$. Furthermore, the function $g : h(\mathbb{R}) \to \mathbb{R}$ defined by

$$g = f \circ h^{-1} \tag{9.4.4}$$

is continuous and satisfies $xg(x) > 0$ for $x \neq 0$.

Theorem 9.4.1. Suppose $\psi(x) > 0$ for $x \neq 0$. If $x = \phi(t)$ is a solution of (9.4.1) on some interval I, then $z = h \circ \phi(t)$ is a solution of the equation

$$(a(t)z'(t))' + q(t)g(z(t)) = e(t), \qquad (9.4.5)$$

where h and g are as in (9.4.3) and (9.4.4).

Conversely, if $z = \xi(t)$ is a nontrivial solution of (9.4.5) on some interval I, then $x = h^{-1} \circ \xi(t)$ is a nontrivial solution of equation (9.4.1) on some interval $J \subset I$. If in addition $\psi(0) \neq 0$, or $z(t) \neq 0$ for all $t \in I$, then $J = I$.

Proof. It is easy to verify, by using (9.4.3) and (9.4.4), that z and x defined above are respectively solutions of equations (9.4.5) and (9.4.1) and that $z' = \psi(x)x'$ and $J \subset I$. In fact, this inequality together with (9.4.3) shows that if $\psi(0) = 0$ and z vanishes at some $t_1 \in I$, then x' may not exist at t_1; in this case, J is a proper subset of I. The last statement of the theorem also follows at once. ∎

Corollary 9.4.1. Every oscillatory solution of (9.4.1) generates an oscillatory solution of equation (9.4.5).

Corollary 9.4.2. There is one–to–one correspondence between the non–oscillatory solutions of equations (9.4.1) and (9.4.5).

Corollary 9.4.3. Suppose $e(t) = 0$ and $\psi(0) = 0$. If the solution $z(t)$ of (9.4.5) with $z(t_0) = z'(t_0) = 0$ for every $t_0 \geq 0$ is unique, then equation (9.4.1) has no nontrivial oscillatory solutions. If in addition (9.4.5) is oscillatory, then equation (9.4.1) has no nontrivial continuable solutions.

Proof. Suppose $x(t)$ is a nontrivial solution of (9.4.1), then by Theorem 9.4.1 and (9.4.3), $z(t) = h(x(t))$ is a nontrivial solution of equation (9.4.5) such that

$$z'(t) = \psi(x(t))x'(t). \qquad (9.4.6)$$

By (9.4.3), $x(t)$ vanishes if and only if $z(t)$ vanishes. Since $\psi(0) = 0$ it follows from (9.4.6) that if $z(t_1) = 0$ for some $t_1 \geq t_0 \geq 0$, then $z'(t_1) = 0$ and hence from the uniqueness assumption $z(t) \equiv 0$ and so is $x(t)$, which is a contradiction. ∎

Example 9.4.1. Consider the differential equation

$$\left(x^2(t)x'(t)\right)' + \frac{1}{3}x^3(t) = 0 \qquad (9.4.7)$$

and let $x(t)$ be its solution, then $z(t) = x^3(t)/3$ is a solution of the linear equation $z''(t) + z(t) = 0$. Thus, $z(t) = A\sin(t + \theta)$ for some constants A and θ, and hence $x(t) = C\sin^{1/3}(t + \theta)$ where $C = (3A)^{1/3}$. As $x'(t)$ does not exist for $t = k\pi - \theta$, $k = 1, 2, \cdots$ equation (9.4.7) has no nontrivial continuable solution.

Example 9.4.2. Consider the differential equation

$$\left(x^{2n}(t)x'(t)\right)' + \frac{k}{t^2}x^{2n+1}(t) = 0, \tag{9.4.8}$$

where n is a positive integer and k is a constant. Here, $h(x) = x^{2n+1}/(2n+1)$ and the associated equation is

$$z''(t) + (2n+1)\left(\frac{k}{t^2}\right)z(t) = 0. \tag{9.4.9}$$

As equation (9.4.9) is oscillatory for $k > 1/(8n+4)$ and nonoscillatory for $k \le 1/(8n+4)$, by Corollary 9.4.3, equation (9.4.8) can have continuable solutions only when $k \le 1/(8n+4)$, and hence by Corollary 9.4.2 no nontrivial solution of (9.4.8) is oscillatory.

We now consider the unforced differential equation

$$(a(t)\psi(x(t))x'(t))' + q(t)f(x(t)) = 0 \tag{9.4.10}$$

subject to the additional condition $\psi(x) > 0$ for $x \ne 0$. The transformation h in (9.4.3) reduces (9.4.10) to

$$(a(t)z'(t))' + q(t)g(z(t)) = 0, \tag{9.4.11}$$

where g is defined in (9.4.4). If we now let

$$s = R(t) = \int_{t_0}^{t} \frac{du}{a(u)} \tag{9.4.12}$$

equation (9.4.11) is reduced to

$$\ddot{y} + R^*(s)Q(s)g(y) = 0, \quad (\cdot = d/ds) \tag{9.4.13}$$

where $R^*(s) = a[t(s)]$, $Q(s) = q[t(s)]$ and $y(s) = z[t(s)]$.

We now state the following interesting results.

Theorem 9.4.2. Suppose $\psi(x) > 0$ for $x \ne 0$, $q(t) < 0$ on $[t_1, t_2]$, $t_1 \ge 0$ and $x(t)$ is a solution of equation (9.4.10) on $[t_1, t_2]$ such that $x(t_1) = x'(t_1) = 0$. Then, $x(t) = x'(t) = 0$ for $t \in [t_1, t_2]$ if and only if

$$\int_{0+}^{1} \frac{\psi(u)}{\sqrt{F(u)}}du = \infty \quad \text{and} \quad \int_{0-}^{-1} \frac{\psi(u)}{\sqrt{F(u)}}du = -\infty,$$

where $F(x) = \int_0^x \psi(u)f(u)du$.

Theorem 9.4.3. Suppose $\psi(x) > 0$ for $x \neq 0$, $\int_0^{\pm\infty} \psi(u)du = \pm\infty$, and $q(t) < 0$ on $[t_1, t_2]$, $t_1 \geq 0$. Then equation (9.4.10) has a solution $x(t)$ such that $\lim_{t \to T} |x(t)| = \infty$ for some $T \in (t_1, t_2)$ if and only if

$$\int_0^\infty \frac{\psi(u)}{\sqrt{1 + F(u)}} du < \infty \quad \text{or} \quad \int_0^{-\infty} \frac{\psi(u)}{\sqrt{1 + F(u)}} du > -\infty,$$

where $F(x) = \int_0^x \psi(u)f(u)du$.

Next we state two simple oscillation criteria for equation (9.4.10).

Theorem 9.4.4. Suppose $\psi(x) > 0$ for $x \neq 0$ and the following conditions are satisfied

$$f'(x) \geq 0 \quad \text{for} \quad x \neq 0, \qquad \int^\infty \frac{du}{a(u)} = \infty, \qquad (9.4.14)$$

$$\left| \int_0^{\pm\infty} \psi(u)du \right| < \infty \quad \text{and} \quad \int^\infty q(s) \left(\int_{t_0}^s \frac{du}{a(u)} \right) ds = \infty. \qquad (9.4.15)$$

Then equation (9.4.10) is oscillatory.

Theorem 9.4.5. Suppose $\psi(x) > 0$ for $x \neq 0$, conditions (9.4.14) and (9.4.15) hold, and $\int_{\pm 1}^{\pm\infty} (\psi(u)/f(u))du < \infty$. Then equation (9.4.10) is oscillatory.

The next theorem describes the behavior of oscillatory solutions of equation (9.4.10) when $\psi(0) = 0$.

Theorem 9.4.6. Suppose $x(t)$ is a solution of equation (9.4.10) on $[t_1, t_2]$ such that $x(t_1) = x(t_2) = 0$. If $\psi(0) = 0$ and $q(t)$ does not change sign on $[t_1, t_2]$, then $x(t) = 0$ on $[t_1, t_2]$.

Proof. Suppose there exists $t_3 \in (t_1, t_2)$ such that $x(t_3) \neq 0$, then there exist $T_1, T_2 \in [t_1, t_2]$ such that $x(T_1) = x(T_2) = 0$ and $x(t) \not\equiv 0$ on (T_1, T_2). Integrating equation (9.4.10) from T_1 to T_2, we obtain

$$\int_{T_1}^{T_2} (a(s)\psi(x(s))x'(s))'ds + \int_{T_1}^{T_2} q(s)f(x(s))ds = 0.$$

As $x(T_1) = x(T_2) = 0$ and $\psi(0) = 0$, the first integral is zero, and hence $\int_{T_1}^{T_2} q(s)f(x(s))ds = 0$. As the integrand is of one sign and a continuous function of t, it follows that $q(t)f(x(t)) = 0$ for all $t \in [T_1, T_2]$, and hence $x(t) \equiv 0$ on $[T_1, T_2]$, which is a contradiction. ■

Corollary 9.4.4. Suppose $\psi(0) = 0$ and $q(t)$ does not change sign. Then the only oscillatory solution of equation (9.4.10) is the eventually identically zero solution.

9.5. Notes and General Discussions

1. Lemma 9.1.1 is taken from Leighton [21] and Swanson [26], Theorems 9.1.1 – 9.1.3 are due to Komkov [19]. Theorems 9.1.4 – 9.1.8 are borrowed from Graef and Spikes [13]. For related works we refer to Graef and Spikes [12], Kreith [20] and Swanson [26,27].

2. Lemma 9.2.1 is due to Chanturija [7]. Theorems 9.2.1 – 9.2.4 are extracted from the work of Kawane et. al. [17]. For more on this topic, we refer to Kartsatos [15].

3. The results of Section 9.3 are taken from Graef [11]. For the related subject we refer to Atkinson [1,2], Bellman [3], Burlak [4], Burton et. al. [5,6], Hinton [14], Kauffman et. al. [16], Knowles [18], Patula et. al. [23,24], Spikes [25], Weyl [28], and Wong et. al. [29,30].

4. The results of Section 9.4 are due to Mahfoud and Rankin [22]. For more general results we refer to Grace et. al. [8–10].

9.6. References

1. **F.V. Atkinson,** Nonlinear extensions of limit point criteria, *Math. Z.* **130**(1973), 297–312.

2. **F.V. Atkinson,** On second order differential inequalities, *Proc. Roy. Soc. Edinburgh Sec. A* **72**(1974), 109–127.

3. **R. Bellman,** *Stability Theory of Differential Equations,* McGraw Hill, New York, 1953.

4. **J. Burlak,** On the nonexistence of L_2–solutions of nonlinear differential equations, *Proc. Edinburgh Math. Soc.* **14**(1965), 257–268.

5. **T.A. Burton and R. Grimmer,** On the asymptotic behavior of solutions of $x'' + a(t)f(x) = e(t)$, *Pacific J. Math.* **41**(1972), 43–55.

6. **T.A. Burton and W.T. Patula,** Limit circle results for second order equations, *Monatsh. Math.* **81**(1976), 185–194.

7. **T.A. Chanturija,** Some comparison theorems for higher order ordinary differential equations, *Bull. Acad. Polon. Sci. Ser. Sci. Math. Astronom. Phys.* **25**(1977), 749–756 (in Russian).

8. **S.R. Grace,** Oscillation theorems for nonlinear differential equations of second order, *J. Math. Anal. Appl.* **171**(1992), 220–241.

9. **S.R. Grace and B.S. Lalli,** Integral averaging techniques for the oscillation of second order nonlinear differential equations, *J. Math. Anal. Appl.* **149**(1990), 277–311.

10. **S.R. Grace, B.S. Lalli and C.C. Yeh,** Oscillation theorems for nonlinear second order differential equations with a nonlinear damping term, *SIAM J. Math. Anal.* **15**(1984), 1082–1093.

11. **J.R. Graef,** Limit circle criteria and related properties for nonlinear equations, *J. Differential Equations* **35**(1980), 319–338.

12. **J.R. Graef and P.W. Spikes,** Sufficient conditions for nonoscillation of a second order nonlinear differential equation, *Proc. Amer. Math. Soc.* **50**(1975), 289–292.

13. **J.R. Graef and P.W. Spikes,** Comparison and nonoscillation results for perturbed nonlinear differential equations, *Ann. Mat. Pura Appl.* **116**(1978), 135–142.

14. **D. Hinton,** Limit point – Limit circle criteria for $(py')' + qy = \lambda ky$, in *Ordinary and Partial Differential Equations, Lecture Notes in Math.* **415**, Springer–Verlag, New York, 1974, 173–183.

15. **A.G. Kartsatos,** On the maintenance of oscillation of nth order equations under the effect of small forcing term, *J. Differential Equations* **10**(1971), 355–363.

16. **R.M. Kauffman, T.T Read and A. Zettl,** *The Deficiency Index Problem for Powers of Ordinary Differential Expressions, Lecture Notes in Math.* **621**, Springer–Verlag, New York, 1977.

17. **N. Kawano, T. Kusano and M. Naito,** Nonoscillatory solutions of forced differential equations of the second order, *J. Math. Anal. Appl.* **90**(1982), 323–342.

18. **I. Knowles,** On a limit–circle criterion for second order differential operators, *Quart. J. Math.* **24**(1973), 451–455.

19. **V. Komkov,** A note on a generalization of the Sturm–Picone theorem, *Colloq. Math.* **XXXIX**(1978), 173–176.

20. **K. Kreith,** *Oscillation Theory, Lecture Notes in Math.* **324**, Springer–Verlag, New York, 1973.

21. **W. Leighton,** Comparison theorems for linear differential equations of second order, *Proc. Amer. Math. Soc.* **13**(1962), 603–610.

22. **W.E. Mahfoud and S.M. Rankin,** Some properties of solutions of $(r(t)\Psi(x)x')' + a(t)f(x) = 0$, *SIAM J. Math. Anal.* **10**(1979), 49–54.

23. **W.T. Patula and P. Waltman,** Limit point classification of second order linear differential equations, *J. London Math. Soc.* **8**(1974), 209–216.

24. **W.T. Patula and J.S.W. Wong,** An L^p analog of the Weyl alternative, *Math. Ann.* **197**(1972), 9–28.

25. **P.W. Spikes,** On the integrability of solutions of perturbed nonlinear differential equations, *Proc. Roy. Soc. Edinburgh Sec. A* **77**(1977), 309–318.

26. **C.A. Swanson,** *Comparison and Oscillation Theory of Linear Differential Equations*, Academic Press, New York, 1968.

27. **C.A. Swanson,** Picone's identity, *Rend. Mat.* **8**(1975), 373–397.

28. **H. Weyl,** Über gewöhnliche Differentialgleichungen mit Singularitäten und die augehörige Entwicklung willkürlicher Funktionen, *Math. Ann.* **68**(1910), 220–269.

29. **J.S.W. Wong,** Remarks on the limit–circle classification of second order differential operators, *Quart. J. Math.* **24**(1973), 423–425.

30. **J.S.W. Wong and A. Zettl,** On the limit point classification of second order differential equations, *Math. Z.* **132**(1973), 297–304.

Chapter 10

Nonoscillation Theory for Multivalued Differential Equations

10.0. Introduction

In our previous chapters, we have presented several nonoscillation criteria for second order differential equations. In the present chapter, we shall introduce nonoscillatory theory for second order differential and neutral inclusions. Our results rely on fixed point theorems for multivalued maps, and on a compactness criterion.

10.1. Preliminaries

In this chapter, we shall provide nonoscillatory results for the second order differential inclusions

$$(a(t)x'(t))' \in e(t) + F(t, x(t)), \quad t \geq t_0 \geq 0 \qquad (10.1.1)$$

and the neutral inclusions

$$\frac{d}{dt}\left(a(t)\frac{d}{dt}(x(t) + px[t - \tau])\right) \in F(t, x(t)), \quad t \geq t_0 \geq 0. \qquad (10.1.2)$$

A nontrivial solution of (10.1.1), or (10.1.2) is called oscillatory if it has arbitrarily large zeros, otherwise it is called nonoscillatory. To present our nonoscillation criteria we shall need the following two fixed point theorems for multivalued maps.

Theorem 10.1.1 (Ky–Fan's Fixed Point Theorem [4,10,11]). Let Q be a nonempty, closed, convex subset of a Banach space E and $F : Q \to CK(Q)$ a upper semicontinuous, compact map; here $CK(Q)$ denotes the family of nonempty convex compact subsets of Q. Then, there exists $x \in Q$ with $x \in F(x)$.

Theorem 10.1.2 (Fitzpatrick–Petryshyn Fixed Point Theorem [9]). Let Q be a nonempty, closed, convex subset of a Banach space E and $F : Q \to CK(Q)$ a upper semicontinuous, condensing map with $F(Q)$ bounded. Then, there exists $x \in Q$ with $x \in F(x)$.

We shall also need the following compactness criterion in $B[T, \infty)$ (the Banach space of all continuous, bounded real valued functions on $[T, \infty)$ endowed with the usual supremum norm, i.e., $\|u\|_\infty = \sup_{t \in [T,\infty)} |u(t)|$ for $u \in B[T, \infty)$.)

Theorem 10.1.3 [7]. Let E be an equicontinuous and uniformly bounded subset of the Banach space $B[T, \infty)$. If E is equiconvergent at ∞, then it is relatively compact.

10.2. Differential Inclusions

In what follows with respect to the differential inclusion (10.1.1), we shall assume that the functions a and e are single valued and F is a multifunction. Furthermore, the following conditions hold:

(i) $a \in C([t_0, \infty), \mathbb{R}^+)$,

(ii) $e \in L^1([t_0, \infty), \mathbb{R}^+)$,

(iii) there exists an $\eta \in C([t_0, \infty), \mathbb{R})$ such that $(a(t)\, \eta'(t))' = e(t)$, $t \geq t_0$,

(iv) $F : [t_0, \infty) \times \mathbb{R} \to CK(\mathbb{R})$ is a L^1–Carathéodory multifunction: by this we mean

(a). for each measurable $u : [t_0, \infty) \to \mathbb{R}$ the map $t \mapsto F(t, u(t))$ has measurable single valued selections,

(b). for a.e. $t \in [t_0, \infty)$ the map $u \mapsto F(t, u)$ is upper semicontinuous,

(c). for each $r > 0$ there exists $h_r \in L^1[t_0, \infty)$ with $|F(t, u)| \leq h_r(t)$ for a.e. $t \in [t_0, \infty)$ and all $u \in \mathbb{R}$ with $|u| \leq r$; here $|F(t, u)| = \sup\{|z| : z \in F(t, u)\}$,

(v) $\inf_{t \in [t_0, \infty)} \eta(t) > -\infty$.

Remark 10.2.1. In (iv), part (a) could be replaced by: the map $t \mapsto F(t, u)$ is measurable for all $u \in \mathbb{R}$.

Now, let $\beta \in \mathbb{R}$ be such that

$$\beta > - \inf_{t \in [t_0, \infty)} \eta(t), \tag{10.2.1}$$

and let $d > 0$ be such that

$$\beta + \inf_{t \in [t_0, \infty)} \eta(t) \geq d. \tag{10.2.2}$$

Theorem 10.2.1. Suppose (i) – (v) hold, and let β (respectively d) be chosen as in (10.2.1) (respectively (10.2.2)). Also, assume that the following three conditions are satisfied

$$F : [t_0, \infty) \times (0, \infty) \rightarrow CK([0, \infty)), \tag{10.2.3}$$

$$\text{there exists} \quad M > d \quad \text{with} \quad M > \beta + \sup_{t \geq t_0} \eta(t) \tag{10.2.4}$$

and

$$\int^{\infty} \frac{1}{a(s)} \int_s^{\infty} \sup_{w \in [d,M]} |F(t,w)| dt ds \ < \ \infty. \tag{10.2.5}$$

Then there exists a nonoscillatory solution x of (10.1.1) for a.e. $t \geq T$ with

$$\lim_{t \to \infty} (x(t) - \eta(t)) \ = \ \beta \quad \text{and} \quad \lim_{t \to \infty} a(t)(x(t) - \eta(t))' = 0;$$

here T is chosen as in (10.2.6).

Proof. From (10.2.5) there exists $T \geq t_0$ such that

$$\int_T^{\infty} \frac{1}{a(s)} \int_s^{\infty} \sup_{w \in [d,M]} |F(t,w)| dt ds \ \leq \ M - \left[\beta + \sup_{t \geq T} \eta(t)\right]. \tag{10.2.6}$$

We wish to apply Theorem 10.1.1 with $E = (B[T, \infty), \| \cdot \|_\infty)$ and

$$Q \ = \ \{x \in B[T, \infty) : \ d \leq x(t) \leq M \quad \text{for} \quad t \geq T\}.$$

Clearly, Q is closed and convex. Also, if $x \in Q$ then for $t \geq T$, we have from (10.2.3) and the definition of Q that $0 \leq u(t) \leq \sup_{w \in [d,M]} |F(t,w)|$ for each $u(t) \in F(t, x(t))$. Define a mapping $N : Q \to \mathcal{P}(E)$ (the power set of E) by (here $x \in Q$),

$$N\,x(t) \ = \ \beta + \eta(t) + \int_t^{\infty} \frac{1}{a(s)} \int_s^{\infty} F(v, x(v)) dv ds \quad \text{for} \quad t \geq T.$$

The Proposition 1.1 in [6, pp. 777] guarantees that $N : Q \to C(E)$; here $C(E)$ denotes the family of nonempty, convex subsets of E. We shall first show that

$$N : Q \ \rightarrow \ C(Q). \tag{10.2.7}$$

For notational purposes for any $x \in Q$ let

$$\mathcal{F}(x) \ = \ \{u \in L^1[T, \infty) : \ u(t) \in F(t, x(t)) \text{ for a.e. } t \in [T, \infty)\}.$$

Let $x \in Q$ and take $w \in N\,x$. Then there exists $\tau \in \mathcal{F}(x)$ with

$$w(t) \;=\; \beta + \eta(t) + \int_t^\infty \frac{1}{a(s)} \int_s^\infty \tau(v)\,dv\,ds \quad \text{for } t \geq T.$$

Then for $t \geq T$, we have from (10.2.6) that

$$
\begin{aligned}
w(t) &\leq \beta + \sup_{t \geq T} \eta(t) + \int_t^\infty \frac{1}{a(s)} \int_s^\infty \sup_{w \in [d,M]} |F(v,w)|\,dv\,ds \\
&\leq \beta + \sup_{t \geq T} \eta(t) + \left[M - \left\{ \beta + \sup_{t \geq T} \eta(t) \right\} \right] \;=\; M.
\end{aligned}
$$

As a result $w(t) \leq M$ for $t \geq T$ for each $w \in N\,x$. On the other hand if $t \geq T$, we have

$$w(t) \;\geq\; \beta + \eta(t) \;\geq\; \beta + \inf_{t \geq t_0} \eta(t) \;\geq\; d.$$

As a result $w(t) \geq d$ for $t \geq T$ for each $w \in N\,x$. Thus (10.2.7) holds.

Next, we shall show that

$$N : Q \;\to\; C(Q) \quad \text{is a compact map.} \tag{10.2.8}$$

For this, we will use Theorem 10.1.3. For $x \in Q$ let $G(x) = N\,x - \eta$. Take any $x \in Q$ and $w \in G(x)$. Then there exists $\tau \in \mathcal{F}(x)$ such that

$$w(t) \;=\; \beta + \int_t^\infty \frac{1}{a(s)} \int_s^\infty \tau(v)\,dv\,ds \quad \text{for } t \geq T.$$

Now, since $N : Q \to C(Q)$, we have for $t \geq T$ that $w(t) \leq M + \max\{|d - \beta|, |M - \beta|\}$, and hence for each $w \in G\,x$ it follows that

$$\|w\|_\infty \;=\; \sup_{t \in [T,\infty)} |w(t)| \;\leq\; M + \max\{|d - \beta|, |M - \beta|\} \quad \text{for each } x \in Q.$$

Thus, the set $Y = \{N\,x - \theta : x \in Q\}$ is a uniformly bounded subset of $B[T, \infty)$. Also, for each $t \geq T$, we have

$$|w(t) - \beta| \;\leq\; \int_t^\infty \frac{1}{a(s)} \int_s^\infty \sup_{w \in [d,M]} |F(v,w)|\,dv\,ds \tag{10.2.9}$$

for $w \in G\,x$, $x \in Q$. Now (10.2.5) and (10.2.9) guarantee that the set Y is equiconvergent at ∞. Also, for t_1, t_2 with $T \leq t_1 \leq t_2$, we have

$$|w(t_2) - w(t_1)| \;\leq\; \int_{t_1}^{t_2} \frac{1}{a(s)} \int_s^\infty \sup_{w \in [d,M]} |F(v,w)|\,dv\,ds$$

for $w \in G\,x$, $x \in Q$. Now Theorem 10.1.3 guarantees that Y is relatively compact in $B[T, \infty)$, and as a result (10.2.8) holds.

It remains to show

$$N : Q \;\to\; CK(Q) \quad \text{is a upper semicontinuous map.} \qquad (10.2.10)$$

From (10.2.10) and [4, pp. 465] it is sufficient to show that the graph of N, $\mathcal{G}(N)$, is closed. Consider $(\mu_n, x_n) \in \mathcal{G}(N)$ with $(\mu_n, x_n) \to (\mu, x)$; here $n \in \mathbb{N}_1 = \{1, 2, \cdots\}$. We must show $\mu \in N\,x$. Fix $t \in [T, \infty)$. Note that $\|x\|_\infty \leq M$, $\|x_n\|_\infty \leq M$ for $n \in \mathbb{N}_1$, since x, $x_n \in Q$ for $n \in \mathbb{N}_1$. Also there exists $z_n \in \mathcal{F}(x_n)$ with

$$
\begin{aligned}
\mu_n(t) \;&=\; \beta + \eta(t) + \int_t^\infty \frac{1}{a(s)} \int_s^\infty z_n(v)\,dv\,ds \\
&=\; \beta + \eta(t) + \int_t^\infty z_n(v) \int_t^v \frac{1}{a(s)}\,ds\,dv.
\end{aligned}
\qquad (10.2.11)
$$

We must show that there exists $u \in \mathcal{F}(x)$ with

$$\mu(t) \;=\; \beta + \eta(t) + \int_t^\infty \frac{1}{a(s)} \int_s^\infty u(v)\,dv\,ds.$$

Notice (iv) guarantees that there exists a $h_M \in L^1[T, \infty)$ with $|z_n(s)| \leq h_M(s)$ for a.e. $s \in [T, \infty)$. Consider $\{z_n\}_{n \in \mathbb{N}_1}$. Take $k \in \mathbb{N}_1$ and $k > t$. From (10.2.11), we have

$$\left| \mu_n(t) - \beta - \eta(t) - \int_t^k z_n(v) \int_t^v \frac{1}{a(s)}\,ds\,dv \right| \;\leq\; \int_k^\infty h_M(v) \int_t^v \frac{1}{a(s)}\,ds\,dv.$$
$$(10.2.12)$$

Now, a standard result from the literature [14, Proposition 1.4] guarantees that $\mathcal{F}_k : B[T, k] = C[T, k] \to L^1[T, k]$ is upper semicontinuous with respect to the weak topology (w–u.s.c.) and also weakly completely continuous; here $\mathcal{F}_k$ is given by

$$\mathcal{F}_k(w) \;=\; \left\{ u \in L^1[T, k] : \; u(t) \in F(t, w(t)) \;\; \text{for a.e. } t \in [T, k] \right\}.$$

Since $z_n \in \mathcal{F}_k(x_n)$ for $n \in \mathbb{N}_1$, there exists a $u_k \in L^1[T, k]$ and a subsequence of S of $\mathbb{N}_1$ with z_n converging weakly in $L^1[T, k]$ to u_k (i.e. $z_n \rightharpoonup u_k$ in $L^1[T, k]$) as $n \to \infty$ in S. Now $x_n \to x$ in $C[T, k]$ and $z_n \rightharpoonup u_k$ in $L^1[T, k]$ as $n \to \infty$ in S, together with $z_n \in \mathcal{F}_k(x_n)$ for $n \in S$ and $\mathcal{F}_k : C[T, k] \to L^1[T, k]$ w–u.s.c., implies that

$$u_k \;\in\; \mathcal{F}_k(x). \qquad (10.2.13)$$

Note as well that $\|x\|_k = \sup_{s \in [T, k]} |x(s)| \leq M$, $\|x_n\|_k \leq M$ for $n \in S$, and $|u_k(v)| \leq h_M(v)$ for a.e. $v \in [T, k]$. Let $n \to \infty$ through S in

(10.2.12), to obtain

$$\left| \mu(t) - \beta - \eta(t) - \int_t^k u_k(v) \int_t^v \frac{1}{a(s)} ds dv \right| \le \int_k^\infty h_M(v) \int_t^v \frac{1}{a(s)} ds dv.$$
(10.2.14)

Similarly, we can show that there exists $u_{k+1} \in L^1[T, k+1]$ and a subsequence of S, say S_1, with $z_n \rightharpoonup u_{k+1}$ in $L^1[T, k+1]$ as $n \to \infty$ in S_1 and with $u_{k+1} \in \mathcal{F}_{k+1}(x)$. Of course this implies $z_n \rightharpoonup u_{k+1}$ in $L^1[T, k]$ as $n \to \infty$ in S_1 so $u_{k+1}(v) = u_k(v)$ for a.e. $v \in [T, k]$. In addition note that $|u_{k+1}(v)| \le h_M(v)$ for a.e. $v \in [T, k+1]$. Continue and construct $u_{k+2}, u_{k+3}, \cdots$. For $l \in \{k, k+1, \cdots\} = \mathbb{N}_k$ let $u_l^*(v)$ be any extension to $[T, \infty)$ of u_l with $|u_l^*(v)| \le h_M(v)$ for a.e. $v \in (l, \infty)$. Also, let

$$\mathcal{F}_l^*(w) = \{z \in L^1[T, \infty) : \quad z(v) \in F(v, w(v)) \text{ for a.e. } v \in [T, l],$$
$$|z(v)| \le h_M(v) \text{ for a.e. } v \in [T, \infty)\}.$$

Now, $\{u_l^*\}_{l \in P}$ is a weakly compact sequence in $L^1[T, \infty)$, (see [5 or 7]), so there exists a subsequence which converges weakly to a function $u \in L^1[T, \infty)$. Note $u(v) = u_k(v)$ for a.e. $v \in [T, k]$ since $u_{k+m}(v) = u_k(v)$ for a.e. $v \in [T, k]$; here $m \in \mathbb{N}_1$. This together with (10.2.14) yields

$$\left| \mu(t) - \beta - \eta(t) - \int_t^k u(v) \int_t^v \frac{1}{a(s)} ds dv \right| \le \int_k^\infty h_M(v) \int_t^v \frac{1}{a(s)} ds dv.$$
(10.2.15)

We next claim that

$$\mathcal{F}(x) = \bigcap_{l \in K} \mathcal{F}_l^*(x) \quad \text{(here } K = \{[T]+1, [T]+2, \cdots\}) \quad (10.2.16)$$

(and $\mathcal{F}(w)$ is nonempty, closed and convex). Note first that $\|x\|_\infty \le M$ so $|F(v, x(v))| \le h_M(v)$ for a.e. $v \in [T, \infty)$. Let x_k be the restriction to the interval $[T, k]$, $k \in \mathbb{N}_1$, of x. Note [10,11 or 12] that

$$\mathcal{F}_k(x_k) = \{z \in L^1[T, k] : z(v) \in F(v, x_k(v)) \text{ for a.e. } v \in [T, k]\}$$

is closed in $L^1[T, k]$ for all $k \in K$. Let

$$\mathcal{F}_k^*(x_k) = \{z \in L^1[T, \infty) : \quad z \in \mathcal{F}_k(x_k) \text{ for } v \in [T, k]$$
$$\text{and } z(v) = 0 \text{ for } v > k\}.$$

It is immediate that $\mathcal{F}_k^*(x_k)$ is a closed set in $L^1[T, \infty)$ for each $k \in K$. Let

$$R_k = \{z \in L^1[T, \infty) : \quad z(v) = 0 \text{ for } v \in [T, k],$$
$$|z(v)| \le h_M(v) \text{ for a.e. } v \in (k, \infty)\}$$

and notice it is clear that $\mathcal{F}_k^*(x) = \mathcal{F}_k^*(x_k) \oplus R_k$. It is clear that $\mathcal{F}_k^*(x)$ is a closed set in $L^1[T,\infty)$. Also, for each $k \in K$, we have $\mathcal{F}(x) \subseteq \mathcal{F}_k^*(x)$, and so $\mathcal{F}(x) \subseteq \cap_{l \in K} \mathcal{F}_l^*(x)$. On the other hand if $z \in \mathcal{F}_l^*(x)$ for each $l \in K$, then $z(v) \in F(v, x(v))$ for a.e. $v \in [T, \infty)$, and so $\cap_{l \in K} \mathcal{F}_l^*(x) \subseteq \mathcal{F}(x)$. Thus, (10.2.16) holds and also $\mathcal{F}(x)$ is a closed subset of $L^1[T, \infty)$. Now, since u belongs to $\cap_{l \in K} \mathcal{F}_l^*(x)$ (note for each $l \in K$ that $u \in \mathcal{F}_l^*(x)$), we have $u \in \mathcal{F}(x)$. Let $k \to \infty$ in (10.2.15), to obtain

$$\mu(t) - \beta - \eta(t) - \int_t^\infty u(v) \int_t^v \frac{1}{a(s)} ds\, dv = 0$$

and hence

$$\mu(t) = \beta + \eta(t) + \int_t^\infty \frac{1}{a(s)} \int_s^\infty u(v) dv\, ds.$$

Consequently, $\mathcal{G}(N)$ is closed, so $N : Q \to C(Q)$ is a upper semicontinuous map (see [4, pp. 465]); in fact (10.2.8) guarantees that $N : Q \to CK(Q)$.

Theorem 10.1.1 now guarantees that there exists $x \in Q$ with $x \in N\,x$, i.e., for every $t \geq T$, we have

$$x(t) \in \beta + \eta(t) + \int_t^\infty \frac{1}{a(s)} \int_s^\infty F(v, x(v)) dv\, ds,$$

and so

$$(a(t)x'(t))' \in e(t) + F(t, x(t)) \quad \text{for a.e. } t \geq T.$$

In addition

$$|x(t) - \eta(t) - \beta| \leq \int_t^\infty \frac{1}{a(s)} \int_s^\infty \sup_{w \in [d, M]} |F(v, w)| dv\, ds$$

$$\to 0 \quad \text{as } t \to \infty$$

and

$$|a(t)[x(t) - \eta(t)]'| \leq \int_t^\infty \sup_{w \in [d, M]} |F(v, w)| dv \to 0 \quad \text{as } t \to \infty. \quad \blacksquare$$

We shall next show that an existence result for (10.1.1) can be obtained if (10.2.3) is replaced by

$$F : [t_0, \infty) \times (0, \infty) \to CK((-\infty, 0]). \tag{10.2.17}$$

Theorem 10.2.2. Suppose (i) – (v) and (10.2.17) hold, and let β (respectively d) be chosen as in (10.2.1) (respectively (10.2.2)). Also, assume that the following two conditions are satisfied

there exist $K > 1$ and $M > \dfrac{d}{K}$ with $M \geq \beta + \sup_{t \geq t_0} \eta(t)$ (10.2.18)

and

$$\int^{\infty} \frac{1}{a(s)} \int_s^{\infty} \sup_{w \in [d/K, M]} |F(t, w)| dt ds < \infty. \tag{10.2.19}$$

Then there exists a nonoscillatory solution x of (10.1.1) for a.e. $t \geq T$ with $\lim_{t \to \infty} (x(t) - \eta(t)) = \beta$ and $\lim_{t \to \infty} a(t)(x(t) - \eta(t))' = 0$; here T is chosen as in (10.2.20).

Proof. From (10.2.19) there exists $T \geq t_0$ with

$$\int_T^{\infty} \frac{1}{a(s)} \int_s^{\infty} \sup_{w \in [d/K, M]} |F(t, w)| dt ds \leq \left(\frac{K-1}{K}\right) d. \tag{10.2.20}$$

We will apply Theorem 10.1.1 with $E = (B[T, \infty), \| \cdot \|_{\infty})$ and

$$Q = \{x \in B[T, \infty) : d/K \leq x(t) \leq M \text{ for } t \geq T\}.$$

Let N and $\mathcal{F}$ be as in Theorem 10.2.1. First, we shall show that

$$N : Q \to C(Q). \tag{10.2.21}$$

Let $x \in Q$ and take $w \in N x$. Then there exists $\tau \in \mathcal{F}(x)$ such that

$$w(t) = \beta + \eta(t) + \int_t^{\infty} \frac{1}{a(s)} \int_s^{\infty} \tau(v) dv ds \quad \text{for } t \geq T.$$

If $t \geq T$, we also have

$$\begin{aligned} w(t) &\geq \beta + \eta(t) - \int_T^{\infty} \frac{1}{a(s)} \int_s^{\infty} \sup_{w \in [d/K, M]} |F(t, w)| dt ds \\ &\geq \beta + \inf_{t \geq T} \eta(t) - \left(\frac{K-1}{K}\right) d \geq d - \left(\frac{K-1}{K}\right) d = \frac{d}{K}. \end{aligned}$$

As a result $w(t) \geq d/K$ for $t \geq T$ for each $w \in N x$. On the other hand for $t \geq T$, we have $w(t) \leq \beta + \eta(t) \leq \beta + \sup_{t \geq T} \eta(t) \leq M$. Thus, $w(t) \leq M$ for $t \geq T$ for each $w \in N x$. Hence, (10.2.21) holds. Essentially the same reasoning as in Theorem 10.2.1 guarantees that $N : Q \to CK(Q)$ is a upper semicontinuous, compact map. Now, Theorem 10.1.1 proves the result. ∎

In our next result, we will show that the assumption (10.2.17) (respectively, (10.2.3)) in Theorem 10.2.2 (respectively, Theorem 10.2.1) can be removed if we combine the analysis of both the theorems. In this case (10.2.18) has to be adjusted slightly.

Theorem 10.2.3. Suppose (i) – (v) hold, and let β (respectively d) be chosen as in (10.2.1) (respectively, (10.2.2)). Also, assume that the following two conditions are satisfied

$$\text{there exist } K > 1 \text{ and } M > \frac{d}{K} \text{ with } M > \beta + \sup_{t \geq t_0} \eta(t) \tag{10.2.22}$$

and

$$\int^\infty \frac{1}{a(s)} \int_s^\infty \sup_{w\in[d/K,M]} |F(t,w)|dtds < \infty. \tag{10.2.23}$$

Then there exists a nonoscillatory solution x of (10.1.1) for a.e. $t \geq T$ with $\lim_{t\to\infty} (x(t) - \eta(t)) = \beta$ and $\lim_{t\to\infty} a(t)(x(t) - \eta(t))' = 0$; here T is chosen as in (10.2.24).

Proof. From (10.2.23) there exists $T \geq t_0$ such that

$$\int_T^\infty \frac{1}{a(s)} \int_s^\infty \sup_{w\in[d/K,M]} |F(t,w)|dtds$$
$$\leq \min\left\{\left(\frac{K-1}{K}\right)d,\ M - \left[\beta + \sup_{t\geq T} \eta(t)\right]\right\}. \tag{10.2.24}$$

Let Q and N be as in Theorem 10.2.2. Let $x \in Q$ and take $w \in N\,x$. As in Theorem 10.2.2, we find $w(t) \geq d/K$ for $t \geq T$ and as in Theorem 10.2.1, we have $w(t) \leq M$ for $t \geq T$. Thus, $N: Q \to C(Q)$. ∎

Remark 10.2.2. Minor adjustments in the analysis of this section would enable us to discuss the more general differential inclusion

$$(a(t)(b(t)x(t))')' \in e(t) + F(t, x(t)), \quad t \geq t_0 \geq 0.$$

10.3. Neutral Inclusions

In this section, we shall discuss the neutral inclusion (10.1.2), where the function a is single valued, p and τ are constants, and F is a multifunction. In what follows, we shall assume that the following conditions hold:

(vi) $\tau \in \mathbb{R}^+$,

(vii) $a \in C([t_0, \infty), \mathbb{R}^+)$,

(viii) $F: [t_0, \infty) \times \mathbb{R} \to CK(\mathbb{R})$ is a L^1–Carathéodory multifunction.

Theorem 10.3.1. Suppose (vi) – (viii) hold. Also, assume that the following three conditions are satisfied

$$F: [t_0, \infty) \times (0, \infty) \to CK((-\infty, 0]) \tag{10.3.1}$$

$$|p| \neq 1 \tag{10.3.2}$$

and

there exists $K > 0$ such that $\int^\infty \frac{1}{a(s)} \int_s^\infty \sup_{w\in[K/2,K]} |F(t,w)|dtds < \infty.$

$$\tag{10.3.3}$$

Then there exists a bounded nonoscillatory solution x of (10.1.2) for a.e. $t \geq T \geq t_0$ with T suitably chosen (see the proof of the theorem).

Remark 10.3.1 One cannot expect an analog of Theorem 10.3.1 for the case $|p| = 1$ even in the neutral equation case [8, Chapter 5] (see also [2, Chapter 3] for some partial results when $|p| = 1$).

Proof. We need to consider the two cases $|p| < 1$ and $|p| > 1$ separately.

Case (I). $|p| < 1$. Choose $T \geq t_0$ so that $t - \tau > t_0$ for $t \geq T$ and

$$\int_T^\infty \frac{1}{a(s)} \int_s^\infty \sup_{w \in [K/2, K]} |F(t, w)| dt ds \leq \frac{1}{4}(1 - |p|)K. \qquad (10.3.4)$$

Let $T_1 = T - \tau$. We wish to apply Theorem 10.1.2 with $E = (B[T_1, \infty), \| \cdot \|_\infty)$ and

$$Q = \left\{ x \in B[T_1, \infty) : \frac{K}{2} \leq x(t) \leq K \text{ for } t \geq T_1 \right\}.$$

Define the single valued map $N_1 : Q \to E$ and the multivalued map $N_2 : Q \to \mathcal{P}(E)$ as follows (here $x \in Q$):

$$N_1 x(t) = \begin{cases} \frac{3}{4}(1+p)K - px[T - \tau], & T_1 \leq t \leq T \\ \frac{3}{4}(1+p)K - px[t - \tau], & t \geq T \end{cases}$$

and

$$N_2 x(t) = \begin{cases} \{0\}, & T_1 \leq t \leq T \\ -\int_T^t \frac{1}{a(s)} \int_s^\infty F(v, x(v)) dv ds, & t \geq T. \end{cases}$$

First, we shall show that

$$N = N_1 + N_2 : Q \to C(Q). \qquad (10.3.5)$$

For notational purposes for any $x \in Q$ let

$$\mathcal{F}(x) = \left\{ u \in L^1[T_1, \infty) : u(t) \in F(t, x(t)) \text{ for a.e. } t \in [T_1, \infty) \right\}.$$

Let $x \in Q$ so that $K/2 \leq x(t) \leq K$ for $t \in [T_1, \infty)$, and take $w \in N_2 x$. Then there exists $\tau_0 \in \mathcal{F}(x)$ with

$$w(t) = \begin{cases} \{0\}, & T_1 \leq t \leq T \\ -\int_T^t \frac{1}{a(s)} \int_s^\infty \tau_0(v) dv ds, & t \geq T \end{cases}$$

Now, we need to consider two subcases, namely, $0 \le p < 1$ and $-1 < p < 0$.

Subcase (1). $0 \le p < 1$. If $T_1 \le t \le T$, then clearly

$$
\begin{aligned}
N_1\, x(t) + w(t) &= \frac{3}{4}(1+p)K - px[T - \tau] \\
&\ge \frac{3}{4}(1+p)K - pK = \left(\frac{3}{4} - \frac{1}{4}p\right) K \ge \frac{K}{2}
\end{aligned}
$$

and

$$
N_1\, x(t) + w(t) \le \frac{3}{4}(1+p)K - p\frac{K}{2} = \left(\frac{3}{4} + \frac{1}{4}p\right) K \le K.
$$

If $t \ge T$, then (10.3.1) implies

$$
N_1\, x(t) + w(t) \ge \frac{3}{4}(1+p)K - px[t - \tau] \ge \frac{3}{4}(1+p)K - pK \ge \frac{K}{2},
$$

and (10.3.4) implies

$$
\begin{aligned}
N_1\, x(t) + w(t) &\le \frac{3}{4}(1+p)K - px[t - \tau] + \frac{1}{4}(1-p)K \\
&\le \frac{3}{4}(1+p)K - p\frac{K}{2} + \frac{1}{4}(1-p)K = K.
\end{aligned}
$$

As a result $K/2 \le N_1\, x(t) + w(t) \le K$ for $t \ge T_1$ for each $w \in N_2\, x$. Thus, (10.3.5) holds in this case.

Subcase (2). $-1 < p < 0$. If $T_1 \le t \le T$, then clearly

$$
\begin{aligned}
N_1\, x(t) + w(t) &= \frac{3}{4}(1+p)K - px[T - \tau] \\
&\ge \frac{3}{4}(1+p)K - p\frac{K}{2} = \left(\frac{3}{4} + \frac{1}{4}p\right) K \ge \frac{K}{2}
\end{aligned}
$$

and

$$
N_1\, x(t) + w(t) \le \frac{3}{4}(1+p)K - pK = \left(\frac{3}{4} - \frac{1}{4}p\right) K \le K.
$$

If $t \ge T$, then (10.3.1) implies

$$
N_1\, x(t) + w(t) \ge \frac{3}{4}(1+p)K - px[t - \tau] \ge \left(\frac{3}{4} + \frac{1}{4}p\right) K \ge \frac{K}{2},
$$

and (10.3.4) implies

$$
N_1\, x(t) + w(t) \le \frac{3}{4}(1+p)K - pK + \frac{1}{4}(1+p)K = K.
$$

As a result $K/2 \leq N_1 x(t) + w(t) \leq K$ for $t \geq T_1$ for each $w \in N_2 x$. Thus, (10.3.5) holds in this case also.

Essentially the same reasoning as in Theorem 10.2.1 guarantees that

$$N_2 : Q \rightarrow C(E) \quad \text{is a upper semicontinuous, compact map.} \quad (10.3.6)$$

Next, we claim that

$$N_1 : Q \rightarrow E \quad \text{is a contractive map.} \quad (10.3.7)$$

To see this notice for $x_1, x_2 \in Q$ and $T_1 \leq t \leq T$, we have

$$|N_1 x_1(t) - N_1 x_2(t)| = |p\{x_1[T - \tau] - x_2[T - \tau]\}| \leq |p| \|x_1 - x_2\|_\infty,$$

whereas if $t \geq T$, we have

$$|N_1 x_1(t) - N_1 x_2(t)| = |p\{x_1[t - \tau] - x_2[t - \tau]\}| \leq |p| \|x_1 - x_2\|_\infty.$$

Thus, it follows that

$$\|N_1 x_1 - N_1 x_2\|_\infty \leq |p| \|x_1 - x_2\|_\infty,$$

so in view of $|p| < 1$, (10.3.7) holds.

Now, (10.3.5), (10.3.6) and (10.3.7) imply that

$$N : Q \rightarrow CK(Q) \quad \text{is a upper semicontinuous, condensing map.} \quad (10.3.8)$$

Thus, Theorem 10.1.2 implies that there exists $x \in Q$ with $x \in N_1 x + N_2 x$. Hence, for $t \geq T$, we have

$$x(t) \in \frac{3}{4}(1 + p)K - px[t - \tau] - \int_T^t \frac{1}{a(s)} \int_s^\infty F(v, x(v)) dv ds.$$

This completes the proof of Case (I).

Case (II). $|p| > 1$. Choose $T \geq t_0$ so that $t - \tau > t_0$ for $t \geq T$ and

$$\int_T^\infty \frac{1}{a(s)} \int_s^\infty \sup_{w \in [K/2, K]} |F(t, w)| dt ds \leq \frac{1}{4}(|p| - 1)K.$$

Let $T_1 = T - \tau$, and E and Q be as in Case (I). Define $N_1 : Q \rightarrow E$ and $N_2 : Q \rightarrow \mathcal{P}(E)$ as follows (here $x \in Q$):

$$N_1 x(t) = \begin{cases} \dfrac{3}{4}\left(\dfrac{1+p}{p}\right)K - \dfrac{1}{p}x[T + \tau], & T_1 \leq t \leq T \\[3mm] \dfrac{3}{4}\left(\dfrac{1+p}{p}\right)K - \dfrac{1}{p}x[t + \tau], & t \geq T \end{cases}$$

and

$$
N_2\, x(t) \;=\; \begin{cases} \{0\}, & T_1 \le t \le T \\[2mm] -\dfrac{1}{p}\displaystyle\int_{T+\tau}^{t+\tau}\dfrac{1}{a(s)}\int_{s}^{\infty} F(v,x(v))\,dv\,ds, & t \ge T. \end{cases}
$$

Now, a slight modification of the argument in Case (I) guarantees that $N : Q \to CK(Q)$ is a upper semicontinuous, condensing map. Thus, Theorem 10.1.2 is applicable. This completes the proof. ∎

10.4. Notes and General Discussions

1. All the results of this chapter are based on the recent work of Agarwal, Grace and O'Regan [1]. These results improve the corresponding ones for the single valued case presented in earlier chapters and in [2]. Further applications of the fixed point theorems used in this chapter are available in [3,13].

10.5. References

1. **R.P. Agarwal, S.R. Grace and D. O'Regan,** On nonoscillatory solutions of differential inclusions, *Proc. Amer. Math. Soc.*, to appear.
2. **R.P.Agarwal, S.R. Grace and D. O'Regan,** *Oscillation Theory for Second Order Dynamic Equations,* to appear.
3. **R.P. Agarwal, M. Meehan and D. O'Regan,** *Fixed Point Theory and Applications,* Cambridge University Press, Cambridge, 2001.
4. **C.D. Aliprantis and K.C. Border,** *Infinite Dimensional Analysis,* Springer, Berlin, 1994.
5. **J.P. Aubin and A. Cellina,** *Differential Inclusions,* Springer, Berlin, 1984.
6. **M. Cecchi, M. Marini and P. Zecca,** Existence of bounded solutions for multivalued differential systems, *Nonlinear Analysis* **9**(1985), 775–786.
7. **N. Dunford and J. Schwartz,** *Linear Operators,* Interscience, New York, 1958.
8. **L.H. Erbe, Q.K. Kong and B.G. Zhang,** *Oscillation Theory for Functional Differential Equations,* Marcel Dekker, New York, 1995.
9. **P.M. Fitzpatrick and W.V. Petryshyn,** Fixed point theorems for multivalued noncompact acyclic mappings, *Pacific Jour. Math.* **54**(1974), 17–23.
10. **M. Frigon,** Théorèmes d'existence de solutions d'inclusions différentielles, in *Topological Methods in Differential Equations and Inclusions* (edited by A. Granas and M. Frigon), NATO ASI Series C, **Vol 472**, Kluwer, Dordrecht, 1995, 51–87.

11. **A. Lasota and Z. Opial**, An application of the Kututani–Ky Fan theorem in the theory of ordinary differential equations, *Bull. Acad. Polon. Sci. Ser. Sci. Math. Astron. Phys.* **13**(1965), 781–786.

12. **D. O'Regan,** Integral inclusions of upper semi–continuous or lower semi–continuous type, *Proc. Amer. Math. Soc.* **124**(1996), 2391–2399.

13. **D. O'Regan and R. Precup,** *Theorems of Leray–Schauder Type and Applications*, Gordon & Breach, Amsterdam, 2001.

14. **T. Pruszko,** Topological degree methods in multivalued boundary value problems, *Nonlinear Analysis* **5**(1981), 953–973.

Subject Index